T0176912

Cosmochemistry

Cosmochemistry is a rapidly evolving field of planetary science and the second edition of this classic text reflects the exciting discoveries made over the past decade from new spacecraft missions. Topics covered include the synthesis of elements in stars, behavior of elements and isotopes in the early solar nebula and planetary bodies, and compositions of extra-terrestrial materials. Radioisotope chronology of the early Solar System is also discussed, as well as geochemical exploration of planets by spacecraft, and cosmochemical constraints on the formation of solar systems. Thoroughly updated throughout, this new edition features significantly expanded coverage of chemical fractionation and isotopic analyses; focus boxes covering basic definitions and essential background material on mineralogy, organic chemistry and quantitative topics; and a comprehensive glossary. An appendix of analytical techniques and end-of-chapter review questions, with solutions available at www.cambridge.org/cosmochemistry2e, also contribute to making this the ideal teaching resource for courses on the Solar System's composition as well as a valuable reference for early career researchers.

Harry Y. McSween, Jr. is Chancellor Professor Emeritus at the University of Tennessee. His research on meteorites and Mars has resulted in hundreds of scientific papers. He has authored/co-authored six books on planetary science, including the textbook *Planetary Geoscience* (Cambridge, 2019) and was co-investigator on four NASA spacecraft missions. He has received awards from the US National Academy of Sciences, Meteoritical Society, and American Geophysical Union, and is the namesake for an asteroid.

Gary R. Huss is Research Professor and Director of the W. M. Keck Cosmochemistry Laboratory, University of Hawai'i. He is grandson of H. H. Nininger, the father of modern meteoritics, and has 50 years of experience collecting and carrying out research on meteorites. He has published approximately 130 papers on cosmochemistry. He is a Fellow of, and has served as President of, the Meteoritical Society. He also has an asteroid named after him.

Cosmochemistry

Harry Y. McSween Jr.
University of Tennessee

Gary R. Huss
University of Hawai'i

CAMBRIDGE
UNIVERSITY PRESS

University Printing House, Cambridge CB2 8BS, United Kingdom

One Liberty Plaza, 20th Floor, New York, NY 10006, USA

477 Williamstown Road, Port Melbourne, VIC 3207, Australia

314–321, 3rd Floor, Plot 3, Splendor Forum, Jasola District Centre, New Delhi – 110025, India

103 Penang Road, #05–06/07, Visioncrest Commercial, Singapore 238467

Cambridge University Press is part of the University of Cambridge.

It furthers the University's mission by disseminating knowledge in the pursuit of education, learning, and research at the highest international levels of excellence.

www.cambridge.org
Information on this title: www.cambridge.org/9781108839839
DOI: 10.1017/9781108885263

First published 2021

Printed in the United Kingdom by TJ Books Limited, Padstow Cornwall

A catalogue record for this publication is available from the British Library.

Library of Congress Cataloging-in-Publication Data
Names: McSween, Harry Y., author. | Huss, Gary R., author.
Title: Cosmochemistry / Harry McSween Jr., University of Tennessee, Knoxville, Gary Huss, University of Hawaii, Manoa.
Description: Second edition. | New York : Cambridge University Press, 2021. | Includes bibliographical references and index.
Identifiers: LCCN 2021024813 (print) | LCCN 2021024814 (ebook) | ISBN 9781108839839 (hardback) | ISBN 9781108885263 (epub)
Subjects: LCSH: Cosmochemistry. | BISAC: SCIENCE / Earth Sciences / General
Classification: LCC QB450 .M37 2021 (print) | LCC QB450 (ebook) | DDC 523/.02–dc23
LC record available at https://lccn.loc.gov/2021024813
LC ebook record available at https://lccn.loc.gov/2021024814

ISBN 978-1-108-83983-9 Hardback

For Sue and Jackie

Contents

Preface

Cosmochemistry provides critical insights into the workings of our local star and its stellar companions throughout the galaxy, the origin and timing of our solar system's birth, and the complex processes inside planetesimals and planets (including our own) as they evolve. Much of the database of cosmochemistry comes from laboratory analyses of elements, compounds, and isotopes in our modest collections of extraterrestrial samples. A growing part of the cosmochemistry database is gleaned from remote-sensing measurements by spacecraft instruments, which provide chemical analyses and geologic context for other planets, moons, asteroids, and comets. Because the samples analyzed by cosmochemists are typically so small and valuable, or must be analyzed on bodies many millions of miles distant, this discipline leads in the development of new analytical technologies for use in the laboratory or on spacecraft. These technologies then spread to geochemistry, materials science, and other fields where precise analyses of tiny samples are important.

Despite its cutting-edge qualities and often newsworthy discoveries, cosmochemistry is an orphan. It does not fall clearly within the purview of chemistry, physics, geology, astronomy, or biochemistry but is rather an amalgam of parts of these disciplines. Because it has no natural home or constituency, cosmochemistry is usually taught (if it is taught at all) directly from its scientific literature (admittedly challenging reading) or from specialized books on meteorites and planetary science. In crafting the original edition of this textbook, we attempted to remedy that shortcoming. In this thoroughly revised second edition, we have incorporated new discoveries made and novel insights gleaned during the last decade. We have tried to make this subject accessible to advanced undergraduate and graduate students with diverse academic backgrounds, although we do presume some prior exposure to basic chemistry. This goal may sometimes lead to uneven treatment of some subjects, and our readers should understand that our intended audience is broad.

Cosmochemistry is advancing so rapidly that we can only hope to provide a snapshot of the discipline as it is currently understood and practiced. We have found even that to be a challenge because we could not hope to possess expertise in all the subjects encompassed by this discipline. We have drawn heavily on the contributions of many colleagues, especially those who educate by writing thoughtful reviews. That assistance is gratefully acknowledged through our annotated suggestions for further reading at the end of each chapter.

The topics covered in the chapters of this book include the following, in this order:

- Introduction to how cosmochemistry developed, and to how it differs from geochemistry
- Basic review of the characteristics and behaviors of elements and isotopes
- Discussion of how elements are synthesized within stars, and how the chemistry of the galaxy has evolved over time
- Assessment of the abundances of elements and isotopes in the solar system, and how they are measured
- Description of presolar grains found in meteorites, and how they constrain nucleosynthesis in stars and processes in interstellar space
- Introduction to meteorites, interplanetary dust particles, and lunar samples
- Consideration of processes that have fractionated elements in interstellar space, in the solar nebula, and within planetary bodies
- Consideration of processes that fractionate stable isotopes, as well as isotopic anomalies inherited from the Sun's parent molecular cloud
- Explanation of how long-lived and short-lived radioactive isotopes are used to quantify solar system history
- Synthesis of the radiometric age of the solar system and the ages of its constituents
- Assessment of the solar system's most volatile materials: ices, noble gases, and organic matter
- Survey of planetesimals to provide context on planetary building blocks
- Assessment of the chemistry of asteroids and comets, based on the samples we have of them and on spacecraft remote sensing

- Examples of modern geochemical exploration of solar system bodies: the Moon and Mars
- Synthesis and review of the formation of solar systems, from the perspective of cosmochemistry
- Appendix describing some important analytical methods used in cosmochemistry

More established disciplines are taught using tried-and-true methods and examples, the results of generations of pedagogical experimentation. Cosmochemistry does not yet offer that. Most of those who dare to teach cosmochemistry, including the authors of this book, have never actually been students in a cosmochemistry course. In the authors' case, we have learned directly from a handful of scientists who have guided our introduction to the field, including Calvin Alexander, Bob Pepin, Ed Anders, Jim Hays, Dick Holland, Ian Hutcheon, Klaus Keil, Roy Lewis, Dimitri Papanastassiou, Jerry Wasserburg, and John Wood, and indirectly from many professional colleagues and our own students. We hope that this introduction to cosmochemistry will guide other students and their teachers as they explore together this exciting, interdisciplinary subject, and that they will enjoy the experience as we have.

1 Introduction to Cosmochemistry

Overview

Cosmochemistry is defined, and its relationship to geochemistry is explained. We describe the historical beginnings of cosmochemistry, and the lines of research that coalesced into the field of cosmochemistry are discussed. We then briefly introduce the tools of cosmochemistry and the datasets that have been produced by these tools. The relationships between cosmochemistry and geochemistry on the one hand and astronomy, astrophysics, and geology on the other are considered.

1.1 What Is Cosmochemistry?

A significant portion of the universe is comprised of elements, ions, and the compounds formed by their combinations – in effect, chemistry on the grandest scale possible. These chemical components can occur as gases or superheated plasmas, less commonly as solids, and very rarely as liquids.

Cosmochemistry is the study of the chemical compositions of the universe and its constituents, and the processes that produced those compositions. This is a tall order, to be sure. Understandably, cosmochemistry focuses primarily on the objects in our own solar system, because that is where we have direct access to the most chemical information. That part of cosmochemistry encompasses the compositions of the Sun, its retinue of planets and their satellites, the almost innumerable asteroids and comets, and the smaller samples (meteorites, interplanetary dust particles or "IDPs," returned lunar samples) derived from them. From their chemistry, determined by laboratory measurements of samples or by various remote-sensing techniques, cosmochemists try to unravel the processes that formed or affected them and to fix the chronology of these events. Meteorites offer a unique window on the *solar nebula* – the disk-shaped cocoon of gas and dust that enveloped the early Sun some ~4.57 billion years ago, and from which planetesimals and planets accreted (Fig. 1.1).

Within some meteorites are also found minuscule *presolar grains*, which provide us with an opportunity to analyze directly the chemistry of interstellar matter. Some of these tiny grains are pure samples of the matter ejected from dying stars, and they provide constraints on our understanding of how elements were forged inside stars before the Sun's birth. Once formed, these grains were released into the *interstellar medium* (ISM), the space between the stars. The ISM is filled primarily with diffuse gases, mostly hydrogen and helium, but with oxygen, carbon, and nitrogen contributing about 1% by mass and all the other elements existing mostly as micrometer-sized dust motes. Much of the chemistry in the ISM occurs within relatively dense *molecular clouds*, where gas densities can reach 10^3–10^6 particles/cm^3, high by interstellar standards (but not by our everyday experience: Earth's atmosphere has ~3×10^{19} atoms/cm^3 at sea level). These clouds are very cold, with temperatures ranging from 10 to 100 K, so interstellar grains become coated with ices. Reactions between ice mantles and gas molecules produce organic compounds that can be extracted from meteorites and identified by their bizarre isotopic compositions. Many dust grains were undoubtedly destroyed in the ISM, but some hardy survivors were incorporated into the nebula when the molecular cloud collapsed, and thence were accreted into meteorites.

Processes that occur inside stars, in interstellar space, and within the solar nebula have no counterparts in our terrestrial experience. They can be studied or inferred from astronomical observations and astrophysical theory, but cosmochemical analyses of materials actually formed or affected by these processes provide constraints and insights that remote sensing and theory cannot. Our

Figure 1.1 An artist's conception of the solar nebula, surrounding the violent young Sun. Figure courtesy of NASA.

terrestrial experience places us on firmer ground in deciphering the geologic processes occurring on the Earth's Moon. In studying lunar rocks and soils, we can use familiar geochemical tools developed for understanding the Earth. We have also measured the chemical compositions of some other planetary bodies and their smaller cousins, geologically processed planetesimals, using telescopes and instruments on spacecraft. In some cases, we even have meteorites ejected during impacts onto these bodies. Chemical measurements (whether from laboratory analyses of samples or *in situ* analyses of rocks and soils by orbiting or landed spacecraft) add quantitative dimensions to our understanding of planetary science. All extraterrestrial materials are fair game for cosmochemistry.

1.2 Geochemistry versus Cosmochemistry

Traditionally, cosmochemistry has been treated as a branch of *geochemistry* – usually defined as the study of the chemical composition of the Earth. Geochemistry focuses on the chemical analysis of terrestrial materials, as implied by the prefix "geo," and geochemistry textbooks commonly devote only a single chapter to cosmochemistry, if the subject is introduced at all. However, the line between geochemistry and cosmochemistry has always been somewhat fuzzy. The most prominent technical journal in this discipline, *Geochimica et Cosmochimica Acta*, has carried both names since its inception in 1950. The burgeoning field of planetary geochemistry appropriates the "geo" prefix, even though its subject is not Earth. A broader and more appropriate definition of geochemistry might be the study of element and isotope behavior during geologic processes, such as occur on and within the Earth and other planets, moons, and planetesimals. Using this definition, we will include *planetary geochemistry* as an essential part of our treatment of cosmochemistry.

It is worth noting, though, that the geochemical and cosmochemical behaviors of elements do show some significant differences. A geochemical perspective of the Periodic Table is illustrated in Figure 1.2 (adapted from Railsback, 2003). As depicted, this diagram is decidedly Earth-centric, but the controls on element behavior during geologic processes apply to other bodies as well. Determining relative elemental abundances is an important part of geochemistry, and the relative abundances of elements in the Earth's crust vary over many orders of magnitude. Crustal abundances are illustrated in Figure 1.2, because most geochemical data are based on readily accessible samples of the crust. Geochemistry is also concerned with determining the composition of the Earth's interior – its mantle and core – and a more comprehensive figure would include those abundances as well. Very few native elements (pure elements not chemically bound to any others) occur naturally in the Earth, so Figure 1.2 distinguishes elements that occur commonly as cations or anions (positively and negatively charged particles, respectively). The electrical properties of elements control how they combine into compounds (minerals), dissolve in natural fluids, or are concentrated into melts (magmas) at high temperatures. The elements in Figure 1.2 are also grouped by their so-called geochemical affinities: *lithophile* (rock-loving) elements tend to form silicates or oxides (the constituents of most rocks), *siderophile* (iron-loving) elements combine with iron into metal alloys, *chalcophile* (sulfur-loving) elements react with sulfur to form sulfides, and *atmophile* elements tend to form gases and reside in the atmosphere. Many elements exhibit several affinities, depending on conditions, so the assignments illustrated in Figure 1.2 offer only a rough approximation of the complexity of element geochemical behavior. Finally, an important part of geochemistry takes advantage of the fact that most elements have more than one *isotope*. Measuring isotopic abundances has great value as a geochemical tool, and the most commonly used isotope systems are illustrated by boxes with heavy lines in Figure 1.2. Stable isotopes of some light elements provide information on sources of elements, the conditions under which minerals form, and the processes that separate isotopes from each other. Unstable (*radioactive*) *nuclides* and their decay products (*radiogenic* nuclides) similarly constrain element sources and geologic processes, as well as permit the ages of rocks

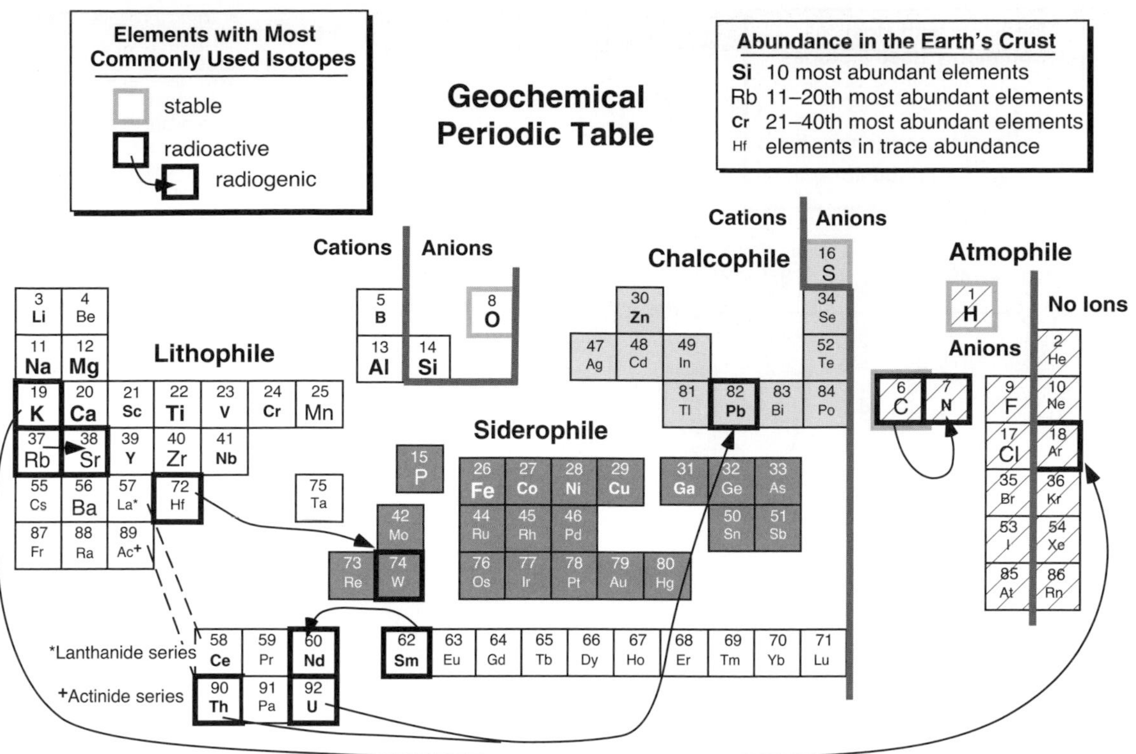

Figure 1.2 A geochemical Periodic Table, illustrating controls on element behavior during geologic processes. Relative abundances of elements in the Earth's crust are indicated by symbol sizes. Cations and anions are usually combined into minerals. Elements having affinities for silicate or oxide minerals (lithophile), metal (siderophile), sulfide minerals (chalcophile), and gas (atmophile) phases are distinguished. Elements having stable isotopes that are commonly used in geochemistry are shown as boxes with bold gray outlines. Radioactive and radiogenic isotopes used for chronology are shown by boxes with bold black outlines and arrows showing decay relationships.

and events to be determined. The isotopic compositions of many other elements in terrestrial materials are now being analyzed, and a future Figure 1.2 will certainly expand the list of commonly used isotopic systems.

By way of contrast, Figure 1.3 illustrates a cosmochemical perspective of the Periodic Table. The element abundances shown in this figure are atomic concentrations in the Sun (relative to the abundance of silicon), as best we can determine them. The Sun comprises >99.8% of the mass of solar system matter, so solar composition is approximately equivalent to the average solar system (often incorrectly called *cosmic*) composition. The behavior of elements in space is governed largely by their *volatility*, which we quantify by specifying the temperature interval at which elements change state from a gas to a solid on cooling. (The liquid state is not generally encountered at the very low pressures of space; liquids tend to be more common in geochemistry than cosmochemistry.) All elements occur as gases at high enough temperatures, and they either condense at lower temperatures to form solid minerals or ices, or react with already condensed phases to form other solid phases. Some elements condense at such low temperatures that they effectively remain as gases. Thermodynamic data can be used to predict the temperatures at which solid phases become more stable than their components in a gas of solar composition. Assignment of elements to the various *refractory* and *volatile* groups in Figure 1.3 is based on the temperature at which 50% of each element has condensed into solid phases. It is convenient in cosmochemistry to identify elements according to the kinds of minerals into which they condense – lithophile, siderophile, and chalcophile. Some volatile elements only condense at very low temperatures to form ices, or do not condense at all. Also illustrated in Figure 1.3 are the most commonly used isotope systems in cosmochemistry; the complete list is considerably longer than for geochemistry, and would include stable isotopes measured in

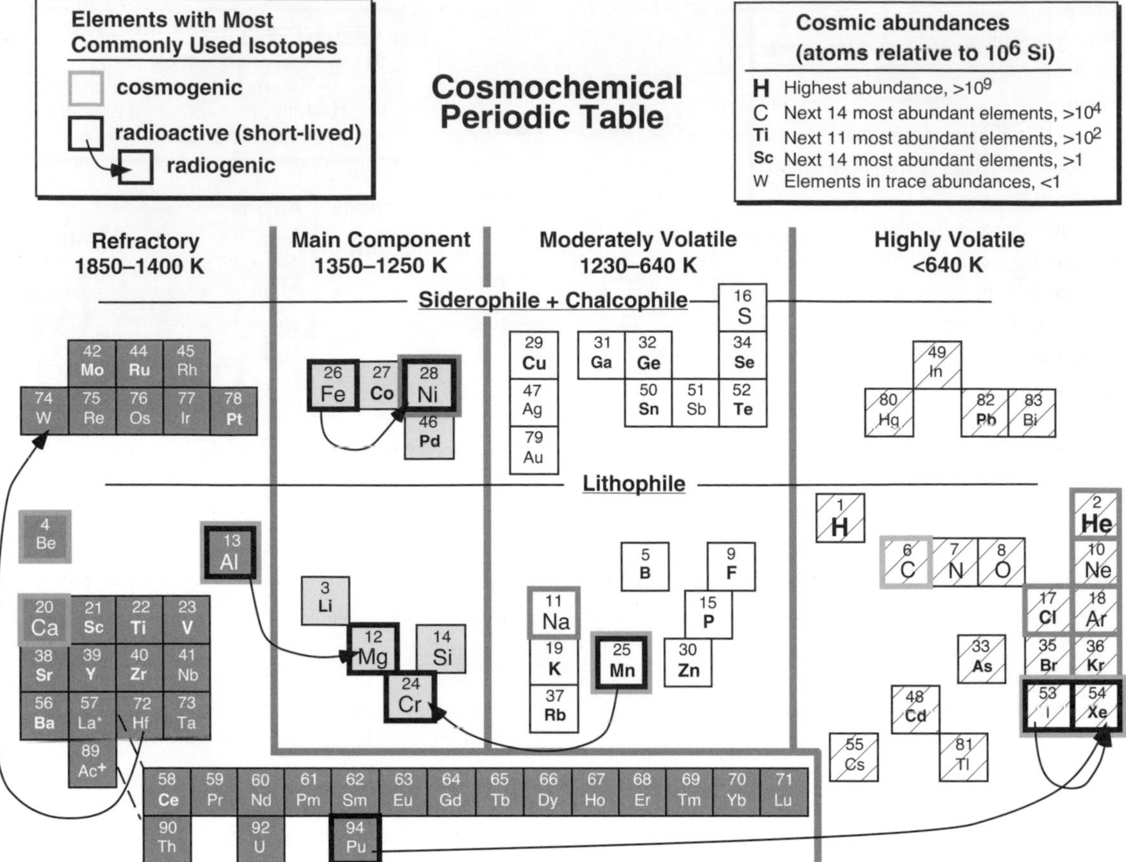

Figure 1.3 A cosmochemical Periodic Table, illustrating the behavior of elements in chondritic meteorites. Relative solar system abundances are indicated by symbol sizes. Volatilities of elements reflect the temperatures at which 50% of each element would condense into a solid phase from a gas of solar composition. As in Figure 1.2, the chemical affinities of each element – lithophile for silicates and oxides, siderophile for metals, and chalcophile for sulfides – are indicated. Some of the most highly volatile phases may have remained uncondensed in the nebula. Stable, radioactive, and radiogenic isotopes used in cosmochemistry are indicated by bold outlines, as in Figure 1.2. Abundances and 50% condensation temperatures come from tabulations by Lodders and Fegley (1998).

presolar grains in meteorites, cosmogenic nuclides formed by interaction with cosmic rays in space, and now-extinct radioactive isotopes that existed in the early solar system.

Comparison of Figures 1.2 and 1.3 reveals that the chemical behavior of an element may differ depending on whether it is in a geochemical or cosmochemical environment. This book's topics will refer to both figures in understanding the compositions of extraterrestrial materials. In cosmochemistry we are concerned with the origin and behavior of elements in space, whereas in planetary geochemistry we focus on their behavior once they are accreted into bodies that undergo geologic (usually thermal) processing. Planetary geochemistry follows more or less the same rules as on the Earth, although these rules must be modified to accommodate different geologic conditions or starting compositions. And the geochemical consequences of biology, so important on Earth, do not apply on other worlds, so far as we can determine presently.

1.3 Beginnings of Cosmochemistry (and Geochemistry)

1.3.1 Philosophical Foundations

The philosophical foundations of cosmochemistry date to the last half of the eighteenth century when Immanuel Kant (1724–1804) and Pierre-Simon Marquis de Laplace (1749–1827) put forward comprehensive models for the origin of the solar system. Kant's model, published in 1755, started with the Sun at the center of a gaseous nebula. In order for this cocoon of gas and dust to be stable in the gravitational field of the Sun, the nebula had to rotate about the Sun. Kant suggested that the matter in the disk would segregate into large bodies that would become the planets. This segregation would take place slowly, with each body developing into a miniature version of the solar system. Kant showed that the rotation of the planets and their satellites would be in the same sense as their revolution around the Sun. In 1796, Laplace published a model that started with the primordial Sun occupying the entire volume now occupied by the planetary orbits. This hot, luminous "solar nebula" rotated as a rigid body so that linear velocity was greatest at the outer edge. As the nebula cooled and contracted, it rotated faster to preserve angular momentum. When centrifugal force exceeded gravitational attraction, a ring was left behind. This process was repeated many times and the rings contracted to form planets. During the nineteenth century, these two models became intertwined into a "nebular hypothesis" that was generally accepted in some form until the beginning of the twentieth century. Ideas based on these models, such as a hot solar nebula, have remained part of mainstream cosmochemical thought until very recently.

1.3.2 Meteorites and Microscopy

Meteorites (Fig. 1.4) are central to cosmochemistry, because they are our most accessible source of extraterrestrial samples. Though people have seen stones falling from the sky for thousands of years, the fact that meteorites actually fall was not acknowledged by the European and American scientific establishments until early in the nineteenth century. Credit for putting meteorites on the scientific map generally goes to Ernst Chladni (1756–1827). In a 63-page book with the long title (translated from German) *On the Origin of the Mass of Iron Found by Pallas and of Other Similar Iron Masses, and on a Few Natural Phenomena Connected Therewith*, published in 1794, Chladni laid out a case based on historical records of observed falls that stone and iron masses enter the Earth's atmosphere from space and form fireballs as they plunge through the atmosphere. These ideas contradicted two beliefs that were strongly held by his scientific contemporaries: rocks and masses of metal do not fall from the sky, and no small bodies exist in space beyond the Moon. However, during the next five years, four falls of stony meteorites were witnessed and widely reported in Europe. Chemist Edward Howard (1774–1816) and mineralogist Jacques-Louis de Bournon (1751–1825) carried out a series of chemical and mineralogical analyses of stones said to have fallen from the sky and found that they were similar in texture and composition, and significantly different from terrestrial rocks. The publication of these findings in early 1802 was followed by the fall in 1803 of nearly 3000 stones at L'Aigle in Normandy, France. These events provided evidence to support Chladni's claims, and meteorites entered the realm of scientific study.

Figure 1.4 Broken surface of the Allende chondritic meteorite. Note the abundant round or broken chondrules and irregular white calcium-aluminum inclusions. Centimeter scale at the bottom.

A major step in understanding meteorites came with Henry Clifton Sorby's (1826–1908) development of the petrographic microscope in the mid-1800s. Using this instrument, thin sections (paper-thin slices of rock, mounted on glass slides) are observed by passing polarized light through them from below, providing a means of identifying minerals and observing the textures of rocks. Sorby soon turned his attention to a type of meteorite called *chondrites*, describing the round droplets of solidified melt in them (called *chondrules*, after the Greek "chondros" for "grains" or "seeds") as drops of a fiery rain (Fig. 1.5). Chondrites will be described in detail in Chapter 6.

A significant part of meteoritics literature focuses on petrographic description and classification. This does not usually make for exciting reading, but an orderly classification is essential for interpreting the chemical compositions of meteorites and recognizing relationships among them. Beginning in the 1860s, Gustav Rose (1798–1873) at the University Museum of Berlin and Nevil Story-Maskelyne (1823–1911) at the British Museum developed meteorite taxonomies based on microscope observations. Gustav Tschermak (1836–1927) later refined Rose's classification, and Aristides Brezina (1848–1909) refined the Rose–Tschermak classification,

Figure 1.5 Transmitted-light photomicrograph of the Tieschitz chondritic meteorite. The rounded, millimeter-sized chondrules contain crystals of olivine and pyroxene, and the chondrules are set in a fine-grained, opaque matrix. Horizontal field of view is ~2.3 mm.

which reached its final form in 1904. This classification was based on mineralogy, because at the time there were few chemical analyses of meteorites, and those that existed were of uneven quality. George Prior (1862–1936) devised a simpler mineralogical classification for chondrites in 1920. Prior's major mineralogical subdivisions for meteorites are still used today, but his system has been supplanted by one devised by Randall Van Schmus and John Wood (1967) that separates primary characteristics of chondrites, such as bulk composition, from secondary characteristics, such as degree of metamorphic or aqueous alteration. We will discuss meteorite classification in detail in Chapter 6.

1.3.3 Spectroscopy and the Compositions of Stars

In the early nineteenth century, determining the compositions of the Sun and other stars posed a fundamental hurdle for astronomy. The French philosopher Auguste Compte (1798–1857) confidently asserted that never, by any means, would we be able to study the chemical compositions of celestial bodies. But *spectroscopy* soon proved him wrong. Spectroscopes attached to telescopes were used to spread out starlight into its component wavelengths. The spectra of stars and of the Sun showed numerous narrow, dark gaps where particular wavelengths were missing. These gaps (absorption lines) are due to the various chemical elements in a star's outer layers absorbing light emanating from the hotter interior. Each element absorbs (or emits) light at specific wavelengths characteristic of its electronic structure.

Box 1.1 Lockyer and the Discovery of Helium in the Sun

Joseph Lockyer (1836–1920) was one of the pioneers of solar spectroscopy. In examining the spectra of solar prominences in 1869, Lockyer noticed an absorption line that he could not identify. Reasoning that it represented an element not present on Earth, he proposed a new element – helium, from the Greek word "helios" for "Sun." This idea failed to achieve acceptance by Lockyer's scientific colleagues, until a gas having the same mysterious spectral line was found 25 years later in rocks. The helium in terrestrial uranium ore formed as a decay product of radioactive uranium. Thus, this abundant element was first discovered in the Sun rather than in the laboratory. Lockyer's cosmochemical discovery was recognized by the British government, which created a solar physics laboratory for him. Lockyer also founded the scientific journal *Nature*, which he edited for 50 years.

In the late 1800s, after decades of work on the spectroscopy of stars, Lockyer developed his "meteoritic hypothesis." According to this idea, meteorites were the primary dust of the universe. Nebulae observed by astronomers were interpreted as swarms of meteorites bound together through gravitation and interacting much like atoms in a gas. Lockyer postulated that the solar system and other objects had formed from these meteorite swarms (Lockyer, 1890). Although the original hypothesis was soon abandoned, the idea that meteorites might be chemically primitive materials that sample the cosmos was not far off the mark.

Identifying the elements present in the Sun and stars from their spectra was one thing, but determining their relative abundances was quite another. The solar absorption lines for iron are particularly prominent, leading astronomers to believe that iron was the most abundant element in the Sun, as it is in the Earth and in many meteorites. Princeton astronomer Henry Russell (1877–1957) even conjectured that if the Earth's crust were heated to the temperature of the Sun, its spectrum would resemble the solar spectrum. It took until the 1920s before a clear understanding of how spectra arise was established, permitting evaluation of the true compositions of the Sun and stars. The key to understanding stellar spectra was discovered in 1925 by Cecilia Payne (1900–1979). She showed that the spectral lines arose from the excitation of the electrons surrounding the atomic nucleus and that the energy levels of the electrons were a function of stellar temperature. When temperature was taken into account, the abundances of elements in stars were shown to be nearly the same in a variety of stars, in spite of them having different spectra. Her work also showed that hydrogen and helium are the most abundant elements in the Sun and other stars. This last result was not widely immediately accepted and was downplayed in her published thesis. But by 1930, her work had completely supplanted previous interpretations and modern spectroscopy was born.

Figure 1.6 Victor Goldschmidt, as pictured on a Norwegian postage stamp issued in 1974.

1.3.4 Solar System Element Abundances

The term "cosmochemistry" apparently derives from the work of Victor Goldschmidt (Fig. 1.6), who is often described as the father of geochemistry. This is yet another crossover and, in truth, Goldschmidt also established cosmochemistry as a discipline. In 1938, he published a cosmic abundance table based on the proportions of elements in meteorites. He used the term "cosmic" because, like his contemporaries, he believed that meteorites were interstellar matter. Chemist William Harkins (1873–1951) had formulated an earlier (1917) table of elemental abundances – arguably the first cosmochemistry paper, although he did not use that term. As explained in Chapter 4, *solar system abundance* is now preferred over *cosmic abundance*, although the terms are often used interchangeably.

Goldschmidt and his colleagues in Germany, and later in Norway (where he escaped the grasp of the Nazis in World War II), analyzed and compiled a wealth of chemical data on terrestrial rocks and meteorites. The compositions of terrestrial rocks have been modified by partial melting (leaving some components behind as residues) and by fractional crystallization (where crystals are segregated from the melt, causing the liquid's composition to change). However, Goldschmidt recognized that chondrites have not experienced wholesale melting and have thereby escaped geologic processing. They are basically cosmic sediments – physical mixtures of nebular matter whose chemical abundances have remained unchanged since they formed. To obtain accurate compositions using the then-new analytical technique of emission spectroscopy, Goldschmidt separated chondrites into their more readily measurable silicate, metal, and sulfide components and analyzed each in turn. Consequently, he was able to determine how various elements were partitioned among these coexisting phases (thereby inventing the terms "lithophile," "siderophile," and "chalcophile" to describe their geochemical affinities). He then calculated what he called the "cosmic abundances" of 66 elements by using the weighted means of element concentrations in meteorite silicate (10 parts), metal (2 parts), and sulfide (1 part). At about the same time, astronomers began using the Sun's spectra to estimate elemental abundances. It soon became apparent that solar elemental abundances were similar to Goldschmidt's cosmic (chondritic) abundances, except that the meteorites were depleted in the most volatile elements like hydrogen and helium. Cosmic abundances (more appropriately called solar system abundances) are a cornerstone of cosmochemistry because they represent the raw material from which the solar system formed.

1.3.5 Isotopes and Nuclear Physics

Isotopes were recognized at the beginning of the twentieth century as a result of studies of radioactivity. Careful studies found three naturally occurring radioactive decay series – the thorium, uranium, and actinium series – all of which ended with stable lead. The existence of isotopes was confirmed in 1911 when Fredrick Soddy (1877–1956) measured the atomic masses of lead obtained from uranium-rich and thorium-rich ores and showed that they were different. During this same period, J. J. Thompson (1856–1940) discovered that ions accelerated through an electric field would adopt different parabolic trajectories depending upon their masses. Thompson's student, Francis Aston (1877–1945), used this principle to design several different mass spectrographs, which separated particles by mass/charge ratio and recorded the output on photographic plates. By 1920, Aston was reporting the presence and relative abundances of isotopes of numerous elements, including oxygen, neon, argon, krypton, xenon, and mercury (Aston, 1920). In 1922, he received the Nobel Prize in Chemistry for his work on isotopes using a mass spectrograph. However, an understanding of the structure of the atom had to await the discovery of the neutron in 1930.

During the 1930s, further technological advances permitted detailed studies of the masses and relative abundances of isotopes. Modern mass spectrometers designed by Alfred Nier (1911–1994) at the University of Minnesota had greater mass-resolving power and were more sensitive than any built previously. Nier made accurate measurements of the isotopic abundances of argon, potassium, zinc, rubidium, and cadmium and, in the process, discovered ^{40}K, which would later become an important isotope for dating rocks. He also pioneered the development of uranium–lead and thorium–lead dating.

In 1940, Nier successfully separated ^{235}U from ^{238}U using a mass spectrometer, providing an enabling technology for the Manhattan Project. During World War II, many of the top nuclear physicists worked on the development of the atomic bomb. The fascinating story of these years and their effect upon the participants is beyond the scope of this book. But as the war ended, many of them turned their attention away from the tools of war toward understanding our planet and universe, and their knowledge became available to cosmochemistry.

The leading figure in cosmochemistry during the 1950s and 1960s was Harold Urey (1893–1981). Urey (Fig. 1.7) was one of the first practitioners of cosmochemistry as we understand it today. He was awarded the Nobel Prize in Chemistry in 1934 for his work on deuterium and heavy water. During the war, he and his colleagues developed the gaseous diffusion method for separating ^{235}U from ^{238}U. After the war, he became a professor at the University of Chicago, where he did pioneering work using the ^{18}O/^{16}O ratio in paleoclimate research, developed theories about the origin of the elements and their abundances in stars, pointed out the importance of short-lived radionuclides such as ^{26}Al, investigated the origins of life on Earth, and made many other contributions. He was a leader in developing the scientific rationale for returning samples from the Moon.

Figure 1.7 Harold Urey, one of the fathers of cosmochemistry. NASA image.

During World War II, Hans Suess (1909–1993) was part of a team of German scientists working on heavy water. In 1950, he emigrated to the United States. His work with Urey on *nucleosynthesis* and the abundances of the elements is a cornerstone of cosmochemistry. In 1965, along with Heinrich Wänke (1928–2015), he proposed that the extremely high noble gas contents in some chondritic meteorites were due to the implantation of solar wind. He also worked on climate research and ^{14}C dating. Together, Suess and Urey (1956) published the first table of cosmic abundances to include the abundances of the isotopes.

The new knowledge of nuclear physics affected cosmochemistry in another way. A classic paper by astrophysicists Margaret and Geoffrey Burbidge (husband and wife), William Fowler, and Fred Hoyle (this paper was so influential that it has come to be known by scientists simply as "B^2FH"), and a similar contribution by Alastair Cameron, both published in 1957, provided the theoretical basis for understanding how elements are produced in stars, as described in Chapter 3.

Radiometric dating using long-lived radionuclides came into its own in the 1940s and 1950s with the advent of better mass spectrometers. The uranium-isotope decay scheme was first shown to be useful as a geochronometer by Fritz Houtermans and Arthur Holmes in 1946. The first accurate determination of the age of the Earth was made in 1956 by Clair Patterson, who used the uranium–lead method to date meteorites. The ^{40}K–^{40}Ar decay scheme was shown to be a useful chronometer for meteorites by Gerald Wasserburg in his doctoral thesis, completed in 1954. The first age determination by the ^{87}Rb–^{87}Sr method was published by Hahn et al. (1943), and the method came into wide use in the 1950s. Its application to meteorites peaked in the late 1960s and 1970s, in conjunction with work on the lunar samples.

The first *short-lived radionuclide* (one whose primordial abundance has decayed away) was shown to have been present in meteorites by John Reynolds in 1960. Reynolds found large excesses of ^{129}Xe, the decay product of short-lived ^{129}I, in chondritic meteorites. This discovery showed that elements had been synthesized in stars shortly before the formation of the solar system. A more important short-lived radionuclide, ^{26}Al, was demonstrated to have been present in meteorites by Typhoon Lee et al. in 1977. This isotope is particularly significant, as it is thought to have been a potent source of heating for asteroids and planets early in solar system history. A variety of other short-lived isotopes have now been confirmed in meteorites and are the basis for high-resolution chronometry of the early solar system.

Nuclides formed by nuclear reactions induced by high-energy cosmic rays are called *cosmogenic*. Cosmogenic isotopes are more common in meteorites than on the Earth, because our planet's atmosphere screens out most cosmic rays. However, meteorites traveling in space are heavily irradiated by cosmic rays, and the production of cosmogenic isotopes can be used to estimate the times since the meteorites were liberated from their parent asteroids (these times are called *cosmic-ray exposure ages*). The first cosmic-ray exposure ages were measured in the late 1950s. Since then, thousands of cosmic-ray exposure ages have been estimated using a variety of cosmogenic nuclides, and new modeling techniques have allowed the interpretation of complex irradiation histories.

In 1956, John Reynolds pioneered a new and highly sensitive method for measuring noble gases, which effectively created the field of noble gas geochemistry and cosmochemistry. Noble gases have many isotopes and, because they do not bond with rock-forming elements, they have very low abundances in most materials. Thus, additions from the decay of radioactive nuclides or cosmic-ray interactions are easy to detect. Noble gases also exhibit different elemental and isotopic ratios in meteorites, reflecting different processes operating in the early solar system. In addition, isotopic anomalies that could not be explained by any processes known to be operating in the solar system were found in xenon (discovered by John Reynolds and Grenville Turner in 1964) and neon (discovered by David Black and Robert Pepin in 1969) extracted from meteorites. These noble gases provided the first hints that presolar grains might have survived in the nebula, although they were not widely recognized at the time.

The pursuit of the carriers of exotic noble gas components by Edward Anders and colleagues at the University of Chicago and in other laboratories eventually led to the isolation of presolar grains from meteorites. The approach used by Anders involved laborious tracking of exotic noble gas carriers through steps of increasingly harsh chemical dissolutions and physical separations. He was rewarded with the discovery of presolar diamond (the first isolated presolar grain), silicon carbide, and graphite, the carriers of the three main exotic noble gas components. These three materials are all carbon rich, but subsequent work has identified presolar oxides, nitrides, and silicates, as discussed in Chapter 5.

1.3.6 Spacecraft, Returned Samples, and Remote Chemical Analyses

The launch of Sputnik by the Soviet Union in 1957 changed the world forever. The immediate impact was to change the nature of the Cold War by demonstrating the feasibility of intercontinental ballistic missiles. But Sputnik also raised the curtain on the scientific exploration of space and on visiting and obtaining samples from other solar system bodies. The first target was the Moon. Close-up images of the Moon were provided by the United States' Ranger missions, which

Table 1.1 Samples returned by spacecraft missions

Sample source	Mass	Date Returned to Earth	Mission
Moon: Mare Tranquilitatis	21.55 kg	Jul 24, 1969	Apollo 11 (NASA)
Moon: Oceanus Procellarum	34.30 kg	Nov 24, 1969	Apollo 12 (NASA)
Moon: Mare Fecunditatis	101 g	Sep 24, 1970	Luna 16 (USSR)
Moon: Fra Mauro Highlands	42.80 kg	Feb 9, 1971	Apollo 14 (NASA)
Moon: Hadley–Apennine	76.70 kg	Aug 7, 1971	Apollo 15 (NASA)
Moon: Apollonius Highlands	55 g	Feb 25, 1972	Luna 20 (USSR)
Moon: Descartes Highlands	95.20 kg	Apr 27, 1972	Apollo 16 (NASA)
Moon: Taurus–Littrow	110.40 kg	Dec 19, 1972	Apollo 17 (NASA)
Moon: Mare Crisium	170.1 g	Aug 22, 1976	Luna 24 (USSR)
Earth–Sun Lagrange 1	Solar wind atoms	Sep 8, 2004	Genesis (NASA)
Comet Wild 2	1000s of dust particles	Jan 15, 2006	Stardust (NASA)
Asteroid 25143 Itokawa	1000s of particles	Jun 13, 2010	Hayabusa (JAXA)
Asteroid 162173 Ryugu	~5.4 g	Dec 5, 2020	Hayabusa2 (JAXA)
Moon: South Pole–Aitken	~2 kg	Dec 2020	Chang'e 5 (CNSA)
Asteroid 101955 Bennu	~250 g?	Sep 2023 (planned)	OSIRIS-REx (NASA)

Figure 1.8 Apollo astronaut on the Moon. NASA image.

impacted the Moon in 1964 and 1965. The first lunar chemical data were provided by the Soviet Luna and American Surveyor spacecraft in 1966. The first manned landing on the Moon was the Apollo 11 mission in 1969. Five more successful Apollo missions followed, before the program was abruptly terminated.

The return of 381 kg of lunar rocks and soils from six sites on the Moon's nearside by Apollo astronauts (Fig. 1.8) and 326 g from three sites by Soviet Luna robotic landers in the 1970s (Table 1.1) provided a bonanza of new extraterrestrial materials for cosmochemistry. The intense interest in these samples encouraged a considerable expansion of laboratory techniques and capabilities. Fortuitously, two large meteorites (the Allende and Murchison chondrites) fell to Earth in 1969, just as the new laboratories were gearing up for lunar sample return. The new analytical techniques were applied to these meteorites, initially as a means of demonstrating capability, but the two chondrites turned out to be incredibly interesting in their own right and provided a new impetus for the study of all types of extraterrestrial materials. Because lunar samples were so precious, many groups simultaneously analyzed the same rocks, and competition forced the quality of the analyses to new heights. Lunar rocks were especially useful because mapping of the Moon by telescopes and orbiting spacecraft provided the geologic context for these samples. Lunar soils allowed for wider sampling of rock types, because the soils consist of rock particles thrown tremendous distances by impacts. In addition, lunar soils contain implanted solar wind particles, providing a window onto solar element abundances. A decade after the Apollo program ended, the first lunar meteorites were recognized in Antarctica. Their discovery, made possible by comparing them with Apollo samples, proved that rocks could be ejected from one body and travel to another. For the first time, rocks delivered by natural processes from another planet were available for direct geochemical analysis. While the Apollo and Luna samples were collected from a geographically restricted area ($<5\%$ of the surface area) of the Moon, the nearly 150 distinct lunar meteorites (as of time of writing) are thought to provide a more representative sample. In 2020, China's Chang'e 5 spacecraft returned ~2 kg of lunar samples from a young volcanic mound, the only returned lunar samples in the last four decades (Table 1.1).

Remote sensing of the Moon by spacecraft complemented the return of samples. Using X-ray fluorescence spectrometers, the Apollo 15 and 16 service modules obtained orbital measurements of aluminum, silicon, and magnesium at a spatial resolution of ~50 km. In 1994, the Clementine orbiter mapped the surface reflectance of the Moon at 11 wavelengths with spatial resolutions of 100–300 m. These multispectral images have been used to infer mineralogy as well as the abundances of iron and titanium. In 1997–1998, the Lunar Prospector orbiter used a gamma-ray spectrometer (GRS) to obtain global maps of the abundances of iron and thorium at 15 km spatial resolution and potassium and titanium at 60 km resolution. Additional GRS data for oxygen, magnesium, aluminum, silicon, calcium, and uranium were obtained at 150 km resolution, and a neutron spectrometer measured hydrogen concentrations in lunar soils. In 2009, the Lunar Reconnaissance Orbiter and its LCROSS impactor characterized the lunar radiation environment and found water ice at one of the poles. In 2011, the GRAIL mission, consisting of two orbiters, refined our understanding of lunar gravity, and LADEE studied the atmosphere and airborne dust in 2013. Besides the United States and Russia, other space-faring nations (the European Union, Japan, China, and India) have sent probes to the Moon that have made remote-sensing measurements.

The first close-up images of the planet Mars were returned by the Mariner 4 spacecraft, which flew by the planet in 1964. Mariners 6 and 7 flew past Mars in 1969, returning more images, and Mariner 8 was the first spacecraft to go into orbit around Mars in 1971. The Viking missions landed two spacecraft on Mars in 1976. The Viking landers took high-resolution images of their landing sites and gathered data on the composition of the atmosphere and surface soils.

In 1979, two unusual meteorites, both igneous rocks, were hypothesized to be Mars samples. This controversial idea was based primarily on their young crystallization ages (several hundred million years), because it was difficult to envision how melting could occur on small asteroids so late in solar system history. The idea found wide acceptance several years later when trapped gases in these meteorites were found to have chemical compositions identical to the martian atmosphere, as measured by the Viking landers (Fig. 1.9). Three decades later, nearly 150 separate martian meteorites have been recognized.

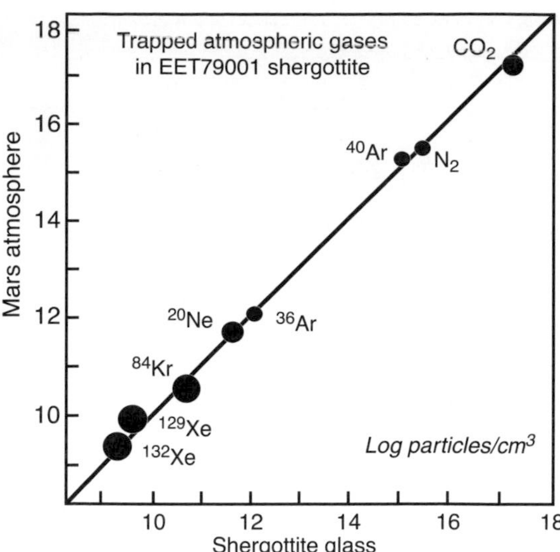

Figure 1.9 Comparison of trapped gases in impact-melt glass in the EET79001 martian meteorite with the composition of the martian atmosphere as measured by the Viking landers. This remarkable agreement is the evidence that convinced most planetary scientists that SNC meteorites came from Mars.

The chemical compositions of these meteorites – shergottites, nakhlites, chassignites, commonly abbreviated to SNCs – have provided a wealth of new insights into geochemical processes on Mars.

More recently, chemical analyses of martian rocks and soils were obtained using α-particle X-ray spectrometers (APXS) on the Mars Pathfinder rover (Sojourner) in 1997, two long-lived Mars Exploration Rovers (Spirit and Opportunity) beginning in 2004, and the Mars Science Laboratory rover (Curiosity) beginning in 2012 (Fig. 1.10). The mineralogy of martian surface materials has been determined by a thermal emission spectrometer (TES) on the Mars Global Surveyor orbiting spacecraft at spatial resolutions of a few tens of kilometers, and higher spatial resolution mineralogical mapping has been provided by a thermal emission mapping spectrometer on Mars Odyssey and visible/near-infrared spectrometers on Mars Express (OMEGA) and the Mars Reconnaissance Orbiter (CRISM). Abundances and global distributions of iron, potassium, thorium, silicon, calcium, and chlorine have also been determined by a GRS on the Mars Odyssey orbiter. A neutron spectrometer determined concentrations of hydrogen in the martian subsurface. In 2005, Phoenix landed near the northern martian pole and probed for ice. The Curiosity rover, carrying a variety of chemical and mineralogical tools, has explored an ancient martian lake bed since 2012. MAVEN and the Trace Gas Orbiter are analyzing

Figure 1.10 A family portrait of Mars rovers. Sojourner (Mars Pathfinder rover) is in the foreground, Spirit (Mars Exploration Rover, MER) is on the back left, and Curiosity (Mars Science Laboratory, MSL) is on the right. Perseverance (Mars 2020 rover, not pictured) is approximately the size of Curiosity. Image from NASA and JPL. *For the colour version, refer to the plate section.*

the martian atmosphere, and the InSight lander studied Mars' interior structure beginning in 2018. The Mars 2020 rover (Perseverance) was launched in 2020 and will explore Jezero crater beginning in 2021. This rover will analyze samples using X-ray fluorescence (PIXL) and ultraviolet native fluorescence and Raman spectra (SHERLOCK), as well as cache samples for possible future return to Earth. The United Arab Emirates and China have also launched spacecraft intending to perform remote-sensing measurements on Mars.

Several Soviet Venera and Vega spacecraft landed on the surface of Venus in the early 1980s and survived for a few minutes before succumbing to the stifling heat and pressure. X-ray fluorescence chemical analyses for a number of major elements in surface samples were reported. Chemical and isotopic analyses of the Venus atmosphere were made by Pioneer Venus, Venera, and Venus Express. In 1990, Magellan provided global radar maps of the surface.

The Mariner 10 spacecraft flew by Mercury and photographed one side of the planet in 1975. Mercury was not visited again until the MESSENGER spacecraft flew past the planet in 2008 and 2009 before achieving orbit in 2011. In addition to high-resolution imagery of the entire planet, MESSENGER acquired chemical and mineralogical information about the planet, studied its core and magnetic field, and investigated the composition of its very thin atmosphere.

Aspects of the chemical composition of the atmospheres of Jupiter, Saturn, Uranus, and Neptune were measured by the Voyager and Galileo spacecraft in the 1980s and 1990s, respectively. The Juno orbiter arrived at Jupiter in 2016 and is now measuring its composition. The atmospheres of Saturn and of Titan, Saturn's largest satellite, were analyzed by the Cassini spacecraft and its Huygens probe (Fig. 1.11). Much of what is known or inferred about the compositions of satellites of the giant plants is based on the spectra of sunlight reflected from their cloud tops.

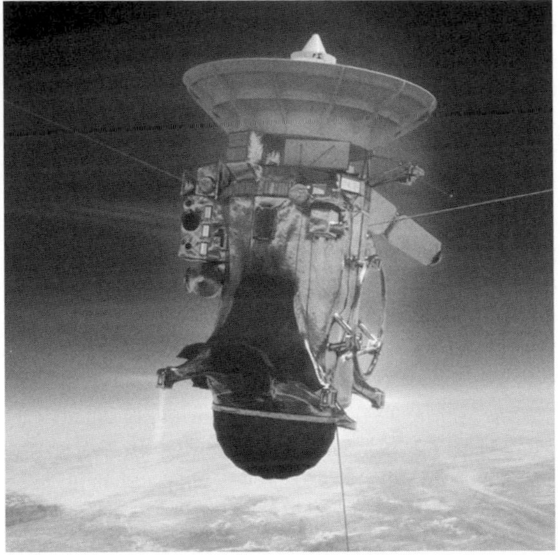

Figure 1.11 The nuclear-powered Cassini spacecraft, launched towards Saturn in 1997, was active for nearly 20 years. The ring on the right side held the Huygens probe, which parachuted onto Titan's surface. NASA image. *For the colour version, refer to the plate section.*

Several spacecraft have obtained *in situ* chemical data for asteroids. The NEAR Shoemaker spacecraft, which orbited asteroid Eros in 2000–2001, made compositional measurements of the surface using X-ray fluorescence (XRF) and gamma-ray spectra. The Japanese Hayabusa spacecraft, which visited asteroid Itokawa in 2003, also carried an XRF spectrometer and measured the surface composition. Both asteroids were found to have chondritic compositions. The Dawn spacecraft orbited asteroid Vesta in 2011, characterizing its mineralogy with visible/near-infrared spectra and its chemistry with a gamma-ray and neutron spectrometer. Vesta, the second most massive body in the asteroid belt, is thought to be the source for howardite, eucrite, and diogenite (HED) meteorites, and that idea was confirmed by Dawn. The spacecraft then traveled to the *dwarf planet* (formerly asteroid) Ceres, where it orbited and analyzed its surface composition and geology from 2015 to 2018 and discovered that it is a partly differentiated icy body.

The Hayabusa mission returned thousands of tiny particles from Itokawa to Earth (Table 1.1), which confirmed its classification as chondritic. Hayabusa2 rendezvoused with asteroid Ryugu in 2018. The spacecraft collected two samples totaling 5.4 g and returned them to Earth in 2020. OSIRIS-REx arrived in orbit around asteroid Bennu in 2018; it collected abundant surface samples in 2020 and will return them to Earth in 2023 (Table 1.1).

Spacecraft missions have visited five comets and have provided some spectacular images of the comet nuclei and data on their chemical and mineralogical compositions. In 1985, NASA's ICE spacecraft passed through the tail of comet Giacobini–Zinner. In 1986, five spacecraft visited comet 1P/Halley (Halley's Comet). The Soviet Vega 1 and Vega 2 probes sent images of the nucleus, and the European Giotto probe made a close pass by the nucleus. All three of these probes had mass spectrometers that returned data on the composition of dust expelled from the comet. Two Japanese probes also visited comet 1P/Halley. Comet Grigg–Skjellerup was imaged by the Giotto spacecraft in 1989. NASA's demonstration spacecraft Deep Space 1 imaged the nucleus of comet Borrelly in 2001.

More ambitious comet missions have flown subsequently. The Deep Impact mission went to comet Temple 1, arriving in 2005. Deep Impact released a large impactor that crashed into the comet, and then studied the material that was released using spectrometers and cameras. Information on the chemical composition and mineralogy of the comet showed that it contains both low-temperature and high-temperature materials. The Stardust mission visited comet Wild 2 in 2004. In addition to taking some spectacular images, the spacecraft brought samples of comet dust back to Earth in 2006 (Table 1.1). Rosetta orbited comet Churyumov–Gerasimenko in 2014, conducting the most detailed comet observations to date. Its Philae lander detected organic compounds on the comet's surface.

In 2016, New Horizons flew by Pluto, now demoted from planetary status to that of dwarf planet, and its moon Charon. New Horizons returned images and compositional analyses. Pluto is the innermost object of the Kuiper belt, which contains thousands of icy worlds that orbit beyond the outermost planet, Neptune. Continuing its traverse outward, the spacecraft flew by and imaged Arrokoth, a small Kuiper belt object, in 2019.

1.3.7 Multiple Sources of Extraterrestrial Materials

For many years, cosmochemistry depended on the chance discovery of meteorites – either witnessed "falls" and serendipitous "finds," or the dogged determination of a few private collectors who systematically searched for them. That changed in 1969, when Japanese explorers in Antarctica led by Masaru Yoshida stumbled onto meteorites exposed on bare ice. American geologist William Cassidy immediately recognized an opportunity, and with support from the National Science Foundation he mounted a joint expedition with the Japanese to the Allan Hills region of Antarctica in 1977 to recover meteorites. This was the first of many expeditions, sponsored by the National Science Foundation and headed first by Cassidy and later by Ralph Harvey, that have returned to Antarctica every year to collect meteorites (Fig. 1.12). The Japanese have operated a parallel field program in the Yamato Mountains region of Antarctica. Other countries, including the countries that comprise the European Space Agency, China, and Russia have also sent meteorite-collecting expeditions to Antarctica. Altogether, these

Figure 1.12 A frozen meteorite exposed on Antarctic ice. The numbers at the bottom of the identifying counter at the right are a centimeter scale. NASA image. *For the colour version, refer to the plate section.*

expeditions have recovered well over 55,000 meteorites (including paired specimens), which have dramatically increased the number of meteorites available to science. Among the meteorites returned by these expeditions were the first lunar meteorites and several other previously unknown meteorite types.

Meteorites fall onto the Antarctic continent and become frozen into the ice. Glaciers, carrying this treasure trove of meteorites, constantly creep downslope towards the edges of the continent. Wherever a moving glacier is obstructed by a mountain, its flow stagnates and layers of ice are removed by wind ablation, exposing meteorites in great concentrations. American expeditions collect the meteorites and send them to NASA's Johnson Space Center in Houston, where they are cut into pieces and allocated to scientists all over the world. Japanese expeditions return their samples to the National Institute for Polar Research in Tokyo for curation and allocation.

About the same time that meteorites were found in Antarctica, an important collection of meteorites was being put together in Roosevelt County, New Mexico. From 1966 to 1972, several meteorite hunters collected ~140 meteorite specimens, representing about 100 separate fall events. This collection demonstrated another way in which nature concentrates meteorites. The meteorites in Roosevelt County were found in "blowout" areas, where up to a meter of soil had been blown away by the wind, leaving meteorites in plain view on the hardpan surface. Based on this experience, systematic and successful searches of desert areas in Western Australia and the southwestern United States have been carried out. Subsequently, the deserts of North Africa have turned out to be especially prolific sources of meteorites. The shifting desert sands expose meteorites that have accumulated over thousands of years. The meteorites are collected by nomads and sold to Western collectors. Although most desert meteorites are weathered to some degree, new and rare meteorite classes have been discovered this way.

In 1974, astronomer Donald Brownlee began collecting IDPs in the stratosphere using collectors on the leading edges of airfoils on high-altitude aircraft. The smallest or fluffiest of these tiny particles (e.g. Fig. 1.13), those with the highest surface-to-volume ratio, are decelerated in the upper atmosphere without being substantially heated. The drifting particles can be either captured in the stratosphere or recovered from low-dust environments on the Earth's surface for laboratory studies. Similar particles and slightly larger *micrometeorites* have also been collected frozen in polar ice. IDPs appear to be materials from asteroids and comets.

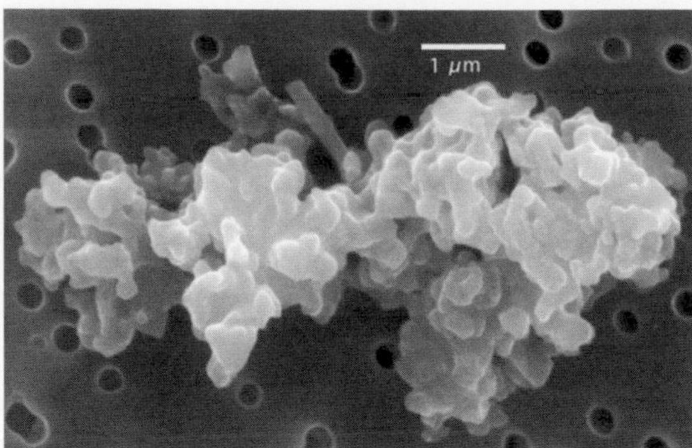

Figure 1.13 Scanning electron micrograph of an interplanetary dust particle. Image courtesy of D. Brownlee.

The Genesis spacecraft collected solar wind particles during two years in orbit. The solar collectors were returned to Earth in 2005, but unfortunately a parachute failed to open and the sample container ruptured during its hard landing. Although the collector devices were damaged by the impact and there was significant contamination, valuable data on the Sun's composition have been, and continue to be, obtained from the damaged collectors.

In 2006, the Stardust spacecraft returned samples of comet Wild 2 to Earth. These samples are tiny particles expelled from the comet nucleus and captured in aerogel (a highly porous medium that can snag speeding cometary grains) when the spacecraft flew through the coma at high speed. Even though the deceleration caused the particles to disaggregate and portions of them to melt, some mineral grains were preserved intact, and elemental and isotopic signatures have been measured along the tracks in the aerogel and in particles at the ends of the tracks.

1.3.8 Organic Matter and Extraterrestrial Life?

Some meteorites, known as carbonaceous chondrites, contain several percent of carbon-bearing compounds, primarily organic molecules. The earliest characterizations of organic matter in chondritic meteorites were attempted in the late 1800s. Researchers speculated that the organic material was probably extraterrestrial in origin. However, interest soon waned and, curiously, not a single paper appeared on meteoritic organic matter between 1899 and 1953. However, interest in extraterrestrial organic compounds intensified in the 1960s, when Bartholomew Nagy and his colleagues published a series of papers claiming that the Orgueil carbonaceous chondrite contained biogenic hydrocarbons and "organized elements" that resembled fossilized algae. However, it was soon shown that the organic material came from terrestrial contamination, a problem that continues to plague organic cosmochemistry. The large amounts of meteoritic material available from the falls of the Murchison chondrite in 1969 and the Tagish Lake chondrite in 2000 generated considerable activity in determining both the structures and relative abundances of organic compounds and their stable-isotope compositions. Subsequent work has provided important new constraints on the processes that formed organic compounds in the ISM, the solar nebula, and on meteorite parent bodies.

Since ancient times, philosophers have wondered whether life on Earth was seeded from elsewhere in the universe, an idea known as "panspermia." It is generally agreed that meteorites and comets contain prebiotic compounds that could have provided the raw material for life on Earth. A more contentious question is whether the comets might actually contain microscopic organisms, as proposed by Hoyle and Wickramasinghe (1981). This hypothesis has not been taken seriously by most scientists, although it is hard to disprove.

In 1996, David McKay et al. published a paper that claimed to show evidence for possible extraterrestrial life in the martian meteorite ALH 84001 (Fig. 1.14). One of the lines of evidence involved organic compounds found in this meteorite. The paper generated tremendous interest and controversy and resulted in the creation of a new NASA research initiative to study astrobiology. Intense study has provided nonbiological explanations for all the original claims. Nonetheless, the intense astrobiology research effort generated by the 1996 study is still going strong.

Figure 1.14 A piece of the ALH 84001 martian meteorite, ~8 cm across. Dark material on the surface is fusion crust, formed during passage through the Earth's atmosphere. This sample created a stir when it was proposed to contain evidence for extraterrestrial life. Image courtesy of the Smithsonian Institution. *For the colour version, refer to the plate section.*

1.4 The Tools of Cosmochemistry

In this section, we briefly review the tools used by cosmochemists and the kinds of datasets that have been acquired over the years. The tools identified below are described in more detail in the Appendix at the end of this book. Drawing on the data gained from these tools and on the intellectual framework of cosmochemistry, we can formulate questions and define research protocols to answer key questions.

1.4.1 Laboratory and Spacecraft Analyses

1.4.1.1 *Mineral Identification and Petrography*

The most basic information about any object is obtained by examining it with our own senses. What does it look like? How heavy is it? Is it solid or crumbly? A powerful extension of these senses is the petrographic microscope, which is used to examine slices of rocks and meteorites that are thin enough to transmit light. A skilled petrographer can identify minerals based on their interactions with polarized light. Meteorite classification is heavily based on the mineralogical and textural characteristics of the samples. Petrography has also played a major role in inferring the geologic history of the Moon from lunar samples and of Mars from martian meteorites. Electron-beam instruments (electron microprobe, scanning electron microscope, transmission electron microscope) extend the petrographer's senses by providing information on mineralogy and crystal structure at a much higher spatial resolution.

Minerals can also be identified by remote sensing, using spectroscopy. Reflectance spectra have absorption bands in the visible/near-infrared region caused by electronic transitions, commonly of iron in certain crystal sites. Thermal emission spectra are produced when solar energy is absorbed by minerals and re-emitted at thermal infrared wavelengths. Both types of spectra can be diagnostic for specific minerals, and spectroscopic measurements are made using telescopes and instruments on orbiting or landed spacecraft. A recent flight instrument is the ChemCam instrument on the Curiosity rover on Mars. A laser beam vaporizes grains in a target rock, and spectrographs identify elements in the plasma. Specific minerals can sometimes be identified if the analyzed spot is only one phase. A Raman spectrometer is included on the Perseverance Mars rover. Raman and luminescence spectroscopes are being developed for future Mars rover missions.

X-ray diffraction (XRD), a mainstay in mineralogical identification on Earth, has been applied on Mars by the CheMin instrument on the Curiosity rover. This tool has provided the most definitive identification of minerals that are present in low abundance.

1.4.1.2 *Mineral Chemistry*

Prior to the 1960s, mineral compositions were determined either by wet chemical analyses of mineral separates or by reference to an extensive database of the optical properties of minerals as functions of their compositions. With the development of the electron microprobe, which uses electrons to excite characteristic X-rays, it became possible to measure directly the chemical composition of each mineral in a thin section. The energy-dispersive X-ray detector available on most scanning electron microscopes can provide *in situ* chemical analyses of minerals. Certain mineral compositions can also be determined from spectroscopy.

1.4.1.3 *Bulk-Rock Chemistry (Major and Minor Elements)*

During the nineteenth century and the first half of the twentieth century, bulk chemical compositions of meteorites were determined by wet chemistry techniques. The most reliable bulk chemical datasets were compiled by Hugo Wiik and Eugene Jarosewich. During the second half of the twentieth century, new methods were established to determine chemical compositions. XRF analysis was used extensively during the 1950s and 1960s. In this technique, a beam of X-rays excites electrons in the sample, which then lose this energy by re-emitting X-rays that are unique to each element. XRF has also been used to analyze a few elements on the lunar surface from orbit, and was used by the Viking landers to carry out more complete analyses of soils on Mars in 1975.

During the 1960s, this technique was largely supplanted by neutron activation analysis. In this technique, small samples are placed into a nuclear reactor, and neutrons from the reactor interact with the elements in the sample to produce radioactive isotopes. The sample is then removed from the reactor and the radioactive isotopes decay, emitting characteristic gamma rays, which are then counted to determine the chemical composition of the sample. Neutron activation analysis can detect 74 elements, but unfortunately silicon, a dominant element in most cosmochemical samples, cannot be directly measured. Today, datasets based on this technique provide a cornerstone for ideas about meteorites and lunar samples.

Gamma-ray and neutron spectrometers on orbiting spacecraft have been used to determine elemental abundances in surface samples from the Moon, Mars, and several asteroids. Gamma rays are produced by nuclear transitions, such as those that occur during interactions of surface materials with solar cosmic rays. The energy of gamma rays identifies specific nuclei, which can be translated into abundances for a handful of elements. Accurate major and minor element abundances in Mars rocks and soils have also been determined by APXS on four rovers. Alpha particles (^{4}He nuclei) produced by radioactive decay bombard a sample, producing X-rays that are characteristic of specific elements. A microfocusing XRF spectrometer (PIXL) is part of the instrument package on the most recent Mars rover, Perseverance.

1.4.1.4 *Trace-Element Abundances*

The distributions of trace elements between minerals and within a suite of related rocks provide powerful tools for constraining the origin and history of rocks and meteorites. Trace-element abundances for rocks are typically part of the dataset collected when determining bulk compositions. Trace-element compositions of minerals require more powerful techniques such as the ion microprobe (secondary ion mass spectrometer, or SIMS) or laser-ablation inductively coupled plasma mass spectrometry (LA-ICPMS).

Potassium and thorium, which normally occur in minor or trace abundances, have been determined in surface materials on the Moon, Mars, and an asteroid by gamma-ray spectroscopy from orbiting spacecraft.

1.4.1.5 *Radiogenic Isotopes and Chronology*

Radiometric dating utilizes the constant decay rates of radioactive (unstable) isotopes to determine how much time has passed since an object formed. Long-lived radionuclides (those for which some of the primordial abundance is still present in the solar system) provide absolute ages for objects relative to today. Short-lived, now-extinct radionuclides provide high-resolution relative ages for objects that formed in the earliest epochs of solar system history. Mass spectrometry is the main tool for determining ages by radiometric dating.

Radiometric dating is not the only method for determining age relationships. Relative ages can be determined by examining the textural relationships between various objects. For example, an object enclosed by another object must have formed earlier. Some complex relative chronologies can be developed based on textural observations. A melt that contains a mineral grain unrelated to the melt must have formed after the mineral grain that it contains. On a larger scale, a rock unit that stratigraphically overlies or crosscuts another rock formation must have formed after the underlying formation. The relative age of a planetary surface can also be estimated from the density of impact craters on that surface. Cratering rates can be calibrated by radiometric dating of rocks from that surface or by comparison with other surfaces that have been dated. Using all these techniques, detailed chronologies can and have been developed for solar system materials.

Cosmic-ray exposure dating is used to determine how long a meteorite has traveled through space as a small object on the way to Earth from its parent asteroid, or how long a material was exposed to cosmic rays on the surface of a planet (or the Moon) during its history. Exposure to high-energy cosmic rays alters the isotopic chemistry of the target material, and measurement of the products of these interactions can be used to determine cosmic-ray exposure ages.

1.4.1.6 *Stable-Isotope Compositions*

Chemical and physical processes affect the stable isotopes of an element differently because of slight variations in the energy levels of the electrons caused by different numbers of neutrons in the nucleus of each isotope. These slight differences result in mass-dependent fractionations (changes in relative abundances of the isotopes) that are characteristic of these processes. For example, during precipitation of a mineral from a fluid, the isotopic composition of the mineral will differ slightly from that of the fluid. The magnitude of the difference is typically temperature dependent, so the compositions of the liquid and the mineral can be used to define the temperature at which the mineral formed. This is an example of mass fractionation during an equilibrium process. As another example, if a melt evaporates into empty space, the isotopic composition of the melt becomes heavier as lighter isotopes are preferentially lost. The isotopic composition of the residual melt will be proportional to the amount of evaporation. This is an example of kinetic isotope fractionation. As with radiometric dating, the primary tool for this type of analysis is the mass spectrometer.

The Curiosity rover carried the Sample Analysis at Mars (SAM) instrument – actually a suite of instruments: mass spectrometer, gas chromatograph, and tunable laser spectrometer – that measured the abundance of light elements and their stable-isotope compositions.

1.4.1.7 *Organic Compounds*

Organic molecules are identified by a variety of techniques, after they have been extracted from chondrites using liquid solvents or stepwise heating. Mass spectrometers can separate many compounds and identify them by their masses. The SAM instrument on the Curiosity rover has identified organic molecules on Mars. Other techniques can identify reactive functional groups or specific structural types of molecules. A visible/near-infrared spectrometer on the Dawn spacecraft used this approach to identify organic molecules on Ceres.

1.4.2 Cosmochemical Theory

A powerful theoretical tool for cosmochemical models is *thermodynamics*. This formalism considers a system in a state of *equilibrium*, one consequence of which is that observable properties of the system undergo no net change with time. (We offer a somewhat more rigorous discussion of thermodynamics in Chapter 7.) The tools of thermodynamics are not useful for asking questions about a system's evolutionary history. However, with the appropriate equations, we are able to estimate what a system at equilibrium would look like under any environmental conditions. The methods of thermodynamics allow for the use of temperature and pressure, plus the system's bulk composition, to predict which minerals will be stable and in what relative amounts they will be present. In this way, the thermodynamic approach to a cosmochemical system can help us measure its stability and predict the direction in which it will change if environmental parameters change.

Cosmochemistry literature often contains discussions of equilibrium condensation from a gas of solar composition. During the 1960s through the 1980s, the intellectual framework for these discussions was a hot, monotonically cooling solar nebula (recall the Laplace model mentioned earlier). However, the discovery in chondrites of isotopic anomalies that could not have survived in the gaseous state and grains that existed in interstellar space prior to the formation of the solar system showed that this simple model for the early solar system was not strictly correct. Cosmochemists still refer to condensates, and the previous work still provides important constraints for understanding solar system bodies. However, the application of these principles is generally in reference to a local environment or specific situation rather than a global environment like a hot solar nebula. We are also now more aware that thermodynamics reveals nothing about the pathways to the final states that we observe, and that similar compositions can result from condensation or evaporation.

In the planetary geochemical literature, there also are extensive discussions about the thermodynamics of crystallizing melts. Sophisticated programs have been developed to model the compositions of the minerals that crystallize from a cooling liquid, as well as changes in the residual melt.

A third way in which thermodynamics can be used is to invert the measured major, minor, and trace-element compositions of coexisting minerals to estimate the temperature, pressure, and oxidation state under which they formed. Valid results can only be obtained if the system was truly in equilibrium at the inferred conditions and the minerals did not subsequently change composition. Nevertheless, this is a powerful tool to infer the formation conditions of various solar system materials.

An alternative way to study a system is to examine its *kinetics*. Using kinetics, we can study the pathways along which a system may evolve between states of thermodynamic equilibrium and determine the rates of change of system properties along those pathways. Cosmochemical systems can evolve along a variety of pathways, some of which are more efficient than others. In a kinetic study, the task is often to determine which of the competing pathways is dominant.

Cosmochemists are interested in determining not only what should have happened in a real system (the thermodynamic answer) but also how it is most likely to have happened (the kinetic answer). The two are intimately linked. While the focus of thermodynamics is on the end states of a system – before and after a change has taken place – a kinetic treatment sheds light on what happens between the end states. Where possible, it is advantageous to keep both approaches in mind.

Theories of the origin of the solar system and of the formation and evolution of the bodies within it provide the intellectual framework for detailed investigations of processes and events in the early solar system. These theories are often based primarily on work in fields other than cosmochemistry as we have defined it. For example, observations made by astronomers tell us about the environments in which stars form. Astrophysics modelers describe the gravitational collapse of dense cores in molecular clouds to form protostars and accretion disks. Dynamicists model the evolution of planetary systems starting with the solar nebula and continuing through the formation of the planets and the evolution of the distribution of asteroids and icy bodies as we see them today. Input from these and other disciplines generates global models that provide the often-unstated framework for cosmochemistry. Cosmochemists must be familiar with these broad ideas, but must also test them against observations from their own research. Observations are the ultimate arbiter of a successful theory.

1.5 Relationship of Cosmochemistry to Other Disciplines

It should be clear from the above discussions that cosmochemistry is a highly interdisciplinary subject that draws on aspects of chemistry, physics, geology, astronomy, astrophysics, and perhaps eventually even biology. Understanding the formation of the elements requires consideration of the astrophysical settings in which nucleosynthesis occurs – stars, supernovae, and the ISM. Astronomy provides direct observations of these settings, as well as spectroscopic data on chemical abundances that complement chemical analyses of presolar grains in meteorites. Astrophysics provides the theoretical predictions of relative abundances of nuclides produced in various stellar settings that can be tested by the chemical and isotopic compositions of presolar grains. Hypotheses about the formation of organic molecules in interstellar space can be evaluated by studying organic matter in meteorites. Spectroscopic elemental abundances measured in the Sun can be compared with element ratios in bulk chondritic meteorites, and the agreement allows measured chondritic abundances to be substituted for those elements without measurable solar spectral lines. Astrophysical models of changes in the composition of the Sun over time can be tested by analyzing solar wind gases implanted into soils on the Moon and asteroids. Nowhere in science is there a better example of synergy between disparate disciplines.

Likewise, planetary geochemistry cannot be interpreted without understanding its geologic context. Geologic processes on Mars, the Moon, and other planets can be inferred from spacecraft imagery, and the conditions under which they occur can be determined by *in situ* measurements by orbiters or landers. Element behavior during melting and crystallization can be inferred by analyzing martian and lunar meteorites and returned lunar samples. The timing of major geologic events and the ages of mantle reservoirs can be defined by radiogenic isotope measurements. Geochemistry and geology provide complementary ways of assessing how planets work.

Small bodies such as asteroids and comets contain primordial material but may also have experienced heating and impacts. Metamorphism (mineralogical and textural changes wrought by heat and shock) can blur or even destroy the cosmochemical and chronological records in meteorites, and it is essential to catalog these geologic processes so that we can interpret the records correctly. Melting and sequestering of siderophile elements in some asteroid cores produced iron meteorites, analyses of which shed light on an inaccessible region of our own planet. Astronomical observations and spectroscopy of asteroids complement direct measurements of their chemical compositions by analyses of meteorites.

In this book, we will explore cosmochemistry and planetary geochemistry, usually in the context of astronomy, geology, and physics. In this way we hope to illuminate how the abundances and chemical behaviors of elements and isotopes are critical to understanding the formation and inner workings of the Earth's neighboring planets, the processes that shaped the early solar system, the chronology of major events, and the origins of the elemental building blocks of the universe.

Questions

1. How does cosmochemistry relate to other, more well-known disciplines such as geology, chemistry, and astronomy?
2. Distinguish between the following terms: lithophile, siderophile, chalcophile, atmophile.
3. How do element behaviors in the solar nebula or in interstellar space differ from those in planets?
4. Who was Victor Goldschmidt and what role did he play in the development of geochemistry and cosmochemistry?
5. What are some of the sample sets that are available to cosmochemists?
6. List five techniques by which elements and/or isotopes can be measured.

Suggestions for Further Reading

Davis A. M., editor (2014) *Treatise on Geochemistry, 2nd Edition, Vol. 1: Meteorites and Cosmochemical Processes*, and *Vol. 2: Planets, Asteroids, Comets and the Solar System*, 454 pp. and 415 pp., respectively, Elsevier, Oxford. Comprehensive review chapters of cosmochemistry and planetary geochemistry topics by authorities in the field; these volumes cover many of the subjects of the present book, but generally at a more advanced level.

Gregory T. (2020) *Meteorite: How Stones from Outer Space Made Our World*, 299 pp., Hachette Book Group, New York. A popular account of meteorites, with especially well-researched historical accounts of meteorite falls and finds.

Lauretta D. S., and McSween H. Y., editors (2006) *Meteorites and the Early Solar System II*, 943 pp., University of Arizona Press, Tucson. Another comprehensive, modern collection of reviews by prominent meteoriticists. Everything you want to know about meteorites is here, but the chapters are at an advanced level.

Papike J. J., editor (1998) *Planetary Materials, Reviews in Mineralogy*, **36**, Mineralogical Society of America, Washington. Chapters on IDPs; chondritic and nonchondritic meteorites; lunar samples; and martian meteorites provide exhaustive references on these materials.

Other References

Aston F. W. (1920) Isotopes and atomic weights. *Nature*, **105**, 617–619.

Black D. C., and Pepin R. O. (1969) Trapped neon in meteorites II. *Earth & Planetary Science Letters*, **6**, 395–405.

Burbidge E. M., Burbidge G. R., Fowler W. A., and Hoyle F. (1957) Synthesis of elements in stars. *Reviews of Modern Physics*, **29**, 547–650.

Cameron A. G. W. (1957) *Stellar Evolution, Nuclear Astrophysics, and Nucleogenesis* (CRL-41; AECL-454). Atomic Energy of Canada, Ltd., Chalk River, Ontario.

Hahn O., Strassman F., Mattauch J., and Ewald H. (1943) Geologische Altersbestimmungen mit der Strontiummethode. *Chemische Zeitung*, **67**, 55–56.

Holmes A. (1946) An estimate of the age of the Earth. *Nature*, **157**, 680–684.

Houtermans F. G. (1946) Die Isotopenhäufigkeiten im Natürlichen Blei und das Alter den Urans. *Naturwissenschaften*, **33**, 185–187.

Hoyle F., and Wickramasinghe C. (1981) Where microbes boldly went. *New Scientist*, **91**, 412–415.

Lauretta D. S., and Killgore M. (2005) *A Color Atlas of Meteorites in Thin Section*. Golden Retriever Press, Tucson, 301 pp.

Lee T., Papanastassiou D. A., and Wasserburg G. J. (1977) Aluminum-26 in the early solar system: Fossil or fuel? *Astrophysical Journal Letters*, **211**, L107–L110.

Lockyer J. N. (1890) *The Meteoritic Hypothesis*. Macmillan & Co., New York, 560 pp.

Lodders K., and Fegley B. Jr. (1998). *The Planetary Scientist's Companion*. Oxford University Press, New York, 371 pp.

McKay D. S., Gibson E. K., Thomas-Keprta K. L., et al. (1996) Search for past life on Mars: Possible relic biogenic activity in Martian meteorite ALH 84001. *Science*, **273**, 924–930.

Patterson C. C. (1956) Age of meteorites and the Earth. *Geochimica et Cosmochimica Acta*, **10**, 230.

Railsback L. B. (2003) An earth scientist's periodic table of the elements and their ions. *Geology*, **31**, 737–740. An updated version (available in six languages) was released in 2012 and can be found online: railsback.org/PT.html.

Reynolds J. H. (1960) Determination of the age of the elements. *Physical Reviews Letters*, **4**, 8–10.

Reynolds J. H., and Turner G. (1964) Rare gases in the chondrite Renazzo. *Journal of Geophysical Research*, **49**, 3263–3281.

Suess H. E., and Urey H. C. (1956) Abundances of the elements. *Reviews of Modern Physics*, **28**, 53–74.

Van Schmus W. R., and Wood J. A. (1967) A chemical-petrologic classification for the chondritic meteorites. *Geochimica et Cosmochimica Acta*, **31**, 747–765.

Wasserburg G. J., and Hayden R. J. (1955) Age of meteorites by the ^{40}Ar–^{40}K method. *Physics Review*, **97**, 86–87.

2 Nuclides and Elements

The Building Blocks of Matter

Overview

In order to understand the origin of the elements, their relative abundances, their interactions, and the processes that produced the objects we observe today in the solar system and in the universe as a whole, it is necessary to have a basic knowledge of the building blocks and structure of matter. Cosmochemists operate at the boundaries between traditional fields of physics, chemistry, geology, and astronomy. In this chapter, we will review some basic nuclear physics to provide a background for understanding the origin and abundances of the elements. Fundamentals of radioactive decay and isotope cosmochemistry will be presented. We will also review some basic chemistry to provide a background for discussions of chemical processes in later chapters.

2.1 Elementary Particles, Isotopes, and Elements

At the simplest level, an *atom* consists of a nucleus composed of protons and neutrons surrounded by a cloud of electrons. A neutral atom consists of an equal number of *protons* (positively charged) and *electrons* (negatively charged) and a similar number of *neutrons*. Protons and neutrons have nearly equivalent masses, while electrons are much smaller, 1/1836 times the mass of a proton (Table 2.1). An *ion* is an atom that has either lost or gained one or more electrons, giving it a positive or negative charge. The chemical properties of an element are governed by the number of electrons it has and the ease with which they are lost or gained. The atomic weights and abundances of the elements are governed by the composition and physics of the nucleus.

Each element consists of between one and 10 stable *isotopes* and many unstable (*radioactive*) ones, most with very short *half-lives* (the time it takes for half of the atoms present to decay). Isotopes of an element have the same number of protons, which define the element, but differ in the number of neutrons in the nucleus. For example, oxygen has eight protons and either eight, nine, or 10 neutrons. The isotopes of oxygen are ^{16}O, ^{17}O, and ^{18}O; the numbers refer to the total number of protons + neutrons in the nucleus. The *atomic number*, Z, which is unique to each element, is the number of protons in the nucleus (equal to the number of electrons in the neutral atom). The *mass number*, A, is the number of protons plus the number of neutrons, N, in the nucleus ($A = Z + N$). Isotopes can either be stable or radioactive, and the isotopes resulting from radioactive decay are said to be *radiogenic*. Natural samples contain 266 stable isotopes comprising 81 elements, along with 65 radioactive isotopes of these and nine additional elements. (When discussing the isotopes of more than one element, the term *nuclide* is typically used.) In addition, more than 1650 nuclides and 22 elements not found on Earth have been created in the laboratory with nuclear reactors and particle accelerators. Most of the man-made isotopes are very short-lived. Most probably exist briefly in special natural settings, such as in an exploding supernova, but we typically have no way of detecting them. Radioactive technetium, which does not exist naturally in the solar system, has been identified in the spectra of highly evolved low-mass stars.

The notation used for isotopes, ions, and chemical compounds is illustrated by this example: $^{16}O_2^0$. The number to the upper left of the element symbol is the mass number (A). The number to the upper right of the element symbol is its charge, with 0 referring to a neutral atom, and +, 2+, 3+, −, 2−, and 3− referring to its degree and sign of ionization. The number to the lower right is the number of atoms present in one molecule of this element. When reading the symbol for an isotope, one should always say the element name first, then the isotope

Table 2.1 Constituents of an atom and other particles

Particle	Symbol	Location in Atom	Mass	Charge
Proton	p	Nucleus	938.256 MeV = 1.673×10^{-24} g = 1.00728 u	+1
Neutron	n	Nucleus	939.550 MeV = 1.675×10^{-24} g = 1.00866 u	0
Electron	β^-	Cloud around nucleus	0.511 MeV = 9.110×10^{-28} g	−1
Gamma ray	γ	Emitted in nuclear reaction	0	0
Neutrino	η	Emitted in nuclear reaction	Very small	0

number. For example, say "oxygen-16," not "16-O." A concern when this convention was adopted was that people might say the isotope number first, rather than the element. As you will see as you gain experience in cosmochemistry, this concern was well founded.

The masses of the isotopes are described in cosmochemistry using atomic mass units, or *amu*. The official unit of measure, the *unified atomic mass unit* (abbreviated *u*) or *Dalton* (abbreviated *Da*), is defined as 1/12 of the mass of one atom of ^{12}C, and thus is approximately equal to the mass of a proton or a neutron. Thus, the mass of ^{16}O, which consists of eight protons and eight neutrons, is approximated as 16 u, and that of ^{18}O, which has two additional neutrons, is approximated as 18 u. The exact masses are slightly less than the sum of the masses of the individual protons and neutrons. This is because the potential energy of a bound stable nucleus is lower than that of its constituent parts. Albert Einstein showed that mass (m) and energy (E) are related by the famous equation, $E = mc^2$, where c is the velocity of light. When a nucleus forms, the potential energy of the dispersed components is converted to heat, which can be lost from the system as the system cools. As a result, the cooled system has a lower mass than the constituent protons and neutrons did before the nucleus was bound together. Thus the atomic mass of the ^{16}O atom is 15.9949 u rather than 16.12752 u, which is the sum of eight protons and eight neutrons. This *mass defect* is the *nuclear binding energy.*

Figure 2.1 shows the nuclear binding energy per nucleon (the protons and neutrons in the nucleus) as a function of mass number (A) for the stable isotopes. Notice that the binding energy per nucleon increases rapidly from H to D (^{2}H) to ^{3}He to ^{4}He. This is because the attractive force between the nucleus and each additional proton or neutron exceeds the additional electrostatic repulsion between the positively charged protons. This trend generally continues through ^{23}Na, although some nuclear configurations between mass 4 and mass 12 are not energetically favored. For elements between ^{24}Mg and Xe (masses 124–136), the binding energy per nucleon is approximately constant, although it continues to increase slowly to ^{62}Ni, the most tightly bound nucleus. This is because the nucleus has become large enough that the nuclear force associated with a proton or neutron at one edge of the nucleus no longer extends completely across its width. For all nuclei with masses $\leq^{62}$Ni, the addition of a proton or neutron releases energy, the energy of fusion. For example, combining four protons to form a ^{4}He nucleus releases about 7 MeV of energy per nucleon, or a total of 28 MeV.

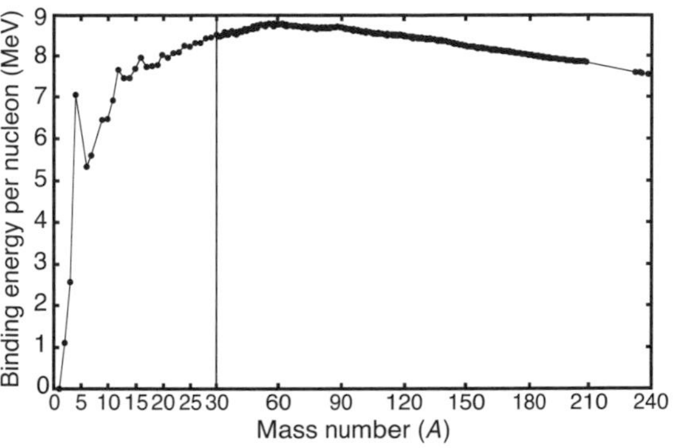

Figure 2.1 Binding energy per nucleon (total nuclear binding energy divided by the number of protons and neutrons in the nucleus) plotted against atomic mass number, A. The highest binding energy per nucleon (= highest stability) is found for elements near iron (A = ~56). Nuclear binding energy can be released by combining light nuclei into heavier ones (fusion) or by splitting heavier nuclei into lighter ones (fission). Note the local peaks in binding energy among the light nuclei. The region of lower binding energy per nucleon between mass numbers 4 and 12 helps to explain the low abundance of lithium, beryllium, and boron.

For isotopes with masses $>^{62}$Ni, the *total* binding energy of the nucleus continues to increase, but the binding energy *per nucleon* decreases slowly with the addition of a proton or neutron until Xe. Thereafter, adding a proton or neutron significantly reduces the binding energy per nucleon. The heaviest stable nuclide is ^{209}Bi. All heavier nuclides are unstable and decay by emission of an *α-particle* (a ^{4}He nucleus consisting of two neutrons and two protons) or by fission into smaller nuclides, releasing energy as the system achieves a more stable configuration. For example, the splitting of a ^{238}U nucleus into two ^{119}Pd nuclei releases a total energy of 240 MeV. In some cases, the decay occurs almost instantaneously after the nuclide forms, while in others, such as ^{232}Th, ^{235}U, and ^{238}U, the nuclei are almost stable and their half-lives against decay are measured in hundreds of millions to tens of billions of years. The energy released when a heavy isotope fissions is the energy that powers nuclear power stations. The awesome destructive power of an atomic bomb comes from the instantaneous release of the binding energy of uranium or plutonium through a neutron-induced fission chain reaction. The even-more-powerful hydrogen bomb uses the energy of fusion.

The *atomic mass of an isotope* (*A*) should not be confused with the *atomic weight of an element*, which is the abundance-weighted average value of the various stable-isotope masses of an element. For example, oxygen, which is composed of 99.757% ^{16}O (15.9949 *u*), 0.038% ^{17}O (16.9991 *u*), and 0.205% ^{18}O (17.9992 *u*), has an atomic weight of 15.9994, and chlorine, which is composed of 75.78% ^{35}Cl (34.9689 *u*) and 24.22% ^{37}Cl (36.9659 *u*), has an atomic weight of 35.453. The atomic weights of most elements on Earth are constant to a very high precision in natural materials because the relative abundances of the isotopes are always the same. Exceptions to this rule are elements with radioactive or radiogenic isotopes, whose isotopic abundances change with time, and very small mass-dependent isotope fractionations caused by physical and biological processes. The latter are the basis for stable-isotope geochemistry. Why should the mixture of isotopes always be the same for each of the elements? The answer to this question has profound implications for how one views the origin of the solar system. It turns out that samples from meteorites do occasionally show measurable variations in isotopic compositions of constituent elements. We will discuss how isotopic compositions provide information about solar system formation in later chapters. There is one caveat to this general rule about elements on Earth, however. Samples of elements that have been processed by humans can have highly anomalous isotopic compositions and thus unusual atomic weights.

2.2 Chart of the Nuclides: Organizing Elements by Their Nuclear Properties

The Chart of the Nuclides ©™ provides a very useful way to organize the large number of different nuclides. The Chart, a publication of Bechtel Marine Propulsion Corporation, is available through their website at www.nuclidechart.com. We recommend the 17th edition *Chart of the Nuclides* booklet as a supplement to this textbook. It contains a thorough but very readable discussion of nuclear structure and radioactive decay. Analogous compilations are also available from other sources. A schematic version of the Chart of the Nuclides is shown in Figure 2.2. The Chart plots the atomic number, *Z*, against the number of neutrons, *N*. For each nuclide, the Chart gives the symbol and mass number, the isotopic abundance (or the half-life if it is radioactive), the atomic mass, and a number of other nuclear properties, some of which will be discussed below. Naturally occurring radioactive isotopes are identified, as are members of natural radioactive decay chains. Especially stable nuclides, such as those with closed neutron or proton shells – the so-called *magic numbers* – are identified.

The Chart of the Nuclides reveals several important features of matter. The stable nuclides lie along a trend that starts at close to 45 degrees, corresponding to approximately equal numbers of protons and neutrons, but becomes shallower as *N* and *Z* increase. This reflects the extra neutrons required to stabilize the nucleus against the electrostatic repulsion between the positively charged protons. When the nuclides become too large, the forces holding them together can no longer overcome the electrostatic forces trying to drive them apart, and they are unstable. On either side of the array of stable isotopes are many unstable isotopes that have either too many protons (those on the left) or too many neutrons (those on the right) (Fig. 2.2). These nuclides must eliminate their excess protons and neutrons. They eliminate neutrons by emitting an electron from the nucleus (β^- decay), and they eliminate excess protons by either emitting a positron (β^+ decay) or by a proton capturing an electron to become a neutron (electron capture). The nuclides move toward the so-called *Valley of β Stability*, where the nuclides are stable against β^- decay, β^+ decay, and electron capture.

Look closely at the distribution of stable elements in Figure 2.2. For some values of *Z* there is only one stable isotope, whereas for others there are many isotopes. In fact, there are 20 stable elements that have only one isotope, such as ^{9}Be, ^{19}F, ^{23}Na, ^{27}Al, ^{31}P, ^{45}Sc, ^{51}V, and ^{55}Mn. For all of these, *Z* is an odd number. A detailed view of a portion of the Chart of the Nuclides (Fig. 2.3) shows that for even-*Z* elements with several isotopes, the even-numbered isotopes are more abundant than the odd-numbered isotopes. In fact, many of the odd-numbered isotopes are unstable. For odd-*Z* elements with multiple isotopes, the even-numbered isotopes are typically

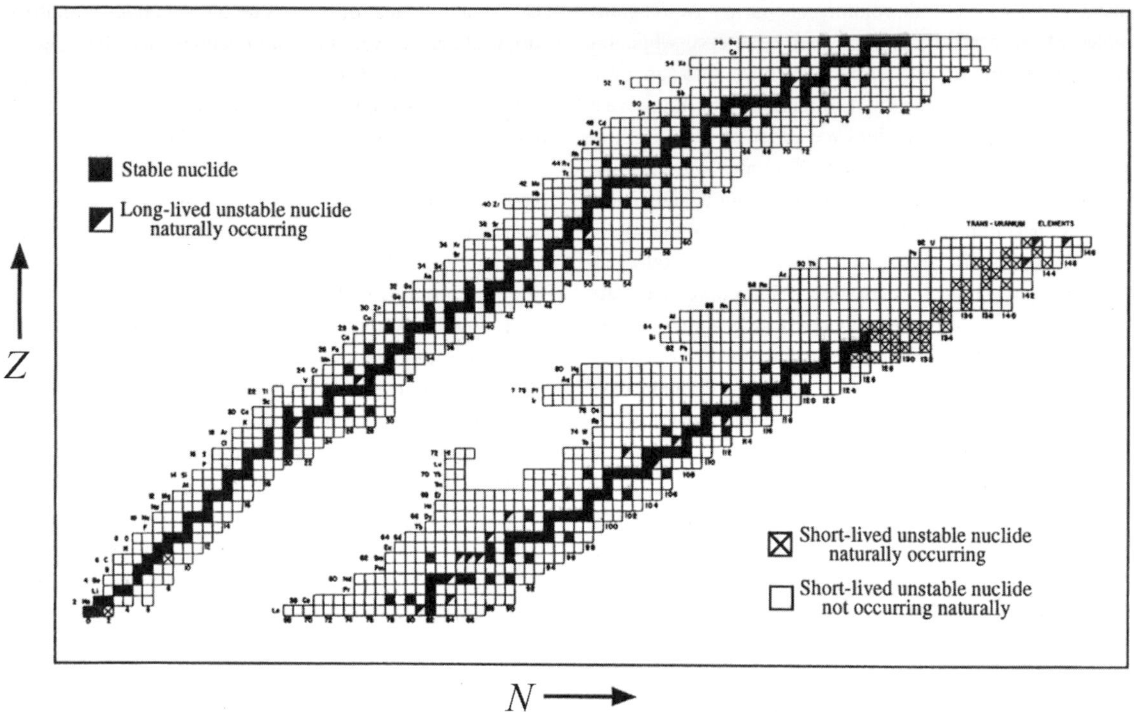

Figure 2.2 Schematic presentation of the Chart of the Nuclides, which plots *Z* (number of protons) versus *N* (number of neutrons). Stable isotopes, shown in black, define a narrow band within a wider band of unstable nuclides. In general, elements with even *Z* are more abundant and have more isotopes than elements with odd *Z*. Among the isotopes of a given element, those with even *N* are more abundant than those with odd *N*.

Atomic number (Z)														
29			Cu56 93 ms β^+	Cu57 196 ms β^+	Cu58 3.21 s β^+, E.C.	Cu59 1.36 m β^+	Cu60 23.7 m β^+, E.C.	Cu61 3.35 h β^+, E.C.	Cu62 9.74 m β^+, E.C.	**Cu63** **69.17** 62.929601	Cu64 12.70 h E.C., β^-, β^+	**Cu65** **30.83** 64.927794	Cu66 5.10 m β^-	Cu67 2.58 d β^-
	Ni53 0.05 s β^+	Ni54 0.10 s β^+	Ni55 202 ms β^+	Ni56 5.9 d E.C.	Ni57 35.6 h E.C., β^+	**Ni58** **68.0769** 57.935348	Ni59 7.6e4 a E.C., β^+	**Ni60** **26.2231** 59.930791	**Ni61** **1.1399** 60.931060	**Ni62** **3.6345** 61.928349	Ni63 101 a β^-	**Ni64** **0.9256** 63.927970	Ni65 2.52 h β^-	Ni66 2.28 d β^-
27	Co52 0.12 s β^+	Co53 0.25 s β^+	Co54 1.46 m, 193 ms β^+	Co55 17.53 h β^+	Co56 77.3 d E.C., β^+	Co57 271.8 d β^+	Co58 9.1 h, 70.9 d E.C., β^+	**Co59** **100** 58.933200	Co60 10.5 m, 5.3 a β^-	Co61 1.65 h β^-	Co62 13.9 m, 1.5 m β^-	Co63 27.5 s β^-	Co64 0.30 s β^-	Co65 1.17 s β^-
	Fe51 0.31 s β^+	Fe52 46 s, 8.3 h β^+	Fe53 2.6 m, 8.5 m β^+	**Fe54** **5.845** 53.939615	Fe55 2.73 a E.C.	**Fe56** **91.754** 55.934942	**Fe57** **2.119** 56.935399	**Fe58** **0.282** 57.933280	Fe59 44.5 d β^-	Fe60 2.6e6 a β^-	Fe61 6.0 m β^-	Fe62 68 s β^-	Fe63 6 s β^-	Fe64 2.0 s β^-
25	Mn50 1.74 m, 283 ms β^+, E.C.	Mn51 46.2 m β^+, E.C.	Mn52 21.1 m, 5.6 d E.C., β^+	Mn53 3.7e6 a E.C.	Mn54 312.1 d E.C., β^+	**Mn55** **100** 54.938050	Mn56 2.578 h β^-	Mn57 1.45 m β^-	Mn58 65 s, 3.0 s β^-	Mn59 4.6 s β^-	Mn60 1.77 s, 50 s β^-	Mn61 0.67 s β^-	Mn62 0.67 s β^-	Mn63 276 ms β^-
	Cr49 42.3 m β^+, E.C.	**Cr50** **4.345** 49.946050	Cr51 27.702 d E.C.	**Cr52** **83.789** 51.940512	**Cr53** **9.501** 52.940654	**Cr54** **2.365** 53.938885	Cr55 3.497 m β^-	Cr56 5.9 m β^-	Cr57 21 s β^-	Cr58 7.0 s β^-	Cr59 0.5 s β^-	Cr60 0.6 s β^-	Cr61 0.26 s β^-	Cr62 0.19 s β^-
23	V48 15.98 d E.C., β^+	V49 331 d E.C.	**V50** **0.250** 1.4e17 a E.C., β^+ 49.947163	**V51** **99.750** 50.943964	V52 3.76 m β^-	V53 1.54 m β^-	V54 49.8 s β^-	V55 6.5 s β^-	V56 0.23 s β^-	V57 0.33 s β^-	V58 0.21 s β^-	V59 0.1 s β^-	V60 0.1 s β^-	V61 ~43 ms β^-
Neutron number (N)		26		28		30		32		34		36		38

Figure 2.3 A detailed view of a portion of the Chart of the Nuclides. Stable isotopes are shaded. For stable nuclides, isotopic abundances are given below the element symbol and isotopic masses are given at the bottom of the square. Half-lives of unstable nuclides are given along with their decay modes.

unstable. Now look at the columns of constant N. For some values of N there is only one stable isotope, while for others there are several. Again, nuclides with an even number of neutrons are more common, and Figure 2.3 shows that the odd-N nuclides are either unstable or have low abundances relative to other isotopes of the same element. The Chart of the Nuclides does not contain sufficient information to evaluate the relative abundances of isotopes with odd N and variable Z because only the relative abundances of the isotopes within an element are shown, not the absolute abundances. Nuclides whose boxes join at the corners to make diagonal lines running from the upper left to the lower right on the diagram have the same mass number and are called *isobars*. In Figure 2.3, there are several cases where two stable nuclides have the same mass number: ^{54}Cr and ^{54}Fe, ^{58}Fe and ^{58}Ni, ^{64}Ni and ^{64}Zn. Nuclides that are the only isotope of an element never have another stable nuclide on their isobar. These patterns can be understood in terms of how the protons and neutrons in the nucleus are arranged.

Neutrons and protons reside in shells in the nucleus, each of which contains a specific number of protons or neutrons. This structure is broadly analogous to the shells that control the distribution of electrons, which will be discussed in some detail below. When protons or neutrons in a shell are paired, the structure is more energetically favored. This is why nuclei with even numbers of protons or neutrons are more stable than those with odd numbers. Table 2.2 summarizes the abundances of nuclides with different nuclear configurations. As one might suspect from looking at the Chart of the Nuclides, the most stable configuration is to have even numbers of both protons and neutrons. Having an even number of either protons or neutrons is also an energetic advantage. But for nuclides with both unpaired protons and unpaired neutrons, the nucleus is typically unstable. The four stable nuclides with unpaired protons and neutrons are all among the lightest nuclides.

A nucleus is even more stable when the proton or neutron shells are full, again analogous to the electron shell structure we will discuss below. Nuclides with these configurations are more tightly bound than their neighbors and are said to be "magic." These magic numbers occur at 2, 8, 20, 28, 50, 82, and 126. Because neutrons and protons have separate shells, a nuclide can have a full proton shell and a partially filled neutron shell, or vice versa. If neutrons and protons both have filled shells, the nuclide is said to be "doubly magic." Stable nuclei that are doubly magic are ^{4}He, ^{16}O, ^{40}Ca, ^{48}Ca, and ^{208}Pb. There are other doubly magic nuclei that are not stable because of the imbalance in the number of protons and neutrons (e.g. ^{56}Ni, ^{78}Ni, ^{100}Sn, ^{132}Sn). One version of the shell theory predicts that a nuclide with 114 protons and 184 neutrons should also be doubly magic, producing a so-called *island of stability* well beyond the heaviest stable nuclide. Nuclides of this mass would not be stable, but would exist for longer than neighboring configurations – perhaps long enough to be detected. Although nuclides with 114–118 protons have been produced, they had too few neutrons to be magic (174–177 neutrons versus 184 required to be magic), so as yet there is no experimental verification of the island of stability (e.g. Hofmann and Münzenberg, 2000; Kallunkathariyil et al., 2020).

The stable nuclides are flanked by unstable (radioactive) nuclides (Figs. 2.2 and 2.3). For these nuclides, the binding energy per nucleon is not sufficient to hold the nucleus together. The lower binding energy can also be seen as a *mass excess* relative to the most stable nuclide with the same atomic mass (A). Figure 2.4 plots the masses of the nuclides of constant A for three values of A. Figure 2.4a shows the masses of the nuclides with $A = 61$. Only one of these nuclides is stable, ^{61}Ni, for which $Z = 28$. The stable nuclide has the lowest mass, which means that is has the largest binding energy for this mass number. Nuclides with higher Z (to the left on Fig. 2.4a) have too many protons and are unstable. These nuclides decay by positron emission (β^+ decay) or in some cases by electron capture, changing a proton into a neutron, reducing the proton number by one, decreasing the mass excess, and increasing the binding energy. Nuclides with too many neutrons (those on the right in Fig. 2.4a) decay by emitting an electron (β^- decay), changing a neutron into a proton. This decreases mass excess and increases the binding energy. The larger the excess of protons or neutrons compared to the stable isobar, the more unstable the nuclide is. Nuclides with odd A typically have only one stable isobar.

Figure 2.4b shows the masses of the nuclides with $A = 40$. For even values of A, two or three stable isobars typically exist. In this case, both ^{40}Ca and ^{40}Ar exist in

Table 2.2 Numbers of stable nuclides as a function of nuclear configuration

A	Z	N	Number of Stable Nuclides
Even	Even	Even	159
Odd	Even	Odd	53
Odd	Odd	Even	50
Even	Odd	Odd	4

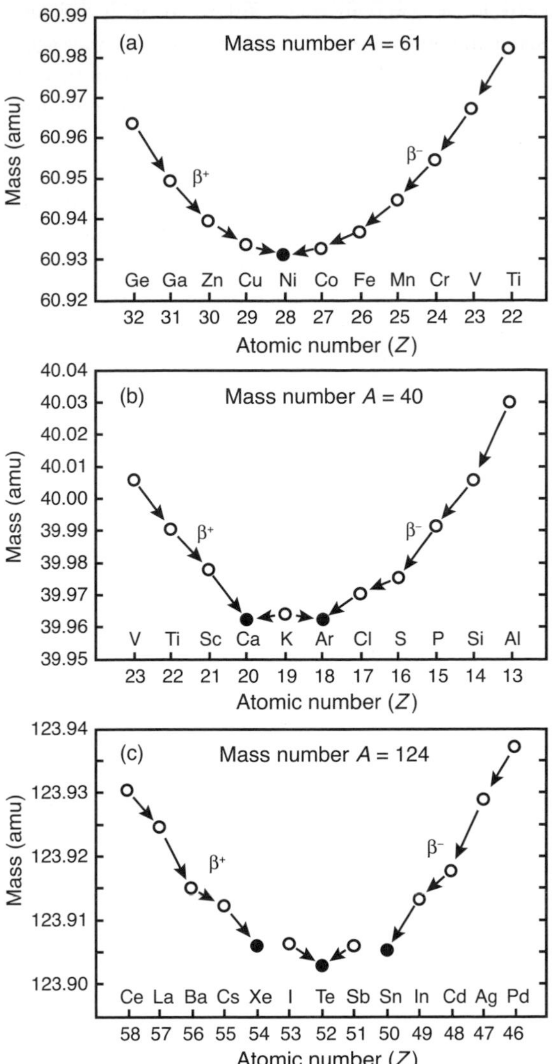

Figure 2.4 Nuclear mass versus atomic number (Z) for nuclides with constant mass number (A). The three panels show data for three different values of A. Stable isotopes are shown as filled circles, and unstable isotopes are open circles. Radioactive isotopes decay by either β^+ decay, on the left, or β^- decay, on the right, until they reach the stable nuclides in the Valley of β Stability.

nature. Both are even-even nuclides. ^{40}K is an odd-odd nuclide and has a larger mass and lower binding energy. It is unstable and decays by β^- decay to form ^{40}Ca and by electron capture to form ^{40}Ar. The ^{40}K–^{40}Ar decay is the basis for one of the most widely used radiochronometers in geo- and cosmochemistry (see Chapter 9). The Chart of the Nuclides lists 33 nuclides that decay by both β^- decay and β^+ decay or electron capture.

A few mass numbers have three stable isobars (e.g. A = 96, 124, 136). Figure 2.4c shows the masses and decay paths of nuclides with A = 124. There are three stable isobars with mass 124: ^{124}Xe, ^{124}Te, and ^{124}Sn. It is not possible for ^{124}Xe and ^{124}Sn to decay by either β^+ or β^- decay because the adjacent nuclides have lower binding energy and are less stable. ^{124}I and ^{124}Sb can decay to ^{124}Te.

Figure 2.4 illustrates the Valley of β Stability. When plotted as in Figure 2.4, each mass number A shows a similar shape, with the stable isotopes at the bottom of the generally parabolic distribution. Nuclides on either end of the distribution decay by β^- emission, β^+ emission, or neutron capture and move down the slope to the floor of the Valley.

2.3 Radioactive Elements and Their Modes of Decay

The 65 naturally occurring radioactive isotopes generally fall into three categories, depending upon their half-lives:

1. Many of the longer-lived radioactive isotopes were produced during stellar nucleosynthesis and have sufficiently long half-lives that all of the atoms have not yet decayed. Examples include ^{40}K, ^{87}Sr, ^{232}Th, ^{235}U, and ^{238}U.
2. Some radioactive isotopes are continuously produced in the Earth's atmosphere, in rocks at the surface of planetary bodies, or in meteoroids and small grains orbiting in space by collisions between high-energy cosmic rays and stable isotopes. The collisions break the stable isotope into smaller fragments, some of which are radioactive. Such nuclides are said to be *cosmogenic*, and the reactions that produce them are called *spallation* reactions. An example is the cosmogenic isotope ^{14}C, which is used to date archeological finds. It is produced from ^{14}N in the Earth's atmosphere.
3. The third category of radioactive nuclide consists of isotopes produced by naturally occurring radioactive chains. For example, when an atom of ^{238}U decays, it passes through radioactive ^{234}Th, ^{234}Pa, ^{234}U, ^{230}Th, ^{226}Ra, ^{222}Rn, ^{218}Po, and several other isotopes before it finally ends up as stable ^{206}Pb. The intermediate isotopes have half-lives ranging from a few hours to almost 10^5 years – too short for there to be any contribution from stellar nucleosynthesis in solar system materials. These species are also too heavy to be produced in significant amounts by cosmic-ray-induced spallation reactions.

Radioactive decay occurs by several pathways, as illustrated in Figure 2.5. Beta decay (β^-), emission of an electron by the nucleus, converts a neutron to a proton and increases Z by one without changing A. Isotopes of interest to cosmochemistry that decay only by β^- decay

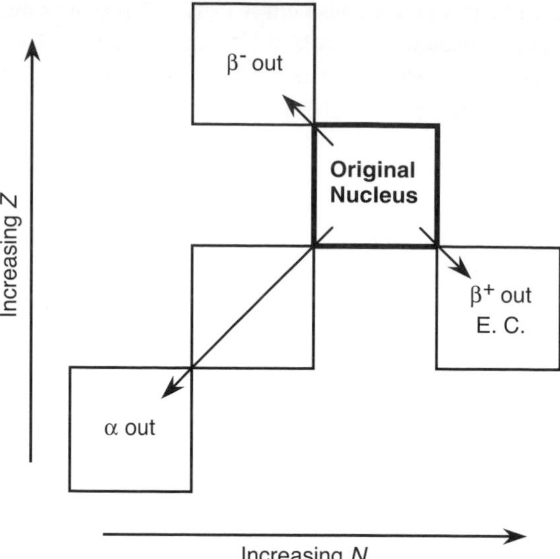

Figure 2.5 The effects of different modes of radioactive decay on the position of an isotope on the Chart of the Nuclides. Beta decay, which changes a neutron to a proton, moves the nuclide up and left. Positron decay or electron capture, which change a proton to a neutron, move the nuclide down and right. And α decay, which is the emission of a ^{4}He nucleus, moves the nuclide down and left.

include ^{3}H, ^{10}Be, ^{14}C, ^{36}Cl, ^{60}Fe, and ^{87}Rb. Positron decay (β^+) and electron capture both convert a proton to a neutron. Although the two processes differ in nuclear detail, both reduce Z by one without changing A (Fig. 2.5). Isotopes that decay by one or both of these processes include ^{22}Na, ^{26}Al, ^{41}Ca, and ^{53}Mn. The isotope ^{40}K, which is important for radiometric dating, decays by β^- decay to ^{40}Ca and by β^+ decay to ^{40}Ar. Radioactive isotopes can also decay by emission of an α-particle, which consists of two proton and two neutrons (α decay). This decreases Z by two and A by four (Fig. 2.5). Alpha decay is the primary mode of decay for heavy radioactive isotopes. The decay series for actinides such as ^{232}Th, ^{235}U, and ^{238}U consist primarily of α decays, punctuated by occasional β decay to keep the resulting daughter isotopes close to the Valley of β Stability.

Isotopes heavier than about 230 u can also decay by spontaneous fission. In spontaneous fission, the nucleus splits into two smaller nuclei. The split is typically asymmetrical, leaving fragments of different Z. Spontaneous fission is a minor decay pathway for naturally occurring isotopes such as ^{232}Th, ^{235}U, and ^{238}U. A more efficient pathway for fission to occur is for the nucleus first to absorb a neutron. The extra neutron triggers the nucleus to split. This process is called neutron-induced fission and it is utilized for nuclear reactors and weapons. It is possible for neutron-induced fission to occur naturally. Spontaneous fission typically releases neutrons, which can then be absorbed by other atoms. If the concentration of fissionable nuclei is high enough, the released neutrons can trigger additional fission reactions. The process can be further enhanced if a moderating material such as water is present to slow down the ejected neutrons, making them easier to capture. In 1956, P. K. Kuroda speculated that early in Earth's history, when the relative abundance of ^{235}U was higher, a deposit of natural uranium and water might generate a self-sustaining nuclear chain reaction – a natural nuclear reactor. Such a deposit, called Oklo, was found in the African Gabon Republic in 1972 (Cowan, 1976). It was recognized because the uranium in the deposit had a very low abundance of ^{235}U, which had been destroyed by neutron-induced fission in the self-sustaining reactor.

The decay of radioactive nuclides is used to determine the chronology of events from the earliest epochs of the solar system to recent times. The usefulness of radiometric dating depends critically on the constancy of radioactive decay rates. There are a number of observations that confirm that decay rates do not change with time. For example, radionuclides used to date rocks and minerals that formed throughout solar system history decay by all of the above mechanisms (e.g. ^{40}K by β^- decay and electron capture, ^{87}Rb by electron capture, ^{147}Sm and ^{232}Th by α decay, ^{187}Re by β^- decay, ^{235}U and ^{238}U by α decay and spontaneous fission). In spite of these differences, dates obtained from different decay systems are consistent for rocks and minerals formed at various times over the ~4.567 billion-year history of the solar system. For decay rates to vary and produce this result, they would have to vary in concert, which would require several different nuclear properties to vary in such a way as to maintain the illusion of constancy. Astronomy provides another line of evidence. Supernovae are known to produce large quantities of radioactive isotopes, the decay of which drives their light curves as observed from Earth. Supernovae that exploded billions of years ago, but whose light is only now reaching Earth, show the same characteristic light curves as supernovae that exploded a few million years ago. Thus, the decay of newly formed radioactive isotopes has followed the same laws for billions of years. These and other lines of direct evidence show that the laws of nuclear physics and thus the decay constants of radioactive isotopes have remained constant within tight limits since the universe formed.

As the ability to measure decay rates and as the understanding of the laws of physics have improved, a few exceptions to the rule of constant decay rates have been identified. However, these exceptions are effectively irrelevant to cosmochronology. For example, in the interiors of stars, where temperatures exceed 100 million degrees and pressures reach millions of atmospheres, the stabilities of atomic nuclei change dramatically. For example, the half-life of ^{60}Fe can decrease from 2.6 million years in normal space to a few minutes in the interior of a massive star (Limongi and Chieffi, 2006). If the temperatures and pressures become high enough, the atomic structure breaks down completely and the material degenerates into a tightly packed ball of neutrons: a neutron star. But because matter that can be dated by radioactive decay must exist in the same form from the time of its origin until it is dated, we need not worry about what would happen if it found itself in the interior of a star.

One radioactive decay pathway, electron capture, has been found to have a minor dependence on external conditions that are of relevance to cosmochemistry. The rate at which an electron in the cloud surrounding the nucleus is captured by a proton to make a neutron is proportional to the probability of an electron ending up close to the nucleus. Changes in the electron distribution can potentially change the electron capture rate. Such changes can be produced by chemical bonding, high pressure, low temperature, superconductivity, and strong electric and magnetic fields (Huh, 1999; Liu and Huh, 2000). However, *ab initio* calculations of the decay rate of ^{40}K under high pressure (25 GPa) predict an increase in the ^{40}K decay rate of only ~0.025% (2.5 parts in 10^4) (Lee et al., 2008), and a measurement performed on ^{83}Rb, which, like ^{87}Rb decays by electron capture, showed no measurable change in decay rate at 42 GPa. The extremely small changes in electron capture rate that occur under cosmochemically relevant conditions have no effect on the ages obtained using radiochronometers.

2.4 The Periodic Table: Organizing Elements by Their Chemical Properties

Credit for constructing the Periodic Table of the Elements (Fig. 2.6) is generally given to Dmitri Mendeleev, a

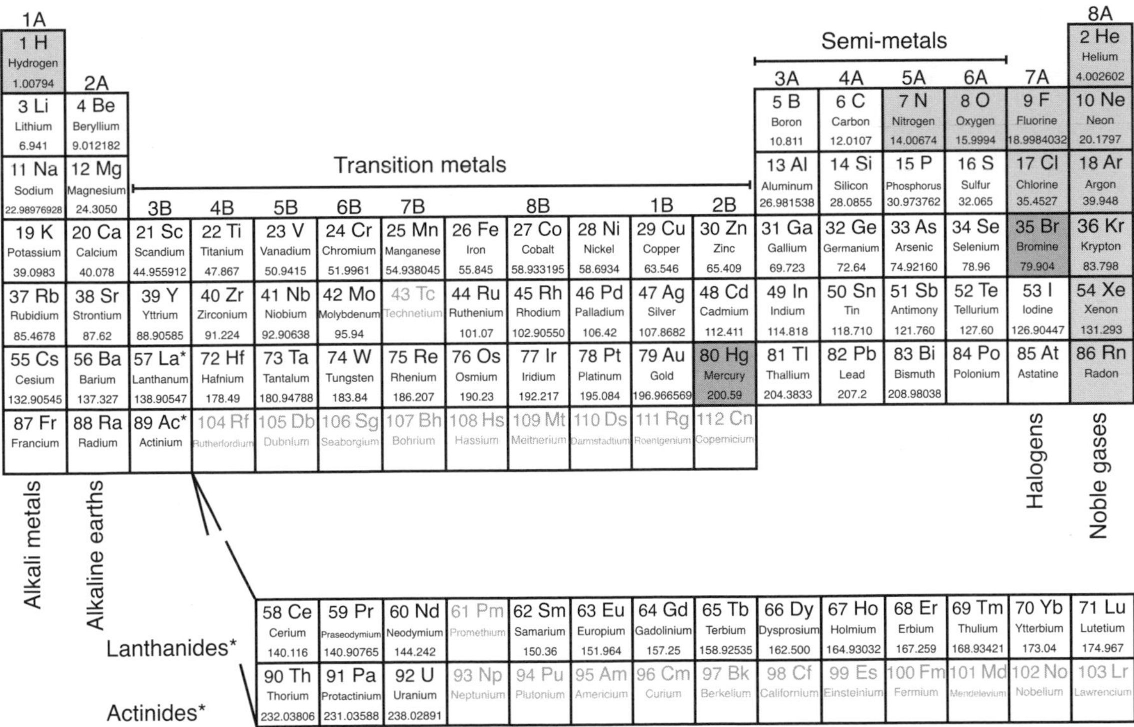

Figure 2.6 The Periodic Table of the Elements. Elements are designated by their atomic number and symbol. The atomic weights are given for the stable elements and those with long-lived radioactive isotopes. Elements consisting of short-lived, naturally occurring radioactive isotopes are shown without atomic weights. Elements that are not naturally found on Earth, but which have been made in accelerators, are shown in gray type. The elements in white boxes are solids at room temperature, those in dark gray boxes are liquids, and those in light gray boxes are gases.

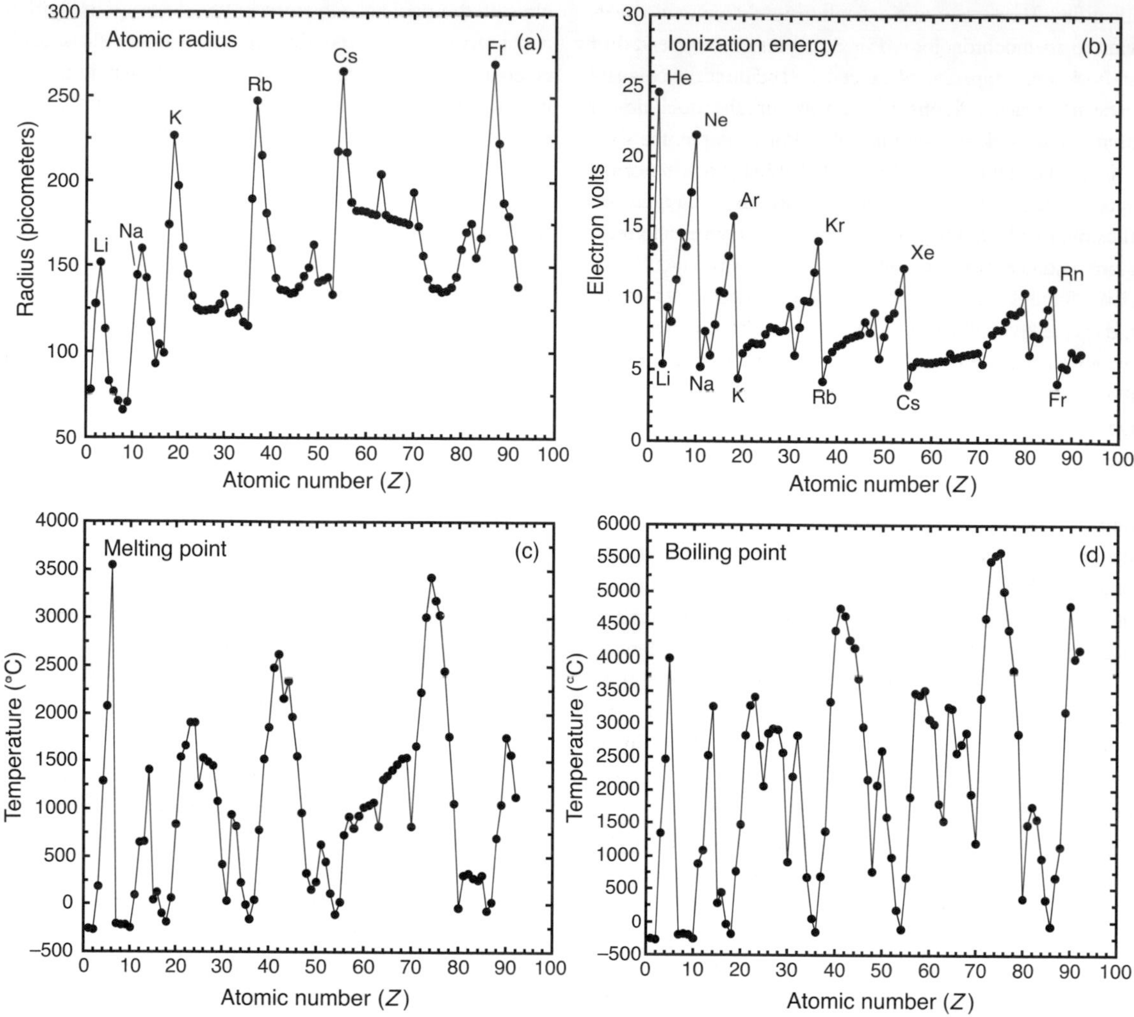

Figure 2.7 Properties of the elements as a function of atomic number (Z). Atomic radius (a) and ionization energy (b) directly reflect the atomic structure of the atoms. Other properties, such as melting point (c) and boiling point (d), are controlled indirectly by the atomic structure through the bonding between atoms.

professor at St. Petersburg University. While writing a textbook on inorganic chemistry, he noticed patterns in the properties and atomic weights of sets of elements such as Cl–K–Ca, Br–Rb–Sr, and I–Cs–Ba. He then attempted to identify similar patterns for all of the 63 known elements. The result of his work was his statement of the periodic law, and his classic paper, *On the Relationship of the Properties of the Elements to their Atomic Weights*, was published in 1869. At the time that Mendeleev developed his Periodic Table, the experimentally determined atomic weights of the elements were not always accurate. He found that 17 elements had to be moved to new positions in the Table from those indicated by their accepted atomic weights. In most cases, the new positions were confirmed by subsequent measurements of atomic weight. There were also some gaps in his Table, and from those gaps Mendeleev predicted the existence of additional elements. For example, he predicted elements that he called eka-aluminum, eka-boron, and eka-silicon, which were eventually found and renamed gallium, scandium, and germanium. The discovery of the element argon by Lord Rayleigh in 1895 was not predicted by the Periodic Table, and in fact there was no place for it in the Table. A few years later, William Ramsey suggested that argon be placed between chlorine and potassium in a family with helium. Based on this suggestion, he accurately predicted the existence and properties of neon.

In the modern Periodic Table (Fig. 2.6), the elements are arranged in order of increasing atomic number and then in horizontal rows called "periods" and vertical columns called "groups." Members of each group have similar physical and chemical properties. Group 1A elements are called the alkali metals, Group 2A elements are the alkaline earths, Group 7A elements are the halogens, etc. Figure 2.7 illustrates some of the properties that vary periodically with atomic number. For example, atomic radius peaks at $Z = 3$ (Li), 12 (Mg), 19 (K), 37 (Rb), 55 (Cs), and 87 (Fr). With the exception of magnesium, all are members of Group 1A of the Periodic Table. Ionization energy peaks at 2 (He), 10 (Ne), 18 (Ar), 36 (Kr), 54 (Xe), and 86 (Rn), all of which are noble gases and members of Group 8A. Likewise, the melting points and boiling points of the elements vary periodically with atomic number (Fig. 2.7c, d). As we have already noted in the case of magnesium, the periodicity of the properties of the elements is not always perfectly correlated with the atomic number. This reflects the details of the forces that hold the atoms together and the order in which the electron shells are filled.

The use of atomic number rather than atomic weight, as in Mendeleev's original Table, dates to the discovery that the characteristic X-ray lines emitted by the elements are a strict function of the atomic number, *Z*. Electrons of an element that have been excited to a higher energy level lose the excess energy by emitting X-rays of specific frequency (= energy). The most energetic of these lines are called the K-series, the next the L-series, then the M-series, and so on. Measuring the intensities of these lines is the basis for a variety of techniques to determine the compositions of rocks and minerals (see Appendix). In 1913–1914, Henry Mosley studied these X-ray lines and showed that the square root of the X-ray energy ($= h\nu$, Planck's constant multiplied by frequency) is proportional to *Z* (Fig. 2.8). He concluded that the order of the elements in the Periodic Table should be based on *Z*. This insight changed the positions of some elements in the Table, because atomic weights depend on the number and relative abundances of the different isotopes of an element, whereas the periodic properties of the elements and the X-ray frequencies depend on the number of protons and electrons in the atom. Note for example that tellurium, which has eight stable isotopes, has a higher atomic weight than the element with the next higher atomic number, iodine, which has a single isotope. With the new ordering, the elements still missing from the Periodic Table could now be predicted with certainty. In 1913, seven gaps still existed in the Periodic Table among the elements between H ($Z = 1$) and U ($Z = 92$). These gaps were at $Z = 43$, 61, 72, 75, 85, 87, and 91. All have now been identified. Two of them, hafnium (72) and rhenium (75), are stable. The rest are radioactive and rare on Earth, existing only as decay products of heavier radioactive nuclides.

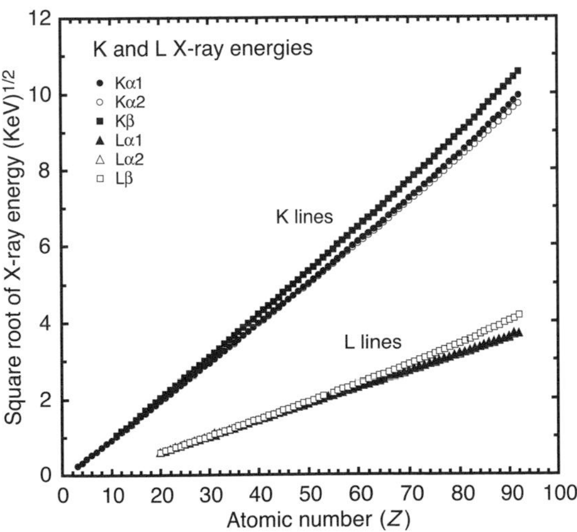

Figure 2.8 The K and L X-ray energies of the elements plotted as a function of atomic number (*Z*). The tight relationship between the square root of the X-ray energy (or frequency) and atomic number led to the modern ordering of the elements in the Periodic Table according to *Z* and shows that the Table is now complete.

The chemical properties of an element are functions of the number and distribution of its electrons around the nucleus. In 1913, Niels Bohr devised a model for the atom that successfully explained why atomic spectra consist of discrete lines, not only in X-rays, as discussed above, but also in visible light. Building on the ideas of quantum mechanics, he postulated that the angular momentum of an orbiting electron can only have certain fixed values. If so, then the orbital energy associated with any electron cannot vary continuously, but can only have discrete quantum values. He described a series of spherical shells at fixed distances from the nucleus. However, chemical physicists soon created a better model that uses the language of wave equations. In 1926, Erwin Schrödinger proposed a model incorporating not only a single quantum number, as Bohr had suggested, but three others as well. Schrödinger's equations describe geometrical charge distributions (atomic orbitals) that are unique for each combination of quantum numbers. The electrical charge that corresponds to each electron is not ascribed to an orbiting particle, as in Bohr's model. Instead, you can think of each electron as being spread over the entire orbital, which is a complex three-dimensional figure.

The *principle quantum number*, n, which can take integral values from 1 to infinity, describes the effective volume of an orbital. Chemists commonly use the word *shell* to refer to all orbitals with the same value of n, because each increasing value of n defines a layer of electron density that is farther from the nucleus.

The orbital *angular momentum quantum number*, l, describes the shape of the region occupied by an electron. Depending on the value of n, a particular electron can have values of l that range from 0 to $n - 1$. If $l = 0$, the orbital described by Schrödinger's equations is spherical and is given the shorthand symbol s. If $l = 1$, the orbital looks like one of those in Figure 2.9 and is called a p orbital. Orbitals with $l = 2$ (d orbitals) are shown in Figure 2.10, and $l = 3$ orbitals (f orbitals) have shapes that are not easy to illustrate in a flat drawing.

The third quantum number, m_l, describes the orientation of the electron orbital relative to an arbitrary direction. Because an external magnetic field (such as might be induced by a neighboring atom) provides a convenient reference direction, m_l is usually called the *magnetic orbital quantum number*. It can take an integral valuc from $-l$ to l.

The fourth quantum number, m_s, does not describe an orbital itself, but imagines the electron as a particle within the orbital spinning around its own polar axis. In doing so, the electron becomes a tiny magnet with a north and south pole. The *magnetic spin quantum number*, m_s, can be either positive or negative depending on whether the electron's magnetic north pole points up or down relative to an outside magnetic reference.

Each orbital can therefore contain no more than two electrons, with opposite spin quantum numbers. This rule, which affects the order in which electrons may fill orbitals, is known as the *Pauli exclusion principle.* Table 2.3 summarizes the configuration of electron orbitals for the first three shells. The orbitals are labeled with the numerical value of n and a letter corresponding to the value of l (s, p, d, f . . .). As you can see from Table 2.3, the $n = 1$ shell can hold up to two electrons, both in the s orbital, the $n = 2$ shell can hold up to eight electrons (2 in the s and 6 in the p orbital), the $n = 3$ can hold up to 18 electrons (2 s, 6 p, and 10 d), and the $n = 4$ shell can hold up to 32 electrons (2 s, 6 p, 10 d, and 14 f). The lowest-energy orbitals are occupied first. So, for hydrogen, which has one electron, the electron resides in the 1s orbital. For lithium, which has three electrons, two are in the 1s orbital and the third is in the 2s orbital. For silicon ($Z = 14$), there are two electrons in 1s, two electrons in 2s, six electrons in 2p, two electrons in 3s, and two electrons in 3p.

The layout of the Periodic Table (Fig. 2.5) reflects the shell structure of the electrons. Hydrogen and helium have only K-shell electrons. The elements in Period 2 have K- and L-shell electrons, with the 1s orbitals always filled and the 2s and 2p orbitals filled in succession. Those in Period 3 have K- and L-shell electrons, with 1s, 2s, and 2p orbitals filled and the 3s and 3p orbitals filled in succession. Elements in Period 4 have K-, L-, and M-shell electrons, with the 1s, 2s, 3s, 2p, and 3p orbitals completely filled. After the 4s orbitals are filled, the 3d orbitals are filled, giving the transition metals. Then come the 4p orbitals. Period 5 is filled in an analogous fashion. In Period 6, the lanthanides, which fit between lanthanum and hafnium, reflect the appearance of the N-shell electrons, which fill the f orbitals. Period 7, which contains the actinides, also has K-, L-, M-, and N-shell electrons.

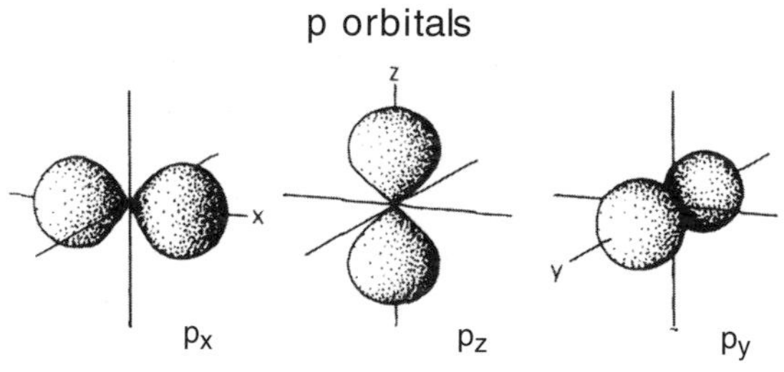

Figure 2.9 The three p orbitals in any shell have figure-eight shapes, oriented along Cartesian x-, y-, and z-axes centered on the atomic nucleus.

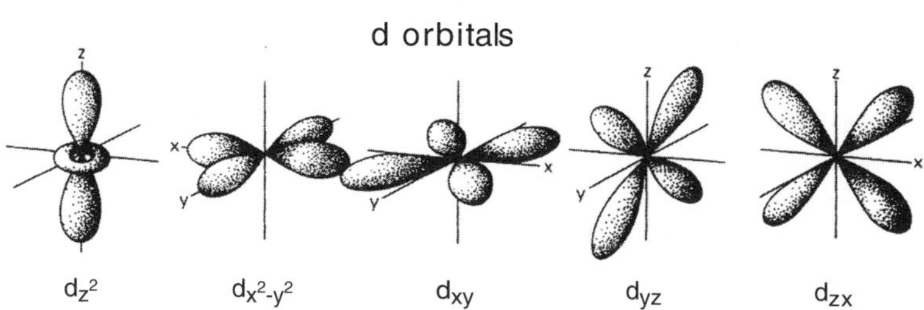

Figure 2.10 Three of the d orbitals (d_{xy}, d_{yz}, and d_{zx}) in any electron shell have four lobes, oriented between the Cartesian axes. A fourth ($d_{x^2-y^2}$) also has four lobes, which lie along the x- and y-axes. The fifth (d_{z^2}) has lobes parallel to the z-axis and a ring of charge density in the x–y plane.

Table 2.3 Configuration of electrons in orbitals

		n	l	m_l	m_s
K shell	1s orbital	1	0	0	+
2 s electrons		1	0	0	−
L shell	2s orbital	2	0	0	+
2 s, 6 p electrons		2	0	0	−
	2p orbitals	2	1	−1	+
		2	1	−1	−
		2	1	0	+
		2	1	0	−
		2	1	1	+
		2	1	1	−
M shell	3s orbital	3	0	0	+
2 s, 6 p, 10 d electrons		3	0	0	−
	3p orbitals	3	1	−1	+
		3	1	−1	−
		3	1	0	+
		3	1	0	−
		3	1	1	+
		3	1	1	−
	3d orbitals	3	2	−2	+
		3	2	−2	−
		3	2	−1	+
		3	2	−1	−
		3	2	0	+
		3	2	0	−
		3	2	1	+
		3	2	1	−
		3	2	2	+
		3	2	2	−
N shell	4s, 4p, 4d, 4f orbitals				
2 s, 6 p, 10 d, 14 f electrons					

The periodic behavior of the elements now makes sense. The noble gases, which are positioned at the right side of the Periodic Table, have completely filled electron shells and thus have little or no tendency to form chemical bonds. The alkali metals, which have only one s electron in the outermost shell, can easily lose that electron and become a positive ion. In contrast, the halogens are only one electron short of having filled-out shells, and so they often acquire an extra electron and become a negative ion. Note that the ionization energy, the energy required to remove one electron from an atom, increases from left to right across the Periodic Table, peaking with the noble gases (Fig. 2.7b). Electronegativity, which is the tendency to acquire another electron, also increases across the Periodic Table, but peaks with the halogens. The noble gases have very low electronegativities because they have filled electron shells.

2.5 Chemical Bonding

The interactions between elements in nature and in the laboratory are manifestations of chemical bonding between the elements. The nature of chemical bonds between elements is a direct result of their electron configurations. The outermost unfilled shells of electrons surrounding the nucleus generate the various types of bonds, and the electrons that create the bonds are known as the *valence electrons*. The valence of an atom is defined as the number of hydrogen-atom equivalents that could combine in that molecule. For example, the sodium in NaCl has a valence of +1, while the magnesium in MgO has a valence of +2. The chlorine and oxygen in these compounds have valences of −1 and −2, respectively. In this section we briefly review the different types of chemical bonding to provide the basics for later discussions. Detailed discussions of chemical bonding are available elsewhere (e.g. Roher, 2001).

An *ionic bond* is formed between a metal and a nonmetal by the electrostatic attraction between two oppositely charged ions. A metal donates one or more electrons, forming a positive ion (cation). The donated electrons enter the nonmetal, filling vacancies in the outer electron shell and creating a negative ion (anion). The resulting electrostatic attraction between the positively and negatively charged ions forms the bond. An example is common table salt, sodium chloride. Sodium, an alkali metal, has one 3s electron in its outermost (valence) shell, while chlorine, a halogen, has one empty spot in its 3p shell. Sodium donates an electron to chlorine, giving both of them closed-shell electron configurations. The positively charged sodium ion then bonds to the negatively charged chlorine ion in an ionic bond. The charge of an ion in a stable ionic compound is governed by the balance between the energy required to remove electrons from the metal (ionization energy) and the energy released when the electron is added to the nonmetal, plus the energy released by bringing the ions together to form a bond. In general, the ionization energy is lower when removing electrons from an outer shell that has only one electron (e.g. Na^+, K^+, Ag^+) or when the resulting ion has a filled outer shell (e.g. Mg^{2+}, Ca^{2+}, Zn^{2+}), although other configurations do form ionic bonds.

Box 2.1 (cont.)

Water has high surface tension. It tends to form drops rather than to spread out as a thin film over a surface. This is because the attractive forces between water molecules are larger than the attraction between the water and the surface in most cases. Molecules inside a drop are equally attracted to surrounding water molecules in all directions. However, at the surface there are no water molecules on one side, so a molecule experiences attractions on the sides and toward the interior of the drop. This tends to minimize the surface area of a drop and creates an effect like an elastic membrane on the water's surface. The distinction between surface molecules and interior molecules is particularly great in water because of the hydrogen bonding. High surface tension allows water drops to cling to surfaces, allows large waves to form, and can help move water and the nutrients dissolved in it from plant roots to leaves.

Water exists on the surface of the Earth as a solid, a liquid, and a gas. This permits the hydrologic cycle to operate and move water from the oceans to the tops of the highest mountains, where it is stored as snow and ice to be released later in the year. Having a surface temperature in the range where water can exist in all three forms may be a key ingredient for a planet to have life. On the present-day surface of Mars, water exists only in solid and vapor forms.

the other. At that moment, the molecule is a dipole and can attract another molecule. Furthermore, the attraction of the temporary dipole can cause the electrons in the other molecule to move toward the positive end of the temporary dipole, creating a second dipolar molecule. The electrons continue to move, so the polarity of the dipole is constantly changing. The electrons in the nearby molecules follow the dance of the strongest dipole, and the synchronous temporary dipoles result in a permanent attraction between the molecules. This latter type of dipole attraction is known as "dispersion forces."

These two types of interactions, dipole-dipole and dispersion forces, determine properties such as the melting and boiling points of molecular solids and of atoms such as the noble gases. In general, the melting and boiling points increase with increasing size of the molecule or atom because there are more electrons in the atom, resulting in a stronger dipole attraction.

Hydrogen bonds are intermolecular bonds that form between molecules consisting of hydrogen and some of the most electronegative elements (e.g. N, O, F, Cl). These electronegative elements have one or two pairs of electrons in the outermost shell that are not involved in the bonding within the molecule. These lone pairs of electrons align themselves around the nucleus on the side opposite to the hydrogen atoms. The shared electrons from the hydrogen atoms are strongly attracted to the electronegative atom, so the net result is a highly polar molecule. The positively charged hydrogen end of the molecule is strongly attracted to the lone pairs of electrons around the electronegative atoms in the adjacent molecules, making a hydrogen bond. The attraction is not as strong as in a true covalent bond. Hydrogen bonds are about one-tenth the strength of the average covalent bond, but they are significantly stronger than the dipole-dipole attraction of van der Waals forces (Jeffrey, 1997).

2.6 Chemical and Physical Processes Relevant to Cosmochemistry

The basic chemical properties of elements described above play key roles in the physical and chemical properties that cosmochemists investigate in their research.

2.6.1 Evaporation and Condensation

When the elements are ejected from the stars where they were produced, they are in the gas phase. Subsequently, they combine in various chemical compounds and most condense as solids. The nature of those compounds and their behavior in the various environments encountered on their way to becoming part of the solar system can, in principle, be determined from the basic chemical properties of the elements. Evaporation and condensation are also important in the solar system and have played a defining role in determining the properties of planets, moons, asteroids and the meteorites derived from them, comets, dust grains, and all other solar system materials, as discussed in Chapter 7. Cosmochemists have spent a lot of effort quantifying the evaporation and condensation behavior of the elements and the compounds that they form.

2.6.2 Melting and Crystallization

The interiors of planets, moons, and many asteroids either are, or have in the past been, molten. The behavior of molten silicates and metal is important in understanding how a planet or small body evolved from an undifferentiated collection of presolar materials into the differentiated object we see today. Basaltic volcanism is ubiquitous on the terrestrial planets and many asteroids. A knowledge of atomic structure and chemical bonding is necessary to understand how basaltic melts are generated and how they crystallize. Melting and crystallization are also important processes in the formation of chondrules, tiny millimeter-sized spherical objects that give chondritic meteorites their name. The melting, crystallization, and sublimation of ices are dominant processes in the histories of the moons of the outer planets, comets, many asteroids, and the Earth.

2.6.3 Dissolution and Precipitation of Solids from Solution

When liquid water, the universal solvent, is present on a planet or small body, a wide variety of chemical reactions may take place that can completely alter the mineralogy and chemistry of the body. Some meteorites show extensive evidence of aqueous alteration. To understand the conditions under which the alteration occurred, one must be able to infer the amount, composition, and temperature of the fluids from the minerals that they produced.

2.6.4 Solid-State Diffusion

Elements also move around within solids. This motion is typically very slow, but over time it turns a rock composed of a disequilibrium assemblage of materials produced at high and low temperatures into an assemblage where all phases are in equilibrium. Understanding the nature and extent of these solid-state processes is critical to understanding and quantifying the histories of meteorites, asteroids, comets, and planets.

Summary

In this chapter, we introduced the constituents and structure of the atom and showed that elements typically have several isotopes (same number of protons but different numbers of neutrons). Using the Chart of the Nuclides, we briefly discussed the distribution and stability of the isotopes. Radioactive isotopes were introduced and we mentioned that they can be used for dating geologic and cosmochemical events. We then discussed the Periodic Table and showed that the physical and chemical properties of the elements are derived from the number and arrangement of electrons. The different types of chemical bonding were introduced. The range of chemical and physical processes that can be understood from basic chemical principles was discussed.

The contents of this chapter will be the foundation of all subsequent discussions. Your depth of understanding of each of these subjects will grow as you continue through the book, but it is critical to have a grasp of these basic ideas before moving ahead. In the next chapter, we will discuss the origin of the elements via nucleosynthesis in stars.

Questions

1. What is nuclear binding energy? What are the two main ways it can be released? Which of these is currently powering our Sun?
2. When discussing stable and radioactive isotopes, what is meant by the "Valley of β Stability"?
3. What are four different modes of radioactive decay?
4. In order to use radioactive decay of an element for a chronometer, the decay constant must not have changed with time. How do we know this is true?
5. What are the five main types of chemical bonds? Give two important properties of each type.
6. Why is water such an important compound in geochemistry and cosmochemistry? What are some of its distinctive properties?

Suggestions for Further Reading

Baum E. M., Ernesti M. C., Knox H. D., et al. (2010) *Nuclides and Isotopes, Chart of the Nuclides, 17th edition*. Bechtel Marine Propulsion Corporation, Schenectady, NY, 94 pp. Everything you might want to know about nuclides in one convenient place. The first 41 pages of the book edition summarize the basic information needed to make full use of the Chart of the Nuclides.

Other References

Cowan, G. A. (1976) A natural fission reactor. *Scientific American*, **235**, 36–47.

Hofmann S., and Münzenberg G. (2000) The discovery of the heaviest elements. *Reviews of Modern Physics*, **72**, 733–767.

Huh C.-A. (1999) Dependence of the decay rate of ^{7}Be on chemical forms. *Earth & Planetary Science Letters*, **171**, 325–328.

Jeffrey G. A. (1997) *An Introduction to Hydrogen Bonding*. Oxford University Press, New York, 303 pp.

Kallunkathariyil J., et al. (2020) Stability of the heaviest elements: K isomer in ^{250}No. *Physical Review C*, **101**, 011301(R).

Lee K. K. M., and Steinle-Neumann G. (2008) Ab-initio study of the effects of pressure and chemistry on the electron-capture radioactive decay constants of ^{7}Be, ^{22}Na, and ^{40}K. *Earth & Planetary Science Letters*, **267**, 628–626.

Limongi M., and Chieffi A. (2006) The nucleosynthesis of ^{26}Al and ^{60}Fe in solar metallicity stars extending in mass from 11 to 120 $M_{\odot}$: The hydrostatic and explosive contributions. *Astrophysical Journal*, **647**, 483–500.

Liu L., and Huh C.-A. (2000) Effect of pressure on the decay rate of ^{7}Be. *Earth & Planetary Science Letters*, **180**, 163–167.

Roher G. S. (2001) *Structure and Bonding in Crystalline Materials.* Cambridge University Press, Cambridge, 552 pp.

3 Origin of the Elements

Overview

Several processes were responsible for producing the current inventory of elements in the cosmos. Hydrogen, helium, and some lithium were created in the Big Bang, a massive explosion that is thought to have produced the universe. Elements heavier than hydrogen and helium, known as *metals* in astronomy, were produced in stars by processes collectively called stellar *nucleosynthesis*. The chemical elements other than hydrogen and helium in our solar system are the result of nucleosynthesis that occurred in stars that lived and died before the solar system formed. These processes involve fusion of light elements into heavy elements, sometimes at modest rates as stars evolve, and sometimes at furious rates in stellar explosions. Significant amounts of a few rare elements, such as lithium, beryllium, and boron, were created via spallation reactions, in which collisions between highly energetic cosmic rays (typically protons or helium ions) and atoms break up the heavier nuclides into lighter fragments. And, of course, some nuclides have been produced by the decay of radioactive nuclides. In this chapter, we will review these processes and discuss the evolution of the elemental abundances with time in the universe and the galaxy.

3.1 In the Beginning

The cosmological model that best explains the origin of the universe is the *Big Bang*. According to this model, the universe began at a finite time in the past and at a discrete point in space, expanded from a hot, dense initial state of very small size, and continues to expand to this day. The framework for the model comes from Albert Einstein's Theory of General Relativity. In 1922, Alexander Friedman, a Russian cosmologist and mathematician, derived a series of equations based on General Relativity indicating that the universe might be expanding, contrary to the then-popular idea of a static universe. In 1927, Belgian physicist and Catholic priest Georges Lemaître independently derived similar equations. In 1929, American astronomer Edwin Hubble published the results of a decade of observations showing that the distances to faraway galaxies were proportional to the *red-shift* of the light received from them. The red-shift is the lengthening of the wavelength of light caused by rapid movement away from the observer (the *Doppler shift*). Thus, Hubble's observations imply that galaxies are receding from us at speeds proportional to their distance. Lemaître's "hypothesis of the primeval atom," published in 1931, combined the theoretical and observational work into the first version of the Big Bang model.

3.1.1 The Big Bang Model

The Big Bang model has become much more sophisticated over the ~90 years since it was first articulated. Readable summaries of the main features of the Big Bang model are given by Hawking (1988) and Weinberg (1993). Mathematical models break down when describing the initial state of the universe, so the earliest stages are subject to considerable uncertainty. Immediately after the Big Bang, the universe consisted of an incredibly high energy density of unimaginable temperature and pressure, and it was expanding and cooling very rapidly. Approximately 10^{-35} seconds into the expansion, the universe went through a phase of exponential expansion, known as *cosmic inflation*, driven by a poorly understood negative-pressure vacuum energy density. During this brief phase, the universe expanded faster than the speed of light. Because of inflation, regions on opposite sides of the universe can no longer exchange information, because this exchange is limited by the speed of light. Cosmic inflation explains why the universe seems to have identical physical properties throughout, even though distant parts of the universe cannot exchange information. Cosmic inflation also explains the origin of large-scale structure in the cosmos, which is the manifestation of quantum fluctuations in the microscopic pre-inflationary universe magnified to cosmic size.

After inflation stopped, the universe consisted of an expanding quark–gluon plasma containing all of the elementary particles. From this point on, detailed, self-consistent mathematical models describe the evolution of the universe. Temperatures were so high that motions of particles were near the speed of light, and particle–antiparticle pairs of all kinds were being continuously created and destroyed in collisions. At some point, an unknown process called *baryogenesis* violated the conservation of baryon number (protons and neutrons are baryons), leading to a very small (one part in 30 million) excess of quarks and leptons over antiquarks and antileptons. This led to the predominance of matter over antimatter in the present universe. As the universe continued to expand and cool, various symmetry-breaking phase transitions put the fundamental forces of physics and the parameters of elementary particles into their present form.

After about 10^{-11} seconds, particle energies had dropped to values that can be attained in particle physics experiments. At $\sim 10^{-6}$ seconds, quarks and gluons combined to form baryons (protons and neutrons). The small excess of quarks over antiquarks led to a small excess of baryons over antibaryons. Because the temperature was no longer high enough to create new proton–antiproton pairs, a mass annihilation occurred, leaving just one in 10^{10} of the original protons and neutrons and none of their antiparticles. At about 1 second, a similar process happened for electrons and positrons. After these annihilations, the energy density of the universe was dominated by photons.

A few minutes into the expansion, when the temperature had dropped to $\sim 10^9$ degrees, neutrons combined with protons to form deuterium and helium nuclei. Most protons remained uncombined as hydrogen nuclei. As the universe continued to cool, the rest-mass energy density of matter (gravity) came to exceed the energy density of the photons (electromagnetic radiation). After about 379,000 years, electrons and nuclei combined into atoms, mostly hydrogen, and the universe became transparent to light. Over the next $\sim 10^9$ years, slightly denser regions of the nearly uniformly distributed matter gravitationally attracted nearby matter and began to form gas clouds, stars, galaxies, and other astronomical structures observable today.

Three types of matter exist in the universe today: cold dark matter (matter that cannot be detected by electromagnetic radiation), hot dark matter (dark matter traveling at near the speed of light, often considered to be neutrinos), and baryonic matter (normal chemical elements). The dominant form of matter today appears to be cold dark matter, with the two other types of matter making up less than ~18% of the universe. Observations also indicate that ~72% of the total energy density in today's universe is in the form of dark energy, a mysterious form of energy that cannot be directly detected. Dark energy is postulated to explain an apparent acceleration of the expansion of the universe over the last few billion years.

3.1.2 Observational Evidence

The Big Bang model is based on three lines of observational evidence. The first is Hubble's original observations relating distance to red-shift and to the speed of receding galaxies. This relationship is described by the *Hubble Constant*. The second is the *cosmic microwave background*, which is the electromagnetic radiation left over from the Big Bang that can be measured today. The third is the relative abundances of hydrogen, helium, and lithium in the universe. Their abundances are in good agreement with specific predictions of the Big Bang model.

3.1.2.1 *The Hubble Constant*

Astronomers use a variety of methods to determine the distance to objects in the universe. One of the most effective is the "standard candle" provided by type Ia supernovae. These supernovae originate in binary star systems when a white dwarf star accretes matter from its companion. When the white dwarf reaches the *Chandrasekhar mass* of ~1.44 solar masses, a thermonuclear runaway occurs that completely disrupts the star in a cataclysmic explosion that makes a supernova as bright as an entire galaxy. Because type Ia supernovae occur in stars with similar masses and because the nuclear burning affects the entire star, they all have essentially the same intrinsic brightness, and their apparent brightness observed from Earth can be used to derive the distance to the supernova.

A measure of the speed at which a galaxy is traveling relative to the Earth can be obtained by looking at the spectrum of electromagnetic radiation (including visible light) arriving at Earth from that galaxy. When an observer is moving relative to the source of waves, the wavelength changes, becoming longer if the distance between the observer and the source is increasing and shorter if the distance is decreasing. A similar Doppler shift is commonly observed when a vehicle sounding a siren approaches, passes, and moves away from an observer. A higher pitch is heard as the vehicle approaches and a lower pitch is heard as it moves away. The same phenomenon is observed with electromagnetic radiation, a star emits a characteristic spectrum that depends on the mass and stage of evolution of the star. Lines appear in the spectrum due to absorption by atoms in the stellar atmosphere. When the star is moving away from the Earth, the entire spectrum, including the absorption lines, is shifted to longer (redder) wavelengths; the spectrum is said to be *red-shifted*. A star moving toward the Earth will have a spectrum shifted toward shorter (bluer) wavelengths; such a spectrum is said to be *blue-shifted*. The same is true for galaxies, if they are far enough away. Nearby galaxies have spectra that are dominated by the effects of rotation, with the side of the galaxy where stars are moving toward the Earth being blue-shifted and the side where they are moving away being red-shifted.

Hubble's crucial observation was that, in every direction one looks, the farther away a galaxy is, the more the light from that galaxy is red-shifted. If the red-shift is a Doppler shift, this implies that the farther away the galaxy is, the faster it is moving away from us. The most reasonable explanation for this observation is that the universe is uniformly expanding everywhere. The current best estimate for the Hubble Constant, which describes the rate of expansion, is ~70.8 km/s/Mpc, with an uncertainty of ~5.6%.

3.1.2.2 *Cosmic Microwave Background*

During the initial stages after the Big Bang, all of the particles and photons in the universe were in full thermal equilibrium. Under these conditions, the electromagnetic radiation had a *black-body spectrum*. Electrons and atomic nuclei (mostly protons) were unbound, and photons were continuously scattered by the electrons, making the early universe opaque to light. When the temperature fell due to expansion to a few thousand kelvin, electrons and nuclei began to combine to form atoms. Scattering of photons ceased, and the radiation became decoupled from the matter. Photons could travel unimpeded. As the universe expanded and cooled, the energy of the photons was red-shifted. Now, this primordial radiation falls into the microwave region of the electromagnetic spectrum and corresponds to a temperature of ~3 K. This radiation, which permeates the universe, was accidentally discovered in 1964 by Arno Penzias and Robert Wilson, who were conducting diagnostic observations at Bell Laboratories using a new microwave receiver. Penzias and Wilson were awarded the Nobel Prize for their discovery.

3.1.2.3 *Big Bang Nucleosynthesis*

The outlines of Big Bang nucleosynthesis were laid out in the calculations of Ralph Alpher and George Gamow in the 1940s. Together with Hans Bethe, they published a seminal paper (Alpher et al., 1948) outlining the theory of light-element production in the early universe. Big Bang nucleosynthesis began about three minutes after the Big Bang, when the universe had cooled sufficiently to permit protons and neutrons to combine into stable nuclei. The nuclear physics of Big Bang nucleosynthesis is well understood because the temperatures involved (~10^9 K) are found in the interiors of stars. As long as the universe was hot enough, protons and neutrons could easily transform into each other. As the temperature dropped, the equilibrium shifted in favor of protons because of their slightly lower mass, and nucleosynthesis ended with about one neutron for every seven protons. Helium-4 is more stable than free neutrons and protons and so has a strong tendency to form. However, formation of ^{4}He requires an intermediate step, which is the formation of deuterium (^{2}H), consisting of a proton and a neutron. Deuterium is only marginally stable and two deuterium atoms will easily combine to form ^{4}He. But until the universe became cool enough that the mean energy per particle was lower than the binding energy of deuterium, significant amounts of deuterium could not build up. This is known as the "deuterium bottleneck." Production of significant ^{4}He had to wait until the temperature of the universe became low enough for deuterium to be stable ($T = 10^9$ K). At this point a burst of nucleosynthesis occurred, but soon thereafter – at about twenty minutes after the Big Bang – the universe became too cool for any fusion reactions to occur. This froze in the abundances before the nuclear reactions could go to completion.

From this point on, the abundances were fixed, except for the decay of radioactive nuclides, until stars formed and stellar nucleosynthesis began. Big Bang nucleosynthesis did not produce elements heavier than beryllium because there are no stable nuclides with five or eight nucleons. In stars, this bottleneck is bypassed when three ^{4}He atoms combine to form ^{12}C (the triple-alpha process, see below). But by the time enough ^{4}He had built up, the temperature of the universe was too low for this process to operate efficiently, so it made a negligible contribution to Big Bang nucleosynthesis. The Big Bang model predicts mass abundances of ~75% ^{1}H (91.7% by number), ~25% ^{4}He (~8.3% by number), ~0.01% ^{2}D (~0.005% by number), and trace amounts (~10^{-10}) of ^{3}He, ^{6}Li, ^{7}Li, ^{9}Be, and perhaps ^{10}B and ^{11}B. Radioactive species such as ^{3}H (tritium), ^{7}Be, ^{10}Be, and shorter-lived species were produced but decayed away.

There is good agreement between observations and the predictions of Big Bang nucleosynthesis. The abundances of H, ^{4}He, ^{3}He, and deuterium match the

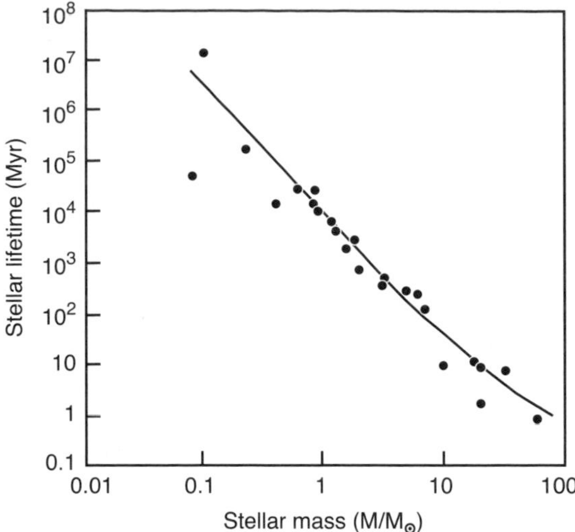

Figure 3.2 Lifetimes of stars are a strong function of stellar mass. The lifetimes of stars with masses <0.9 M$_\odot$ exceed the age of the galaxy. In contrast, stars more massive than 10 M$_\odot$ live fast and die young, expending their nuclear fuel in <10 million years. The change in slope of the trend for high-mass stars is due to mass loss, which significantly lowers their masses and increases their lifetimes.

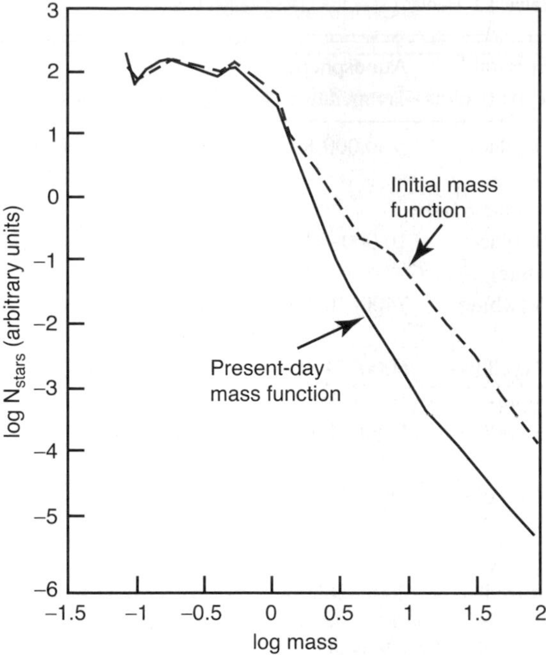

Figure 3.3 Number of stars plotted as a function of stellar mass in units of solar masses. Two distributions are plotted. The present-day mass function (PDMF) illustrates the relative abundances of stars in the galaxy today. The initial mass function (IMF) shows the number of stars of each mass produced in a single episode of star formation. The two distributions can be compared if the IMF is normalized to the same volume of space represented by the PDMF. All of the low-mass stars that have formed in the portion of the galaxy plotted here are still present, but high-mass stars are depleted in the PDMF. The difference in the two curves at high stellar mass reflects the absence from the PDMF of the stars that have exhausted their nuclear fuel and died over galactic history. After Basu and Rana (1992).

distributions can be compared (Fig. 3.3). Note that the two curves are the same for the lowest-mass stars, for which even the oldest have not yet exhausted their nuclear fuel. But for high-mass stars, the IMF is significantly higher than the PDMF, reflecting the fact that most of the high-mass stars that formed in the galaxy have long ago exhausted their fuel and died. These missing stars are the ones that have contributed the heavy elements to the galaxy. The IMF has been very reproducible over the history of the galaxy, and it provides the basis for modeling galactic chemical evolution.

3.2.2 The Life Cycles of Stars

The life cycle of a star begins in a cold, dark molecular cloud. A particularly dense region in the cloud may become gravitationally unstable and collapse. As the density increases, the temperature rises due to the conversion of gravitational potential energy to thermal energy. The central region quickly contracts to form a *protostar*. Differential motions cause the cloud core to rotate, and material arriving later joins the star through an accretion disk. Processes within the disk and ejection of material via polar jets (Fig. 3.4) are mechanisms by which angular momentum is removed from the system, aiding accretion of the protostar. As the protostar contracts and heats up, hydrogen and helium are ionized and the protostar evolves toward hydrostatic equilibrium. During this period, the protostar becomes completely convective and has a very high luminosity. As low-mass protostars evolve, luminosity decreases substantially, but surface temperature remains about the same. Low-mass protostars follow a nearly vertical track on the H–R diagram, called the Hayashi track, toward the main sequence (Fig. 3.5). Higher-mass stars move down and/or to the left to join the main sequence (Fig. 3.5).

A protostar first begins to shine by radiating gravitational potential energy, with half of the released energy going into increasing the temperature in the star and half being radiated into space. But, as the star approaches the main sequence, fusion reactions begin. When the core temperature reaches ~10^6 degrees, hydrogen begins to combine with deuterium to make ^{3}He. The reaction is $^1H + ^2D \Rightarrow ^3He + \gamma$, where γ is a gamma ray (reactions

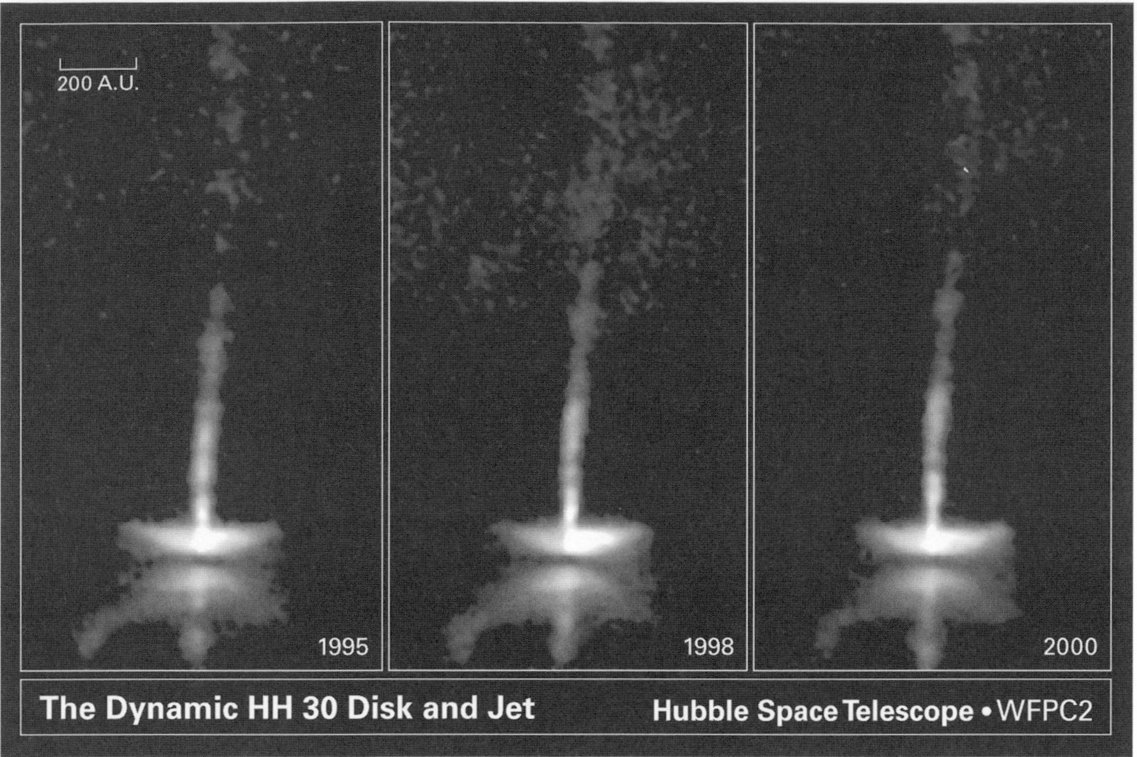

Figure 3.4 Images of a bipolar jet from a protostar taken by the Hubble Space Telescope. Evolution of the jet with time is visible across the three images.

such as this are generally written ^{2}D(p,γ)^{3}He; see Box 3.1). But the deuterium is soon exhausted because of its low initial abundance in the star, so deuterium burning only slows the contraction somewhat. Proton reactions also occur with lithium, beryllium, and boron as the temperature passes through a few million degrees. But as with deuterium, the fuel is soon exhausted and the protostar continues to contract. Fusion of hydrogen must begin before enough energy is released to stabilize the star against further contraction. As the protostar approaches hydrostatic equilibrium, its position on the H–R diagram drops to slightly below the main sequence (Fig. 3.5).

The nuclear reaction that finally stabilizes the structure of the protostar is the fusion of two protons to form a deuterium atom, a positron, and a neutrino: ^{1}H(p,$\beta^+\nu$)^{2}D. This reaction becomes important at a temperature of a few million degrees. The newly produced deuterium quickly reacts with a proton to form ^{3}He, which quickly reacts through several channels to form ^{4}He. The *proton–proton chains* (see below) are the main

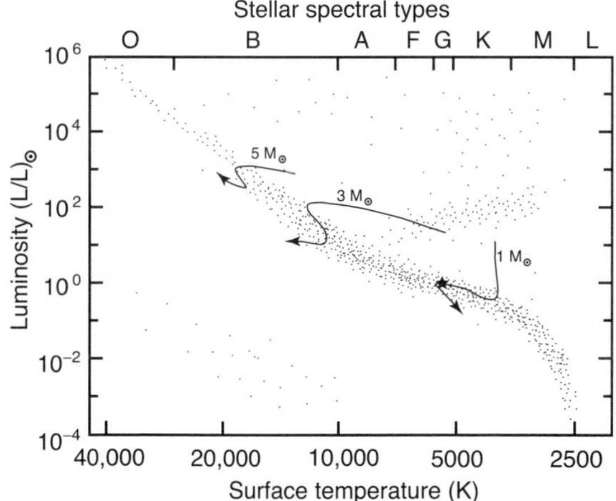

Figure 3.5 Tracks followed by 1, 3, and 5 $M_\odot$ pre-main sequence stars as they evolve toward the main sequence, as shown on an H–R diagram. Pre-main sequence stars shine primarily by conversion of gravitational potential energy to heat, although energy released by burning of deuterium and other elements also plays a role.

Box 3.1 Convention for Writing Nuclear Reactions

Nuclear reactions can be described quickly and clearly using the following convention. A reaction of the form

$$A + B \rightarrow C + D \quad \text{is typically written} \quad A(B,C)D$$

where A and D are nuclides and B and C are typically light particles such as a neutron (n), proton (p), α-particle, electron or positron (β^- or β^+), photon (γ), or neutrino (ν). As an example, neutron-capture reactions such as

$$^{48}\mathrm{Ti} + \mathrm{n} \rightarrow {}^{49}\mathrm{Ti} + \gamma \quad \text{would be written} \quad {}^{48}\mathrm{Ti}(\mathrm{n},\gamma)^{49}\mathrm{Ti}$$

Reactions involving protons such as

$$^{25}\mathrm{Mg} + \mathrm{p} \rightarrow {}^{26}\mathrm{Al} + \gamma \quad \text{would be written} \quad {}^{25}\mathrm{Mg}(\mathrm{p},\gamma)^{26}\mathrm{Al}$$

Reactions with α-particles (helium nuclei) such as

$$^{12}\mathrm{C} + \alpha \rightarrow {}^{16}\mathrm{O} + \gamma \quad \text{would be written} \quad {}^{12}\mathrm{C}(\alpha,\gamma)^{16}\mathrm{O}$$

More complicated reactions can also be written using this formalism. For example, the reactions of carbon burning, where two carbon atoms fuse to make heavier elements, are written

$$^{12}\mathrm{C}(^{12}\mathrm{C},\alpha)^{20}\mathrm{Ne} \text{ and } {}^{12}\mathrm{C}(^{12}\mathrm{C},\gamma)^{24}\mathrm{Mg}$$

A chain of reactions can also be written using this convention. For example, the reactions that make up the CNO cycle can be written

$$^{12}\mathrm{C}(\mathrm{p},\gamma)^{13}\mathrm{N}(,\beta^+\nu)^{13}\mathrm{C}(\mathrm{p},\gamma)^{14}\mathrm{N}(\mathrm{p},\gamma)^{15}\mathrm{O}(,\beta^+\nu)^{15}\mathrm{N}(\mathrm{p},\alpha)^{12}\mathrm{C}$$

In this chain, ^{12}C captures a proton and emits a gamma ray to become ^{13}N, which spontaneously decays by emitting a positron and a neutrino to become ^{13}C, which captures a proton and emits a gamma ray to become ^{14}N, which captures a proton and emits a gamma ray to become ^{15}O, which spontaneously decays to ^{15}N, which captures a proton and emits an α-particle to become ^{12}C.

Using this convention, almost any nuclear reaction can be written clearly and concisely.

source of nuclear energy in the Sun. In stars more massive than our Sun, catalyzed hydrogen burning can proceed via the *CNO cycles* (see below). With the initiation of hydrogen burning in the stellar core, a star settles into a stable configuration, in which gravitational contraction is balanced by the thermal energy of hydrogen fusion, and joins the main sequence. The star will spend most of its life on the main sequence, during which time it will slowly increase in luminosity and cross the main sequence on the H–R diagram (Figs. 3.1 and 3.5). The implications of the Sun's evolution for the planets are considered in Box 3.2.

The end of hydrogen burning in the core marks the end of a star's life on the main sequence. Figure 3.6 summarizes the stages of evolution of low- to intermediate-mass stars from the main sequence to the end of nuclear burning. When the hydrogen in the core is exhausted, the helium core contracts because it is no longer supported by sufficient nuclear energy, and heats up due to compression. The thermal instability introduced by the cessation of core hydrogen burning causes the outer 70–80% of the star to become convective. The convection mixes throughout the envelope the products of partial hydrogen burning that took place at the base of the stratified envelope during the main sequence. This mixing event is known as the *first dredge-up*, because this is the first time that nucleosynthesis products appear at the stellar surface. Because of the increasing temperature and pressure in the hydrogen-rich zone just outside of the helium core, hydrogen burning resumes in a shell around the hydrogen-depleted core. The shell gradually moves outward, consuming the hydrogen as it goes, and the star heats up. The envelope expands greatly during this period due to increasing thermal pressure and the outer envelope of the star cools. The star is now a red giant and on the H–R diagram has moved up and to the right of its position on the main sequence (Fig. 3.7).

Box 3.2 A Faint Young Sun

Ever since the nuclear source for stars' energy was first understood (Burbidge et al., 1957), it has been recognized that stellar luminosity increases over time. This is a natural consequence of the fusion ("burning") of hydrogen atoms into helium. Our Sun is no exception. The Sun initially had ~71% hydrogen in its core, but it now has ~34%. The production and accumulation of helium in its core increases the core density, thereby causing gravitational contraction and adiabatic heating, which in turn increase core temperature and luminosity (Bahcall et al., 2001). The solar luminosity at the Sun's birth is calculated to have been only ~70% of its current value.

A cooler Sun should have equated to colder planetary surface temperatures. In fact, the early Earth is estimated to have been frigid enough to be covered with ice. However, there is ample evidence that the Archean Earth (4000–2500 million years ago) had liquid water on its surface. This conundrum is known as "the faint young Sun problem." Mars, too, was warmer and wetter in its early history than predicted by solar heating alone. Several hypotheses have been advanced to try to get around this problem.

One idea is that the Sun was more massive and thus more luminous in its youth than it is today. The Sun then blew off mass through solar wind and coronal mass ejections. Such mass loss is plausible, because rotating stars gradually lose angular momentum and, in the process, eject mass. However, astrophysical observations indicate that spindown, and therefore mass loss, occurs during the first few hundred million years of a star's lifetime, far too quickly to explain an unfrozen early Earth and Mars half a billion years later.

A more likely scenario is atmospheric heating due to greenhouse gases. The dominant gases in the early Earth's atmosphere were CO_2 and N_2 (Catling and Kasting, 2017), and the carbon dioxide, perhaps aided by some methane, could have absorbed solar infrared radiation and moderated the climate. Other factors, such as a lower cloud albedo and a low ratio of land to water, may have also played a role. Mars is a little more complicated and uncertain, but a thicker early atmosphere with abundant CO_2, produced through serpentinization of olivine and pyroxene by hydrothermal fluids in the martian crust, may have helped promote at least transient warming episodes.

For a more complete discussion of solar energy production over time and the evidence for a warm and wet early Earth and Mars, see Spencer (2019).

Eventually, the core becomes hot enough for helium to begin to fuse into carbon. In low-mass stars, core helium burning begins with a helium flash. After the helium flash, fusion in the core settles down to a balance between thermal pressure and gravitational compression. The star contracts, its surface temperature increases, and it moves back toward the main sequence in the H–R diagram (Fig. 3.7). For intermediate-mass stars, this track is called the horizontal branch. This stage, during which helium is converted to carbon and oxygen in the core via the *triple-alpha process* and hydrogen fuses to helium in the hydrogen shell outside of the core, lasts about 10% as long as the main sequence.

The red giant stage ends when helium in the core is exhausted. Again, the core contracts and the thermal structure of the star becomes unstable. Convective mixing again reaches down toward the layers that have experienced nuclear burning. This mixing event is known as the *second dredge-up*. In low-mass stars, the convection is not deep enough to bring newly synthesized material to the surface, but in stars of greater than 4–5 $M_{\odot}$, the mixing encounters the hydrogen shell and nucleosynthesis products are again dredged up. As the mixing subsides, the hydrogen shell is again established and the core slowly contracts and heats up. A zone of helium builds up between the hydrogen shell and the carbon-rich core.

When the temperature gets high enough, the helium outside the core ignites and burns in a runaway "flash," creating a thermal pulse that mixes the material from the helium shell and the intershell region between the hydrogen and helium shells. The star then readjusts its structure to radiate the energy produced, and the convective envelope reaches down into the material mixed by the thermal pulse and brings the products of helium burning and *s*-process nucleosynthesis to the surface. This mixing event is known as the *third dredge-up*. The helium flash terminates quickly due to the exhaustion of helium and the star settles down again, with only the hydrogen shell burning to supply energy. The hydrogen shell burns outward, leaving behind a hydrogen-depleted, helium-rich region. The interior of the star contracts and heats up, eventually triggering another helium-shell flash and another third dredge-up mixing event. This sequence repeats 10–100 times depending on the mass of the star. On an H–R diagram, low- and intermediate-mass stars move back toward the red giant branch, approaching it asymptotically (Fig. 3.7).

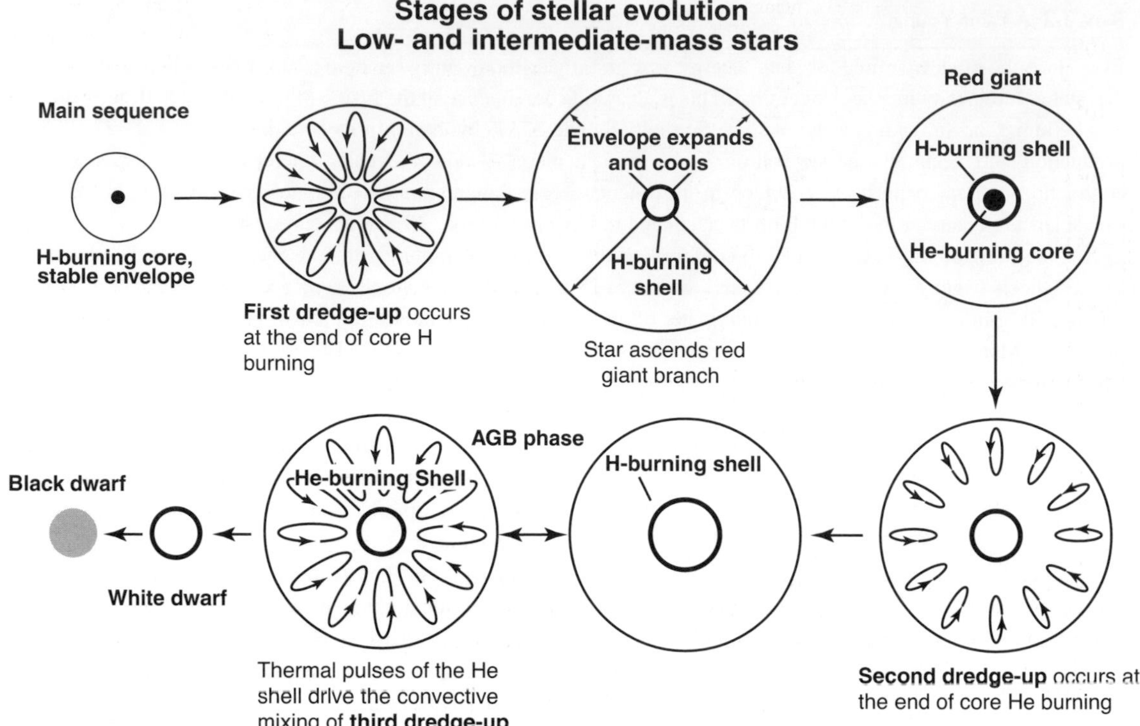

Figure 3.6 Stages of stellar evolution for low- and intermediate-mass stars. As the star exhausts its hydrogen fuel and leaves the main sequence, the helium core contracts and heats up, and the resulting thermal instability mixes the products of partial hydrogen burning throughout the stellar envelope in the first dredge-up. The star expands and the surface cools. Hydrogen burning continues in a shell around the core, and the star ascends the red giant branch (see Fig. 3.7). When the core is hot enough, helium burning is initiated and the star returns to a stable configuration: the red giant stage. When helium is exhausted in the core, the core again contracts. The envelope becomes convective again in the second dredge-up. Core contraction continues, along with shell hydrogen burning, which leaves behind a region dominated by helium outside of the core. When the temperature gets high enough, the helium ignites and burns in a runaway "flash" that again disrupts the thermal structure of the star. Convection in the envelope brings products of helium burning and *s*-process nucleosynthesis to the surface in one of many cycles of the third dredge-up. The star is now an asymptotic giant branch (AGB) star. The surface of the star heats up as the hydrogen shell burns toward the surface, and toward the end of the AGB phase, a planetary nebula is emitted. Then the nuclear fuel runs out and the star cools and contracts to become a white dwarf and eventually a black dwarf.

Such stars are known as *asymptotic giant branch* (AGB) stars. Herwig (2005) gives a detailed review of the evolution of AGB stars, and Busso et al. (1999) gives a good review of nucleosynthesis in these stars.

During the AGB phase, the stars eject matter in strong stellar winds that increase in intensity with time. Towards the end of the AGB stage, the surface temperature of the star begins to increase as the hydrogen shell gets closer to the surface. The star moves to the left across the H–R diagram (Fig. 3.7). The surface temperature of the star can exceed 25,000 K and the brightness can be several thousand times the solar luminosity. The flux of ionizing photons increases with the surface temperature and begins to excite the dust that condensed in the earlier winds. The radiation flux illuminates the surrounding dust shell, making a *planetary nebula* that is visible from Earth. The surface temperature of the star continues to rise, reaching a maximum of >100,000 K, and the planetary nebula reaches peak illumination. This planetary nebula stage lasts about 10,000 years, after which the nuclear fuel is exhausted. The luminosity and temperature of the central star decrease and the star contracts, shining from gravitational and stored energy only and eventually becoming a *white dwarf*. The core continues to contract until it is

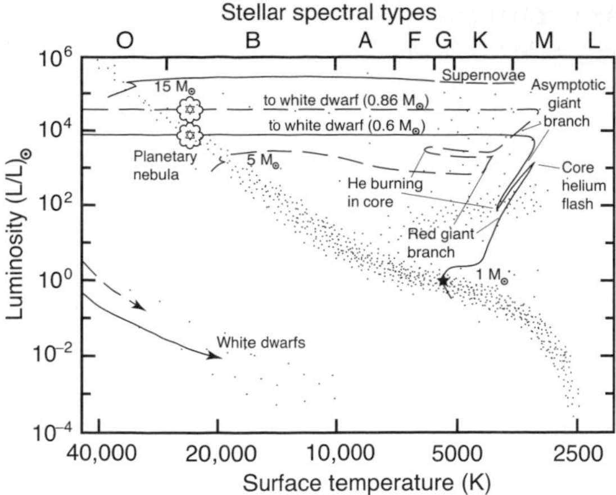

Figure 3.7 H–R diagram with tracks followed by 1, 5, and 15 $M_\odot$ stars as they leave the main sequence after hydrogen burning ends in their cores. Energy is provided by a series of different nuclear fuels, with more elements involved in nuclear reactions in higher-mass stars (see text). Low-mass stars ascend the red giant branch until core helium burning begins. The star heats up and moves back toward the main sequence. At the end of core helium burning, the star again moves upward along the asymptotic giant branch. As the hydrogen shell burns toward the stellar surface, the star heats up and moves to the left across the upper part of the diagram, ejecting its envelope via a strong wind. The ejected gas and dust eventually are illuminated as a planetary nebula. Shortly thereafter, the star exhausts its fuel and burning ceases. The star moves down the diagram to become a white dwarf. Intermediate-mass stars follow a similar progression, but they tend to cool more during the red giant stage, moving right across the diagram. Massive stars move right across the top of the diagram, becoming supergiants before ending their lives with type II supernova explosions.

supported by *electron degeneracy*. In a normal gas, compression to a higher density occurs by squeezing out the space between the atoms. In a degenerate gas, this process has reached its limit and the electrons reside in the lowest available energy states. The matter can contract no further. White dwarfs are very hot, very small, faint stars. They form a diagonal band across the lower left of the H–R diagram, roughly parallel to the main sequence (Fig. 3.7). White dwarfs have masses ranging from slightly larger than the Sun to considerably smaller, but they have radii ranging from about 0.01 to 0.001 times that of the Sun (roughly the size of the Earth). They have exhausted all of their nuclear fuel and are no longer generating energy. They are cooling and contracting and after several hundred billion years they will become cold *black dwarfs*.

For stars between ~3 $M_\odot$ and ~8 $M_\odot$, the core temperature is high enough to initiate carbon burning. An inert core of oxygen, magnesium, and neon accumulates. After ~1000 years, the star runs out of carbon and the core cools and contracts. The carbon-burning stage occurs at the end of the AGB stage at a point when the star is emitting massive winds. These stars also generate planetary nebulae before cooling and contracting to become O–Ne–Mg white dwarfs.

Figure 3.8 summarizes the stages of the post-main-sequence evolution of stars with masses >10 $M_\odot$. The early part of this evolution is similar to that for low- and intermediate-mass stars, but in stars with masses >10 $M_\odot$, the nuclear burning does not stop with hydrogen, helium, and carbon. In these stars, once the carbon in the core is exhausted, gravity causes the core to contract and heat up sufficiently to initiate neon burning, followed in turn by oxygen burning and silicon burning (Fig. 3.8). As each fuel is exhausted in the core, a surrounding shell burning that fuel is ignited, creating an onion-shell strcture. These stages are increasingly rapid because less and less energy is being released, until a core of iron is produced. Fusion reactions cannot proceed further because nuclear reactions no longer supply energy; instead they require energy to proceed (cf. Fig. 2.1). The iron core of the star builds up very quickly to the ~1.44 $M_\odot$ Chandrasekhar mass, the maximum mass that can be supported by electron degeneracy. This happens in about a day in a 25 $M_\odot$ star. The temperature in the core reaches 10 billion degrees. The iron nuclei disintegrate into protons and neutrons. This requires energy, so the core cools. The very high pressure in the core causes the electrons and protons to combine into neutrons – a process that releases a blast of neutrinos. The neutrinos carry away a tremendous amount of energy from the core and the neutrons collapse to nuclear density (>10^{17} g/cm^3), shrinking the neutron core to a tiny fraction of its previous size almost instantaneously. The neutrons now form a degenerate gas (neutron degeneracy is broadly analogous to electron degeneracy; the neutrons cannot move in spite of their high energy, and they behave like matter at near 0 K). The surrounding layers are no longer supported and collapse onto the neutron ball at supersonic speed. The infalling material bounces off the neutron core and a shock wave is created, which along with the neutrinos released from the core, ejects the outer layers of the star at speeds of 10,000–20,000 km/s. The star explodes as a type II core-collapse supernova, with a luminosity of 10^8 times the solar luminosity. Most supernovae leave behind a neutron star with a radius of a few kilometers and a mass similar to that of the Sun, rotating as fast as 30 times per second (these are observed as pulsars). If the mass left behind by the explosion is high enough, the remaining material may collapse into a black hole. The mass cut that seems to separate these two outcomes is ~25 $M_\odot$.

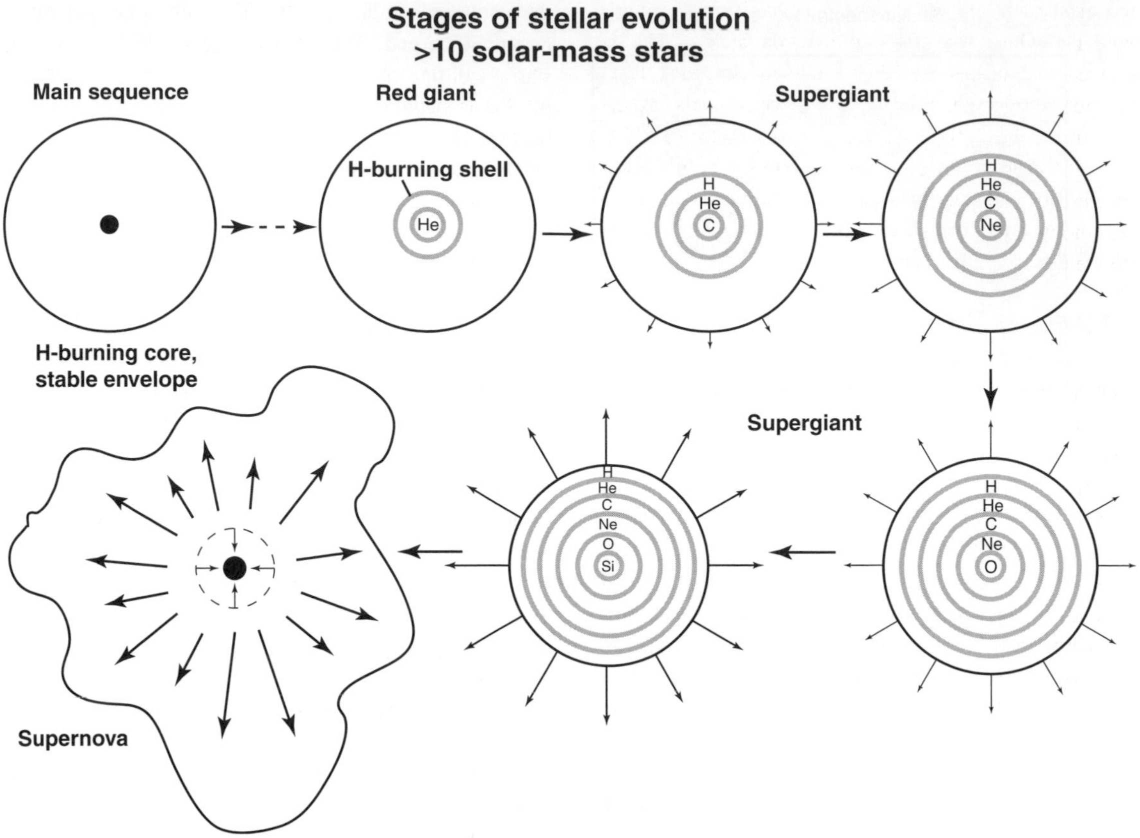

Stages of stellar evolution
>10 solar-mass stars

Main sequence
H-burning core, stable envelope
Red giant
H-burning shell
He
Supergiant
H
He
C
Ne
O
Si
Supernova

Figure 3.8 Stages of stellar evolution for stars with masses >10 $M_{\odot}$. The initial stages of burning in a massive star are analogous to those in low- and intermediate-mass stars (Fig. 3.6). But after helium is exhausted in the core, the star burns a series of new fuels – predominantly carbon, neon, oxygen, and silicon – in the core. As each fuel is exhausted in the core, it begins to burn in a shell around the core, creating an onion-shell structure, with each layer containing products of different nucleosynthetic burning. When silicon has been consumed in the core, there is no more nuclear fuel to support the star against gravity and it collapses catastrophically (see text). The result is a type II supernova explosion that ejects the outer and intermediate layers of the star, returning newly synthesized elements to the galaxy. The interior zones collapse back on the remnant, which can become either a neutron star or, if the progenitor was >25 $M_{\odot}$, a black hole.

The stars destined to become type II supernovae are the *supergiants* that occupy the top of the H–R diagram (Fig. 3.7). Supergiants have very high luminosities and appear in all spectral classes. These are all very massive stars that have two or more shells of nuclear burning. These stars are often losing mass at a tremendous rate. Stars of more than ~33 $M_{\odot}$ may lose their entire envelopes through massive stellar winds and become Wolf–Rayet stars. These unusual stars have newly synthesized material at their surfaces. They also eventually become core-collapse supernovae. Supernovae are classified spectroscopically. Those without hydrogen in their spectra are type I supernovae and those with hydrogen are type II supernovae. Because stars of <33 $M_{\odot}$ retain their hydrogen envelope, they form type II supernovae. Wolf–Rayet stars explode as type Ib (no hydrogen but helium in the spectrum) or Ic (no hydrogen or helium in the spectrum) core-collapse supernovae. Supernova explosions are major contributors to the elements heavier than hydrogen and helium in the galaxy.

Roughly 40% of the objects in the galaxy are binary star systems (or stated another way, ~60% of stars have companions). In many cases one member of the system is significantly larger than the other and evolves more quickly. The more-evolved star becomes a white dwarf, and when the envelope of the other star expands during the red giant stage, the outer portion is gravitationally pulled onto the white dwarf. The transferred material is compressed and heated and eventually an explosive fusion reaction is triggered. If the explosive reaction only involves the surface layers of the star, a *nova* results. But if the accretion triggers explosive nuclear burning of the carbon-oxygen core, the star is completely disrupted as a type Ia supernova explosion. Novae and type Ia supernovae contribute newly synthesized material to the galaxy.

3.2.3 Stellar Nucleosynthesis: Quiescent Processes

Nuclear fusion is the engine that makes stars shine, and it is also a dominant pathway for synthesizing new elements. In order for nuclear fusion to occur, two nuclei, which are positively charged due to the protons in them, first have to overcome the repulsive electrostatic force between them – the Coulomb barrier. This repulsive force acts over a long range and falls off with distance, r, as $1/r^2$. When the two nuclei get close enough together (within about the diameter of an atomic nucleus), the attractive, short-range strong nuclear force takes over and the nuclei fuse. Figure 3.9a illustrates the competition between electrostatic repulsion and the strong nuclear attractive force.

In classical physics, the only way to overcome the Coulomb barrier is to increase the kinetic energy of the incoming particle. To do this by heating a gas requires a temperature many orders of magnitude higher than that in the core of the Sun. So how does fusion happen? Quantum mechanics comes to the rescue. According to quantum mechanics, there is a finite probability that an incoming particle, for a very short period of time, will have enough energy to overcome the Coulomb barrier and fuse with the nucleus. This is known as quantum tunneling. A particle in the high-energy tail of the Maxwell–Boltzmann thermal distribution that is at the same time existing in a high-energy quantum state will have a higher probability of fusing with the nucleus, if it is close enough. The overlap of the tunneling probability and the Maxwell–Boltzmann distribution is called the Gamow peak. The Gamow peak allows fusion reactions to occur at the relatively low temperatures that are found in the interiors of stars. Note that the larger the nucleus, the larger the repulsive electrostatic force, so the more massive the nuclei, the higher the temperature that is required for nuclear fusion to occur.

3.2.3.1 *Hydrogen Burning*

The dominant source of nuclear energy powering a star is the conversion of hydrogen to helium. There are two main ways to do this: 1) proton–proton chains, and 2) CNO cycles. Both methods have the net result of converting four hydrogen atoms into a ^{4}He atom, which releases 26.73 MeV of energy per ^{4}He atom. The reactions must

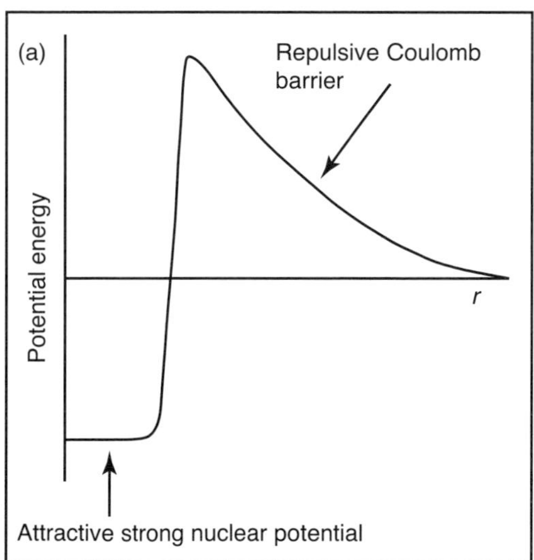

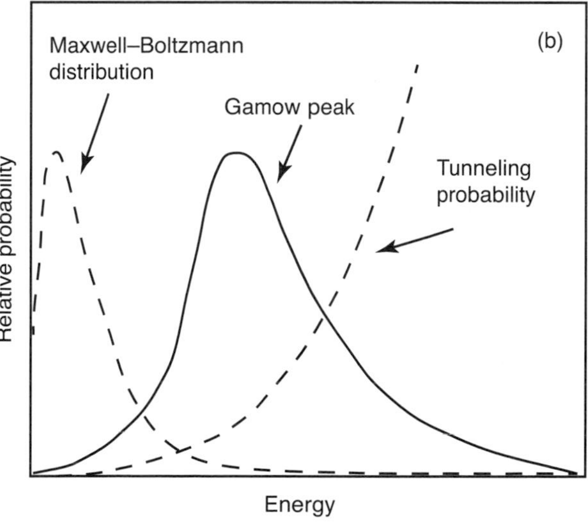

Figure 3.9 Energetics of nuclear fusion. (a) Schematic plot of potential energy as a function of separation distance (r) as two nuclei approach one another. For separation greater than about the diameter of an atomic nucleus, the electrostatic repulsion due to the positive charges on the nuclei dominates. If the nuclei can get within about a nuclear diameter, the strong nuclear force takes over and the nuclei can fuse. (b) The Gamow peak (solid curve) is the product of the Maxwell–Boltzmann distribution and the tunneling probability. Fusion takes place between particles with relative energies within the Gamow peak.

change two protons into two neutrons through either electron capture or positron emission, so two neutrinos are emitted for each ^{4}He created.

In first-generation stars, those that consist almost entirely of hydrogen and helium, and in stars of later generations where the core temperature is less than ~20 million degrees ($M \leq M_{\odot}$), hydrogen burns to helium via *proton–proton chains.* The simplest chain, and the one activated at the lowest temperature (~5 million degrees), is called the PP I chain. Three reactions are involved. In the first step, two protons fuse to form a ^{2}He atom, which immediately emits a positron to form a deuterium atom and a neutrino [^{1}H(p,$\beta^+\nu$)^{2}D]. This reaction is very slow and controls the rate of the PP I chain. Deuterium then quickly combines with a proton to form ^{3}He [^{2}D (p,γ)^{3}He], and two ^{3}He nuclei combine to form ^{4}He and two hydrogen atoms [^{3}He(^{3}He,2p)^{4}He].

Helium-3 can also react with ^{4}He to make ^{7}Be, which then captures an electron to form ^{7}Li and a neutrino [^{3}He (α,γ)^{7}Be(β^-,ν)^{7}Li]. Lithium-7 then captures a proton and the resulting ^{8}Be nucleus breaks up into two ^{4}He atoms – one more than was present initially [^{7}Li(p,α)^{4}He]. This is the PP II chain, and it becomes more important as the star ages and ^{4}He becomes more abundant, and as the temperature increases beyond ~15 million degrees inside the star.

Beryllium-7 also reacts via two branches. Instead of capturing an electron to become ^{7}Li, ^{7}Be can react with a proton to form ^{8}B plus a gamma ray. ^{8}B emits a positron and a neutrino to form ^{8}Be, which then breaks up into two ^{4}He atoms [^{7}Be(p,γ)^{8}B(,$\beta^+\nu$)^{8}Be*(,α)^{4}He]. This is the PP III chain. PP III becomes important at temperatures above ~25 million degrees.

Most stars are second-, third-, or higher-generation stars that contain elements heavier than hydrogen and helium produced by the earlier generations of stars. In these stars, hydrogen is converted to helium by several reactions that use atoms of carbon, nitrogen, and oxygen as catalysts: the *CNO cycles*. A schematic representation of the three main CNO cycles is shown in Figure 3.10. The CN I cycle is the first to operate efficiently as the temperature rises. It consists of the following sequence of reactions (see Box 3.1):

$$^{12}\mathrm{C(p,\gamma)}^{13}\mathrm{N}(,\beta^{+}\nu)^{13}\mathrm{C(p,\gamma)}^{14}\mathrm{N}\ldots$$

$$\ldots^{14}\mathrm{N(p,\gamma)}^{15}\mathrm{O}(,\beta^{+}\nu)^{15}\mathrm{N(p,\alpha)}^{12}\mathrm{C}$$

The (,$\beta^+\nu$) reactions are essentially instantaneous and so are much faster than the other reactions at the temperatures in low- and intermediate-mass stars. The fastest proton reaction in this series is ^{15}N(p,α)^{12}C, while the slowest reaction is ^{14}N(p,γ)^{15}O. As a result, extensive CN cycling converts much of the ^{15}N, ^{12}C, and ^{13}C into ^{14}N. In the CN cycle, ^{13}C is destroyed more rapidly than ^{12}C by about a factor of three. But in the solar composition, ^{12}C is 89 times more abundant than ^{13}C, so initially more ^{13}C is produced from the destruction of ^{12}C than is destroyed by proton reactions. The ^{12}C/^{13}C ratio decreases until it reaches an equilibrium value equal to the inverse of the reaction rates, where as much ^{13}C is being destroyed as is being produced. From this point on, the ^{12}C/^{13}C ratio remains the same as carbon is gradually converted to ^{14}N.

The CNO bicycle, consisting of the CN I and CNO II loops in Figure 3.10, becomes important as the temperature increases. A small fraction of the ^{15}N is lost from the CN cycle via ^{15}N(p,γ)^{16}O. The relative importance of the two cycles is governed by the factor of ~1000 difference in the rates of the two ^{15}N reactions. The rate of the ^{16}O (p,γ)^{17}F reaction is very much slower than the reactions in the CN cycle, but ^{16}O is a factor of ~5000 more abundant than ^{15}N, so the rates at which the ^{15}N and ^{16}O reactions process nuclei through the CNO II cycle are similar. However, because the CNO II cycle is a factor of ~1000 slower than the CN cycle, it contributes little to the energy budget of the star.

There are additional cycles involving the CNO isotopes that also produce ^{4}He. ^{18}O is rapidly converted to ^{15}N via ^{18}O(p,α)^{15}N. Initially this dumps most of the original ^{18}O nuclei into the CN cycle. But the ^{18}O is slowly replaced via the CNO III cycle from ^{16}O via ^{16}O (p,γ)^{17}F(,$\beta^+\nu$)^{17}O(p,γ)^{18}F(,$\beta^+\nu$)^{18}O, and the ^{16}O nuclei are largely replaced through two pathways: ^{18}O(p,γ)^{19}F (p,α)^{16}O and ^{18}O(p,α)^{15}N(p,γ)^{16}O. A few nuclei escape from the CNO tricycle through ^{19}F to be synthesized into ^{20}Ne and heavier elements.

Figure 3.11 shows the abundances of hydrogen, helium, and the CNO isotopes as a function of stellar radius in a 1.5 $M_{\odot}$ star at the end of the main sequence. The horizontal axis represents the position within the star (in units of mass, a rather quirky astrophysics convention). The center of the star is at the left and the surface is at the right. The lines on the diagram show the mass fraction of each isotope at each position inside the star. For example, at the surface of the star, most of the mass consists of hydrogen, helium is the next most abundant, followed by ^{16}O, ^{12}C, ^{14}N, and so forth. The starting composition of this hypothetical star, and the composition that remains at the stellar surface, is the same as that of our Sun.

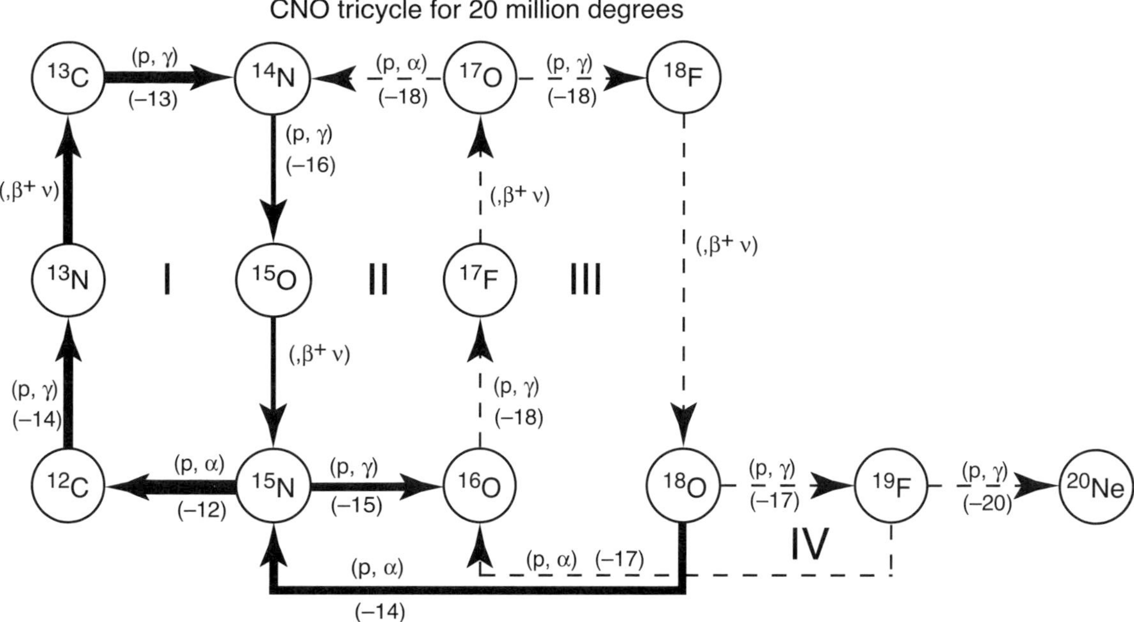

Figure 3.10 Schematic representation of catalytic hydrogen burning via the CNO tricycle. The arrows show the paths that catalyst nuclei follow. The sizes of the arrows are proportional to the reaction rates, so large arrows represent fast reactions with a high throughput. The type of reaction and the exponent of the reaction rate at 20 million degrees are shown for each reaction. At this relatively low temperature, the $(,\beta^+\nu)$ reactions are effectively instantaneous (to avoid cluttering the diagram, they are shown with the same size arrow as the reaction feeding them). The cycles labeled I, II, and III each consume four protons and produce an α-particle (^{4}He nucleus), with the original species returned to start a new cycle. The net result of extensive CNO cycling, in addition to converting hydrogen to helium, is to convert most of the CNO nuclei to ^{14}N. After Rolfs and Rodney (1988), with reaction rates from Caughlan and Fowler (1988).

In the center of the star, all of the hydrogen has been converted to helium, and most of the CNO nuclei have been converted to ^{14}N. The small amount of carbon that still remains has a ^{12}C/^{13}C ratio of ~3, and oxygen has a ^{17}O/^{16}O ratio almost 1000 times higher than in the starting composition. However, this portion of the star contains only ~15% of the mass. Outside of this helium core, hydrogen burning has not gone to completion. The temperature inside the star increases as a function of depth, and the nuclear reaction rates increase very strongly as a function of temperature. Between ~0.2 and 0.9 $M_\odot$, partial hydrogen burning has modified the initial composition of the star. With increasing depth, the first effects are seen in ^{15}N, which has the fastest reaction rate at all temperatures. Below 0.9 $M_\odot$, ^{15}N is depleted relative to the starting composition, and below 0.7 $M_\odot$, its abundance is a tiny fraction of its initial abundance, sustained only by CNO cycling. In contrast, ^{13}C, which is fed by the high-abundance isotope ^{12}C, initially builds up in abundance until the destruction rate of ^{13}C in CNO cycling matches that of ^{12}C. Below this point, the ^{12}C/^{13}C ratio stabilizes at ~3 and both isotopes are converted to ^{14}N. At a depth of ~0.5 $M_\odot$, the ^{18}O abundance drops as it is burned to ^{15}N in the CNO III cycle. But note that the ^{16}O abundance remains essentially unchanged until just outside the helium core. Compare the depths at which the abundance of each isotope begins to change with the relative reaction rates shown in Figure 3.10.

As hydrogen is exhausted in the stellar core, the energy source supporting the star is removed and the core contracts. Because the thermal balance has been disturbed, the thin convective layer at the surface of the star reaches down and mixes the products of partial hydrogen burning throughout the envelope. This mixing is the first dredge-up, described earlier. At the end of the first dredge-up, the envelope is enriched in ^{14}N and ^{17}O compared to the starting composition, and is depleted in ^{15}N and ^{18}O.

processes: the *s-process* and the *r-process*. The *s*-process is neutron capture on a timescale that is slow compared to the rate of β^- decay. The *r*-process is neutron capture on such a rapid timescale that many neutrons can be captured before β^- decay occurs.

The *s*-process occurs during the AGB stage in low- and intermediate-mass stars, when the hydrogen shell is burning outward from the core and the helium shell repeatedly ignites, generating third-dredge-up mixing events. It can also occur in massive stars during analogous stages. There are two main reactions that provide the neutrons to drive the *s*-process: $^{13}C(\alpha,n)^{16}O$ and $^{22}Ne(\alpha,n)^{25}Mg$. Both reactions require α-particles (^4He nuclei). The $^{13}C(\alpha,n)^{16}O$ reaction operates primarily in the inter-shell region between the hydrogen and helium shells while the helium shell is quiescent. The $^{22}Ne(\alpha,n)^{25}Mg$ reaction is marginally activated during the helium-shell flash. Because the neutrons from the two reactions are released at different temperatures, the details of the *s*-process nucleosynthesis that results are different and provide a probe of nucleosynthesis in stars.

The *s*-process is responsible for approximately half of the isotopes heavier than iron. Figure 3.13 is a portion of the Chart of the Nuclides with the *s*-process path highlighted. The *s*-process path defines the center of the Valley of β Stability on the Chart of the Nuclides. When the *s*-process is activated, the stable isotopes acquire neutrons and move to the right (constant *Z*) until the resulting nuclide is unstable and β decays to become an isotope of the next heavier element. As the process continues, the isotopes that do not capture neutrons very efficiently (those said to have low *neutron-capture cross sections*) build up, while those with high neutron-capture cross sections are depleted. If the neutron fluence (total number of neutrons available) is high enough, the abundances of the isotopes approach a steady state, where the abundances of isotopes of similar mass are proportional to the inverse of their neutron-capture cross sections.

3.2.4 Stellar Nucleosynthesis: Explosive Processes

Explosive nucleosynthesis occurs under conditions where temperature and density are changing rapidly with time, either due to the passage of a shock wave or because of a thermonuclear runaway. We highlight some of the more important processes here.

3.2.4.1 *Explosive Hydrogen Burning*

Explosive hydrogen burning occurs via the hot CNO cycle in nova explosions when the hydrogen accreted onto a white dwarf or neutron star from a binary companion reaches a critical density of ~10^3 g/cm^3. The resulting

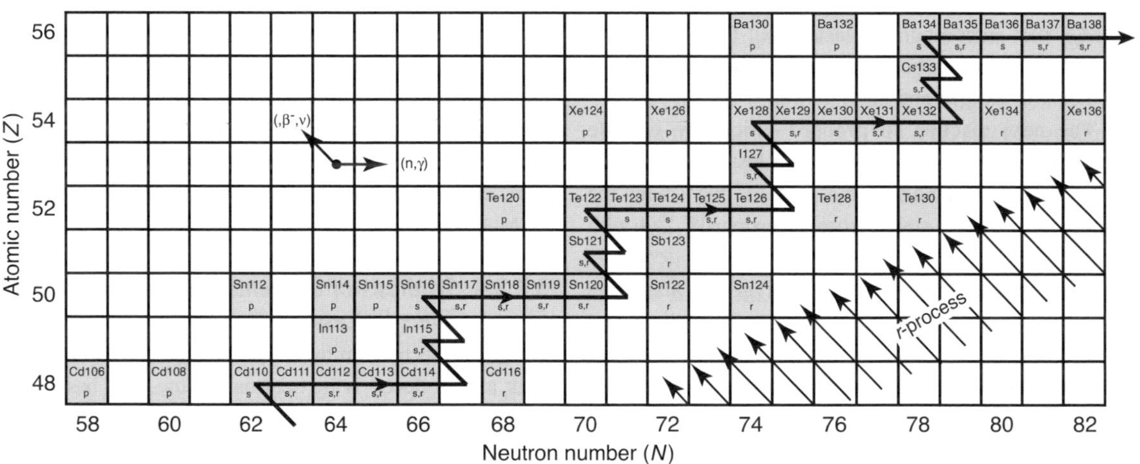

Figure 3.13 A portion of the Chart of the Nuclides showing the *s*-process and *r*-process pathways. The *s*-process pathway, shown by the dark line in the center of the Valley of β Stability, shows how a nuclide that successively captures individual neutrons would evolve. Each added neutron moves the nuclide to the right on the diagram, until it reaches an unstable nuclide, in which case it β decays to the next stable nuclide with a higher *Z*. In contrast, in situations where nuclides capture neutrons very rapidly (the *r*-process), they will be driven far to the right of the Valley of β Stability until the timescale for neutron capture matches that for β decay. They will then move to higher *Z* and capture more neutrons until they either reach a size that causes them to fission (break) into smaller nuclei (which can then capture more neutrons) or until the neutrons disappear, in which case they will β decay back to the first stable isotope along paths of constant *A* (arrows).

thermonuclear runaway powers the nova explosion. Explosive hydrogen burning also occurs in core-collapse supernovae as the shock passes through the hydrogen-rich envelope.

3.2.4.2 *Explosive Helium Burning*

Explosive helium burning occurs in supernova explosions when the shock wave passes through helium-rich layers. If the layer has a sufficiently high ^{22}Ne abundance, the passage of the shock wave triggers the ^{22}Ne(α,n)^{25}Mg reaction, which releases a burst of neutrons. This burst drives the stable seed nuclei to the neutron-rich side of the Valley of β Stability.

3.2.4.3 *Explosive Oxygen, Carbon, and Neon Burning*

Explosive oxygen, carbon, and neon burning occur in the oxygen-rich layers of massive stars during the passage of a supernova shock wave. The passage of the wave generates temperatures near 3 billion degrees. A key product is ^{28}Si, but some iron-group elements are also produced. In layers that are heated to 2–3 billion degrees, photodisintegration reactions [(γ,p), (γ,n), (γ,α)] on stable nuclei produce nuclei on the neutron-deficient side of the Valley of β Stability. Outside the layer that experiences explosive oxygen burning, conditions are appropriate for explosive carbon and neon burning. Explosive carbon burning via the reactions ^{12}C(^{12}C,2α)^{16}O, ^{12}C(^{12}C,α)^{20}Ne, ^{12}C(^{12}C,γ)^{24}Mg, and ^{12}C(^{12}C,p)^{23}Na is also the underlying energy source for type Ia supernovae.

3.2.4.4 *The r-Process*

The *r*-process is the process of rapid-neutron-capture nucleosynthesis and is responsible for about half of the isotopes heavier than iron. It occurs in explosive environments when heavy seed nuclei are exposed to a high density of neutrons. In core-collapse supernovae, the seed nuclei are the products of explosive nuclear statistical equilibrium, which obtains until the temperature drops sufficiently for charged-particle freeze-out, leaving the nuclei in a sea of free neutrons. The seed nuclei then rapidly capture the neutrons, making the heavy *r*-process nuclides. The timescale for neutron capture in the *r*-process is very short, and many neutrons are captured by each nuclide before it can β^- decay. The nuclides are driven far to the neutron-rich side of the Valley of β Stability (Fig. 3.13). When the neutrons are all consumed, the isotopes β decay along paths of constant *A* back toward the Valley of β Stability, following the arrows on Figure 3.13, until they reach a stable configuration. An analogous *p*-process, known as the *rp*-process, may also operate in which a nucleus captures protons and moves up and to the left on Figure 3.13 before decaying back toward the Valley of β Stability. The low-abundance isotopes to the left of the Valley of β Stability that are bypassed by the *s*-process may be produced either by the supposed *rp*-process or by the photodisintegration reactions described above.

There are two peaks in the solar system's *r*-process abundances: one at $A = {\sim}130$ and one at $A = {\sim}195$. A single *r*-process cannot produce this distribution. The abundances of short-lived radionuclides ^{129}I and ^{182}Hf in the early solar system also appear to require production in different sites. Hafnium-182 ($t_{1/2} = 8.9$ Myr) is significantly more abundant relative to stable hafnium isotopes in the early solar system than is ^{129}I ($t_{1/2} = 17$ Myr), in spite of having a much shorter half-life. Spectroscopic studies of very metal-poor stars, which formed during the first billion years after the Big Bang, also show evidence of more than one *r*-process site. Together these observations suggest that the light *r*-process nuclei (${\sim}110 < A < {\sim}130$) have a different source than the heavy *r*-process elements ($A > {\sim}130$).

The elements from ${\sim}90 < A < {\sim}110$ (sometimes called the CPR elements) are thought to be produced efficiently by charged-particle reactions in the neutrino-driven winds generated by the formation of a neutron star during a core-collapse supernova explosion (Woosley and Hoffman, 1992). These CPR elements are produced in an approximately fixed yield pattern in all supernovae from progenitors of $\leq 25\ M_\odot$. Although the CPR nuclei serve as seeds for the *r*-process, their abundances are not strongly coupled to heavy *r*-process elements, and the current understanding indicates that there are too few neutrons in the neutrino-driven wind to drive a true *r*-process (Qian and Woosley, 1996).

The gravity-wave observations from the merging of two neutron stars (GW170817) in August 2017 by the LIGO and Virgo detectors, followed by an associated gamma-ray burst (GRB170817A) and other electromagnetic effects, brought another potential location for the *r*-process back into the spotlight. Merging neutron stars had been discussed off and on since the late 1970s as a potential *r*-process site, but their occurrence rates and the mass yields of *r*-process isotopes were considered to be too low for them to play a major role in *r*-process synthesis. Work since the discovery of GW170817 changed the picture completely. The pattern of *r*-process isotopes obtained in model calculations is apparently quite robust and is consistent with the observed pattern in the solar system and in low-metallicity stars (e.g. Korobkin et al., 2012). The *r*-process yield of an

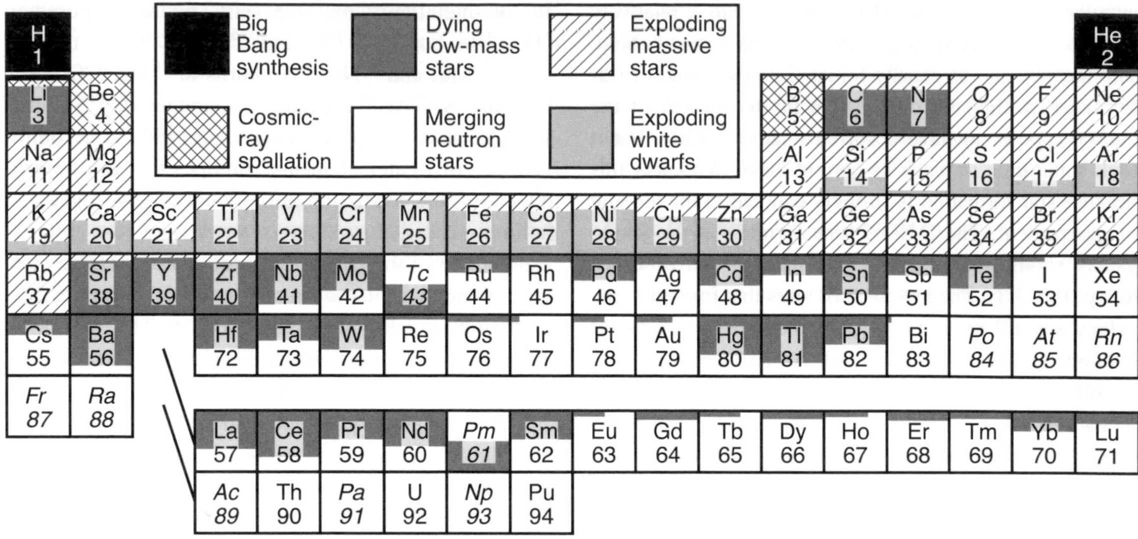

Figure 3.14 Nucleosynthetic sources of the elements in the solar system. Each element in the Periodic Table is shaded to show the approximate percentages contributed by each of six nucleosynthetic sources. The elements that have no surviving stable isotopes from the time of solar system formation are shown in italics. Not all possible sources are included and some simplifications have been made, but the diagram gives a good general picture of the results of galactic chemical evolution. Modified from Johnson (2019).

indicates that we understand the basic principles of galactic chemical evolution.

Summary

In this chapter, we reviewed the broad outlines of the Big Bang model for the origin of the universe and discussed some of the supporting observations. We showed that the Big Bang gave rise to hydrogen, helium, and some lithium, beryllium, and boron, but that other elements were produced primarily in stars. To understand stellar nucleosynthesis, it is necessary to understand the characteristics of stars. Astronomers use the H–R diagram (Fig. 3.1) and the color spectrum of the light emitted by stars (Table 3.1) to classify and characterize stars. Stars range in mass from 0.08 to ~120 $M_{\odot}$. There are many more low-mass stars than intermediate-mass stars, and many more intermediate-mass stars than massive stars (Fig. 3.3). The rate of nuclear burning within a star, and thus the lifetime of the star, is a strong function of stellar mass. Low-mass stars the size of the Sun or smaller shine for in excess of 10 billion years. Massive stars live fast and die young, expending their nuclear fuel in as little as a few million years.

For most of their lifetimes, stars are powered by hydrogen burning, which turns hydrogen into helium. Hydrogen burning can take place through either the proton–proton chains or through CNO burning. Once hydrogen has been exhausted in the stellar core, the star passes through a series of stages, during which power generation moves to helium burning and, in more massive stars, to carbon burning, neon burning, oxygen burning, and silicon burning. Each of these processes generates new elements. Synthesis of elements heavier than iron requires energy, so the buildup of heavier elements occurs either through the *s*-process, which takes place during alternate hydrogen-shell and helium-shell burning, or through explosive nucleosynthesis, which includes the *r*-process. Newly synthesized elements are returned to the interstellar medium to be incorporated into the next generation of stars through massive stellar winds during the late stages of the stars' lives, or through stellar explosions such as supernovae and novae. One of the great triumphs of human intellectual endeavor is the level of detail that we have reached in our understanding of these processes.

In the final section of this chapter, we discussed the formation of galaxies and the formation and chemical evolution of the Milky Way. This sequence of events set the stage for the formation of the solar system. In Chapter 4, we will look at the resulting abundances of the elements and isotopes, both in the solar system and in the galaxy. The solar system abundances of the elements are a fundamental constraint for understanding the Sun, the planets, and the smaller bodies in the solar system.

Questions

1. What was the Big Bang? What is some of the observational evidence that supports this model?
2. What is a Herzsprung–Russell diagram and how is it used?
3. What is the initial mass function (IMF) of stars and how and why is it different from the present-day mass function (PDMF)?
4. What are the main nuclear reactions that power main sequence stars? What happens when this energy source is exhausted?
5. What are the dominant processes that produced the elements heavier than iron? Give the basic outlines of how each one works.
6. How would you identify a star that formed early in galactic history? List two important characteristics of such a star.
7. Why did massive stars dominate nucleosynthesis early in the history of the galaxy?

Suggestions for Further Reading

Busso M., Gallino R., and Wasserburg G. J. (1999) Nucleosynthesis in asymptotic giant branch stars: Relevance for galactic enrichment and solar system formation. *Annual Reviews of Astronomy & Astrophysics*, **37**, 239–309. A good review of nucleosynthesis in low- and intermediate-mass stars.

Hawking S. W. (1988) *A Brief History of Time*. Bantam Books, New York, 198 pp. A readable but authoritative book about cosmology and the Big Bang, written by one of the leading scientists in the field.

Herwig F. (2005) Evolution of asymptotic giant branch stars. *Annual Reviews of Astronomy & Astrophysics*, **43**, 435–479. An excellent discussion of the evolution of AGB stars.

Johnson J. A. (2019) Populating the period table: Nucleosynthesis of the elements. *Science*, **363**, 474–478. An accessible discussion of all of the various processes that produced the elements that make up the universe.

Weinberg S. (1993) *The First Three Minutes*. Basic Books, New York, 191 pp. Nobel Prize-winning physicist Weinberg discusses the very beginnings of the universe in this influential book. It is a bit more detailed and challenging than Hawking's book but provides more detail about the Big Bang.

Other References

Alpher R. A., Bethe H., and Gamow G. (1948) The origin of chemical elements. *Physical Review*, **73**, 803–804.

Bahcall J. N., Pinsonneault M. H., and Basu S. (2001) Solar models: Current epoch and time dependences, neutrinos, and helioseismological properties. *Astrophysical Journal*, **555**, 990–1012.

Basu S., and Rana N. C. (1992) Multiplicity-corrected mass function of main-sequence stars in the solar neighborhood. *Astrophysical Journal*, **393**, 373–384.

Burbidge E. M., Burbidge G. R., Fowler W. A., and Hoyle F. (1957) Synthesis of elements in stars. *Reviews of Modern Physics*, **29**, 547–650.

Caughlan G. R., and Fowler W. A. (1988) *Thermonuclear reaction rates V. Cosmic Data and Nuclear Data Tables*, **40**, 283–334.

Catling D. C., and Kasting J. F. (2017) *Atmospheric Evolution on Inhabited and Lifeless Worlds*. Cambridge University Press, New York, 579 pp.

Champagne A. E., and Wiescher M. (1992) Explosive hydrogen burning. *Annual Reviews of Nuclear and Particle Science*, **42**, 39–76.

Dearborn D. S. P. (1992) Diagnostics of stellar evolution: The oxygen isotopes. *Physics Reports*, **210**, 367–382.

Kobayashi C., Umeda H., Nomoto K., et al. (2006) Galactic chemical evolution: Carbon through zinc. *Astrophysical Journal*, **653**, 1145–1171.

Korobkin O., Rosswog S., Arcones A., and Winteler C. (2012) On the astrophysical robustness of the neutron star merger r-process. *Monthly Notices of the Royal Astronomical Society*, **426**, 1940–1949.

Kroupa P. (2002) The initial mass function of stars: Evidence for uniformity in variable systems. *Science*, **295**, 82–91.

Qian Y.-Z., and Woosley S. E. (1996) Nucleosynthesis in neutrino-driven winds. I. The physical conditions. *Astrophysical Journal*, **471**, 331–351.

Qian Y.-Z., and Wasserburg G. J. (2007) Where, oh where has the *r*-process gone? *Physics Reports*, **442**, 237–268.

Rolfs C. E., and Rodney W. S (1988) *Cauldrons in the Cosmos*. University of Chicago Press, Chicago, 561 pp.

Spencer J. (2019) The faint young sun problem revisited. *GSA Today*, **29**, 4–10.

Other ReferencesThielemann F.-K., Eichler M., Panov I.V., and Wehmeyer B. (2017) Neutron star mergers and nucleosynthesis of heavy elements. *Annual Reviews of Nuclear & Particle Science*, **67**, 24.

Other Referencesvan de Voort, F., Quataert E., Hopkins P. F., et al. (2015) Galactic r-process enrichment by neutron star mergers in cosmological simulations of a Milky Way-mass galaxy. *Monthly Notices of the Royal Astronomical Society*, **447**, 140–148.

Woosley S. E., and Hoffman R. D. (1992) The α-process and the *r*-process. *Astrophysical Journal*, **395**, 202–239.

4 Solar System and Cosmic Abundances
Elements and Isotopes

Overview

In this chapter, we discuss the abundances of the elements and isotopes in the solar system. First, we look at the techniques used to determine solar system abundances, including spectroscopy of the stellar photosphere, measurements of solar wind, and analyses of chondritic meteorites. The solar system abundances of the elements and isotopes are then presented. These abundances are then compared to the abundances in the solar neighborhood of the galaxy and elsewhere. Finally, we introduce how solar system abundances provide a basis for much of what we do in cosmochemistry.

4.1 Chemistry on a Grand Scale

The bulk chemical composition of the solar system is an important cornerstone of our ideas about how the Sun formed and how the various planets, asteroids, meteorites, and comets came to be as we see them today. In older literature, you will typically see the term *cosmic abundances*. In more recent literature, this term is being replaced by *solar system abundances*. This shift reflects an evolution of our understanding of the composition of the solar system and how it relates to that of the surrounding neighborhood and the galaxy as a whole. The terms have been interchangeable for most of the past century, and for most purposes in cosmochemistry they still are. However, as we continue to learn about how the solar system came to be, it will become important to be specific about which composition we want to discuss. We will therefore discuss solar system abundances when talking about our solar system, and use cosmic abundances when talking about the abundances in a more general sense.

4.2 Historical Perspective

Of what is the universe made? This question has been the subject of serious scientific study for more than a century. One of the first papers on the subject, entitled "The Relative Abundances of the Chemical Elements," was presented in 1889 before the Philosophical Society of Washington by Frank W. Clarke. In it, the following sentences appear: "An attempt was made in the course of this investigation to represent the relative abundances of the elements by a curve, taking their atomic weight for one set of ordinates. It was hoped that some sort of periodicity might be evident, but no such regularity appeared." Clarke was in the early stages of a career-long study of the chemical compositions of all types of terrestrial rocks and of the Earth's crust. By the end of the 1800s, the chemical compositions of meteorites were well known, and spectroscopy had been used to infer (incorrectly, as it turns out) the compositions of the Sun, stars, and comets. Based on these observations, Norman Lockyer published "The Meteoritic Hypothesis" in 1890, in which he proposed that "all self-luminous bodies in the celestial spaces are composed either of swarms of meteorites or of masses of meteoritic vapor produced by heat." While this sounds naïve today, one must remember that, at the time, scientists did not know that the extended "nebulae" that they observed with telescopes were, in fact, galaxies rather than diffuse gas, that comets were composed of ice, that stars were powered by nuclear reactions, and many of the other ideas that are the basis for our understanding of the cosmos. However, Lockyer's idea that the celestial space around the solar system consists of material with a composition similar to that of the solar system was profound and became one of these basic ideas. Lockyer had, in effect, proposed that the composition of the visible universe could be determined by measuring the compositions of meteorites.

The first half of the twentieth century saw major advances in our understanding of the visible universe and in gathering detailed information about chemical abundances. On the chemistry side, Giuseppe Oddo (1914) and William Harkins (1917) independently showed that elements with even atomic numbers are more abundant than the adjacent odd-numbered elements on either side (see the zigzag abundance pattern in Figure 4.1). This fundamentally important observation, which reflects the nuclear shell structure discussed in Chapter 2, became known as the *Oddo–Harkins rule*. In the 1920s and 1930s, Victor Goldschmidt compiled a huge database of chemical data on terrestrial rocks, meteorites, and their constituents. In 1937, he published a table of element abundances based on his meteorite data and a recipe of 10 parts silicates, 2 parts metal, and 1 part sulfide, as noted earlier in Chapter 1. He called his table the "cosmic abundances," because some contemporary astronomers believed that meteorites came from interstellar space. He also recognized that chondritic meteorites had not been melted (although their constituent chondrules had) and thus had not experienced the chemical fractionations associated with partial melting and crystallization. Goldschmidt's table provided the cosmic abundances of 66 elements.

The next major step forward was taken by Suess and Urey (1956), who published a new table of cosmic abundances based on meteorite data, solar abundances, and theoretical arguments based on the growing understanding of nucleosynthesis. These authors found regularities in the elemental abundances that had eluded Clarke. But these regularities had nothing to do with the Periodic Table, which organizes the elements based on chemical behavior. Instead, the observed abundance patterns reflect the nucleosynthetic processes that created the elements. The odd-even effect could now be explained in terms of nuclear physics (see Chapter 2). Suess and Urey were able to interpolate abundances of elements that could not be directly measured (e.g. noble gases) based on the observation that the abundances of odd-mass-number nuclides vary smoothly with increasing mass number. Their table included isotopic abundances as well as elemental abundances. Since the Suess and Urey table was published, subsequent work has primarily refined the determinations of the cosmic abundances through improved measurements of meteorites, a better understanding of which meteorites should be considered for this work, improved measurements of the solar composition, and a better understanding of nuclear physics.

In recent decades, spectroscopy has revealed that the elemental and isotopic abundances in the galaxy vary

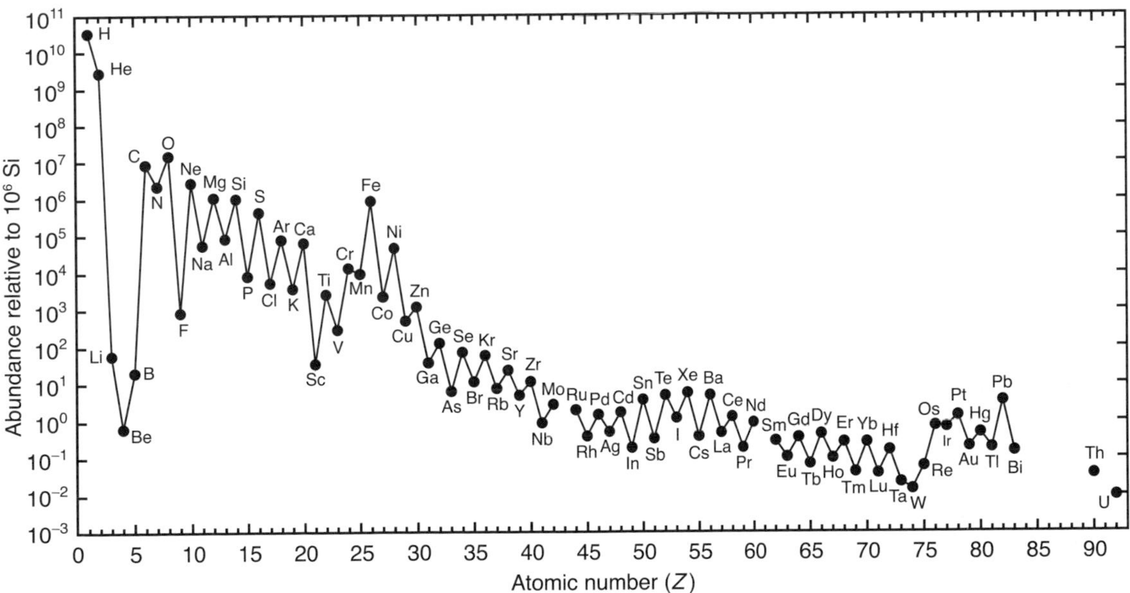

Figure 4.1 Solar system abundances of the elements (relative to 10^6 Si atoms), as a function of atomic number. (These are modern abundances, not those available to Harkins in 1917.) The zigzag pattern illustrates the higher abundance of even relative to adjacent odd atomic numbers. Data from Lodders et al. (2009), Scott et al. (2015a, 2015b), and Grevesse et al. (2015).

with radial position and that the Sun has a somewhat different composition than the molecular clouds and diffuse interstellar medium in the solar neighborhood. For this reason, we can no longer think of the solar system abundances as truly "cosmic" abundances.

4.3 How Are Solar System Abundances Determined?

Determining the composition of the solar system is a complicated undertaking. Our goal is to determine the original composition of the solar system: the composition of the material from which the solar system formed. As you will see in later chapters, if we know the current composition of a meteorite, planet, or other solar system object, and we know the composition of the material from which it was derived, we have a good chance of understanding the processes that have operated to produce the current object. Over the ~4.57 billion years of solar system history, chemical and physical processes have segregated the elements from one another so that it is very difficult to obtain a representative sample of the solar system to measure. The challenge is to find material that has survived until today without significant modification.

The Sun contains >99.8% of the mass in the solar system, so to a very good first approximation, the composition of the Sun is the composition of the solar system. However, in addition to being rather hard to determine, the composition of the Sun has been evolving since the solar system formed. The nuclear reactions that power the Sun are transmuting elements into other elements, modifying the initial composition. Fortunately, most (but not all; see below) of these reactions are occurring deep in the Sun and the evidence has not yet appeared at the surface. Physical and chemical processes operating in the outer parts of the Sun have also caused the compositions of the outer layers to change somewhat relative to the starting composition. These changes are relatively small, however, and good theoretical models have been developed to account for them. Consequently, the Sun is a reasonable object to use to analyze the original composition of the solar system. The Sun's composition can be measured spectroscopically, and the solar wind can be captured and analyzed. These techniques, and their advantages and disadvantages, will be discussed below.

The Earth and other planetary bodies have been heavily modified by planetary-scale differentiation, smaller-scale melting and the resulting chemical fractionations, collisions that mix material with different histories, and other processes. Samples of planetary materials are thus not suitable for determining the solar system composition. More-primitive objects, such as comets and chondritic meteorites, have compositions that are more similar to the composition of the Sun. Until very recently, comets could only be measured spectroscopically when they passed close to the Sun and the signature tails of gas and dust formed. However, the process of forming the tail can fractionate the chemical elements from one another. We also have samples of comet dust in the form of *interplanetary dust particles* (IDPs) captured in the upper stratosphere. These will be described in Chapter 6. But IDPs are too small to be chemically representative of their cometary sources by themselves. They come from many different comets, each perhaps with a different history and composition, and we do not know how to reassemble them to obtain the original composition of even one comet. NASA's Stardust mission, which returned samples from comet Wild 2, solves some of these problems, but while we have learned and will continue to learn a tremendous amount about this comet, it is not clear that an accurate bulk composition of the comet will be one of the results of this work. Chondritic meteorites are much more promising. They are larger than comet particles and we can measure their compositions very precisely in the laboratory. However, over the years, it has become apparent that most meteorites are made up of material that has evolved chemically from the original solar system composition. Only one rare group of chondritic meteorites, the CI carbonaceous chondrites, have compositions that appear to be essentially unmodified from the original solar system composition. This is surprising, considering that the CI chondrites have experienced extensive mineralogical alteration on their parent body.

Our current level of understanding about the original composition of the solar system is the result of a bootstrap effort involving more than a century of gathering compositional data about the Sun, stars, other astronomical objects, the Earth, other planets, and meteorites; establishing relationships between objects; and developing increasingly sophisticated models of stellar nucleosynthesis and the chemical and physical processes that operated in the early solar system.

4.4 Determining Chemical Abundances in the Sun

4.4.1 Spectroscopic Observations of the Sun

Atoms and molecules emit and absorb photons with characteristic energies. As a result, measurements of stars, comets, and other luminous bodies with a spectrograph, which permits the output to be measured as a function of

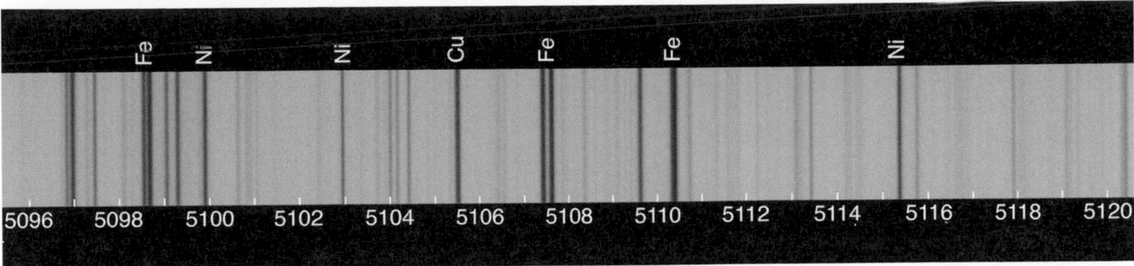

Figure 4.2 Portion of the high-spectral-resolution solar spectrum (2 milli-Ångstrom resolution), recorded by Louis Delbouille in the 1970s at the Jungfraujoch Observatory. The spectrum contains thousands of absorption lines that are used to determine the composition of the solar photosphere. The elements responsible for some of the strongest lines are noted at the top of the figure. The numbers at the bottom give wavelength in Ångstrom. Image courtesy of J.-M. Malherbe.

wavelength, reveal numerous emission or absorption lines (Fig. 4.2). These lines can be used to infer the composition of the objects. The first spectroscopic measurements of the Sun, stars, and other luminous objects were made in the last half of the nineteenth century. However, it wasn't until the late 1920s that relatively accurate elemental abundances for the Sun and the stars were determined (see Box 4.1).

It is now relatively easy to obtain detailed spectra of the Sun, but determining the Sun's chemical composition from such spectra requires a realistic model of the stellar atmosphere and a detailed understanding of how the lines are formed and interact to give the observed spectrum. The elemental abundances are not "observed" but are derived from the observations, and the uncertainties in models of the stellar atmosphere and of line formation are much greater than the uncertainties in the spectral measurements (Hill, 2001).

The last ~15 years have seen a minor revolution in determining solar and stellar abundances (Grevesse et al., 2007, 2015; Asplund et al., 2009; Lodders et al., 2009; Scott et al. 2015a, 2015b). Much of the previous work assumed that the spectral lines originate in local thermodynamic equilibrium (LTE), and the stellar atmosphere had been modeled in a single dimension. Since 2000, improved computing power has permitted three-dimensional modeling of the Sun's atmosphere and non-LTE treatment of line formation. The result has been significant shifts in inferred solar abundances.

In LTE modeling of line formation, the strength of a line can be straightforwardly predicted from a few properties of the line and the species that generate it, once a model atmosphere and continuous opacity are known everywhere. Non-LTE modeling does not assume anything, but calculates line strengths from first principles. In non-LTE modeling, everything depends on everything else, everywhere else; thus, many more inputs into the model and much more computational power are required. However, the resulting models of line formation are much more realistic, and presumably more accurate.

The three-dimensional modeling of the solar atmosphere permits an accurate representation of the *granularity* of the Sun's surface. "Granularity" is the term used to describe areas of convective upwelling of hot material and narrower regions of down-going cooler material (Fig. 4.3). Temperature modeling of the Sun's atmosphere must account for this granularity. The three-dimensional modeling of the Sun's granularity has resulted in much better matches of line widths and related properties, while at the same time showing that absorption lines should typically be stronger than those modeled in the one-dimensional case (consequently, fewer atoms are required to achieve the same absorption). The new modeling thus results in lower overall abundances of elements heavier than hydrogen and helium. The resulting inferred solar *metallicity*, the mass fraction of all elements other than hydrogen (X) and helium (Y), is $Z = 0.012$, rather than earlier values of $Z = 0.017$–0.0189. Table 4.1 gives the solar abundances as determined by this new model. As of time of writing, these new lower abundances for elements like carbon, nitrogen, and oxygen do not agree with the results of *helioseismology*, which match the old abundances quite well (Basu and Antia, 2004). See Christensen-Dalsgaard (2002) for a review of helioseismology.

There has been a large volume of literature attempting to resolve the discrepancy between solar spectroscopy and helioseismology. One suggestion was that the inferred neon abundance in the Sun, which cannot be determined spectroscopically, is too low (Bahcall et al., 2005). But new determinations of the solar neon abundance based on Genesis mission results ruled this out. Another idea is that coronal-hole solar wind, which gives a higher metallicity,

Box 4.1 Cecilia Payne Cracks the Spectral Code

"The most brilliant Ph.D. thesis ever written in astronomy." This high praise was given by Otto Struve, the director of the Yerkes Observatory, to the thesis of Cecilia Payne, entitled "Stellar Atmospheres." In her thesis, published in 1925, Payne provided the key to unlocking the mysteries of stellar spectroscopy, for the first time allowing accurate determinations of the compositions of stars from their spectral lines. Her methodology and conclusions were so ingenious and well founded that they were rapidly accepted and completely changed the nature of stellar spectroscopy. But history does not give her the acclaim that her discoveries deserve.

Payne was born in 1900 in Wendover, England, to an upper-class family. Her early education was in botany, physics, and chemistry, but her passion was astronomy. In 1922, she heard a lecture by Harlow Shapley, the director of the Harvard College Observatory. The lecture inspired her to seek admission to Harvard, and with strong recommendations from her mentors in England, including Sir Arthur Eddington, she was admitted and arrived on campus in 1923. Two years later, in 1925, she became the first student, male or female, to earn a Ph.D. from the Harvard College Observatory.

When Payne began her work in the 1920s, stellar spectroscopy was a very active area of research. Numerous elemental and molecular lines had been identified in stellar spectra. The lines observed in each star varied with the inferred temperature of the star, which was understood to mean that the elemental abundances varied with temperature. This body of data was the basis for the spectral typing of stars (O, B, A, F, G, K, M, L). However, the power source for stars was not understood, and it was not clear why the composition of a star should be related to its temperature. In the 1920s, it was also widely believed that the Sun had the same composition as the Earth; models considered the Earth to have formed from the outer layers of the Sun. Payne used the new quantum mechanical understanding of atomic structure to show how and why the spectral lines of the different elements varied as a function of stellar spectral type. She demonstrated how the temperature of the stellar surface controls the spectral lines that are observed. Her analysis led to the conclusion that the chemical composition of the starry universe is homogeneous, and that stars are composed almost entirely of hydrogen and helium. The latter result was not so readily accepted, as it went directly against conventional wisdom, and she downplayed it in her thesis. However, in 1929, Henry Russell published a paper entitled "On the Composition of the Sun's Atmosphere," in which he argued on the basis of a new theoretical understanding of the physics of the hydrogen atom that the Sun's atmosphere must be hydrogen rich. He did give credit to Payne's conclusions based on spectroscopy, but history credits Russell with showing that the Sun is mostly hydrogen.

Why did Cecilia Payne not get credit for her groundbreaking work? It was not easy for women to work in science in the first half of the twentieth century. Those with exceptional talent were able to make exceptional contributions, but they often worked in obscurity. But this was only one factor – timing was also an issue. A person making a profound discovery is often well ahead of his or her time, and the importance of the new result is not recognized right away. The person who gets credit is the person who convinces the world of the correctness and importance of the result. In the case of the Sun being made of hydrogen, that person was Russell. Although Payne's career was marked by slow promotions and low salaries, she valued immensely the joy of discovery. She encouraged the same single-minded purpose in her students, telling them, "Your reward will be the widening of the horizon as you climb. If you achieve that reward, you will ask no other."

Payne (later Payne-Gaposchkin) went on to have a brilliant academic career. Although she suffered years of gender discrimination, she eventually became Harvard's first woman full professor and chair of the Astronomy Department.

is more representative of the bulk Sun than the photosphere. But this leads to too-high abundances of magnesium, silicon, sulfur, and iron to be consistent with the flux of solar neutrinos. Another suggestion is that the surface convective zone in the Sun has lower metallicity than the bulk Sun because it accreted hydrogen-rich gas from the nebula after the heavy elements had segregated into planets. However, as of time of writing, neither these nor any of a large number of proposed models resolve the disagreement between the new lower solar abundances of carbon, nitrogen, and oxygen and the results of helioseismology (Vagnozzi et al., 2017).

Other considerations must also be kept in mind when using the current composition of the Sun to represent the

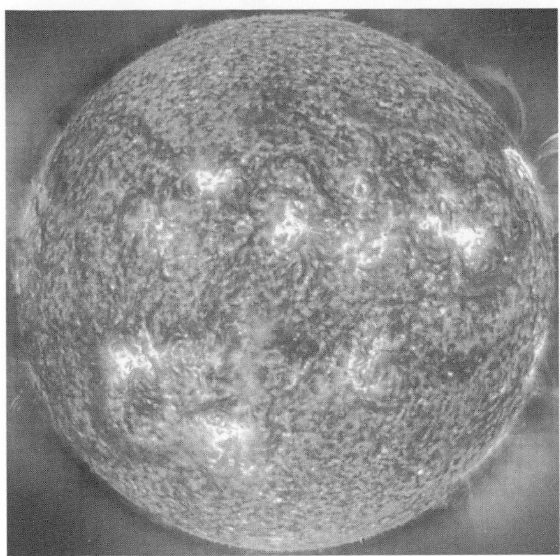

Figure 4.3 Image of the Sun taken by the SOHO spacecraft, showing the turbulent convection in its surface layers. Lighter areas are hotter and darker areas are cooler. Image courtesy of SOHO (ESA and NASA).

bulk solar system. The composition of the Sun is not the same as it was when the constituent materials first came together to form a protostar. The initial stages of nuclear burning, which took place while the protostar was completely convective, burned ^{2}D to form ^{3}He, lowering the D/H ratio and raising the $^{3}He/^{4}He$ ratio. Lithium is depleted by a factor of ~160 in the Sun compared to meteoritic abundances, probably also due to nuclear reactions in the solar envelope, but the details of how this happened are not understood. Other than these changes, main sequence nucleosynthesis in the solar core and at the base of the solar envelope is burning hydrogen to form helium and modifying the abundances and isotopic compositions of carbon, nitrogen, and oxygen. These effects will not appear at the surface until the Sun leaves the main sequence.

Although the Sun may have accreted as a homogeneous body, elemental settling has occurred in the nonconvecting region below the surface convection zone (Vauclair, 1998). Element segregation is the result of two kinds of processes. First, individual atoms want to move in the presence of a gravitational field, a pressure or thermal gradient, radiative acceleration, or a concentration gradient. Through these motions, heavier atoms tend to migrate toward the center of a star. Second, the motions of the atoms are slowed down due to collisions with other atoms. As a result, they share angular momentum in a random way. In our Sun, these processes result in compositional variations of about 10%. Variations of this magnitude are easily detected by helioseismology, and theoretical modeling matches the observations very well. It is not clear how the segregation of elements in the Sun beneath the surface convection zone affects surface abundances.

We also know that processes that eject solar wind fractionate elements. Over time, ejection of solar wind that is fractionated relative to the bulk Sun changes the surface composition of the Sun – the part that we measure spectroscopically. There are theoretical models to account for this effect, but they introduce uncertainties into the inferred composition of the bulk Sun.

4.4.2 Collecting and Analyzing the Solar Wind

The *solar wind* consists of charged particles emitted from the Sun's upper atmosphere. The hot ionized plasma in the corona is accelerated by the Sun's magnetic field. Like the Sun itself, the wind consists primarily of protons (ionized hydrogen atoms), but it also contains all the other elements in the Sun. Thus, if one could measure the composition of the solar wind, one might get an independent measurement of the composition of the Sun. However, this turns out to be harder than it might initially appear. The density of the solar wind (mostly hydrogen) is not very high, only about 3 atoms/cm^3 at the Earth's orbit. A solar wind collector must be exposed for a long time to gather enough atoms to measure. In addition, the process that accelerates the solar wind and ejects it from the solar surface fractionates the elements. The biggest fractionations follow the first ionization potential, the energy necessary to ionize the elements. Those elements that are easy to ionize are preferentially incorporated into the wind over those that are hard to ionize. Because it is so difficult to quantify these effects, measurements of the solar wind have been used primarily to determine the isotopic composition of the Sun (all isotopes of an element have the same ionization potential), and by extension the solar system. However, recent studies of solar wind samples brought back by the Genesis mission (see Box 4.2) suggest that isotopes may also be fractionated in the solar wind by a process called "Coulomb drag." This fractionation would limit the usefulness of solar wind isotopic data in determining solar system abundances.

To measure the solar wind, one must find or make suitable solar wind collectors. The Earth is protected from the solar wind by its magnetic field, so the collectors must be outside of the influence of the planet's magnetic field. The first measurements of the noble gases in the solar wind were made in a subclass of chondritic meteorites (originally called "gas rich"), which are actually samples of the surface regolith on the parent asteroid. The implanted solar noble gases in these meteorites show a

Table 4.1 Solar system abundances of the elements

	Mean CI Chondrites		Photospheric Abundances	Solar System Abundances	
Element	LPG* (2009) ppm by Weight	Atoms/10^6 Si	SSGASP† Atoms/10^6 Si	Atoms/10^6 Si	
1 H	19,700 ± 1970	5.130 × 10^6 (10)	3.090 × 10^{10} (7)	3.090 × 10^{10} (7)	P
2 He	9.17 × 10^{-3}	0.6013	[2.630 × 10^9] (2)	2.630 × 10^9 (2)	P
3 Li	1.47 ± 0.19	55.59 (13)	0.3467 (26)	55.59 (13)	M
4 Be	0.0210 ± 0.0015	0.6116 (7)	0.7413 (23)	0.6116 (7)	M
5 B	0.775 ± 0.078	18.816 (10)	15.488 (58)	18.816 (10)	M
6 C	34,800 ± 3480	7.605 × 10^5 (10)	8.318 × 10^6 (12)	8.318 × 10^6 (12)	P
7 N	2950 ± 443	5.528 × 10^4 (15)	2.089 × 10^6 (12)	2.089 × 10^6 (12)	P
8 O	459,000 ± 4590	7.630 × 10^6 (10)	1.514 × 10^7 (12)	1.514 × 10^7 (12)	P
9 F	58.2 ± 8.7	804.09 (15)	776.25 (78)	804.09 (15)	M
10 Ne	1.8 × 10^{-4}	2.341 × 10^{-3}	[2.630 × 10^6] (26)	2.630 × 10^6 (26)	T
11 Na	4990 ± 250	5.697 × 10^4 (5)	5.012 × 10^4 (10)	5.530 × 10^4 (4)	A
12 Mg	95,800 ± 2874	1.035 × 10^6 (3)	1.202 × 10^6 (10)	1.045 × 10^6 (3)	A
13 Al	8500 ± 255	8.269 × 10^4 (3)	8.318 × 10^4 (10)	8.273 × 10^4 (3)	A
14 Si	107,000 ± 3210	≡1.000 × 10^6	≡1.000 × 10^6	≡1.000 × 10^6	
15 P	967 ± 97	8195 (10)	7943 (7)	8022 (6)	A
16 S	53,500 ± 2675	4.379 × 10^5 (5)	4.074 × 10^5 (7)	4.266 × 10^5 (4)	A
17 Cl	698 ± 105	5168 (15)	9772 (100)	5168 (15)	M
18 Ar	1.33 × 10^{-3}	9.622 × 10^{-3}	[7.762 × 10^4] (35)	7.762 × 10^4 (35)	T
19 K	544 ± 27	3652 (5)	3388 (12)	3607 (5)	A
20 Ca	9220 ± 461	6.038 × 10^4 (5)	6.457 × 10^4 (7)	6.167 × 10^4 (4)	A
21 Sc	5.9 ± 0.30	34.45 (5)	44.668 (10)	34.45 (5)	M
22 Ti	451 ± 36	2473 (8)	2630 (10)	2575 (8)	A
23 V	54.3 ± 2.7	279.8 (5)	239.88 (20)	276.7 (5)	A
24 Cr	2650 ± 80	1.338 × 10^4 (3)	1.288 × 10^4 (10)	1.334 × 10^4 (3)	A
25 Mn	1930 ± 58	9221 (3)	8128 (10)	9107 (3)	A
26 Fe	185,000 ± 5550	8.695 × 10^5 (3)	9.120 × 10^5 (10)	8.727 × 10^5 (3)	A
27 Co	506 ± 15	2254 (3)	2630 (12)	2270 (3)	A
28 Ni	10,800 ± 324	4.830 × 10^4 (3)	4.898 × 10^4 (10)	4.835 × 10^4 (3)	A
29 Cu	131 ± 13	541.1 (10)	467.74 (12)	505.76 (8)	A
30 Zn	323 ± 32	1296 (10)	1122 (12)	1212 (8)	A
31 Ga	9.71 ± 0.49	36.56 (5)	32.359 (12)	35.80 (5)	A
32 Ge	32.6 ± 3.3	117.80 (10)	131.83 (17)	121.41 (8)	A
33 As	1.74 ± 0.16	6.096 (9)		6.096 (9)	M
34 Se	20.3 ± 1.4	67.48 (7)		67.48 (7)	M
35 Br	3.26 ± 0.49	10.71 (15)		10.71 (15)	M
36 Kr	5.22 × 10^{-5}	1.635 × 10^{-4}	[54.954] (15)	54.954 (15)	T
37 Rb	2.31 ± 0.16	7.094 (7)	9.120 (17)	7.283 (6)	A
38 Sr	7.81 ± 0.55	23.40 (7)	20.89 (15)	22.50 (8)	A
39 Y	1.53 ± 0.15	4.517 (10)	5.012 (12)	4.696 (8)	A
40 Zr	3.62 ± 0.36	10.416 (10)	12.023 (10)	11.10 (7)	A
41 Nb	0.279 ± 0.028	0.7882 (10)	0.9120 (15)	0.8190 (8)	A
42 Mo	0.973 ± 0.097	2.662 (10)	2.344 (23)	2.600 (9)	A
44 Ru	0.686 ± 0.041	1.782 (6)	1.738 (20)	1.778 (6)	A
45 Rh	0.139 ± 0.014	0.3545 (10)	0.2399 (20)	0.3545 (10)	M
46 Pd	0.558 ± 0.028	1.376 (5)	1.097 (15)	1.376 (5)	M

Table 4.1 *(cont.)*

	Mean CI Chondrites		Photospheric Abundances	Solar System Abundances	
Element	LPG* (2009) ppm by Weight	Atoms/10^6 Si	SSGASP† Atoms/10^6 Si	Atoms/10^6 Si	
47 Ag	0.201 ± 0.010	0.4891 (5)	0.2818 (26)	0.4891 (5)	M
48 Cd	0.674 ± 0.047	1.574 (7)	1.820 (41)	1.574 (7)	M
49 In	0.0788 ± 0.0055	0.1779 (7)	0.1950 (58)	0.1779 (7)	M
50 Sn	1.63 ± 0.24	3.604 (15)	3.236 (26)	3.604 (15)	M
51 Sb	0.145 ± 0.022	0.3126 (15)		0.3126 (15)	M
52 Te	2.28 ± 0.16	4.690 (7)		4.690 (7)	M
53 I	0.530 ± 0.106	1.096 (20)		1.096 (20)	M
54 Xe	1.74×10^{-4}	3.48×10^{-4}	[5.370] (15)	5.370 (15)	T
55 Cs	0.188 ± 0.009	0.3713 (5)		0.3713 (5)	M
56 Ba	2.41 ± 0.14	4.606 (6)	5.495 (17)	4.677 (6)	A
57 La	0.242 ± 0.012	0.4573 (5)	0.3981 (10)	0.4425 (5)	A
58 Ce	0.622 ± 0.031	1.165 (5)	1.175 (10)	1.167 (4)	A
59 Pr	0.0946 ± 0.0066	0.1762 (7)	0.1622 (10)	0.1711 (6)	A
60 Nd	0.471 ± 0.024	0.8571 (5)	0.8128 (10)	0.8475 (4)	A
62 Sm	0.152 ± 0.008	0.2653 (5)	0.2754 (10)	0.2672 (4)	A
63 Eu	0.0578 ± 0.0029	0.0998 (5)	0.1023 (10)	0.1003 (4)	A
64 Gd	0.205 ± 0.010	0.3422 (5)	0.3715 (10)	0.3473 (4)	A
65 Tb	0.0384 ± 0.0027	0.0634 (7)	0.0631 (26)	0.0634 (7)	A
66 Dy	0.255 ± 0.013	0.4119 (5)	0.3890 (10)	0.4069 (4)	A
67 Ho	0.0572 ±0.0040	0.0910 (7)	0.0933 (29)	0.0911 (7)	A
68 Er	0.163 ± 0.008	0.2558 (5)	0.2630 (12)	0.2568 (5)	A
69 Tm	0.0261 ± 0.0018	0.0406 (7)	0.0398 (10)	0.0403 (6)	A
70 Yb	0.169 ± 0.008	0.2564 (5)	0.2188 (29)	0.2564 (5)	M
71 Lu	0.0253 ± 0.0013	0.0380 (5)	0.0389 (23)	0.0380 (5)	A
72 Hf	0.106 ± 0.005	0.1559 (5)	0.2188 (12)	0.1559 (5)	M
73 Ta	0.0145 ± 0.0015	0.0210 (10)		0.0210 (10)	M
74 W	0.096 ± 0.010	0.1371 (10)	0.2089 (29)	0.1371 (10)	M
75 Re	0.0393 ± 0.0039	0.0554 (10)		0.0554 (10)	M
76 Os	0.493 ± 0.039	0.6802 (8)	0.7762 (12)	0.7046 (7)	A
77 Ir	0.469 ± 0.023	0.6404 (5)	0.8128 (17)	0.6404 (5)	M
78 Pt	0.947 ± 0.076	1.274 (8)		1.274 (8)	M
79 Au	0.146 ± 0.015	0.1946 (10)	0.2512 (20)	0.1946 (10)	M
80 Hg	0.350 ± 0.070	0.4580 (20)		0.4580 (20)	M
81 Tl	0.142 ± 0.011	0.1824 (8)	0.2455 (58)	0.1824 (8)	M
82 Pb	2.63 ± 0.18	3.332 (7)	2.570 (20)	3.332 (7)	M
83 Bi	0.110 ± 0.010	0.1382 (9)		0.1382 (9)	M
90 Th	0.0310 ± 0.0025	0.0351 (8)	0.0331 (26)	0.0349 (8)	A
92 U	0.00810 ± 0.00065	0.00893 (8)		0.00893 (8)	M

* LPG = Lodders, Palma, and Gail (2009); † SSGASP = Scott et al. (2015a, 2015b), Grevesse et al. (2015).

[ ] Indirect measurements: He from inversion of helioseismology; Ne and Ar from coronal Ne/O and Ar/O ratios; Kr and Xe from interpolation of theoretical *s*-process production rates.

Photospheric abundances of Li through F and noble gases are from (Asplund et al., 2009).

( ) Uncertainties given in %.

P: photosphere, M: meteorites, A: average of photosphere and meteorites, T: theoretical abundance.

Convert from 10^{12} H astronomical scale to 10^6 Si cosmochemistry scale with a factor of 0.03090.

Box 4.2 The Genesis Mission

The Genesis spacecraft was launched on August 8, 2001 with the goals of obtaining precise measurements of the solar isotopic abundances and improving the determination of elemental abundances. The spacecraft contained five arrays of passive collectors and an electrostatic concentrator (Fig. 4.4). The collectors had to be made from very pure materials because the concentrations of elements from the solar wind captured by the collectors were expected to be very low. Collector materials included metals (gold, germanium, and silicon), as well as sapphire (Al_2O_3) and diamond-like hydrocarbon. These different materials permitted all of the major elements of the solar wind to be collected and analyzed without interference from the collector materials.

The solar wind is ejected from the Sun in several ways. The normal solar wind travels at speeds of ~300 km/s and has a density of ~3 particles/cm^3 at the Earth's orbit. High-speed, lower-density, coronal-hole winds travel at ~800 km/s, and coronal mass ejections are violent, sporadic events that eject much more material at speeds of 1000–1500 km/s. The mission was designed to sample all three types of solar wind.

An electrostatic concentrator was specially designed to obtain a good measurement of the oxygen isotopic composition of the Sun. Oxygen has distinct isotopic compositions in different solar system materials, but without knowing the overall isotopic composition of oxygen in the solar system, it is very difficult to infer the origin and

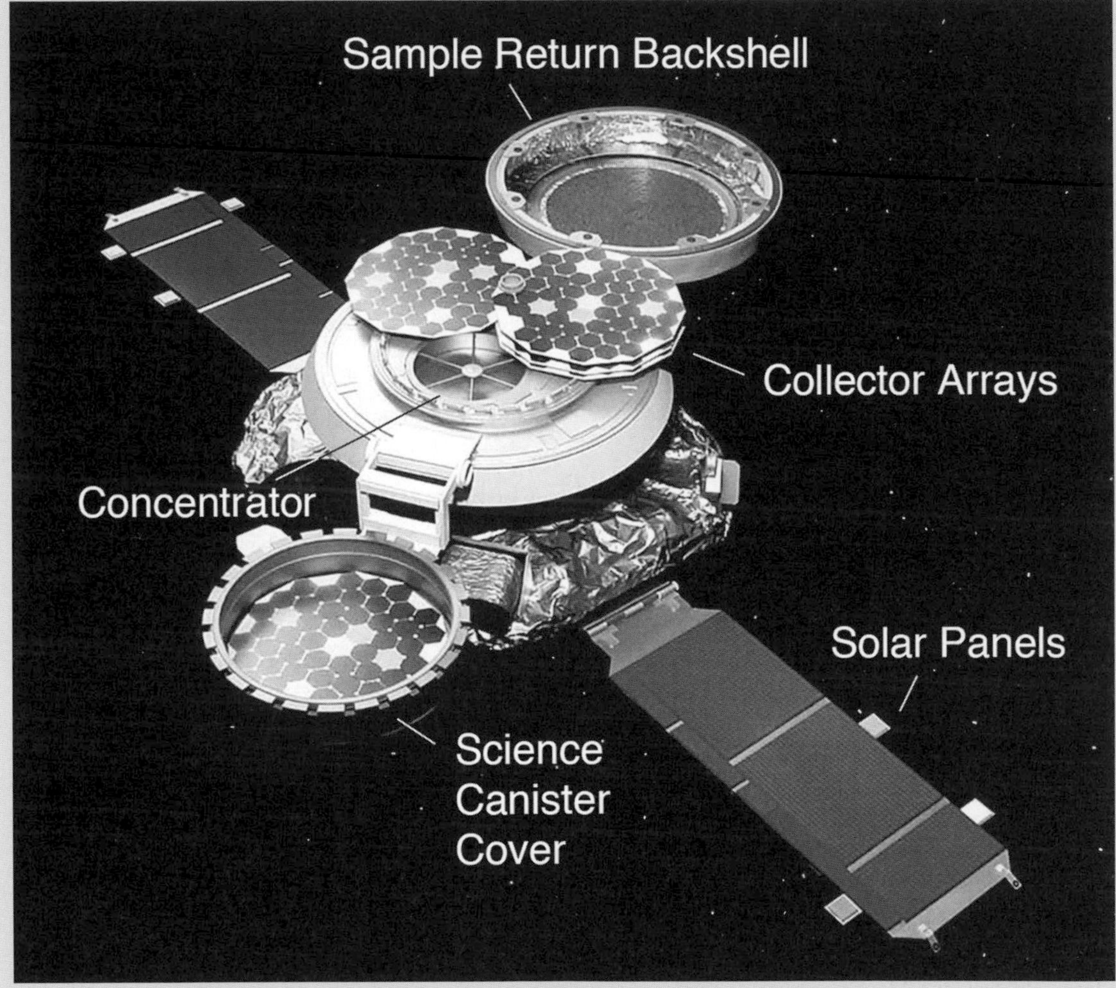

Figure 4.4 Genesis spacecraft with the various solar wind collectors deployed. NASA/JPL image.

Box 4.2 (cont.)

Figure 4.5 Genesis spacecraft shortly after its abrupt return to Earth. NASA/JPL image.

meaning of these isotopic differences. A concentrator was necessary because the integrated oxygen fluence expected in a passive collector over the duration of the mission would have been insufficient to provide enough oxygen atoms for a precise isotopic measurement. The concentrator increased the oxygen concentration in the target material by a factor of ~20.

The spacecraft traveled to a point outside the Earth's magnetic field. It was stationed at the L1 Lagrange Point, between the Earth and Sun, for 884 days while it collected solar wind. The spacecraft then returned to Earth on September 8, 2004. The mission plan was for a helicopter to snag the spacecraft out of the air while it descended on a parachute. Unfortunately, the parachute failed to open and the spacecraft plummeted into the Utah desert (Fig. 4.5).

The crash shattered many of the detectors into pieces just a few millimeters across, and Utah dirt coated the pieces. But while the crash of the spacecraft was a setback, all was not lost. The spacecraft and all of its pieces were carefully collected and returned to the Johnson Space Center in Houston. There, all the tools and skills of ground-based laboratories and scientists were applied to the problem of cleaning the collectors and making them suitable for analysis.

A solar wind mission has a major advantage over many other types of missions. The solar wind is implanted into the collectors, beneath the surface, while the dirt and contamination are confined to the collector surfaces. After careful study, cleaning procedures were established that removed the surface contamination, and within a year after the landing, data on the isotopic composition of the solar wind were being published. The main effect of the crash was simply to delay the publication of the results. One can even find a silver lining to the crash: sample curators no longer have to worry about subdividing the collectors to provide material to different labs. They have thousands of pieces that can be sent out for analysis. An important lesson from this mission is that if one can bring the samples back to Earth, the resources and ingenuity of dedicated people can perform miracles.

much smaller elemental fractionation relative to the expected composition of the Sun than do other noble gas components trapped in meteorites and noble gases in the Earth's atmosphere. The gas-rich meteorites gave the first data on the isotopic compositions of the noble gases in the solar wind.

The Moon, which lies outside of the influence of the Earth's magnetic field, is another excellent solar wind collector. In fact, the Moon's surface is saturated with solar wind hydrogen and helium, which means that every time a new solar wind atom is implanted into the surface, it kicks out an atom that was previously implanted. The concentration of solar wind helium in the lunar surface is high enough to be considered as a potential resource for future lunar mining. Unfortunately, because the Moon's surface consists of rocks and minerals that contain lots of elements, these rocks are not suitable collectors for measuring the solar wind because, except for noble gases and nitrogen, one cannot distinguish the solar wind atoms from the atoms that originally made up the rock. Measurements of lunar soils give estimates of the isotopic compositions of nitrogen and noble gases in the Sun, but provide little information on the relative abundances of the elements. This is because acceleration of solar wind from the Sun, trapping it in the lunar soils, and degassing it from the soil over time as the soil is thermally cycled on the lunar surface all modify the elemental composition of trapped solar wind.

The Apollo missions attempted to take advantage of the Moon's location outside of the Earth's magnetic field to capture solar wind in aluminum collectors. The collectors were deployed for the duration of each mission (from 10 to 45 hours) and provided information on the helium, neon, and argon isotopic compositions of the solar wind.

The NASA Genesis mission (see Box 4.2), launched in 2001, was designed to collect solar wind over a ~2.5 year period. Special collector materials were designed to be free of contaminants so that clean measurements of the isotopic compositions of trapped solar wind elements would be possible. The spacecraft was stationed between the Earth and the Sun at a point where their gravities balance. The mission returned $\sim 10^{20}$ atoms (0.4 mg) of solar wind to Earth. In spite of an unfortunate crash when the spacecraft returned to Earth, most of the main science objectives of the mission have been met.

Results from analyses of Genesis solar wind particles were summarized by Burnett (2013) and Burnett et al. (2019). High-precision abundance and isotopic measurements of noble gases in the solar wind have been obtained, as have measurements of the abundances of magnesium, iron, potassium, sodium, calcium, aluminum, and chromium. Another major discovery was that solar and terrestrial oxygen isotopic compositions are very different, as described later in this chapter in Box 4.3. Prior to Genesis, wide variations in nitrogen isotopes were known in solar system materials; the solar wind value of $^{15}N/^{14}N$ is lower than other planetary and meteorite measurements, except for Jupiter. Measurement of the isotopic composition of magnesium in the solar wind has also been attempted, and measurements of other isotopes are ongoing.

4.4.3 Chemical Abundances in CI Chondritic Meteorites

For over a century, it has been widely accepted that chondrites have primitive compositions, meaning that the physical and chemical processes that formed the meteorites did not significantly change their elemental compositions. However, as analytical techniques improved, it became clear that all chondrites are not the same. In fact, at least 15 compositional groups can be identified among the chondrites (see Chapter 6). Then the question became: which group of chondrites is most representative of the bulk composition of the solar system? By the 1960s, it was clear that a small group of meteorites, the CI chondrites (Fig. 4.6), have the most primitive compositions. This conclusion comes from plots like Figure 4.7, which shows the photospheric abundances of the elements plotted against the abundances in CI chondrites. Elements that plot on the solid line on Figure 4.7 are present in the same relative abundance in the Sun and in CI chondrites. Except for those volatile elements that exist in the gas phase at room temperature (hydrogen, helium, oxygen, carbon, nitrogen, noble gases), and elements that were destroyed in the Sun by the first stages of nuclear burning (e.g. lithium), essentially all elements are present in the same relative abundance in the Sun and the CI chondrites. Figure 4.7 uses the best data available today, but plots based on earlier data have given the same basic picture since the 1960s.

Several techniques have been used over the years to measure the bulk chemical compositions of CI chondrites and other meteorites. Wet chemistry was used in the late nineteenth through the mid-twentieth century to analyze ~15 major and minor elements. X-ray fluorescence spectroscopy was popular in the 1950s and 1960s. This method is easier to apply than wet chemical analysis, but some of the measurements in the literature had problems with alkali elements. Neutron activation analysis

Figure 4.6 Sample of the Orgueil CI carbonaceous chondrite in the collection of the Muséum d'histoire naturelle de Paris.

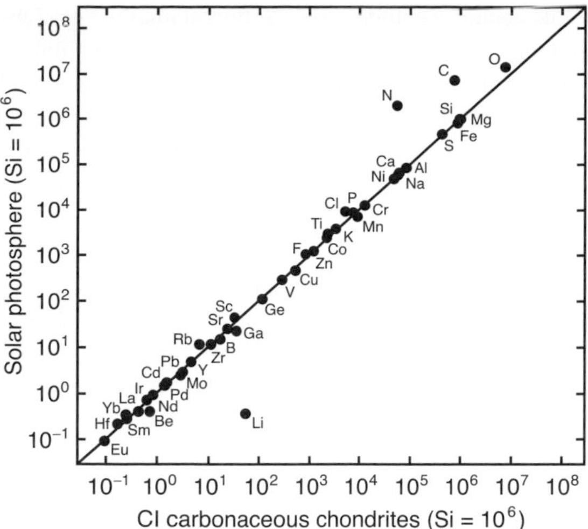

Figure 4.7 Abundances of elements as determined in the solar photosphere versus abundances in CI carbonaceous chondrites (both normalized to 10^6 silicon atoms). The 45-degree line is the 1:1 correlation describing equal relative abundances. Almost all the elements fall within a few percent of the 1:1 line. The exceptions are elements that are gases or form gaseous compounds at room temperature (hydrogen, helium, nitrogen, carbon, oxygen, neon, argon, krypton, xenon), which are not effectively retained in the CI chondrites, and elements like lithium that have been destroyed by nucleosynthesis processes in the Sun over its lifetime. The close agreement between the two compositions indicates that CI chondrites acquired a representative sample of bulk solar system material.

took over in the 1970s as the technique of choice for measuring the compositions of lunar samples and meteorites. Neutron activation analysis can measure up to 74 of the 92 elements with high precision. Unfortunately, silicon, a key major element, is not one of them. Neutron activation analysis can measure very small samples, and care must be taken to measure samples that are large enough to provide a representative analysis. Detailed descriptions of each of these techniques can be found in the Appendix.

4.4.4 Indirect Methods of Estimating Abundances

As our understanding of nuclear physics and stellar nucleosynthesis has improved, it has become possible to use that knowledge along with measured abundances to infer the solar system abundances of elements that cannot be directly measured, either by solar spectroscopy because their ionization potentials are too high, or in meteorites because they were not efficiently incorporated into the chondrites. In particular, the noble gases cannot be directly measured and must be inferred from theoretical considerations or by other indirect means.

Helium: Although helium was first discovered through spectral observations of the Sun, the lines that were observed appear in the high-temperature corona. Helium does not appear in the spectrum of the solar photosphere, and it is very inefficiently retained in meteorites. Solar wind and solar-energetic particles show highly variable helium abundances, and these are rather low when compared with helium values inferred from hot stars or from the H II regions in the interstellar medium (characterized by ionized hydrogen). The best estimates for the helium abundance in the Sun come from two sources: (1) solar standard and non-standard models and (2) inversion of helioseismic data. Standard models give a helium abundance by mass in the protosolar cloud of $Y = 0.27 \pm 0.01$ (Christensen-Dalsgaard, 1998). Non-standard models, which take into account elemental migration over the age of the Sun (see below), give a solar helium mass fraction of $Y = 0.275$ (Gabriel, 1997). Inversion of helioseismic data gives a smaller value, $Y = 0.2485$, as the helium mass fraction in the outer convection zone of the Sun (Basu and Antia, 2004). The best estimate for the helium mass fraction in the outer layers is thus $Y = 0.2485$ (He/H = 0.0851 by number). This is the value given in Table 4.1. The helium mass fraction of the bulk Sun at the time it formed is estimated to have been $Y = 0.275 \pm 0.01$ (He/H = 0.098) (Grevesse and Sauval, 1998). The difference between this number and the current inferred mass fraction in the convection zone is due to elemental segregation over the age of the Sun.

Neon and argon: No spectral lines of neon and argon appear in the photospheric spectrum. The abundances of these elements come from measurements of the Ne/O and Ar/O ratios in coronal material and thus are relatively uncertain. Unfortunately, nucleosynthesis theory does not provide much guidance for either neon or argon.

Krypton and xenon: The abundances of krypton and xenon are determined exclusively from nucleosynthesis theory. They can be interpolated from the abundances of neighboring elements based on the observation that abundances of odd-mass-number nuclides vary smoothly with increasing mass number (Suess and Urey, 1956). The regular behavior of the *s*-process also provides a constraint (see Chapter 3). In a mature *s*-process, the relative abundances of the stable nuclides are governed by the inverse of their neutron-capture cross sections. Isotopes with large cross sections have low abundance because they are easily destroyed, while the abundances of those with small cross sections build up. Thus, one can estimate the abundances of krypton and xenon from the abundances of *s*-only isotopes of neighboring elements (selenium, bromine, rubidium, and strontium for krypton; tellurium, iodine, cesium, and barium for xenon).

4.5 Solar System Abundances of the Elements

Table 4.1 and Figure 4.1 show the solar system abundances of the elements as determined by the methods discussed above. For some elements, the photospheric abundances provide the best estimate, whereas for others, the meteorite data must be used. In some cases, the data are equally reliable and an average of the values determined from the solar photosphere and CI chondrites is used. The abundances of the noble gases come from indirect measurements or theoretical considerations. The method for determining the abundance of each element is indicated in the far-right column of Table 4.1.

Two conventions for scaling abundance data are used in the literature. Astronomers like to calculate the abundances relative to 10^{12} atoms of hydrogen, whereas cosmochemists like to calculate abundances relative to 10^6 atoms of silicon. The two scales differ by a factor of 32.36. In this book, we present the abundances using the cosmochemical scale (normalized to 10^6 silicon atoms).

There are several basic features to note about the chemical abundances of the solar system. First, the Sun, and thus the solar system, consists predominantly of hydrogen and helium, with these two elements making up >98% of the mass of the solar system. Outside of the Sun, hydrogen and helium are found primarily in the gas-giant planets.

Second, lithium, beryllium, and boron have very low abundances (Fig. 4.1). These elements are, for the most part, not made in stars and were not made efficiently in the Big Bang. They are produced via cosmic-ray interactions. Nuclei of heavier atoms, when hit by fast-moving protons or other nuclei, can break into pieces, including protons, neutrons, alpha particles, and heavier fragments. Some of these fragments are lithium, beryllium, and boron nuclei.

Third, carbon, nitrogen, and oxygen are relatively abundant in the solar system (Fig. 4.1). Oxygen is somewhat more abundant than carbon; the C/O ratio is ~0.54. Because CO is a very stable molecule, most of the carbon and oxygen in the early solar system were tied up in CO. The excess oxygen that remained controlled the chemical environment of the early solar system.

Fourth, among the elements that make up rocks and minerals, silicon, magnesium, and iron are of almost equal abundance, followed by sulfur, aluminum, calcium, sodium, nickel, and chromium. Two of the most common minerals in meteorites and in the terrestrial planets are olivine ($(Mg,Fe)Si_2O_4$) and pyroxene ($(Mg,Fe,Ca)SiO_3$). The composition obtained by averaging these two minerals is very similar to the bulk solar system composition, so it is really no surprise that they are so abundant.

Fifth, for elements heavier than iron, the abundances decrease in a regular way with increasing mass (the zigzag pattern superimposed upon this decrease reflects nuclear properties, cf. the Oddo–Harkins rule). This is because the elements of the iron peak (iron, nickel, cobalt) have the highest binding energies per nucleon of all nuclides (cf. Fig. 2.1). For elements below iron, energy was released as protons and neutrons were added to smaller nuclei. However, for elements above iron, energy is required to force another proton or neutron into the nucleus. As nuclear mass increases, the binding energy per nucleon holding the nucleus together decreases until, at a nuclear mass of ~200, many nuclear configurations are not stable at all. The heaviest naturally occurring nuclide in the solar system is ^{238}U, and the abundance of uranium is about one part in 10^8 of that of silicon (Table 4.1).

4.6 Solar System Abundances of the Isotopes

Table 4.2 gives current estimates for the relative abundances of the isotopes in the solar system. The isotopic compositions of most elements, especially those that exist as solids, come from measurements of terrestrial materials. Because the Earth has experienced extensive melting and differentiation, it can be considered a homogeneous isotopic reservoir. However, each of the elements can experience both equilibrium and kinetically based isotopic fractionations during igneous, evaporative, and aqueous processes. The range of compositions introduced by such processes is small for most elements and so does not obscure the overall picture.

Why can one assume that the isotopic compositions of elements on the Earth are representative of the bulk solar system? This was not obvious in the beginning. Isotope-ratio measurements of meteorites made in the late 1950s appeared to show significant variations. However, within a few years, it was shown that those variations were due almost entirely to problems with the measurements, not to real variations in the samples. By the early 1960s, enough measurements had been made to convince most scientists that in the material to which they had access, terrestrial rocks and meteorites, each element always had the same isotopic composition (except for

Table 4.2 Abundances of the isotopes in the solar system

Element	*A*	Atom %	Abundance	Process
1 H	1	99.99806	2.59×10^{10}	
	2	0.00194	5.03×10^{5}	U
		100	2.590×10^{10}	
2 He	3	0.0166	1.03×10^{6}	U,h?
	4	99.9834	2.51×10^{9}	U,h
		100	2.51×10^{9}	
3 Li	6	7.589	4.2	X
	7	92.411	51.4	U,x,h
		100	55.6	
4 Be	9	100	0.612	X
5 B	10	19.820	3.7	X
	11	80.180	15.1	X
		100	18.8	
6 C	12	98.889	7.11×10^{6}	He
	13	1.111	79,900	H,N
		100	7.19×10^{6}	
7 N	14	99.775	2.12×10^{6}	H
	15	0.225	4791	H,N
		100	2.12×10^{6}	
8 O	16	99.774	1.57×10^{7}	He
	17	0.0358	5635	N,H
	18	0.1900	29,899	He,N
		100	1.57×10^{7}	
9 F	19	100	804	N
10 Ne	20	92.9431	3.06×10^{6}	C
	21	0.2228	7330	C,Ex
	22	6.8341	2.25×10^{5}	He,N
		100	3.29×10^{6}	
11 Na	23	100	57,700	C,Ne,Ex
12 Mg	24	78.992	8.10×10^{5}	N,Ex
	25	10.003	1.03×10^{5}	Ne,Ex,C
	26	11.005	1.13×10^{5}	Ne,Ex,C
		100	1.026×10^{6}	
13 Al	27	100	84,600	Ne,Ex
14 Si	28	92.230	9.22×10^{5}	O,Ex
	29	4.683	46,800	Ne,Ex
	30	3.087	30,900	Ne,Ex
		100	1.00×10^{6}	
15 P	31	100	8300	Ne,Ex
16 S	32	95.018	4.0026×10^{5}	O,Ex
	33	0.75	3160	Ex
	34	4.215	17,800	O,Ex
	36	0.017	72	Ex,Ne,S
		100	4.213×10^{5}	
17 Cl	35	75.771	3920	Ex
	37	24.229	1250	Ex,C,S
		100	5170	
18 Ar	36	84.5945	78,438	Ex
	38	15.3809	14,262	O,Ex
	40	0.0246	22.81	S,Ne
	*40**		*22*	
		100	92,723	
19 K	39	93.2582	3501	Ex
	40	0.0117	0.439	S,Ex,Ne
	*40**		*5.53*	
	41	6.7301	252.6	Ex
		100	3754	
20 Ca	40	96.941	58,500	Ex
	42	0.647	391	Ex,O
	43	0.135	82	Ex,C,S
	44	2.086	1260	Ex,S
	46	0.004	2	Ex,C,Ne
	48	0.187	113	E,Ex
		100	60,348	
21 Sc	45	100	34.4	Ex,Ne,E
22 Ti	46	8.249	204	Ex
	47	7.437	184	Ex
	48	73.72	1820	Ex
	49	5.409	134	Ex
	50	5.185	128	E
		100	2470	
23 V	50	0.2497	0.7	Ex,E
	51	99.7503	285.7	Ex
		100	286.4	
24 Cr	50	4.3452	569	Ex
	52	83.7895	11,000	Ex
	53	9.5006	1240	Ex
	54	2.3647	309	E
		100	13,118	
25 Mn	55	100	9220	Ex,E
26 Fe	54	5.845	49,600	Ex
	56	91.754	7.78×10^{5}	Ex,E
	57	2.119	18,000	E,Ex
	58	0.282	2390	He,E,C
		100	8.48×10^{5}	
27 Co	59	100	2350	E,C
28 Ni	58	68.0769	33,400	E,Ex
	60	26.2231	12,900	E
	61	1.1399	559	E,Ex,C
	62	3.6345	1780	E,Ex,O
	64	0.9256	454	Ex
		100	49,093	
29 Cu	63	69.174	374	Ex,C
	65	30.826	167	Ex
		100	541	

Table 4.2 *(cont.)*

Element	*A*	Atom %	Abundance	Process
30 Zn	64	48.63	630	Ex,E
	66	27.9	362	E
	67	4.10	53	E,S
	68	18.75	243	E,S
	70	0.62	8	E,S
		100	1296	
31 Ga	69	60.108	22.0	S,e,r
	71	39.892	14.6	S,e,r
		100	36.6	
32 Ge	70	21.234	24.3	S,e
	72	27.662	31.7	S,e,r
	73	7.717	8.8	e,s,r
	74	35.943	41.2	e,s,r
	76	7.444	8.5	E
		100	114.5	
33 As	75	100	6.10	R,s
34 Se	74	0.89	0.60	P
	76	9.37	6.32	S,p
	77	7.64	5.15	R,s
	78	23.77	16.04	R,s
	80	49.61	33.48	R,s
	82	8.73	5.89	R
		100	67.48	
35 Br	79	50.686	5.43	R,s
	81	49.314	5.28	R,s
		100	10.71	
36 Kr	78	0.362	0.20	P
	80	2.326	1.30	S,p
	82	11.655	6.51	S
	83	11.546	6.45	R,s
	84	56.903	31.78	R,S
	86	17.208	9.61	S,r
		100	55.85	
37 Rb	85	72.0569	5.121	R,s
	87	27.9431	1.986	S
	87*		*2.108*	
		100	7.107	
38 Sr	84	0.5551	0.1296	P
	86	9.8162	2.2923	S
	87	7.3829	1.72405	S
	87*		*1.60196*	
	88	82.2458	19.2061	S,r
		100	23.3521	
39 Y	89	100	4.63	S
40 Zr	90	51.452	5.546	S
	91	11.223	1.210	S
	92	17.146	1.848	S
	94	17.38	1.873	S
	96	2.799	0.302	R
		100	10.779	
41 Nb	93	100	0.780	S
42 Mo	92	14.525	0.370	P
	94	9.151	0.233	P
	95	15.838	0.404	R,s
	96	16.672	0.425	S
	97	9.599	0.245	R,s
	98	24.391	0.622	R,s
	100	9.824	0.250	R
		100	2.549	
44 Ru	96	5.542	0.099	P
	98	1.869	0.033	P
	99	12.758	0.227	R,s
	100	12.599	0.224	S
	101	17.060	0.304	R,s
	102	31.552	0.562	R,S
	104	18.621	0.332	R
		100	1.781	
45 Rh	103	100	0.370	R,s
46 Pd	102	1.02	0.0139	P
	104	11.14	0.1513	S
	105	22.33	0.3032	R,s
	106	27.33	0.371	R,S
	108	26.46	0.359	R,S
	110	11.72	0.159	R
		100	1.357	
47 Ag	107	51.839	0.254	R,s
	109	48.161	0.236	R,s
		100	0.490	
48 Cd	106	1.25	0.020	P
	108	0.89	0.014	P
	110	12.49	0.197	S
	111	12.8	0.201	R,S
	112	24.13	0.380	S,R
	113	12.22	0.192	R,S
	114	28.73	0.452	S,R
	116	7.49	0.118	R
		100	1.574	
49 In	113	4.288	0.008	p,s,r
	115	95.712	0.170	R,S
		100	0.178	
50 Sn	112	0.971	0.035	P
	114	0.659	0.024	P,s
	115	0.339	0.012	p,s,r
	116	14.536	0.524	S,r
	117	7.676	0.277	R,S
	118	24.223	0.873	S,r

Table 4.2 (*cont.*)

Element	*A*	Atom %	Abundance	Process
	119	8.585	0.309	S,R
	120	32.593	1.175	S,R
	122	4.629	0.167	R
	124	5.789	0.209	R
		100	3.605	
51 Sb	121	57.213	0.179	R,s
	123	42.787	0.134	R
		100	0.313	
52 Te	120	0.096	0.005	P
	122	2.603	0.122	S
	123	0.908	0.043	S
	124	4.816	0.226	S
	125	7.139	0.335	R,s
	126	18.952	0.889	R,S
	128	31.687	1.486	R
	130	33.799	1.585	R
		100	4.691	
53 I	127	100	1.10	R
54 Xe	124	0.129	0.007	P
	126	0.112	0.006	P
	128	2.23	0.122	S
	129	27.46	1.499	R
	130	4.38	0.239	S
	131	21.80	1.190	R
	132	26.36	1.438	R,s
	134	9.66	0.527	R
	136	7.87	0.429	R
		100	5.457	
55 Cs	133	100	0.371	R,s
56 Ba	130	0.106	0.005	P
	132	0.101	0.005	P
	134	2.417	0.108	S
	135	6.592	0.295	R,s
	136	7.853	0.351	S
	137	11.232	0.502	S,r
	138	71.699	3.205	S
		100	4.471	
57 La	138	0.09013	0.000412	P
	138*		*0.000416*	
	139	99.90987	0.456584	S,r
		100	0.456996	
58 Ce	136	0.1860	0.002193	P
	138	0.2503	0.002952	P
	138*		*0.002948*	
	140	88.4497	1.0428	S,r
	142	11.1140	0.1310	R
		100	1.1790	
59 Pr	141	100	0.172	R,S
60 Nd	142	27.007	0.23150	S
	143	12.141	0.10407	R,S
	143*		*0.10292*	
	144	23.697	0.20312	S,R
	145	8.751	0.07501	R,s
	146	17.107	0.14663	R,S
	148	5.708	0.04893	R
	150	5.588	0.04790	R
		100	0.85716	
62 Sm	144	3.060	0.00820	P
	147	15.358	0.04118	R,s
	147*		*0.04003*	
	148	11.193	0.03001	S
	149	13.760	0.03690	R,S
	150	7.348	0.01970	S
	152	26.627	0.07140	R,S
	154	22.654	0.06075	R
		100	0.26815	
63 Eu	151	47.81	0.0471	R,s
	153	52.19	0.0514	R,s
		100	0.0985	
64 Gd	152	0.203	0.0007	P,s
	154	2.181	0.0078	S
	155	14.800	0.0533	R,s
	156	20.466	0.0736	R,s
	157	15.652	0.0563	R,s
	158	24.835	0.0894	R,s
	160	21.864	0.0787	R
		100	0.3598	
65 Tb	159	100	0.0634	R
66 Dy	156	0.056	0.0002	P
	158	0.095	0.0004	P
	160	2.329	0.0094	S
	161	18.889	0.0762	R
	162	25.475	0.1028	R,s
	163	24.896	0.1005	R
	164	28.260	0.1141	R,S
		100	0.4036	
67 Ho	165	100	0.0910	R
68 Er	162	0.139	0.0004	P
	164	1.601	0.0042	P,S
	166	33.503	0.088	R,s
	167	22.869	0.060	R
	168	26.978	0.071	R,S
	170	14.910	0.039	R
		100	0.2626	
69 Tm	169	100	0.0406	R,s

Table 4.2 (*cont.*)

Element	A	Atom %	Abundance	Process
70 Yb	168	0.12	0.0003	P
	170	2.98	0.0076	S
	171	14.09	0.0361	R,s
	172	21.69	0.0556	R,S
	173	16.10	0.0413	R,s
	174	32.03	0.0821	S,R
	176	13.00	0.0333	R
		100	0.2563	
71 Lu	175	97.397	0.037025	R,s
	176	2.603	0.000990	S
	176*		*0.001075*	
		100	0.038015	
72 Hf	174	0.162	0.000252	P
	176	5.258	0.008196	S
	176*		*0.008111*	
	177	18.596	0.02899	R,s
	178	27.282	0.04253	R,S
	179	13.622	0.02123	R,s
	180	35.081	0.05469	S,R
		100	0.15589	
73 Ta	180	0.0123	0.0000026	p,s,r
	181	99.9877	0.020987	R,S
		100	0.02099	
74 W	180	0.120	0.0002	P
	182	26.499	0.0363	R,s
	183	14.314	0.0196	R,s
	184	30.642	0.0420	R,s
	186	28.426	0.0390	R
		100	0.1371	
75 Re	185	37.298	0.020720	R,s
	187	62.702	0.034832	R
	187*		*0.037380*	
		100	0.055552	
76 Os	184	0.0199	0.000135	P
	186	1.5920	0.010817	S
	187	1.6412	0.011151	S
	187*		*0.008603*	
	188	13.2869	0.090278	R,s
	189	16.1999	0.110071	R
	190	26.3446	0.178999	R
	192	40.9156	0.278003	R
		100	0.679454	
77 Ir	191	37.272	0.250	R
	193	62.728	0.421	R
		100	0.671	
78 Pt	190	0.01363	0.000174	P
	192	0.78266	0.009972	S
	194	32.967	0.42007	R
	195	33.83156	0.43108	R
	196	25.24166	0.32163	R
	198	7.16349	0.09128	R
		100	1.2742	
79 Au	197	100	0.195	R
80 Hg	196	0.15	0.001	P
	198	9.97	0.046	S
	199	16.87	0.077	R,S
	200	23.10	0.106	S,r
	201	13.18	0.060	S,r
	202	29.86	0.137	S,r
	204	6.87	0.031	R
		100	0.458	
81 Tl	203	29.524	0.054	R,S
	205	70.476	0.129	S,R
		100	0.183	
82 Pb	204	1.983	0.0660	S
	206	18.724	0.6234	R,S
	206*		*0.6143*	
	207	20.588	0.6855	R,S
	207*		*0.6798*	
	208	58.705	1.9547	R,s
	208*		*1.9458*	
		100	3.3297	
83 Bi	209	100	0.1382	R,s
90 Th	232	100	0.03512	RA
	232*		*0.04399*	
92 U	235	0.7303	0.000065	RA
	235*		*0.005780*	
	238	99.2697	0.008896	RA
	238*		*0.018019*	
		100	0.008961	

Abundances from Lodders, Palma, and Gail (2009) and Lodders (2003). O and N isotopes from Genesis results. Nucleosynthetic processes from Anders and Grevesse (1989). Processes are listed in order of importance, with minor processes (10–30%) shown in lower case.
U: cosmological nucleosynthesis, H: hydrogen burning, N: hot or explosive hydrogen burning, He: helium burning, C: carbon burning, O: oxygen burning, Ne: neon burning, Ex: explosive nucleosynthesis, E: nuclear statistical equilibrium, S: *s*-process, R: *r*-process, RA: *r*-process producing actinides, P: *p*-process, X: cosmic-ray spallation.
Italicized values refer to abundances 4.55×10^9 years ago.
* Indicates isotopes of particular importance to radiochronology.

small fractionations of stable isotopes and shifts due to decay of radioactive isotopes, as discussed in Chapter 8). At the same time, models for the formation of the solar system envisioned that it passed through a hot gaseous state where all pre-existing solids were evaporated. Had such a state existed, it would have isotopically homogenized the solar system. So observations and experiments pointed to a single, well-defined composition that could be measured, and the idea of a single homogenized solar system composition became a cornerstone of cosmochemistry. The "fly in the ointment" was the noble gases, which exhibited a range of isotopic compositions in different materials, well outside the range to be expected from known processes in the solar system.

Starting in 1973, evidence began to accumulate from elements other than noble gases that the isotopic composition of the solar system was not homogeneous. As you will see in later chapters, isotopic variations, though typically small, are ubiquitous in solar system materials. These isotopic "anomalies" appear in the most primitive solar system materials, the components of chondritic meteorites. With the exception of oxygen, the anomalies either disappear or are much smaller in objects with more complicated histories involving processes that mix isotopes, such as evaporation and condensation, melting and crystallization, and aqueous processes. Products of these processes typically have very similar isotopic compositions. Thus, although the solar system was not homogenized isotopically by passing through a hot gas phase, the mixture of gas and dust that made up the solar system was apparently well mixed and the processes that acted on that material served to homogenize it at the atomic scale. Small isotopic differences have been measured in different kinds of meteorites that are inferred to reflect differences in source materials (see Chapter 8), but these isotopic variations are measured in parts per 10,000 and do not negate the general concept of a solar system isotopic composition (Table 4.2).

NASA's Genesis mission (Box 4.2) produced a significant change in the inferred isotopic compositions of oxygen and nitrogen in the solar system. Both elements are significantly more abundant in the Sun than in the terrestrial samples that are typically measured for a table like Table 4.2. Oxygen in the Sun is enriched in ^{16}O by almost 6% relative to standard mean ocean water, and solar nitrogen is enriched in ^{14}N by almost 40% compared to terrestrial nitrogen. We report the Genesis ratios for oxygen and nitrogen in Table 4.2.

Along with the abundances of the isotopes, Table 4.2 also provides information on the processes that synthesized the isotopes. These processes are listed in order of importance, with minor processes shown in lower case. Note that most isotopes are produced by more than one process, and that isotopes of elements heavier than iron are produced predominantly by the *s*- and *r*-processes. *P*-process isotopes invariably have quite low abundances.

The isotopic compositions of several elements have changed since the formation of the solar system. Radioactive isotopes have decayed, and the abundances of their daughter isotopes have increased over time. For isotopes of particular importance to radiochronology (flagged with an asterisk), Table 4.2 gives both the present abundance and the abundance at the time the solar system formed (italicized values).

Table 4.2 and similar tables should *not* be used as the source of isotopic compositions of terrestrial standards for high-precision isotopic measurements. While Table 4.2 gives a good overview of the abundances of the isotopes, the values in the table do not always represent state-of-the-art isotopic measurements. Nor do they attempt to take into account the isotopic fractionations that occur in terrestrial materials. Typically, an isotope system that has been measured for many years has a convention about how isotopic ratios are reported. Research papers should state how their measurements were standardized and what composition is being used for the standards. Those who develop measurement techniques for a new isotope system must address standardization before they can publish their data.

4.7 How Did Solar System Abundances Arise?

As discussed in Chapter 3, the abundances inherited by the solar system are the result of a long history of nucleosynthesis in the galaxy. By the time the solar system formed, many generations of massive stars and several generations of intermediate-mass stars had contributed material to the galaxy over the billions of years leading up to the formation of the solar system. Low-mass stars were just beginning to return newly synthesized material to the galaxy when the solar system formed. The elements produced by all these different sources were mixed through the galaxy by turbulence and thermal mixing driven primarily by stellar explosions. The same types of stars live and die throughout the galaxy, but there are more of them in the galactic center than on the edges. Over timescales of a few hundred million years, rough chemical and isotopic homogeneity can be attained over an annulus of the galactic disk, although radial gradients

Box 4.3 The Oxygen Isotopic Composition of the Sun

The oxygen isotopic composition of the solar system is a key constraint for understanding the environment in which the Sun formed, the environment in the early solar system, and the processes that operated in the early solar system. Oxygen isotopic composition is also a key property by which we classify meteorites. Yet until very recently, we have not been able to make any measurement that could unambiguously tell us what the oxygen isotopic composition of the Sun, and by extension the solar system, really is.

Previous measurements: Meteorites show a very large range in oxygen isotopic compositions. Unlike the range in oxygen compositions observed on Earth, which can be explained by mass-dependent processes operating on a single terrestrial composition, the meteorite compositions fall well off the terrestrial mass fractionation line (Fig. 4.8). This plot compares $^{17}O/^{16}O$ versus $^{18}O/^{16}O$, compared to a terrestrial standard. At one extreme are the calcium-aluminum-rich inclusions (CAIs) and isolated hibonite grains from carbonaceous chondrites, which have compositions that are enriched by up to ~6% in ^{16}O relative to the terrestrial composition. At the other extreme is isotopically heavy water from some ordinary chondrites that is depleted by ~15% in ^{16}O compared to the terrestrial composition. Although both extremes have been proposed as the true composition of the solar system, historically the terrestrial composition has been viewed as a reasonable approximation to the solar composition.

Attempts have been made to measure the oxygen-isotopic composition of the solar photosphere with spectroscopy. Unfortunately, the photosphere is too hot to look at the best lines for this work and the results to date have uncertainties that span the range of compositions determined for meteorites and planetary bodies. The Genesis mission, described in Box 4.2, was designed in large part to get a precise measurement of the oxygen composition of the solar wind. A special concentrator was designed so that enough atoms could be collected to make a measurement with high precision. A special ion microprobe, called MegaSims, was designed with some of the features of an accelerator mass spectrometer in order to measure the solar wind oxygen composition. Three of the four targets in the concentrator survived the crash intact. The open circle in Figure 4.8 is the measured solar wind value (McKeegan et al., 2011); as with other elements, oxygen can also be mass fractionated in the solar wind (along the dotted line), and the most plausible solar composition ($\Delta^{17}O = -28.4 \pm 3.6‰$) lies on the CAI line (star). Thus, the Sun is more ^{16}O-rich than almost anything else that has been measured so far in the solar system. Because the Sun contains the vast majority of the mass of the solar system, we have used the Genesis results for oxygen (and nitrogen) in our table of solar system abundances (Table 4.2).

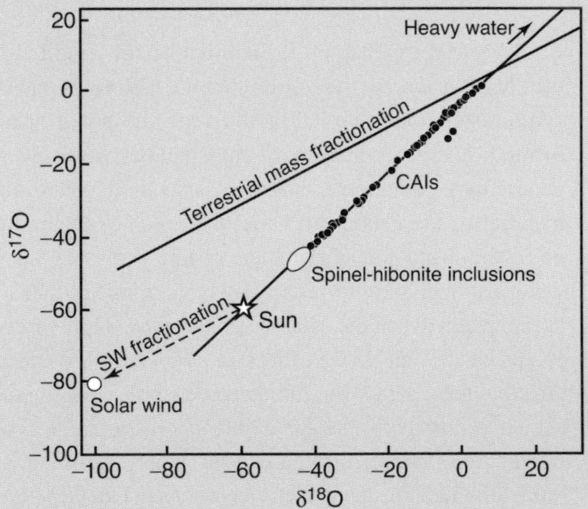

Figure 4.8 Oxygen isotopic compositions of the solar wind, Sun, Earth, and meteorite components. Earth samples lie along the terrestrial mass fractionation line. CAIs and spinel-hibonite inclusions in carbonaceous chondrites and heavy water in ordinary chondrites define another fractionation line with a slope of ~1. The open circle gives the solar wind composition measured by McKeegan et al. (2011). Like other elements, oxygen in the solar wind has been mass fractionated, and the Sun's true composition (star) likely lies on the CAI line. By convention, $^{17}O/^{16}O$ and $^{18}O/^{16}O$ ratios are plotted as deviations from the composition of standard mean ocean water (SMOW) in units of parts per thousand (permil or ‰). δ values are calculated as follows: $\delta^{17}O = (((^{17}O/^{16}O)_{sample}/(^{17}O/^{16}O)_{SMOW})-1) \times 1000$, and similarly for $\delta^{18}O$. Another parameter, $\Delta^{17}O$, defined as $\Delta^{17}O = \delta^{17}O - 0.52 \times \delta^{18}O$, gives the distance above or below the terrestrial fractionation line on a plot like Figure 4.8. Figure modified from McKeegan et al. (2011).

Box 4.3 (cont.)

Recently, a refractory inclusion with a ^{16}O-rich, solar-like oxygen isotopic composition ($\Delta^{17}O = -29.1 \pm 0.7‰$) was discovered (Kööp et al., 2020). This unusual CAI has no evidence of ^{26}Al and may have formed before that short-lived isotope was incorporated into the nebula. Only a handful of other objects are similarly ^{16}O rich, and they are typically also ^{26}Al poor. Most unaltered CAIs have $\Delta^{17}O = \sim 23.5‰$ and do show evidence for ^{26}Al. It is not clear why most CAIs have compositions that are not as extreme as the Sun. If most CAIs were fully equilibrated with the gas in which they formed, as most workers believe, and most of the original accretion-disk solids were less ^{16}O rich than the solar gas, the addition of evaporated solids to the gas could give most CAIs a less ^{16}O-rich composition. The solar-like CAIs would then be those that did not equilibrate with the modified gas reservoir.

remain (see below). Therefore, it is no accident that the first measurements of the compositions of the Sun, stars, galaxies, comets, and other celestial objects gave compositions that are to first order rather similar.

On a local scale, gas and dust are regularly forced together by expanding gas from supernova explosions into molecular clouds, where material cools and densities increase until especially dense regions collapse to initiate star formation. Most stars form in giant molecular clouds out of which hundreds or thousands of stars may form in what is often called "cluster star formation." The solar system probably formed in such a region. The solar system primarily incorporated material inherited by the molecular cloud, but depending on the timing of solar system formation, it may also have incorporated newly synthesized material ejected by the earliest-formed massive stars in the cloud. There is good evidence from short-lived radionuclides that this happened in the solar system, as we will discuss in Chapters 10 and 15. An active area of research is trying to infer the details of the environment in which the solar system formed, and the materials from which it formed.

The solar system abundances of the elements are the result of the Big Bang, which produced hydrogen and helium, ~7.5 billion years of stellar nucleosynthesis and cosmic-ray interactions, which produced most of the rest of the elements, and the physical processes that mixed the materials together to form the Sun's parent molecular cloud. The unique features of the solar system composition may also reflect the stochastic events that occurred in the region where the Sun formed just prior to solar system formation.

4.8 Differences between Solar System and Cosmic Abundances

As mentioned in the introduction to this chapter, the terms "solar system abundances" and "cosmic abundances" have been used interchangeably over the last century. This was justified because the available evidence suggested that other stars had essentially the same composition as the Sun. However, with the new observational capability provided by state-of-the-art telescopes and instrumentation, we are learning that the composition of the solar system is not the same as that of most of the rest of the galaxy. There is really no reason to expect that the compositions should be the same. The chemical and isotopic composition of the galaxy has been evolving continuously since it formed. Stellar nucleosynthesis has added elements heavier than helium to the mix, and the galaxy has been growing through accretion of metal-poor gas and assimilation of smaller galaxies. When the Sun formed ~4.57 billion years ago, it captured a snapshot of this evolution in the region where it formed. But the galaxy has continued to evolve. It is also likely that the Sun no longer resides in the same region of the galaxy where it formed due to dynamical scattering caused by close passages with other stars and the mass of the galactic arms. To further complicate matters, self-enrichment of the cloud from which the solar system formed by massive stars that formed and died in the same star-forming region, or mechanical separation of gas from dust in the Sun's parent molecular cloud could have caused the Sun to acquire a composition different from that of the galaxy in the solar neighborhood.

It is not straightforward to determine the current chemical composition of the galaxy in the neighborhood of the Sun. Although characteristic absorption lines are generated by cold matter in interstellar space, their strengths are not necessarily proportional to the amounts of the elements present. Elements in the gas phase absorb efficiently, while those in dust do not. Consequently, absorption lines cannot be used directly to provide details of elemental abundances. Stars produce emission lines in their photospheres, but low-mass stars live billions of years and the surrounding galaxy can evolve considerably during that time. Low- and intermediate-mass stars provide a way to look back at the chemical composition of the galaxy billions of years ago. The best way to estimate the current composition of the galaxy in the solar neighborhood is to look at the photospheres of B stars. These massive stars are short-lived and thus formed recently, within the last ~10 million years, so there has not been time for the composition of the galaxy to evolve since they formed. Nor has there been time for the stars to migrate from where they formed.

Systematic observations of B stars show that their heavy-element abundances relative to hydrogen are only 50–70% of those in the Sun (the new lower abundances derived from three-dimensional solar models do not eliminate this difference). Theory predicts that the metallicity of the galaxy has been increasing with time, and a systematic study of stars formed over most of the history of the galaxy shows this to be true. In addition, there can be a spread of almost an order of magnitude in the Fe/H ratios of stars that formed at roughly the same time. Relative to nearby solar-type stars of similar age, the Sun has a ~50% higher Fe/H ratio than average.

The Sun might have acquired a higher Fe/H ratio than many stars of similar age because of where the solar system formed. If the Sun formed closer to the galactic center, where star formation and galactic evolution proceed more rapidly, it would have acquired a higher abundance of heavy elements from the average material in that location. After formation, the Sun might then have been gravitationally scattered outward to where it now resides. On the other hand, the galaxy is not a closed system, and new heavy-element-poor gas has been falling onto the galaxy from surrounding intergalactic space. If enough metal-poor gas fell into the solar neighborhood, more-recently formed stars might have lower metallicity than the Sun.

The solar system is different than the rest of the galaxy in other ways as well. Although one cannot use spectral observations of cold matter to reliably determine elemental abundances, it is possible to measure isotopic abundances using absorption lines. This is because the different isotopes of most elements do not partition significantly between gas and solid phases in the interstellar medium. However, because absorption lines can become saturated if the column density between the source and the observer is too high (no light gets through, so, for example, doubling the amount of material makes no measurable difference), it is often easier to measure ratios of isotopes that have similar abundances. For example, $^{16}O/^{17}O$ in solar system material is ~2630, whereas $^{16}O/^{18}O$ is ~490. The ^{16}O line often saturates, resulting in spurious results. However, $^{17}O/^{18}O$ is ~0.19 and thus is relatively easy to measure. Table 4.3 shows isotopic data for carbon, nitrogen, and oxygen for molecular clouds in the galactic center and the outer galaxy compared to the solar system ratios.

The $^{13}C/^{12}C$ ratios for molecular clouds throughout the galaxy are higher than in the solar system, with ratios increasing toward the galactic center. The $^{15}N/^{14}N$ ratios in molecular clouds throughout the galaxy are lower than in the solar system and the ratios decrease toward the galactic center. The carbon and nitrogen ratios apparently reflect the addition of the products of CNO-catalyzed hydrogen burning to the interstellar medium. The oxygen ratios are harder to understand. CNO cycling should produce higher $^{17}O/^{16}O$ ratios and lower $^{18}O/^{16}O$ ratios. Yet $^{18}O/^{16}O$ increases toward the galactic center, while $^{17}O/^{18}O$ ratios, which are everywhere higher than in the solar system, are nearly the same across the galaxy. An additional source of ^{18}O is required – one that produces enough ^{18}O to balance its destruction by CNO burning in the outer galaxy, but which is operating much more efficiently in the galactic core. Massive stars may be that source.

Another possible way to reconcile some of the chemical and isotopic differences between the solar system and the solar neighborhood is to postulate that the solar system formed in a region of active star formation where the abundances of material from massive stars were enhanced over the galactic average due

Table 4.3 Isotopic ratios of C, N, and O in molecular clouds relative to solar ratios

Material	$^{13}C/^{12}C$	$^{15}N/^{14}N$	$^{18}O/^{16}O$	$^{17}O/^{18}O$
Solar system	0.01124	0.003676	0.00204	0.1861
Outer galaxy	0.0157 ± 0.0034	0.00291 ± 0.00034	0.00204 ± 0.00041	0.275 ± 0.009
Galactic center	0.0360 ± 0.0090	0.00147 ± 0.00034	0.0041 ± 0.0016	0.296 ± 0.19

Data from Wannier (1989).

to supernova explosions just before the solar system formed. These models are currently poorly constrained and to date have not been very successful, but the concept behind them may well be true, as indicated by the presence of short-lived radionuclides in our solar system (see Chapter 10).

4.9 How Are Solar System Abundances Used in Cosmochemistry?

We have now considered how the solar system abundances are determined and have discussed the uncertainties in the abundance numbers. Here we briefly discuss how these abundances are used in cosmochemistry. These uses will be discussed further in later chapters.

4.9.1 Precursor Materials

Solar system abundances help to determine the nature of the materials from which the solar system formed. As discussed above, the solar system abundances arose through a long history of star formation, stellar nucleosynthesis, and ejection of newly formed elements into the interstellar medium. If we know the isotopic compositions of materials formed within the solar system, we can recognize materials that were formed outside of the solar system. These "presolar grains," discussed in Chapter 5, provide ground truth for models of stellar nucleosynthesis. By studying solar system abundances in the context of models of stellar nucleosynthesis, cosmochemists can infer a great deal about the types of stars that contributed to the solar system mixture and the environment in which the solar system formed. Comparison of solar system abundances with the elemental and isotopic compositions of nearby and distant stars and molecular clouds, and with the isotopic compositions of presolar grains, gives insight into galactic chemical evolution. For this kind of work, the solar system abundances are known with sufficient accuracy and precision to serve as a firm foundation.

As models for the origin of the solar system become more sophisticated, they make increasingly precise predictions about the evolution of the chemical and isotopic composition of solar system materials. For example, oxygen isotopes are used to classify meteorites (discussed in Chapter 6) and to investigate relationships between different bodies in the early solar system. But the origin of the oxygen isotopic variations is currently not understood. Several models have been proposed. A key to understanding solar system oxygen is the isotopic composition of the Sun, which is basically equivalent to the solar system. That is why so much effort has been expended to determine the oxygen isotopic composition of the Sun (see Box 4.3). This single data point has eliminated several of the proposed models, although the distribution of oxygen isotopes among solar system bodies has not yet been explained. Determining the origin of the Earth's water is a problem for which the bulk composition of the hydrogen in the solar system provides critical input, as discussed in Chapter 15. Various proxies have been identified and measured to infer the solar system composition (e.g. Jupiter's atmosphere, cometary ices), and depending on the composition that is chosen, different models for the origin of the Earth's water can be constructed. There is also an isotopic dichotomy in several rock-forming elements between carbonaceous chondrites and related meteorites and most other types of chondrites (discussed in Chapter 8). The evidence for the dichotomy is clear, but the explanation is not, in part because we do not know the bulk isotopic compositions of the elements that define the dichotomy for the solar system. An area of continuing research will be to improve our understanding of the bulk chemical and isotopic composition of the solar system so that better models of early solar system processes can be produced.

4.9.2 Fractionations

The processes that produced the planets, their moons, meteorites, and comets from the raw materials from which the solar system formed resulted in changes in the chemical compositions relative to the starting material. We refer to these changes as "chemical fractionations," and they will be discussed at length in Chapter 7. In order to have a sensible discussion about how the elements have been fractionated by a given process, it is necessary to know the starting composition. For discussions of processes in the early solar system, the solar system abundances give the starting composition, although in some cases the compositions of the CI carbonaceous chondrites (upon which solar system abundances are based) are used for that purpose.

4.9.3 Normalizing to Solar System Abundances

In the cosmochemistry literature, you will often see data normalized to (that is, divided by) solar system abundances (most commonly those of CI chondrites). An important reason for doing this is illustrated in Figure 4.9. The top panel (a) plots the composition of a chondrule with the elements arranged in order of their volatility, from most volatile on the left to most refractory on the right. This plot reveals little about the extent to which the chondrule differs from the composition of the

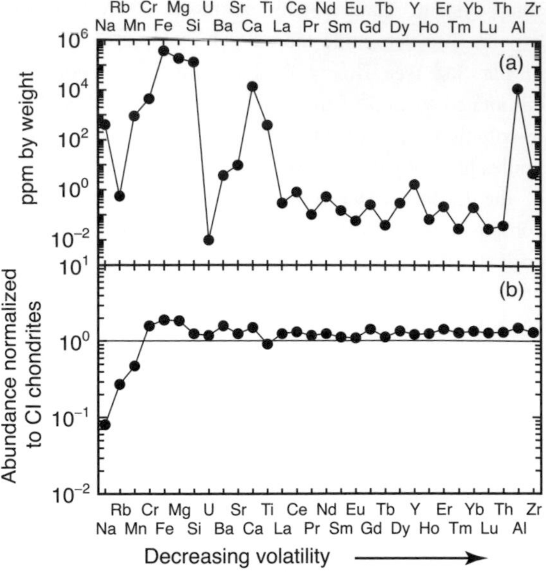

Figure 4.9 (a) Composition of a chondrule from a CR chondrite, plotted as ppm by weight. Elements are arranged in order of decreasing volatility. (b) The same data normalized to the abundances in CI chondrites (Table 4.1). A composition that is identical to CI chondrites will plot on the horizontal line at a value of 1 (10^0). In the normalized plot, depletions of the most volatile elements in the chondrule relative to solar system abundances are now evident. Elements more refractory than chromium have the same relative abundances as CI chondrites, although the absolute abundances are higher due to depletion of volatile elements.

bulk solar system. However, if the composition is normalized to the solar system abundances, as in the bottom panel (b), the pattern of chemical fractionation is immediately evident. This normalization eliminates the zigzag pattern due to the Oddo–Harkins rule and puts abundant and trace elements onto the same scale. The normalized plot shows that this chondrule is depleted in volatile elements relative to solar system abundances and is enriched in elements more refractory than chromium. (Note that in a chemical plot such as this, which considers the weight percent or the atom percent of each constituent, if some elements are depleted, the remaining elements in the solid must have higher concentrations because the elements must total 100%. Some of it plus the rest of it must equal all of it!)

In geochemical literature, as well as in the cosmochemical literature, you will often see compositions normalized to carbonaceous chondrites (meaning CI chondrites). In many cases, this normalization is chosen because it is "conventional," not because the normalization is particularly instructive for the problem being investigated. It is not necessary to normalize data in this way. You can normalize trace-element compositions of minerals to the bulk composition of the system to illustrate fractionation processes during formation of the minerals, or you can normalize everything, including the bulk composition, to CI chondrites. When reading the literature, make sure you understand how data have been normalized, and when you report your data, make sure that you clearly explain any normalizations. Normalizing data is a powerful tool, but if the wrong normalizations are used, they can generate artifacts in the data that can result in spurious interpretations.

Summary

In this chapter, we have introduced solar system abundances and cosmic abundances of the elements and isotopes and briefly discussed their importance in cosmochemistry. The idea of cosmic abundances has been around for over a century and is rooted in measurements of many different celestial objects. Only in the last few decades has our knowledge of the compositions of the solar system, of stars and molecular clouds in our galaxy, and of other galaxies grown sufficiently that we can now detect clear compositional differences between our solar system and other objects in the galaxy and the universe.

Solar system abundances are determined by a combination of spectroscopy of the solar photosphere, detailed chemical measurements of CI chondrites and the solar wind, and in the cases of elements not amenable to direct measurement, theoretical considerations. The solar system and galactic abundances arose through a combination of Big Bang nucleosynthesis, nucleosynthesis in stars over the history of the universe, and cosmic-ray-induced spallation reactions that fragment pre-existing nuclei. Solar system abundances differ from abundances in other parts of the galaxy and the universe because the details of the processes and rates of nucleosynthesis vary from place to place. Thus, it is no longer appropriate to use the terms "cosmic abundances" and "solar system abundances" interchangeably.

Solar system abundances are a cornerstone of our understanding of the origin and evolution of the solar system and the elements from which it formed. To understand how physical and chemical processes produced the current planets, asteroids, comets, and other solar system bodies, it is necessary to know the original composition of the material upon which those processes operated. Knowledge of the chemical and isotopic composition of the solar system is required to recognize objects that formed outside of the solar system. To decipher the role of recent stellar nucleosynthesis in the solar system and in the solar neighborhood, it is necessary to know what the chemical and isotopic composition of the material was before nucleosynthesis began to operate.

In the next few chapters, we will investigate these topics further. For example, in Chapter 5, we will introduce presolar grains and show how they can be used to investigate stellar nucleosynthesis and processes operating in interstellar space and in the early solar system, and in Chapters 7 and 8, we will discuss chemical and isotopic fractionation processes that operate in the solar system.

Questions

1. What are the three main ways of obtaining information about solar system abundances?
2. What are the advantages and limitations of spectroscopic measurements of the solar photosphere for determining solar system abundances?
3. Why are CI chondrites so important to cosmochemistry?
4. What are some of the differences between the composition of the solar system and the composition of the galaxy? How might these differences have arisen?
5. Why would you want to normalize chemical data obtained for a chondrule or a meteorite? What would you normalize the data to and why?

Suggestions for Further Reading

Asplund M., Grevesse N., Sauval A. J., and Scott P. (2009) The chemical composition of the Sun. *Annual Review of Astronomy & Astrophysics*, **47**, 481–522. A comprehensive discussion of the issues around converting stellar spectra to solar abundances.

Palme H., Lodders K., and Jones A. (2014) Solar system abundances of the elements. In *Treatise on Geochemistry, 2nd edition, Vol. 2: Planets, Asteroids, Comets and the Solar System*, Davis A. M., editor, pp. 15–36, Elsevier, Oxford. An excellent summary of element abundances and how they are determined.

Other References

Anders E., and Grevesse N. (1989) Abundances of the elements: Meteoritic and solar. *Geochimica et Cosmochimica Acta*, **53**, 197–214.

Bahcall J. N., Basu S., and Serenelli A. M. (2005) What is the neon abundance of the Sun? *Astrophysical Journal*, **63**, 1281–1285.

Basu S., and Antia H. M. (2004) Constraining solar abundances using helioseismology. *Astrophysical Journal*, **606**, L85–L88.

Burnett D. S. (2013) The Genesis solar wind sample return mission: Past, present, and future. *Meteoritics & Planetary Science*, **48**, 2351–2370.

Burnett D. S., Jurewicz A. J. G., and Woolum D. S. (2019) The future of Genesis science. *Meteoritics & Planetary Science*, **54**, 1092–1114.

Christensen-Dalsgaard J. (1998) The 'standard Sun'. Modelling and helioseismology. *Space Science Reviews*, **85**, 19–36.

Christensen-Dalsgaard J. (2002) Helioseismology. *Reviews of Modern Physics*, **74**, 1073–1129.

Clarke F. W. (1889) The relative abundances of the chemical elements. *Bulletin of the Philosophical Society of Washington*, **11**, 131–142.

Gabriel M. (1997) Influence of heavy element and rotationally induced diffusions on solar models. *Astronomy & Astrophysics*, **327**, 771–778.

Grevesse N., and Sauval A. J. (1998) Standard solar composition. *Space Science Reviews*, **85**, 161–174.

Grevesse N., Asplund M., and Sauval A. J. (2007) The solar chemical composition. *Space Science Reviews*, **130**, 105–114.

Grevesse N., Scott P., Asplund M., and Sauval A. J. (2015) The elemental composition of the Sun III. The heavy elements Cu to Th. *Astronomy & Astrophysics*, **573**, A27.

Harkins W. D. (1917) The structure of atoms and the evolution of the elements as related to the composition of the nuclei of atoms II. *Science*, **46**, 443–448.

Hill V. (2001) From stellar spectra to abundances. *Astrophysics & Space Science*, **277** (Supplement), 137–146.

Kööp L., Nagashima K., Davis A. M., and Krot A. N. (2020) A refractory inclusion with solar oxygen isotopes and the rarity of such objects in the meteorite record. *Meteoritics & Planetary Science*, **55**, 524–534.

Lockyer J. N. (1890) *The Meteoritic Hypothesis*. Macmillan & Co., New York, 560 pp.

Lodders K. (2003) Solar system abundances and condensation temperatures of the elements. *Astrophysical Journal*, 591, 1220–1247.

Lodders K., Palme H., and Gail H. P. (2009) Abundances of the elements in the solar system. In *Landolt-Börnstein, New Series, Vol. VI/4B*, Trümper J. E., editor, pp. 560–630, Springer-Verlag, New York.

McKeegan K. D., Kallio A. P. A., Heber V. S., et al. (2011) The oxygen isotopic composition of the Sun inferred from captured solar wind. *Science*, **332**, 1528–1532.

Oddo G. (1914) Die Molekularstruktur der radioaktiven Atome. *Zeitschrift für Anorganische Chemie*, **87**, 253–268.

Payne C. H. (1925) *Stellar Atmospheres; a Contribution to the Observational Study of High Temperature in the Reversing Layers of Stars*, Ph.D. Thesis, Radcliffe College.

Russell H. N. (1929) On the composition of the Sun's atmosphere. *Astrophysical Journal*, **70**, 11–82.

Scott P., Grevesse N., Asplund M., et al. (2015a) The elemental composition of the Sun I. The intermediate mass elements Na to Ca. *Astronomy & Astrophysics*, **573**, A25.

Scott P., Asplund M., Grevesse N., et al. (2015b) The elemental composition of the Sun II. The iron group elements Sc to Ni. *Astronomy & Astrophysics*, **573**, A26.

Suess H. E., and Urey H. C. (1956) Abundances of the elements. *Reviews of Modern Physics*, **28**, 53–74.

Vagnozzi S., Freese K., and Zurbuchen T. H. (2017) Solar models in the light of new high metallicity measurement from solar wind data. *Astrophysical Journal*, **839**, 55.

Vauclair S. (1998) Elemental settling in the solar interior. *Space Science Reviews*, **85**, 71–78.

Wannier P. G. (1989) Abundances in the galactic center. In *The Center of the Galaxy*, Morris M., editor, pp. 107–119, IAU Symposium #136.

5 Presolar Grains

A Record of Stellar Nucleosynthesis and Processes in Interstellar Space

Overview

Presolar grains give us a direct window into stellar nucleosynthesis and provide probes of processes in interstellar space and in the solar nebula. Known types of presolar grains originated in the winds or ejecta of stars that lived and died before the solar system formed. After presenting a short history of how presolar grains came to be recognized, we describe how to identify presolar grains, the techniques used to study them, and the various types of currently recognized grains. We then review what presolar grains can tell us about stellar nucleosynthesis, the environments around evolved stars and in the interstellar medium, and how the grains can be used as probes of conditions in the early solar nebula.

5.1 Grains that Predate the Solar System

In recent years, a new source of information about stellar nucleosynthesis and the history of the elements between their ejection from stars and their incorporation into the solar system has become available. This source is tiny dust grains that condensed from gas ejected from stars at the end of their lives and that survived unaltered to be incorporated into solar system materials. These *presolar grains* (Fig. 5.1) originated before the solar system formed and were part of the raw materials that built the Sun, the planets, and other solar system objects. They survived the collapse of the Sun's parent molecular cloud and the formation of the accretion disk and were incorporated essentially unchanged into the parent bodies of chondritic meteorites. They are found in the fine-grained matrix of the least-metamorphosed chondrites and in *interplanetary dust particles* (IDPs), materials that were not processed by high-temperature events in the solar system.

There are two types of presolar grains. *Circumstellar condensates* condensed from hot gas ejected from dying stars in the immediate vicinity of their parent stars. They are also sometimes called *stardust*. Circumstellar condensates give snapshots of the life histories of the stars from which they formed. Although the stars themselves no longer exist, the grains can tell us the mass and metallicity of the stars and the stages of their evolution when the grains formed. Most of the known types of presolar grains are circumstellar condensates.

Interstellar grains are those that formed in interstellar space. These grains have no direct association with a specific stellar source. Their constituents started out in stars, just like those that make up circumstellar condensates, but either they did not condense into grains immediately after being ejected from the stars, or the grains into which they were originally incorporated were vaporized in interstellar space by supernova shocks or intense radiation. The atoms subsequently recondensed into grains in dense molecular clouds. Interstellar grains have poorly defined compositions and structures. They are unstable and were easily transformed into crystalline grains in high-energy environments such as the solar nebula. Most of the material that was to become the solar system likely resided in interstellar dust grains, but these are hard to recognize and only a handful of candidates are currently known.

The potential of presolar grains to provide information about nucleosynthesis, stellar evolution, galactic chemical evolution, interstellar processes, and nebular processes is only beginning to be tapped. But as new instrumentation is developed, more and more of the information that they carry will be extracted.

5.2 A Cosmochemical Detective Story

By the 1960s, mass spectrometers were evolving from special tools to investigate fundamental physics questions to more general analytical tools that could be used by

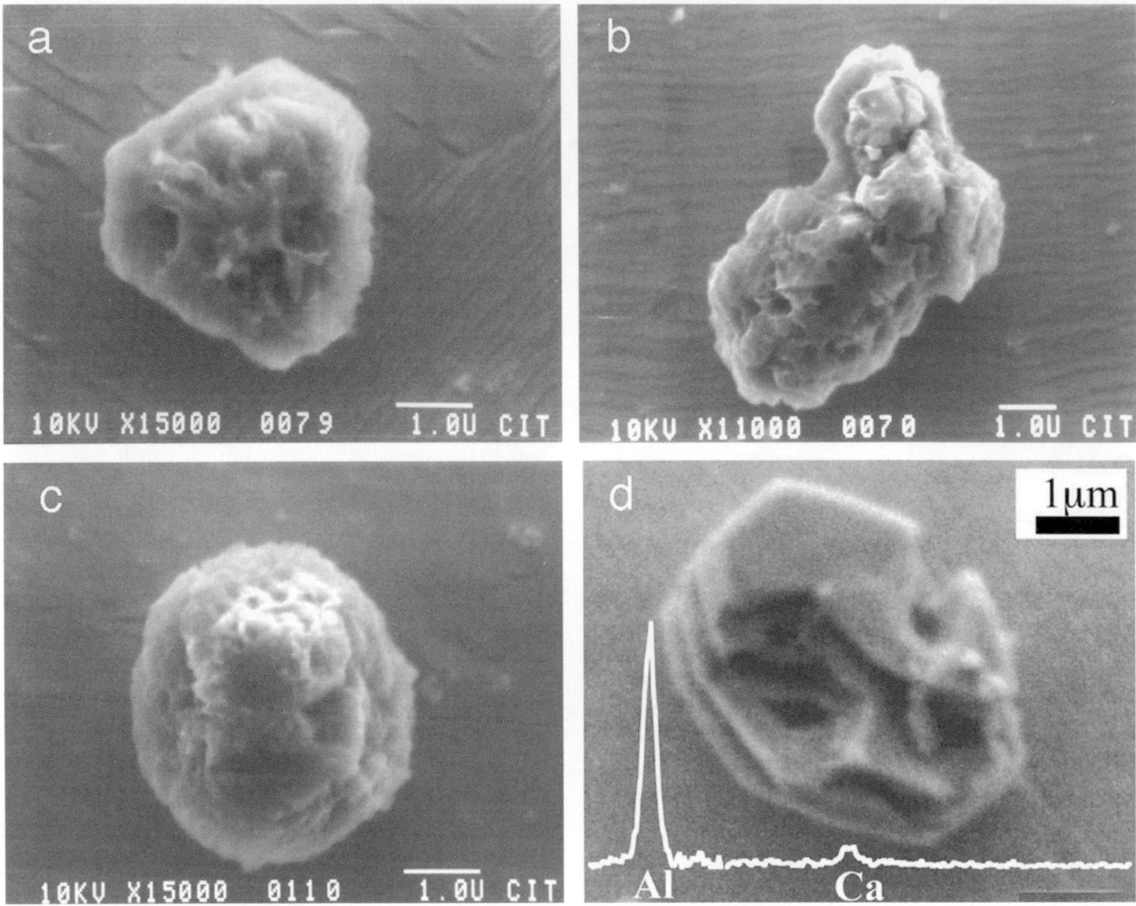

Figure 5.1 Examples of presolar silicon carbide from the Orgueil meteorite (a–c) and hibonite from the Semarkona meteorite (d). These examples are relatively large for presolar grains. Note the geometric outlines of crystal faces in images (a) and (d). Image (d) from Choi et al. (1999), reproduced with permission of the American Astronomical Society.

geochemists and cosmochemists. After some false starts, where reported isotopic anomalies in meteorites turned out to be experimental artifacts, most workers came to the conclusion that the isotopic compositions of the elements in the solar system were consistent with a single homogeneous mixture. The only known isotopic variations could be explained either as the result of radioactive decay or *spallation* reactions, or of processes that resulted in mass-dependent fractionation of isotopes. At about the same time, Cameron (1962) published a theoretical model that indicated that the solar system must have passed through a stage where temperatures were so high that all solids evaporated. The solar system would have been completely homogenized isotopically in passing through this stage, so the apparent absence of isotopic anomalies was consistent with theory. But even at this time, there was an exception to this picture, although it was not recognized as such. Reynolds and Turner (1964) reported two types of isotopic anomalies in the noble gas xenon from the Renazzo chondrite. There were clear excesses of ^{129}Xe, which were attributed to the decay of extinct radioactive ^{129}I (discussed more fully in Chapter 9). But there was also a general anomaly in the form of excesses of the heavy isotopes of xenon (^{134}Xe, ^{136}Xe) relative to the underlying meteoritic composition. The authors did not make much of these excesses and attributed them to some form of nuclear fission. This was the first evidence of presolar grains in meteorites.

The story of how the xenon anomalies led to the discovery of presolar grains is full of twists, turns, dead ends, and luck. During the 1960s, the excesses of heavy xenon isotopes were found in a number of carbonaceous chondrites. The pattern of the excesses is similar to that expected from fission, but it does not match the pattern

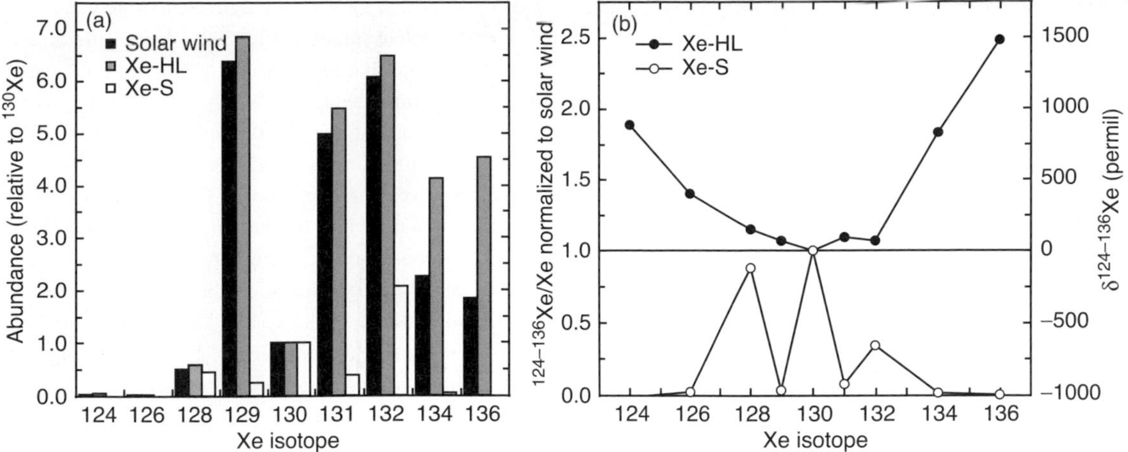

Figure 5.2 Isotopic compositions of Xe-HL (formerly called CCFXe or Xe-X) and Xe-S compared to the isotopic composition of solar wind xenon. In (a), the isotopic abundances of all three compositions are shown as histograms, normalized so that ^{130}Xe has the same abundance in all of them. In (b), the Xe-HL and Xe-S compositions have been normalized to the composition of solar wind xenon, which now plots as a flat line at $y = 1$ (0 in delta notation). This plot highlights the deviations from the solar wind composition. Xe-HL has large excesses of the heaviest (H) and lightest (L) isotopes. Xe-S consists of only the middle five xenon isotopes. Delta notation, shown on the right axis, is given by $\delta^{xxx}\text{Xe}/^{130}\text{Xe} = (((^{xxx}\text{Xe}/^{130}\text{Xe})_{HL}/(^{xxx}\text{Xe}/^{130}\text{Xe})_{solar})-1) \times 1000$. This notation was devised to facilitate discussions of isotopic shifts relative to a standard composition and it works well for shifts of a few parts in 10^2–10^3. However, with large shifts it becomes very asymmetrical, with the maximum positive value of infinity and a maximum negative value of −1000.

produced by uranium fission. This xenon component became known as *CCFXe*, for carbonaceous chondrite fission xenon. Manuel et al. (1972) showed that the supposed fission xenon is accompanied by excesses of the lightest xenon isotopes (Fig. 5.2). They proposed that the excesses resulted from the release of an "isotopically anomalous xenon" component that they called *Xe-X*, which contained excesses of both heavy and light isotopes. This interpretation was not immediately accepted, however. A more popular explanation of the Xe-X pattern was that it was due to severe mass-dependent fractionation favoring the light isotopes plus fission xenon.

Meanwhile, work was being carried out to determine the parent of the supposed fission xenon component. The composition of plutonium fission xenon was measured by Alexander (1971). A sample of plutonium from a nuclear reactor was heated to release all trapped gases and then left on a shelf for 23 months to allow pure plutonium fission xenon to accumulate. The resulting precise composition of plutonium fission xenon eliminated plutonium as the parent of CCFXe. Another hypothesis, championed by Edward Anders of the University of Chicago, was that the CCFXe had been produced by fission of a super-heavy element. Nuclear physicists had presented theoretical arguments that nuclei near proton number 114 and neutron number 184, heavier than any known natural nuclide, might form an "island of stability" and that nuclides in this mass range might be stable enough to survive for some time in nature.

While searching for the supposed carrier phase of the super-heavy element, Lewis et al. (1975) made a remarkable discovery. To isolate the carrier of CCFXe, they dissolved a sample of the Allende carbonaceous chondrite in a caustic acid mixture. They found that a residue containing <1% of the starting mass contained most of the noble gases in the bulk meteorite. They further discovered that when they treated this residue with oxidizing nitric acid, the major noble gas component in the meteorite was removed, leaving behind the CCFXe in almost pure form. The mass loss involved in this "etching" was very small. Based on this work, these authors named the carrier of the major noble gas component in carbonaceous chondrites Q, for quintessence. The remaining residue that carried the CCFXe was dominated by chromite, which they assumed to be the carrier.

For the next decade, the Berkeley group under John Reynolds and the Chicago group under Edward Anders struggled to identify the carrier of CCFXe. The Berkeley group focused on carbonaceous material, while the Chicago group designed their work around finding the mineral that would have carried the super-heavy element postulated to be the parent for CCFXe.

As the search for the carrier of CCFXe was going on, several important related discoveries were made. Black and Pepin (1969) reported extreme isotopic variations in neon from the Orgueil carbonaceous chondrite. They defined five neon components, which they labeled A through E. *Neon-E* was inferred to be almost pure ^{22}Ne. The only plausible source that these authors could come up with for pure ^{22}Ne was the decay of very short-lived ^{22}Na (half-life = 2.6 years). These authors stated that the carrier of the Ne-E was likely to be presolar material, having survived from a stellar source where ^{22}Na was produced. This interpretation did not attract much attention at the time.

A University of Chicago group then reported the discovery of isotopic anomalies in oxygen that could not be explained by known processes operating in the solar system (Clayton et al., 1973). Analyzing calcium-aluminum-rich inclusions (CAIs), these authors found an array of oxygen compositions that varied in the amount of ^{16}O present and that plotted along a line with slope ~1 on an oxygen three-isotope plot (Fig. 5.3). These anomalies were interpreted as reflecting the addition of material from a supernova that exploded shortly before the solar system formed.

Srinivasan and Anders (1978) reported that a highly processed acid residue from the Murchison meteorite had a xenon composition enriched in ^{128}Xe and ^{130}Xe and depleted in ^{129}Xe, ^{131}Xe, ^{134}Xe, and ^{136}Xe. When they calculated the pure end-member composition, they found

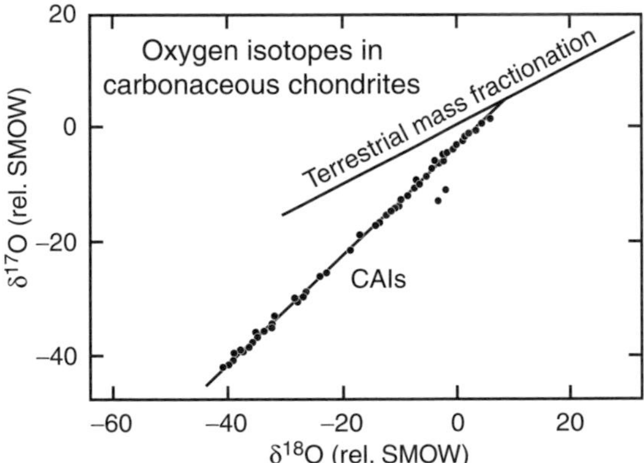

Figure 5.3 Oxygen three-isotope plot showing excesses of ^{16}O in minerals from CAIs. All samples from Earth rocks plot along the terrestrial mass fractionation line. Mass-dependent fractionation processes cannot move a composition off a line with slope 1/2, so the excesses of ^{16}O in CAIs are clearly isotopic anomalies. After Clayton et al. (1977).

a component that almost exactly matched predictions for *s*-process nucleosynthesis (Fig. 5.2). Again, the authors had to postulate that presolar dust grains were the carrier of this new component, which they called *Xe-S*.

By the end of the 1970s, support for the fission hypothesis for the origin of CCFXe was waning, and over the next few years the component's name was changed to *Xe-HL*, which describes the enrichments in heavy (H) and light (L) xenon isotopes without ascribing a mode of origin. Searches continued for the carriers of Xe-HL, Xe-S, and Ne-E at Berkeley and Chicago, and in Bern, Switzerland, under the direction of Peter Eberhardt.

Starting with the discovery of oxygen isotopic anomalies in Allende refractory inclusions and continuing at an increasing pace throughout the early 1980s, a stream of isotopic anomalies in many elements (strontium, neodymium, titanium, calcium, etc.) were reported (summarized by Clayton et al., 1988). These discoveries posed a serious challenge to the idea that the solar system was isotopically homogeneous. In addition, by the late 1980s, many theoretical models were also showing that the accretion disks around newly forming stars are not hot enough to vaporize pre-existing dust except very close to the star. These observations and models implied that the solar nebula did not pass through a hot gaseous stage where all dust was evaporated. Thus, by the late 1980s, the stage was set for the discovery of surviving presolar grains.

The discovery happened by accident. Lewis and Anders were frustrated by their failure to find the carrier of anomalous xenon in carbonaceous chondrites. They decided to try an extreme treatment to see if they could dissolve the carrier. They treated a sample of the colloidal fraction of an Allende residue with the harshest chemical oxidant known: hot perchloric acid. The black residue turned white, and to their surprise, when they measured it, the anomalous xenon was still there! The residue consisted entirely of carbon, and when they performed electron diffraction measurements, they found that it consisted of tiny (nanometer-sized) diamonds. After a detailed characterization that included chemical, structural, and isotopic studies, they reported the discovery of presolar diamond (Lewis et al., 1987). The 23-year search for the carrier of CCFXe (Xe-HL) was over, and the study of presolar grains had begun.

The carriers of anomalous Ne-E (two forms of which were now known) and Xe-S were quickly identified. Neon-E(H), which is released at temperatures above ~1200°C in stepped heating experiments, and Xe-S were found to be carried in presolar silicon carbide (Tang and Anders, 1988). Neon-E(L), which is released below

~900°C, was found to be carried by presolar graphite (Amari ct al., 1990). Once these presolar compounds were shown to be present in meteorites, studies were carried out to identify all of the different types of meteorites that carry presolar grains. Concentrated searches for other presolar phases also were initiated, and many new types of presolar grains have since been found. However, we cannot yet account for the majority of the presolar components that must have been present in the Sun's parent molecular cloud.

5.3 Recognizing Presolar Grains in Meteorites

At the present time, the only way to definitively identify a grain as presolar is by its exotic isotopic composition. The logic is based on the old idea that the solar system passed through a hot gaseous stage during which all pre-existing solids evaporated, producing a single homogeneous isotopic composition. Any grain that has an isotopic composition that cannot be explained in terms of known processes (radioactive decay, spallation, mass-dependent fractionation) acting on this solar system composition must have originated elsewhere. Although few believe any longer that the solar system was homogenized in this way, this reasoning is still basically valid. Most processes acting in the solar system and on planetary bodies (evaporation, melting, dissolution by a fluid, etc.) tend to homogenize the isotopic composition of the starting materials, so an object with an anomalous isotopic composition must have survived homogenizing processes. A stronger case for a presolar origin can be made if the observed isotopic composition can be explained by a process operating elsewhere, such as nucleosynthesis in stars. We will present examples of such compositions later in this chapter.

In practice, it is not sufficient that an object has an isotopic composition that cannot be explained by radioactive decay or mass-dependent fractionation effects. The object must also have *physical and chemical* characteristics that make it unlikely to be a product of solar system processes. For example, millimeter- to centimeter-sized CAIs from primitive chondrites have been shown to contain small (parts in 10^3–10^4) isotopic anomalies in many elements. However, based on the size, composition, physical characteristics, and abundance of the inclusions, it is generally believed that these objects formed within the solar system. They preserve small isotopic anomalies because they did not form from a representative sample of the bulk solar system (see Chapters 8 and 15). As mass spectrometry has become more precise, even smaller isotopic differences have been found between bulk meteorite samples, which implies that different meteorites incorporated slightly different samples of solar system materials. So, isotopic anomalies can indicate either that an object is itself presolar or that it formed in the solar system from precursor material that was not fully homogenized in the solar system. As we discuss below and in Chapter 8, these anomalies and bona fide presolar grains can be used as probes of processes in the early solar system.

As noted previously, most of the presolar grains so far identified are circumstellar condensates (stardust), but some grains formed in interstellar space. The interstellar grains are not likely to contain large isotopic anomalies. So how can we recognize these interstellar grains in meteorites?

Although we have not found other characteristics that are as definitive as anomalous isotopic compositions, there are a number of ways to potentially recognize interstellar grains. A general idea of the nature of most interstellar dust comes from astronomical observations, which give information on the size, shape, composition, and structure of dust particles. These observations indicate that interstellar dust has a power-law size distribution that peaks at around 0.1 μm. Polarization studies indicate that interstellar grains tend to be elongated rather than equant. The abundances of elements in the gas phase in the diffuse interstellar medium are inversely proportional to their condensation temperature. Volatile elements tend to remain in the gas phase, whereas moderately volatile and refractory elements are located mostly in dust. Structurally, most of the dust appears to be amorphous. Upper limits on the crystalline fraction are on the order of a few percent. This is consistent with formation via condensation in a cold environment where kinetics inhibits the formation of well-crystallized mineral phases. Based on these observations, much of the dust that went into making the solar system consisted of <1 μm-sized grains of amorphous material containing moderately volatile and refractory elements in approximately bulk solar system abundances. Unfortunately, material like this was very susceptible to being altered or destroyed by processes in the accretion disk or within meteorite parent bodies, where temperatures were higher and liquid water may have been present. However, material like that expected from astronomical observations has been found in a handful of the least altered chondrites.

Some of the common processes that operate in interstellar space should leave recognizable effects in grains. For example, high-energy galactic *cosmic rays* permeate the galaxy. These cosmic rays are strongly attenuated in the inner solar system by the Sun's magnetic field, but dust grains in interstellar space are subject to constant bombardment. This bombardment disrupts the structure of silicate grains and can turn crystals into amorphous grains. Atoms can be ejected from the grains and atomic nuclei can break into smaller pieces (*spallation*). The crystalline structure of a silicate subject to tens of millions of years of cosmic-ray bombardment should have been almost completely destroyed and its composition modified. Bradley (1994) found a large number of tiny amorphous grains within IDPs (to be discussed in Chapter 6). He named these "**g**lass with **e**mbedded **m**etal and **s**ulfide" (GEMS) and proposed that they might be presolar grains. But without isotopic data to back up his proposal, this idea has remained controversial. With the invention of a new generation of ion microprobes (secondary ion mass spectrometers or SIMS) in the early 2000s, particularly the NanoSIMS, it has become possible to measure the isotopic compositions of the GEMS in IDPs. Most have isotopic compositions indistinguishable from solar system values (although the uncertainties are large), but a few have exhibited isotopic anomalies that label them as true presolar grains. Now the debate centers on the isotopically "normal" GEMS. Are they presolar as well?

Cold molecular clouds exhibit deuterium/hydrogen (D/H) ratios up to ~125 times the value for the average interstellar medium. It is believed that high D/H ratios in molecular clouds result from ion–molecule reactions involving organic molecules on grain surfaces (see Chapter 8). Given enough time, extreme enrichments are possible. Meteorites and IDPs also contain organic material with high D/H ratios, up to ~25 times the value in the local interstellar medium. The molecules that carry these isotopic anomalies may be derived from the Sun's parent molecular cloud and thus may be presolar components. The isotopic compositions of organic matter in meteorites are discussed in Chapters 8 and 11.

5.4 Known Types of Presolar Grains

Approximately 20 different types of material have been identified as presolar so far (Table 5.1). The carbon-rich types were the first to be recognized. These were the phases that survived harsh chemical treatments and had anomalous noble gases to serve as beacons to guide the search. Several kinds of carbides and Fe–Ni were found inside presolar graphite or silicon carbide grains (Table 5.1). Many of these materials have chemical compositions that are alien to the solar system (oxides are typically more stable in our system, due to the abundance of oxygen). Presolar oxides and nitrides were the next types to be found. They survived the harsh chemical treatments and were found primarily by measuring their oxygen isotopic compositions by ion microprobe. Silicates (forsterite and enstatite) were not identified until later because they are destroyed by the chemical treatments used to separate carbon-rich grains and oxides. *In situ* measurements of chondrite thin sections using isotope-ratio mapping located these phases. Deuterium-rich organic matter probably contains a presolar component, but the organic fraction consists of many different

Table 5.1 Known types of presolar grains

Name	Composition	Size	Matrix-Normalized Abundance
Diamond	C	1–2 nm	1500 ppm
Silicon carbide	SiC	0.1–10 μm	10–15 ppm
Graphite	C	0.1–20 μm	5–10 ppm
Spinel	$MgAl_2O_4$	0.1–3 μm	10–20 ppb
Corundum	Al_2O_3	0.2–3 μm	100 ppb
Hibonite	$CaAl_{12}O_{19}$	1–5 μm	20 ppb
TiO_2	TiO_2	n.d.	$<$10 ppb
Si_3N_4	Si_3N_4	1–5 μm	1–20 ppb
Forsterite	Mg_2SiO_4	0.2–0.5 μm	10–1800 ppm
Enstatite	$MgSiO_3$	0.2–0.5 μm	10–1800 ppm
Amorphous silicate	Variable	0.2–0.5 μm	20–3600 ppm
D-rich organics	HCNO	n.d.	n.d.
Carrier of P1 (= Q) noble gases	Probably organic compounds	n.d.	n.d.
TiC, MoC, ZrC, RuC, FeC, FeNi	Inside other minerals	1–25 nm	n.d.

compounds with a variety of different histories. Techniques are not yet available to identify individual presolar organic compounds. The carrier of the dominant heavy noble gas component in chondrites has not been identified, but it is likely to be some kind of organic material. An origin in cold interstellar molecular clouds has been proposed to explain the high abundance of the heavy noble gases, the low abundances of helium and neon, and the close association of the carrier with presolar diamond, silicon carbide, and graphite.

5.5 Locating and Identifying Presolar Grains

Historically, there have been two basic approaches to locating and identifying presolar grains for further study. One approach – the one used to find the first presolar grains – can be likened to burning down the haystack to find the needle. Meteorite samples are progressively dissolved in a series of strong acids, leaving behind only the most acid-resistant phases (spinel, corundum, hibonite, silicon carbide, diamond, graphite; see Table 5.1), many of which are presolar. Details of this chemistry can be found in Amari et al. (1994) and Huss and Lewis (1995). This method is used now primarily to generate large collections of presolar grains for survey work, or to investigate in detail a particular aspect of stellar nucleosynthesis (see below).

We now know that the haystack is important as well. Therefore, searches are now carried out on material that has been either minimally processed or not processed at all. The ion microprobe (SIMS) is the instrument used for this work; it uses a focused beam of high-energy ions to sputter and ionize atoms from the sample. These secondary ions are analyzed using a high-performance mass spectrometer. Modern ion microprobes such as the NanoSIMS or CAMECA IMS 1270 or 1280 equipped with a solid-state imaging detector called SCAPS (see Appendix) have spatial resolutions of 100–500 nanometers, similar to or smaller than the sizes of most presolar grains, so these instruments can be used to identify presolar grains *in situ* in meteorite matrix or in IDPs. To search for presolar grains, the ion microprobe is used in ion-imaging mode to make isotope-ratio maps. Presolar grains show up as "hot spots" on these maps. The isotopes of oxygen, a major element in many types of grains, are commonly used in this work. The technique can also be applied to separates from meteorites that are enriched in presolar grains relative to the bulk meteorite. Isotope-ratio mapping by ion probe resulted in identification of the first presolar silicates (forsterite, enstatite, amorphous silicate) in meteorites and IDPs (Messenger et al., 2003; Nagashima et al., 2004; Nguyen and Zinner, 2004).

A third method to locate and identify presolar grains has been developing slowly. The main problem with ion probe screening is that identifying the presolar grains is, itself, destructive. Ideally, we would like to identify the presolar grains without damaging them. Alexander et al. (1990) pointed out that the silicon X-ray yield is higher from presolar silicon carbide, where half of the atoms are silicon, than from typical silicates, where the silicon content is significantly lower. They used this property to identify silicon carbide grains in thin section. Bernatowicz et al. (2003) applied an enhanced version of this technique to a disaggregated matrix sample from the Murchison CM2 chondrite. They mapped the sample with energy-dispersive spectrometer (EDS) on a scanning electron microscope (SEM) and, using a combination of screening thresholds (e.g. silicon counts greater than a predetermined amount and magnesium less than a certain amount), were able to identify 81 presolar silicon carbide grains. Liu et al. (2017) took this approach further by using a combination of backscattered electron and SEM-EDS imaging and Raman spectroscopy to screen a silicon carbide separate from the Murchison meteorite. Based on distortions of the silicon carbide crystal structure produced by different minor- and trace-element abundances, they were able to identify rare supernova grains and other unusual grains in a completely nondestructive way. These and new nondestructive techniques will permit more information to be extracted from each presolar grain.

5.6 Characterizing Presolar Grains

Once the grains are identified, the next task is to characterize them. Properties of interest include size, shape, crystal structure, major-, minor-, and trace-element composition, and isotopic composition. When presolar grains were first discovered, scientists were only able to determine the basic mineralogy and show that the grain was presolar by measuring the isotopic compositions of one or two elements. However, it soon became clear that more information on each grain was needed to fully understand their sources and history. It is now possible to gather detailed structural and chemical information and to measure isotopic compositions of five or six elements in the same 2–3 μm grain. Some techniques are destructive, while others are not (more information on those described below is given in the Appendix). To get the most information from a grain, it makes sense to use the nondestructive techniques first. Nondestructive techniques can provide chemical and, occasionally, structural data on a presolar grain, but they cannot unequivocally identify a grain as presolar. Therefore, as described above, the ion microprobe is commonly used to identify the grains, particularly silicates, and then a combination of destructive and nondestructive techniques can be applied to characterize them.

Grains found *in situ* pose a special problem. Because other grains surround them in a close-packed arrangement, clean measurements of their chemical and isotopic compositions are difficult to make. The grains can be removed from the surrounding material using a focused ion beam (FIB) lift-out technique, but this is very time consuming. The most-detailed information comes from grains that have been separated by physical and chemical means from the surrounding material.

5.6.1 Nondestructive Characterization

Morphological information and major-element compositions can be obtained using a scanning electron microscope (SEM). The SEM uses a focused electron beam to scan the sample and creates an image from either secondary electrons or backscattered electrons. Features as small as several tens of nanometers can be resolved on the surfaces of presolar grains with a field emission SEM. The chemical composition can be determined by SEM-EDS, which measures characteristic X-rays excited in the sample by the electron beam of the SEM. The spatial resolution of the SEM-EDS combination is ~1 μm. The auger microprobe, which studies electrons emitted at characteristic energies, can analyze the composition of a thin layer on the surface of a grain and has a spatial resolution of $\leq$20 nanometers. This instrument is thus very useful in gathering chemical information on small presolar grains *in situ* in meteorite matrix or in an IDP. Raman spectroscopy, which measures the inelastic (Raman) scattering of laser light, can identify the mineralogy of grains and can provide basic structural information for some materials. It is particularly useful to study the structural characteristics of organic materials. Modern micro-Raman systems have spatial resolutions approaching 1 μm. Electron backscatter diffraction (EBSD) examines the diffraction patterns of electrons scattered at low angles from the surface layers of a sample. The diffraction patterns are characteristic of the internal structure of a mineral grain. This technique is normally applied to carefully prepared polished sections, but in some instances it can produce a pattern from an isolated presolar grain. All of these techniques are nondestructive; the sample remains for further analysis.

The transmission electron microscope (TEM) is a powerful tool for studying the internal structure and composition of a presolar grain. The TEM can obtain images and chemical compositions of objects as small as a few tens of nanometers across (Fig. 5.4) and can resolve the atomic structure of a grain from lattice fringe effects. The TEM requires a very thin (tens of nanometers thick) sample to get good electron diffraction and electron-energy-loss spectral (EELS) data. TEM measurements require careful sample preparation, which is, in part, destructive. Presolar grains can be sliced in a microtome and the slices are then mounted on porous grids for analysis. Exquisitely detailed information can be obtained from these sections, but one has little control over which section will be in the right place on the sample mount to

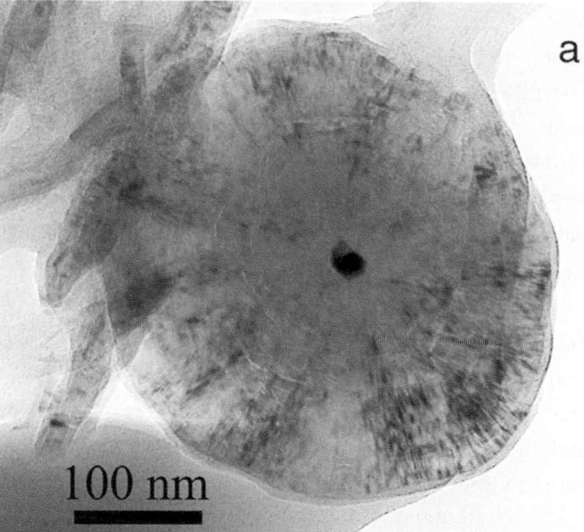

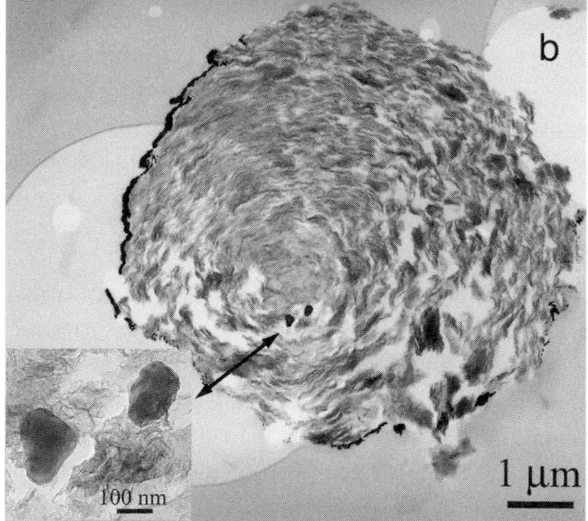

Figure 5.4 (a) TEM images of slices through presolar graphite grains. (a) An "onion-type" presolar graphite grain containing a central titanium-vanadium carbide grain, which apparently served as a nucleation site for the graphite. (b) A "cauliflower-type" graphite showing concentric layers of graphite and two (Ti,Zr,Mo) C crystals. Images courtesy of K. Croat.

be studied. A FIB can also prepare samples for TEM work. This technique uses the ion beam to remove all of the material of no interest, leaving behind a well-characterized sample for TEM analysis.

Recently, synchrotron techniques have been applied in presolar grains research. X-ray diffraction (XRD) can be used to give structural information on very small particles. X-ray fluorescence (XRF) gives composition information at spatial scales as small as 50–100 nm. X-ray absorption spectroscopy (XAS) can give information about atomic coordination in a structure. X-ray absorption near-edge spectroscopy (XANES) uses hard X-rays to investigate the electronic structure of target materials and can investigate the element composition, oxidation state, and symmetry of a material. Scanning transmission X-ray spectroscopy (STXM) allows maps to be generated by scanning the X-ray beam over the sample. Phase-contrast maps can be generated from X-ray fluorescence maps, and three-dimensional tomography can be done. Synchrotron techniques are nominally nondestructive, although X-ray beams carry a lot of energy and can disrupt the structures of some materials.

5.6.2 Destructive Techniques

In order to get isotopic information about a grain, it is necessary to count the atoms of which it is composed, which means that the grains must be at least partially destroyed. SIMS can measure the major-, minor-, and trace-element abundances and isotopic compositions of presolar grains. Several types of ion probes are used. With the earlier generation of ion probes, it was possible to measure the isotopic compositions of silicon, carbon, nitrogen, magnesium, and titanium in individual 4–8 μm-sized grains (e.g. Fig. 5.1). With the newer, more sensitive instruments, multiple elements can be measured in considerably smaller grains. But even the most sensitive ion probes only put about one atom in 10^2–10^4 into the detector to be counted.

Resonance ionization mass spectrometry (RIMS) is much more efficient at getting ions into the detectors. First, atoms are removed from the sample surface with a pulsed laser, which releases most atoms without ionizing them. The ionized atoms are removed using a pulsed electric field. Then, by using carefully tuned lasers, the element(s) of interest in the neutral gas plume are ionized and extracted into a time-of-flight mass spectrometer. RIMS can put several percent to several tens of percent of the target atoms into the detectors, while other elements are not analyzed at all, making this technique 10–1000 times more sensitive than SIMS. With the completion of the new Chicago Instrument for Laser Ionization (CHILI) at the University of Chicago (Stephan et al., 2016), and a similar instrument (Laser Ionization of Neutrals or LION) at Lawrence Livermore National Laboratory, RIMS has joined the arsenal of analytical tools that can be applied to presolar grains. Results from the RIMS technique are discussed later in this chapter.

Bulk techniques such as noble gas mass spectrometry, thermal-ionization mass spectrometry (TIMS), and inductively coupled plasma mass spectrometry (ICPMS) can be used to measure trace elements in presolar grains. These techniques can give precise isotopic results, but to utilize this capability one must have enough atoms. For many elements, there are not enough atoms in a single grain to provide a meaningful isotopic composition. Consider xenon, which has nine isotopes that vary in abundance by a factor of ~200. In order to get a meaningful isotopic composition, ~30,000 atoms must be measured. But only about one in a million presolar diamonds has a xenon atom in it, and a single 1 μm presolar silicon carbide grain only has ~1000 xenon atoms. Thus, large collections of grains must be measured at one time, and the resulting compositions are averages from many stellar sources. Such compositions do still provide important information, and bulk measurements of noble gases led to the discovery of presolar grains in the first place.

Bulk techniques are still used to follow isotopic tracers for presolar materials. For example, chromium enriched in ^{54}Cr has been identified in acid residues of the most primitive chondrites. As with the exotic noble gases, the ^{54}Cr excesses were targets of various chemical experiments and bulk isotopic measurements. Eventually, these studies led to the identification of tiny <100 nm oxide grains (e.g. Dauphas et al., 2010; Qin et al., 2011) as the carrier of the anomalies, and work is proceeding to better characterize the carrier (e.g. Nittler et al., 2018). In Chapter 8, we will discuss an isotopic dichotomy that is observed in several elements among bulk meteorite samples that implies either an inhomogeneous original distribution of presolar components or a separation of those components in the early solar system.

5.7 Identification of Stellar Sources

The isotopic characteristics of circumstellar condensates can be used to identify the types of source stars. This information also provides "ground truth" that helps evaluate the details of nucleosynthesis models.

5.7.1 Grains from AGB Stars

Most presolar silicon carbide and oxide grains, and a significant fraction of presolar silicate grains found in meteorites, come from low- to intermediate-mass stars in the asymptotic giant branch (AGB) phase (see Chapter 3). Evidence for this conclusion derives from two sources: 1) spectroscopic observations of the envelopes of these stars and 2) comparisons of the isotopic compositions of the grains with theoretical models of stellar nucleosynthesis.

Spectroscopic observations have provided important information about nucleosynthesis in low- and intermediate-mass stars. AGB stars of low to intermediate mass often show high abundances of *s*-process elements in their envelopes. These elements are brought to the surface during third dredge-up mixing events (Chapter 3). Although these observations do not tell us anything about the isotopic compositions of the *s*-process elements, their high abundances in stellar envelopes show that *s*-process synthesis is going on. One of the first indications that presolar grains were present in meteorites was the discovery of Xe-S in an acid residue of the Murchison meteorite (Fig. 5.2). Presolar silicon carbide grains carry *s*-process signatures in krypton, xenon, strontium, molybdenum, barium, neodymium, samarium, and other elements.

Spectroscopic observations have also identified a class of AGB stars in which carbon is more abundant than oxygen. These are known as carbon stars. Before the discovery of presolar grains, theoretical modeling had shown that the envelopes of low-mass stars could become carbon rich through the dredge-up of ^{12}C from the helium shell during the AGB phase (Iben and Renzini, 1983). Carbon and oxygen combine to form CO before either combines in significant amounts with other elements. When oxygen is more abundant than carbon, oxide minerals will form using the leftover oxygen. When carbon is more abundant than oxygen, carbon-rich compounds will form from the excess carbon. Carbon stars provide an ideal environment for the formation of carbon-rich presolar grains such as silicon carbide. A strong case for the idea that presolar silicon carbide comes from carbon stars can be made from observations of silicon carbide spectral features in the outflows of some carbon-rich AGB stars. The case is strengthened further by the close similarity between the $^{12}C/^{13}C$ ratios measured in silicon carbide grains and the $^{12}C/^{13}C$ ratios measured spectroscopically in carbon stars (Fig. 5.5). This similarity not only indicates that presolar silicon carbide came from carbon stars, but also that a population of stars much like those in the galaxy today, rather than a single carbon star, contributed silicon carbide grains to the early solar system.

Spectroscopic observations show that red giant stars typically have higher $^{17}O/^{16}O$ ratios than the solar system, and have $^{18}O/^{16}O$ ratios that are near to or slightly lower than solar. This is the basic pattern expected in the envelopes of stars where the products of partial hydrogen burning have been mixed into the envelope. Most of the measured red giant stars are in the mass range where they should evolve into AGB stars. Measurements from presolar oxide grains match the stellar observations fairly well (Fig. 5.6), in spite of rather large errors in the stellar measurements. But as the number of measurements of presolar oxide grains increased, and as biases in the screening methods to locate the grains were recognized and accounted for, new groups of oxide grains were recognized that were not obvious in the stellar data.

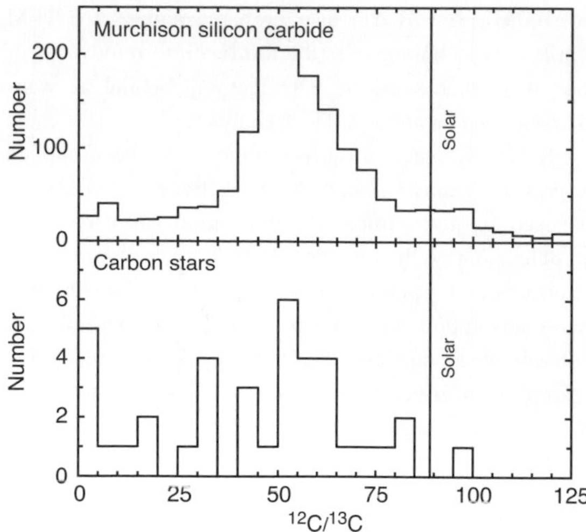

Figure 5.5 Carbon isotopic compositions of presolar silicon carbide grains from the Murchison meteorite compared with the carbon isotopic compositions of carbon stars (low- to intermediate-mass AGB stars). The composition of carbon in the solar system is indicated by the vertical (solar) line. The similarity in the distributions of compositions in the two plots indicates that silicon carbide in the Murchison meteorite came from a population of carbon stars very similar to that in the galaxy today. Data for carbon stars from Smith and Lambert (1990).

Theoretical modeling provides strong evidence that most presolar silicon carbide grains come from 1.5 to 3 $M_\odot$ stars. As discussed in Chapter 3, stellar modeling of the evolution of the CNO isotopes in the envelopes of these stars makes clear predictions about the $^{12}C/^{13}C$, $^{14}N/^{15}N$, $^{17}O/^{16}O$, and $^{18}O/^{16}O$ ratios as a star evolves. For example, in the envelopes of low- to intermediate-mass stars of solar composition, the $^{12}C/^{13}C$ ratio drops to ~40 (from a starting value of 89), and $^{14}N/^{15}N$ increases by a factor of six as carbon and nitrogen processed by hydrogen burning via the CNO cycle during the main sequence are mixed into the envelope by first dredge-up. When the third dredge-up thermal pulses start, newly synthesized ^{12}C from the helium shell is added to the

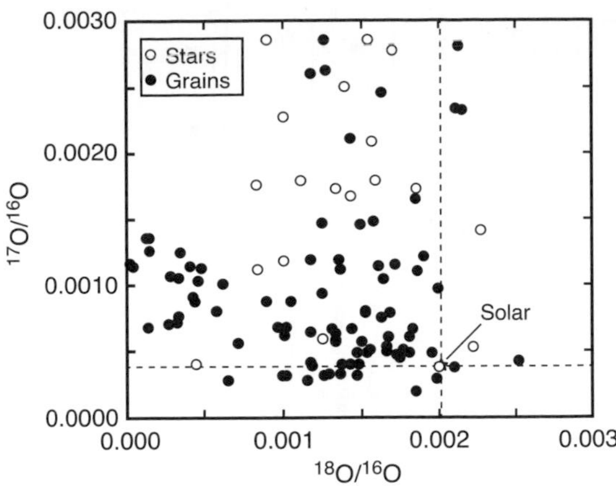

Figure 5.6 Oxygen isotopic compositions of presolar oxide grains are compared with those of red giant stars. Stellar data are shown without error bars, which are large on this scale. Both data sets are characterized by higher $^{17}O/^{16}O$ ratios and lower $^{18}O/^{16}O$ ratios compared to solar oxygen. Stellar data from Smith and Lambert (1990).

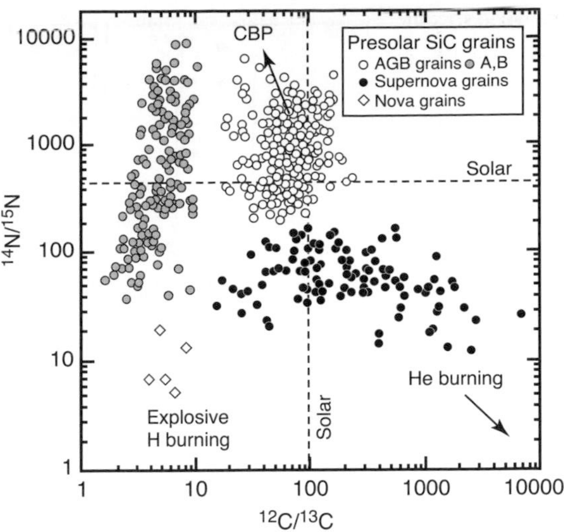

Figure 5.7 Carbon and nitrogen isotopic ratios for presolar silicon carbide grains. The vast majority of grains (AGB grains) plot just above and to the left of the solar composition (indicated by the intersecting dashed lines). Grains with $^{14}N/^{15}N$ ratios >2000 and $^{12}C/^{13}C$ ratios <40 are not predicted by standard AGB models. But if extra mixing (cool-bottom processing, CBP) is put into the models, compositions move in the direction of the arrow. Supernova grains, which are rich in ^{12}C from helium burning, plot mostly in the lower right quadrant. Rare grains thought to come from novae, which are powered by explosive hydrogen burning, plot in the lower left quadrant. The sources for the A and B grains are not currently known. Modified from Zinner (2014).

envelope and the $^{12}C/^{13}C$ ratio rises, while nitrogen is unaffected. After several pulses, the actual number of which depends on the stellar mass, the envelope becomes carbon rich and silicon carbide grains can form from the star's ejecta. Without extra hydrogen burning in the envelope, the $^{12}C/^{13}C$ ratios can rise to >150. But in stars of greater than ~4 $M_\odot$, hot-bottom burning at the base of the envelope processes the new ^{12}C to ^{13}C and on to ^{14}N. The $^{12}C/^{13}C$ ratio continues to drop, approaching the CNO equilibrium value of ~2.5. Such stars probably do not become carbon stars, because carbon is converted to nitrogen more quickly than oxygen is and does not accumulate in the envelope. The distributions of $^{12}C/^{13}C$ and $^{14}N/^{15}N$ ratios predicted from models of carbon-rich AGB stars of less than ~4 $M_\odot$ are similar to the distribution exhibited by silicon carbide grains (Fig. 5.7). This similarity supports the idea that silicon carbide grains come from carbon stars. Figure 5.7 also shows silicon carbide grains with $^{12}C/^{13}C$ ratios below ~40 and/or $^{14}N/^{15}N$ ratios above 2000. These grains are not accounted for by standard models, but as we discuss below, the existence of these grains has led to new insight into the structure of low-mass stars.

Oxide grains can form in the winds ejected from red giant stars or from AGB stars before they become carbon rich (Chapter 3). Theoretical modeling predicts that the stellar envelopes present at the time these oxide grains might form should be depleted in ^{18}O and enriched in ^{17}O relative to the starting material, with ^{16}O largely unaffected. Results of such models are summarized on an oxygen three-isotope plot in Figure 5.8a, along with measured compositions of presolar oxides. For a given initial composition, the envelopes of stars of different mass at the end of first dredge-up define an array of compositions. For all but the least massive stars, the $^{18}O/^{16}O$ ratio achieves a near-constant value controlled primarily by dilution of processed material by unprocessed material. In contrast, the $^{17}O/^{16}O$ ratio increases with mass to a maximum value for stars near ~4 $M_\odot$, and declines again in stars of higher mass. Different initial compositions produce different arrays. The initial compositions are thought generally to reflect the metallicity of the star, with $^{18}O/^{16}O$ and $^{17}O/^{16}O$ increasing with increasing metallicity. Standard models can explain approximately 75% of the oxide grains that have been measured (Fig. 5.8a), indicating that the grains come primarily from low- to intermediate-mass stars. We will return to the other grains later in this chapter.

Theoretical modeling of the *s*-process is well developed and makes specific predictions that can be tested against the compositions of presolar grains. Noble gas data for presolar silicon carbide show clear

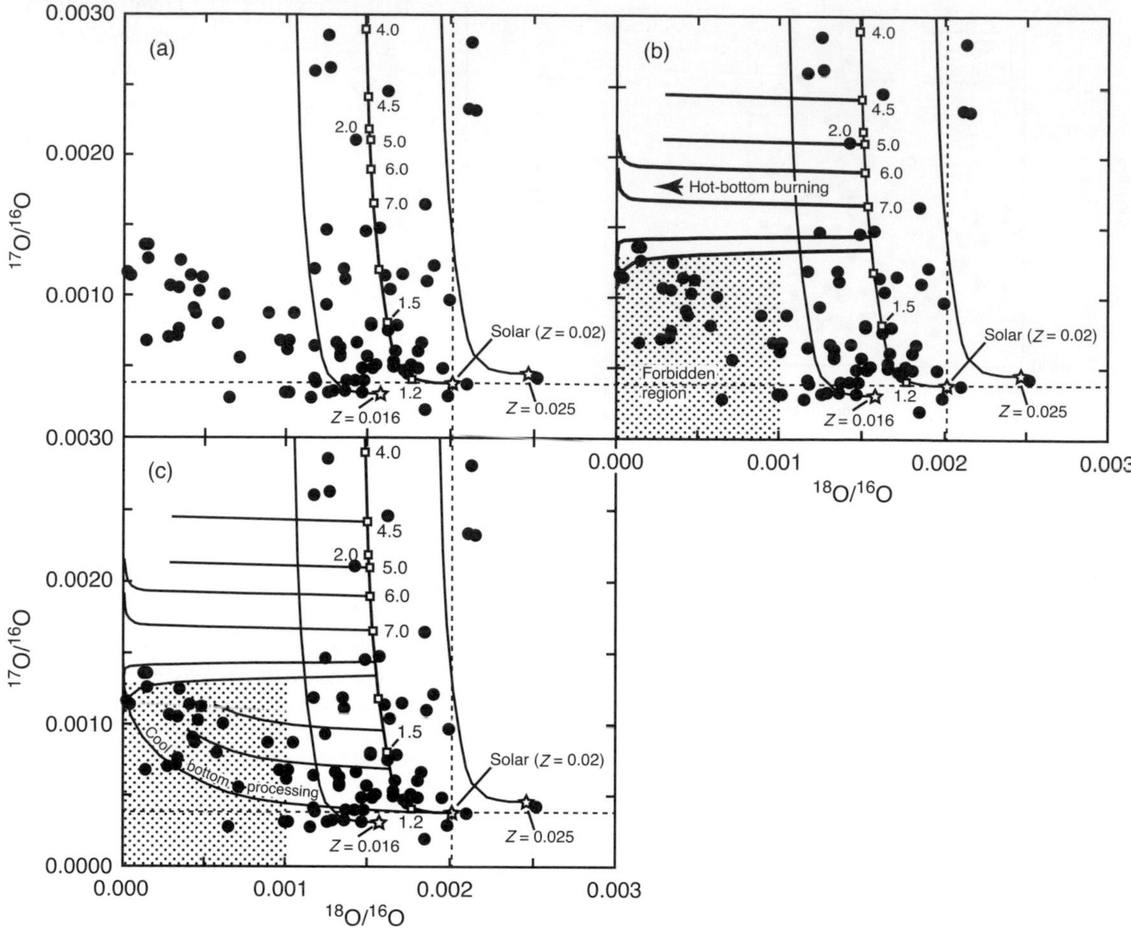

Figure 5.8 Oxygen three-isotope plots comparing predictions of stellar models with measurements of presolar oxide grains (solid circles). (a) Solid curves are arrays of final compositions of the stellar atmospheres of stars of masses 1.2–7 $M_\odot$ (small open squares). The three curves represent stars of different metallicities: the middle line is solar metallicity, the one on the left is for metallicity 25% less than solar, and the one on the right is for metallicity 25% more than solar. Oxygen compositions of many grains fall within the area covered by these curves, so the models can successfully explain these grains. But many grains fall to the left of these curves. (b) If hot-bottom burning at the base of the stellar envelope is permitted in the models, the envelopes of stars of mass $\geq$4.5 $M_\odot$ evolve to the left from their positions on the curves in panel (a). Stars that experienced hot-bottom burning can explain grains in the upper left quadrant of the plot, but this quadrant has almost no grains. Grains in the lower left quadrant cannot be explained by this process. (c) Models of low-mass stars that include a parameterized prescription for rotational mixing at the base of the envelope (known as cool-bottom processing) can produce compositions at the lower left of the plot. The match between the grain compositions and these models strongly indicates that three-dimensional models that correctly describe the mixing in a rotating star are required to adequately model the oxygen isotopic compositions of low-mass stars.

s-process compositions that agree well with detailed theoretical models. Data for different size fractions show clear evidence for a predicted branching in the krypton *s*-process chain. Unstable ^{85}Kr has a half-life that is long enough for it to capture another neutron before it can decay and become stable ^{86}Kr, if the neutron density is high enough. The abundance of ^{86}Kr varies dramatically among silicon carbide size separates (Fig. 5.9), demonstrating that, on average, larger silicon carbide grains contain krypton processed at higher neutron densities than smaller grains. Isotopic measurements of barium, strontium, neodymium, and samarium in bulk samples of silicon carbide by thermal ionization mass spectrometry match general expectations for the *s*-process. Barium

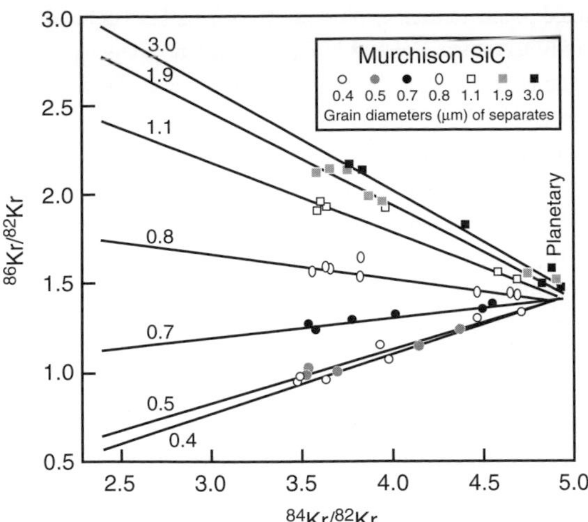

Figure 5.9 Krypton three-isotope plot for presolar silicon carbide from the Murchison meteorite, showing that $^{86}Kr/^{82}Kr$ varies with grain size, while $^{84}Kr/^{82}Kr$ is essentially unchanged. The labeled lines are mixing lines between the *s*-process end member estimated by Gallino et al. (1990) on the left and normal solar system krypton on the right. The variation in $^{86}Kr/^{82}Kr$ is due to a branch point in the krypton *s*-process chain that permits the production of ^{86}Kr if the neutron density is high enough. Because neutron density correlates with stellar mass, these data indicate that many stars contributed to the silicon carbide found in the Murchison meteorite. After Lewis et al. (1994).

and strontium isotope ratios in bulk samples show a grain-size effect similar to that in krypton, but the relationship between neutron exposure and grain size is opposite to that in krypton. The data described above were obtained from samples containing many silicon carbide grains, and thus give only a general picture of the *s*-process. Measurements of individual silicon carbide grains by RIMS allow direct comparisons of the compositions of molybdenum, zirconium, strontium, and barium from discrete times in the lives of individual stars with predictions of theoretical models. Not only are the agreements remarkable, confirming that 1.5–3 $M_\odot$ stars are the source of the silicon carbide grains, but these data also provide specific tests of nucleosynthesis models (see below).

5.7.2 Supernova Grains

Not all presolar grains are derived from AGB stars. Some come from type II supernovae. Massive stars that have exhausted their nuclear fuel experience core collapse, which in turn results in a supernova explosion (Chapter 3). The nucleosynthetic products of a type II supernova are the result of the regular sequence of shell burning in the presupernova massive star overprinted by explosive nucleosynthesis. Detailed predictions of the compositions of supernova grains are difficult to make because mixing and the details of the boundary between ejected material and material that cannot escape from the star are not understood. However, in general, supernova ejecta should be enriched in ^{12}C, ^{16}O, ^{20}Ne, ^{24}Mg, ^{28}Si, ^{32}S, etc. – isotopes that are produced from helium, carbon, and oxygen burning. About 1% of the silicon carbide grains in meteorites are highly enriched in ^{12}C and ^{28}Si compared to the AGB grains (Figs. 5.7 and 5.10; such grains are overabundant in these figures because special efforts have been made to find and study them). Immediately upon their discovery, these so-called "X" grains were proposed to have come from supernovae. Many low-density graphite grains have similar isotopic characteristics. The case for a supernova origin was dramatically strengthened when large excesses of ^{44}Ca, the decay product of ^{44}Ti ($t_{1/2} = 59$ years), were found in the X grains. One graphite grain had 138 times more ^{44}Ca than it should have had based on the other isotopic abundances (Nittler et al., 1996); the calcium was almost pure ^{44}Ca! Other isotopic signatures attributed to type II supernovae include excesses of ^{15}N and ^{18}O relative to solar system nitrogen and oxygen, very high inferred initial

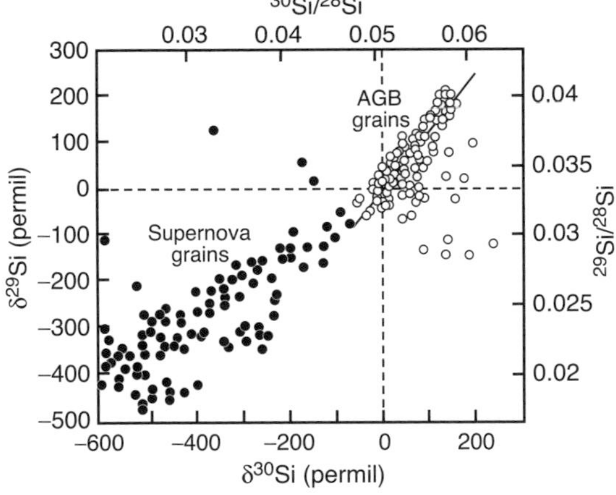

Figure 5.10 Silicon isotopic compositions for presolar silicon carbide grains. The top and right axes give the isotope ratios and the left and bottom axes give the ratios in terms of delta notation. Grains from AGB stars tend to fall in the upper right quadrant, with most grains falling along a steep array that passes slightly to the right of the solar composition (dashed lines). This "mainstream" array primarily reflects galactic chemical evolution, although the slight shift to the right of solar, and the AGB data significantly to the right of the array, probably reflect *s*-process nucleosynthesis in the AGB stars. Grains from supernovae are typically strongly enriched in ^{28}Si and ^{12}C relative to the starting compositions of the parent stars. Modified from Zinner (2014).

abundances of the short-lived radionuclides ^{26}Al ($t_{1/2}$ = 730,000 years) and ^{41}Ca ($t_{1/2}$ = 350,000 years), and excesses of ^{42}Ca, ^{43}Ca, and ^{49}Ti. Very rare type C silicon carbide grains have large excesses in heavy silicon isotopes ($^{29,30}Si$) and isotopically light ^{32}S, probably from the decay of ^{32}Si ($t_{1/2}$ = 150 years). Supernova grains make up a small percentage of presolar graphite, silicon carbide, and oxide grains and are the dominant source of Si_3N_4 grains.

5.7.3 Nova Grains

About 30–40% of classical novae (Chapter 3) have been observed to produce dust 20–100 days after the outburst. A few silicon carbide and graphite grains appear to have come from novae. They carry the signatures of explosive hydrogen burning, although the magnitudes of the isotopic effects are not as extreme as models predict. The silicon carbide grains have low $^{12}C/^{13}C$ ratios and very low $^{14}N/^{15}N$ ratios, large ^{30}Si excesses, and high inferred $^{26}Al/^{27}Al$ ratios. Graphite grains of putative nova origin have low $^{12}C/^{13}C$ ratios and $^{20}Ne/^{22}Ne$ ratios lower than predicted from models of AGB stars. Very low $^{20}Ne/^{22}Ne$ ratios imply that the ^{22}Ne came from the decay of ^{22}Na ($t_{1/2}$ = 2.5 years). Very low ^{4}He abundances accompanying the low $^{20}Ne/^{22}Ne$ ratios imply that the grains are almost devoid of implanted noble gases. Recent modeling of CO novae has produced better matches with the isotopic compositions of some candidate nova grains without requiring any dilution of the ejecta (e.g. Iliadis et al., 2018).

5.7.4 Other Stellar Sources

The stellar sources of a relatively small fraction of presolar grains, such as the A and B silicon carbide grains (Fig. 5.7), remain unknown. The isotopic compositions of these grains do not match expectations for grains from the sources discussed above. Possible sources for ^{14}N-rich AB SiC grains include J-type carbon stars and "born-again" AGB stars (stars that have left the AGB but still have enough helium outside the core to support a late helium flash and a temporary return to the AGB). Suggested sources for ^{15}N-rich AB SiC grains include supernovae with explosive hydrogen burning. AB grains may also come from massive Wolf–Rayet stars (Chapter 3). Better modeling, better stellar observations, and more grain data will be required to sort out the sources of these unusual grains.

5.8 Presolar Grains as Probes of Stellar Nucleosynthesis

Presolar grains provide precise isotopic compositions that give snapshots of the envelope compositions of stars at fixed points in their evolution. But we will never be able to study those stars directly because they no longer exist. In contrast, observational astronomy provides relatively low-precision isotopic measurements of the envelopes of specific stars, which can be studied for many different characteristics. The two datasets complement each other, with the presolar grains providing the detailed information and the astronomical observations providing the context. Data from presolar grains have resulted in a resurgence of the study of stellar nucleosynthesis. The detailed information from presolar grains points out problems with theoretical models or their input data. The new constraints also permit modelers to investigate processes for which there were previously no observational constraints.

5.8.1 Input Data for Stellar Models

Nuclear reaction rates and neutron-capture cross sections are key inputs to stellar models. Numerical values for these rates and cross sections come from laboratory experiments or from theoretical estimates. Some are well determined, while others have large uncertainties. The isotopic compositions of presolar grains have pointed out cases where reaction rates or cross sections were significantly in error. One spectacular example concerns barium. Early attempts to model the isotopic composition of *s*-process barium in silicon carbide grains produced relatively poor agreement. However, good agreement could be achieved if the neutron-capture cross sections were adjusted by factors of 30% to more than a factor of two. Gallino et al. (1993) proposed that the experimental values were incorrect and made specific predictions for what the correct values should be. Shortly afterward, these predictions were confirmed by new experimental determinations (e.g. Voss et al., 1994; Beer et al., 1997). Similar predictions by Gallino et al. (1993) regarding the neutron-capture cross sections for ^{142}Nd and ^{144}Nd were also confirmed. Other suggestions for reactions and cross sections that need re-evaluation have been made and are being tested. As models become better and measurements of presolar grains become more precise, additional predictions for improvements in reaction-rate determinations will undoubtedly result.

5.8.2 Internal Stellar Structure

Presolar grains have also provided examples of where theoretical models are not adequate to describe the internal structure of real stars. Oxygen isotopic data for presolar oxide grains are broadly consistent with low- and intermediate-mass stars as their sources. As already discussed, most grains are significantly enriched in ^{17}O and slightly depleted in ^{18}O relative to the starting compositions of the stars. However, the data for some grains are

Box 5.1 Can Presolar Grains Constrain the Age of the Galaxy?

Do presolar grains provide a constraint on the age of the galaxy? Nittler and Cowsik (1997) proposed that they do. Their logic was as follows.

The solar system formed ~4.57 billion years ago (Chapter 10). In order for a star to contribute grains to the early solar system, it must have completed its life cycle by that time. The lifetimes of stars are a strong function of stellar mass (see Fig. 3.2), with lifetimes of low-mass stars exceeding 10 Gyr. Silicon carbide grains apparently come mostly from AGB stars of 1.5–3 $M_\odot$. The lifetimes of such stars range from a few hundred million years to ~1.5 Gyr. Oxide grains apparently come from stars that range from 1.1 to 1.3 $M_\odot$ up to several solar masses. The low-mass end of this range corresponds to stars that live 3–9 Gyr. If the longest-lived star that contributed material to the solar system lived for ~9 Gyr, then it must have formed $9 + 4.57 = 13.57$ billion years ago. This calculation provides a lower limit on the age of the galaxy.

The stars that provided oxide grains to the solar system apparently had metallicities between half of solar and somewhat greater than solar (similar to the case for silicon carbide grains). According to models of galactic chemical evolution, it took a billion years or longer for the average metallicity to reach half the Sun's value. This time must be added to the minimum age estimated above. Nittler and Cowsik (1997) used theoretical curves such as those in Figure 5.8 to estimate the mass and metallicity of the parent stars for a number of grains. This gave them the lifetimes of the parent stars. They then used a galactic chemical evolution model to estimate the time it took for the galaxy to achieve the estimated metallicity. They added this time to the lifetime of the parent star and to the age of the solar system to get an estimate for the galactic age from each grain. They then took an average of the eight oldest determinations to get an estimate for the age of the galaxy of 14.4 ± 1.3 billion years.

Nittler and Cowsik (1997) recognized that their estimate of the galactic age has large systematic uncertainties. The greatest uncertainty comes from the mass estimate for the parent star. For example, increasing the mass from 1.1 to 1.15 $M_\odot$ can decrease its lifetime by a billion years. The masses and metallicities of the parent stars are inferred from models of stellar nucleosynthesis, and these models still have significant uncertainties. The theoretical estimates of stellar lifetimes as a function of mass also have uncertainties. Errors can be introduced from the estimate of metallicity, which depends on the accuracy of stellar models. There are also large uncertainties in the galactic chemical evolution models, which means that the time estimated for the galaxy to reach the metallicities inferred for the parent stars of the grains is quite uncertain. In fact, the only number in this calculation that does not have a large uncertainty is the age of the solar system.

In spite of these uncertainties, the resulting estimate of galactic age seems reasonable. The age of the universe is estimated at 13.7 ± 0.2 billion years, and the galaxy must be younger than the universe. Ballpark estimates for the time necessary for galaxies to form are roughly a billion years. So from these considerations, the galaxy should have formed 12.7 billion years ago. This is just outside of the quoted statistical uncertainty in the Nittler and Cowsik (1997) estimate. This is a reasonable agreement and suggests that as we continue to improve stellar models and our understanding of galactic chemical evolution, presolar grains might provide an independent constraint on the age of the galaxy.

not consistent with the model predictions. Figure 5.8a shows data for presolar oxide grains compared to the arrays of envelope compositions as a function of mass and initial composition predicted by standard stellar models. Many grains plot well away from the region predicted by the standard models. In intermediate-mass stars, hydrogen burning via the CNO cycle takes place at the base of the convecting envelope after first dredge-up, destroying ^{18}O and producing ^{17}O, while leaving ^{16}O largely unchanged. This hot-bottom burning shifts the predicted oxygen compositions as shown in Figure 5.8b. However, the region explained by hot-bottom burning has essentially no oxide grains, and most of the oxide grains fall in a "forbidden region" that was not explained by any of the models.

A significant problem with stellar nucleosynthesis models is that they do not adequately describe the *physical* state of the star. To decrease the amount of computer time required to run the models to a manageable level, most models have been one-dimensional, meaning that the star was assumed to be spherically symmetrical. Convective mixing can be modeled based on density considerations, but the effects of mixing generated by stellar rotation cannot. Even before the data for presolar oxides were available, there were observations of low-mass red giant stars that showed lower $^{12}C/^{13}C$ and higher $^{14}N/^{15}N$ ratios in the envelopes than predicted by the models. Extra mixing of envelope material into nuclear burning zones was postulated as an explanation. When put into the models in a parameterized way, this "cool-bottom processing" can explain not only the oxygen isotopic compositions of the presolar grains (Fig. 5.8c), but also the nitrogen isotopic compositions of silicon carbide grains, the ^{26}Al abundances observed in presolar grains, and the ^{3}He budget of low-mass stars. Although it is not yet possible to model this extra mixing from first principles in three-dimensional models, it is clear what the nature of the mixing must be and the extent to which it modifies the envelope compositions of red giant and AGB stars.

5.8.3 The Neutron Source(s) for the *s*-Process

Silicon carbide grains provide a means to study the details of the *s*-process in low-mass stars. The structures of these grains accommodate a wide variety of minor and trace elements. The neutrons that drive the *s*-process come from two sources. One source is the reaction $^{13}C(\alpha, n)^{16}O$, which occurs in the intershell region between the hydrogen shell and the helium shell while the hydrogen shell is burning and the helium shell is quiescent. This reaction is most likely the dominant source of neutrons in low-mass stars. Unfortunately, it is not yet possible to determine the strength of this neutron source from first principles in stellar models. The other neutron source is the $^{22}Ne(\alpha,n)^{25}Mg$ reaction, which occurs at higher temperatures in the helium shell when it ignites (see Chapter 3). This short-duration source can be modeled from first principles. The relative strengths of these two sources can be evaluated by looking at the details of the isotopic compositions of *s*-process elements such as strontium, zirconium, molybdenum, ruthenium, and barium. There are a number of branch points where, at low neutron density, a nucleus β decays to the element with a higher Z, but if the neutron density is high enough, the nucleus can capture a neutron before it decays to form a neutron-rich isotope. Because of these branch points, the isotopic compositions of *s*-process elements vary as a function of neutron density. The neutron density is a function of the temperature and the neutron source.

Comparisons of the calculated compositions from models of stellar nucleosynthesis and the measured compositions of strontium, zirconium, molybdenum, ruthenium, and barium in the most common type of AGB silicon carbide grains indicate that the ^{13}C source is the most important and that the parent stars of these grains were mostly 1.5–2 $M_{\odot}$ stars, with a few 3 $M_{\odot}$ stars also contributing. A similar analysis done with silicon and titanium isotopes seems to indicate stars of slightly higher mass, 2–3 $M_{\odot}$. The discrepancy between the results of these two studies may indicate either that the grains that were measured come from different populations of stars, or that the models or their input data need to be improved. As discrepancies like this are resolved, our understanding of stellar nucleosynthesis will continue to improve.

5.8.4 Constraining Supernova Models

It is important to understand the nucleosynthesis and dynamics of core-collapse supernovae for many reasons. These stars produce a large fraction of the heavy elements in the universe. Because of the power of these supernovae and the short lifetimes of the presupernova stars, they play a significant role in the formation, evolution, and destruction of star-forming regions. A supernova may have been responsible for triggering the gravitational collapse of the dust cloud that formed the solar system and probably provided the short-lived radionuclides that are the major tools of early solar system chronology (discussed in Chapters 9, 10, and 15). Prior to the discovery of presolar grains, supernova models generally started with an onion-shell structure for the presupernova star (see Chapter 3) and then propagated a shock wave through this spherically symmetrical (one-dimensional) model. Such a model does a reasonable job of describing nucleosynthesis in the supernova. However, when presolar grains from supernovae were identified and studied, some interesting problems appeared. Many of the grains contained newly synthesized matter from several different zones in the supernova. According to the models, these materials should not have come into contact with one another, but clearly they did. High-resolution observations of recent supernova remnants also show clear evidence of mixing in the ejecta. But until the grains were studied, the extent and spatial scale of the mixing were unknown. It will be a huge task to generate models that accurately describe the mixing that takes place during supernova explosions, but with the grains providing ground truth, such modeling may eventually be possible.

5.8.5 Galactic Chemical Evolution

Presolar grains from low- and intermediate-mass stars also carry a record of the initial compositions of the parent stars before any stellar nucleosynthesis took place. Elements such as silicon and titanium are only slightly affected by nuclear burning in the hydrogen and helium shells. For example, the most common kind of AGB silicon carbide grains fall along a rather steep array on a silicon three-isotope plot (Fig. 5.10). Grains that fall along this array are often called "mainstream" grains in the literature. Theoretical modeling shows that stellar nucleosynthesis in low- and intermediate-mass stars cannot produce this array. Instead, silicon compositions evolve along much shallower lines. Thus, the array defined by the grains must dominantly represent the compositions of the parent stars, with only a small overprint of *s*-process nucleosynthesis. Nucleosynthesis within the parent stars is probably responsible for shifting the array to the right of the solar composition. Similar plots show that titanium isotopes in silicon carbide generally track the silicon isotopes (Fig. 5.11). The *s*-process has the greatest effect on ^{50}Ti, but the data arrays for silicon carbide show large spreads for ^{46}Ti that are almost as great as those observed for ^{50}Ti. This indicates that the titanium isotope variations among the grains were also probably inherited in large part from the parent stars.

The general expectation is that secondary isotopes, such as ^{17}O, ^{18}O, ^{29}Si, ^{30}Si, ^{46}Ti, ^{47}Ti, ^{49}Ti, and ^{50}Ti, increase in relative abundance compared to primary isotopes, such as ^{16}O, ^{28}Si, and ^{48}Ti, with time and overall metallicity in the universe (Chapter 3). The first relatively simplistic attempts to model galactic chemical evolution based on nucleosynthetic yields for massive stars confirm this expectation. But the models are not able to reproduce in detail the data arrays derived from the presolar grains.

From the basic trend of galactic chemical evolution, we would expect the grains at the lower left of the silicon-isotope mainstream array to be from stars of lower metallicity than those at the upper right (Fig. 5.10). Thus, the Sun appears to lie at the low-metallicity end of the trend. This implies that the parent stars of the silicon carbide grains, especially the larger ones, had metallicities higher than that of the Sun even though the stars lived and died before the solar system formed. There is a gradient of increasing metallicity from the edge to the center of the galaxy due to the higher star-formation rate near the center, where the density of matter is higher. One idea to explain the apparent SiC contradiction is that the stars that provided ^{29,30}Si-rich silicon carbide to the solar system formed closer to the galactic center than where the Sun formed and migrated outward during their lifetimes. These stars deposited "high-metallicity" dust in the Sun's parent molecular cloud (Clayton, 1997). On the other hand, metallicity need not closely track isotopic composition. Stochastic variation in the metallicity of the gas going into new stars can result in a spread of metallicities and isotopic compositions in a relatively limited region of the galaxy. But there is

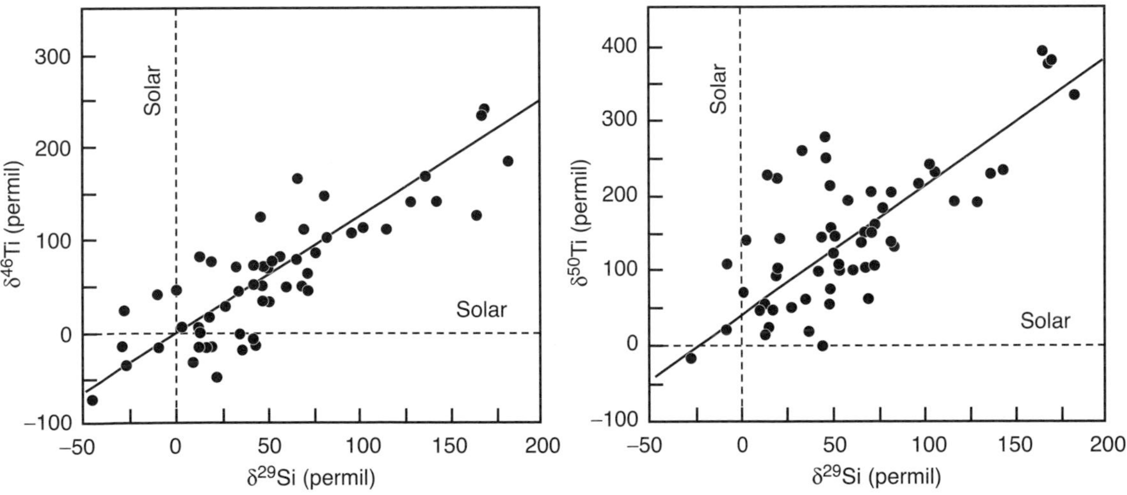

Figure 5.11 Titanium isotope data for mainstream silicon carbide grains versus δ^{29}Si. The correlation between excesses of minor titanium isotopes and minor silicon isotopes most likely reflects galactic chemical evolution. The offset of the δ^{50}Ti trend to pass above the solar composition probably reflects *s*-process nucleosynthesis in the parent stars, which most strongly affects ^{50}Ti. Data from Huss and Smith (2007) and references therein.

something else to consider. Isotopic measurements of presolar grains are easiest on larger grains, and the compositions of mainstream silicon carbide grains plotted in Figures 5.10 and 5.11 come from the high-mass tail of the distribution of presolar grains. Larger grains form more easily in ejecta from high-metallicity stars, which have higher concentrations of their constituent elements. If our current datasets are biased toward grains from high-metallicity stars, the Sun might only *appear* to be at the low-metallicity end of the array of stars that supplied silicon carbide grains to the solar system.

Variations in the oxygen isotopic compositions of presolar oxides also reflect galactic chemical evolution overprinted by stellar nucleosynthesis (Nittler et al., 1997). While the range in $^{17}O/^{16}O$ is governed mostly by stellar nucleosynthesis, the range in $^{18}O/^{16}O$ can only partially be explained by stellar processing. The data clearly point to a range of initial compositions for the parent stars of the measured grains, but with oxygen it is less clear where the solar system fits relative to this range than it is in the cases of silicon and titanium.

The discussions above show that presolar grains can give considerable information about galactic chemical evolution. However, the theoretical framework within which to interpret that information is not as mature at that for stellar nucleosynthesis.

5.9 Presolar Grains as Tracers of Circumstellar and Interstellar Environments

Circumstellar condensates carry information about the environments in which they formed and about the environments though which they passed on the way to being incorporated into meteorites and IDPs.

5.9.1 Silicon Carbide

The structure of presolar silicon carbide grains can provide information about the conditions of formation. Crystalline silicon carbide is known to form about 100 different polytypes, including cubic, hexagonal, and rhombohedral structures. Presolar silicon carbide exists in only two of these, a cubic (β-SiC) polytype and a hexagonal (α-SiC) polytype (Daulton et al., 2003). These two types are the lowest-temperature forms of silicon carbide. The temperatures of formation are broadly consistent with expectations for conditions in circumstellar outflows.

Silicon carbide grains are known to contain subgrains of titanium carbide. Equilibrium thermodynamics predicts that titanium carbide will condense before silicon carbide at relatively high pressures and low C/O ratios (Fig. 5.12). The titanium carbide grains were apparently accreted by the growing silicon carbide grains and were enclosed as the silicon carbide grain continued to grow.

The external morphology of presolar silicon carbide grains also provides useful information. Comparing grains that have experienced the chemical treatments required to produce silicon carbide separates with grains that have been identified *in situ* ("pristine grains") helps to identify the features that reflect their formation and their journey through the interstellar medium and the solar nebula. Most pristine grains are bounded by one or more crystal faces, and more than half of these crystal faces exhibit polygonal depressions with symmetries consistent with the internal crystal structure. On chemically processed grains, the polygonal depressions are accompanied by numerous surface pits, probably etched defects in the crystal structure. These observations indicate that the polygonal depressions are primary features. Together, the polygonal depressions and the high density of defects indicate that the growth of the grains was initially very rapid and was quenched when the gas pressure dropped.

5.9.2 Graphite Grains from AGB Stars

Presolar graphite grains are inviting targets for study because of their larger sizes (up to 20 μm in diameter) and because they are relatively soft and easy to slice for study. Graphite grains from AGB stars and from supernovae (identified by their isotopic compositions) have been extensively studied. AGB graphite grains exhibit two morphologies that have been labeled "onion-type" and "cauliflower-type." The outer portions of the onion-type grains consist of well-crystallized layers that form concentric shells (Fig. 5.4a). The cores of many of these grains consist of nanocrystalline carbon comprised of randomly oriented graphene sheets (one-atom-thick planar sheets of carbon rings densely packed in a honeycomb lattice; graphite consists of many graphene sheets stacked together in a regular pattern). As much as one-quarter of the core material may be in the form of polycyclic aromatic hydrocarbons (PAHs). Onion-type graphite grains apparently grew rapidly from supersaturated gas during the initial stages, producing the nanocrystalline cores. Subsequently, heterogeneous nucleation on these cores produced the concentric graphite sheets. Cauliflower-type graphites consist of turbostratic graphite sheets that are wavy and contorted. Some show roughly concentric structures with short-range coherency (<50 nanometers thick). The structures tend to be loose-packed with visible gaps in cross section (Fig. 5.4b). Many contain tiny carbide crystals. It is not clear whether or not the

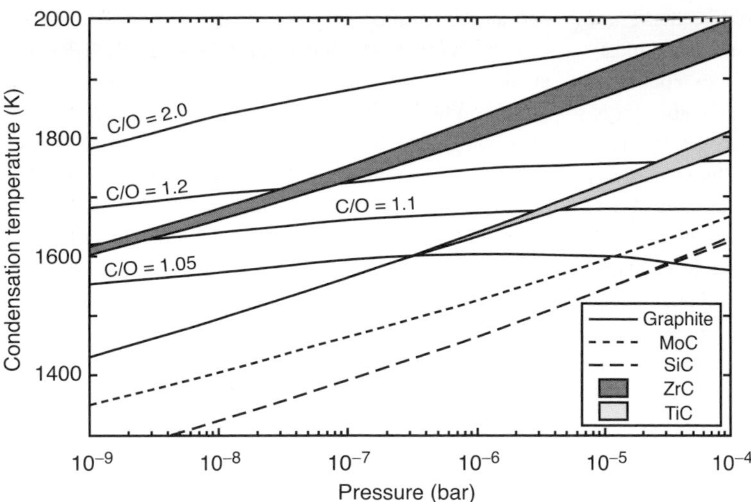

Figure 5.12 Condensation curves for graphite, silicon carbide, titanium carbide, molybdenum carbide, and zirconium carbide as a function of temperature, gas pressure, and C/O ratio. The curves represent the conditions at which the phases condense from a gas of solar composition, except for C/O ratio. Condensation temperatures for silicon carbide, zirconium carbide, and titanium carbide vary slightly with C/O ratio and are shown as envelopes, with the lower boundary corresponding to C/O = 1.05 and the upper boundary to C/O = 2.0. The carbide condensation curves are roughly parallel and all crosscut the graphite condensation curve. At low pressures, graphite condenses before most of the carbides, particularly at high C/O ratios. At higher pressures, the carbides condense before graphite. Zirconium carbide would have been the easiest to incorporate into presolar graphite because it condenses at higher temperatures than graphite over most of the pressure range illustrated and under C/O ratios up to almost 2. Presolar graphite grains that contain subgrains of titanium carbide would have condensed at pressures between ~3×10^{-7} and 10^{-4} bars, depending on C/O ratio. Silicon carbide is not observed in graphite grains and it is difficult to incorporate into graphite because it condenses after graphite under most conditions. After Bernatowicz et al. (1996).

onion-type and cauliflower-type graphite formed in the same astrophysical environment.

Both types of AGB graphite contain small inclusions of refractory carbides. Some grains have cores of refractory carbide (usually titanium carbide) that served as nucleation sites for the graphite (Fig. 5.4a). Titanium carbide is the most abundant type, and many carbides have substantial enrichments in *s*-process elements (zirconium, molybdenum, ruthenium). In some cases, the *s*-process elements can dominate the carbides, with only minor amounts of titanium. The presence of these carbides places relatively strong constraints on the environment of formation for the graphite. Carbides of molybdenum, zirconium, and titanium always condense before silicon carbide over a wide range of temperature and pressure conditions (Fig. 5.12). Unlike for the carbides, whose appearance depends on both nebular pressure and temperature, the appearance of graphite depends little on pressure, but depends strongly on the C/O ratio. This means that the lines of first appearance for graphite on a stability diagram cross those for the carbides. Observations indicate that the carbides formed before graphite. Figure 5.12 indicates that titanium carbide will condense before graphite, at $P \geq 3 \times 10^{-7}$ atmospheres for C/O = 1.05 and at $P \geq 3 \times 10^{-5}$ atmospheres for C/O = 1.2; low C/O ratios and relatively high pressures were the rule where graphite condensed. Astronomical observations of carbon stars have a mean C/O ratio of 1.15 ± 0.16. This compares very favorably to the conditions indicated in Figure 5.12 for the condensation of graphite grains with inclusions of refractory carbides.

5.9.3 Graphite Grains from Supernovae

Graphite grains from supernovae are identified by isotopic excesses of ^{18}O and ^{28}Si. The grains that have been studied in detail are 4–10 μm in diameter. Structurally, they consist of concentric turbostratic layers with a degree of graphitization intermediate between the onion- and cauliflower-types from AGB stars. Supernova graphite grains have high abundances of tiny internal crystals of titanium carbide. These crystals differ from those in AGB graphite grains in that they do not have significant amounts of *s*-process elements such as molybdenum, zirconium, and ruthenium. Supernova graphite grains often contain metallic iron–nickel, either as discrete grains or epitaxially grown onto the crystal faces of the titanium carbide crystals. The sequence of condensation indicated by these grains is titanium carbide first, followed by iron–nickel metal, followed in turn by graphite.

Both the number and composition of the internal titanium carbide grains can vary as a function of distance from the center of the graphite grains. The vanadium/titanium ratio in the titanium carbide grains tends to increase from the core to the rim of the graphite grains, while the abundance of the titanium carbide grains per volume of graphite typically decreases significantly. Vanadium carbide condenses at substantially lower temperatures than titanium carbide and thus serves as a monitor of gas-phase temperature during the growth of the graphite grains. The abundance differences from core to rim could either reflect a changing growth rate for the graphite grains or a decrease with time in the abundance of titanium carbide grains (Croat et al., 2003).

The morphologies of the titanium carbide grains in supernova graphite show evidence of "weathering" in the supernova outflows. Most grains show some erosion of the crystal faces, but the faces are still clearly defined. Some titanium carbides and some metal grains have rims of amorphous or nearly amorphous material 3–15 nm thick (Croat et al., 2003). These rims resemble the ~50 nm amorphous rims observed on grains in lunar soils. The lunar rims were caused by irradiation with solar wind protons and helium ions, which hit the grains at several hundred kilometers per hour. The rims on the inclusions in supernova graphite were likely created by the grains traveling at high speed and impacting slower-moving parcels of gas. It is unlikely that the rims are due to chemical reactions because they appear on two different mineral phases, and gas that is condensing TiC, FeNi, and graphite is unlikely to be corrosive to them.

The observations summarized above provide some basic information about the supernova ejecta in which the graphite grains formed. The high abundances of titanium carbide grains imply a very dusty environment. Gas pressures necessary to supply enough titanium to produce the high abundance of titanium carbide grains are in the range of a few 10^{-7} to a few 10^{-6} atmospheres. As we learn more about supernova graphite grains, it may be possible to make detailed models of the conditions in supernova ejecta based on the grains produced there.

5.9.4 Aluminum Oxide Grains

Presolar aluminum oxide grains are found in three structural forms: corundum (α-Al_2O_3), amorphous alumina, and an unidentified phase of Al_2O_3 that contains significant magnesium (Stroud et al., 2004, 2007; Takigawa et al., 2014). Seven of nine presolar aluminum oxide grains studied by Takigawa et al. (2014) had rough surfaces with a distinctive platy morphology, consistent with presolar aluminum oxide grains studied by Choi et al. (1998), while two had smooth surfaces. Two of the presolar grains were single-crystal corundum and a third was polycrystalline corundum. Amorphous surfaces on aluminum oxide grains could have been produced by irradiation, although the dose required to amorphize corundum is about 10 times higher than the dose that amorphizes silicates. Dissolution experiments show that, while corundum is unaffected by chemical treatments, other aluminum oxide phases are attacked by acids. The rough surfaces on the grains probably formed by secondary processes such as irradiation, which damaged the crystal structure, followed by chemical etching.

5.9.5 Interstellar Grains

For the most part, the work that has been done on presolar grains has focused on circumstellar condensates because they have large isotopic anomalies that make them easy to identify. As we discussed earlier in this chapter, much of the material in interstellar space has probably evaporated and recondensed many times because of the passage of shocks generated by strong stellar winds or stellar explosions. The resulting interstellar dust may consist of chemically unstable material that was easily converted to more-stable compounds in the solar nebula. But grains from the more stable tail of this material may have survived somewhere in primitive meteorites or in comets. If we can find and recognize this material, we will have a means of probing the details of the environment in interstellar space. GEMS, found in IDPs (Bradley, 1994), may be some of this material (see Chapter 6). They have experienced extreme radiation damage and perhaps chemical fractionation. The site of this processing is not clear. Some GEMS have anomalous isotopic compositions, suggesting that they either started out as circumstellar condensates or incorporated such material at some point in their history.

5.10 Presolar Grains as Probes of the Early Solar System

The solar system inherited a complex mixture of different types of presolar grains (Table 5.1). The known types of presolar grains exhibit a wide range of thermal and chemical stabilities. There are highly refractory components, such as diamond, Al_2O_3, and hibonite. There are components that are outgassed or destroyed at relatively low temperatures, such as graphite and the low-temperature P3 noble gas component in diamond. Some components are stable in a gas with C/O>1, and others are stable in a gas with C/O<1. Because of this range of stability, the relative abundances of the different types of presolar grains in a meteorite give direct information about the processes and environments that affected the meteorite and its precursor material.

Almost immediately after the discovery of presolar grains, it was clear that they could only be found in the most primitive chondrites, i.e. those that had suffered the lowest degree of thermal metamorphism. Further work showed that the abundances of presolar grains, when normalized to the content of fine-grained matrix where the grains reside, correlate inversely with the metamorphic grade of the host meteorite (Huss and Lewis, 1995). Figure 5.13 shows this correlation for ordinary chondrites, but similar patterns are seen in all classes in which enough meteorites have been studied to evaluate the trend. Note that thc *relative* abundances of the different types of presolar components also change with metamorphic grade. Thermally resistant components survive best, while thermally labile components are destroyed at low grade. The survival rate also changes with the chemical environment. Diamond and silicon carbide survive to higher metamorphic grades in the highly reduced enstatite chondrites, relative to the more-oxidized ordinary and carbonaceous chondrites.

The characteristics of the presolar diamonds also change with the metamorphic grade of the host meteorite. Figure 5.14 shows the typical bimodal release of heavy noble gases (here illustrated by xenon) in Orgueil, an unheated chondrite. This pattern is compared to the xenon-release patterns of diamonds from two ordinary chondrites that have experienced different degrees of mild metamorphism. The amount of low-temperature gas, labeled P3 for historical reasons, is a sensitive function

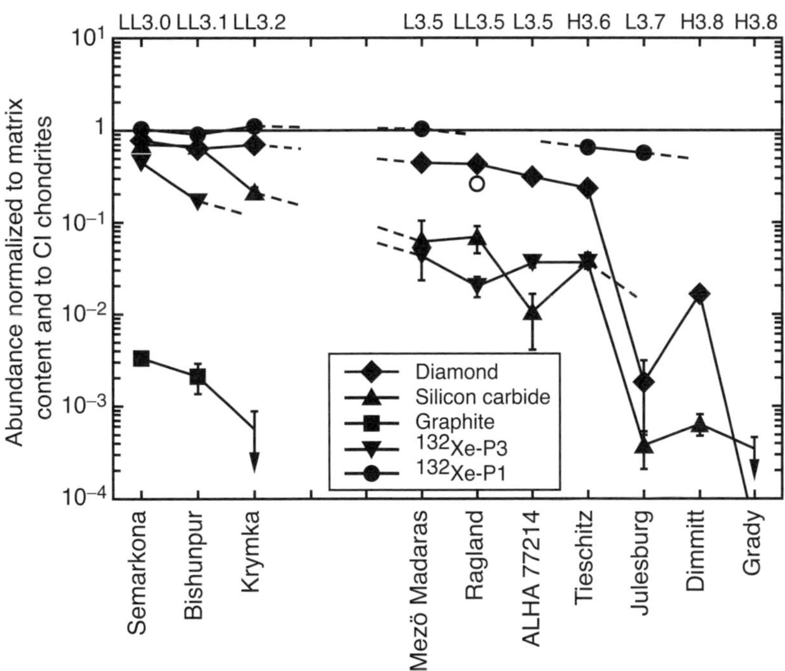

Figure 5.13 Abundances of presolar diamond, silicon carbide, graphite, and two noble gas components (P3 sited in diamonds, and P1 sited in the elusive phase Q) as a function of metamorphic grade for type 3 ordinary chondrites. The abundances are normalized to the matrix content of the host meteorite and to the abundances in the essentially unheated CI chondrites in order to illustrate the differences in abundance as a function of metamorphic grade. The line at unity represents abundances equal to those in CI chondrites. Decreases in the abundances with metamorphic grade reflect different thermal and chemical stability (metamorphism in type 3 chondrites is discussed in Chapter 6). For example, meteorites of grades 3.0–3.5 retain their full complement of 132Xe-P1; however, only types 3.0–3.1 show any presolar graphite. Types 3.7 and higher have less than half of the 132Xe-P1 and only traces of the other presolar components. Adapted from Huss et al. (2006).

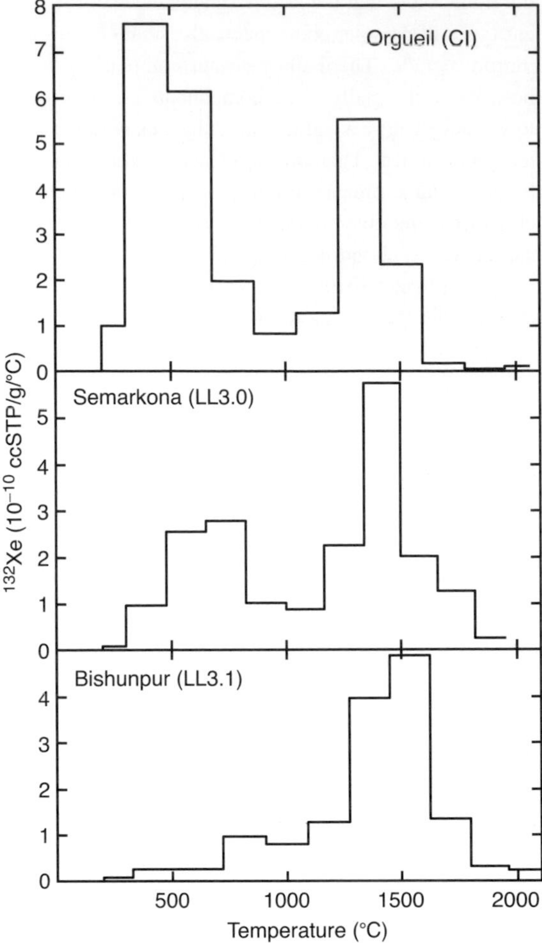

Figure 5.14 Release for ^{132}Xe in presolar diamonds as a function of temperature during stepped heating experiments. Data are normalized to °C so that the areas under the histograms accurately represent gas amounts. Orgueil, which can be considered unheated, has the largest release of low-temperature P3 xenon. The low-temperature peak decreases systematically with increasing metamorphic grade from unheated Orgueil to slightly more heated Semarkona to Bishunpur. Higher petrologic types show even smaller low-temperature gas releases. Data from Huss and Lewis (1994).

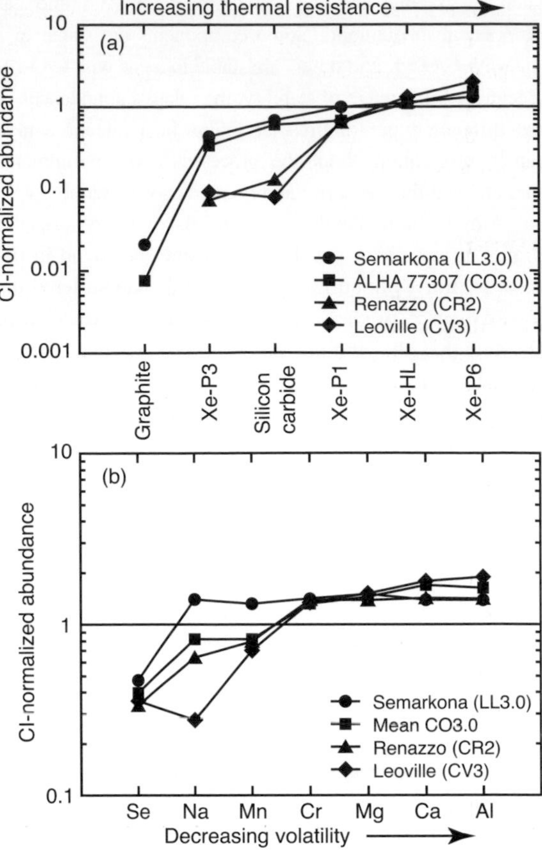

Figure 5.15 (a) Abundances of six presolar components with different thermal stabilities, normalized to the abundances in CI chondrites, for the least metamorphosed members of the LL, CO, CR, and CV chondrites. They are arranged in order of increasing thermal resistance. Peak metamorphic temperatures for these meteorites are constrained to be less than 200°C. (b) Abundances of seven elements with volatilities ranging from volatile to refractory are plotted as a function of their volatility for the same meteorites, normalized to the abundances in CI chondrites. Note that the LL chondrite, Semarkona, which has the smallest depletions of easily destroyed presolar grains, is depleted only in the most volatile element plotted, selenium. The other elements are present in CI relative abundances, although their absolute abundances are higher than in CI due to depletion of the most volatile elements. Depletions in fragile presolar grains and in volatile and moderately volatile elements increase from LL to CO to CR to CV chondrites. These correlated depletions indicate that the same process was responsible for the loss of easily destroyed presolar grains and the volatile elements.

of temperature. Its abundance correlates well with other indicators of metamorphism (Chapter 6). A very rough temperature scale can be derived for the metamorphism in primitive chondrites from these different features.

The abundances of presolar grains in even unmetamorphosed chondrites seem to reflect some thermal processing in the nebula prior to accretion (Huss et al., 2003). The differences in thermal and chemical resistance among grain types mean that mixtures of grains exposed to different conditions in the nebula have been modified in different ways. Chondritic material exposed to high temperatures in the nebula loses volatile elements through evaporation. If one looks at the least-metamorphosed members of each chondrite class, the abundances of presolar grains show patterns that correlate with the abundances of the chemical elements (Fig. 5.15). The presolar components that are easiest to destroy are most depleted in the meteorites that have low abundances of volatile elements. The correlations suggest that the same thermal processing was responsible for both sets of depletions. These depletions and their possible causes will be discussed in more detail in Chapter 7.

Summary

As we complete revisions to this book, it has been just over three decades since the discovery of presolar grains. These grains are unambiguous, though highly biased samples of the material that made up the Sun's parent molecular cloud. Presolar grains have revitalized the study of stellar nucleosynthesis by providing ground truth against which to check models of stellar nucleosynthesis. They provide the first direct look at some of the types of dust present in interstellar space. They also provide a means to investigate the environments around dying stars, conditions in interstellar space, and conditions in the early solar system. In addition, they have fundamentally changed the way we think about the formation of the solar system. The widespread existence of presolar grains in primitive materials shows that the solar system did not pass through a hot nebular stage in which all pre-existing solids were vaporized. All of this has occurred in spite of not yet having a complete census of the types of presolar material and without having fully characterized the presolar materials that we currently recognize. Studying presolar grains will be a rich area for research for many years to come.

Presolar grains are very difficult to study because of their small size. The technical requirements for studying presolar grains have led to the development of a new generation of analytical equipment that has made analysis of particles down to a few tenths of a μm in diameter feasible. Further improvements in analytical capability may eventually reach the point where the precision of our measurements is limited by the number of atoms present in a grain, a limit that cannot be overcome by any analytical technique.

We have so far discussed the origin of the elements and their abundances and isotopic compositions in the solar system and in the galaxy. We have investigated the grains that carry the elements from their stellar sources to the solar system. In the next chapter, we examine the suites of cosmochemical samples available for study – the many types of meteorites, as well as their smaller relatives, the IDPs.

Questions

1. What are presolar grains and why are they important?
2. How are presolar grains recognized?
3. What are the important differences between circumstellar condensates and interstellar grains?
4. How can presolar grains tell us about stellar nucleosynthesis?
5. What can presolar grains tell us about the environment of their formation?
6. What can presolar grains tell us about the environment and processes in interstellar space and in the early solar system?

Suggestions for Further Reading

Lewis R. S., Tang M., Wacker J. F., et al. (1987) Interstellar diamonds in meteorites. *Nature*, **326**, 160–162. The paper describing the discovery of the first presolar grains.

Bernatowicz T. J., and Zinner E. (1997) *Astrophysical implications of the laboratory study of presolar materials*. In AIP Conference Proceedings Vol. 402. American Institute of Physics, Woodbury, New York, 750 pp. An important volume describing the first decade of research on presolar grains.

Nittler L. R., and Ciesla F. (2016) Astrophysics with extraterrestrial materials. *Annual Reviews of Astronomy & Astrophysics*, **54**, 53–93. A recent overview of the use of presolar grains to investigate nucleosynthesis, galactic chemical evolution, and the origin of the solar system.

Floss C., and Haenecour P. (2016) Presolar silicate grains: Abundances, isotopic and elemental compositions, and the effects of secondary processing. *Geochemical Journal*, **50**, 3–25. This paper discussed presolar silicates in meteorites and the extent to which they survived solar system processing.

Other References

Alexander E. C., Jr., Lewis R. S., Reynolds J. H., and Michel M. C. (1971) Plutonium-244: Confirmation as an extinct radioactivity. *Science*, **172**, 837–840.

Alexander C. M. O'D., Swan P., and Walker R. M. (1990) In situ measurement of interstellar silicon carbide in two CM chondrite meteorites. *Nature*, **348**, 715–717.

Amari S., Anders E., Virag A., and Zinner E. (1990) Interstellar graphite in meteorites. *Nature*, **345**, 238–240.

Amari S., Lewis R. S., and Anders E. (1994) Interstellar grains in meteorites. I. Isolation of SiC, graphite, and diamond; size distributions of graphite and SiC. *Geochimica et Cosmochimica Acta*, **58**, 459–470.

Beer H., Corvi F., and Mutti P. (1997) Neutron capture of the bottleneck isotopes ^{138}Ba and ^{208}Pb, *s*-process studies, and the *r*-process abundance distribution. *Astrophysical Journal*, **474**, 843–861.

Bernatowicz T. J., Cowsik R., Gibbons P. E., et al. (1996) Constraints on stellar grain formation from presolar graphite in the Murchison meteorite. *Astrophysical Journal*, **472**, 760–782.

Bernatowicz T. J., Messenger S., Pravdivtseva O., et al. (2003) Pristine presolar silicon carbide. *Geochimica et Cosmochimica Acta*, **67**, 4679–4691.

Black D. C., and Pepin R. O. (1969) Trapped neon in meteorites II. *Earth & Planetary Science Letters*, **6**, 395–405.

Bradley J. P. (1994). Chemically anomalous, preaccretionally irradiated grains in interplanetary dust particles from comets. *Science*, **265**, 925–929.

Cameron A. G. W. (1962) The formation of the Sun and planets. *Icarus*, **1**, 13–69.

Choi B.-G., Huss G. R., Wasserburg G. J., and Gallino R. (1998) Presolar corundum and spinel in ordinary chondrites: Origins from AGB stars and supernova. *Science*, **282**, 1284–1289.

Clayton D. D. (1997) Placing the Sun and mainstream SiC particles in galactic chemodynamic evolution. *Astrophysical Journal*, **484**, L67–L70.

Clayton R. N., Grossman L., and Mayeda T. K. (1973) A component of primitive nuclear composition in carbonaceous meteorites. *Science*, **182**, 485–487.

Clayton R. N., Onuma N., Grossman L., and Mayeda T. K. (1977) Distribution of the pre-solar component in Allende and other carbonaceous chondrites. *Earth & Planetary Science Letters*, **34**, 209–224.

Clayton R. N., Hinton R. W., and Davis A. M. (1988) Isotopic variations in the rock-forming elements in meteorites. *Philosophical Transactions of the Royal Society of London*, A**325**, 483–501.

Croat T. K., Bernatowicz T., Amari S., et al. (2003) Structural, chemical, and isotopic microanalytical investigations of graphite from supernova. *Geochimica et Cosmochimica Acta*, **67**, 4705–4725.

Daulton T. L., Bernatowicz T. J., Lewis R. S., et al. (2003) Polytype distribution in circumstellar silicon carbide: Microstructural characterization by transmission electron microscopy. *Geochimica et Cosmochimica Acta*, **67**, 4743–4767.

Dauphas N., Remusat L., Chen J. H., et al. (2010) Neutron-rich chromium isotope anomalies in supernova nanoparticles. *Astrophysical Journal*, **720**, 1577–1591.

Gallino R., Busso M., Picchio G., and Raiteri C. M. (1990) On the astrophysical interpretation of isotope anomalies in meteoritic SiC grains. *Nature*, **348**, 298–302.

Gallino R., Raiteri C. M., and Busso M. (1993) Carbon stars and isotopic Ba anomalies in meteoritic SiC grains. *Astrophysical Journal*, **410**, 400–411.

Huss G. R., and Lewis R. S. (1994) Noble gases in presolar diamonds II: Component abundances reflect thermal processing. *Meteoritics*, **28**, 811–829.

Huss G. R., and Lewis R. S. (1995) Presolar diamond, SiC, and graphite in primitive chondrites: Abundances as a function of meteorite class and petrologic type. *Geochimica et Cosmochimica Acta*, **59**, 115–160.

Huss G. R., and Smith J. A. (2007) Titanium isotopes in isotopically characterized silicon carbide grains from the Orgueil CI chondrite. *Meteoritics & Planetary Science*, **42**, 1055–1075.

Huss G. R., Meshik A. P., Smith J. B., and Hohenberg C. M. (2003) Presolar diamond, silicon carbide, and graphite in carbonaceous chondrites: Implications for thermal processing in the solar nebula. *Geochimica et Cosmochimica Acta*, **67**, 4823–4848.

Huss G. R., Rubin A. E., and Grossman J. N. (2006) Thermal metamorphism in chondrites. In *Meteorites and the Early Solar System II*, Lauretta D. S., and McSween H. Y., editors, pp. 567–586, University of Arizona Press, Tucson.

Iben I. Jr., and Renzini A. (1983) Asymptotic giant branch evolution and beyond. *Annual Reviews of Astronomy & Astrophysics*, **21**, 271–342.

Iliadis C., Downen L. N., José J., et al. (2018) On presolar stardust grains from CO classical novae. *Astrophysical Journal*, **855**, 76.

Lewis R. S., Srinivasan B., and Anders E. (1975) Host phase of a strange xenon component in Allende. *Science*, **190**, 1251–1262.

Lewis R. S., Amari S., and Anders E. (1994) Interstellar grains in meteorites. II. SiC and its noble gases. *Geochimica et Cosmochimica Acta*, **58**, 471–494.

Liu N., Steele A., Nittler L. R., et al. (2017) Coordinated EDX and micro-Raman analysis of presolar silicon carbide: A novel, nondestructive method to identify rare subgroup SiC. *Meteoritics & Planetary Science*, **52**, 2550–2569.

Manuel O. K., Hennecke E. W,. and Sabu D. D. (1972) Xenon in carbonaceous chondrites. *Nature*, **240**, 99–101.

Messenger S., Keller L. P., Stadermann F. J., et al. (2003) Samples of stars beyond the solar system: Silicate grains in interplanetary dust. *Science*, **300**, 105–108.

Nagashima K., Krot A. N., and Yurimoto H. (2004) Stardust silicates from primitive meteorites. *Nature*, **428**, 921–924.

Nguyen A. N., and Zinner E. (2004) Discovery of ancient silicate stardust in a meteorite. *Science*, **303**, 1496–1499.

Nittler L. R., and Cowsik R. (1997) Galactic age estimates from O-rich stardust in meteorites. *Physical Review Letters*, **78**, 175–178.

Nittler L. R., Amari S., Zinner E., et al. (1996) Extinct ^{44}Ti in presolar graphite and SiC: Proof of a supernova origin. *Astrophysical Journal Letters*, **462**, L31–L34.

Nittler L. R., Alexander C. M. O'D., Gao X., et al. (1997) Stellar sapphires: The properties and origins of presolar Al_2O_3 in meteorites. *Astrophysical Journal*, **483**, 475–495.

Nittler L. R., Alexander C. M. O'D., Liu N., and Wang J. (2018) Extremely ^{54}Cr- and ^{50}Ti-rich presolar oxide grains in a primitive meteorite: Formation in rare types of supernovae and implications for the astrophysical context of solar system birth. *Astrophysical Journal Letters*, **856**, L24.

Qin L, Nittler L. R., Alexander C. M. O'D., et al. (2011) Extreme ^{54}Cr-rich nano-oxides in the CI chondrite Orgueil: Implications for a late supernova injection into the solar system. *Geochimica et Cosmochimica Acta*, **75**, 629–644.

Reynolds J. H., and Turner G. (1964) Rare gases in the chondrite Renazzo. *Journal of Geophysical Research*, **49**, 3263–3281.

Smith V. V., and Lambert D. L. (1990) The chemical composition of red giants. III. Further CNO isotopic and *s*-process abundances in thermally pulsing asymptotic giant branch stars. *Astrophysical Journal Supplement*, **72**, 387–416.

Srinivasan B., and Anders E. (1978) Noble gases in the Murchison meteorite: Possible relics of s-process nucleosynthesis. *Science*, **201**, 51–56.

Stephan T., Trappitsch R., Davis A. M., et al. (2016) CHILI – the Chicago Instrument for Laser Ionization – a new tool for isotope measurement in cosmochemistry. *International Journal of Mass Spectrometry*, **407**, 1–15.

Stroud R. M., Nittler L. R., and Alexander C. M. O'D. (2004) Polymorphism in presolar Al_2O_3 grains from asymptotic giant branch stars. *Science*, **305**, 1455–1457.

Stroud R. M., Nittler L. R., Alexander C. M. O'D., and Zinner E. (2007) Transmission electron microscopy and secondary ion mass spectrometry of an unusual Mg-rich presolar Al_2O_3 grain. *Lunar and Planetary Science* **38**, abstract #2203.

Takigawa A., Tachibana S., Huss G. R., et al. (2014) Morphology and crystal structures of solar and presolar Al_2O_3 in unequilibrated ordinary chondrites. *Geochimica et Cosmochimica Acta*, **124**, 309–327.

Tang M., and Anders E. (1988) Isotopic anomalies of Ne, Xe, and C in meteorites. II. Interstellar diamond and SiC: Carriers of exotic noble gases. *Geochimica et Cosmochimica Acta*, **52**, 1235–1244.

Voss F., Wisshak K., Guber K., and Kappler F. (1994) Stellar neutron capture cross sections of the Ba isotopes. *Physical Review C*, **50**, 2582–2601.

Zinner E. (2014) Presolar grains. In *Treatise on Geochemistry, 2nd Edition, Vol. 1: Meteorites and Cosmochemical Processes*, Davis A. M., editor, pp. 181–213, Elsevier, Oxford.

6 Meteorites, Interplanetary Dust, and Lunar Samples

Overview

Most meteorites (~86%) are *chondrites*, which are primitive rocks that have elemental compositions similar to that of the Sun. They are named for the millimeter-sized droplets of quenched silicate melt, called *chondrules*, that are particularly abundant in these meteorites. Chondrites come from the asteroid belt. They are physical mixtures of accreted components whose characteristics and mineralogy we will describe. The classification of chondrites utilizes a combination of primary features (bulk chemical composition, or oxygen isotopes) and secondary features (produced by thermal metamorphism or aqueous alteration). About 14% of meteorites arriving on Earth consist of differentiated materials. Their compositions have been affected by melting and the crystallization of melts, and they include various kinds of achondritic meteorites, irons, and stony irons from the asteroid belt, along with lunar samples and martian meteorites. Achondrites, so named because they do not contain chondrules, are divided into two types. Primitive achondrites are the solid residues left behind after extraction of partial melts from chondritic material. Magmatic achondrites crystallized from the extracted melts or from completely melted and differentiated chondritic material. Irons and stony irons formed by separation of molten metal/sulfide from silicates. Lunar rocks include cumulates from the ancient highlands and mare basalts that flooded huge craters. Martian meteorites include basalts and cumulates formed from basaltic magmas. *Interplanetary dust particles* (IDPs) are another class of extraterrestrial materials. They come in two morphological varieties: smooth IDPs, which are similar to aqueously altered carbonaceous chondrites and represent asteroidal dust, and porous IDPs, which consist of unaltered minerals and glass and are likely derived from comets.

6.1 Primitive versus Differentiated

Cosmochemistry focuses on the study of the chemical compositions of various solar system materials. Chondrites are the most abundant primitive samples. They are essentially sedimentary rocks composed of mechanical mixtures of materials with different origins (chondrules, refractory inclusions, metal and sulfide grains, matrix), which we will refer to as "components." Chondrites formed by the accretion of solid particles within the solar nebula or onto the surfaces of growing planetesimals. They are very old (>4.5 billion years, as measured by radioactive chronometers) and contain some of the earliest formed solids in the solar system. Chondrites have bulk chemical compositions very similar to the solar photosphere, except that they lack the gaseous elements and are often depleted in volatile and moderately volatile elements. CI chondrites (see Section 6.3) are not as depleted in volatile and moderately volatile elements and are representative of the bulk solar system composition (see Chapter 4).

Chondrites have never been melted, although some of their components melted and solidified before being accreted. Where melting occurs in nature, it is commonly incomplete. Partial melting selectively affects minerals that liquefy at lower temperatures, so the melt fraction and the solid residue have complementary chemical compositions. Melts ("magmas" in geologic parlance) tend to be mobile, and segregation of a melt from its residue occurs readily once a critical amount of melt is present. Consequently, the solid residue and the rocks formed by crystallization of the magma have different chemical compositions, a result described as *differentiation*. Other changes in chemical composition can occur during crystallization of the magma. Early-formed crystals may segregate because of density differences between crystals and magma or because they become concentrated in certain regions by convecting melt. Separation of crystals from the melt is called *fractional crystallization*. Because the early crystals and the melt have different compositions, this process also leads to differentiation.

Some meteorites, and all planetary samples, have undergone melting and differentiation at some stage. Hence, the compositions of differentiated materials do not resemble solar system abundances. These samples can, however, tell us about various geochemical processes within asteroids and planets.

We will now describe each of the various kinds of meteoritic samples available for cosmochemical investigation, progressing from primitive materials (chondrites and IDPs) to samples from differentiated bodies. Presolar grains extracted from meteorites have already been described in Chapter 5.

6.2 Components of Chondrites

Chondrites are physical aggregates of various components, most of which formed in the nebula. Before describing the chondrites themselves, let's take a look at their components.

6.2.1 Chondrules

Chondrules, the predominant constituent of most chondrites, were first named by the German mineralogist Gustav Rose in 1864, although references in the literature to "curious globules" in meteorites appeared as early as 1802. These droplets were formed by melting at temperatures of 1770–2120 K, and they subsequently experienced rapid cooling and solidification in minutes to hours. The heating mechanism for chondrules remains a contentious subject (Connolly and Jones, 2016). The conventional view is that clumps of nebular dust were flash melted, possibly by the passage of shock waves through the protoplanetary disk (Desch et al., 2012). An alternative view is that sprays of droplets were released from low-velocity impacts into already molten planetesimals (Asphaug et al., 2011). Chondrule cooling rates are fairly well constrained by laboratory experiments that reproduce the distinctive igneous textures of chondrules. Not all chondrules were completely melted, and relict grains of olivine can sometimes be recognized in them. The relict grains appear to be igneous crystals themselves, suggesting that chondrules may have been repeatedly cycled through the chondrule furnace. Many chondrules have abraded surfaces and some have rims of melted material.

Chondrules exhibit a bewildering variety of compositions and textures (Fig. 6.1a,b). Most are composed primarily of olivine and/or pyroxene, commonly with some glass. (For a crash course in mineralogy, see Box 6.1.) If melt solidifies so quickly that its atoms cannot organize into crystalline minerals, it quenches into glass. Iron–nickel metal and iron sulfide occur in many chondrules, often clustered near the peripheries. The textures of most chondrules are porphyritic, meaning that they contain large crystals set in a finer-grained or glassy background ("mesostasis"). Porphyritic chondrules are sometimes subdivided into type I chondrules, which contain MgO-rich olivine and pyroxene, and type II chondrules, which have FeO-rich silicates. Barred olivine chondrules consist of parallel plates of olivine with interstitial glass, sometimes armored by a surface layer of olivine. Excentroradial chondrules contain thin pyroxene blades that radiate from a point of nucleation. Other chondrules are aluminum rich and contain plagioclase, spinel, and sometimes silica.

The chemical compositions of individual chondrules have been determined by neutron activation of extracted samples or by electron microprobe analyses of chondrules *in situ*. Some (but not all) chondrules are depleted in moderately volatile elements. There is a compositional continuum between the olivine-rich and aluminum-rich chondrules. Original concentrations of the short-lived radionuclide ^{26}Al in chondrules suggest they formed very early, before all of this isotope decayed, but as much as 2–5 million years after the formation of CAIs (see Section 6.2.2).

Chondrules are volumetrically abundant in most chondrites, comprising up to 70% of some chondrite groups. However, chondrule abundance varies among different groups, and one kind of chondrite, CI, contains no chondrules at all. Occasionally, compound chondrules consisting of several objects stuck together are observed. The chondrules must have been hot when they touched, implying that the number densities of melt droplets in the regions where chondrules formed were high enough that molten droplets encountered each other.

6.2.2 Refractory Inclusions

Two types of refractory-element-rich inclusions are found in chondrites: *calcium-aluminum-rich inclusions* (CAIs, Fig. 6.1c) and *amoeboid olivine aggregates* (AOAs, Fig. 6.1d). CAIs were first described in the newly fallen Allende chondrite in 1970, and AOAs were first described in 1976, also in Allende. CAIs elicited considerable excitement when they were first discovered because the minerals that comprise them were the very phases predicted to condense from a cooling gas of solar composition. In the early 1970s, the paradigm for the early solar system was that all solids evaporated during the

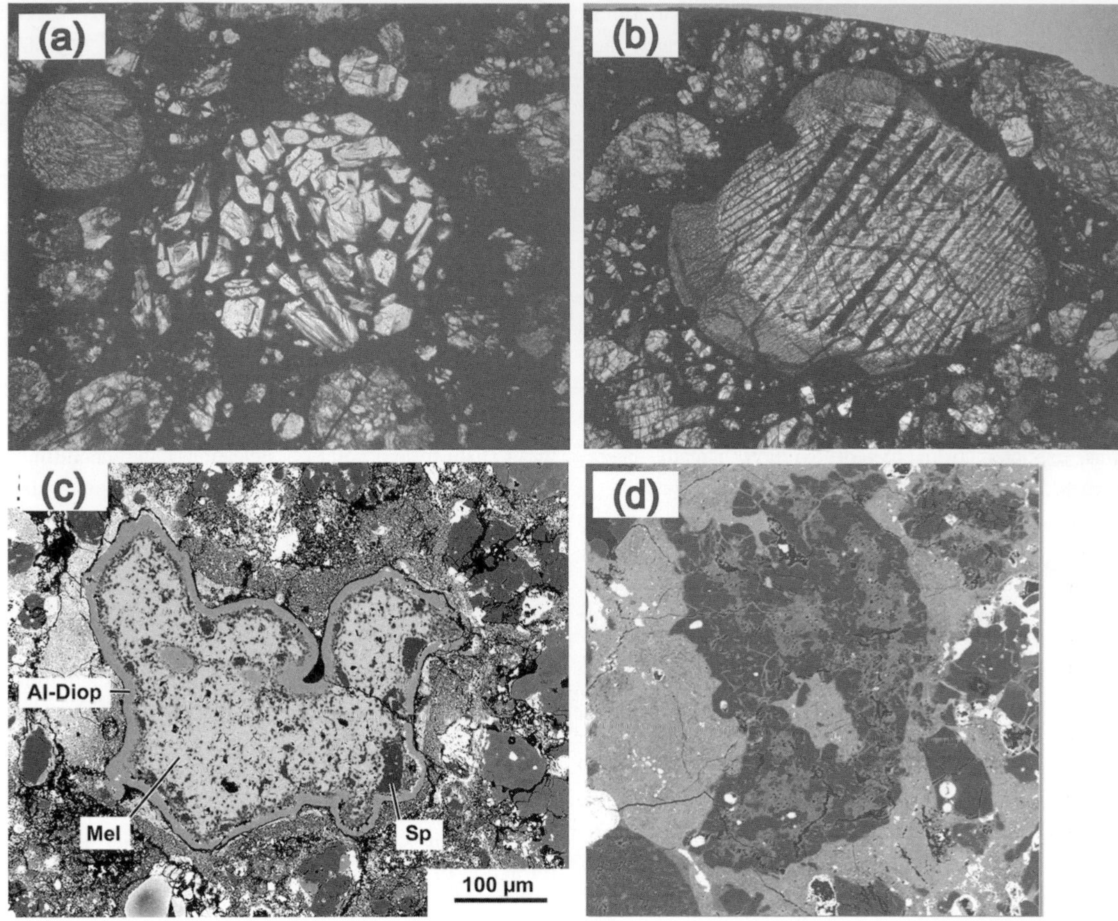

Figure 6.1 Components of chondrites. (a) Photomicrograph of a porphyritic chondrule showing olivine crystals in glassy mesostasis (dark), in the Clovis ordinary chondrite, field of view (FOV) ~1.8 mm. (b) Photomicrograph of a barred chondrule, containing oriented plates of olivine in glass (dark), in the Bishunpur ordinary chondrite, FOV ~2.7 mm. (c) Backscattered electron image of a CAI in the MAC 87300 carbonaceous chondrite; mineral abbreviations are Mel (melilite), Sp (spinel), and Al-Diop (aluminum-rich diopside). (d) Backscattered electron image of an AOA in the Y-81020 carbonaceous chondrite, containing olivine (dark gray) and clots of spinel-anorthite-diopside (lighter gray), FOV = 300 μm. (a) and (b) reprinted from Lauretta et al. (2006), with permission; (c) reprinted from Russell et al. (2000), with permission; (d) image courtesy of L. Chizmadia.

gravitational collapse of the Sun's parent molecular cloud and that the solids we now observe condensed from the cooling gas. Although the "hot nebula" paradigm has been supplanted by models involving local evaporation and condensation, the refractory mineralogy of CAIs makes them unique among cosmochemical samples (Krot, 2019).

CAIs are composed of a variety of minerals, primarily hibonite, perovskite, melilite, spinel, aluminum- and titanium-rich diopside, anorthite, forsterite, and occasionally corundum or grossite. They also show significant enrichments in refractory trace elements. CAIs exhibit a host of isotopic anomalies inherited from incorporated presolar grains or from the early nebula itself.

Evidence for the former existence of the short-lived radionuclide ^{26}Al was first documented by analyzing CAIs. The high abundance of ^{26}Al, as well as ages determined from long-lived isotope chronometers, reveal that CAIs are the most ancient components of chondrites (excluding presolar grains). In fact, they appear to be the oldest surviving objects that formed within the solar system.

About a third of CAIs are thought to have condensed directly from the gas phase. These have irregular shapes, fine-grained, fluffy textures, and mineralogies that match

Box 6.1 A Brief Review of Mineralogy

The documented mineralogy of meteorites is quite varied; at least 435 different meteoritic minerals have been reported (tabulated by Rubin and Ma, 2017, and further described in a series of papers by Hazen and Morrison, 2020). In the list below, we focus only on those minerals that are especially important, either because they are volumetrically abundant or because they provide important constraints on meteorite formation.

By definition, a *mineral* must possess a unique crystal structure (a regular internal arrangement of atoms or ions) and a specific chemical composition or limited range of compositions. Minerals having a range of compositions are called "solid solutions," and allowed chemical substitutions are shown in chemical formulae by elements in parentheses. For example, olivine [$(Mg,Fe)_2SiO_4$] represents solid solution compositions between the end members forsterite [Mg_2SiO_4] and fayalite [Fe_2SiO_4]. Different minerals having the same chemical composition but distinct crystal structures are called "polymorphs."

Although the crystallography of minerals is an inherent part of their definition, here we will focus on the chemistry of minerals, given the subject of this book. Readers interested in crystallography are referred to any number of texts in mineralogy. Materials without ordered crystal structures are either glass (quenched from high temperatures so quickly that crystals did not form), amorphous (poorly crystalline phases usually formed at low temperatures where crystallization is sluggish), metamict (formerly crystalline minerals whose atomic structures have been destroyed by irradiation), or diaplectic (formerly crystalline minerals whose atomic structures have been randomized by impact shock).

Minerals can be subdivided into the following groups:

Native elements – elements not combined with any other elements

Silicates – SiO_2, or compounds of various cations with various complex ions of silicon and oxygen. Cations can combine with individual SiO_4^{4-} tetrahedra, chains of SiO_3^{2-}, and sheets of $Si_4O_{10}^{4-}$. Silicates are the most abundant minerals in chondrites, IDPs, and most types of differentiated material.

Oxides and hydroxides – compounds of cations (most commonly Fe) with O^{2-} or OH–

Sulfides and phosphides – compounds of cations (usually Fe) with S^{2-} or P^{6-}, respectively

Carbonates – compounds of cations (usually Ca, Mg, or Fe) with CO_3^{2-}

Sulfates – compounds of cations (usually Ca or Mg) with SO_4^{2-}

Phosphates – compounds of cations (usually Ca) with PO_4^{3-}

Halides – compounds of cations (usually Na or K) with Cl^-, Br^-, or F^-

The following minerals are important in meteorites and other extraterrestrial samples. Some important classes of minerals in terrestrial rocks are not observed or are extremely rare in meteorites or planetary samples (e.g. micas, amphiboles) and will not be considered.

Native Elements

Kamacite – Fe alloy with minor Ni

Taenite – Fe alloy with 24–77% Ni

Platinum group alloy – various alloys of Pt, Ir, Os, Ru, Rh with Fe

Graphite, diamond – C polymorphs

Silicates

SiO_2 – tridymite is the most common polymorph in meteorites; quartz is most common on Earth

Zircon – $ZrSiO_4$

Olivine – solid solution between Mg_2SiO_4 (forsterite) and Fe_2SiO_4 (fayalite)

Pyroxenes – can be divided into low-calcium pyroxenes and high-calcium pyroxenes. Low-calcium pyroxenes make up a solid solution between $Mg_2Si_2O_6$ (enstatite) and $Fe_2Si_2O_6$ (ferrosilite) and include minerals such as bronzite and hypersthene. Low-calcium pyroxenes can be orthorhombic or monoclinic. Low-calcium pyroxene in type 3 chondrites is predominantly monoclinic, but becomes orthorhombic in higher petrologic types. High-calcium pyroxenes are monoclinic and make up a solid solution between $CaMgSi_2O_6$ (diopside) and $CaFeSi_2O_6$ (hedenbergite), including $Ca(Mg,Fe)Si_2O_6$ (augite). Extensive substitution of aluminum and titanium into high-calcium pyroxenes produces aluminous diopside and a titanium-rich pyroxene called grossmanite (formerly

Box 6.1 (cont.)

known as fassaite). Pigeonite is intermediate in composition between clinoenstatite and diopside and is monoclinic.

Melilite – solid solution between $Ca_2Al_2SiO_7$ (gehlenite) and $Ca_2MgSi_2O_7$ (akermanite)

Plagioclase – feldspar solid solution between $CaAl_2Si_2O_8$ (anorthite) and $NaAlSi_3O_8$ (albite). Maskelynite is a diaplectic glass of plagioclase composition.

Orthoclase – $KAlSi_3O_8$ feldspar

Nepheline – $NaAlSiO_4$ feldspathoid

Serpentine – $(Mg,Fe)_6Si_4O_{10}(OH)_8$ phyllosilicate. Although Fe-rich varieties are uncommon on Earth, they are abundant in some meteorites.

Phyllosilicates (clay minerals) – layered silicates such as $Al_4Si_4O_{10}(OH)_8$ (kaolinite) and $(1/2Ca,Na)_{0.7}(Al,Mg,Fe)_4(Si,Al)_8O_{20}(OH)_4.nH_2O$ (montmorillonite, also called smectite)

Oxides

Magnetite – Fe_3O_4

Spinel – $MgAl_2O_4$

Chromite – $(Mg,Fe)Cr_2O_4$

Ilmenite – $FeTiO_3$

Hematite – Fe_2O_3

Perovskite – $CaTiO_3$

Corundum – Al_2O_3

Grossite – $CaAl_4O_7$

Hibonite – $CaAl_{12}O_{19}$

Sulfides, Phosphides, Carbides

Troilite – FeS (stoichiometric)

Pyrrhotite – $Fe_{1-X}S$ (nonstoichiometric)

Pyrite – FeS_2

Pentlandite – $(Fe,Ni)_9S_8$

Oldhamite – CaS

Alabandite – (Mn,Fe)S

Niningerite – MgS

Keilite – (Fe,Mg)S

Schreibersite – $(Fe,Ni)_3P$

Cohenite – Fe_3C

Moisanite – SiC (artificial SiC is called carborundum)

Carbonates, Sulfates, Phosphates, Halides

Calcite – $CaCO_3$

Magnesite – $MgCO_3$

Dolomite – $CaMg(CO_3)_2$

Siderite – $FeCO_3$

Breunnerite – $(MgFe)CO_3$

Gypsum – $CaSO_4.2H_2O$

Anhydrite – CaSO

Kieserite – $MgSO_4.H_2O$ (one of many magnesium sulfates with varying H_2O)

Jarosite – $KFe_3(SO_4)_2(OH)_6$

Merrillite (whitlockite) – $Ca_3(PO_4)_2$

Apatite – $Ca_5(PO_4)_3(F,Cl,OH)$

Halite – NaCl

Sylvite – KCl

condensation calculations, as discussed in Chapter 7. Other CAIs are coarser-grained and more compact, and their textures show that they crystallized from melts. Many melted CAIs show evidence of evaporation of more-volatile elements during the melting event. The melted CAIs may be refractory residues formed by heating and partial evaporation of clumps of refractory dust. Many CAIs are surrounded by delicate rims composed of forsterite and other minerals that probably condensed in the nebula. CAIs can be subdivided into various types, based on their mineralogy.

CAIs are especially abundant in carbonaceous chondrites, but they occur in lesser abundance in other chondrite groups as well. Most types of refractory inclusions occur in all groups, but their relative proportions and sizes vary.

The "amoeboid" descriptor for AOAs refers to their irregular shapes. AOAs tend to be fine-grained and porous, and have comparable sizes to CAIs in the same meteorite. They consist mostly of forsterite and lesser amounts of iron–nickel metal, with a refractory component composed of anorthite, spinel, aluminum-rich diopside, and (rarely) melilite. The refractory component is sometimes recognizable as a CAI embedded within the AOA. The AOAs show no evidence of having been melted, but some contain CAIs that have melted.

AOAs, like CAIs, are depleted in volatile and moderately volatile elements, but to a lesser degree. AOAs are generally interpreted as aggregates of grains that condensed from nebula gas. Forsterite and metal are predicted to have condensed after the minerals that comprise CAIs, and their lower abundances of ^{26}Al suggest that AOAs formed after CAIs (typically by a few hundred thousand years). Mineralogical similarities between AOAs and the forsterite-rich rims on some CAIs may suggest that they formed contemporaneously.

6.2.3 Metals and Sulfides

Two different kinds of *metal* are found in chondrites. Small nuggets composed of highly refractory siderophile elements (iridium, osmium, ruthenium, molybdenum, tungsten, rhenium) occur within CAIs. Such refractory alloys have been predicted to condense at temperatures above 1600 K from a gas of solar composition. However, except for tungsten, they are also the expected residues of CAI oxidation.

More commonly, larger and much more abundant grains of metallic iron, or more properly iron–nickel–cobalt, occur within and as rims on chondrules and dispersed throughout the matrix. The metal grains outside chondrules may have originally formed within chondrules and been expelled. On slow cooling, the solidified metal separates into low-nickel (kamacite) and high-nickel (taenite) phases. Most metal compositions have been modified by later heating events, so it is difficult to find metal grains that retain their original compositions. However, primitive iron–nickel–cobalt grains found in one chondrite group (CH) have compositions consistent with condensation at about 1400 K.

Calculations predict that metallic iron should react with sulfur in nebular gas when temperatures drop below 650 K to produce iron sulfide FeS (troilite). Indeed, brassy troilite grains are commonly observed in association with metal in chondrites. However, sulfur is so readily mobilized during later heating that it is doubtful that troilite grains formed by nebular reactions have been preserved in their original form.

6.2.4 Matrix

Another chondrite component is an optically opaque (in thin section) assortment of very fine-grained minerals that fills the spaces between the larger chondrules, refractory inclusions, and metal grains. This material is called *matrix*. Characterization of matrix minerals is hampered by their tiny particle sizes (as small as 50–100 nm). Moreover, the fine grain sizes, high porosity, and permeability of matrix make it especially susceptible to alteration during later heating or exposure to aqueous fluids.

Matrix minerals are complex mixtures of silicates (especially olivine and pyroxene), oxides, sulfides, metal, and – in meteorites that have suffered aqueous alteration – phyllosilicates and carbonates. The bulk chemical composition of matrix is broadly chondritic and richer in volatile elements than the other chondrite components. Some chondrules have rims of adhering matrix that appear to have been accreted onto them prior to final assembly of the meteorite. Small lumps of matrix also occur in many chondrites. Presolar grains, described in Chapter 5, occur in matrix.

The origin of matrix materials is uncertain. Matrices are probably complex mixtures of presolar grains, nebular condensates formed at various temperatures, and disaggregated chondrules and refractory inclusions. Matrix material has almost always been altered to varying degrees after accretion of the meteorite parent bodies. In thermally metamorphosed chondrites, the original matrix has been recrystallized.

6.3 Chondrite Classification

Chondrules, refractory inclusions, metal, troilite, and matrix have been assembled in varying proportions to

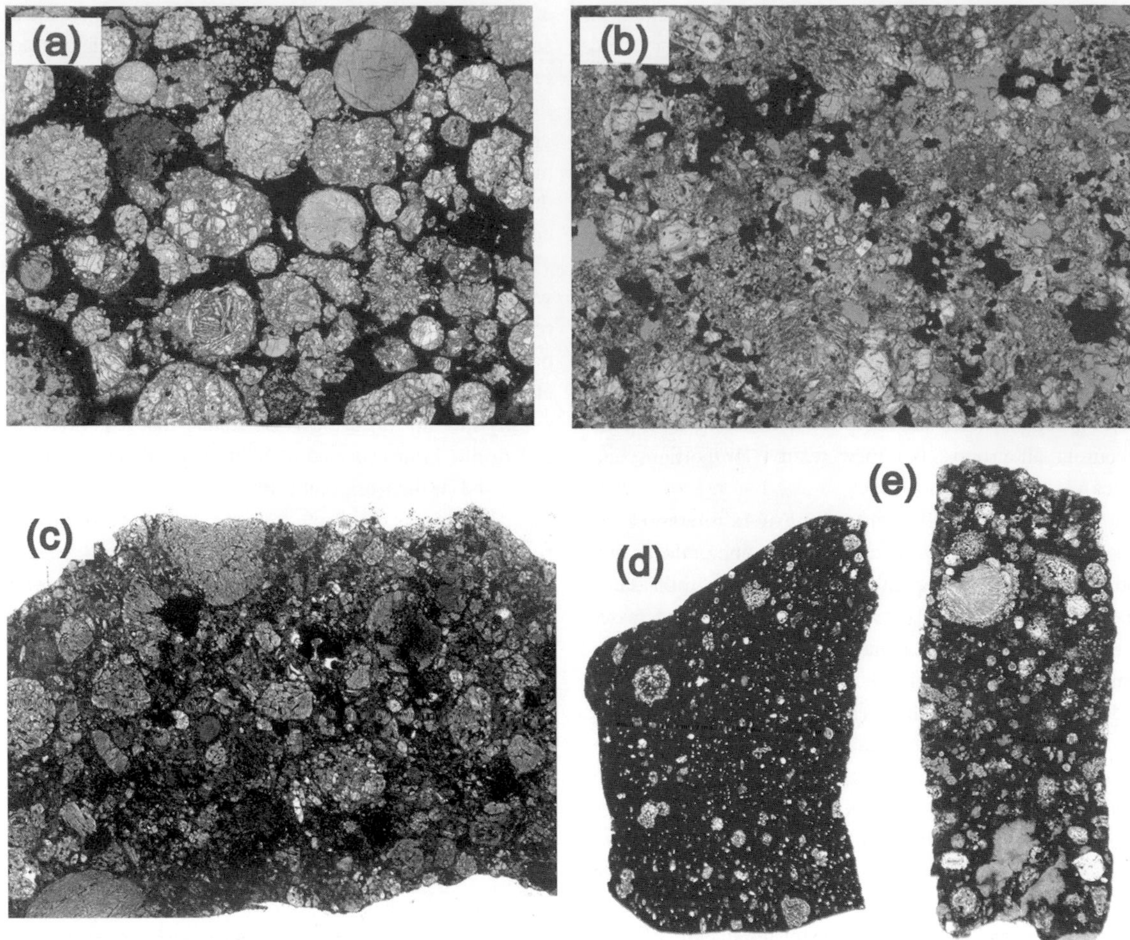

Figure 6.2 Photomicrographs of chondrites, all in plane-polarized light. (a) Unmetamorphosed Semarkona LL3 ordinary chondrite, containing abundant chondrules and dark matrix, FOV = 5.4 mm. (b) Metamorphosed Mt. Tazerzait L5 ordinary chondrite, with original chondrule outlines blurred because of recrystallization, black patches are metal and troilite, FOV = 5.4 mm. (c) KLE 98300 EH3 enstatite chondrite, showing chondrules in dark matrix, FOV = 5.4 mm. (d) Murchison CM2 carbonaceous chondrite, in which dark matrix is particularly abundant, FOV = 1 cm. (e) Allende CV3 carbonaceous chondrite, with CAIs (large white objects) and chondrules in dark matrix, FOV = 1 cm. All images from Lauretta and Killgore (2005), with permission.

form chondrites. Some microscopic images of chondrite thin sections are illustrated in Figure 6.2.

A serviceable classification scheme for chondrites has taken many decades to develop, because earlier schemes confused primary and secondary properties. A primary property is one resulting from the chondrite's original formation as an accreted rock, and a secondary property is one acquired at some later time by another process. The bulk chemical composition of a chondrite is a primary property, as are the relative abundances of chondrules, CAIs, metal, and matrix. Most chondrites have suffered some kind of alteration after accretion – either thermal metamorphism (changes brought on by heating) or aqueous alteration (changes resulting from interaction with fluids, produced by melting of ice that was accreted along with the rock). Recrystallization occurs during thermal metamorphism, blurring original chondrite textures (compare Figs. 6.2a and 6.2b) and equilibrating the mineral compositions. Aqueous alteration produces pervasive changes in mineralogy, forming hydrated minerals (phyllosilicates) from olivine and pyroxene, and precipitating salts (carbonates, sulfates, and halides). Many chondrites have also experienced shock metamorphism (changes wrought by the high pressures achieved during impacts). Shock produces new minerals and causes fracturing or melting of target rocks. Thermal metamorphism, aqueous alteration, and shock metamorphism are examples of secondary processes. The current chondrite classification scheme utilizes both primary and secondary properties and is based on a taxonomy originally formulated by Van Schmus and Wood (1967).

6.3.1 Primary Characteristics: Chemical Compositions

Using bulk chemical compositions, chondrites are divided into clans and subdivided into groups, identified by a letter or combination of letters. The *ordinary* (O) *chondrite* clan, so called because these chondrites are the most abundant types of meteorites, is separated into the H, L, and LL groups. H chondrites have the highest concentration of iron (hence the H), occurring mostly as metal, and L chondrites have lower iron. The LL group was subsequently recognized to have an even lower abundance of iron, occurring mostly as ferrous iron in silicates. Since the group name L was already taken, an extra L was added. Ordinary chondrites are shown in Figures 6.2a and 6.2b.

The *enstatite* (E) *chondrites*, another clan, are strongly reduced, with virtually all the iron in them occurring as metal. Unusual sulfide minerals result from changes in the behavior of some elements, like calcium and manganese, from lithophile to siderophile or chalcophile, at these low oxidation states. Two groups, EH and EL, are distinguished based on their high (H) and low (L) iron metal contents. An EH3 chondrite is shown in Figure 6.2c.

The *carbonaceous* (C) *chondrites* are traditionally considered to be a large clan with many chemical groups: CI, CM, CR, CB, CO, CV, CK, and CH. The second letter refers to a type specimen for each meteorite group: Ivuna (I), Mighei (M), Renazzo (R), Bencubbin (B), Ornans (O), Vigarano (V), Karoonda (K), and ALH 85085 (H; this last one violates the first-letter convention – the H refers to its high iron concentration and metal abundance). These groups reflect a wide variety of compositions, oxidation states, and petrography. In fact, the carbonaceous chondrites range from a group with essentially unfractionated elemental abundances, a high proportion of presolar grains, and no chondrules (CI chondrites) to other groups with compositions highly depleted in volatile elements and enriched in refractory elements, with moderate numbers of chondrules and low proportions of presolar grains (CV and CO chondrites). Unlike ordinary and enstatite chondrites, the carbonaceous chondrites cannot be thought of as a clan consisting of closely related meteorites. In the following discussions, we will refer to individual chondrite groups rather than the carbonaceous chondrite clan as a whole. A CM and a CV chondrite are shown in Figures 6.2d and 6.2e, respectively.

The *Rumuruti* (R) group is highly oxidized and has a high abundance of matrix (~50%), but is otherwise similar in many respects to ordinary chondrites.

These chemical groups constitute one (vertical) axis of a diagram defining the Van Schmus and Wood classification system, as illustrated in Figure 6.3. The other (horizontal) axis is based on secondary characteristics.

6.3.2 Secondary Characteristics: Petrologic Types

Chondrites are divided into various *petrologic types* based on the degree to which they have been thermally metamorphosed or aqueously altered. In the original Van Schmus and Wood (1967) classification, CI chondrites were considered to be the most primitive meteorites, because their compositions provide the best match to

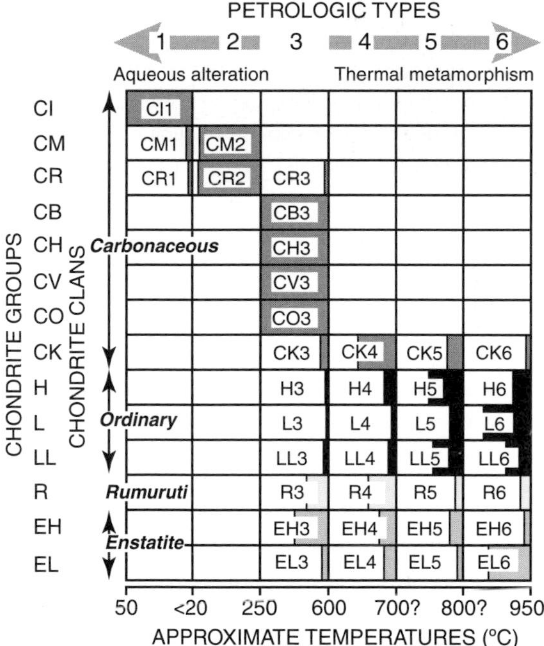

Figure 6.3 Classification system for chondrites, adapted from Van Schmus and Wood (1967). A meteorite is classified by identifying its chemical group and petrologic type. The chondrite groups are collected into related clans, shown by the two-headed arrows. Approximate temperatures for metamorphism (petrologic types 4–6) or aqueous alteration (petrologic types 1–2) are shown at the bottom. Shaded areas illustrate approximate relative proportions of each petrologic type in the world's meteorite collections.

solar elemental abundances. They were classified as petrologic type 1, some other carbonaceous chondrites were type 2, relatively unmetamorphosed ordinary and enstatite chondrites were type 3, and petrologic types 4, 5, and 6 indicated progressively higher degrees of thermal metamorphism. A few ordinary chondrites of type 7 have been recognized, but this type is now generally attributed to shock metamorphism. Van Schmus and Wood (1967) provided mineralogical and textural criteria to recognize each petrologic type.

Although this serviceable chondrite taxonomy remains in use, its original interpretation was partly flawed. As currently used, petrologic types 2 and 1 are reinterpreted to reflect increasing degrees of aqueous alteration (McSween, 1979), while types 3–6 continue to be interpreted as increasing degrees of thermal metamorphism (arrows in Fig. 6.3). Thus, the most primitive chondrites are actually petrologic type 3, at least from the perspective of secondary characteristics. Carbonaceous chondrites of types 1 and 2 have altered matrix material composed primarily of serpentine, a clear indication of aqueous alteration. Many of these meteorites also contain veins of sulfate and carbonate that precipitated from aqueous solutions. The iron in most type 1 and 2 chondrites is completely oxidized; much of it is present as magnetite. CR chondrites, which contain abundant metal, are an exception. Much of the oxidation in carbonaceous chondrites may also result from aqueous alteration processes that converted metal to magnetite. The CI chondrites, although type 1 and highly aqueously altered, remain the most primitive meteorites in terms of solar system abundances, so the alteration that affected them is inferred to have occurred with little or no chemical change.

Sears et al. (1980) subdivided type 3 chondrites into types 3.0–3.9, based on their thermoluminescence (TL) characteristics. TL sensitivity is controlled by the degree of crystallization of chondrule mesostasis (the very fine-grained minerals or quenched glass between the larger crystals). Chondrules in type 3.0 chondrites have a glassy mesostasis, which progressively crystallizes into feldspar during mild metamorphism. Subtypes 3.0–3.9 can also be distinguished from a plot of CaO versus FeO concentrations in olivine chondrules (Scott et al., 1994). Detailed studies have revealed subtle changes in the compositions of minerals (Grossman and Brearley, 2005) and the structure of organic matter in type 3.0–3.9 chondrites (Bonal et al., 2006). These studies have resulted in further subdivision of the type 3.0–3.2 chondrites. The subdivision of type 3 chondrites has proven very useful in identifying the most primitive chondritic materials (sometimes called *unequilibrated chondrites* because their minerals have not been heated sufficiently for equilibration to occur). A recent compilation of the diagnostic characteristics of petrologic types is summarized in Table 6.1.

The type 3 CV chondrites show a diversity of secondary characteristics that have resulted in their subdivision into two subgroups, the oxidized CVs (CV_{ox}) and the reduced CVs (CV_{red}), primarily on the basis of metal content (McSween, 1977). The oxidized subgroup has been subsequently divided into the Allende-like (CV_{oxA}) and Bali-like (CV_{oxB}) subgroups (Krot et al., 1998). The CV_{oxB} chondrites experienced aqueous alteration and contain abundant phyllosilicates, magnetite, fayalite, and other alteration products. The matrices of CV_{oxA} chondrites consist largely of lath-shaped, Fe-rich olivine, Ca, Fe pyroxene, and nepheline. These meteorites may have been reheated and dehydrated after experiencing aqueous alteration. The reduced CV3 chondrites experienced only very little aqueous alteration. Subtyping of CV chondrites based on thermal metamorphism has been provided by Raman spectroscopy (Bonal et al., 2006).

6.3.3 Chondrite Taxonomy

As seen in Figure 6.3, the taxonomy used for chondrites is a combination of chemical group and petrologic type, as in H4 or CM2. This system avoids the problems of previous chondrite classifications by cleanly separating primary and secondary characteristics. The placement of unaltered (type 3) chondrites near the middle of the numerical sequence is an historical artifact, as the original Van Schmus and Wood (1967) classification scheme did not recognize that the intensity of aqueous alteration proceeds to lower numbers.

Temperatures corresponding to the various petrologic types are difficult to determine. At high temperatures, they can be roughly constrained using "geothermometers" – coexisting mineral pairs that exchange chemical components by reactions that are dependent on temperature. At lower temperatures, estimates come from nonreversible transitions in the structure of compounds or from the destruction of certain phases. Estimates of the temperatures corresponding to the various petrologic types are shown at the bottom of Figure 6.3. Also illustrated in this figure are the relative proportions of meteorites in each group that fall into the various petrologic types (indicated by the proportion of each box that is shaded). From this diagram and the relative abundances of the different classes of chondrites, it should be obvious that most chondrites have suffered high degrees of thermal metamorphism or aqueous alteration. This makes the least altered (type 3.0–3.1) chondrites especially

Table 6.1 Criteria for classifying chondrites according to petrologic type (Weisberg et al., 2006)

Criterion	1	2	3	4	5	6	7
Homogeneity of olivine compositions		>5 % mean deviation		≤5 %	Homogeneous		
Structural state of low-Ca pyroxene		Predominantly monoclinic		>20 % monoclinic	≤20 % monoclinic	Orthorhombic	
Feldspar		Minor primary grains		Secondary <2 μm grains	Secondary 2–50 μm grains	Secondary >50 μm grains	
Chondrule glass	Altered or absent	Mostly altered, some preserved	Clear, isotropic	Devitrified	Absent		
Metal: maximum Ni (wt. %)		<20 taenite minor or absent	>20 kamacite and taenite in exsolution relationship				
Sulfides: mean Ni (wt. %)		>0.5	<0.5				
Matrix	Fine-grained opaque	Mostly fine-grained opaque	Opaque to transparent	Transparent, recrystallized			
Chondrule-matrix integration	No chondrules	Sharp chondrule boundaries		Some chondrules can be discerned, fewer sharp edges		Chondrules poorly delineated	Primary textures destroyed
Carbon (wt. %)	3–5	0.8–2.6	0.2–1	<0.2			
Water (wt. %)	18–22	2–16	0.3–3	<1.5			

important for cosmochemistry, although as noted earlier, CI chondrites (we typically do not write CI1 because all CI chondrites are type 1) provide the best match to solar composition.

This is also a very useful scheme for recognizing whether other characteristics are primary or secondary. For example, noble gases and carbon were driven out of ordinary chondrites during thermal metamorphism, as revealed by an inverse correlation of noble gas and carbon abundances with increasing petrologic type. In fact, a characteristic that historically defined carbonaceous chondrites was the relatively high carbon content, along with the opaque matrix. The vast majority of ordinary chondrites are metamorphosed and so do not contain significant carbon. However, type 3.0–3.5 ordinary chondrites are as carbon rich as many carbonaceous chondrites. Thus, the historical distinction between carbonaceous and ordinary chondrites is in part a reflection of metamorphism, a secondary property.

6.3.4 Other Classification Parameters: Shock and Weathering

Another important secondary process affecting many chondrites is shock metamorphism. Stöffler et al. (1991) quantified the shock effects observed in olivine and plagioclase, and used these for a shock classification of ordinary chondrites. Scott et al. (1992) extended this shock classification to carbonaceous chondrites, and

Rubin et al. (1997) developed a scheme for enstatite chondrites by including shock effects in orthopyroxene. The shock classification system for chondrites has been subsequently updated by Stöffler et al. (2018). The shock classification recognizes shock stages S1 to S6, representing increasing degrees of shock pressure, finally culminating in completely melted rocks (S7).

Many shocked chondrites are *breccias*, formed from fragments of rocks that have been cemented together to form new rocks. A chondritic breccia in which all the fragments are of the same chemical group (but usually different petrologic types) is called *genomict*. A breccia containing fragments of different meteorite groups is called *polymict*.

Terrestrial weathering has significantly altered the chemistry of some chondrites. Weathering leaches out certain mobile elements and introduces others, as well as oxidizing metals to form rust. Meteorites that are recovered after having been on Earth for some time ("finds") are commonly weathered, with effects ranging from mild to severe enough to make the meteorite hard to recognize. Terrestrial weathering completely destroys meteorites on timescales of a few thousand to tens of thousands of years (barring extreme circumstances of preservation). Even "falls" (meteorites observed to fall and recovered almost immediately), however, can experience some alteration during long and sometimes unclean storage in museums. A weathering classification scheme has been proposed by Wlotzka (1993) for ordinary chondrites from temperate climates. A different weathering classification is used for Antarctic meteorites.

6.4 Oxygen Isotopes in Chondrites

Because of the isotopic variability and the high solar system abundance of oxygen, oxygen isotopes are very useful for meteorite classification. Below the condensation temperature of silicates and above the condensation temperature of ices, approximately 25% of the oxygen in the solar nebula is predicted to have occurred in condensed solids, with the remainder in gaseous molecules. Chondrites provide samples of the condensed oxygen in the early solar system.

Each meteorite class has a unique oxygen composition. The origin of these unique compositions is not fully understood, but it is known to reflect two types of isotopic variations. *Mass-dependent isotope variations* result from physical and chemical processes that slightly favor one isotope over another due to the difference in mass. *Mass-independent variations* probably reflect mixing of components with isotopic compositions that originated either through nucleosynthesis or through processes that do not depend on mass. We will discuss these processes in detail in Chapter 8.

Figure 6.4 shows the measured whole-rock isotopic compositions for various chondrite groups on a three-isotope plot (Clayton, 2004), which we previously introduced in Chapter 4. This figure uses the delta (δ) notation, where measured isotopic ratios of ^{17}O and ^{18}O to ^{16}O are reported as deviations from a terrestrial standard. Terrestrial samples all lie along a single line with a slope of approximately 1/2 on an oxygen three-isotope plot. This reflects the larger shift of $\delta^{18}O$ relative to $\delta^{17}O$ due to the two-times-larger mass difference between ^{18}O–^{16}O compared to ^{17}O–^{16}O. Except for the enstatite chondrites, the different chrondrite groups plot in various positions off the terrestrial mass fractionation line, which indicates that they formed from different mixtures of oxygen components. The carbonaceous chondrites are arrayed along a mass-independent mixing line with a slope of approximately unity in Figure 6.4, and the ranges for individual carbonaceous chondrite groups tend to be significantly greater than for other chondrite groups. The large variation among CV chondrites has been shown to result from varying amounts of ^{16}O-rich refractory inclusions. Ordinary chondrite groups (H, L, LL) and R chondrites have small but clearly resolvable differences in oxygen isotopes.

The systematic variations in oxygen isotopes provide an independent means of classifying chondrites that generates the same groups as the bulk chemical compositions.

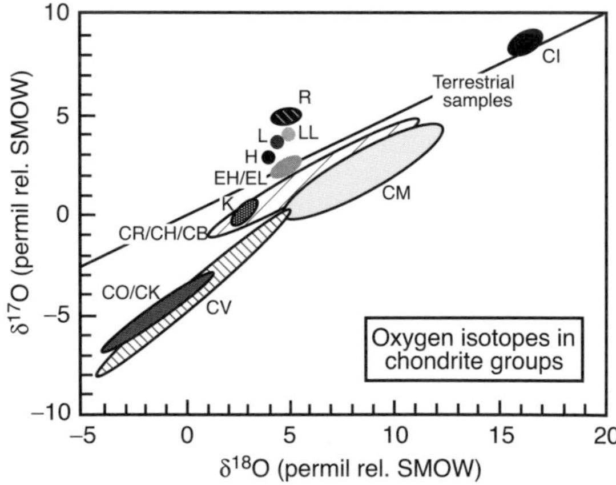

Figure 6.4 Oxygen isotopic compositions for bulk chondrites, after Clayton (2004). By convention, $^{17}O/^{16}O$ and $^{18}O/^{16}O$ ratios are plotted as deviations from the composition of standard mean ocean water (SMOW) in units of parts per thousand (permil). The δ values are calculated as follows: $\delta^{17}O = ((^{17}O/^{16}O)_{sample}/(^{17}O/^{16}O)_{SMOW}) - 1) \times 1000$, and similarly for $\delta^{18}O$.

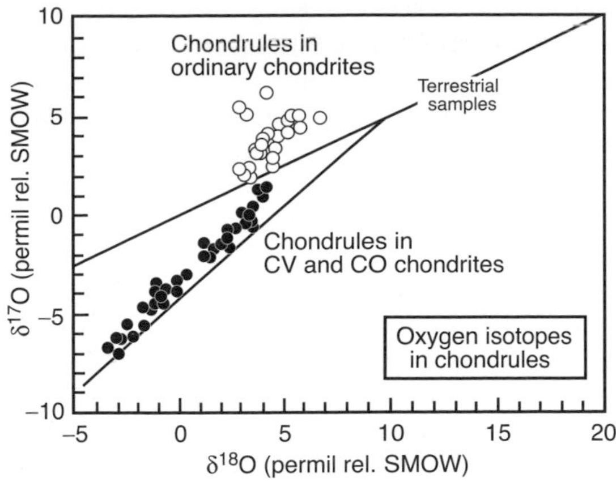

Figure 6.5 Oxygen isotopic compositions of individual chondrules in ordinary and carbonaceous chondrites, after Clayton (2004).

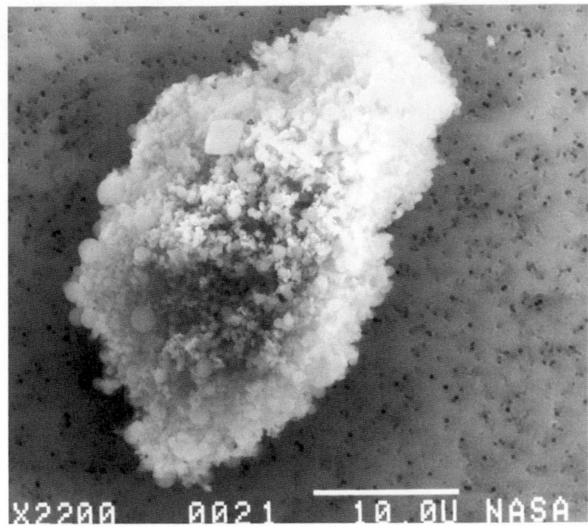

Figure 6.6 Electron microscope image of a chondritic porous IDP. Scale bar is 10 μm. Image courtesy of D. Brownlee.

The oxygen isotopes also work for classifying nonchondritic meteorites. Oxygen isotopic compositions are somewhat easier to obtain than detailed chemical data and so are often used to nail down a classification.

Because chondrites formed from components with diverse origins and histories, these components may have different oxygen isotopic compositions (unless they have been equilibrated during metamorphism). Figure 6.5 shows the compositions of individual chondrules from ordinary and carbonaceous chondrites (Clayton, 2004). These chondrules show mass-independent variations. Interestingly, the ranges of isotopic compositions of chondrules from H, L, and LL chondrites are indistinguishable, even though the bulk meteorite compositions are distinct. The CAIs in carbonaceous chondrites plot on the same slope-1 line as their chondrules, but with even more extreme isotopic compositions (refer back to Figure 4.8, where CAIs are compared to the Sun's oxygen isotopic composition). The various minerals in some individual CAIs may also have different oxygen isotopic compositions, all aligned on the three-isotope plot. The sequence, in terms of decreasing ^{16}O enrichment, is typically spinel, pyroxene, olivine, melilite, and anorthite. The linearity suggests mixing of two components.

6.5 Interplanetary Dust Particles

Approximately 40,000 tons of microscopic particles enter the Earth's atmosphere each year. These IDPs (Fig. 6.6) are derived from both asteroids and comets.

Tiny particles less than ~100 μm in diameter in the atmosphere are decelerated at high altitudes, and because of their large surface areas relative to their masses, they decelerate slowly enough that they are not heated to the point where they would melt. IDPs have been collected in the stratosphere (at altitudes of ~20 km), adhering to collection pads on the wings of U2 aircraft. They occur in sizes ranging from 1 to >50 μm, although most are 5–15 μm. The larger IDPs, which commonly fragment on collection, often consist of many grains and are referred to as "cluster" particles. Asteroidal dust dominates the larger size fractions, which are called *micrometeorites*. Larger particles from ~100 μm to ~2 mm in diameter melt during atmospheric entry (Folco and Cordier, 2015). These are mostly glassy or cryptocrystalline spheres, although partly melted micrometeorites tend to be angular.

There are two distinct morphological classes of IDPs: smooth and porous. Because both classes have approximately chondritic bulk compositions, they are referred to as *chondritic smooth* (CS) and *chondritic porous* (CP). CS IDPs have compact textures and are thought to be particles from asteroids. CP IDPs are fluffy (porosities commonly exceed 30%, and values as high as 70% have been measured) and some texturally resemble bunches of grapes. Their densities vary considerably but can be as low as 0.3 g/cm^2. It is not obvious how grains having such high porosities could have survived compaction within kilometer-sized parent bodies unless ices were present, filling the volume between grains.

Advances in techniques for handling and analyzing very small particles have allowed detailed examination and characterization of IDPs. Especially useful instruments include the transmission electron microscope, synchrotron facilities, and the ion microprobe. As implied by their names, the bulk compositions of chondritic IDPs, in aggregate, have roughly CI bulk compositions (Schramm et al., 1989). Enrichment in carbon and volatile elements suggest that IDPs are very primitive materials.

The CP and CS IDPs are mineralogically distinct (Rietmeijer, 1998; Jessberger et al., 2001; Bradley, 2013). CS IDPs (Fig. 6.7a) are characterized by phyllosilicates (saponite clay) and carbonates. Their mineralogical similarity to CI and CM chondrites suggests derivation from carbonaceous chondrite asteroids that experienced aqueous alteration. In contrast, CP IDPs (Fig. 6.7b) are anhydrous, consisting mostly of olivine, pyroxene, iron sulfides, FeNi metal, and glass. These phases, which occur as very small (generally submicron-sized) grains, are mixed in various proportions to form heterogeneous aggregates. Olivine and pyroxene are usually magnesium rich (forsterite and enstatite). Enstatite crystals have unusual platelet or whisker morphologies (Fig. 6.7c), consistent with growth from a vapor phase. A unique component in CP IDPs is referred to as "kool" grains because they consist of **ko**smochloric pyroxene (Na- and Cr-rich augite) and FeO-rich **ol**ivine, and sometimes sodic plagioclase and sulfides (Joswiak et al., 2009). Sulfides are mostly pyrrhotite, but other iron, nickel, and zinc sulfide minerals also occur. Some pyrrhotite has an unusual cubic

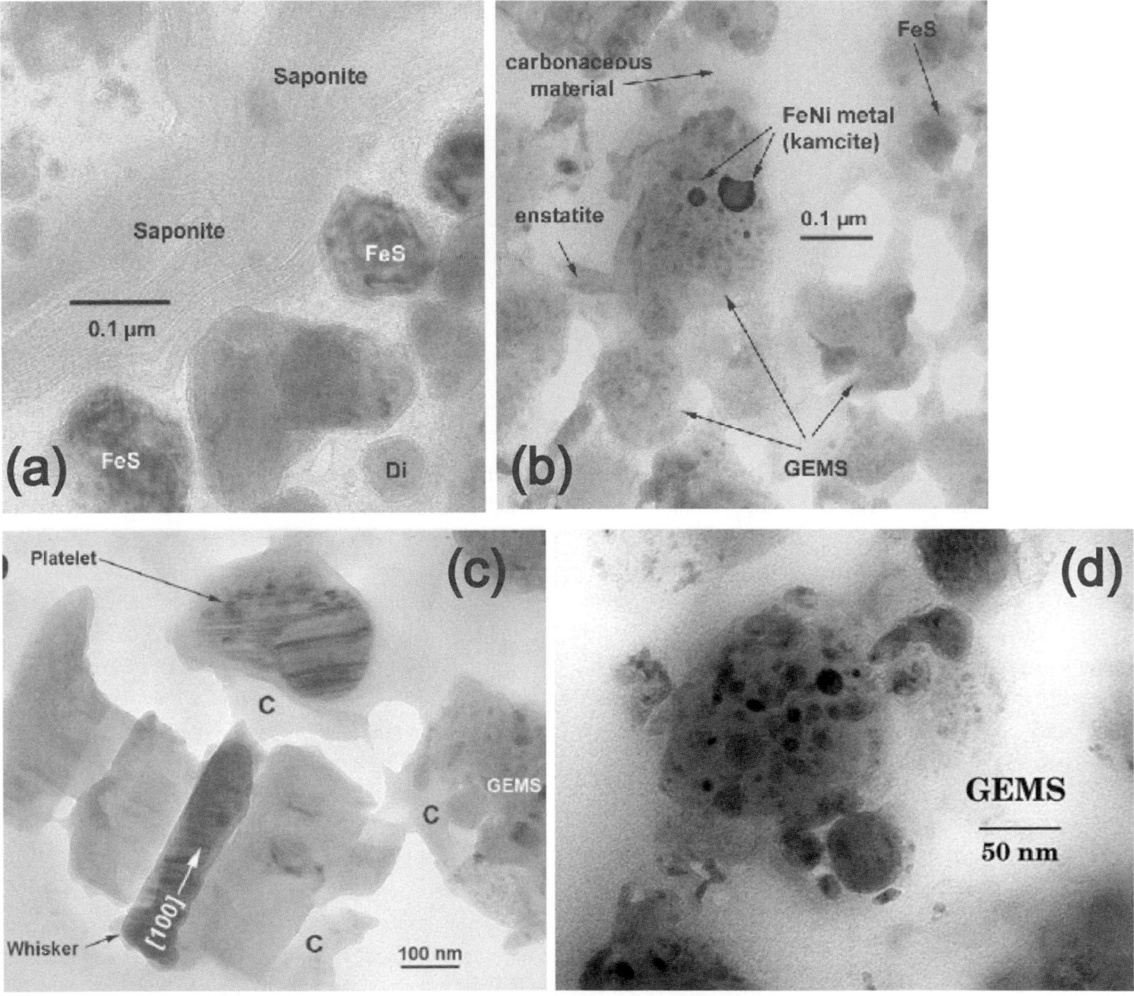

Figure 6.7 Transmission electron microscope images of IDPs. (a) Chondritic porous IDP, showing phyllosilicate (saponite), iron sulfide, and diopside (Di). (b) Chondritic porous IDP, containing pyroxene (enstatite), FeNi metal, iron sulfide, GEMS, and organic matter. (c) Whisker and platelet morphologies of enstatite. (d) GEMS, composed of rounded metal and sulfide particles embedded in glass. Images courtesy of J. P. Bradley.

crystal structure, unlike the more common hexagonal form. The cubic form, which has not been previously reported in nature, is metastable and transforms to hexagonal pyrrhotite when heated. A similar mineral has been produced in the laboratory by vapor-phase growth, suggesting that the IDP sulfides may have formed by condensation from a gas.

Among the most exotic materials in CP IDPs are particles of **g**lass with **e**mbedded **m**etal and **s**ulfide, identified by the acronym GEMS (Bradley, 1994; Ishii et al., 2018; Fig. 6.7d). Although GEMS are commonly described as having chondritic bulk compositions, that description applies only for aggregates of these submicron particles. The nanometer-sized FeNi metal (kamacite) inclusions and pyrrhotite are clearly out of equilibrium with the enclosing magnesium-rich silicate glass. The eroded textures of mineral inclusions in GEMS suggest that they were exposed to ionizing radiation, as would be expected for interstellar dust. Some (but not all) GEMS have exotic oxygen isotopic compositions. In terms of their physical properties and compositions, GEMS are similar to interstellar silicates inferred from astronomical observations, and they were briefly mentioned when we discussed interstellar grains in Chapter 5.

Carbonaceous material (Fig. 6.7b) is intimately mixed with silicates and is very abundant (carbon abundance averages 13% and varies up to ~50%) in CP IDPs. Some carbon is elemental (graphite), but C–H stretching resonances in infrared spectra show that aliphatic hydrocarbons are also present. Polycyclic aromatic hydrocarbons also occur. Presolar nanodiamonds have been identified in cluster IDPs, but not in smaller CP IDPs. Enormous D/H and $^{15}N/^{14}N$ anomalies have been measured in bulk IDPs, and the hydrogen isotopic anomalies are correlated with organic-rich domains. Ratios of D/H as high as 25 times the solar ratio suggest the presence of molecular cloud materials.

6.6 Nonchondritic Meteorites

Differentiated meteorites do not have chondritic compositions. *Achondrites* are stony meteorites that have undergone melting and differentiation. Their precursors are commonly assumed to have had chondritic compositions, although no chondrules survived the melting process. Transitional meteorites that share some characteristics with both chondrites and achondrites are called *primitive achondrites* (an unfortunately confusing term, but widely used in the meteoritics literature). Primitive achondrites are the solid residues left when partial melts of rocks have been extracted. If only a small amount of melting occurred, the composition of the residue may be difficult to distinguish from the precursor chondrite, hence the prefix "primitive." The magmatic achondrites (usually just called "achondrites," but we use the "magmatic" prefix to distinguish them) are igneous rocks, formed directly from melts – they are either solidified magmas or "cumulates" formed when crystals accumulate in magma chambers. Cumulate rocks no longer have the compositions of their parent melts.

Stony irons are nonchondritic meteorites that contain roughly equal proportions of silicate minerals and metal. Two types of stony irons – pallasites and mesosiderites – are distinguished. Pallasites consist of approximately equal amounts of metal and olivine (one small group contains pyroxene as well). Mesosiderites also have approximately equal proportions of metal and silicate, but the silicate fraction is basalt.

Irons are nonchondritic meteorites that are predominantly metal. Iron meteorites formed by melting of, most likely, chondritic material and segregation of metal melt from silicate. Many apparently represent asteroidal cores, although some may have formed as dispersed metal pockets in the parent asteroids.

Partial melting of chondritic materials can produce a variety of melt compositions. With increasing temperature, the first melt to form is a eutectic metal-sulfide liquid. Eutectic melts are those that form at the lowest temperatures. Segregation of dense metallic liquids produced iron and stony-iron meteorites. With further heating of chondrites not significantly depleted in volatile elements, such as carbonaceous and ordinary chondrites, the first silicate minerals will be rich in silica and alkalis, having "andesitic" compositions (Collinet and Grove, 2020a). Continued partial melting produces silicate magmas richer in magnesium, calcium, and aluminum. These magmas are termed "basaltic," as they are similar in composition to the common basalts produced by partial melting of the Earth's mantle rock.

The behavior of elements during partial melting and crystallization of the resulting magma can vary greatly, depending on how the elements are partitioned between crystals and coexisting liquid. The size and charge of an ion determine whether it will fit easily into sites in common mineral structures. For example, Ni^{2+} is similar in size and has the same charge as Mg^{2+}, so it substitutes readily into magnesium-rich olivine. Nickel is said to be "compatible." On the other hand, U^{3+} is a large ion with high charge, and most common minerals exclude it. Only minerals containing other large, high-charge cations as major elements can easily incorporate "incompatible" uranium. Such minerals usually crystallize late after most of the liquid has solidified, causing incompatible element concentrations to increase in the residual liquid. Compatibility also determines how elements partition during partial melting, as incompatible elements prefer the melt phase and thus are concentrated in magmas. Partitioning of elements between crystals and magma is important for understanding differentiated meteorites; it will be described more rigorously in Chapter 7.

Possible heat sources for melting in meteorite parent bodies are discussed in Chapter 12. Differentiated meteorites generally have old radiometric ages (Chapter 10), which indicate that igneous activity began and ended soon after their parent bodies accreted.

A classification of nonchondritic meteorites is shown in Table 6.2. These meteorites will be described in the following sections, and their chemistry will be considered in Chapter 13.

Table 6.2 Classification of differentiated meteorites

Primitive Achondrites – residues from partial melting
Acapulcoites – Lodranites
Brachinites
Ureilites
Winonaites
Magmatic Achondrites – crystallized from silicate magmas
Aubrites
Howardites – Eucrites – Diogenites (HEDs)
Angrites
Stony Irons
Mesosiderites
Pallasites – main group, Eagle Station group, pyroxene group
Irons
IAB
IC
IIAB
IIC
IID
IIE
IIIAB
IIICD
IIIE
IIIF
IVA
IVB
Martian Meteorites (SNCs)
Shergottites – basaltic, olivine-phyric, lherzolitic
Nakhlites
Chassignites
Orthopyroxenite (ALH 84001)
Breccia (NWA 7034 and paired meteorites)
Lunar Meteorites
Highlands breccias
Mare basalts

6.7 Primitive Achondrites

Most primitive achondrites exhibit metamorphic textures, as appropriate for the solid residues from which melts were extracted. In a few cases, some primitive achondrites have recognizable chondritic textures, but often they are so thoroughly recrystallized that chondrules are not identifiable. Primitive achondrites represent an extension of the highly metamorphosed type 6 chondrites, which generally show no eutectic melting of metal and sulfide (occurring at ~975°C, although the onset temperature of melting varies with composition). Tompkins et al. (2020) proposed that the transition from chondrites to primitive achondrites be based on the onset of silicate melting (~1050°C, with little compositional variation). This definition would preclude the need to reclassify some type 6 chondrites that experienced metal-troilite melting.

Goodrich and Delaney (2000) noticed that primitive achondrites could be readily distinguished from magmatic achondrites using a diagram of molar Fe/Mn versus Fe/Mg (Fig. 6.8). The primitive achondrites plot nearly on the diagonal line representing chondritic Mn/Mg values, whereas achondrites crystallized from magmas plot to the right of that line. Partial melts of chondrites have higher Fe/Mg ratios than the original chondritic source, and the corresponding solid residues have correspondingly lower Fe/Mg, as shown by the arrows in Figure 6.8. The residue composition moves to the left on this diagram as the degree of partial melting increases. However, primitive achondrites have nearly chondritic ratios of iron, manganese, and magnesium, because they commonly represent residues from very low degrees of partial melting. Extraction of small amounts of melt from residues can also be recognized by depletion in highly incompatible lithophile trace elements and in siderophile trace elements if eutectic melting of metal occurred. Once a magma forms and begins to crystallize, its Fe/Mg ratio will change as it accumulates or loses crystals, as shown in Figure 6.8.

Here we describe the most important groups of primitive achondrites.

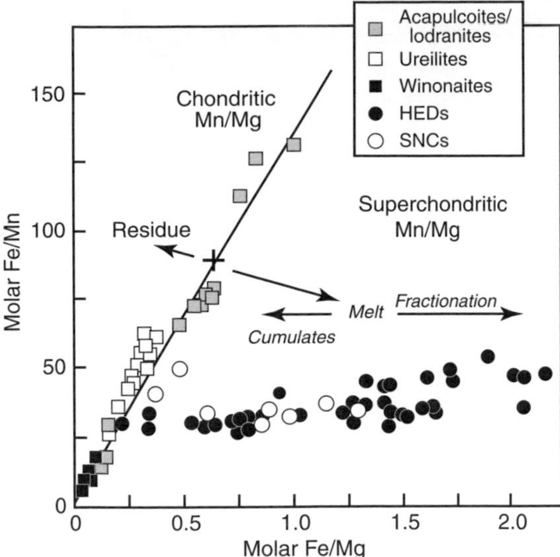

Figure 6.8 Diagram used to distinguish magmatic achondrites that crystallized from magmas from primitive achondrites that represent the solid residue once melt has been extracted. Primitive achondrites (acapulcoites/lodranites, ureilites, winonaites) plot on or to the left of the chondritic Mn/Mg ratio line, whereas magmatic achondrites (HEDs, SNCs) plot to the right. These differences occur because a chondrite (symbolized schematically by the cross; most chondrites actually plot closer to the origin, but this is offset for clarity) and the partial melts formed from it have different Mn/Mg values, as shown by arrows. The length of the arrow for residue represents extraction of 30% melt, and residues from less complete melt extraction lie along that arrow. The complementary melt composition arrow represents 2% melting, and melts formed by higher degrees of melting lie along the arrow joining the 2% melt to its chondritic source. The melt composition can be modified by fractional crystallization, forming cumulates and fractionated liquids, as shown by horizontal arrows. Modified from Goodrich and Delaney (2000).

6.7.1 Acapulcoites and Lodranites

The acapulcoites (Fig. 6.9a) have recrystallized textures (Keil and McCoy, 2018), but a few relict chondrules have been reported. The relative proportions of the minerals (olivine, orthopyroxene, diopside, plagioclase, metal) in acapulcoites are similar to those in ordinary chondrites, and the abundances of the major elements are nearly chondritic. However, elements that have siderophile or chalcophile affinities are depleted, reflecting extraction of a eutectic metal-sulfide melt. Some acapulcoites exhibit tiny metal-troilite veins that represent pathways for melts migrating out of the rocks. Interconnected plagioclase-pyroxene networks provide textural evidence for silicate melting (Tompkins et al., 2020).

The closely related lodranites (Fig. 6.9b) are more thoroughly recrystallized and show no vestige of chondrite textures (Keil and McCoy, 2018). Their bulk chemical compositions are no longer chondritic, because a greater degree of partial melting produced a silicate liquid of basaltic composition which migrated out of the rocks. The compositions of lodranites are depleted in the incompatible trace elements that were concentrated in basaltic melt. Siderophile and chalcophile elements are also depleted, as in acapulcoites.

6.7.2 Ureilites

The ureilites (Fig. 6.9c) constitute the second largest group of nonchondritic meteorites (Goodrich, 1992), comprising more than 550 samples. They are composed primarily of olivine and pyroxene, with interstitial material that is rich in elemental carbon (mostly graphite). The textures of silicate minerals are variable but highly recrystallized – so much so that they resemble igneous rocks, which led to past controversy about whether ureilites are magmatic rocks or residues. Graphite was apparently the original carbon mineral in ureilites, but intense shock pressures converted some of it into its dense polymorph, diamond, in some meteorites. The chemical compositions of ureilites indicate depletion in many siderophile and lithophile elements, reflecting loss of FeNi sulfide and basaltic melts. Their high carbon contents and variable oxygen isotopic compositions (discussed later) suggest that ureilites formed as residues from melting a carbonaceous chondrite-like precursor. The ureilite parent asteroid melted incrementally, over a range of oxidation states, preserving its isotopic heterogeneity (Collinet and Grove, 2020b).

6.7.3 Brachinites

The brachinites are ultramafic rocks composed mostly of olivine, with lesser clinopyroxene and small to trace amounts of orthopyroxene, plagioclase, iron sulfide, FeNi metal, chromite, and phosphate. The olivine is iron rich relative to other primitive achondrites, indicating a high degree of oxidation. They are mostly highly recrystallized, with olivine showing triple junctions, although a few have olivine fabrics that suggest accumulation from magmas (Hasegawa et al., 2019). Brachinites are interpreted as residues from low degrees of partial melting of precursor rocks similar to R chondrites (Keil, 2014). Several unusual achondrites (GRA 06128 and 06129) have andesitic compositions and may represent partial melts extracted from the brachinites.

6.7.4 Winonaites and IAB Silicate Inclusions

The winonaites are compositionally similar to silicate inclusions in some IAB irons (described below). They

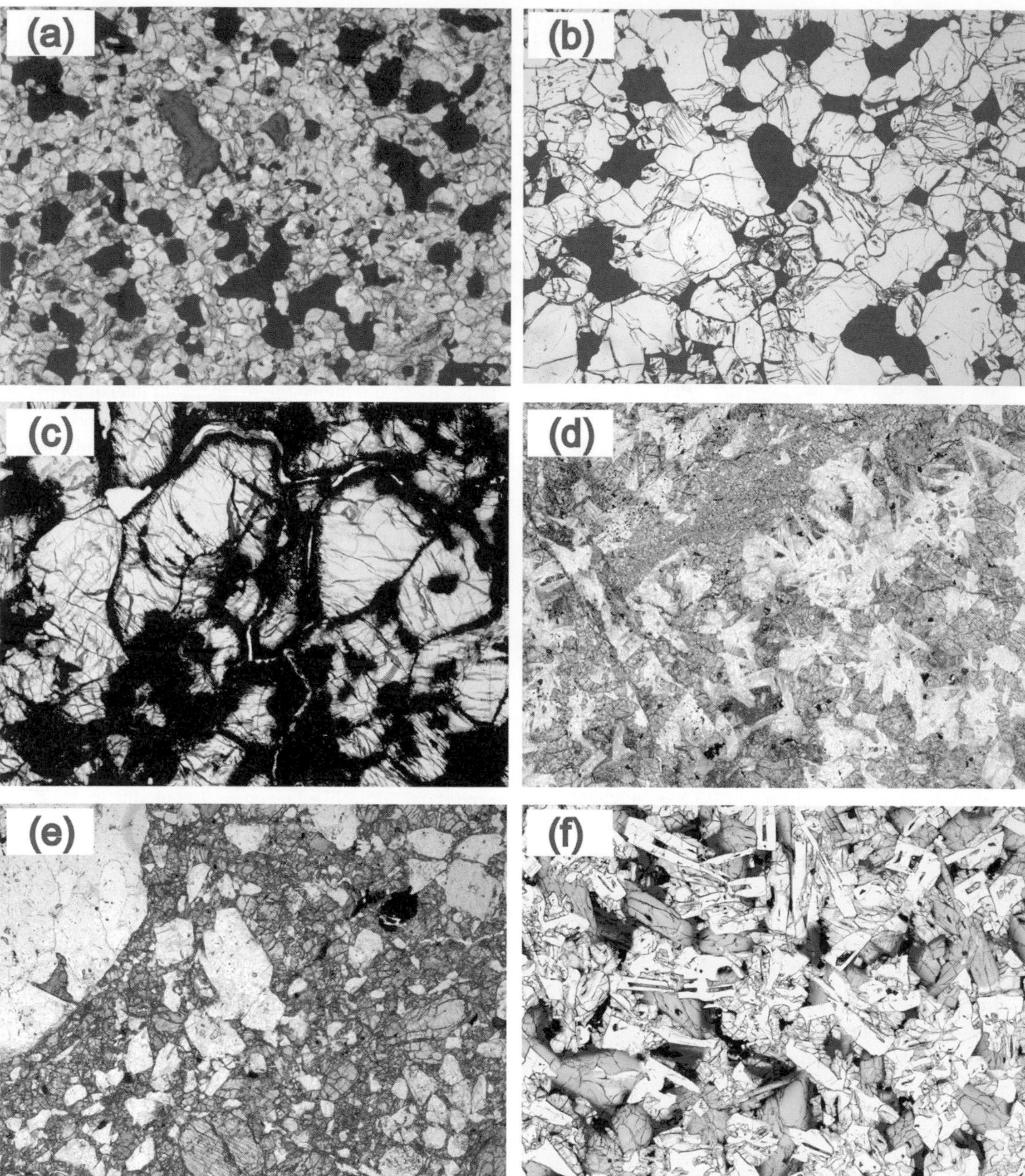

Figure 6.9 Photomicrographs of primitive and magmatic achondrites, all in plane-polarized light with field of view = 5.4 mm. (a) Acapulco and (b) Lodran, the type specimens for acapulcoite and lodranite; light grains are olivine and pyroxene, and black grains are metal. (c) NWA 942 ureililte, showing olivine crystals (white) in dark, carbonaceous mesostasis. (d) Millbillillie eucrite, consisting of pyroxene (gray) and plagioclase (white), and (e) NWA 1401 diogenite, consisting of brecciated pyroxene – both members of the HED class. (f) D'Orbigny angrite, consisting of pyroxene and olivine (gray), plagioclase (white), and mesostasis (black). All images from Lauretta and Killgore (2005), with permission.

have chondritic compositions, and relict chondrules have been found in some meteorites. They consist of olivine, pyroxenes, plagioclase, metal, troilite, and other minor minerals (Benedix et al., 1998), and most have been recrystallized. Like the acapulcoites, they have experienced only small degrees of melting.

6.8 Magmatic Achondrites

The various groups of magmatic achondrites formed as either solidified melts or cumulates from a crystallizing magma, as noted earlier. Although they are thought to have formed by melting of chondritic sources, their elemental abundances are far from chondritic, owing to the differentiation resulting from partial melting and fractional crystallization. Partial melts of chondritic material are in most cases basaltic, so achondrites are mostly basalts or cumulates derived from basaltic magmas. However, they exhibit surprising variability in composition and oxidation state, reflecting differences in their chondrite precursors. Here we describe several of the more prominent achondrite groups.

6.8.1 Aubrites

The aubrites are the most reduced magmatic achondrites (Keil, 2010). Their silicates are essentially free of iron and they contain minor metallic iron. A variety of unusual sulfides of calcium, chromium, manganese, titanium, and sodium – all usually lithophile elements – occur in aubrites. These unusual sulfides also characterize the highly reduced enstatite chondrites, which may have been precursors for these rocks.

The most abundant mineral in aubrites is coarse-grained orthopyroxene (as it is in enstatite chondrites), and only a small amount of plagioclase is present. Aubrites are commonly brecciated, and several clasts of related basalt have been observed. The aubrites are depleted in siderophile and chalcophile elements relative to chondritic abundances, indicating fractionation of metal and sulfide, and their concentrations of incompatible elements are varied (McCoy et al., 1999).

6.8.2 Howardites, Eucrites, and Diogenites

The howardite–eucrite–diogenite group, commonly abbreviated as HEDs, is the largest suite of differentiated meteorites (Mittlefehldt et al., 1998; McSween et al., 2011), represented by more than a thousand meteorites. This group consists of the following rock types: lavas (basaltic eucrites); coarser-grained gabbros that contain some accumulated crystals (cumulate eucrites); cumulates of orthopyroxene, with varying amounts of olivine (diogenites); and brecciated mixtures of eucrite and diogenite clasts (howardites). Most of these meteorites are breccias and many have been thermally metamorphosed.

The basaltic eucrites (Fig. 6.9d) are fine-grained rocks with textures formed by rapid cooling of magmas on or near the surface of their parent asteroid. However, the igneous textures of many eucrites are obscured by brecciation and metamorphism. They are composed of pigeonite and plagioclase, with minor tridymite, ilmenite, chromite, and other accessory minerals. The mineralogy of cumulate eucrites is similar, but their coarse grain sizes attest to slower cooling in intrusions at depth. Diogenites (Fig. 6.9e) consist mostly of orthopyroxene, with small but variable amounts of olivine and minor chromite. These rocks are cumulates related to eucrite magmas. Howardites were formed from eucrites and diogenites that were pulverized and mixed by impacts on the parent body surface.

The geochemistry of HED meteorites reflects pervasive melting and differentiation of the parent asteroid, Vesta (McSween et al., 2013). They are depleted in siderophile elements, lost when a metallic core segregated during differentiation. Partial melting and fractional crystallization induced changes in incompatible lithophile elements in the HED magmas. This igneous activity occurred very early in solar system history (ending by about 4.4 billion years ago), as judged from radiometric dating.

6.8.3 Angrites

The angrites (Fig. 6.9f) are basaltic rocks with unusual mineralogy (Mittlefehldt and Lindstrom, 1990). They contain aluminum- and titanium-rich diopside (a variety called grossmanite), calcium-rich olivine (kirschsteinite), plagioclase, and nepheline, as well as accessory spinel, troilite, whitlockite, magnetite, and other minerals. Their textures vary: some have quenched textures, indicating rapid solidification from a melt; some are porphyritic, suggesting slower crystal growth; and one (Angra dos Reis) contains coarse crystals that join at triple junctions, indicating metamorphic textural equilibration.

The geochemistry of angrites is characterized by strong silica undersaturation, by which we mean that there is not enough SiO_2 to combine with various cations to form common silicate minerals. The result is the formation of silica-poor minerals like kirschsteinite and nepheline. These meteorites also show strong depletions in moderately volatile elements. They are thought to have formed as partial melts of a chondritic source under oxidizing conditions.

6.9 Irons and Stony Irons

Iron meteorites offer the unique opportunity to examine metallic cores from deep within differentiated bodies. Most of these samples were exposed and dislodged when asteroids collided and fragmented. Although irons constitute only about 6% of meteorite falls, they are well represented in museum collections. Most iron meteorites show wide variations in siderophile element abundances, which can be explained by processes like fractional crystallization in cores that mimic those in achondrites. However, a few show perplexing chemical trends that may be inconsistent with their formation as asteroid cores.

6.9.1 Classification and Compositions of Iron Meteorites

Iron meteorites consist primarily of iron–nickel metal that commonly occurs as intergrowths of kamacite and taenite (Haack and McCoy, 2004). Historically, irons were classified according to the geometries of their metallic structures into hexahedrites, octahedrites, and ataxites. Octahedrites exhibit a "Widmanstätten pattern" (Fig. 6.10), developed as an intergrowth of kamacite and taenite during slow cooling. Hexahedrites have lower nickel contents and consist entirely of kamacite. Ataxites, whose name means "without structure," are nickel rich and do not display a Widmanstätten pattern. They consist of a very fine-grained intergrowth of kamacite and taenite known as plessite. Common accessory minerals include troilite, schreibersite, graphite, chromite, and cohenite. Sulfur- and phosphorus-rich minerals, such as troilite and schreibersite, vary in abundance among groups of irons. Also, differences in oxidation state are reflected in accessory minerals, as typified by the presence of graphite and cohenite in more-reduced irons and phosphates in more-oxidized irons.

Figure 6.10 A cube of the Muonionalusta iron meteorite, an octahedrite showing a three-dimensional Widmanstätten structure composed of intergrown plates of kamacite and taenite.

Once the chemical compositions of enough irons had been measured, it became clear that, although the various structural classes were reflected in chemical differences, chemical composition cannot be deduced reliably from structure. Using instrumental neutron activation analysis, John Wasson and his colleagues have analyzed hundreds of iron meteorites for siderophile elements over the past four decades. Compositional clusters (Fig. 6.11) based on the abundances of nickel, cobalt, germanium, gallium, iridium, gold, tungsten, and other elements define samples from different asteroid cores.

The taxonomy for irons is complicated, reflecting the tortured historical evolution of their classification. Originally, groups I–IV were defined by their gallium and germanium concentrations. As more data became available, some groups were subdivided (e.g. group III was separated into IIIA and IIIB and then recombined into IIIAB as meteorites with intervening compositions were found). Twelve major groups are now recognized, as shown in Table 6.2. Numerous ungrouped irons, representing as many as 50 distinct classes, are also known.

Many of the compositional trends seen in irons can be explained by fractional crystallization within metallic liquid cores. Various iron groups have different concentrations of these elements, but their interelement trends are parallel, suggesting common magmatic processes (Fig. 6.11a). These meteorites are called the "magmatic irons." A few iron groups show large elemental variations that cannot be explained by fractional crystallization. These iron groups (IAB, IIICD, and IIE; Fig. 6.11b) also contain abundant silicate inclusions, often having chondritic compositions. They are commonly labeled "silicate-bearing" or "nonmagmatic" irons (the latter term is unfortunate, because all irons were once molten; however, they apparently formed by different processes, perhaps within dispersed metal pools rather than large, central cores). The iron meteorite groups also show a wide range of gallium and germanium contents (Fig. 11b). Gallium and germanium are moderately volatile elements and their abundances reflect nebula history. The IIIF, IVA, and IVB irons show extreme depletions in gallium and germanium (Fig. 11b), suggesting loss of these elements prior to accretion of the parent asteroids.

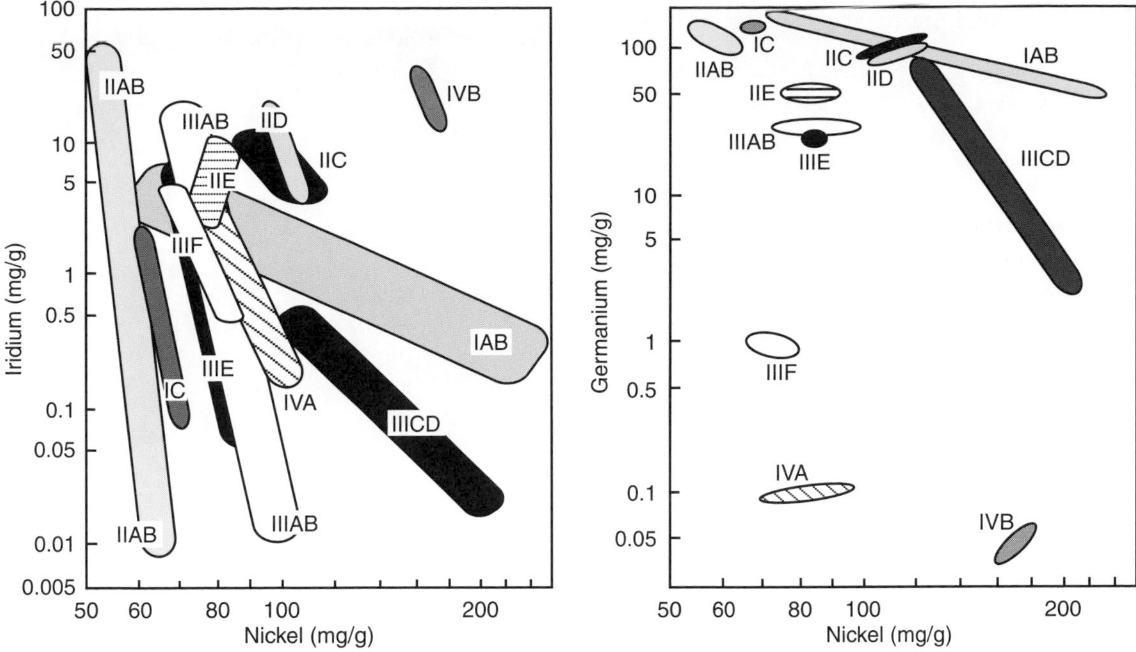

Figure 6.11 Siderophile element (iridium, nickel, germanium) plots used to classify iron meteorites into groups. Modified from Scott and Wasson (1975).

Another, rather broad classification system for irons distinguishes them based on whether they formed by extraction of melted iron from carbonaceous or noncarbonaceous chondrite precursors (Rubin, 2018). The carbonaceous "CC irons" (groups IIC, IID, IIF, IIIF, IVB) tend to have higher nickel and iridium contents and higher refractory siderophile element contents; the latter might result from refractory metal nuggets in CAIs. The noncarbonaceous "NC irons" are groups IC, IIAB, IIIAB, IIIE, and IVA. Ungrouped iron meteorites are mostly CC irons.

6.9.2 Pallasites

More than 100 known pallasites can be divided into "main-group" pallasites, the Eagle Station grouplet (currently with five members), and a handful of ungrouped and anomalous pallasites (McKibbin et al., 2019). Main-group pallasites consist mostly of magnesium-rich olivine and FeNi metal (Fig. 6.12). The olivine is sometimes dispersed and sometimes clustered, and grains are either rounded or angular. Rounded grains developed by residence in, and reaction with, FeNi-FeS liquid. Large areas of the Brenham pallasite contain no olivine at all. The metal composition in main-group pallasites is similar to that in IIIAB irons and, like those irons, has a Widmanstätten pattern. The oxygen isotope composition of silicates and the sulfur isotope composition of sulfides are also the same. The Eagle Station grouplet, in addition to olivine and metal, contains pyroxenes, chromite, and phosphates. The olivine is more iron rich than in main-group pallasites, and the metal has higher iridium and nickel contents. Oxygen isotopes in the Eagle Station pallasites are more ^{16}O rich than in the main-group pallasites. Ungrouped pallasites differ in mineralogy and appear to represent distinct parent bodies.

Olivine and metal are expected to be in contact at the boundary between a core and its overlying silicate mantle. That is the traditionally favored location for the formation of pallasites. Core–mantle models vary from the slow displacement of silicate melt in an olivine cumulate or residue by molten metal, to violent mixing of core metal and olivine by deep impact (Boesenberg et al., 2012). Surface mixing of metal and olivine has also been proposed (Bryson et al., 2015). Although concentration profiles for minor elements in olivine have been used extensively to derive rapid cooling rates for pallasites at high temperatures, recent experimental work has shown that the concentration profiles likely result from the nucleation of new phases such as chromite or schreibersite during cooling. A model based on this work suggests cooling rates of 100–300°C/Myr at high temperatures, decreasing to ~1°C/Myr at the low temperatures recorded by the metal phase (Donohue et al., 2018).

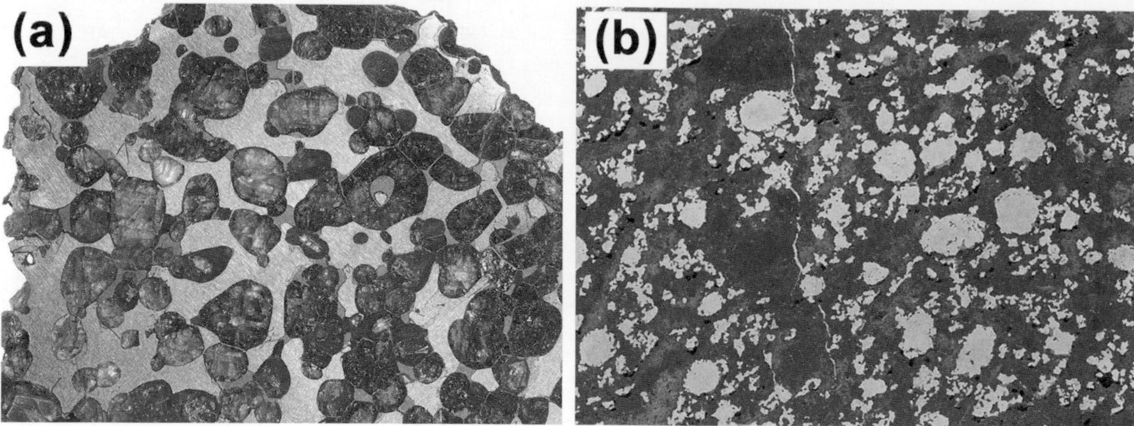

Figure 6.12 (a) Brenam pallasite, showing rounded olivine grains embedded in FeNi metal. Sample width ~8 cm. (b) Estherville mesosiderite, composed of subequal amounts of metal and silicate clasts. Sample width ~4 cm. Photograph (a) by G. Notkin and (b) courtesy of the Natural History Museum, London.

6.9.3 Mesosiderites

Mesosiderites are an enigmatic group of differentiated meteorites. They are breccias composed of FeNi metal and silicate in roughly equal proportions (Benedix et al., 2014). The metal is commonly thought to represent molten material from the deep interior of an asteroid, whereas the silicate fraction consists of basalts and pyroxene cumulates similar to HED meteorites that formed near the surface. It is difficult to construct models that allow mixing of such diverse materials, but these disparate materials are generally thought to have been violently mixed by impact and then cooled slowly in a large body (Scott et al., 2001). Metal in the least metamorphosed mesosiderite, NWA 1878, occurs as clusters of centimeter-sized spherules (Kimura et al., 2020), distinctly different from the coarse metal grains in other mesosiderites, and suggests that current models for the formation of these meteorites are incorrect. Although suggested to have formed by an impact on Vesta (Haba et al., 2019), exploration by the Dawn spacecraft did not observe any metal-rich areas on the asteroid.

6.10 Lunar Samples and Meteorites

The Apollo astronauts returned 382 kg of lunar sample to Earth, and this collection was supplemented by 326 g of soil samples collected by the Soviet Luna landers. The first lunar meteorite was found in 1982 in Antarctica. As of time of writing, over 415 lunar meteorites representing about 150 different fall events have been collected. The total mass of these meteorites is ~670 kg. Most of these meteorites were recovered in Antarctica by American and Japanese teams, and most of the rest were recovered in the deserts of North Africa and Oman. The lunar meteorites have significantly expanded the areas of the Moon from which we have samples.

Lunar rocks (Fig. 6.13) represent samples from the ancient *highlands* and the younger *maria* (Papike et al., 1998; Warren and Taylor, 2014). Highlands rocks are readily distinguished by having high Al/Si ratios (reflecting their high abundance of aluminum-rich plagioclase) and lower TiO_2 contents (which are higher in ilmenite-rich mare basalts). Lunar samples also include volcanic glasses and soils.

Almost all highlands rocks are breccias composed of clasts fragmented by impacts. "Pristine" highlands rocks (Warren and Wasson, 1977) have chemical compositions corresponding to igneous rocks, reflecting the fact that many are brecciated samples of a single rock type (monomict). Nonpristine rocks are breccias composed of multiple rocks (polymict), and their compositions reflect their mixed parentage and/or contamination by the impacting meteorite. Pristine highlands samples fall into three categories: ferroan anorthosites, magnesium-rich rocks, and alkali-rich rocks. The anorthosites are coarse-grained cumulates of plagioclase with only minor amounts of pyroxenes and olivine. These rocks represent the most ancient lunar crust, with radiometric ages of ~4.35 Ga, although they are probably older. The rocks of the magnesium-rich suite are comparable in age and consist

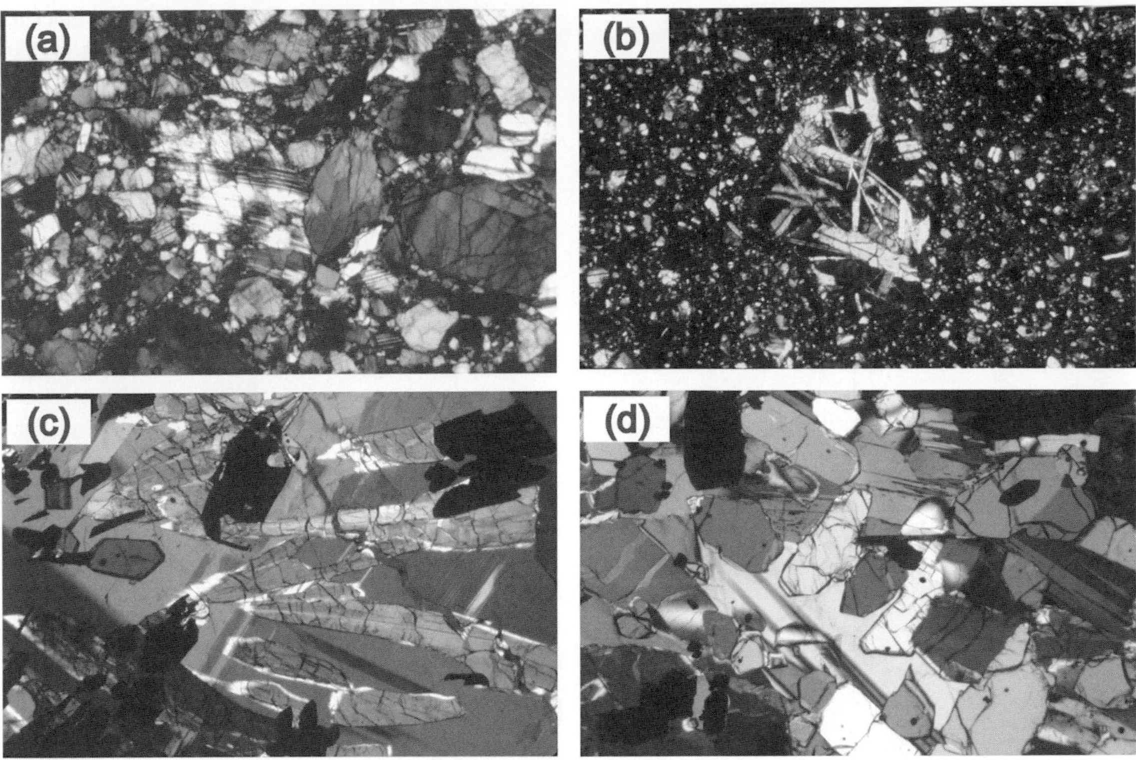

Figure 6.13 Photomicrographs of lunar rocks, all viewed under crossed nicols with field of view = 5.4 mm. (a) Ferroan anorthosite 60025, a breccia composed almost entirely of plagioclase. (b) KREEP-rich breccia 70126, exhibiting mineral and basaltic rock clasts. (c) High-Ti mare basalt 70017, composed of pyroxene, plagioclase, and opaque minerals (elongated grains are ilmenite). (d) Low-Ti mare basalt 12005, composed of pyroxene, plagioclase, and opaque oxides. Images courtesy of D. Kring.

of coarse-grained rocks composed of olivine, pyroxenes, and plagioclase. The alkali suite includes basalts, anorthosites and gabbros with higher sodium contents, and "granites" containing K-feldspar. Nonpristine breccias exhibit a wide variety of textures, reflecting the complex processes that produced them. These polymict breccias typically contain highly variable amounts of KREEP (an acronym of the symbols for potassium, rare earth elements, phosphorus), an incompatible-element-rich component.

Mare basalts are composed of pyroxenes, plagioclase, sometimes with olivine or silica, and a variety of minor minerals. The most prominent accessory mineral is ilmenite, which is commonly present in higher abundances than in terrestrial basalts. Ilmenite concentration, which controls TiO_2 abundance, is used to subdivide mare basalts into high-titanium (>6 wt.%), low-titanium (1.5–6%), and very low-titanium (<1.5%) varieties. (These are relative terms; it is somewhat of a misnomer to call a rock with 5% TiO_2 "low titanium.") A secondary classification is based on abundances of K_2O and Al_2O_3. Mare basalts are younger than highlands rocks, with radiometric ages varying between approximately 3.9 and 2.0 billion years.

Several kinds of volcanic glass beads were collected by Apollo astronauts. Numerous chemical classifications have been devised. The best known are the Apollo 15 green glasses and the Apollo 17 orange glasses. Each of these glasses is compositionally uniform in terms of major elements, but the beads have coatings that are enriched in highly volatile elements formed during fire-fountain eruptions.

Lunar soil (more properly called *regolith*) is a collection of highlands and mare rocks pulverized by impacts. These soils consist of recognizable mineral grains and rock fragments, often bonded by impact-melted glass into small aggregates ("agglutinates"). Because of impact gardening over long periods of time, the soils were repeatedly exposed on the lunar surface and provide chemical evidence of irradiation by the solar wind. Soils that have been lithified into coherent rocks are regolith breccias.

The chemistry of lunar materials will be considered in much greater detail in Chapter 14.

6.11 Martian Meteorites

Martian meteorites (SNCs, for shergottite, nakhlite, chassignite) comprise a diverse suite of igneous rocks (McSween and McLennan 2014; McSween, 2015; Udry et al., 2020). The SNC moniker is outdated now, as a few samples do not fall into those groups. Presently, more than 250 of these meteorites are recognized, although many are paired, and ~140 of these are separate samples. Most of these meteorites have been recovered from Antarctica and North African deserts (Udry et al., 2020). With only two exceptions, they have relatively young crystallization ages.

Shergottites are the most abundant type of martian meteorite, comprising 83% of known samples. Basaltic shergottites (Fig. 6.14a) are lavas composed of pyroxenes (pigeonite and augite), plagioclase (converted to maskelynite by shock), and a variety of accessory minerals including iron-titanium oxides, sulfides, and whitlockite. Gabbroic shergottites are coarser-grained and contain cumulus pyroxene and plagioclase; they represent the plutonic equivalents of basaltic shergottites. Olivine-phyric shergottites are texturally and mineralogically similar to basaltic shergottites, except that they also contain olivine. Most of the large olivine crystals ("phenocrysts") crystallized from the shergottite liquid, although some grains may be foreign crystals incorporated into the magma ("xenocrysts"). The unusual augite-phyric shergottites contain phenocrysts of augite, plagioclase, and olivine, but lack pigeonite, and have distinctly older ages than other shergottites. Poikilitic shergottites are cumulate rocks composed of olivine, pyroxenes, and lesser amounts of maskelynite, formed by slow cooling of shergottite-like magmas at depth; these were formerly called lherzolitic shergottites, but more representative sampling shows that they contain too much plagioclase to be ultramafic

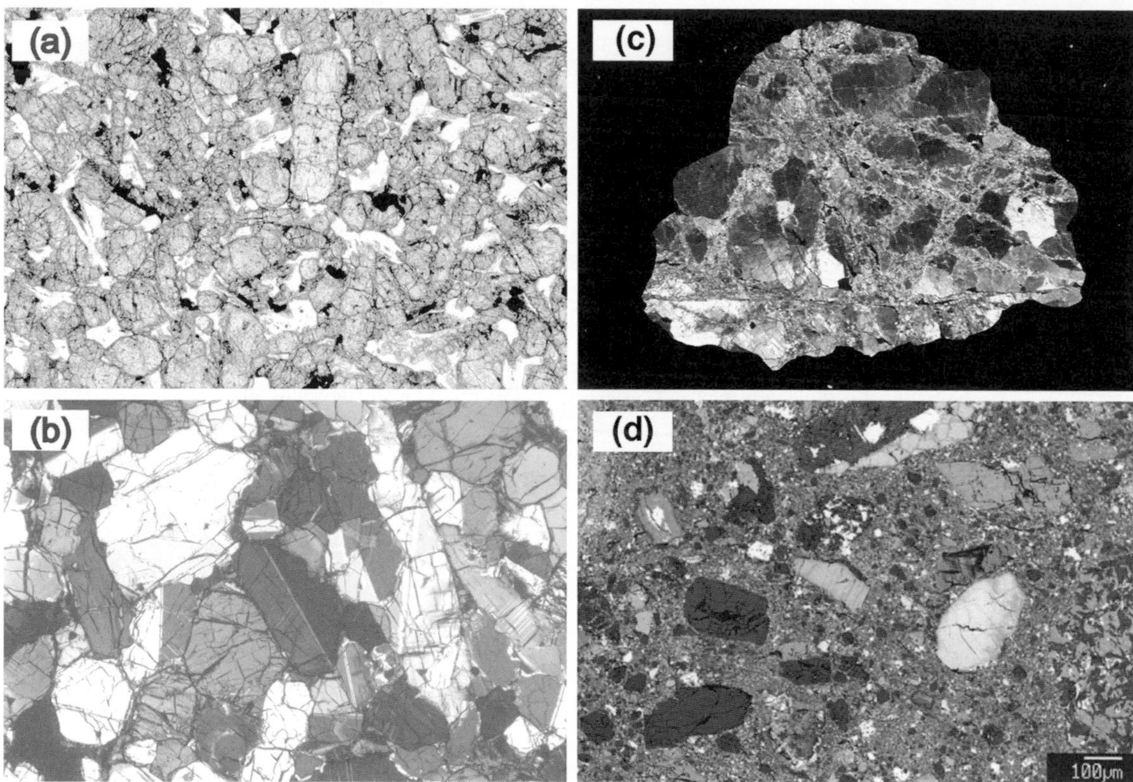

Figure 6.14 Photomicrographs of martian meteorites. (a) Zagami basaltic shergottite, composed of pyroxene (gray) and plagioclase (white), which has been converted to maskelynite by shock, plane-polarized light, field of view (FOV) = 5.4 mm. (b) Lafayette nakhlite, composed mostly of high-calcium pyroxene, plane-polarized light, FOV = 5.4 mm. (c) ALH 84001 orthopyroxenite showing crushed grains, cross-polarized light, FOV = 2 cm. (d) NWA 7034 martian meteorite breccia, backscattered electron image. (a) and (c) from Lauretta and Killgore (2005), with permission; (d) from Agee et al. (2013).

lherzolites. Most shergottites range in age from 575–165 Ma, although the augite-phyric shergottites crystallized at ~2.4 Ga.

Nakhlites (Fig. 6.14b) are augite-rich basaltic rocks (Udry and Day, 2018). They are cumulates formed by concentration of augite and minor olivine. They formed on or close to the surface, as indicated by the fine grain size of their groundmass, which consists of needles of feldspars, delicate spinel crystals, and glass. All the nakhlites have ages of ~1.3 Ga, and compositional similarities indicate that they are related. The two known chassignites are olivine cumulates with minor pyroxene. They also date from 1.3 Ga and are related to nakhlites.

ALH 84001 (Fig. 6.14c) is a unique martian meteorite with an ancient age of ~4.1 Ga. It consists mostly of orthopyroxene crystals that accumulated in basaltic magma (Mittlefehldt, 1994). Its most distinctive feature, however, is the occurrence of small globules of carbonates with unusual compositions and textures in crushed zones. The controversial hypothesis (now largely discredited) that the carbonates contain evidence of extraterrestrial life (McKay et al., 1996) made this the most famous meteorite on Earth. ALH 84001 has also experienced intense shock metamorphism.

NWA 7034 (Agee et al., 2013), NWA 7533 (Humayun et al., 2013), and a handful of other paired meteorites are polymict breccias (Fig. 6.14d) composed of igneous clasts of varying composition (basalt, trachyandesite, mugearite), as well as impact-melted clasts. At ~4.4 Ga, these ancient meteorites provide the only samples of the early martian crust.

We will return to a more detailed consideration of the chronology of martian meteorites in Chapter 10 and of their chemistry in Chapter 14.

6.12 Oxygen Isotopes in Differentiated Meteorites

Many achondrites, lunar samples, and martian (SNC) meteorites can be distinguished by their oxygen isotopic compositions. The data for various groups (Clayton, 2004) are illustrated in the upper part of Figure 6.15. Lunar rocks plot nearly on the terrestrial fractionation line, although minor differences can be resolved (see Box 8.2). The Moon is thought to have formed from a massive impact on the Earth, and the almost identical oxygen isotopic compositions of both bodies suggest that they were compositionally similar or their materials were equilibrated during the collision (Young et al., 2016). The aubrites plot on the terrestrial fractionation line, as do the enstatite chondrites from which they were probably derived. The similarity between the oxygen in these meteorites and the Earth's oxygen composition supports the idea that the Earth may have accreted from enstatite-chondrite-like material. SNCs and HEDs define their own mass fractionation lines.

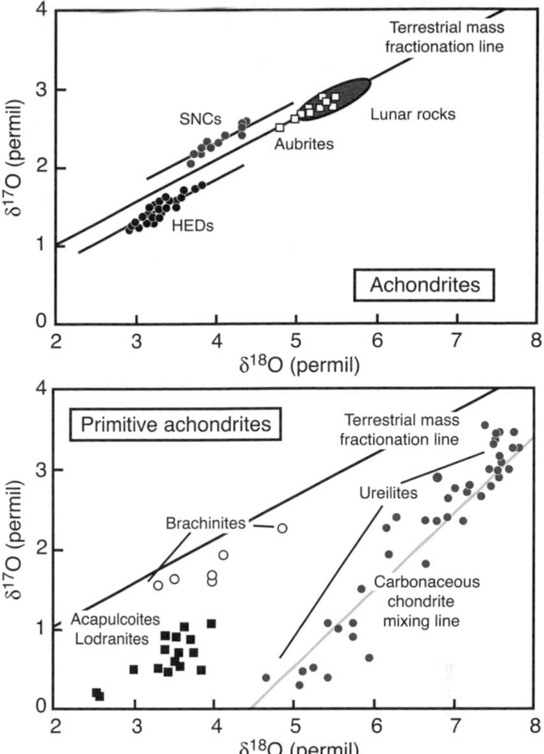

Figure 6.15 Oxygen isotopes in magmatic achondrites (above) and primitive achondrites (below). The δ notation and units are explained in the caption for Figure 6.4. Most achondrites define mass fractionation lines parallel to, but slightly offset from, the terrestrial line. Aubrites and lunar samples plot nearly on the terrestrial line. Primitive achondrites generally do not define oxygen mass fractionation lines, but are scattered and reflect the compositions of their chondrite precursors.

Unlike chondrites, differentiated meteorites define mass-dependent fractionation lines with slope +1/2 parallel to the terrestrial fractionation line. Melting homogenizes oxygen isotopes in the parent body, and differentiation of those magmas smears them along a mass-dependent fractionation line. Primitive achondrites, however, do not always define mass fractionation lines. The partial melts extracted from those meteorites presumably had homogenized oxygen isotopes, but the solid residues could have remained heterogeneous unless metamorphism and melting were extensive. For example, limited melting in the acapulcoite-lodranite parent body did not erase its chondritic isotopic signature (lower part of Fig. 6.15). The ureilites are arrayed along the mixing line for carbonaceous chondrites (Fig. 6.15), consistent with their formation as residues from partial melting of carbonaceous-chondrite-like materials.

Iron meteorites commonly contain oxygen-bearing minerals, such as chromite, phosphates, and sometimes silicates. Oxygen isotopes in such inclusions provide information on possible connections between irons and other meteorite groups. For example, the isotopic fingerprint in inclusions in IIIAB irons is similar to that of the olivines in main-group pallasites, suggesting that these meteorites might be samples of the core and core–mantle boundary within the same asteroid.

6.13 Trading Rocks: Meteorites on Other Worlds

The phenomenon of falling meteorites is, of course, not unique to Earth. Lunar soils contain enhanced abundances of siderophile elements, which have been used to estimate a finely distributed meteoritic component of a few volume percent. Intact meteorite clasts have been recognized in lunar soils and regolith breccias, including carbonaceous, ordinary, and enstatite chondrites, irons and stony irons, and possibly achondrites (Joy et al., 2016). Bench Crater, a type of chondrite not recognized in our meteorite collections, was also found in soil (McSween, 1976). A small clast of granite in a lunar breccia returned by Apollo astronauts has even been suggested to be a terrestrial meteorite (Bellucci et al., 2019). Geochemical analysis suggests that the ancient (~4 Ga) rock formed in a water-rich environment under a high pressure that corresponds to a depth within the Moon that would be unlikely to have been sampled.

Regolith breccias from asteroids commonly contain fragments of other meteorite types, most commonly CM-like carbonaceous chondrites. The Dawn spacecraft measured an area of high hydrogen abundance on asteroid Vesta, which was interpreted as a concentration of carbonaceous chondrite from a large impactor (Prettyman et al., 2012).

Meteorites have also been delivered to (and from) Mars. Martian soils show an elevated abundance of nickel, which reflects the addition of a few volume percent of meteorites. The Spirit, Opportunity, and Curiosity rovers have encountered >50 meteorites during their traverses across the martian surface, including irons (Fig. 6.16) and a variety of other suspected stony meteorites (Ashley, 2015). Alpha particle X-ray spectrometer chemical analyses of nickel, germanium, and gallium are sufficiently precise to classify the iron meteorites.

Figure 6.16 An iron meteorite (31 cm across), officially named Meridiani Planum, encountered and analyzed on the martian surface by the Opportunity rover. NASA image.

Summary

Primitive chondrites provide a record of nebular processes and conditions. In this chapter we have described the components (chondrules, refractory inclusions, metal and sulfide, matrix) that formed in the nebular setting and were accreted together to produce chondrites. Each of these components, in turn, consists of phases whose compositions, mineral associations, and textures constrain their origin. Proper interpretations of cosmochemical analyses of bulk chondrites, their components, and their constituent minerals also require an understanding of their geologic histories after accretion into planetesimals. Chondrite parent bodies experienced a variety of secondary processes, including thermal metamorphism, aqueous alteration, and shock metamorphism. A serviceable classification scheme for chondrites is based on a primary characteristic (bulk chemical compositions) and several secondary characteristics (petrologic type). Oxygen isotopes provide a convenient way to distinguish most chondrite groups.

IDPs are divided into chondritic smooth (CS) and chondritic porous (CP) varieties. The CS IPDs are similar in mineralogy and chemical composition to carbonaceous chondrites and are likely derived from chondritic asteroids. The CP IDPs are more primitive and are thought to derive from comets.

Differentiated meteorites record geologic processes on or within bodies that were heated to the point of

Box 6.2 Meteorite Data Sources and Names

New meteorites are constantly added to the world's collections and keeping track of them is a difficult task. The British Museum's *Catalogue of Meteorites* (Grady, 2000) serves as an authoritative resource, although it is not constantly updated. It is available in printed and electronic form. This reference not only provides classification, mass, and information on circumstances of recovery, but also lists publications on each meteorite. The Meteoritical Society, an international organization devoted to meteorite research, periodically publishes the *Meteoritical Bulletin*, listing newly classified meteorites and approved meteorite names. This database can be accessed online at meteoriticalsociety.org. Also, MetBase, a commercially produced computer database for meteorites, is available for purchase online at www.metbase.de. It includes meteorite descriptions, analyses, and a bibliography of relevant scientific literature.

A meteorite is named after the geographic locality closest to its discovery, whether fall or find. These localities can be political features like a town, railroad station, or historical site, or physiographic features like a mountain or lake. When two meteorites fall or are found at the same locality, they can be distinguished by the date of discovery, as in Wethersfield 1971 and Wethersfield 1982. For dense collection areas, such as Antarctica, a generic prefix conveying geographic information and a number are used, as in Allan Hills (abbreviated ALH) 84001. If the provenance cannot be accurately determined, a region (such as Northwest Africa, abbreviated NWA) is used with a number. Meteorite names are approved by the Nomenclature Committee of the Meteoritical Society. Approved meteorite names require that a type specimen of 20% of the mass or 20 g, whichever is less, be deposited at an approved curation institution so that scientists can access it for study. Detailed guidelines for meteorite nomenclature can be found in the *Meteoritical Bulletin*, No. 94.

melting. Melting of primitive chondritic materials is almost always partial rather than complete, producing a metal-sulfide eutectic liquid and, as melting advances, a silicate magma with the composition of basalt. Fractional crystallization of the resulting liquids produces changes in the chemical composition of the resulting rocks (differentiation). Magmatic achondrites formed by crystallization of silicate magmas, and irons and stony irons formed by solidification of metallic liquids and their commingling with silicates. Samples of the solid residues left behind after extraction of partial melts are called primitive achondrites. Lunar rocks and all planetary samples, such as martian meteorites, are differentiated rocks as well. These rocks formed in most cases by more protracted planetary evolution, encompassing multiple periods of melting and differentiation, so that the mantles that partially melted to produce magmas may no longer have had chondritic compositions. Oxygen isotopes in achondrites generally show mass-dependent fractionations (primitive achondrites are an exception), but their compositions can sometimes be used to recognize differentiated samples from a common parent body.

Now that we have gained some appreciation for the value and limitations of the various kinds of samples available for cosmochemical analysis, we will begin to consider what these materials really tell us. In a previous chapter, we have seen how solar system abundances of elements and isotopes are determined using the most primitive chondrites. In subsequent chapters, we will see how elements are fractionated in space, planetesimals, and planets.

Questions

1. Describe the components of chondritic meteorites.
2. Explain the classification system used for chondrites.
3. What is differentiation and how does it cause chemical changes?
4. Distinguish between the origins of magmatic achondrites, primitive achondrites, and irons.
5. How are oxygen isotopes in chondrites and achondrites illustrated graphically, and how are the isotopic compositions useful in classifying meteorites and recognizing relationships between them?

Suggestions for Further Reading

There are a number of excellent (but rather technical) chapters in recent books that describe the classification of meteorites in greater detail than presented here. The following are highly recommended. All have excellent photographs, but some of these resources probably offer more information than most beginning readers can use.

Davis A. M., editor (2014) *Treatise in Geochemistry, 2nd edition, Vol. 1: Meteorites and Cosmochemical Processes.*

Elsevier, Oxford, 454 pp. A number of chapters in this book provide excellent descriptions of meteorites and IDPs:

Krot A. N., Keil K., Scott E. R. D., et al., Classification of meteorites and their genetic relationships, pp. 1–63.

Scott E. R. D., and Krot A. N., Chondrites and their components, pp. 65–137.

Mittlefehldt D. W., Achondrites, pp. 25–266.

Benedix G. K., Haack H., and McCoy T. J., Iron and stony-iron meteorites, pp. 267–285.

Bradley J. P., Early solar nebula grains – interplanetary dust particles, pp. 287–308.

Grady M. M. (2000) *Catalogue of Meteorites*. Cambridge University Press, Cambridge, 689 pp. plus disk. A compiled list of known meteorites, their classifications and recovery information.

Hazen R. M., and Morrison S. M. (2020) An evolutionary system of mineralogy. Part I: Stellar mineralogy (>13 to 4.6 Ga). *American Mineralogist*, **105**, 626–651. The first in a series of papers detailing the mineralogy of extraterrestrial materials. Other papers in the series, all to be published in *American Mineralogist*, are:

Part II: Interstellar and solar nebula primary condensation mineralogy (>4,565 Ga).

Part III: Primary chondrule mineralogy (4.566 to 4.561 Ga).

Part IV: Planetesimal differentiation and impact mineralization (4.566 to 4.560 Ga).

Part V: Aqueous and thermal alteration of planetesimals (~4.565 to 4550 Ma).

Weisberg M. K., McCoy T. J., and Krot A. N. (2006) Systematics and evaluation of meteorite classification. In *Meteorites and the Early Solar System II*, Lauretta D. S., and McSween H. Y., editors, pp. 19–52, University of Arizona Press, Tucson. Provides a thorough description of meteorite classification.

Brearley A. J., and Jones R. H. (1998) Chondritic meteorites. *Planetary Materials, Reviews in Mineralogy, 36*, Papike J. J., editor, pp. 3-002 to 3-398. The most complete published description of chondrites.

Other References

Agee C. B., Wilson N. V., McCubbin F. M., et al. (2013) Unique meteorite from early Amazonian Mars: Water-rich basaltic breccia Northwest Africa 7034. *Science*, **339**, 780–785.

Ashley J. W. (2015) The study of exogenic rocks on Mars – An evolving subdiscipline in meteoritics. *Elements*, **11**, 10–11.

Asphaug E., Jutzi M., and Movshovitz M. (2011) Chondrule formation during planetesimal accretion. *Earth & Planetary Science Letters*, **308**, 369–379.

Bellucci J. J., Nemchim A. A., Grange M., et al. (2019) Terrestrial-like zircon in a clast from an Apollo 14 breccia. *Earth & Planetary Science Letters*, **510**, 173–185.

Benedix G. K., McCoy T. J., Keil K., et al. (1998) A petrologic and isotopic study of winonaites: Evidence for early partial melting, brecciation, and metamorphism. *Geochimica et Cosmochimica Acta*, **62**, 2535–2554.

Benedix G. K., Haack H., and McCoy T. J. (2014) Iron and stony-iron meteorites. In *Treatise in Geochemistry, 2nd edition, Vol. 1: Meteorites and Cosmochemical Processes*, Davis A. M., editor, pp. 267–285, Elsevier, Oxford.

Boesenberg J. S., Delaney J., and Hewins R. (2012) A petrological and chemical re-examination of main-group pallasite formation. *Geochimica et Cosmochimica Acta*, **89**, 134–158.

Bonal L., Quirico E., Bourot-Denise M., and Montagnac G. (2006) Determination of the petrologic type of CV3 chondrites by Raman spectroscopy of included organic matter. *Geochimica et Cosmochimica Acta*, **70**, 1849–1863.

Bradley J. P. (1994) Chemically anomalous, preaccretionally irradiated grains in interplanetary dust particles from comets. *Science*, **265**, 925–929.

Bryson J. F. J., Nichols C. I. O., Herrero-Albilos J., et al. (2015) Long-lived magnetism from solidification-driven convection on the pallasite parent body. *Nature*, **517**, 472–475.

Clayton R. N. (2004) Oxygen isotopes in meteorites. In *Treatise on Geochemistry, Vol. 1: Meteorites, Comets, and Planets*, Davis A. M., editor, pp. 129–142, Elsevier, Oxford.

Collinet M., and Grove T. L. (2020a) Widespread production of silica- and alkali-rich melts at the onset of planetesimal melting. *Geochemica et Cosmochimica Acta*, 277, 334–357.

Collinet M., and Grove T. L. (2020b) Incremental melting in the ureilite parent body: Initial composition, melting temperatures, and melt compositions. *Meteoritics & Planetary Science*, **55**, 832–856.

Connolly H. C. Jr., and Jones R. H. (2016) Chondrules: The canonical and noncanonical views. *Journal of Geophysical Research, Planets*, **121**, 1885–1899.

Desch S. J., Morris M. A., Connolly H. J., and Boss A. P. (2012) The importance of experiments: Constraints on chondrule formation models. *Meteoritics & Planetary Science*, **47**, 1139–1156.

Donohue P. H., Hill E., and Huss G. R. (2018) Experimentally determined subsolidus metal-olivine element partitioning with applications to pallasites. *Geochemica et Cosmochimica Acta*, 222, 305–318.

Folco L., and Cordier C. (2015) Micrometeorites. *Planetary Mineralogy, EMU Notes in Mineralogy*, **15**, 253–297.

Goodrich C. A. (1992) Ureilites: A critical review. *Meteoritics*, **27**, 327–352.

Goodrich C. A., and Delaney J. S. (2000) Fe/Mg-Fe/Mn relations of meteorites and primary heterogeneity of primitive achondrite parent bodies. *Geochimica et Cosmochimica Acta*, **64**, 149–160.

Grossman J. N., and Brearley A. J. (2005) The onset of metamorphism in ordinary and carbonaceous chondrites. *Meteoritics & Planetary Science*, **40**, 87–122.

Haack H., and McCoy T. J. (2004) Iron and stony-iron meteorites. In *Treatise on Geochemistry, Vol. 1: Meteorites, Comets, and Planets*, Davis A. M., editor, pp. 325–345, Elsevier, Oxford.

Haba M. K., Wotzlaw J.-W., Lai Y.-J., et al. (2019) Mesosiderite formation on asteroid 4 Vesta by a hit-and-run collision. *Nature Geoscience*, **12**, 510–515.

Hasegawa H., Mikouchi T., Yamaguchi A., et al. (2019) Petrological, petrofabric, and oxygen isotopic study of five ungrouped meteorites related to brachinites. *Meteoritics & Planetary Science*, **54**, 752–767.

Humayun M., Nemchin A., Zanda B., et al. (2013) Origin and age of the earliest martian crust from meteorite NWA 7533. *Nature*, **503**, 513–516.

Ishii H. A., Bradley J. P., Bechtel H. A., et al. (2018) Multiple generations of grain aggregation in different environments preceded solar system body formation. *Proceedings of the National Academy of Sciences, USA*, **115**, doi:10.1073/pnas.1720167115.

Jessberger E. K., Stephan T., Rost D., et al. (2001) Properties of interplanetary dust: Information from collected samples. In *Interplanetary Dust*, Grun E., Gustafson B. A. S., Dermott S. F., and Fechteg H., editors, pp. 253–294, Springer, Berlin.

Joswiak D. J., Brownlee D. E., Matrajt G., et al. (2009) Kosmochloric Ca-rich pyroxenes and FeO-rich olivines (Kool grains) and associated phases in Stardust tracks and chondritic porous interplanetary dust particles: Possible precursors to FeO-rich type II chondrules in ordinary chondrites. *Meteoritics & Planetary Sciences*, 44, 1561–1588.

Joy K. H., Crawford I. A., Curran N. M, et al. (2016) The Moon: An archive of small body migration in the Solar System. *Earth Moon Planets*, **118**, 133–158.

Keil K. (2010) Enstatite achondrite meteorites (aubrites) and the histories of their asteroidal parent bodies. *Chemie der Erde*, **70**, 295–317.

Keil K. (2014) Brachinite meteorites: Partial melt residues from an FeO-rich asteroid. *Chemie der Erde*, **74**, 311–329.

Keil K., and McCoy T. J. (2018) Acapulcoite-lodranite meteorites: Ultramafic asteroidal partial melt residues. *Chemie der Erde*, **78**, 153–203.

Kimura M., Sugiura N., Yamaguchi A., and Ichimura K. (2020) The most primitive mesosiderite Northwest Africa 1878, subgroup 0. *Meteoritics & Planetary Science*, **55**, 1116–1127.

Krot A. N. (2019) Refractory inclusions in carbonaceous chondrites: Records of early solar system processes. *Meteoritics & Planetary Science*, **54**, 1647–1691.

Krot A. N., Petaev M. I., Scott E. R. D, et al. (1998) Progressive alteration in CV3 chondrites: More evidence for asteroidal alteration. *Meteoritics & Planetary Science*, **33**, 1065–1085.

Lauretta D. S., and Killgore M. (2005) *A Color Atlas of Meteorites in Thin Section.* Golden Retriever Press, Tucson, 301 pp.

Lauretta D. S., Nagahara H., and Alexander C. M. O'D. (2006) Petrology and origin of ferromagnesian silicate chondrules. In *Meteorites and the Early Solar System II*, Lauretta D. S., and McSween H. Y., editors, pp. 431–462, University of Arizona Press, Tucson.

McCoy T. J., Dickinson T. L., and Lofgren G. E. (1999) Partial melting of the Indarch (EH4) meteorite: A textural, chemical, and phase relations view of melting and melt migration. *Meteoritics & Planetary Science*, **34**, 735–746.

McKay D. S., Gibson E. K., Thomas-Keprta K. L., et al. (1996) Search for past life on Mars: Possible relic biogenic activity in martian meteorite ALH84001. *Science*, **273**, 924–930.

McKibbin S. J., Pittarello L., Makarona C., et al. (2019) Petrogenesis of main group pallasite meteorites based on relationships among texture, mineralogy, and geochemistry. *Meteoritics & Planetary Science*, **54**, 2814–2844.

McSween H. Y. (1976) A new type of chondritic meteorite found in lunar soil. *Earth & Planetary Science Letters*, **31**, 193–199.

McSween H. Y. (1977) Petrographic variations among carbonaceous chondrites of the Vigarano type. *Geochimica et Cosmochimica Acta*, **41**, 1777–1790.

McSween H. Y. (1979) Are carbonaceous chondrites primitive or processed? A review. *Reviews of Geophysics & Space Physics*, **17**, 1059–1078.

McSween H. Y. (2015) Petrology on Mars. *American Mineralogist*, **100**, 2380–2395.

McSween H. Y., and McLennan S. M. (2014) Mars. In *Treatise on Geochemistry, 2nd Edition, Vol. 2: Planets, Asteroids,, Comets and the Solar System*, Davis A. M., editor, pp. 251–300, Elsevier, Oxford.

McSween H. Y., Mittlefehldt D. W., Beck A. W., et al. (2011) HED meteorites and their relationship to the geology of Vesta and the Dawn mission. *Space Science Reviews*, **163**, 141–174.

McSween H. Y., Binzel R. P., DeSanctis M. C., et al. (2013) Dawn; the Vesta-HED connection; and the geologic context for eucrites, diogenites, and howardites. *Meteoritics & Planetary Science*, **48**, 2090–2104.

Mittlefehldt D. W. (1994) ALH 84001, a cumulate orthopyroxenite member of the martian meteorite clan. *Meteoritics*, **29**, 214–221.

Mittlefehldt D. W., and Lindstrom M. M. (1990) Geochemistry and genesis of the angrites. *Geochimica et Cosmochimica Acta*, **54**, 3209–3218.

Mittlefehldt D. W., McCoy T. J., Goodrich C. A., and Kracher A. (1998) Non-chondritic meteorites from asteroidal bodies. *Planetary Materials, Reviews in Mineralogy, 36*, Papike J. J., editor, pp. 4-1 to 4-195.

Papike J. J., Ryder G., and Shearer C. K. (1998) Lunar samples. *Planetary Materials, Reviews in Mineralogy, 36*, Papike J. J., editor, pp. 5-1 to 5-234.

Prettyman T. H., Mittlefehldt D. W., Lawrence D. J., et al. (2012) Elemental mapping by Dawn reveals exogenic H in Vesta's howarditic regolith. *Science*, **338**, 242–246.

Rietmeijer F. J. M. (1998) Interplanetary dust particles. *Planetary Materials, Reviews in Mineralogy, 36*, Papike J. J., editor, pp. 2-1 to 2-95.

Rubin A. E. (2018) Carbonaceous and noncarbonaceous iron meteorites: Differences in chemical, physical, and collective properties. *Meteoritics & Planetary Science*, **53**, 2357–2371.

Rubin A. E., and Ma C. (2017) Meteoritic minerals and their origins. *Chemie der Erde*, **77**, 325–385.

Rubin A. E., Scott E. R. D., and Keil K. (1997) Shock metamorphism of enstatite chondrites. *Geochimica et Cosmochimica Acta*, **61**, 847–858.

Russell S. S., Davis A. M., MacPherson G. J., et al. (2000) Refractory inclusions from the ungrouped carbonaceous chondrites MacAlpine Hills 87300 and 88107. *Meteoritics & Planetary Science*, **35**, 1051–1066.

Schramm L. S., Brownlee D. E., and Wheelock M. M. (1989) Major element composition of stratospheric micrometeorites. *Meteoritics*, **24**, 99–112.

Scott E. R. D., and Wasson J. T. (1975). Classification and properties of iron meteorites. *Reviews of Geophysics & Space Physics*, **13**, 527–546.

Scott E. R. D., and Krot A. N. (2004) Chondrites and their components. In *Treatise on Geochemistry, Vol. 1: Meteorites, Comets, and Planets*, Davis A. M., editor, pp. 143–200, Elsevier, Oxford.

Scott E. R. D., Jones R. H., and Rubin A. E. (1994) Classification, metamorphic history, and pre-metamorphic composition of chondrules. *Geochimica et Cosmochimica Acta*, **58**, 1203–1209.

Scott E. R. D., Keil K., and Stöffler D. (1992) Shock metamorphism of carbonaceous chondrites. *Geochimica et Cosmochimica Acta*, **56**, 4281–4293.

Scott E. R. D., Haack H., and Love S. G. (2001) Formation of mesosiderites by fragmentation and reaccretion of a large differentiated asteroid. *Meteoritics & Planetary Science*, **36**, 869–881.

Sears D. W. G., Grossman J. N., Melcher C. L., et al. (1980) Measuring the metamorphic history of unequilibrated ordinary chondrites. *Nature,* **287**, 791–795.

Stöffler D., Keil K., and Scott E. R. D. (1991) Shock metamorphism of ordinary chondrites. *Geochimica et Cosmochimica Acta*, **55**, 3845–3867.

Stöffler D., Hamann C., and Metzler K. (2018) Shock metamorphism of planetary silicate rocks and sediments: Proposal for an updated classification system. *Meteoritics & Planetary Science*, **53**, 5–49.

Tomkins A. G., Johnson T. E., and Mitchell J. T. (2020) A review of the chondrite-achondrite transition, and a metamorphic facies series for equilibrated primitive stony meteorites. *Meteoritics & Planetary Science*, **55**, 857–885.

Udry A., and Day J. M. D. (2018) 1.34 million-year-old magmatism on Mars evaluated from the co-genetic nakhlite and chassignite meteorites. *Geochimica et Cosmochimica Acta*, **238**, 292–315.

Udry A., Howarth G. H., Herd C. D. K., et al. (2020) What martian meteorites reveal about the interior and surface of Mars. *Journal of Geophysical Research: Planets*, **125**, e2020JE006523.

Van Schmus W. R., and Wood J. A. (1967) A chemical-petrologic classification for the chondritic meteorites. *Geochimica et Cosmochimica Acta*, **31**, 747–765.

Warren P. H., and Taylor G. J. (2014) The Moon. In *Treatise on Geochemistry, Vol. 1: Meteorites, Comets, and Planets*, Davis A. M., editor, pp. 559–599, Elsevier, Oxford.

Warren P. H., and Wasson J. T. (1977) Pristine nonmare rocks and the nature of the lunar crust. *Proceedings of the Lunar and Planetary Science Conference*, **8**, 2215–2235.

Wlotzka F. (1993) A weathering scale for the ordinary chondrites. *Meteoritics*, **28**, 460.

Young E. D., Kohl I. E., Warren P. H., et al. (2016) Oxygen isotopic evidence for vigorous mixing during the Moon-forming giant impact. *Science*, **351**, 493–496.

7 Element Fractionations by Cosmochemical and Geochemical Processes

Overview

In this chapter, cosmochemical (nebular) and geochemical (planetary) processes that cause the separation of elements from each other (fractionations) are considered. We explain the basics of equilibrium condensation and how condensation sequences are calculated, and discuss the applicability of condensation theory to the early solar nebula. Changes in oxidation state can also affect the fractionation of iron and the volatility of some elements. The partitioning of elements during partial melting and fractional crystallization control the compositions of magmas, residues, and crystal cumulates. Alteration by aqueous fluids fractionates elements based on their solubilities. The physical sorting and segregation of chondrite components in the nebula, the ejection of basaltic melts during pyroclastic eruptions on small bodies, aeolian sorting, and the impact removal of the outer layers of differentiated bodies are examples of physical fractionations. Planetary differentiation – the formation of cores, mantles, and crusts on rocky bodies, the separation of hydrogen and helium in gas giants, and the separation of ices and rock in the ice giants and icy moons – fractionates elements at global scales.

7.1 What Are Element Fractionations and Why Are They Important?

The solar system formed from a collection of gas and dust inherited from its parent molecular cloud. The bulk elemental composition of this material, as best we can know it, is given by the solar system abundances of the elements (Table 4.1). From this bulk material, the planets, moons, asteroids, and comets formed, each with its own particular composition. Processes occurring before, during, and after accretion separated, or *fractionated*, elements from one another. By studying these elemental fractionations, we can identify the processes that separated the elements and learn about the physiochemical conditions involved. This is particularly important for understanding the early solar system, because its processes and conditions are not directly observable.

Quantifying the chemical fractionations in planetary samples and in meteorites – the closest analogs we have to the materials that formed planets – is a necessary first step. This involves careful measurement of their elemental compositions and understanding the compositions of the material from which they formed. Once these fractionations have been identified, experiments (and theory, when the relevant experiments cannot be performed) that yield similar fractionations can point to the processes and constrain the conditions that produced them.

A great deal is known about elemental fractionations on or within the Earth; these fractionations are the business of geochemistry. It is important to remember, though, that our planet is a grand experiment carried out under a particular, and possibly unique, set of conditions. Meteorites, which sample the asteroids, the Moon, and Mars, and samples returned by spacecraft missions from various solar system bodies, allow us to see how similar natural experiments proceeded under different conditions. Based on measured element fractionations in meteorites, a rich variety of processes is indicated.

Cosmochemical fractionations may involve vapor–solid and vapor–liquid transformations (*condensation* of a gas, or its reverse, *evaporation*). Each element condenses over a very limited temperature range, so one would expect the compositions of the condensed phase and the vapor phase to change as a function of the ambient temperature. Many of the chemical fractionations that took place in the early solar system were due, in one way or another, to this phenomenon. It is convenient to quantify volatility by use of the *50% condensation temperature* – that is, the temperature by which 50% of the mass of a particular element has condensed from gas of solar composition. Table 7.1 lists the 50%

Table 7.1 Equilibrium 50% condensation temperatures for a gas of solar composition at 10^{-4} atm

Element	T_{cond} (K)	Initial Condensate	Cosmochemical Classification
Li	1225	Li_2SiO_3 in $MgSiO_3$	Mod. volatile
Be	1490	$BeAl_2O_4$ in melilite and spinel	Refr. lithophile
B	964	$CaB_2Si_2O_8$ in feldspar	Mod. volatile
F	736	$Ca_5(PO_4)F$	Mod. volatile
Na	970	$NaAlSi_3O_8$ in feldspar	Mod. volatile
Mg	1340	Mg_2SiO_4	Major element
Al	1650	Al_2O_3	Refr. lithophile
Si	1340	$Ca_2Al_2SiO_7$	Major element
P	1151	Fe_3P	Mod. volatile
S	648	FeS	Mod. volatile
Cl	863	$Na[AlSiO_4]_3Cl$	Mod. volatile
K	1000	$KAlSi_3O_8$ in feldspar	Mod. volatile
Ca	1518	$CaAl_{12}O_{19}$	Refr. lithophile
Sc	1652	Sc_2O_3	Refr. lithophile
Ti	1549	$CaTiO_3$	Refr. lithophile
V	1455	Dissolved in $CaTiO_3$	Refr. lithophile
Cr	1301	Dissolved in Fe alloy	Mod. volatile
Mn	1190	Mn_2SiO_4 in olivine	Mod. volatile
Fe	1337	Fe alloy	Major element
Co	1356	Dissolved in Fe alloy	Refr. siderophile
Ni	1354	Dissolved in Fe alloy	Refr. siderophile
Cu	1170	Dissolved in Fe alloy	Mod. volatile
Zn	684	ZnS dissolved in FeS	Mod. volatile
Ga	918	Dissolved in Fe alloy	Mod. volatile
Ge	825	Dissolved in Fe alloy	Mod. volatile
As	1012	Dissolved in Fe alloy	Mod. volatile
Se	684	FeSe dissolved in FeS	Mod. volatile
Br	~350	$Ca_5(PO_4)_3Br$	Highly volatile
Rb	~1080	Dissolved in feldspar	Mod. volatile
Sr	1217	Dissolved in $CaTiO_3$	Refr. lithophile
Y	1622	Y_2O_3	Refr. lithophile
Zr	1717	ZrO_2	Refr. lithophile
Nb	1517	Dissolved in $CaTiO_3$	Refr. lithophile
Mo	1595	Refractory metal alloy	Refr. siderophile
Ru	1565	Refractory metal alloy	Refr. siderophile
Rh	1392	Refractory metal alloy	Refr. siderophile
Pd	1320	Dissolved in Fe alloy	Mod. volatile
Ag	993	Dissolved in Fe alloy	Mod. volatile
Cd	430*	CdS in FeS	Highly volatile
In	470	InS in FeS	Highly volatile
Sn	720	Dissolved in Fe alloy	Mod. volatile
Sb	912	Dissolved in Fe alloy	Mod. volatile
Te	680	FeTe dissolved in FeS	Mod. volatile
I			Highly/mod. vol.?
Cs			Highly/mod. vol.?
Ba	1162	Dissolved in $CaTiO_3$	Refr. lithophile
La	1544	Dissolved in $CaTiO_3$	Refr. lithophile

Table 7.1 *(cont.)*

Element	T_{cond} (K)	Initial Condensate	Cosmochemical Classification
Ce	1440	Dissolved in $CaTiO_3$	Refr. lithophile
Pr	1557	Dissolved in $CaTiO_3$	Refr. lithophile
Nd	1563	Dissolved in $CaTiO_3$	Refr. lithophile
Sm	1560	Dissolved in $CaTiO_3$	Refr. lithophile
Eu	1338	Dissolved in $CaTiO_3$	Refr. lithophile
Gd	1597	Dissolved in $CaTiO_3$	Refr. lithophile
Tb	1598	Dissolved in $CaTiO_3$	Refr. lithophile
Dy	1598	Dissolved in $CaTiO_3$	Refr. lithophile
Ho	1598	Dissolved in $CaTiO_3$	Refr. lithophile
Er	1598	Dissolved in $CaTiO_3$	Refr. lithophile
Tm	1598	Dissolved in $CaTiO_3$	Refr. lithophile
Yb	1493	Dissolved in $CaTiO_3$	Refr. lithophile
Lu	1598	Dissolved in $CaTiO_3$	Refr. lithophile
Hf	1690	HfO_2	Refr. lithophile
Ta	1543	Dissolved in $CaTiO_3$	Refr. lithophile
W	1794	Refractory metal alloy	Refr. siderophile
Re	1818	Refractory metal alloy	Refr. siderophile
Os	1812	Refractory metal alloy	Refr. siderophile
Ir	1603	Refractory metal alloy	Refr. siderophile
Pt	1411	Refractory metal alloy	Refr. siderophile
Au	1284	Refractory metal alloy	Refr. siderophile
Hg			Highly/mod. vol.?
Tl	448	Dissolved in Fe alloy	Highly volatile
Pb	520	Dissolved in Fe alloy	Highly volatile
Bi	472	Dissolved in Fe alloy	Highly volatile
Th	1598	Dissolved in $CaTiO_3$	Refr. lithophile
U	1580	Dissolved in $CaTiO_3$	Refr. lithophile

*Temperature for 10^{-5} atm.
Data from Wasson (1985) and Lodders and Fegley (1998).

condensation temperatures for elements in a gas of solar composition at a pressure of 10^{-4} atmospheres, thought to be a typical pressure for the solar nebula. Based on 50% condensation temperatures, the elements can be placed into several groups according to their volatilities (comprehensively reviewed by Davis, 2006). Elements that have 50% condensation temperatures higher than that for chromium are considered to be *refractory elements*, while those with condensation temperatures below that of sulfur, which condenses as FeS, are classified as *highly volatile elements*. The *moderately volatile elements* condense at temperatures between these extremes, at roughly 1300–640 K.

Physical processes in the nebula can also lead to elemental fractionations. Mechanical sorting and segregation of solids with different densities or aerodynamic characteristics may explain variations in chondrite compositions. A few elements can sometimes be segregated by redox state, which can affect density or volatility.

Geochemical fractionations occur on planetesimals or planets after their accretion. Partial melting yields melts and residues that differ in composition, as some elements favor the liquid or solid phase. During *fractional crystallization* of magmas, segregation of crystals produces residual liquids and *cumulates* that differ in composition from the parental magma. Alteration by aqueous fluids can fractionate elements by solubility. Catastrophic collisions can remove the outer portions of differentiated bodies, and ejection of partial melts (basalts) during pyroclastic eruptions can alter the compositions of asteroids.

Finally, global planetary *differentiation* to form silicate crusts and mantles and metallic cores on rocky bodies obviously leads to marked element segregations that depend on the extent of melting, internal pressures, and other factors. Differentiation of icy bodies forms fluids (aqueous or otherwise) that solidify as ice shells or persist as subsurface oceans.

7.2 Condensation as a Fractionation Process

Some of the materials comprising the solar system have apparently been processed though a high-temperature stage, where solids were vaporized. As the system cooled, new solids and perhaps liquids condensed. If the condensates were removed from contact with the remaining gas, compositions that are fractionated relative to the solar system composition would occur. Evaporation is essentially the opposite of condensation, if both occur under equilibrium conditions. Thus, partial evaporation might also explain the composition of some nebular materials.

7.2.1 Condensation Sequences

In Chapter 1 and again above, we introduced the cosmochemical classification of elements based on their relative volatilities in a system of cosmic (solar) composition. In a cooling solar gas, elements condense in a certain order, depending on their volatility (Table 7.1). Condensation and evaporation partition elements between coexisting gas and solid (or liquid) phases, and the removal of one or the other of these phases can fractionate the element abundances of the system as a whole from their original cosmic relative proportions.

Condensation under the conditions that obtained in the solar nebula is not really amenable to direct experiment. However, experiments have provided thermodynamic data with which the condensation process can be modeled theoretically.

Condensation calculations were done long before digital computers made complex calculations possible. In the 1950s and 1960s, Harold Urey and other investigators performed relatively simple calculations to model the condensation behavior of elements in astrophysical settings. Grossman (1972) improved these calculations by taking into account the compositional changes in the gas as condensed materials were sequentially removed. Complex calculations now normally consider the behavior of the 23 elements with the highest cosmic abundance: hydrogen, helium, carbon, nitrogen, oxygen, fluorine, neon, sodium, magnesium, aluminum, silicon, phosphorus, sulfur, chlorine, argon, potassium, calcium, titanium, chromium, manganese, iron, cobalt, and nickel. These calculations also consider thousands of compounds that can potentially form from these elements. Quite a number of especially good reviews of condensation calculations are available (e.g. Grossman and Larimer, 1974; Fegley, 1999; Ebel, 2006).

Condensation theory is based on *thermodynamic equilibrium*. More than a century's worth of experiments have yielded thermodynamic data (entropy and enthalpy of formation, plus heat capacity) for elements and compounds. Equations of state describing the stabilities of different compounds under various conditions can be calculated from these data, as briefly described in Box 7.1. Because liquids are not normally stable at the low pressures appropriate for space, the compounds in condensation calculations are generally solid minerals, but liquids can exist at higher pressures (possibly achievable if some areas of the nebula with enhanced dust concentrations relative to gas were vaporized).

Figure 7.1 illustrates the sequence of appearance as a function of temperature of various solid phases from a cooling solar gas. This particular condensation sequence was calculated at a pressure of 10^{-4} atm (thought to be appropriate for parts of the nebula). It is important to note that not every mineral in the sequence actually condenses directly from the vapor. Instead, some minerals form by reaction of previously condensed solids with the vapor. For example, troilite (FeS) forms by reaction of already condensed Fe metal with sulfur in the gas phase, and olivine first condenses as the magnesium end member forsterite and then becomes progressively more iron rich by reaction with vapor as temperature decreases.

Four condensation sequences for a gas of solar composition (all at the same pressure, in this case 10^{-3} atm) calculated by various investigators are compared in Figure 7.2. The differences in these sequences result primarily from the choice of thermodynamic data. These diagrams illustrate the temperatures of appearance (and, for some cases, disappearance) of phases, but not the fraction of each element condensed. There is remarkable agreement, considering that some calculations did not take titanium or noble gases into account. The absence of titanium prevents condensation of perovskite ($CaTiO_3$), which, in turn, affects the stabilities of other minerals, and the presence of noble gases changes total gas pressure in the calculation.

In condensation calculations, internally consistent thermodynamic datasets are preferred (but not always used) over data haphazardly collected from disparate sources. Another problem is determining how solid-solution minerals are modeled. Many of the condensate minerals, such as melilite and pyroxene, are not pure phases, but rather are chemical mixtures of several

Box 7.1 Gibbs Free Energy and Condensation Calculations

Gibbs free energy (G) is a convenient way to describe the energy of a system. More-easily measured thermodynamic quantities are enthalpy (H) and entropy (S), which describe the heat exchanged during a reaction and the subsequent loss of capacity to do work. The change in free energy, ΔG, during a reaction (such as condensation of a solid phase from gaseous species) is related to changes in enthalpy and entropy by

$$\Delta G = \Delta H - T\Delta S \quad (7.1)$$

where T is the absolute temperature in kelvin, and ΔH and ΔS refer to values calculated for reactants minus products. Energy is available to produce a spontaneous change in the system as long as ΔG is negative. Once free energy is minimized ($\Delta G = 0$), the system has attained equilibrium.

Experimentally determined ΔH and ΔS values for the formation of various compounds from their constituent elements are tabulated under standard conditions (normally 298 K and 1 bar). These values are temperature sensitive and can be extrapolated to the temperature of interest using the measured heat capacity, ΔC_P, by integrating the following equation between 298 K and the temperature of interest:

$$\Delta G_T = \Delta H^0 + \int_{298}^{T} \Delta C_P dT - T\left(\Delta S^0 + \int_{298}^{T} (\Delta C_P/T)dT\right) \quad (7.2)$$

where the superscript 0 denotes enthalpy and entropy values in their standard states.

To illustrate how a (simple) condensation calculation works, let's consider the condensation of corundum (Al_2O_3), one of the earliest phases predicted to condense from a solar gas. The partial pressures of aluminum (P_{Al}) and oxygen (P_O) in the gas phase must first be determined, based on their solar abundances, over a range of temperatures. The equilibrium between solid (s) corundum and gas (g) can be represented by this equation:

$$2Al(g) + 3O(g) = Al_2O_3(s) \quad (7.3)$$

The equilibrium constant (K_{eq}) for this reaction takes the form:

$$K_{eq} = \frac{a_{Al_2O_3}}{(P_{Al}{}^2) \times (P_O{}^3)}, \quad (7.4)$$

where $a_{Al_2O_3}$ is the activity of corundum (equal to 1 for a pure phase). Because

$$\Delta G = -RT\ln K_{eq}, \quad (7.5)$$

we can calculate the temperature at which corundum is in equilibrium with the gas (that is, where $\Delta G = 0$), which is the condensation temperature (1758 K) for this mineral. (A note of caution: it is easy to make an error at this step, because tabulated values of free energy of formation from the elements for corundum do not use Al(g) and O(g) as the standard states. NIST-JANAF tables (Chase, 1998) give free energy of formation of gaseous monatomic species.)

In practice, laborious modern calculations find the optimum distribution of elements between vapor and solid (or liquid) phases by minimizing the Gibbs free energy of the entire system. This calculation is accomplished using the following steps (Ebel, 2006):

1. At a particular temperature and pressure, the thermodynamic stability of all potential condensates is calculated from their equations of state.
2. A small mass increment of each stable solid condensate is added to the system, with complementary subtraction of that mass from the gas composition.
3. Matter is then redistributed between all the coexisting condensed phases and gas, until the calculated free energy of the system achieves its lowest value.
4. The thermodynamic stability of each condensate phase is then checked again.

Iteration of this process at successively lower temperatures gives a condensation sequence. These calculations would be extremely tedious if not for the advent of modern computer systems.

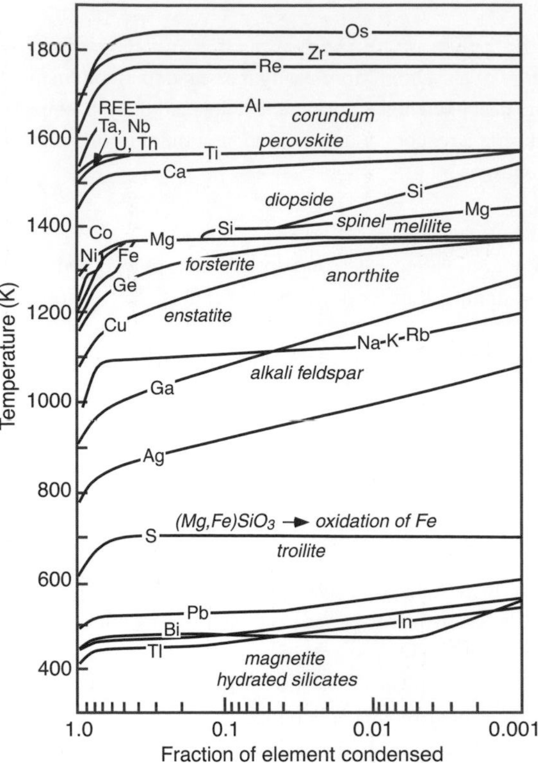

Figure 7.1 Calculated condensation sequence for a gas of solar composition at 10^{-4} atm pressure. Condensed minerals are labeled in italics and curves show the fraction of each element condensed as a function of temperature. Modified from Grossman and Larimer (1974).

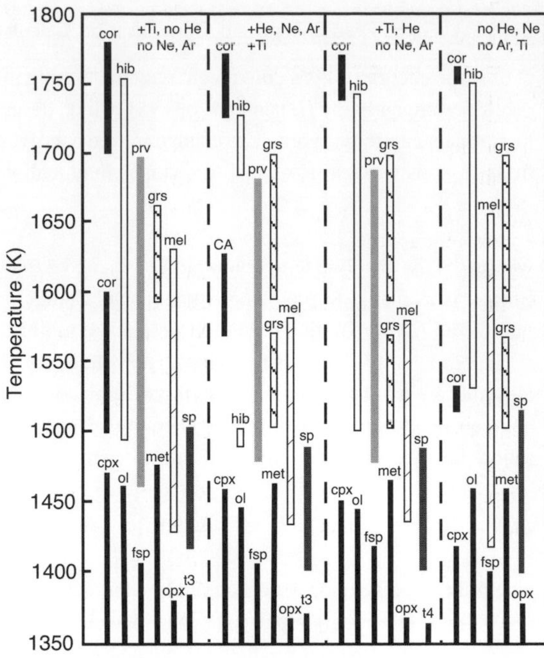

Figure 7.2 Comparison of four condensation sequences, all at 10^{-3} atm, but using different models. These calculations are from (left to right) Petaev and Wood (1998), Ebel and Grossman (2000), Yoneda and Grossman (1995), and Wood and Hashimoto (1993). Mineral abbreviations and formulae are given in Table 7.2. Modified from Ebel (2006).

end-member components. Various mixing models for solid solutions have been devised, which may give different results.

Variations in gas pressure also affect the condensation sequence, as illustrated in Figure 7.3. Except for differing pressures (10^{-3} and 10^{-5} atm), the parameters for these two calculations are identical. The temperatures and orders of appearance of condensed minerals clearly vary with pressure, which could change with position in the nebula.

So far, we have been discussing the condensation of elements with high cosmic abundances. Generally, thermodynamic data for trace elements are lacking, or are only available for pure elements or simple compounds. Chemical analyses of high-temperature minerals in meteorites guide our understanding of the condensation of elements in trace abundances. The analyses indicate that most elements in trace quantities tend to substitute into various host minerals, rather than forming pure phases of their own (see Table 7.1). In calculations, they are allowed to condense as simple metals, oxides, or sulfides, which are then assumed to dissolve into appropriate major phases that are stable under the corresponding conditions. For example, rare earth elements (REEs) condense as oxides and substitute into perovskite or hibonite, and cadmium condenses as CdS and substitutes into FeS. Platinum-group elements (ruthenium, rhodium, palladium, rhenium, osmium, iridium, and platinum) condense as metals at very high temperatures (Fig. 7.1) before Fe metal, so they form a separate refractory metal phase – actually found as tiny nuggets in refractory inclusions in meteorites.

Condensation calculations have also been performed using non-solar starting compositions. The solar composition has more oxygen than carbon, so it is oxidizing. Compositions with more abundant carbon than oxygen tie up oxygen as CO, so they are reducing. Condensation of a carbon-rich gas, which occurs in gases expelled from red giant stars, produces graphite (C) and carbides (SiC and TiC), as found in presolar grains. Portions of the solar nebula may also have been enriched in carbon, yielding the reduced minerals found in enstatite

Table 7.2 Mineral names, abbreviations, and formulae commonly found in condensation sequences

Mineral	Abbreviation	Chemical Formula
Corundum	cor	Al_2O_3
Hibonite	hib	$CaAl_{12}O_{19}$
Grossite	grs	$CaAl_4O_7$
Ca-monoaluminate	CA	$CaAl_2O_4$
Perovskite	prv	$CaTiO_3$
Melilite	mel	$Ca_2(Al_2,MgSi)SiO_7$
Spinel	sp	Al-rich $(Fe,Mg,Cr,Al,Ti)_3O_4$
Olivine	ol	$(Mg,Fe)_2SiO_4$
Clinopyroxene	cpx	$Ca(Mg,Fe)Si_2O_6$
Orthopyroxene	opx	$(Mg,Fe)_2Si_2O_6$
Feldspar	fsp	$(CaAl,NaSi)AlSi_2O_8$
Metal	met	Fe,Ni,Co
	t3	Ti_3O_5
	t4	Ti_4O_7

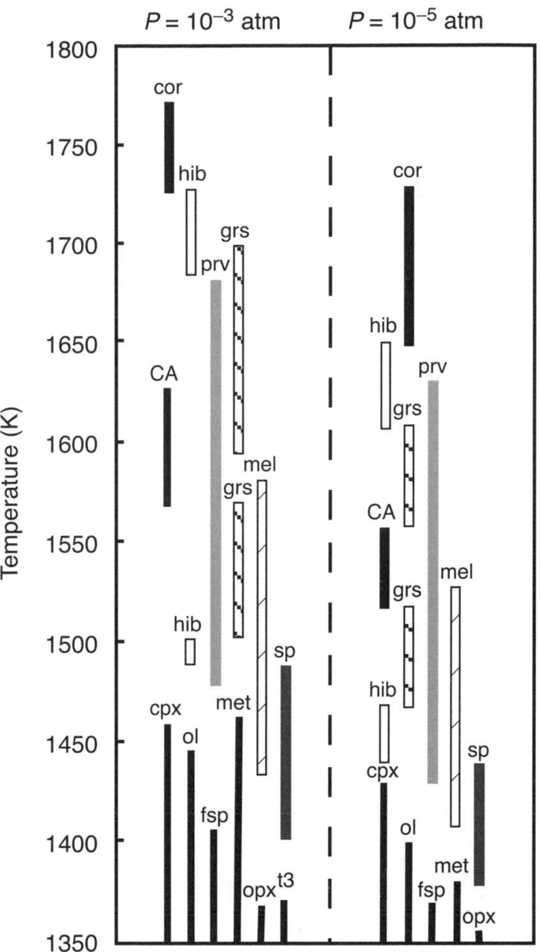

Figure 7.3 Comparison of two condensation sequences calculated at different pressures. Abbreviations and formulae are given in Table 7.2. Modified from Ebel (2006).

chondrites. Local vaporization of dust-enriched areas may also have produced non-solar gas compositions, possibly allowing silicate liquids to condense (perhaps forming some chondrules?).

7.2.2 Applicability of Condensation Calculations to the Early Solar System

During the 1970s and 1980s, equilibrium condensation calculations were thought to be generally applicable to the formation of meteorites, their constituents, and even to planetary bodies. This idea fit well with the hot solar nebula model introduced by Cameron (1962), which proposed that, during formation, the entire inner solar system was heated to high enough temperatures by conversion of gravitational potential energy that all preexisting solids were vaporized. This idea was supported by observations that all materials available for study had apparently sampled the same uniform isotopic reservoir. The calcium-aluminum-rich inclusions (CAIs) found in chondrites were identified in the early 1970s as initial condensates from the solar nebula (Grossman and Larimer, 1974). The bulk compositions of meteorites were interpreted in terms of condensation (Larimer and Anders, 1967, 1970; Wai and Wasson, 1977). Differences in abundances of moderately volatile elements between chondrites were ascribed to differences in their accretion temperatures. Metal was thought to have condensed from the nebula and was accreted in different proportions relative to silicates in the different meteorite classes. CI chondrites were considered to be complete condensates from the solar nebula, with even their mineralogy resulting from gas–solid equilibrium reactions at low temperatures (Larimer and Anders, 1970). However, since that time, improved techniques for measuring isotopic compositions have shown that isotopic variation among samples is the norm, not the exception (to be discussed in Chapter 8), and many different examples of preserved presolar material have been found in chondrites, as discussed in Chapter 5. None of these isotope anomalies or presolar grains could have survived complete evaporation of the nebular solids. So we must ask the question: did equilibrium gas–solid exchange, either by condensation or evaporation, actually take place in the early solar system?

Consider first the CAIs. In general, their bulk compositions are consistent with those calculated for the first 5% of condensable matter (Davis and Richter, 2004). Moreover, many of the minerals that comprise CAIs (hibonite, perovskite, spinel, melilite, diopside, anorthite) are predicted to have been the earliest condensed phases. However, not all CAIs are alike. For example, *type B CAIs* are coarse-grained igneous objects whose mineralogy and mineral chemistry are controlled by igneous processes (Fig. 7.4). Crystallization of a refractory liquid with the composition of CAIs produces many of the same minerals as equilibrium condensation would, but the details of the partitioning of elements between phases and the textural relationships are quite different. Thus, type B CAIs are not condensates, although the bulk compositions of these objects permit their precursors to have been condensates. The isotopic systematics of type B CAIs imply that the precursors were not composed entirely of condensates either. Type B CAIs are characterized by isotopic anomalies in calcium, titanium, and other elements, relative to both the bulk solar system composition and to each other (Niederer et al., 1981; Jungck et al., 1984; Rotaru et al., 1992). Such anomalies could not survive a general evaporation and condensation because the isotopes would have been thoroughly mixed in the gas phase. Local evaporation and condensation in the immediate region where each CAI formed might be permitted. But another possibility is that the immediate precursors to the type B CAIs were collections of refractory grains from a wide variety of sources, some or most of them presolar, that were concentrated by the evaporation of more volatile material, then melted in the solar nebula, and subsequently crystallized to form the objects that we observe.

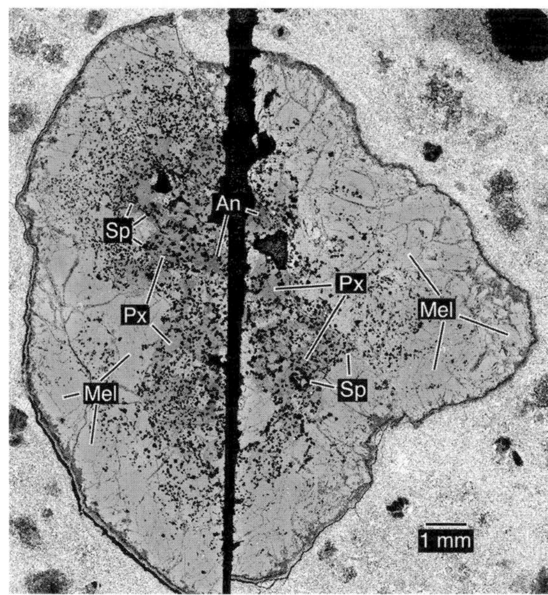

Figure 7.4 Backscattered electron image of a type B CAI in the Allende CV3 chondrite. This object crystallized from a melt and has an outer mantle of melilite (Mel) surrounding an inner zone of pyroxene (Px) and anorthite (An). Tiny grains of spinel (Sp) are abundant in the inner portion of the inclusion. Image courtesy of A. Krot.

On the other hand, fine-grained *type A CAIs* consist of aggregates of nodules with a concentric structure consisting of spinel cores surrounded by melilite, anorthite, and diopside, or melilite cores surrounded by anorthite and diopside (Fig. 7.5). The sequence of minerals in the nodules is consistent with a condensation sequence. However, the textures of the overall objects are complicated and often indicate multistage histories. For these objects, nebular condensation clearly played a major role in their formation, but even in these objects the story is more complicated than envisioned decades ago when they were first described.

Metal, too, has diverse origins. Much of the metal that we currently observe in ordinary chondrites no longer retains its original features, but instead reflects diffusion and grain growth due to metamorphic heating in meteorite parent bodies. In the least metamorphosed chondrites, we get some glimpses of the origin of the metal. Fine-grained matrices of the most primitive chondrites contain tiny micron-sized grains of metal and sulfide. The origin of these grains is not clear: they could be inherited from the

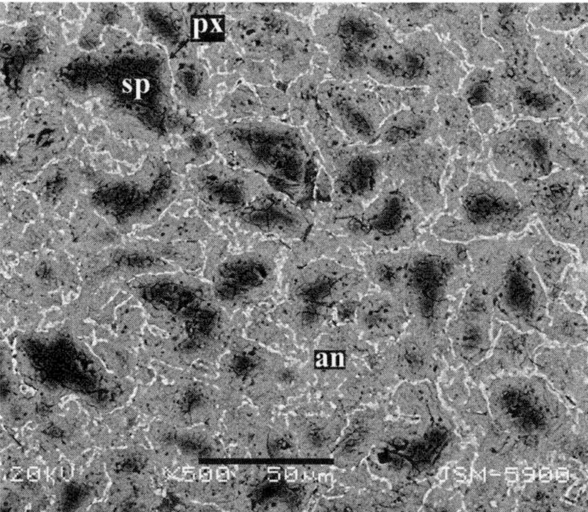

Figure 7.5 Backscattered electron image of a portion of a fine-grained type A CAI in the Efremovka CV3 chondrite. This inclusion consists of nodules of spinel (sp) surrounded by anorthite (an) and thin rims of pyroxene (px). This CAI may be an aggregate of condensates.

interstellar medium or they could be the result of condensation in the nebula. These tiny grains are quickly destroyed by parent-body metamorphism, with their constituent atoms combining into larger metal grains (Huss et al., 2006). Careful study of the trace-element patterns in metal grains inside and outside of chondrules from CR chondrites shows that much of the metal in these meteorites was produced by reduction during chondrule formation and was expelled from the molten chondrules before they solidified (Connolly et al., 2001). Other metal grains in these meteorites, particularly those coating the surfaces of chondrules, show evidence of recondensation of elements that evaporated during the chondrule-heating event. The metal grains in CH chondrites have unusual zoning patterns in siderophile elements consistent with condensation calculations from a gas of solar composition (Campbell et al., 2001). These grains, and some of the chondrules found in CB and CH chondrites, are considered to be equilibrium condensates from a local gas plume produced by a collision between large asteroids.

The volatile element depletions among the various classes of chondrites were once considered to be the result of equilibrium condensation, with accretion of the different classes of meteorites taking place at different temperatures before the missing volatiles could condense. However, evaporation and condensation can both produce solids depleted in volatile elements. Equilibrium evaporation or condensation should produce sharp steps on an abundance versus condensation temperature plot because once the condensed phase becomes stable, the elements will quickly and quantitatively condense into the stable phase, and once the condensed phase becomes unstable, it will quickly and quantitatively evaporate into constituent elements. But this is not what is observed. Figure 7.6 shows the average bulk composition of CR2 chondrites, normalized to CI chondrites. The elements are arranged in order of decreasing volatility from left to right. Volatile and moderately volatile elements (those more volatile than chromium) are depleted, but the depletions increase steadily with increasing volatility over a range in condensation temperatures of 600 K. Refractory elements are enriched relative to CI, reflecting the depletions of volatile elements. Element abundances are not related to lithophile, siderophile, or chalcophile behavior. The easiest way to understand this pattern is if the loss of the volatile elements took place through destruction of various mineral phases formed at different places and times and under different conditions. Evaporation then is a function of the stability of the mineral in which the element is sited, not of the element itself, so the removal

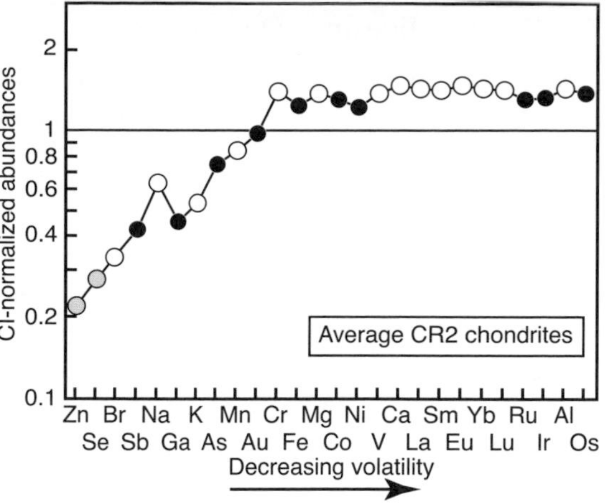

Figure 7.6 Elemental abundances in CR2 chondrites normalized to the CI composition and plotted in order of decreasing volatility from left to right. Lithophile elements are shown with open circles, siderophile elements with black circles, and chalcophile elements with gray circles. Data from Kallemeyn et al. (1994).

of the volatiles from the solid occurs over an extended temperature interval. In Chapter 5, we observed that the depletions of fragile types of presolar grains vary in concert with the volatile element depletions of the host meteorite (see Fig. 5.16). It seems fairly clear, then, that equilibrium condensation did not control the bulk compositions of the chondrites.

We should also address the question of whether CI chondrites represent the complete, low-temperature condensate from the solar nebula. Their bulk composition is consistent with such a model. However, CI chondrites contain among the highest abundances of presolar grains that are not destroyed by aqueous alteration (the mineralogy of CI chondrites is almost entirely due to such alteration). This suggests that it is more likely that CI chondrites formed from representative samples of the dust inherited from the Sun's parent molecular cloud – heavily processed, but neither fractionated nor vaporized in the solar system.

In summary, equilibrium condensation represents one end member of a continuum of gas–solid interactions that took place in the early solar system. Some samples appear to have formed under conditions that closely match those of equilibrium condensation. But the volatility-based fractionations that are widely observed in solar system materials are due to much more complicated processes that involve evaporation (the opposite of condensation) or mixing of materials produced under different conditions, as well as a variety of processes in which kinetic effects played an important role.

7.3 Volatile Element Depletions

In most classes of chondrites, highly volatile and moderately volatile elements are depleted relative to refractory elements when compared to CI chondrites. A plot of volatile element abundances in CV chondrites, normalized to CI chondrites and silicon, versus their 50% condensation temperatures is shown in Figure 7.7. Figure 7.6 shows a similar trend for CR chondrites, in a different format. Like most other chondrite groups, the CV and CR chondrites show a smooth drop-off in the abundances of elements with higher volatility. Different symbols distinguish siderophile, lithophile, and chalcophile elements, illustrating that this pattern is unrelated to geochemical affinity. As discussed above, these depletions were traditionally interpreted in terms of equilibrium condensation from a gas of solar composition. However, there are other ways in which such depletions might arise (Davis, 2006).

7.3.1 Gas–Solid Interactions

The dust that made up the Sun's parent molecular cloud contained mineral grains that condensed in the atmospheres of dying stars (Chapter 5). Stellar atmospheres also likely produced amorphous grains as the temperature and density dropped during expansion of the ejected gas, and some atoms may have reached interstellar space without condensing into grains at all. The journey from the site of nucleosynthesis to the molecular cloud that gave rise to the solar system was a treacherous one. Supernova explosions send shock waves driven by extremely hot gas through the interstellar medium, evaporating all dust for large distances in all directions. Massive stars emit a tremendous flux of UV light that can also evaporate dust. The atoms from these vaporized dust grains were part of a tenuous gas consisting of elements of all volatilities, from the most volatile to the most refractory. A relatively small fraction of the dust that formed in stellar ejecta survived to become part of the Sun's parent molecular cloud. When gas and dust came together to form the cloud, the particle density increased and the temperature dropped to as low as 10 K. Condensable elements in the gas condensed onto pre-existing dust. But this was not equilibrium condensation because the temperatures were too low. Atoms of different volatilities were mixed together without long-range structure. The grains were amorphous and perhaps icy and had a lot of chemical potential energy. So solar system formation began in a cloud consisting of gas composed mostly of hydrogen, helium, and perhaps neon and carbon monoxide, and dust composed mostly of amorphous material plus some mineral grains that condensed in the atmospheres of dying stars.

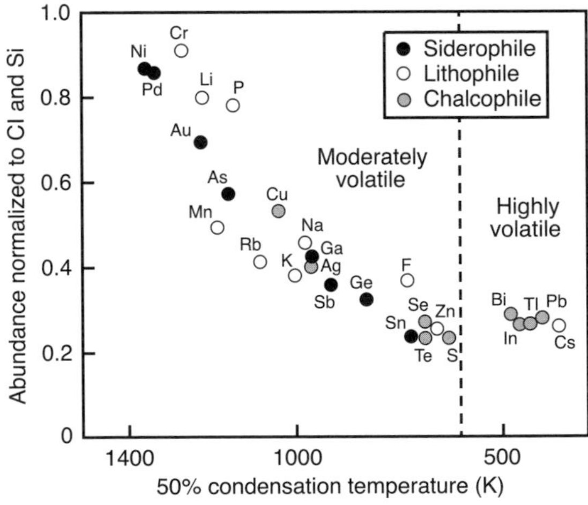

Figure 7.7 Volatile element abundances in CV3 chondrites (normalized to CI chondrites and silicon) lie along a linear array on semi-log plots versus their 50% condensation temperatures. This depletion pattern persists whether the elements are siderophile, lithophile, or chalcophile.

As a portion of the molecular cloud collapsed to form the proto-Sun and its accretion disk, the temperature rose and things began to happen in the dust. Ices evaporated, taking included atoms of refractory elements into the gas phase with them. The refractory elements typically condensed out again on the first grain they encountered. As temperatures rose still further, the kinetic barriers that kept the amorphous grains from crystallizing were overcome and minerals began to crystallize from the amorphous grains. At the same time, volatile elements were lost from the grains. This general process provides a way for volatile elements in the dust to have been fractionated without the system ever achieving chemical equilibrium.

Two types of models have been proposed that use this general picture as the basis for understanding volatile depletions in chondrites. Yin (2005) proposed that the volatile element depletions in chondrites reflect the extent to which these elements were sited in refractory dust in the interstellar medium. Observations show that in the warm interstellar medium, the most-refractory elements are almost entirely in the dust, whereas volatile elements are almost entirely in the gas phase. Moderately volatile elements are partitioned between the two phases. The pattern for the dust is similar to that observed in bulk chondrites. In the Sun's parent molecular cloud, the highly volatile and moderately volatile elements condensed onto the dust grains in ices. Within the solar system, the ices evaporated, putting the volatile elements back into the gas phase, which was separated from the dust.

Thus, in Yin's model, the chondrites inherited their compositions from the interstellar medium. A slightly different model proposes that the fractionated compositions were produced in the solar nebula by thermal processing of gas and dust inherited from the molecular cloud (Huss et al., 2003; Huss, 2004). As the temperature rose in the accretion disk, ices evaporated, releasing trapped refractory elements, as in Yin's model. However, rather than simply being lost from the system, these elements were refluxed back into the dust with an efficiency proportional to their volatility. In this model, the different chondrite groups formed at nebular locations where the ambient temperatures were slightly different. Those with the most volatile-depleted compositions formed at the highest temperatures. In support of this model is a general correlation between the depletion of volatile elements in a chondrite group and the relative abundances of presolar grains in the least metamorphosed meteorites of that group. As shown earlier in Figure 5.16, as the depletions of moderately volatile elements increase from LL to CO to CR to CV, the abundances of the more-fragile presolar grains decrease.

The depletions of volatile elements in achondrites, and in planets, are even more extreme than in chondrites. To see this effect, it is necessary to remove the effects of magmatic fractionations from the data, as all these samples are igneous rocks. This subject is treated in some detail in a later section of this chapter. For our present purposes, it is sufficient to note that some elements behave similarly during melting and crystallization. Examples of elements with similar geochemical behavior are potassium, lanthanum, and uranium. In the plot of uranium versus lanthanum (both refractory elements) in Figure 7.8a, samples from all the bodies plot along the same line, which is the chondritic ratio of these elements. Although changes in their abundances occur during igneous processing, they always travel together. However, in the plot of potassium versus lanthanum (potassium is volatile) in Figure 7.8b, samples from different planets or meteorite parent bodies are arrayed along different lines, all defining volatile-to-refractory element ratios lower than in chondrites. Thus, the various planets and differentiated asteroids show different degrees of volatile element depletion. It is likely that these fractionations reflect processes similar to those that fractionated the volatile elements in chondrites.

7.3.2 Gas–Liquid Interactions

In general, silicate melts are unstable under the conditions that prevailed in the solar nebula. However, chondrules and some CAIs crystallized from partially molten to completely molten droplets. The melting events that produced chondrules and CAIs thus presented an opportunity for element fractionations to occur.

Chondrules: Chondrules formed in the nebula by crystallization of molten droplets produced by some sort

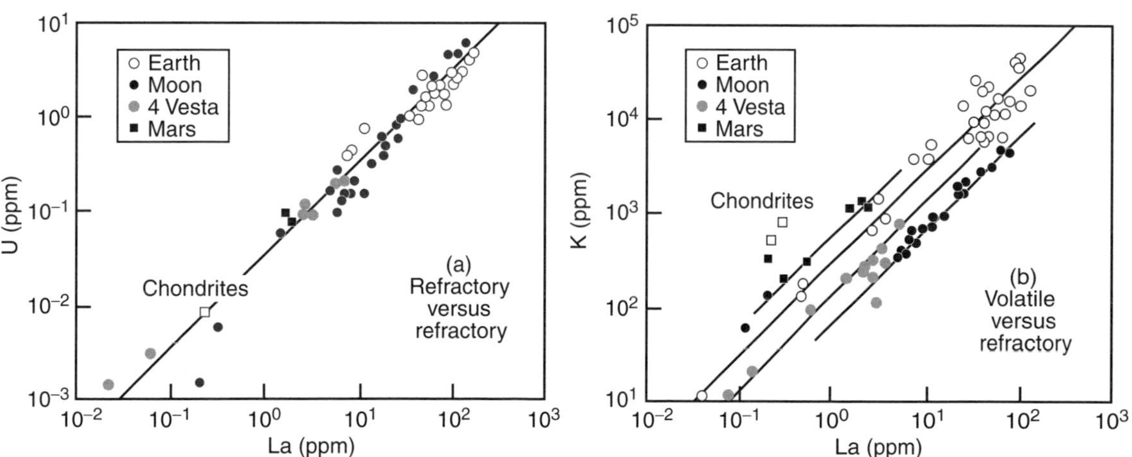

Figure 7.8 Plots of (a) uranium versus lanthanum (two refractory elements), and (b) potassium versus lanthanum (a volatile element and a refractory element) for terrestrial and lunar basalts, HED achondrites (from asteroid Vesta), and martian meteorites. All three elements are incompatible and thus fractionate together during igneous processes, so their ratios remain constant. However, ratios of incompatible elements with different volatilities (K/La) reveal different degrees of volatile element depletion in differentiated bodies. After Wanke and Dreibus (1988).

of flash heating event(s). Heating times are estimated to have been from several seconds to a few minutes to reach peak temperatures between 1770 and 2120 K, followed by rapid but not instantaneous cooling over a few minutes to a few hours. Chondrule melts are unstable in a gas of solar composition and so should have been rapidly losing sulfur, alkalis, iron, and even magnesium. However, evidence that volatile elements were lost from chondrules while they were molten is equivocal. Measurements of the sodium content in crystals from chondrules show no apparent change during crystallization, implying that sodium was not lost from the melt (Alexander et al., 2008). On the other hand, the bulk compositions of chondrites could be interpreted in terms of minor volatile element loss during chondrule formation overprinting the volatility fractionation already in the precursors. One possible explanation for the apparently minor loss of volatiles from chondrules is that many chondrules formed at the same time in close proximity, so that evaporation created a dense atmosphere of evaporated chondrule material that back-reacted with the molten chondrules, minimizing volatile loss and isotopic mass fractionation.

CAIs: Figure 7.9 compares the chemical compositions of CAIs with the MgO and SiO_2 contents expected for condensates from a gas of solar composition. The type A CAIs, interpreted to be condensates, plot close to the condensation trend (heavy black line), but the type B CAIs plot to the left of it. The type B CAIs crystallized from melts during relatively rapid (hours to days) cooling in the solar nebula. Because CAI liquids are unstable under most nebula conditions, it is likely that they were evaporating as they cooled. Using the measured isotopic compositions of silicon and magnesium to infer the amounts of silicon and magnesium that were lost from the molten CAIs by evaporation, one finds that their compositions have moved down and to the left and have lost 10–40% of their original magnesium. Two estimates of the starting composition for each inclusion, determined using two different fractionation factors (squares and diamonds), plot close to the condensation trend, which indicates that the bulk compositions of the precursors were controlled by gas–solid interactions.

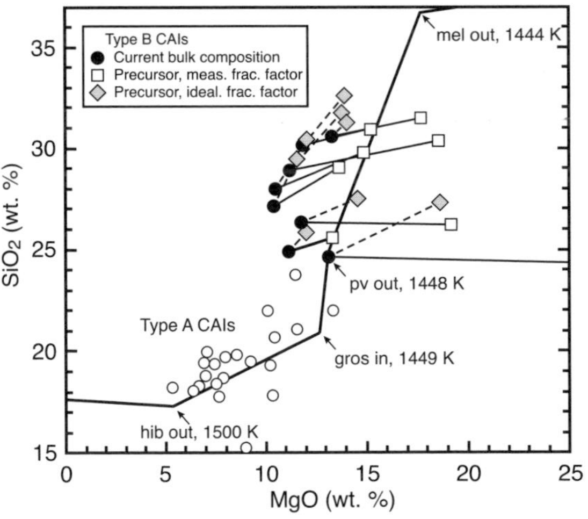

Figure 7.9 Abundances of SiO_2 and MgO in type A and B CAIs compared to the equilibrium condensation trend for a gas of solar composition (heavy line). Unmelted type A CAIs (open circles) generally follow the condensation trend, whereas melted type B CAIs (solid circles) plot to the left due to evaporative loss of both magnesium and silicon. Squares and diamonds represent the initial compositions of the CAIs inferred from the degree of isotopic fractionation in each inclusion (see Chapter 8, Fig. 8.7). Modified from Davis and Richter (2004).

7.4 Physical Fractionations in the Solar Nebula

The solar nebula was a dynamic place, allowing for the separation of different kinds of solid grains by purely physical means. Because the solid grains had different compositions, chondritic planetesimals accreted from them could have distinct element abundances.

7.4.1 Nebular Sorting of Chondrite Components

Several important chemical properties of chondrites likely reflect some sort of physical fractionation process. One example is the mixing of high-temperature components and matrix. CM chondrites provide the clearest example of this. These meteorites consist of CAIs, which are highly depleted in volatile and moderately volatile elements; chondrules, which are depleted in elements more volatile than chromium and gold; and matrix material that is similar to CI chondrites, both in chemical composition and in the abundances of presolar grains. The bulk composition of the Murchison CM chondrite is shown in Figure 7.10. Elements more refractory than gold are enriched relative to CI abundances by about 15%, reflecting the depletion of volatile and moderately volatile elements in chondrules and CAIs. The volatile elements are depleted in bulk Murchison, plotting at ~$0.6 \times$ CI. The volatile elements are present primarily in the matrix, which makes up ~60% of the meteorite. The elements more refractory than uranium are even more enriched relative to CI than the elements between gold and nickel, plotting at ~$1.4 \times$ CI. These refractory elements are

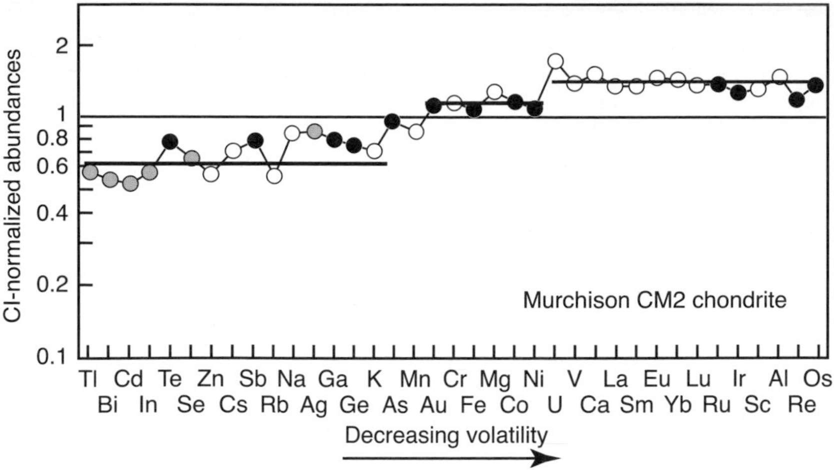

Figure 7.10 Bulk element abundances in the Murchison CM chondrite normalized to CI abundances. Elements are plotted in order of decreasing volatility from left to right. Lithophile elements are shown with open symbols, siderophile elements with black symbols, and chalcophile elements with gray symbols. The Murchison composition is a mixture of three components: a volatile-rich matrix component that contains all elements and makes up ~60% of the meteorite, a volatile-depleted chondrule component that contains elements more refractory than gold and plots at 1.15 × CI, and a refractory component consisting of CAIs that is depleted in elements more volatile than uranium and plots at ~1.4 × CI abundances. Note that there is no fractionation between lithophile, siderophile, and chalcophile elements. Data from Kallemeyn and Wasson (1981) and Krähenbühl et al. (1973).

highly enriched in CAIs, which make up several percent of the meteorite. The observation of fixed levels of enrichment or depletion among elements of similar volatility on diagrams such as Figure 7.10 was critical to the interpretation by Larimer and Anders (1970) that chondrites are mixtures of high-temperature and low-temperature condensates. Although we no longer believe that all the components formed by nebular condensation at different temperatures, the observation that distinct components were mixed to make the chondrites is valid. In most chondrites, the matrix, which varies in abundance from ~60% in CM chondrites to ~15% in ordinary and enstatite chondrites, is the component that carries the majority of volatile elements, although in most chondrites the matrix has also experienced volatile loss relative to CI.

There are also more-subtle examples of physical mixing of components. Figure 7.11 shows the bulk Si/Al and Mg/Al ratios in the various chondrite classes. The ordinary (H, L, LL), R, and enstatite (EH, EL) chondrite classes are arrayed along a roughly linear trend extending from the CI (cosmic) composition that suggests loss of an olivine-rich component, perhaps chondrules. In the same diagram, the carbonaceous (CV, CK, CM, CO, CR, CH) chondrite classes fall along a line pointing from CI toward the origin. Because aluminum is a refractory element, this suggests that CV, CM, and CO chondrites are related to CI by addition of a refractory component. Petrographically, CI chondrites do not appear to contain CAIs, and the abundance of CAIs increases from CO to CM to CV. This strongly indicates that the refractory component reflected in Figure 7.11 is CAIs. Consequently, the sorting and segregation of large particles (chondrules and CAIs) from small grains (matrix) may account for differences in the compositions of chondrites. We will explore further the difference in fractionation patterns between carbonaceous and noncarbonaceous chondrites in Chapter 8.

Another example of fractionation that apparently involved a physical process is the *metal–silicate fractionation* exhibited by chondrites. Figure 7.12 shows the bulk compositions of the ordinary (H, L, and LL) chondrites, with the elements arranged in order of increasing volatility and normalized to the composition of CI chondrites. Lithophile elements are shown as open circles, and siderophile elements are filled black circles. The lithophile and siderophile elements behave as coherent groups, but the proportions of the two groups differ among the three chondrite classes, with the abundance of the siderophile

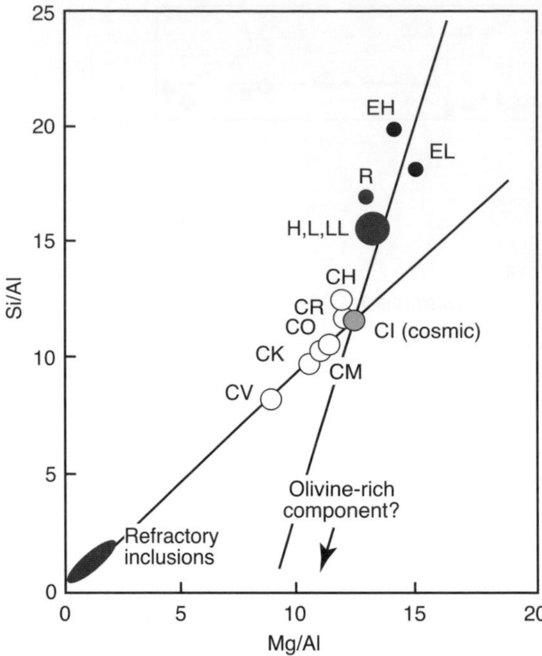

Figure 7.11 Physical fractionation of silicate components (addition of CAIs and loss of an olivine-rich component, perhaps chondrules) from the CI (cosmic) composition are suggested by trends in Si/Al and Mg/Al in bulk chondrites.

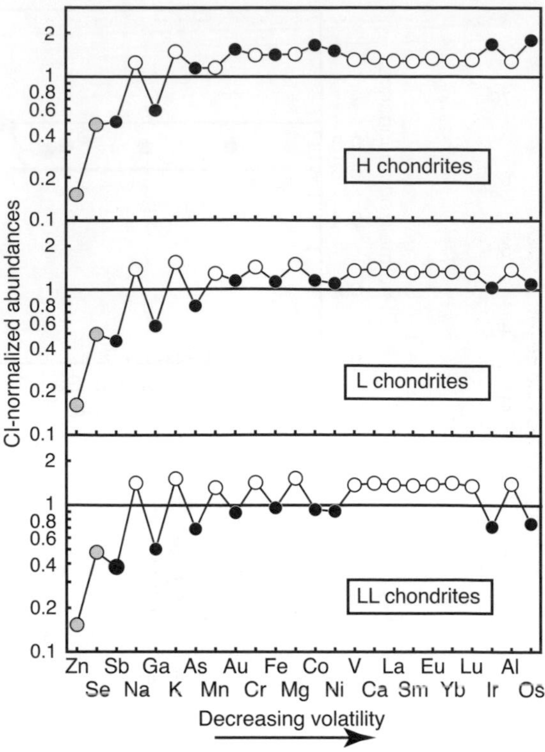

Figure 7.12 Bulk compositions of H, L, and LL chondrites normalized to CI abundances. Elements are plotted in order of decreasing volatility from left to right. Lithophile, siderophile, and chalcophile elements are designated with open, black, and gray symbols, respectively. These diagrams reveal similar fractionations of these element groups among different ordinary chondrite classes. Data from Kallemeyn et al. (1989).

component decreasing from H to L to LL. Petrographically, the abundance of metal follows the same sequence, with H chondrites having the highest metal abundance (hence the H) and LL chondrites having the lowest. Together, these observations suggest that a physical process has fractionated the metal from the silicates and produced different mixtures in these meteorites. It is perhaps telling that metal grains have higher densities and different sizes than chondrules. Gravity sorting of metal grains by mass in a quiescent nebula (the settling of metal to the nebular midplane) or aerodynamic size sorting in turbulent nebular eddies have been proposed (Kuebler et al., 1999).

Fractionation of metal from silicate has also occurred on a much larger scale. Mercury has a much greater proportion of metallic iron than do other planets (Fig. 7.13). This is reflected in the huge size of its core relative to its silicate mantle and crust. The Moon has a much smaller proportion of metal. Some chondrite groups (CB and CH) also show excess metal (Fig. 7.13). The origin of these fractionations is unclear. As noted above, the gravitational or aerodynamic sorting of metal and silicate in the nebula is a plausible explanation for the concentration of metals in chondrites, and might also explain metal–silicate fractionation in planets (Weidenschilling, 1978). Other processes have also been proposed, including photophoresis, whereby intense light preferentially pushes small silicate particles outward (Wurm et al., 2013). Fractionation during large impacts is another possibility and is discussed below as a parent-body rather than a nebula process.

7.4.2 Element Fractionation Resulting from Oxidation/Reduction

Although some kind of physical fractionation of metal and silicate among chondrites is clearly indicated, it is not the entire explanation for the metal abundances of chondrites. It has been known for over a century that, in ordinary chondrites, the abundance of oxidized iron anticorrelates with the abundance of metal. The observations, also known as Prior's rules, are that H chondrites have the most metal, the metal has low nickel content, and the

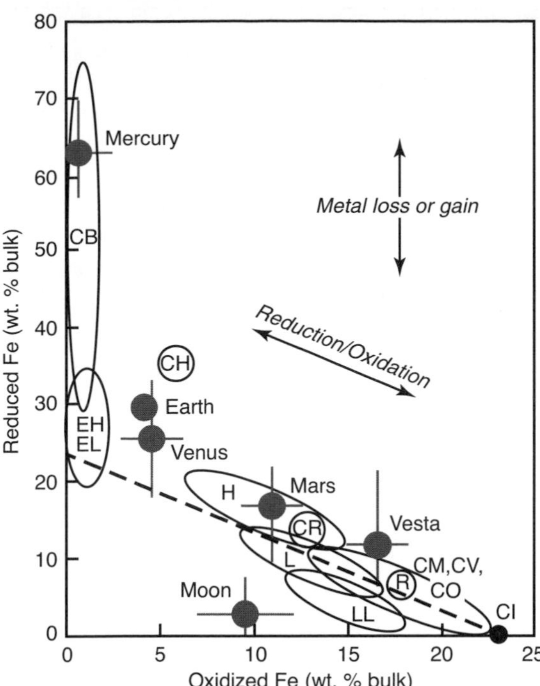

Figure 7.13 Concentrations of reduced iron (as metal or sulfides) and oxidized iron (as silicates or oxides) indicate that both oxidation/reduction and loss of metal are required to explain the compositions of various chondrite classes and planetary bodies. Modified from Cartier and Wood (2019).

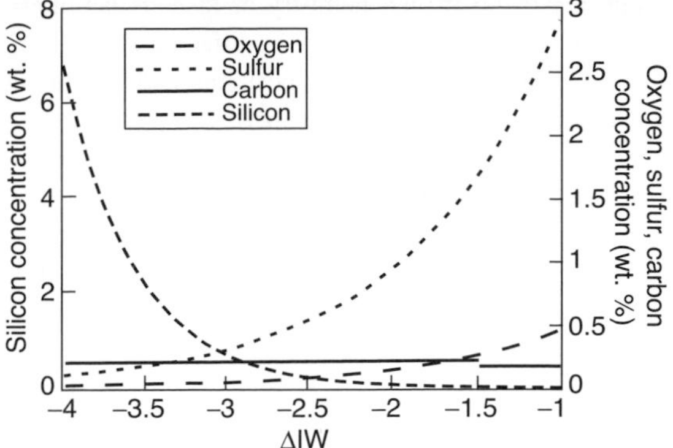

Figure 7.14 Concentrations of light elements in molten iron in equilibrium with silicate vary with oxidation state, expressed as deviations in log fO_2 from the iron-wustite buffer (IW). These curves were calculated for conditions of 5 GPa and 2000 K (Righter et al., 2020).

silicates have low FeO contents, whereas LL chondrites have the least metal, the metal has the highest nickel content, and the silicates have the highest FeO contents. L chondrites are intermediate in all three properties. In a plot of reduced Fe (as metal or sulfides) versus oxidized Fe (as silicates and oxides), loss or gain of metal will move compositions directly downward or upward, respectively (Fig. 7.13). CI chondrites are fully oxidized, and oxidation or reduction will translate materials along or parallel to the heavy dashed line. Most chondrite classes are distributed along this line, reflecting different proportions of reduced and oxidized Fe that are independent of the degree of metal loss. Thus, the cosmochemical behavior of iron is dependent not only on the physical fractionation of metal from silicate, but also on its oxidation or reduction in different parts of the nebula.

Changes in oxidation state can also have a significant effect on the volatility of some elements, including cerium, ytterbium, tungsten, and molybdenum, all of which are refractory in reducing conditions but become increasingly volatile as conditions become more oxidizing. Aluminum and calcium can become volatile if they form hydroxides. Depletions of these elements compared to others with similar volatility in a gas of solar composition can aid in investigating the redox conditions under which certain objects formed.

Redox state also affects the fractionation of elements during core formation. Changes in the concentration of light elements in molten iron equilibrated with silicate under conditions of high pressure and temperature are shown in Figure 7.14. Although carbon shows little fractionation, the partitioning of oxygen, sulfur, and silicon is strongly affected by fO_2. Fractionation of these elements is also sensitive to pressure, and these variables support a sulfur-rich core on Mars and a core containing oxygen and silicon on Earth (Righter et al., 2020).

7.5 Igneous Fractionations

7.5.1 Magmatic Processes that Lead to Fractionations

Melting and crystallization can both produce large chemical fractionations. Any situation in which a melt is separated from the residual solid, or in which crystals forming from a melt are separated from the melt, will produce melts and solids that differ in composition from the starting material. Igneous fractionation processes are important in the histories of achondrites, iron and stony-iron meteorites, and the terrestrial planets. Igneous fractionations are not important in chondrites because these meteorites did not melt and their molten chondrules crystallized without fractionating crystals from melt. Magmatic fractionation is traditionally considered to be within the realm of geochemistry or igneous petrology, but the partitioning of elements in differentiated meteorites has involved magma compositions and conditions outside the range of terrestrial experience.

Partial melting: When a rock is heated and begins to melt, it does not melt all at once. Combinations of minerals with the lowest melting temperatures melt first, and the initial melt typically has a different composition from that of the original rock. *Peridotite*, composed of olivine + pyroxenes + an aluminous phase (garnet, spinel, or plagioclase, depending on the pressure) is a product of the planetary differentiation of chondritic material and is found in the mantles of all the terrestrial planets and differentiated asteroids. Partial melting of peridotite generates basaltic magmas, which are richer in silicon, titanium, iron, calcium, aluminum, and sodium than the starting peridotite. Basalt is invariably the most abundant rock type on the surfaces of the terrestrial planets and differentiated asteroids like Vesta.

Low-degree partial melts of undifferentiated chondritic material generate a range of different compositions. The first melt from a chondritic precursor appears at ~950°C and forms by the melting of FeS, FeNi metal, and phosphates. Crystallized melts of these components are observed in primitive achondrites (acapulcoites and lodranites). As the temperature increases, silicates begin to melt. At 5–10% partial melting, the melt becomes basaltic in composition (McCoy et al., 1997).

Basaltic magmas tend to be buoyant in a gravitational field. Once some critical amount of melt is generated, the magma ascends toward the surface of the body, providing a means of separating liquid from residual solids and thus causing element fractionation.

Fractional crystallization: Crystallization of melts provides another way to fractionate chemical elements. Most crystals are denser than the melt from which they form, so crystal settling can separate the phases. Convection currents in hot magma can also entrain crystals and carry them to the bottom of a magma chamber. Accumulations of crystals are a common occurrence on the floors of terrestrial plutons, and many achondrites are cumulates formed in an analogous way in asteroids. Cumulates and the complementary melt have different elemental compositions. Less commonly, crystals can float to the top of a magma body (this happened on the Moon), also causing element fractionation.

7.5.2 Element Partitioning in Igneous Systems

A magnesium-rich silicate such as forsterite (magnesian olivine, Mg_2SiO_4) is commonly the first mineral to crystallize at high temperatures from a cooling basaltic magma. For this reason, magnesium progressively decreases in residual melt as forsterite crystallizes. Even though forsterite is a silicate, its silica content is actually lower than that of basaltic magma, so silica likewise increases in the liquid during crystallization. If the forsterite is removed from the system as it crystallizes, the remaining melt will be depleted in magnesium and enriched in silica and in other elements relative to the original basaltic composition. As cooling proceeds, different minerals will crystallize and the melt composition will evolve further, becoming more fractionated from the starting composition. We can envision fractional melting and fractional crystallization as opposite processes, both yielding melts and solids of differing composition. The combination of partial melting to produce a basaltic melt followed by fractional crystallization can generate the wide range of igneous compositions seen in many solar system materials.

Petrologic mixing algorithms compute the major- and minor-element compositions of magmas formed by partial melting under various conditions or the evolution of crystallizing magmas. They calculate solid–liquid relations based on thermodynamic properties of the phases and components or from experimentally derived phase relationships. These widely used programs are not considered here but are explained in most igneous petrology and geochemistry texts. Because magma compositions, volatile contents, and redox states vary for different solar system bodies, the applicability of particular mixing programs must be tested. For example, the MELTS program (Ghiorso and Sack, 1995) has been shown to be applicable for martian magmas (Balta and McSween, 2013) and MAGFOX (Longhi, 1991) is usually preferred for lunar magmas (Longhi, 2006).

Elements present in trace concentrations are partitioned in quite different ways. *Trace elements* are not abundant enough to form their own minerals, so they must be incorporated into minerals composed of the major elements. Because they must fit into crystal lattice sites whose sizes and coordination are already fixed, the ease with which they can be included depends on ionic size and charge. Trace ions that fit nicely into sites otherwise occupied by major ions with similar sizes and charges are said to be *compatible*. Conversely, large ions with high charge, referred to as *incompatible elements*, tend to be excluded from most crystal sites.

Element partitioning is expressed in terms of a *distribution coefficient*, *D*, defined as

$$D = C_{i\cdot\text{crystal}}/C_{i\cdot\text{melt}} \qquad 7.6$$

where $C_{i\ \text{crystal}}$ is the concentration of element i in a particular mineral and $C_{i\ \text{melt}}$ is that element's concentration in a coexisting liquid. For compatible elements, D is ≥ 1, whereas for incompatible elements, D is <1 (often far less).

Compatible trace elements are preferentially incorporated into certain sites in crystal structures. For example, Ni^{2+} is partitioned into olivine more strongly than the Mg^{2+} it displaces. Thus, fractional crystallization of olivine produces residual liquids that are increasingly depleted in nickel. This preferential partitioning is a result of crystal field stabilization, whereby certain transition metal ions are bonded more strongly in certain sites than others. An explanation of crystal field theory is beyond the scope of this book, but it can be found in various geochemistry texts; understanding this theory is also important for spectroscopy, since visible/near-infrared spectra are largely determined by transition metals, especially iron.

Incompatible trace element partitioning is illustrated by the REEs. These elements occur in trace amounts because their cosmic abundances are so low. Almost all REEs have ionic radii that are large and charges that are high (trivalent), so they are incompatible in common rock-forming minerals. However, under reducing conditions, europium can exist as a divalent cation, and in this valence it is compatible, substituting comfortably for Ca^{2+} in plagioclase. When REE concentrations are plotted versus atomic number, they exhibit a sawtooth pattern because elements with even atomic numbers have higher cosmic abundances than those with odd atomic numbers (Chapter 4). Therefore, the REE abundances are usually plotted normalized to CI abundances, which removes this effect and produces a smoothly varying pattern, as shown in Figure 7.15. In this figure, europium is clearly enriched relative to the other REEs (a "positive europium anomaly"), indicating that plagioclase has been added as cumulus crystals to the magma that formed this rock. Some rocks also show negative europium anomalies, indicating the removal of plagioclase during fractional crystallization.

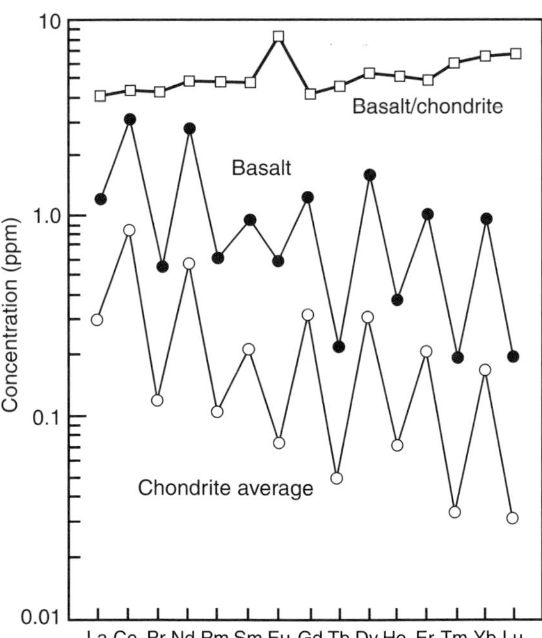

Figure 7.15 REE patterns for basalt and chondrite, and the chondrite-normalized basalt REE pattern. Normalization removes the sawtooth pattern due to differences in odd and even atomic number abundances. The positive europium (Eu) anomaly in the normalized pattern is due to the incorporation of extra plagioclase by fractional crystallization.

Metallic liquids can also experience element fractionations during fractional crystallization. The abundances of trace elements such as gold, gallium, germanium, and iridium and the major element nickel vary among some classes of iron meteorites because of the separation of crystalline metal from molten metal as their parent bodies cooled (discussed earlier in Chapter 6 and Fig. 6.11).

These examples involve the partitioning of trace elements as liquids cooled and crystallized. Partial melting of a solid rock also results in the partitioning of incompatible elements into the liquid phase, which contains no rigid crystalline sites. Separation of the melt then fractionates incompatible elements from the compatible elements left behind in the solid residue.

Trace-element partitioning during melting or crystallization can be modeled if the distribution coefficients between a magma and specific minerals have been measured. These models, described numerically in geochemistry texts, provide quantitative constraints on melting or crystallization processes. Element partitioning can be quite complex, depending on whether multiple minerals are melting or crystallizing together and whether the process occurs under equilibrium or fractional conditions. For example, the equations describing trace element partitioning of element i during crystallization of a melt are:

$C_{i\,\text{melt}}/C^0_{i\,\text{melt}} = 1/(D - FD + F)$ for equilibrium crystallization

and $C_{i\,\text{melt}}/C^0_{i\,\text{melt}} = F^{D-1}$ for fractional crystallization, where $C_{i\,\text{melt}}$ and $C^0_{i\,\text{melt}}$ are the final and original concentrations of element i in the melt, respectively, D is the distribution coefficient, and F is the fraction of liquid remaining. We illustrate a trace-element partitioning calculation in Box 7.2.

Box 7.2 REE Partitioning in Nakhlite Meteorites

Nakhlites are martian cumulate rocks consisting mostly of high-calcium clinopyroxene. MIL 03346 and its pairs, representing the least equilibrated nakhlite, consist of ~70% augite, a few percent olivine, and ~25% intercumulus material (mostly glass) (Udry et al., 2012). The augite is zoned, with homogeneous cores (interpreted to be cumulus) and Fe-rich rims (interpreted to be postcumulus overgrowths). The measured REE abundances in an augite core of MIL 03346 are shown in Figure 7.16a (Day et al., 2006). The element abundances have been normalized to CI chondrite, as is conventional for REE plots.

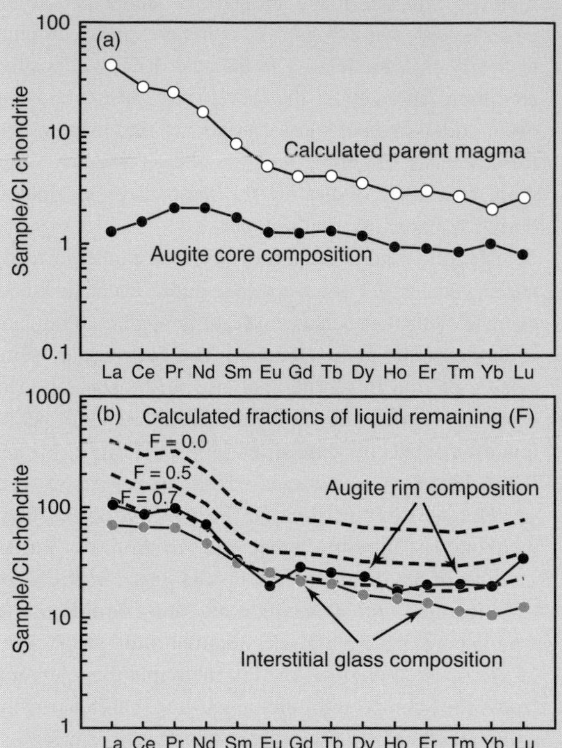

Figure 7.16 Chondrite-normalized REE patterns for the MIL 03346 nakhlite. (a) Measured REE patterns for a cumulus augite core, and for the calculated liquid in equilibrium with pyroxene, interpreted to represent the parent magma. (b) Dashed lines are calculated REE patterns for residual liquids after 30%, 50%, and 100% fractional crystallization of augite (fractions of liquid remaining, $F = 0.7$, 0.5, and 0.0, respectively). 30% fractionation ($F = 0.7$) corresponds approximately to the measured intercumulus liquid (now glass) and augite rim compositions in the meteorite. Modified from Day et al. (2006).

Because augite cores were the earliest crystallizing phase, their composition can be used to estimate the composition of the parent magma. By applying the distribution coefficients for each element in augite, we can calculate the REE pattern in the liquid with which they were in equilibrium, as did Day et al. (2006). We will illustrate this calculation using the measured augite core REE concentrations (Table 8 of Day et al., 2006) and augite distribution coefficients (HB-6 in Table 3 of Dygert et al., 2015) for lanthanum (La), dysprosium (Dy), and lutecium (Lu) in Table 7.3.

Table 7.3 Parameters for calculation of REE abundances in nakhlite magma

Element	$D_{i\text{ augite}}$	$C_{i\text{ augite}}$	$C_{i\text{ liquid}}$	CI Abundance	$C_{i\text{ liquid}}$/CI
La	0.024	0.30 ppm	12.5 ppm	0.320 ppm	39.1
Dy	0.142	0.84 ppm	5.9 ppm	0.238 ppm	24.9
Lu	0.143	0.08 ppm	0.56 ppm	0.0237 ppm	23.6

Substituting the values for each element i into $D = C_{i\text{augite}}/C_{i\text{liquid}}$, solving for $C_{i\text{liquid}}$, and normalizing to CI chondrite gives the liquid concentrations in the right-hand column. The calculated parent magma REE pattern, including the other REE abundances not calculated here, is shown in Figure 7.16a.

We can also calculate the proportion of augite that must be removed from the parent magma to produce the measured residual (interstitial) liquid composition. Substituting the values of $D_{i\text{ augite}}$ and $C_{i\text{ liquid}}$ (as $C^0_{i\text{ melt}}$) into the

Box 7.2 (cont.)

equation for fractional crystallization, $C_{i\ melt}/C^0_{i\ melt} = F^{D-1}$, and solving for $C_{i\ melt}$ for various values of F, the fraction of liquid remaining, gives a variety of REE patterns as the parent melt evolves. These patterns are shown as dashed lines for $F = 0.0$, 0.5, and 0.7 in Figure 7.16b. The pattern for $F = 0.7$ (30% fractional crystallization of augite) compares favorably to the measured patterns for interstitial liquid (now glass) and an augite rim composition in Figure 7.16b.

7.6 Fractionations in Aqueous Systems

Alteration by aqueous fluid is commonly accompanied by the fractionation of soluble elements, which are partitioned into the fluid phase. Because a fluid is so mobile, these elements tend to be lost from the system. How common are aqueous fluids in the solar system?

Liquid water is more common than once thought, following the discovery of subsurface oceans within some icy satellites of the giant planets, and brine eruptions on Enceladus and dwarf planet Ceres. However, we have only limited information on these fluids. We will focus on the two examples that are best understood: carbonaceous chondrite parent bodies and Mars.

7.6.1 Aqueous Alteration of Carbonaceous Chondrites

The CI and CM chondrite parent bodies originally accreted mixtures of anhydrous rock with H_2O ice, which melted when heated by ^{26}Al decay. As long as ice was present, it moderated large temperature increases, so for the most part these aqueous fluids were fairly cold. Alteration of the original anhydrous mineral assemblage produced serpentine, clay minerals, magnetite, sulfates, and carbonates (Brearley, 2006). The association of organic matter with phyllosilicates suggests reactions that formed more-complex molecules accompanied alteration.

Despite these pervasive mineralogical changes, the aqueous alteration in carbonaceous chondrites was nearly isochemical in terms of element abundances. We have already noted repeatedly that CI chondrites have retained their solar-like element abundances, despite having the appearance and mineralogy of congealed mud puddles. However, the abundances of amino acids and insoluble organic matter decrease with increasing degree of alteration, and carbonate content increases (Alexander et al., 2015), so hydrogen, carbon, nitrogen, and oxygen were mobile. The stable-isotope compositions of these elements also changed as they exchanged with fluids (explained further in Chapter 8). The apparent lack of other element fractionations is thought to reflect low ratios of water to rock – the relative proportions of fluid and rock that effectively interact with each other.

7.6.2 Chemical Weathering on Mars

Basaltic rocks and the sediments derived from them cover most of the surface of Mars. Geochemists commonly track the chemical weathering of basalt using the molar A-CNK-FM diagram (Fig. 7.17), where A = Al_2O_3, CNK = $CaO + Na_2O + K_2O$, and FM = $FeO + MgO$. Basalts, composed of pyroxenes + plagioclase ± olivine, generally plot slightly below the FM–feldspar join. Weathering removes soluble CNK components, forming clays and shifting their compositions toward the A–FM join. The Spirit, Opportunity, and Curiosity rovers carried alpha particle X-ray spectrometers (see Appendix) to measure rock chemistry. Rover-analyzed sedimentary rocks from Gusev crater, Meridiani Planum, and Gale crater (Fig. 7.17) are chemically similar to basalts at the same locations, pointing to their volcanic provenance and to limited chemical change. The composition of NWA 7034, the only sedimentary martian meteorite, is also shown. Although mineralogical evidence for chemical weathering is widespread on Mars, bulk geochemical data do not show the leaching of water-soluble elements characteristic of open-system alteration under neutral to slightly basic conditions. The martian sedimentary rocks form a nearly linear array in Figure 7.17, interpreted to reflect either some dissolution of olivine during acidic chemical weathering (Hurowitz and McLennan, 2007) and/or physical sorting of olivine and other minerals during transport (McGlynn et al., 2011).

Despite the evidence for the limited loss of soluble CNK elements from clastic rock compositions at rover landing sites, elsewhere Mars has chemical sediments in the form of evaporites. Evaporation of acidic brines produced widespread sulfates and chlorides, but without the carbonates that are commonly seen in terrestrial deposits. Sulfates also constitute cements in sedimentary rocks at Meridiani Planum, formed by evaporation of ephemeral ponds, and at Gale crater, formed by evaporation of its crater lake. Models for acidic martian brine compositions indicate a different evaporation pathway from terrestrial

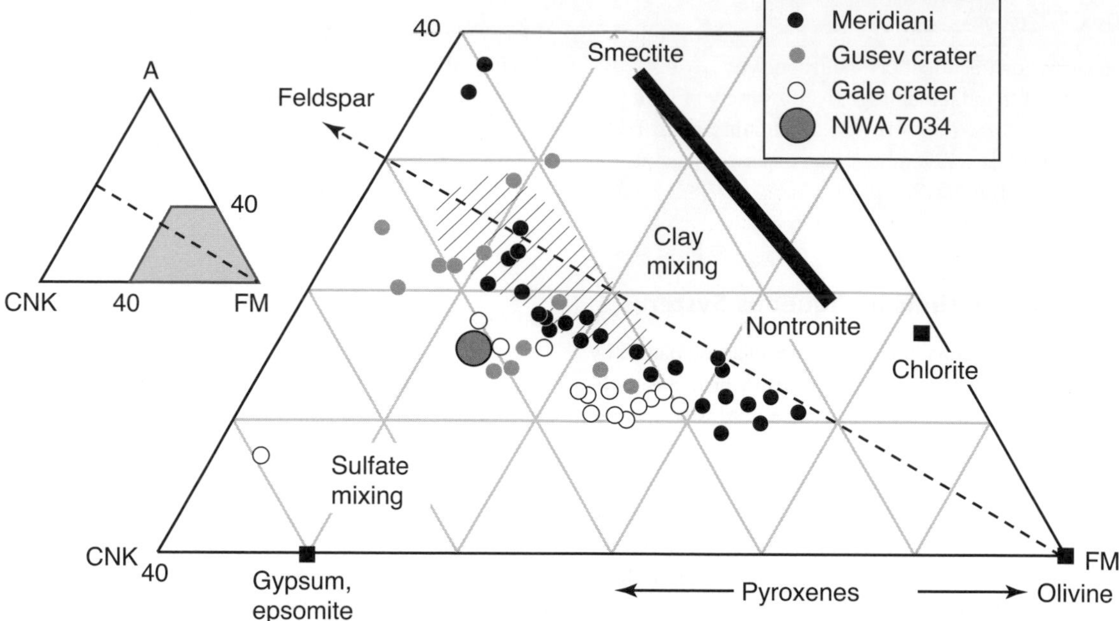

Figure 7.17 Chemical alteration diagram (see text for explanation of the A, CNK, and FM apices) showing the compositions of Mars-rover-analyzed sedimentary rocks from three sites and the NWA 7034 martian meteorite. The rocks have retained their basaltic compositions (compositions of martian basalts are indicated by the hatched field), except for the loss of some olivine, and do not show the loss of soluble CNK elements. Modified from McSween (2015).

seawater, such that magnesium sulfates precipitate before halite and carbonates do not form at all (Tosca et al., 2005). The evaporites clearly demonstrate strong fractionation of water-soluble elements.

In addition to evaporation of aqueous fluids on the martian surface, hydrothermal fluids have caused element fractionation in the subsurface. Opaline silica soils with SiO_2 values as high as 91 wt. % were encountered by the Spirit rover in Gusev crater (Squyres et al., 2008). The soils were also enriched in titanium and chromium, which like silica are insoluble under acidic conditions, leading to the conclusion that the deposits formed through pervasive leaching of other components by hydrothermal fluids. The nearby Home Plate outcrop also exhibits chemical zonation of water-soluble elements like zinc.

7.7 Physical Fractionations on Planetesimals and Planets

Several kinds of physical processes affected planetesimals and planets. These processes could conceivably explain some of the chemical fractionations observed among differentiated meteorites and planets.

7.7.1 Fractionation by Impacts

Collisions between bodies of comparable size are more likely to have been glancing ("hit-and-run") impacts than direct hits (Asphaug et al., 2006). A differentiated or actively differentiating body might lose a significant portion of its silicate crust and mantle in such a collision, with a resulting major change in the bulk composition of the body. Yang et al. (2007) appealed to this process to explain how the cooling rates for IVA iron meteorites were more than 100 times faster than for most iron meteorites. In their model, glancing collisions could produce naked iron cores, which would cool much faster than those insulated by silicate mantles. If such impacts were common, it might explain the metal–silicate fractionations seen in some achondrites.

On a much larger scale, Mercury has a massive core that has been hypothesized to reflect impact stripping of some of its silicate crust and mantle. The ejecta from a glancing impact would not subsequently accrete, thus fractionating metal from silicate (Asphaug and Reufer, 2014). Models of a giant impact on the early Earth that is thought to have formed the Moon also suggest accretion of the impactor's iron core onto the Earth and preferential incorporation of its silicate mantle in the subsequently formed Moon, accounting for its apparent iron loss.

7.7.2 Fractionation during Pyroclastic Eruptions

The production of volatile-charged magmas in planetesimals with high H_2O or CO_2 contents might result in explosive eruptions, producing pyroclastic melt droplets that could escape the gravitational grasp of small bodies (Wilson and Keil, 1991). Such a process might explain the lack of basaltic meteorites that would be expected to accompany the primitive achondrites interpreted as residues of partial melting (e.g. ureilites and aubrites). Dispersal and loss of the basaltic fraction, which would otherwise have comprised the crusts of those asteroids, would result in chemical fractionation of the remaining bodies. The subsequent accretion of such planetesimals into planets would then carry a nonchondritic elemental signature of lost elements that melted at low temperatures or were volatile (Warren, 2008).

Pyroclastic eruptions on planets with substantial gravity fields, like Earth and Mars, also commonly result in separation of ejected crystals, glass shards, and dust during flight. Such components can be deposited at different distances from the source, or at different times, producing stratigraphic fractionation.

7.7.3 Fractionation Due to Aeolian Sorting

Mechanical sorting of rock particles by wind on a planetary surface is analogous to nebular fractionations by gravity or aerodynamic sorting of metal and silicate grains. However, aeolian sorting requires an atmosphere, limiting the locations in the solar system where we can look for this process. Sand transported and deposited into dunes is common on Mars, and dust devils and global dust storms demonstrate the suspension of dust particles. Thus, aeolian sorting of particles by size or density on Mars seems likely. Dune sands composed solely of quartz are common on Earth, but quartz is not found on Mars. The feasibility of sorting olivine and feldspar in basaltic sands under martian conditions has been demonstrated experimentally (Fedo et al., 2015), and chemical (major- and minor-element) changes in physically sorted sediments mimic those measured in martian soils (McGlynn et al., 2011).

Aeolian sorting is an agent for element fractionation on Earth, with its 1 bar atmosphere, and on Mars, which has an atmospheric pressure only 1/100th that of Earth. On planetary bodies with much denser atmospheres, like Venus and Saturn's moon Titan, the wind speeds necessary to entrain grains can be even lower, so mineral and thus element fractionations might be likely there too.

7.8 Differentiation of Rocky Planetesimals and Planets

Planetary differentiation is a fractionation event of the first order. It can be viewed primarily as igneous fractionation. On rocky worlds it involves both metallic and silicate liquids and solids.

As discussed above, the silicate mantles of differentiated bodies consist of peridotite. Partial melting of peridotites produces ubiquitous basaltic liquids that ascend buoyantly to form crusts. Mercury, Venus, and Mars have crusts composed mostly of basalts. Remelting of basaltic crust on the Archean Earth produced magmas richer in silica, the tonalite–trondjemite–granodiorite suite that comprised the Earth's early continental crust (Taylor and McLennan, 2009). Potassium-rich granitic rocks of the post-Archean continental crust formed through water-assisted melting at subduction zones. The occurrence of felsic magmas formed by repeated melting of the crust appears to have been a uniquely terrestrial process. The small amounts of felsic rocks on Mars formed by fractional crystallization of basaltic melts (Udry et al., 2018), as did lunar "granites." The Earth's crust and, indeed, the crusts of all differentiated rocky bodies are enriched in incompatible elements relative to their mantles. This reflects the partial melting of mantle material and extraction of the melts.

The formation of the Moon's crust, composed primarily of feldspar (anorthosite), illustrates how igneous fractionation occurs during differentiation. Early in its history, a significant portion of the Moon was melted to form a magma ocean. The first minerals to crystallize, olivine and pyroxene, sank because of their high densities and formed an ultramafic mantle. Once feldspar began to crystallize, it floated and accumulated near the surface to produce the crust (discussed further in Chapter 14).

The formation of cores, whether in planets or asteroids, is the most spectacular display of physical fractionation – in this case, segregation of metallic liquids from silicate material. The composition of the metallic liquid depends on the degree of melting. Very limited partial melting produces liquids that have the metal-sulfide eutectic composition, which is sulfur rich (~85 wt. % FeS). The eutectic melt is the first melt to form in the Fe–Ni–S system as temperature increases, at ~950°C. As the degree of melting increases, the metallic melt will become progressively richer in iron and nickel. No iron meteorites, nor any planetary cores, have the eutectic composition, although veins of eutectic metal-sulfide formed by low degrees of melting occur in acapulcoites-lodranites.

In experimental studies, metal-sulfide liquids form rounded globules that will sink through a silicate matrix that is itself partly molten. The density difference is sufficient that the segregation of a metallic core though a mush of liquid and crystals should be rapid. However, sinking of metallic liquid through a solid or mostly solid silicate matrix is greatly inhibited. FeNi-rich melts "wet" silicate grains, forming isolated pockets of liquid, and segregation is sluggish or does not occur at all. Only anion-rich (sulfur or perhaps oxygen) metallic melts form interconnected networks and thus can be separated from solid silicate (Haack and McCoy, 2004). However, dynamic experiments indicate that metal-sulfide segregation may be possible at high pressures and strain rates, even in the absence of silicate melting (Rushmer et al., 2000). Also, the difficulties in separating metal droplets from silicate are circumvented if the core grew by the accretion of already differentiated planetesimals with large cores (Wade and Wood, 2005).

7.9 Differentiation of Giant Planets and Icy Satellites

7.9.1 Fractionations within the Giant Planets

As a first approximation, the giant planets are comprised of only hydrogen and helium. The amounts of nebular gases that were incorporated into the giant planets, relative to accreted solids (ices and rock), decrease monotonically from Jupiter outward to Neptune. Understanding the differentiation of these gas and ice giants depends on phase relationships that are unfortunately not well constrained.

To model differentiation of the *gas giants* (Jupiter and Saturn), we must understand the behavior of hydrogen and helium at high pressures and temperatures. The outer regions of Jupiter and Saturn consist of molecular H_2 in fluid form. At temperatures of 3000–4000 K and pressures >1.4 Mbar, the hydrogen molecules dissociate and form a metallic fluid. By "metallic" we mean that this phase is so densely packed that the electron clouds overlap, making it electrically conductive (like a metal). The other important component of the gas giants is helium. Liquid helium and hydrogen become immiscible once metallic hydrogen becomes stable, so they are expected to separate at depth. Helium is denser and rains downward, thus explaining the apparent depletion of helium, relative to the solar H/He ratio, in the atmospheres of Jupiter and Saturn. Various organic molecules (methane, ammonia, and others) and noble gases have been detected in the atmospheres of these planets, and these may be minor components in their interiors as well.

The measured mean densities of Jupiter and Saturn (1.33 and 0.69 g/cm^3, respectively) are averages of molecular and metallic hydrogen plus an estimated 15–30 Earth masses of higher-atomic-weight material. Small, dense, rocky cores are possible but not required by current data; the denser materials could also be partly distributed in a surrounding envelope.

Geothermal gradients in the *ice giants* (Uranus and Neptune) permit molecular H_2 fluids but never reach the stability conditions for metallic hydrogen. The lower pressures allow hydrogen and helium to be miscible, so they do not separate and the H/He ratio in their atmospheres is nearly solar. Their atmospheres contain more methane and ammonia than Jupiter and Saturn, but data on other components is lacking. Unlike the larger gas giants, the interiors of Uranus and Neptune consist mostly of ices, likely made of water, methane, ammonia, and hydrogen sulfide. At low temperatures, H_2O adopts a variety of crystalline forms. The geothermal gradients of

Box 7.3 Another View: Partial Differentiation?

Planets have fully differentiated to form crusts, mantles, and cores, but the fate of smaller planetesimals is less clear. Chondrites and differentiated meteorites (achondrites and irons) are usually thought to have formed on separate parent bodies. This view generally holds, despite evidence of incomplete melting in the form of primitive achondrites, which are chondritic residues from which small amounts of melt were extracted. But could some planestimals have had differentiated interiors and unmelted chondritic crusts?

Paleomagnetic measurements of some chondrites have been interpreted to indicate that magnetic fields originated within the parent bodies themselves, possibly implying that they had molten metal cores that supported dynamos (Weiss and Elkins-Tanton, 2013). Such bodies would also have contained silicate magmas (now achondrites). Other meteorite observations might provide some support for this idea. The IIE iron meteorites are mixtures of formerly molten iron with a variety of silicate rocks, including chondrites and basalts; one interpretation of these meteorites is that they are formed by impact disruption and reassembly of a partly differentiated planetesimal.

Uranus and Neptune pass through the field of hexagonal ice (the familiar form that occurs on Earth) and then into the field of supercritical fluid, with properties very different from liquid water. Phase diagrams for other ice compositions are not well determined at high pressures.

The total mass of higher-atomic-weight materials is roughly the same in all the giant planets, but Uranus and Neptune contain less hydrogen and helium. The mean densities for Uranus and Neptune (1.32 and 1.64 g/cm^3, respectively) are averages for rock (possibly cores), ice mantles, and molecular hydrogen envelopes.

7.9.2 Fractionations in Icy Moons and Dwarf Planets

The icy satellites of the giant planets, as well as dwarf planets Pluto and Ceres, contain crystalline forms of water and lesser amounts of other components composed of carbon, nitrogen, and sulfur. Pressures in the interiors of the largest icy satellites are not high enough to cause phase changes in silicates, but high-pressure forms of water and other ices can occur.

The mean densities of the icy satellites, with the exception of Europa (2.99 g/cm^3), all fall within the range of 1.00–1.94 g/cm^3. These densities are interpreted to indicate cores of rock and outer shells of ices. Thus, the bodies have fractionated water from rock, originally as aqueous fluids, and the brine oceans subsequently froze to produce icy crusts and precipitated salts.

The Galileo spacecraft found magnetic disturbances that are interpreted to indicate the presence of subsurface oceans of liquid water on three moons of Jupiter: Europa, Ganymede, and Callisto. A wobble in the orbit of Enceladus, a moon of Saturn, also requires that part of its icy shell be liquid. Jets of water vapor emanating from fractures on Enceladus are likely fed from this source. On all these satellites around the giant planets, subsurface liquid water is likely maintained by tidal heating, and its stability is increased by dissolved salts. The Dawn spacecraft found recent deposits of salts on Ceres, thought to be precipitates from cryovolcanic eruptions of fluids in pore spaces in the deep interior, as well as a recent cryovolcano. Fractionation of ice and rock is likely continuing on these bodies.

Summary

All of the bodies in the solar system formed from the same mixture of gas and dust inherited from the Sun's parent molecular cloud. The bulk composition of the dust is best approximated by CI chondrites. However, the varying compositions of solar system bodies came about because chemical and physical processes fractionated the elements. Understanding how and why elements fractionated is a central theme of cosmochemistry.

The condensation sequence of a cooling gas of solar composition can be calculated from thermodynamic data for the elements in gaseous or mineral forms. The phases predicted for the high-temperature portion of the condensation sequence are the same minerals found in fine-grained refractory inclusions (type A CAIs) in chondrites. Coarse-grained type B CAIs crystallized from melts, and could be melted condensates or residues from evaporation. Nebular (cosmochemical) fractionations of elements are generally based on their volatility, usually quantified as 50% condensation temperatures. Differences in redox state also affect how some elements (especially iron) are fractionated in the nebula. Physical fractionation of metal and silicate may have resulted from gravitational settling in a quiescent nebula or aerodynamic sorting in turbulent eddies.

Igneous processes (partial melting and fractional crystallization) lead to element fractionations in planetesimals and planets. Incompatible elements tend to be partitioned into melts and compatible elements into certain minerals, depending on ionic size and charge. Segregation of partial melts from residual crystals, or accumulation of crystals from crystallizing magmas, produce rocks with compositions different from the starting materials. Igneous processes account for most of the geochemical fractionations in differentiated meteorites and planetary samples, and they are predictable despite differences in magma compositions and conditions. Aqueous alteration, chemical weathering, and evaporation can fractionate water-soluble elements. Different fluid compositions, water/rock ratios, pH, and redox states complicate these processes.

Fractionation of elements also occurs when solid components are physically sorted and segregated on planets and planetesimals. Glancing collisions between planetesimals can strip silicate mantles from cores, and explosive eruptions on small bodies might disperse basaltic magmas into space. The subsequent accretion of such planetesimals could yield nonchondritic planets. Particles on the surfaces of planets with atmospheres can be sorted by winds, resulting in chemical fractionations.

Differentiation of rocky bodies promotes the segregation of elements according to their geochemical affinities on a planetary scale. Incompatible lithophile elements are concentrated in basaltic liquids during partial melting and thus become especially abundant in crusts. Siderophile and sometimes chalcophile elements are segregated with iron into planetary cores. The gas giants have fractionated hydrogen from helium at high pressures in their deep interiors; metallic hydrogen mantles may overlie rocky cores. The lower pressures inside the ice giants preclude the separation of hydrogen and helium; their interiors are mostly ices overlying rocky cores. The large satellites of the giant planets, as well as dwarf planets Pluto and Ceres, have fractionated ice from rock, and many of these bodies have subsurface oceans of liquid water – evidence of ongoing differentiation.

In the next chapter, we will explore the realm of stable isotopes and isotope anomalies. Just like elements, isotopes are fractionated by various cosmochemical and geochemical processes, and their patterns can be diagnostic of these processes.

Questions

1. What are the major processes that produce chemical fractionations of importance in cosmochemistry?
2. Describe the order of appearance of the important minerals that are calculated to condense from a cooling gas of solar composition. Would an evaporation sequence be similar?
3. How do we quantify the volatility of an element? Where do we see evidence of element fractionation according to volatility?
4. What is the evidence for metal–silicate fractionation? How and where might this have occurred?
5. How and why are elements partitioned during partial melting? During fractional crystallization?
6. What chemical fractionations are produced during the differentiation of rocky planets? Of the giant planets?

Suggestions for Further Reading

Ebel D. S. (2006) Condensation of rocky material in astrophysical environments. In Meteorites and the Early Solar System, II, Lauretta D. S., and McSween H. Y., editors, pp. 253–277, University of Arizona Press, Tucson. A good summary of the modern condensation calculations and modeling of solar system processes.

Wanke H., and Dreibus, G. (1988) Chemical composition and accretion history of terrestrial planets. *Philosophical Transactions of the Royal Society of London*, **A325**, 545–557. This paper describes how chemical fractionations may have resulted from accretion of different materials to form the terrestrial planets.

Other References

Alexander, C. M. O'D., Grossman J. N., Ebel D. S., and Ciesla F. J. (2008) The formation conditions of chondrules and chondrites. *Science*, **320**, 1617–1619.

Alexander C. M. O'D., Bowden R., Fogel M. L., and Howard K. T. (2015) Carbonate abundances and isotopic compositions in chondrites. *Meteoritics & Planetary Science*, 50, 810–833.

Asphaug E., Agnor C. B., and Williams Q. (2006) Hit-and-run planetary collisions. *Nature*, **439**, 155–160.

Asphaug E., and Reufer A. (2014) Mercury and other iron-rich planetary bodies as relicts of inefficient accretion. *Nature Geoscience*, **7**, 564–568.

Balta J. B., and McSween H. Y. (2013) Application of the MELTS algorithm to martian compositions and implications for magma crystallization. *Journal of Geophysical Research, Planets*, **118**, 2502–2519.

Brearley A. J. (2006) The action of water. In *Meteorites and the Early Solar System, II*, Lauretta D. S., and McSween H. Y., editors, pp. 587–624, University of Arizona Press, Tucson.

Cameron A. G. W. (1962) The formation of the sun and planets. *Icarus*, **1**, 13–69.

Campbell A. J., Humayun M., Meibom A., et al. (2001) Origin of zoned metal grains in the QUE94411 chondrite. *Geochimica et Cosmochimica Acta*, **65**, 163–180.

Cartier C., and Wood B. J. (2019) The role of reducing conditions in building Mercury. *Elements*, **15**, 39–45.

Chase M. W. (1998) *NIST-JANAF Thermochemical Tables, 4th edition.* American Institute of Physics, doi:10.18434/T42S31

Connolly H. C., Huss G. R., and Wasserburg G. J. (2001) On the formation of Fe-Ni metal in CR2 meteorites. *Geochimica et Cosmochimica Acta*, **65**, 4567–4588.

Davis A. M. (2006) Volatile element evolution and loss. In *Meteorites and the Early Solar System, II*, Lauretta D. S., and McSween H. Y., editors, pp. 295–307, University of Arizona Press, Tucson.

Davis A. M., and Richter F. M. (2004) Condensation and evaporation of solar system materials. In *Treatise on Geochemistry, Vol. 1: Meteorites, Comets and Planets*, Davis A. M., editor, pp. 407–430, Elsevier, Oxford.

Day J. M. D., Taylor L A., Floss C., and McSween H. Y. (2006) Petrology and chemistry of MIL 03346 and its significance in understanding the petrogenesis of nakhlites on Mars. *Meteoritics & Planetary Science*, **41**, 581–606.

Dygert N., Liang Y., Sun C., and Hess P. (2015) Corrigendum to 'An experimental study of trace element partitioning between augite and Fe-rich basalts.' *Geochimica et Cosmochimica Acta*, **149**, 281–283.

Ebel D. S., and Grossman L. (2000) Condensation in dust-enriched systems. *Geochimica et Cosmochimica Acta*, **65**, 469–477.

Fedo C. M., McGlynn I. O., and McSween H. Y. (2015) Grain size and hydrodynamic sorting controls on the composition of basaltic sediments: Implications for interpreting martian soils. *Earth & Planetary Science Letters*, **423**, 67–77.

Fegley B. Jr. (1999) Chemical and physical processing of presolar materials in the solar nebula and the implications for preservation of presolar materials in comets. *Space Science Reviews*, **72**, 311–326.

Ghiorso M., and Sack R. (1995) Chemical mass transfer in magmatic processes IV. A revised and internally consistent thermodynamic model for the interpolation and extrapolation of liquid-solid equilibria in magmatic systems at elevated temperatures and pressures. *Contributions to Mineralogy & Petrology*, **119**, 197–212.

Grossman L. (1972) Condensation in the primitive solar nebula. *Geochimica et Cosmochimica Acta*, **36**, 597–619.

Grossman L., and Larimer J. W. (1974) Early chemical history of the solar system. *Reviews of Geophysics & Space Physics*, **12**, 71–101.

Haack H., and McCoy T. J. (2004) Iron and stony-iron meteorites. In *Treatise on Geochemistry, Vol. 1: Meteorites, Comets and Planets*, Davis A. M., editor, pp. 325–345, Elsevier, Oxford.

Hurowitz J. A., and McLennan S. M. (2007) A ~3.5 GA record of water-limited acidic weathering conditions on Mars. *Earth & Planetary Science Letters*, **260**, 432–443.

Huss G. R. (2004) Implications of isotopic anomalies and presolar grains for the formation of the solar system. *Antarctic Meteorite Research*, **17**, 132–152.

Huss G. R., Meshik A. P., Smith J. B., and Hohenberg C. M. (2003) Presolar diamond, silicon carbide, and graphite in carbonaceous chondrites: Implications for thermal processing in the solar nebula. *Geochimica et Cosmochimica Acta*, **67**, 4823–4848.

Huss G. R., Rubin A. E., and Grossman J. N. (2006) Thermal metamorphism in chondrites. In *Meteorites and the Early Solar System, II*, Lauretta D. S., and McSween H. Y., editors, pp. 295–307, University of Arizona Press, Tucson.

Jungck M. H. A., Shimamura T., and Lugmair G. W. (1984) Ca isotope variations in Allende. *Geochimica et Cosmochimica Acta*, **48**, 2651–2658.

Kallemeyn G. W., and Wasson J. T. (1981) The compositional classification of chondrites-I. The carbonaceous chondrite groups. *Geochimica et Cosmochimica Acta*, **45**, 1217–1230.

Kallemeyn G. W., Rubin A. E., Wang D., and Wasson J. T. (1989) Ordinary chondrites: Bulk composition, classification, lithophile-element fractionations, and composition-petrographic type relationships. *Geochimica et Cosmochimica Acta*, **53**, 2747–2767.

Kallemeyn G. W., Rubin A. E., and Wasson J. T. (1994) The compositional classification of chondrites: VI. The CR carbonaceous chondrite group. *Geochimica et Cosmochimica Acta*, **58**, 2873–2888.

Krähenbühl U., Morgan J. W., Ganapathy R., and Anders E. (1973) Abundances of 17 trace elements in carbonaceous chondrites. *Geochimica et Cosmochimica Acta*, **37**, 1353–1370.

Kuebler K. E., McSween H. Y., Carlson W. D., and Hirsch D. (1999) Sizes and masses of chondrules and metal-troilite grains in ordinary chondrites: Possible implications for nebular sorting. *Icarus*, **141**, 96–106,

Larimer J. W., and Anders E. (1967) Chemical fractionations in meteorites-II. Abundance patterns and their interpretation. *Geochimica et Cosmochimica Acta*, **31**, 1239–1270.

Larimer J. W., and Anders E. (1970) Chemical fractionations in meteorites-III. Major element fractionations in chondrites. *Geochimica et Cosmochimica Acta*, **34**, 367–387.

Lodders K., and Fegley B. (1998) *The Planetary Scientist's Companion*. Oxford University Press, New York, 371 pp.

Longhi J. (1991) Comparative liquidus equilibria of hypersthene-normative basalts at low pressure. *American Mineralogist*, **76**, 785–800.

Longhi J. (2006) Petrogenesis of picritic mare magmas: Constraints on the extent of early lunar differentiation. *Geochimica et Cosmochimica Acta*, **70**, 5919–5934.

McCoy T. J., Keil K., Muenow D. W., and Wilson L. (1997) Partial melting and melt migration in the acopulcoite-lodranite parent body. *Geochimica et Cosmochimica Acta*, **61**, 639–650.

McGlynn I. O., Fedo C. M., and McSween H. Y. (2011) Origin of basaltic soils at Gusev crater, Mars, by aeolian modification of impact-generated sediment. *Journal of Geophysical Research*, **116**, E00F22.

McSween H. Y. (2015) Petrology on Mars. *American Mineralogist*, **100**, 2380–2395.

McSween H. Y., Moersch J. E., Burr D. M., et al. (2019) *Planetary Geoscience*. Cambridge University Press, Cambridge, 334 pp.

Niederer F. R., Papanastassiou D. A., and Wasserburg G. J. (1981) The isotopic composition of titanium in the Allende and Leoville meteorites. *Geochimica et Cosmochimica Acta*, **45**, 1017–1031.

Petaev M. I., and Wood J. A. (1998) The condensation with partial isolation (CWPI) model of condensation in the solar nebula. *Meteoritics & Planetary Science*, **33**, 1123–1137.

Righter K., Herd C. D. K., and Boujibar A. (2020) Redox processes in early Earth accretion and in terrestrial bodies. *Elements*, **16,** 161–166.

Rotaru M., Birck J. L., and Allegre C. J. (1992) Clues to early solar-system history from chromium isotopic in carbonaceous chondrites. *Nature*, **358**, 465–470.

Rushmer T., Minarik W. G., and Taylor G. J. (2000) Physical processes of core formation. In *Origin of the Earth and Moon*, Canup R. M., and Righter, K., editors, pp. 227–243, University of Arizona Press, Tucson.

Squyres S. W., Arvidson R. E., Ruff S., et al. (2008) Detection of silica-rich deposits on Mars. *Science*, **320**, 1063–1067.

Taylor S. R., and McLennan S. M. (2009) *Planetary Crusts: Their Composition, Origin and Evolution*. Cambridge University Press, Cambridge, 378 pp.

Tosca, N.J., McLennan, S. M., Clark, B. C., et al. (2005) Geochemical modeling of evaporation processes on Mars: Insight from the sedimentary record at Meridiani Planum. *Earth & Planetary Science Letters*, **240**, 122–148.

Udry A., McSween H. Y., Lecumberri-Sanchez P., and Bodnar R. J. (2012) Paired nakhlites MIL 090030, 090032, 090136, and 03346: Insights into the Miller Range parent meteorite. *Meteoritics & Planetary Science*, **47**, 1575–1589.

Udry A., Gazel E., and McSween H. Y. (2018) Formation of evolved rocks at Gale crater by crystal fractionation and implications for Mars crustal composition. *Journal of Geophysical Research, Planets*, **123**, doi:10.1029/2018JE005602.

Wade J., and Wood B. J. (2005) Core formation and the oxidation state of the Earth. *Earth & Planetary Science Letters*, **236**, 78–95.

Wai C. M., and Wasson, J. T. (1977) Nebular condensation of moderately volatile elements and their abundances in ordinary chondrites. *Earth & Planetary Science Letters*, **36**, 1–13.

Warren P. H. (2008) A depleted, not ideally chondritic bulk Earth: The explosive-volcanic basalt loss hypothesis. *Geochimica et Cosmochimica Acta*, **72**, 2217–2235.

Wasson J. T. (1985) *Meteorites*. Freeman and Company, New York, 267 pp.

Weidenschilling S. J. (1978) Iron/silicate fractionation and the origin of Mercury. *Icarus*, **35**, 99–111.

Weiss B. P., and Elkins-Tanton L. T. (2013) Differentiated planetesimals and the parent bodies of chondrites. *Annual Reviews of Earth & Planetary Sciences*, **41**, 529–560.

Wilson L., and Keil K. (1991) Consequences of explosive eruptions on small solar-system bodies – The case of the missing basalts on the aubrite parent body. *Earth & Planetary Science Letters*, **104**, 505–512.

Wood J. A., and Hashimoto A. (1993) Mineral equilibrium in fractionated nebular systems. *Geochimica et Cosmochimica Acta*, **57**, 2377–2388.

Wurm G., Trieloff M., and Rauer H. (2013) Photophoretic separation of metals and silicates: The formation of Mercury-like planets and metal depletion in chondrites. *Astrophysical Journal*, **769**, doi:10.1088/0004-637X/769/1/78.

Yang J., Goldstein J. F., and Scott E. R. D. (2007) Iron meteorite evidence for early catastrophic disruption of protoplanets. *Nature*, **446**, 888–891.

Yin Q.-Z. (2005) From dust to planets: The tale told by moderately volatile elements. In *Chondrites and the Protoplanetary Disk*, Krot A. N., Scott E. R. D., and Reipurth B., editors, pp. 632–644, ASP Conference Series 341, Astronomical Society of the Pacific, San Francisco.

Yoneda S., and Grossman L. (1995) Condensation of CaO-MgO-Al_2O_3-SiO_2 liquids from cosmic gases. *Geochimica et Cosmochimica Acta*, **59**, 3413–3444.

8 Stable-Isotope Fractionations by Cosmochemical and Geochemical Processes

Overview

In this chapter, we discuss isotopic fractionations in extraterrestrial materials and what they can tell us about solar system history. Elemental fractionations (discussed in Chapter 7) depend on the configurations of electrons in the atoms, whereas isotopic fractionations reflect the nuclear structures of the atoms. Because of this difference, isotopic fractionations typically (but not exclusively) probe different processes than elemental fractionations. Here we discuss processes that separate isotopes as a function of mass (mass-dependent fractionation), processes that produce fractionations that do not depend on mass (mass-independent fractionation), and nucleosynthetic isotope signatures that survive in solar system materials. These processes occur in all different environments in the early solar system, and often the observed isotopic effects are a combination of the different types of processes.

8.1 What Are Isotopic Fractionations and Why Are They Important?

Isotopes of an element behave slightly differently in chemical, physical, and biological processes because of differences in their masses. As a consequence, many processes can cause the relative proportions of isotopes to change (to become "fractionated") relative to the starting composition. These are *mass-dependent isotope fractionations*. The field of stable-isotope geochemistry developed around the study of mass-dependent fractionations of hydrogen, carbon, nitrogen, oxygen, sulfur, and chlorine. Stable-isotope cosmochemistry also began with these elements, as the experimental techniques developed for terrestrial isotope studies were adapted to extraterrestrial samples. Detailed discussions of stable-isotope geochemistry can be found in Hoefs (2009) and Sharp (2017). Much of the discussion in this chapter is based on these sources.

The magnitude of the isotopic shifts in most systems is relatively small, so high-precision mass spectrometry has been developed to study them. A special measurement scale was developed to highlight the small isotopic shifts. Isotopic data are typically reported as "delta values," which give the difference between the isotopic composition of the sample and that of a universally accepted standard in parts per thousand. Equation 8.1 gives the definition for $\delta^{18}O$:

$$\delta^{18}O = \left(\frac{\left(\frac{^{18}O}{^{16}O}\right)_{sample} - \left(\frac{^{18}O}{^{16}O}\right)_{standard}}{\left(\frac{^{18}O}{^{16}O}\right)_{standard}} \right) \times 1000$$

$$= \left(\frac{\left(\frac{^{18}O}{^{16}O}\right)_{sample}}{\left(\frac{^{18}O}{^{16}O}\right)_{standard}} - 1 \right) \times 1000 \qquad 8.1$$

(We have already used this notation for oxygen isotopes in the classification of meteorites, as discussed in Chapter 6.) Analogous definitions can be written for δD, $\delta^{13}C$, $\delta^{15}N$, $\delta^{17}O$, and many other isotope ratios. The major isotope is typically used as the denominator in the isotope ratios. For hydrogen and oxygen, the standard is Vienna Standard Mean Ocean Water (VSMOW), the nitrogen standard is the terrestrial atmosphere, and the carbon standard is Vienna PeeDee belemnite (VPDB, a Cretaceous marine fossil from the PeeDee formation). PeeDee belemnite material is no longer available, so other standard materials are used for analysis, but the results are still reported relative to VPDB.

In recent years, *mass-independent isotopic fractionations* have been discovered in oxygen and sulfur (elements with more than two isotopes where mass-dependent and mass-independent effects can be distinguished). The study of mass-independent isotopic variations of oxygen now plays a major role in cosmochemistry. And even more recently, mass-independent isotope effects have been recognized in a wide variety of nonvolatile elements (magnesium, silicon, calcium, titanium, chromium, molybdenum, and many others), and the field has expanded to include studies of these elements as well.

Studies of stable-isotope variations can provide information on processes affecting natural materials at different scales (from minerals to planets and the Sun), on temperatures, on the chemical and oxidation states of various systems, and on material reservoirs and their interactions. In the sections that follow, we will first look at mass-dependent isotope effects. We will examine equilibrium processes and discuss what *equilibrium fractionations* tell us about the state of a cosmochemical system. Then we will discuss systems governed by kinetics and consider what *kinetic fractionations* tell us about processes and rates. Following the discussion of mass-dependent isotope effects, we will discuss mass-independent isotope fractionations that result from various kinds of chemical processes. And finally, we will review what is known about nucleosynthetic isotope effects that were inherited from the solar system's parent molecular cloud. We will not be able to provide an exhaustive treatment of each topic. Instead, we will discuss a few elements as examples of the processes and principles.

The study of isotope fractionations has been facilitated by major improvements in the precision and sensitivity of mass-spectrometric techniques. Techniques such as thermal-ionization mass spectrometry (TIMS), inductively coupled plasma mass spectrometry (ICPMS), and secondary-ion mass spectrometry (SIMS) play major roles in this work (see the Appendix for discussions of these techniques). It is now possible to reliably measure an isotopic difference of ~10 ppm in some systems. The delta scale (δ) defined above for use with traditional stable isotopes has proved to be inadequate to discuss such precise data, so new scales have been developed. The epsilon scale (ε) gives the variations relative to the standard in parts per 10,000. The epsilon equation is like Equation 8.1, except that the factor is 10,000 instead of 1000. This scale is used for titanium, chromium, hafnium, neodymium, tungsten, and similar elements. Magnesium isotopes are sometimes discussed using the μ scale, which gives variations relative to the standard in parts per million. In contrast, osmium isotopic ratios are often reported on the γ scale, which gives variations relative to the standard in parts per 100.

8.2 Mass-Dependent Isotope Fractionations

There are two basic ways to fractionate isotopes mass-dependently. The first is by equilibrium processes, which partition the isotopes among the phases involved without any changes in conditions; there is no net chemical reaction. Equilibrium fractionations reflect the effect of atomic mass on bond energy and can be modeled with the methods of classical thermodynamics. In general, equilibrium fractionation between two phases decreases with increasing temperature, the degree of fractionation is larger for lighter elements because there is a greater relative mass difference between the isotopes, and the heavy isotope is preferentially partitioned into the site with the stiffest (strongest and/or shortest) chemical bonds. Because of the temperature dependence, isotopic shifts between phases at equilibrium can often be used as *geothermometers*.

The second way in which isotopes can be fractionated is via kinetic processes. Kinetic isotope effects are associated with fast, incomplete, or unidirectional processes such as evaporation, diffusion, dissociation reactions, and biological reactions. Isotopic effects produced by diffusion and evaporation are due to differences in translational velocity between different isotopic forms of molecules as they move through a phase or cross a phase boundary. The kinetic energy of a molecule is given by

$$KE = \frac{1}{2}mv^2 \qquad 8.2$$

where m is the mass of the molecule and v is its velocity. Under a given set of conditions, the kinetic energy of molecules of the same compound but with different isotopic configurations will be the same. In order for this to be true, the difference in mass caused by the presence of different isotopes must be counteracted by a difference in velocity – the lighter molecule will move faster. In cosmochemistry, kinetic mass fractionations via evaporation, diffusion, and dissociation reactions are widely observed and are useful for constraining the conditions and processes that produced the objects.

8.2.1 Equilibrium Mass-Dependent Isotope Fractionations

Equilibrium mass-dependent fractionations involve the partitioning of isotopes between phases during *isotope exchange*. Isotope exchange can occur in a wide variety of systems and under a wide variety of conditions, but the common feature of equilibrium isotope exchange is that there is no net chemical reaction. Isotope-exchange reactions can be written as:

$$aA_1 + bB_2 = aA_2 + bB_1 \qquad 8.3$$

where species A and B contain either the light isotope (1) or the heavy isotope (2), and a and b are the coefficients necessary to balance the reaction. For this reaction, the equilibrium constant is given by:

$$K = \frac{\left(\frac{A_2}{A_1}\right)^a}{\left(\frac{B_2}{B_1}\right)^b} \qquad 8.4$$

The isotopic equilibrium constant can be expressed in terms of the partition functions Q of the different species. The equilibrium constant is the quotient of two partition-function ratios, one for the two isotopic species of A and the other for B:

$$K = \frac{\left(\frac{Q_{A2}}{Q_{A1}}\right)}{\left(\frac{Q_{B2}}{Q_{B1}}\right)} \qquad 8.5$$

The partition function is the sum over all of the allowed energy levels, E_i:

$$Q = \Sigma_i\left(g_i e^{-E_i/kT}\right) \qquad 8.6$$

where g_i is a statistical term to account for possible degeneracy, or different states, k is Boltzmann's constant, and T is the temperature. The total partition function can be divided into partition functions that describe the different forms of energy, vibration, rotation, and translation:

$$Q_{total} = Q_{vib}Q_{rot}Q_{trans} \qquad 8.7$$

The translational and rotational energy between isotopes are very similar among compounds in an exchange relationship, leaving vibrational energy as the main source of isotope effects (hydrogen is an exception, where rotational energy must be taken into account).

Figure 8.1 shows how vibrational energy is related to the zero-point energy for three diatomic hydrogen molecules made up of different combinations of isotopes. As two isolated atoms get closer together, an attractive force between electrons and protons pulls the atoms toward each other and the potential energy drops. At very close interatomic distances, electrostatic repulsion between the positively charged nuclei becomes dominant and the potential energy rises again.

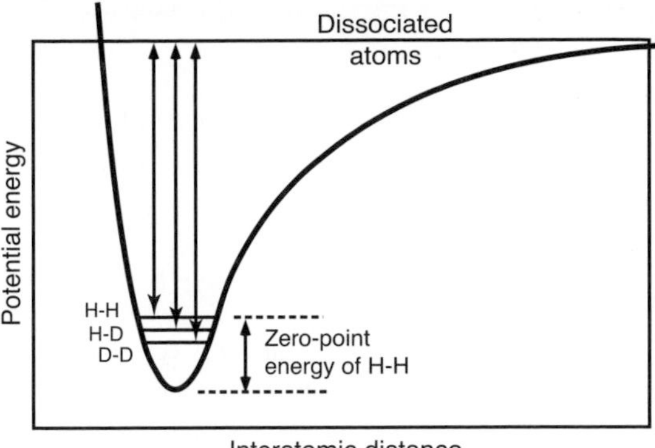

Figure 8.1 Potential energy curve for diatomic hydrogen. Shown are the zero-point energies of the ^1H-^1H, ^1H-^2D, and ^2D-^2D molecules. The zero-point energy of the ^2D-^2D molecule sits deeper in the potential energy well than the other two molecules, so it has a higher dissociation energy and a stronger bond.

The two atoms in a molecule will have an average distance from one another that balances the attractive force between the two atoms as they share electrons and the repulsive force between the two positively charged protons. The atoms behave as a harmonic oscillator and vibrate at a characteristic frequency that depends on the masses of the isotopes. At absolute zero, the energy of the molecule is given by

$$E = \frac{1}{2}hv \qquad 8.8$$

where h is Planck's constant and v is the fundamental vibrational frequency, which depends on the masses of the atoms. This is the *zero-point energy*. Note that the zero-point energy is above the lowest point on the potential energy curve (Fig. 8.1) and is governed by quantum-mechanical effects. When a deuterium atom is substituted for a hydrogen atom, the vibration frequency decreases, producing a corresponding decrease in the zero-point energy of the molecule. This decrease corresponds to an increase in the strength of the bond between the atoms. The D_2 molecule has the strongest bond. At temperatures above absolute zero, the molecules vibrate with more energy, and when the vibrational energy is high enough, the molecules dissociate. Because H_2 is not as deep in the potential well as HD and D_2, it takes less vibrational energy from the environment to dissociate H_2.

As a molecule gains energy, it moves to a higher energy level. Like the zero-point energy, the allowed energy levels are described by quantum mechanics. The general equation is:

$$E = \left(n + \frac{1}{2}\right)hv \quad 8.9$$

where n is the vibrational energy level ($n = 0, 1, 2$, etc.), h is Planck's constant, and v is the vibration frequency (per sec). The higher the vibrational energy level of a molecule, the less additional energy it needs to dissociate. The vibration frequency is a function of the masses of the two atoms.

For isotope-exchange reactions, it is conventional to replace the equilibrium constant, K, with the *fractionation factor*, α. The fractionation factor is defined as the ratio of the numbers of any two isotopes in chemical compound A divided by the corresponding ratio in compound B:

$$\alpha_{A-B} = (A_2/A_1)/(B_2/B_2) \quad 8.10$$

If the isotopes are randomly distributed over all possible positions in compounds A and B, then α is related to K by:

$$\alpha = K^{1/n} \quad 8.11$$

where n is the number of atoms exchanged. For simplicity, isotope-exchange reactions are typically written to describe the exchange of only one atom. In these cases, the equilibrium constant is identical to the fractionation factor.

To illustrate, consider the exchange of ^{18}O and ^{16}O between water and $CaCO_3$:

$$H_2{}^{18}O + 0.33CaC^{16}O_3 \longleftrightarrow H_2{}^{16}O + 0.33CaC^{18}O_3 \quad 8.12$$

The fractionation factor $\alpha_{CaCO_3-H_2O}$ is defined as:

$$\alpha_{CaCO_3-H_2O} = \left({}^{18}O/{}^{16}O\right)_{CaCO_3}/\left({}^{18}O/{}^{16}O\right)_{H_2O}$$
$$= 1.031 \text{ at } 25°C \quad 8.13$$

where the value 1.031 has been determined experimentally. As discussed above, it is common practice to express isotopic compositions in terms of delta values (Eq. 8.1). For two compounds, A and B, the δ values and the fractionation factor α are related by:

$$\delta_A - \delta_B = \Delta_{A-B} \approx 10^3 \ln \alpha_{A-B} \quad 8.14$$

However, if one is trying to compare precisely determined isotope ratios, the fractionation factor, α, determined from the measured ratios (Eq. 8.13) should be used instead of the δ values.

For the purposes of cosmochemistry, the dependence of α on temperature is the most important property. Isotope fractionations tend to go to zero at very high temperatures, but the temperature dependence is not necessarily monotonic. As temperature increases, isotopic fractionations may increase in magnitude or even change sign over a limited temperature range, but they must approach zero at very high temperatures.

For reactions in an ideal gas, there are two temperature regions where the behavior of α is simple. At low temperatures, generally much below room temperature, ln K is proportional to ~$1/T$, where T is the absolute temperature. At room temperature and above, for anhydrous minerals, ln K is proportional to ~$1/T^2$.

The equilibrium partitioning of isotopes between phases is a function of temperature, which means that isotopic compositions of coexisting phases can be used to determine the temperature of a system. Stable isotopes can also be used to trace material reservoirs through various processes. Here we will discuss several applications of equilibrium isotope fractionation to give a feel for the range of applications it has in cosmochemistry.

8.2.1.1 *Isotope Fractionations and Geothermometry in Igneous Minerals*

Oxygen isotopes partition between coexisting minerals and between minerals and melts at temperatures of interest in cosmochemistry. Figure 8.2 shows some experimentally determined mineral–mineral and mineral–melt fractionation factors as a function of temperature ($1/T^2$). Fractionation factors decrease with increasing temperature, going to zero at infinite temperature. There is also a strong dependence on the nature of the chemical bonds, which affect the vibration frequencies. The heavier isotopes of oxygen partition preferentially into minerals with stronger, more covalent Si–O–M bonds (where M is the cation). In Figure 8.2, quartz-magnetite shows the largest fractionation because quartz has only silicate bonds and magnetite has only M–O bonds. Fractionations involving quartz are of limited use in cosmochemistry because quartz is so uncommon, but it is important in geochemistry.

Data like those shown in Figure 8.2 can be used to estimate the temperature in igneous systems. Several conditions must be met to do this. First, the rock or magma should have cooled rapidly (quenched) so that the inferred temperature reflects the temperature in the magma chamber, not the effects of slow cooling when isotope diffusion in minerals can quench at different temperatures. Second, the minerals should be phenocrysts that were in equilibrium with the melt, not foreign xenocrysts with different isotopic compositions. Third, the samples used must be

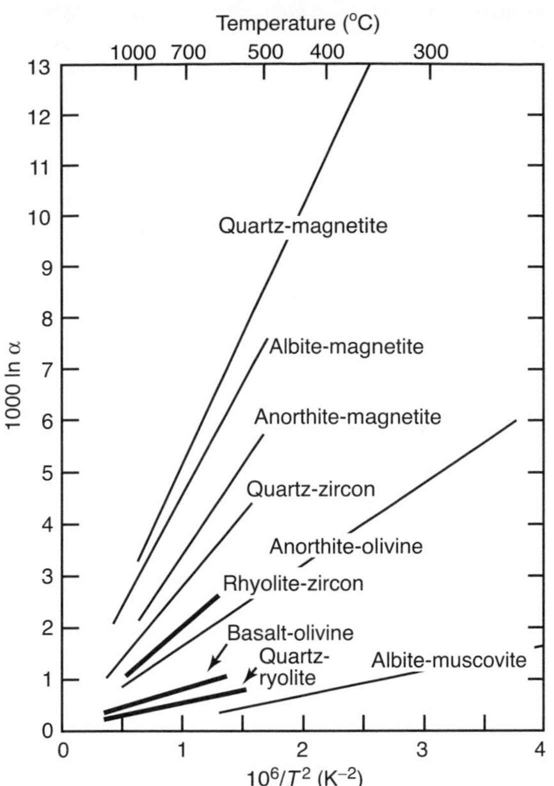

Figure 8.2 Examples of mineral–mineral and mineral–melt isotope fractionation as a function of $1/T^2$. Thin lines are experimental data and thicker lines represent empirical measurements. Adapted from Bindeman (2008), who provides data sources.

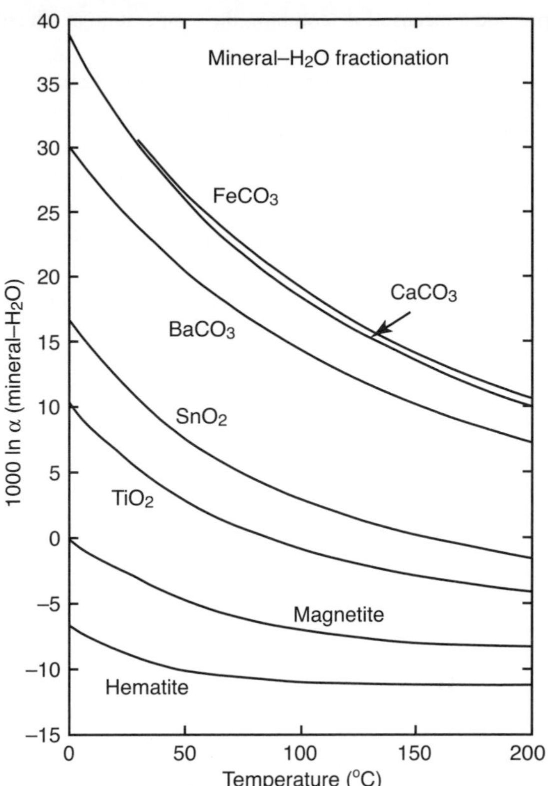

Figure 8.3 Examples of fractionation factors for mineral–H_2O partitioning as a function of temperature. Data are from Zheng (1991, 2011) and Chacko et al. (2001).

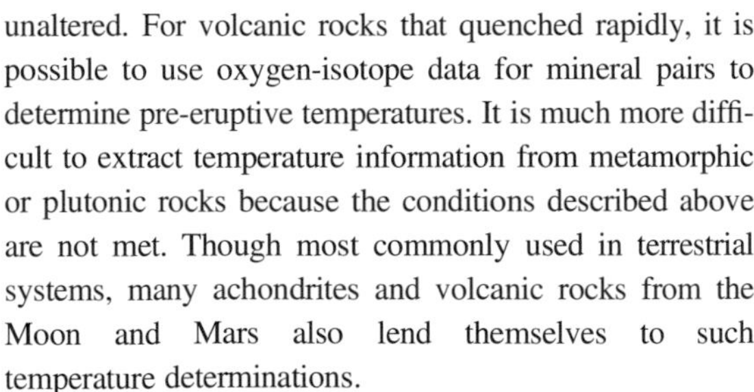

unaltered. For volcanic rocks that quenched rapidly, it is possible to use oxygen-isotope data for mineral pairs to determine pre-eruptive temperatures. It is much more difficult to extract temperature information from metamorphic or plutonic rocks because the conditions described above are not met. Though most commonly used in terrestrial systems, many achondrites and volcanic rocks from the Moon and Mars also lend themselves to such temperature determinations.

Ali et al. (2018) provided an example. They measured the oxygen isotopic compositions of olivine and pyroxene in two martian meteorites, DaG 476 and SaU 008. These two meteorites have porphyritic textures, with large phenocrysts of olivine and pyroxene in fine-grained groundmass. The authors carefully established that olivine and pyroxene were in equilibrium with each other in each meteorite. Then they used the differences in oxygen-isotope composition between olivine and pyroxene to calculate the temperature of the magma chamber where the phenocrysts grew. They found that SaU 008 equilibrated at ~1400°C, while DaG 476 equilibrated at ~1200°C. Combined with other chemical and mineralogical information, these temperatures help to establish the conditions in the Mars mantle where the magmas that produced these meteorites last equilibrated.

8.2.1.2 *Isotope Fractionations during Aqueous Alteration of Carbonaceous Chondrites*

Isotopes also partition during the formation of minerals from aqueous solutions. Figure 8.3 shows some mineral–water fractionation factors as a function of temperature. Carbonates tend to be isotopically heavy compared to the water from which they form, whereas magnetite and hematite tend to be isotopically light. In most mineral–water systems, the isotopic fractionation is a strong function of temperature, so as with high-temperature mineral–melt systems, isotopic fractionations in aqueous systems provide a means of determining the temperature at which a mineral formed.

As discussed in Chapter 6, the CI and CM2 carbonaceous chondrites have experienced extensive aqueous alteration, and some other chondrite classes have experienced minor alteration. Oxygen isotopes provide a means to constrain the conditions under which alteration occurred. Pervasive aqueous alteration results in dissolution of the original minerals and precipitation of new minerals, such as carbonates (calcite, aragonite, dolomite, breunerite, siderite), magnetite, and phyllosilicates. If one knows the oxygen isotopic composition of the fluid from which a mineral formed, the measured oxygen isotopic composition of the mineral can be compared with that of the fluid and with data like those in Figure 8.3 to give the temperature at which the mineral formed. But in cosmochemistry, we generally do not have a sample of the fluid and thus do not know its isotopic composition. The way around this problem is to identify two minerals that formed at the same time from the same fluid. For example, if petrography can demonstrate that calcite and magnetite formed together, then calculated or measured fractionation factors like those shown in Figure 8.3 can be compared and the composition of the water cancels out. The difference in isotopic composition between the two minerals gives the temperature of formation. An example of the use of oxygen isotopes to derive formation temperature is given in Box 8.1.

8.2.1.3 *Evidence for Life on Mars?*

McKay et al. (1996) claimed to have found evidence of ancient life in the martian meteorite ALH 84001 (discussed more fully in Chapter 11, Box 11.3). The paper generated huge interest and launched the new field of astrobiology. However, the evidence presented in the paper was quite controversial. Several lines of evidence depended critically on establishing that carbonate grains in fractures in the meteorite were not due to terrestrial alteration and had formed at low temperatures. Carbon isotopes indicate that the carbonates were unlike terrestrial carbonates and thus formed on Mars (Jull et al., 1995). Oxygen isotopes were used to infer temperatures of formation for the carbonate minerals. Various oxygen isotope measurements have yielded carbonate formation temperatures ranging from 18 to 375°C (Valley et al., 1997; Leshin et al., 1998; Eiler et al., 2002; Halevy et al., 2011), but the consensus now supports low temperatures. Although most of the scientific community do not find the evidence for life in ALH 84001 to be compelling, this has been one of the most important stories in cosmochemistry over the last two decades, and stable-isotope studies have played a key role.

8.2.1.4 *Clumped-Isotope Cosmochemistry*

Clumped isotopes represent a relatively new branch of investigation that studies the ordering of rare isotopes in natural materials (Eiler, 2007). This is an equilibrium technique that uses the full thermodynamic and quantum-mechanical description of the distribution of isotopes in chemical compounds. As discussed above, when a heavy isotope is substituted into a molecule in place of a light isotope, the vibration frequency of the molecule decreases, as does the zero-point energy (Fig. 8.1), making the molecule more stable. It is commonly assumed that the change in bond energy associated with the substitution of two heavy isotopes into a molecule is exactly twice that associated with the substitution of one heavy isotope. This assumption works well for molecules containing the dominant isotopic species, but it is only an approximation to the complete equation for the partition function, which fully describes the vibrational, rotational, and translational energies of molecules (e.g. Equation 8.7). In fact, the decrease in vibrational energy associated with double heavy-isotope substitution often slightly exceeds twice that of a single heavy-isotope substitution. The slight differences between the complete description of the partition function and the standard simplification are the basis for clumped-isotope cosmochemistry.

Application of clumped-isotope theory to cosmochemistry involves high-precision measurements of the abundances of one of the rare *isotopologues*. Table 8.1 shows the isotopologues of CO_2 and their relative abundances. Clumped isotope measurements of CO_2 typically involve mass 47, which is dominated by $^{13}C^{18}O^{16}O$. The measured abundances at mass 47 are typically higher than predicted from a random distribution of the heavy isotopes. The excesses, denoted by Δ_{47}, typically increase with decreasing temperature, making the clumped-isotope approach useful for thermometry. As with any other isotope technique, the results are compared with those for standards carefully prepared to insure isotopic equilibrium at several different temperatures.

A major application of the clumped-isotope technique is in carbonate paleothermometry. As already discussed, conventional carbonate-water paleothermometry depends on knowing $\delta^{18}O$ for the carbonate and for the water from which it formed. In the clumped isotope technique, all of the information needed to reconstruct the equilibrium constant for the formation reaction is preserved within the carbonate mineral itself. And, although the

Box 8.1 An Example of Temperature Determination Using Oxygen Isotopes

Knowing the temperature at which a mineral assemblage formed can be very useful in determining the processes that produced the minerals. Oxygen isotopes can constrain temperature. Jilly-Rehak et al. (2018) measured oxygen isotopes in the magnetite-calcite assemblage found in the matrix of the Al Rais CR2 chondrite. The intimate intergrowth of magnetite and calcite in this meteorite indicates that the minerals coprecipitated from the same isotopically homogeneous fluid. The fractionation factor for mineral and water is given by:

$$\alpha = \left({}^{18}\mathrm{O}/{}^{16}\mathrm{O}\right)_{\mathrm{mineral}}/\left({}^{18}\mathrm{O}/{}^{16}\mathrm{O}\right)_{\mathrm{water}} \quad 8.15$$

Experimental work by Zheng (1991, 2011) found the following relationships describing the fractionation factors as a function of temperature for calcite–water and magnetite–water:

$$\text{Calcite–water: } 10^3 \ln \alpha = \left(4.01 \times 10^6\right)/T^2 - \left(4.66 \times 10^3\right)/T + 1.71 \quad 8.16$$

$$\text{Magnetite–water: } 10^3 \ln \alpha = \left(3.02 \times 10^6\right)/T^2 - \left(12.00 \times 10^3\right)/T + 3.31 \quad 8.17$$

These two fractionation relationships are plotted as a function of temperature in Figure 8.4. To extract temperature information from a single mineral, the fluid composition and mineral composition must be known. If two minerals formed at the same time from the same fluid, the relative equilibrium-fractionation factors can provide temperature information without knowledge of the water composition. The calcite–magnetite fractionation relationship calculated from the above experimental data is also plotted with a solid line in Figure 8.4:

$$\text{Calcite–magnetite: } 10^3 \ln \alpha = \left(0.99 \times 10^6\right)/T^2 - \left(7.34 \times 10^3\right)/T + 1.60 \quad 8.18$$

With these partitioning data, a sample containing an equilibrium mineral assemblage, and precise oxygen-isotope data for the minerals in the sample, a temperature can be calculated. By plotting the relative fractionation measured for calcite and magnetite in Al Rais (29.9 ± 2.2‰), a temperature (converted from kelvin in Fig. 8.4b) of ~60 ± 19°C is inferred for the aqueous alteration of Al Rais.

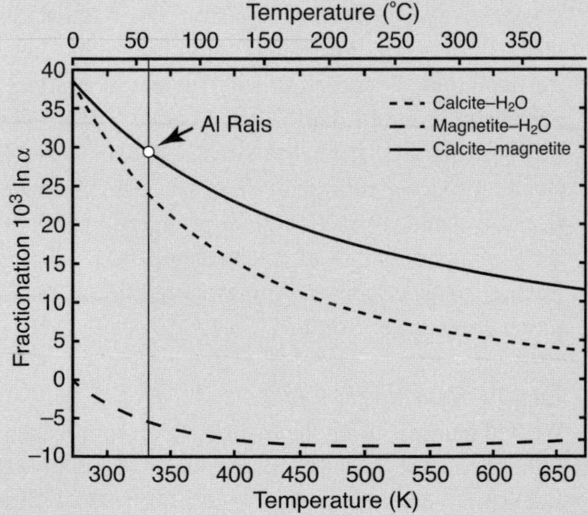

Figure 8.4 Fractionation curves for precipitation of calcite and magnetite from an aqueous solution at various temperatures. The solid black line gives the relative fractionation between calcite and magnetite. The vertical line gives the temperature of formation of the calcite-magnetite assemblage in Al Rais. After Jilly-Rehak et al. (2018).

Table 8.1 Relative abundances of the isotopologues of CO_2

Mass	Isotopologue	Relative Abundance
44	$^{12}C^{16}O_2$	98.40%
45	$^{13}C^{16}O_2$	1.11%
	$^{12}C^{17}O^{16}O$	748 ppm
46	$^{12}C^{18}O^{16}O$	0.40%
	$^{13}C^{17}O^{16}O$	8.4 ppm
	$^{12}C^{17}O_2$	0.142 ppm
47	$^{13}C^{18}O^{16}O$	44.4 ppm
	$^{12}C^{17}O^{18}O$	1.50 ppm
	$^{13}C^{17}O_2$	1.60 ppb
48	$^{12}C^{18}O_2$	3.96 ppm
	$^{13}C^{17}O^{18}O$	16.8 ppb
49	$^{13}C^{18}O_2$	44.5 ppb

clumped-isotope technique does not require $\delta^{18}O$ of water to determine the temperature of formation for a carbonate, one can use the determined temperature, the $\delta^{18}O$ measured in the carbonate at the same time as the clumped isotopes, and the known temperature dependence of $\delta^{18}O$ between carbonate and water to determine the $\delta^{18}O$ of water in the system.

Guo and Eiler (2007) used clumped isotopes to study the temperature of formation of carbonates in several CM chondrites. Prior to this study, temperature estimates of carbonate formation ranged from <20 to <170°C, but problems with these measurements included the likelihood that the oxygen isotopes of the different phases were not in equilibrium. Using the clumped-isotope technique, Guo and Eiler (2007) demonstrated that the carbonates formed at a temperature of ~20 to ~71°C from a fluid with $\delta^{18}O$ of 2.0 to 8.1‰, and $\delta^{17}O$ of –0.7 to 3.0‰. They also concluded that there is evidence for methane formation during the aqueous alteration of CM chondrites.

Halevy et al. (2011) used clumped isotopes to study the carbonates in ALH 84001, the meteorite that was thought to contain evidence for life on Mars (see above and Chapter 11, Box 11.3). Taking advantage of the fact that clumped isotopes can give temperature information from the distribution of isotopes in a single mineral phase, they determined that the carbonates in ALH 84001 formed at 18±4°C.

The clumped-isotope technique presents a number of experimental challenges. Currently, the technique can only be used for elements that are gases at room temperature, and gas-source mass spectrometry has been the tool of choice. Commercial gas-source mass spectrometers have relatively low mass-resolving power ($M/\Delta M =$ ~500), which is insufficient to discriminate among isotopologues with the same mass (e.g. $^{13}C^{18}O^{16}O$ and $^{12}C^{18}O^{17}O$) or to resolve trace contaminants. Also, the clumped-isotope approach requires very precise abundance measurement of isotopologues that are present at ppm to ppb levels (the dominant isotopologue at mass 47 is present at ~44 ppm, Table 8.1). Care must also be taken that sample preparation and mass spectrometry do not modify the molecules that one wants to measure. The clumped-isotope technique has promise, but significant technical advances, theoretical work, and development of standards will be required for clumped-isotope cosmochemistry to become a mainstream technique.

8.2.2 Kinetic Mass-Dependent Isotope Fractionations

Kinetic isotopic fractionations are associated with fast, incomplete, or unidirectional processes such as evaporation, diffusion, dissociation reactions, and most biological processes. Kinetic fractionations are irreversible. In evaporation and diffusion, kinetic fractionation depends on the velocity of the isotopes, which is faster for lighter isotopes in order to maintain constant kinetic energy. The effects are most pronounced for light elements, in which the relative difference in mass is larger.

8.2.2.1 *Isotope Fractionation during Evaporation and Condensation*

Evaporation and condensation can produce large isotopic fractionations and thus are of special interest in geochemistry and cosmochemistry. The fractionations result from a combination of equilibrium and kinetic fractionations. Evaporation and condensation affect us personally through Earth's hydrologic cycle, where water evaporates from the oceans and the resulting clouds produce rain and snow on the continents. Because chemical fractionations produced by volatility are a common feature of cosmochemistry, the isotopic fractionations associated with evaporation and condensation are intensely studied.

Earth's Hydrologic Cycle

We will illustrate using the hydrologic cycle on Earth. At equilibrium, the isotopic compositions of liquid water and water vapor are governed by the fractionation factor, α_{L-V}, defined for this system as the isotope ratio in the liquid divided by the isotope ratio in the vapor ($\alpha_{L-V} = R_L/R_V$). Water molecules pass through the liquid–vapor interface in both directions at the same rate at equilibrium; there is no net flux across the interface.

Condensation is basically an equilibrium process. When a vapor becomes saturated, droplets of water nucleate and grow, taking up the water that exceeds the capacity of the vapor. If condensation takes place in a closed system, the total liquid at any point in the process will be in isotopic equilibrium with the remaining vapor. Figure 8.5a shows the evolution of the isotopic compositions of liquid and vapor during equilibrium condensation as dashed lines. The separation between the two dashed lines comes from the fractionation factor, α_{L-V}. In natural systems, condensation can have a kinetic component if the condensate is removed from contact with the vapor.

In a purely kinetic system, water condensing from a vapor is removed from the system and the isotopic composition of the remaining vapor becomes isotopically lighter (solid lines labeled "open" in Fig. 8.5a). Similarly, if atoms or molecules evaporating from a liquid are immediately removed from contact with the liquid, the compositions of gas and liquid evolve according to a kinetic law. The theoretical foundation for understanding evaporation and condensation under kinetic conditions was presented by Rayleigh (1896); also see Sharp (2017) for a more-detailed discussion. The processes are analogous to fractional crystallization and fractional melting in igneous systems.

The Rayleigh equation for a condensation process is

$$\frac{R_V}{R_{V_i}} = f^{\alpha-1} \tag{8.19}$$

where R_{V_i} is the isotope ratio of the initial bulk composition and R_V is the instantaneous ratio of the remaining vapor (V), f is the fraction of the residual vapor, and the fractionation factor α is given by R_L/R_V. Similarly, the instantaneous isotope ratio of the condensate (R_L) leaving the vapor is given by

$$\frac{R_L}{R_{V_i}} = \alpha f^{\alpha-1} \tag{8.20}$$

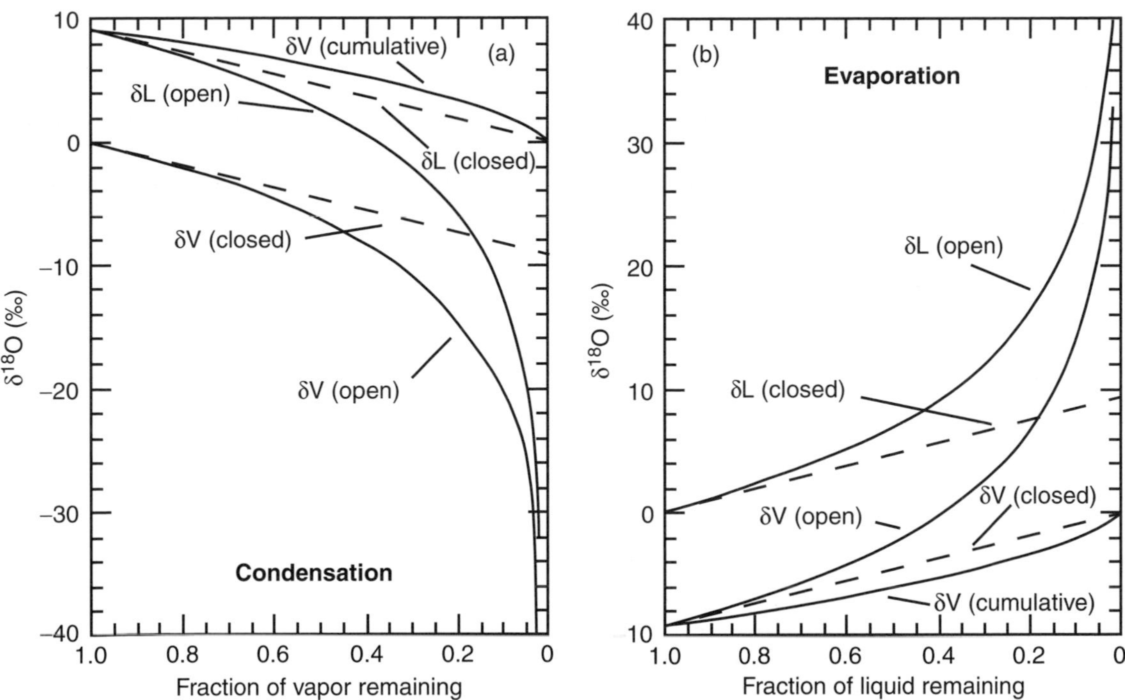

Figure 8.5 Isotopic fractionation during (a) condensation and (b) evaporation in closed and open systems. Equilibrium fractionation between liquid and vapor is assumed and the fractionation factor is $\alpha = 1.0092$ (constant). The solid lines labeled "open" show the instantaneous composition of the liquid and vapor during condensation or evaporation when the condensate or vapor are removed from the system as they are generated (Rayleigh fractionation). The dashed lines show the composition of both phases in a closed system as condensation or evaporation proceeds. The curves marked "cumulative" give the compositions of (a) condensate and (b) vapor generated and removed from the system in a Rayleigh process.

and the average isotope ratio of the separated and accumulated condensate (R_L) at any point during the condensation is given by

$$\frac{\bar{R}_L}{R_{V_i}} = \frac{1-f^{\alpha}}{1-f} \qquad 8.21$$

Figure 8.5a shows the curves described by these equations as solid lines. By the time condensation operating in a purely kinetic mode has removed most of the vapor, the instantaneous condensate would have an extreme isotopic composition. But in a real situation involving condensation of water, the isotopic compositions of vapor and accumulated liquid would closely follow the dashed lines.

In contrast, evaporation is dominantly a kinetic process. Evaporation occurs when the partial pressure of the evaporating compound in the vapor above the liquid is below saturation. When the partial pressure of the evaporating component is low, the number of atoms or molecules leaving the liquid is vastly greater than those returning to the liquid from the vapor. If the partial pressure is high, the flux back into the liquid is more similar to that leaving the liquid and the system behaves more like it was in equilibrium. At equilibrium, the fluxes in and out of the liquid from the vapor would be in balance and there would be no net evaporation.

The behavior of the isotopic ratios of liquid and vapor during equilibrium and kinetic evaporation are shown in Figure 8.5b. At equilibrium, the isotopic compositions of liquid and vapor are related by the fractionation factor, α_{L-V}, and the system evolves along the dashed lines. The Rayleigh equation for the instantaneous isotopic composition of the liquid in an evaporating system is

$$\frac{R_L}{R_{L_i}} = f^{\left(\frac{1}{\alpha}-1\right)} \qquad 8.22$$

and for the vapor leaving the liquid it is

$$\frac{R_V}{R_{L_i}} = \frac{1}{\alpha} f^{\left(\frac{1}{\alpha}-1\right)} \qquad 8.23$$

The average isotope ratio of the separated and accumulated vapor is given by

$$\frac{\bar{R}_V}{R_{L_i}} = \frac{1-f^{1/\alpha}}{1-f} \qquad 8.24$$

where f is the fraction of residual liquid. The isotopic compositions as a function of f are shown by the solid curves in Figure 8.5b.

The interplay between Rayleigh distillation and near-equilibrium condensation can be seen in the isotopic compositions of rain and snow on Earth (Fig. 8.6). Water evaporating from the ocean experiences a large kinetic fractionation and has an isotopic composition that is lighter than that of the ocean. Rain condensing from the clouds is in equilibrium with the water vapor and is isotopically heavier as described by α_{L-V}. The remaining

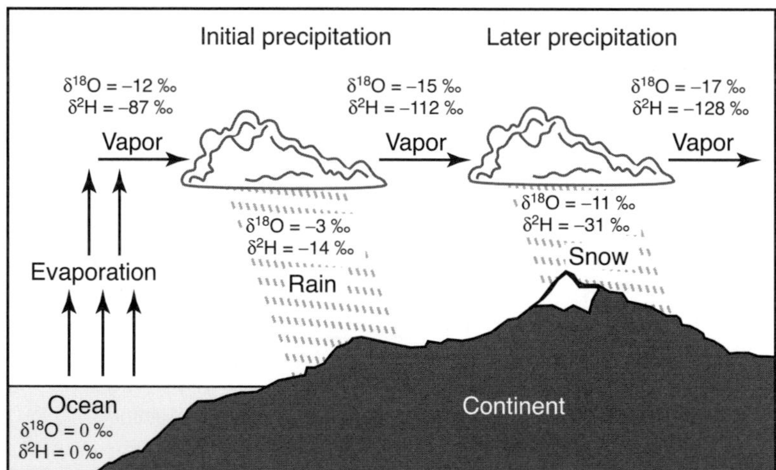

Figure 8.6 Schematic illustration of the isotopic evolution of water vapor as it first evaporates from the ocean, then drops as rain on coastal mountains, then drops as snow on high mountains. The initial evaporation is kinetically controlled and the isotope effect is large, with the vapor isotopically lighter than the ocean. Dropping of rain and snow is a near-equilibrium event, so the isotopic change is significantly smaller. The precipitation is isotopically heavier than the cloud, so the cloud gets a little lighter with each precipitation event. Modified from Sharp (2017).

vapor in the clouds becomes isotopically lighter as a result. When the cloud passes over a high mountain and snow precipitates, the snow will be isotopically heavier than the cloud, but due to the cycles of evaporation and condensation, it will be isotopically lighter than the initial rain that condensed from the ocean vapor. In general, the isotopic composition of rain and snow becomes increasingly light with elevation and distance from the coast. Hydrogen behaves the same way as oxygen, but the isotopic effects are larger due to the larger mass difference between isotopes in hydrogen compared to oxygen.

Temperature affects the amounts and isotopic compositions of liquid and vapor during condensation and evaporation. The amount of water that can exist as a vapor in equilibrium is a function of temperature, with warm gas having a much higher capacity than cold gas. The isotopic fractionation factors generally get larger with decreasing temperature. Consequently, the separation between the two solid curves in Figures 8.5a and 8.5b will increase with decreasing temperature.

Rayleigh-Type Evaporation in CAIs

As discussed in Chapter 7, both evaporation and condensation occurred in the early solar nebula. Isotopic fractionations are powerful tools to understand evaporation and condensation in silicate systems. A silicate melt in the low-pressure environment of the solar nebula would evaporate, producing isotopic fractionations similar to those expected from Rayleigh distillation. In Chapter 7, we discussed how the compositions of type B CAIs, which crystallized from melts in the nebula, evolve chemically during Rayleigh distillation (Fig. 7.9). There are also isotope effects. Type B CAIs are typically enriched in heavy isotopes by several parts per thousand (permil) per amu. Figure 8.7 shows isotopic fractionation in magnesium and silicon for a suite of type B CAIs. The isotopic fractionations for the two elements are correlated, as would be expected.

Although the Rayleigh evaporation model is a powerful tool for understanding the behavior of melts in a vacuum, real systems do not precisely follow the model. A detailed derivation of the Rayleigh equation for evaporation is given by Davis and Richter (2014). The equation describing the relative flux of two isotopes from the liquid to the vapor is

$$\frac{J_{i,2}}{J_{i,1}} = R_{2,1}\frac{\gamma_{i2}}{\gamma_{i1}}\sqrt{\frac{M_1}{M_2}} \qquad 8.25$$

where $R_{2,1} = N_2/N_1$, where N_1 and N_2 are the abundances of isotopes 1 and 2 (defined for magnesium as ^{24}Mg and ^{25}Mg, respectively), γ_{i1} and γ_{i2} are the evaporation coefficients for the two isotopes making up $R_{2,1}$, and M_1 and M_2 are the masses of ^{24}Mg and ^{25}Mg, respectively. The fractionation factor, α, is given by

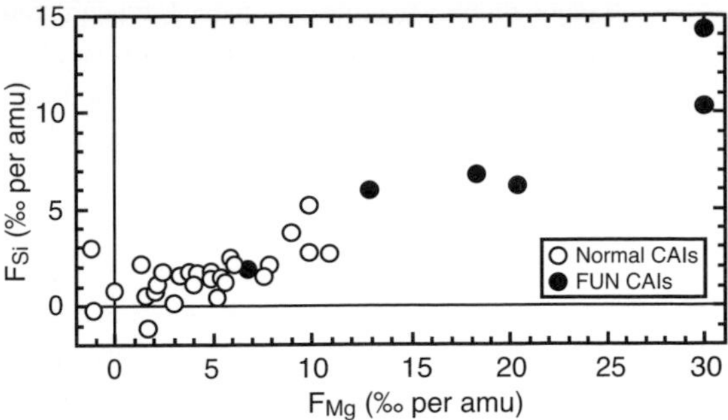

Figure 8.7 Correlated isotopic fractionations in type B CAIs. The FUN inclusions show both much larger isotopic fractionation and large nucleosynthetic isotopic anomalies (Section 8.4). Modified from Davis and Richter (2014).

$$\alpha_{1,2} = \frac{\gamma_{i2}}{\gamma_{i1}}\sqrt{\frac{M_1}{M_2}} \qquad 8.26$$

The γ terms are determined experimentally and are typically very similar. So for now, we will assume that they are unity, giving α as simply the square root of the ratio of the masses (numerically equal to 0.9798 for ^{24}Mg/^{25}Mg):

$$\alpha_{1,2} = \sqrt{\frac{M_1}{M_2}} \qquad 8.27$$

In a system undergoing Rayleigh evaporation, an isotopic ratio in the liquid will evolve according to

$$\frac{R^{res}_{2,1}}{R^{0}_{2,1}} = f_1^{(\alpha-1)} \qquad 8.28$$

where $R^{res}_{2,1}$ is the isotopic composition of the residual liquid, $R^{0}_{2,1}$ is the initial isotopic composition of the liquid, f_1 is the fraction of isotope 1 remaining, and α is the isotopic fractionation factor. Equation 8.28 is valid if α is independent of the evolving composition of the evaporating liquid, and the diffusive transport rate is fast enough to keep the liquid homogeneous. Note that Equation 8.28 is written in terms of α, while the equivalent equation for evaporating water (Equation 8.22) is written in terms of $1/\alpha$. This reflects a difference in the definition of α, which for water is the measured composition of the liquid divided by that of the gas and is >1, and for the silicate melt is $\sqrt{M_{24}/M_{25}} < 1$. Formulated in this way, the two treatments give the same result, namely that the liquid becomes increasingly heavy as Rayleigh evaporation proceeds.

The fractionation of magnesium under Rayleigh conditions using the theoretical fractionation factor based on the square root of the ratio of the masses is compared with experimental data in Figure 8.8. The data follow Rayleigh-like curves, but the apparent isotopic fractionation factor is somewhat closer to 1 than the expected value of $\alpha = \sqrt{24/25}$. An area of active research is to understand the source of this type of deviation. Is there a problem with the experiments, or is the deviation from theory telling us something fundamental about evaporation of silicate melts?

8.2.2.2 *Fractionation Laws*

When studying cosmochemical samples, there are two types of mass-dependent fractionations to worry about. First, all mass spectrometers introduce mass-dependent fractionation during the measurement, and this fractionation must be accounted for. We do this by measuring standards of known isotopic composition. Once the instrumental effects have been accounted for, we can look for the interesting mass fractionations, the ones that are intrinsic to the samples. This is done by comparing the measured composition of the sample to the "solar system composition," which is typically taken to be the isotopic composition of the Earth. Here we will discuss magnesium isotopes in order to make the discussion more specific, but the same principles apply to any element.

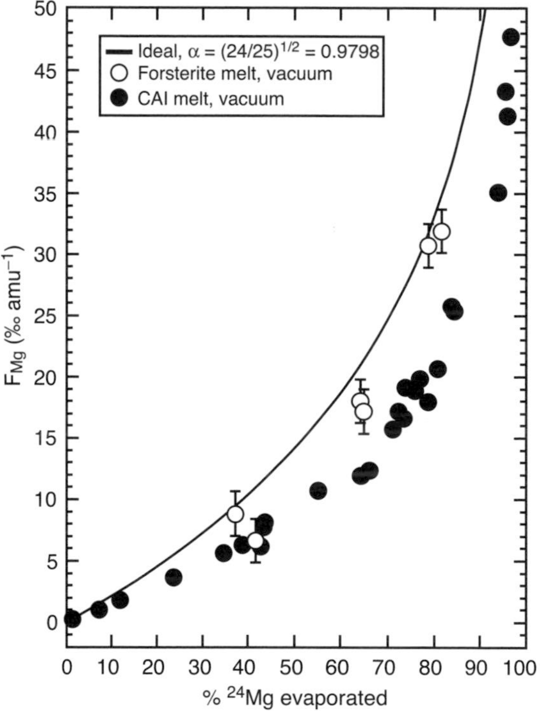

Figure 8.8 Magnesium isotopic composition versus % magnesium evaporated into a vacuum for two different initial compositions. All experiments show similar behavior, with less isotopic fractionation than that expected for evaporation of Mg atoms (the major gas-phase species of magnesium). Forsterite composition melt data are from Davis et al. (1990); CAI-like melt data in vacuum are from Richter et al. (2007). The scatter in the high-precision ICPMS data for vacuum evaporation of CAI-like melt is due to small variations in the isotopic fractionation factor with temperature over the range of 1600–1900°C. The remaining data were collected by ion microprobe; scatter in these data is likely due to the poorer precision of those measurements.

Historically, magnesium isotope ratios are discussed as delta values:

$$\delta^{25}\mathrm{Mg}\left[\frac{\left({}^{25}\mathrm{Mg}/{}^{24}\mathrm{Mg}\right)_{\mathrm{sample}}}{\left({}^{25}\mathrm{Mg}/{}^{24}\mathrm{Mg}\right)_{standard}}-1\right]\times 1000 \qquad 8.29$$

and similarly for $^{26}Mg/^{24}Mg$. A mass fractionation law that depends linearly on mass difference plots as a straight line on three-isotope plots of $\delta^{25}Mg$ versus $\delta^{26}Mg$. Over the years, a variety of mass-fractionation laws have been proposed, based on different understandings of the mass-fractionation process. Many of these laws have an exponential functional form and plot as curves on a $\delta^{25}Mg$ versus $\delta^{26}Mg$ plot. The deviations from a straight line are small, however, and the apparent slopes are close to 0.5, making them difficult to distinguish on a standard $\delta^{25}Mg$ versus $\delta^{26}Mg$ plot. To make discussions easier, Hulston and Thode (1965) introduced the δ' notation, in which the logarithms of the isotopic ratios are plotted against each other:

$$\delta^{25}\mathrm{Mg}' = 1000\times\ln\left[\frac{\left({}^{25}\mathrm{Mg}/{}^{24}\mathrm{Mg}\right)_{\mathrm{sample}}}{\left({}^{25}\mathrm{Mg}/{}^{24}\mathrm{Mg}\right)_{standard}}\right] \qquad 8.30$$

and similarly for $^{26}Mg/^{24}Mg$. It is straightforward to switch between the two systems:

$$\delta^{25}\mathrm{Mg}' = 1000\times\ln\left(\frac{\delta^{25}\mathrm{Mg}}{1000}+1\right) \qquad 8.31$$

and

$$\delta^{25}\mathrm{Mg} = 1000\times\left(e^{\frac{\delta^{25}\mathrm{Mg}'}{1000}}-1\right) \qquad 8.32$$

On a plot of $\delta^{25}Mg'$ versus $\delta^{26}Mg'$, most fractionation laws, except those that depend linearly on mass difference, plot as straight lines, and the linear laws plot as curves. Figure 8.9 plots a variety of fractionation laws that have been used over the years. Figure 8.9a shows the laws on a $\delta^{25}Mg'$ versus $\delta^{26}Mg'$ plot. The slope of a

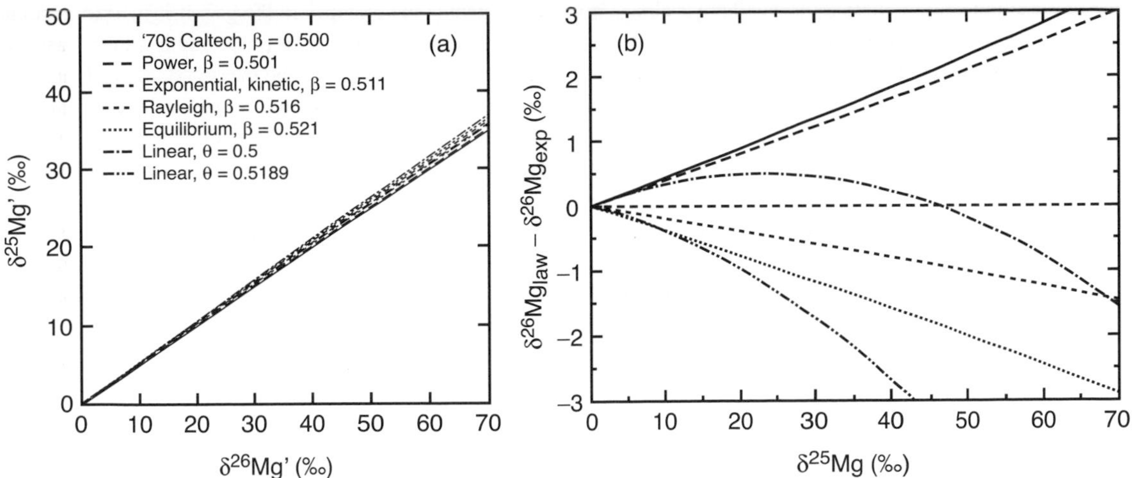

Figure 8.9 Plots of fractionation laws for magnesium isotopes in two different formats. (a) The relationship between $\delta^{25}Mg'$ and $\delta^{26}Mg'$ depends on the fractionation law governing mass fractionation. On this plot, all but the linear laws plot as straight lines. β values are the slopes of the lines on the $\delta^{25}Mg'$ versus $\delta^{26}Mg'$ plot. (b) The differences in the slopes in (a) can be seen more easily when the differences between the fractionation laws are plotted versus the degree of mass fractionation. This plot uses standard delta notation. All laws are plotted as differences from the exponential law. The fractionation laws plot as nearly straight lines, except for the linear laws, which show considerable curvature. Adapted from Davis et al. (2015).

fractionation line on the $\delta^{25}Mg'$ versus $\delta^{26}Mg'$ is defined as β. Figure 8.9b shows the laws as differences compared to the widely used exponential law that was originally defined by Russell et al. (1978) to describe mass fractionation of calcium isotopes in measurements by thermal-ionization mass spectrometry.

Magnesium isotopes are of particular interest in cosmochemistry because magnesium is the daughter element for the ^{26}Al–^{26}Mg short-lived radionuclide system (see Chapter 9). Excesses of ^{26}Mg in an object can be attributed to the decay of ^{26}Al, giving information on the time of formation of an object. But CAIs and chondrules may be mass fractionated relative to bulk solar system magnesium. To get information on the ^{26}Al–^{26}Mg system, this intrinsic mass fractionation must be accounted for. Davis et al. (2015) reported on a series of measurements of both natural CAIs and CAI-analog and forsterite-rich compositions from evaporation experiments. Based on detailed analysis of these natural and experimental samples, they concluded that evaporation of CAI-like melt can be described by a fractionation law that gives a slope of $\beta = 0.5128$ on the $\delta^{25}Mg'$ versus $\delta^{26}Mg'$ plot. These authors recommend this β value to correct for intrinsic mass fractionation in CAI-like and forsterite-rich materials, but caution that β may have a slight temperature dependence.

To use this β value to extract information about the ^{26}Al–^{26}Mg system from measurements of fractionated CAI or forsterite-rich composition, this equation is recommended:

$$\delta^{26}Mg^* = \delta^{26}Mg_{meas} - \left[\left(1 + \frac{\delta^{25}Mg_{meas}}{1000}\right)^{\frac{1}{\beta}} - 1\right] \times 1000 \tag{8.33}$$

where the recommended value of $\beta = 0.5128$ (Davis et al., 2015).

8.2.2.3 *Evaporation under Non-Rayleigh Conditions*

Evaporation does not have to result in measurable isotopic fractionations. Chondrules crystallized from melts in the solar nebula and have variable abundances of volatile elements. They were almost certainly evaporating when they were in the liquid state. But unlike CAIs, they typically show no evidence of isotope fractionation. If evaporation occurs into a gas phase that that has sufficiently high partial pressures of evaporating components, back-reactions between gas and liquid can reduce the isotopic fractionation to near the equilibrium value, which is very small. A gas phase composed predominantly of hydrogen has an interesting effect. Hydrogen in the pressure range of ~10^{-9}–10^{-3} bars enhances the evaporation rate of the evaporating components, but there is almost no effect on the isotopic compositions of the gas and condensed phases. A gas phase with atomic mass more similar to the evaporating species will suppress the isotopic fractionation.

True Rayleigh evaporation can only occur if the condensed phase remains homogeneous and distributes the isotopic fractionation experienced at the surface throughout the body of the condensed phase. Diffusion in molten silicates at temperatures significantly above the liquidus is quite fast, so a homogeneous condensed phase is expected. However, if diffusion in the condensed phase is too slow, it will not be able to keep a melt homogeneous while it loses material from its surface. The most extreme example of this is evaporation from a solid. In this case, diffusion rates can be so slow that only a few atomic layers at the surface are affected by evaporation, and any isotopic fractionation is restricted to that very thin surface layer. From the perspective of the entire solid, there is no significant isotope fractionation associated with evaporating the solid, even though the solid is disappearing. Lack of rapid diffusion in the surrounding gas can also suppress isotopic fractionation because the evaporated material is not removed quickly enough to produce Rayleigh-like evaporation conditions.

In reality, most natural systems operate in a mixed regime. For example, an evaporating chondrule melt may become increasingly isotopically heavy as evaporation occurs, but a mineral grain that forms from the melt early in the process will retain the isotopic composition of the melt at the time it formed because diffusion in the solid is too slow to keep up with the changes in melt composition produced by evaporation. Sulfur isotopes from sulfides inside chondrules from highly unequilibrated ordinary chondrites do not show evidence of isotopic fractionation (Tachibana and Huss, 2005). Because sulfur is so volatile, Rayleigh distillation might be expected. But evaporation of sulfur through a surrounding silicate melt suppresses the isotopic fractionation in sulfur. At low temperatures, before the silicate melt formed, sulfur would evaporate quickly, and the only effective way to suppress sulfur fractionation may be to have had a heating rate that was so high in the low-temperature regime that fractionation did not have time to occur.

8.2.2.4 *Isotope Fractionation during Diffusion*

Diffusion is another important process that has left its isotopic signature in planetary materials. Most discussions are in terms of element diffusion, and isotopic effects of diffusion have received less attention. In most gaseous systems, diffusion rates are sufficiently fast to homogenize the isotopes in the gas. It is hard to generate isotopic gradients in the gas phase and it is even harder to preserve them so they can be measured. In liquids, chemical diffusion occurs whenever there is a concentration gradient, and the chemical diffusion can be accompanied by isotopic effects due to mass-dependent differences in the diffusion rates of the isotopes of an element. Cations that diffuse rapidly through silicate melts relative to silicon show large isotope effects, while those that diffuse slowly tend to have smaller effects. Thermal diffusion can generate isotopic gradients in a homogeneous melt if there is a temperature gradient across the melt (e.g. Richter et al., 2009). Most direct evidence of diffusive isotope fractionation is seen in solids. Details of diffusion in solids are more complicated than diffusion in liquids and gases because of the fixed internal structure. But these details lend themselves to investigation because solids preserve isotopic gradients that can be measured in the laboratory.

Diffusion of an impurity (solute) atom through a solid typically involves vacancies in the solid structure. The solute atom jumps into an adjacent structural vacancy, but in order for it to move further, atoms of the solvent (other than the impurity atom) must exchange sites with the vacancy as well. If no other atoms are involved, the solute atom and the vacancy simply jump back and forth and there is no net diffusion. The details of diffusion of impurities in different materials depend on the relative frequency at which the atoms exchange with vacancies. If the frequency of exchange between and solute and vacancy is small compared to solvent–vacancy exchanges, the vacancy is able to hop around and move away from the impurity, randomizing its position between the infrequent solute–vacancy exchanges. In this case, the overall diffusion of the solute is controlled by the jump frequency of the solute atoms and will depend strongly on the solute's mass. In general, we would anticipate that slowly diffusing impurities will have a large isotope effect, close to the square root of the mass dependence. In contrast, if the solute–vacancy exchange frequency is high relative to the solvent–vacancy exchange frequency, then solute atoms diffuse rapidly compared to solvent elements. In this case, the diffusivity of the solute is controlled by solvent atoms exchanging sites with vacancies, and the mass of the impurity atoms is not an important factor in determining its diffusion coefficient. The resulting isotopic effects are expected to be small.

Details of vacancy diffusion in solids can be complicated. Several types of vacancy diffusion in solids can be defined. *Self-diffusion* of atoms in pure metals by a vacancy mechanism is the simplest case to model. In its pure sense, self-diffusion is a measure of the random walk of constituents in the absence of a chemical gradient. In general, isotopic effects during self-diffusion are expected to be large.

Diffusion of trace elements in a homogeneous medium is similar to self-diffusion. Trace elements that have a large activation energy to jump to an adjacent vacancy and therefore diffuse slowly relative to the major elements can show large isotope effects, approaching square-root-of-the-mass dependence.

Interdiffusion is when two different materials are diffusing in such a way as to homogenize a material. For example, in a diffusion couple made of pure iron and pure nickel, the iron diffuses into the nickel and the nickel diffuses into the iron. Interdiffusion is typically slower than self-diffusion because atoms of both elements must move in order for diffusion to occur. Prediction of isotope effects requires knowledge of the jump frequencies of all components in the system.

Coupled diffusion, where two atoms must move (e.g. the jump of an interstitial atom into an occupied lattice site, displacing the lattice atom into an interstitial site), tends to be slower than self-diffusion and isotope effects are expected to be small.

Interstitial diffusion is when atoms diffuse through interstitial sites rather than via lattice vacancies. Atoms that diffuse this way tend to be smaller than those making up the mineral structure. Because all adjacent interstitial sites of the same type are equivalent, diffusion through interstitial sites occurs by jumps in random directions. The isotope effect for diffusion by purely interstitial mechanisms is expected to be large, close to a square-root-of-the-mass dependence.

Boundary diffusion (i.e. diffusion at grain boundaries) has not been the subject of many experimental studies. In general, isotope effects are expected to be smaller for diffusion along grain boundaries than for diffusion by a vacancy mechanism.

8.2.3 Some Cosmochemical Applications

8.2.3.1 *Mass-Dependent Fractionations in Planetary Atmospheres*

Signatures of mass-dependent isotope fractionations can be found in the atmospheric compositions of the planets. The atmospheres of Venus, Mars, and Titan have significantly elevated D/H ratios compared to Earth's atmosphere (δD = ~125,000‰, ~4,000‰, and ~500‰, respectively). The atmospheres of Mars and Titan also show significant enrichments of ^{15}N compared to that of the Earth (Lammer et al., 2008). Gases can be lost from planetary atmospheres via thermal processes (e.g., Jeans escape) or hydrodynamic escape (discussed further in Chapter 11). At any given temperature, the atoms in a gas have a distribution of velocities: the Maxwell distribution. We experience this distribution as a temperature. The Maxwell distribution is asymmetrical, with a tail that extends to high velocities. At very high altitudes, atmospheric atoms do not collide with each other, so those that have velocities in excess of escape velocity can be lost to space. If the atmosphere is hot enough, the high-energy tail can be lost. Because hydrogen moves faster than deuterium at the same temperature, hydrogen is lost preferentially. Jeans escape can be enhanced by extreme ultraviolet (EUV) or X-ray photons to generate hydrodynamic escape. Examples of nonthermal escape processes include: photochemical escape, where high-energy photons react with and break up molecules, imparting enough energy to the released atoms to cause them to escape; sputtering escape, where solar wind ions can impart enough energy to drive an atom out of the atmosphere; charge exchange escape, which creates fast-moving neutral atoms that can escape the magnetic field lines; and impact erosion, where a meteoroid impact leads to wholesale loss of atmospheric gases.

Venus shows the largest enrichment of deuterium among solar system bodies, but its carbon, nitrogen, and oxygen ratios are similar to those on Earth (Lammer et al., 2008). The extreme deuterium enrichment implies that Venus has lost most of its water. Kasting (1988) suggested that Venus was close enough to the Sun that the CO_2-rich atmosphere reached a temperature high enough to boil the oceans. The resulting steam-rich atmosphere extended to great altitudes. Water was photodissociated in the upper atmosphere, releasing hydrogen, which was lost to space by Jeans escape, leaving deuterium-enriched hydrogen in Venus' atmosphere. Thermal escape was likely enhanced by the much higher flux of X-ray and EUV photons from the young Sun, resulting in removal of almost all of the atmospheric hydrogen. Today, even with an atmospheric temperature at Venus' surface of 467°C, thermal escape is insignificant for hydrogen, carbon, nitrogen, and oxygen, because the much lower EUV flux from the modern Sun results in the base of the exosphere being much lower than it was during the early history of the solar system, making a much deeper gravitational well. Hydrogen loss from Venus today is primarily from photochemical reactions and other processes that are not mass dependent (Lammer et al., 2008).

Earth, being somewhat farther from the Sun, avoided a runaway CO_2-H_2O greenhouse (Kasting, 1988). Liquid oceans formed instead, possibly as early as 4.4 Ga (Valley et al., 2002). Several models suggest that the early Earth's upper atmosphere should have been hydrogen rich, which could have led to relatively rapid loss of hydrogen by thermal escape, perhaps driven by EUV into a hydrodynamic regime. As life arose, hydrogen in the atmosphere was converted to methane and hydrogen loss slowed, but it was still likely to be significant until the rise of oxygen at 2.4 Ga. The Earth's magnetic field had a significant impact on hydrogen loss because it deflected the solar wind, eliminating direct erosion and decreasing upper atmospheric heating. At present, Earth's main escaping ion is O^+, which originates in the ionosphere. Earth probably also avoided solar wind erosion over most of its history because of its magnetic field. Overall the fraction of hydrogen lost on Earth has been lower than on Venus, Mars, or Titan.

Mars' atmosphere has the second highest D/H ratio among the terrestrial planets, estimated at δD = 4000–6000, while its mantle hydrogen has δD similar to Earth's oceans (e.g. Barnes et al., 2020). Mars has a mass that is only about one-tenth the mass of Earth, so it can lose atmospheric gases more easily. Mars' surface geology suggests that it once had an ocean equivalent to 150–200 meters deep, or even more, over the entire planet (Carr and Head, 2003). Estimates of how much water has been lost by Mars over its history range from a layer ~1.5 meters deep to up to 80 meters deep (see Lammer et al., 2008). The primary mechanisms for this loss of hydrogen from Mars are commonly thought to have been Jeans escape and hydrodynamic escape. Another proposed mechanism (Stone et al., 2020) depends on dust storms, which heat the atmosphere so that it can hold more water vapor. Storms carry water vapor upward into the ionosphere, where carbon dioxide is ionized and splits water molecules apart, allowing hydrogen to escape. Data for hydrogen isotopes on Mars come primarily from ion microprobe analyses of martian meteorites (e.g. Leshin, 2000; Barnes et al., 2020).

The nitrogen isotopic composition of Mars' interior, as measured in martian meteorites (Mathew and Marti, 2001), is essentially identical to that of Earth's interior. Atmospheric nitrogen on Mars is significantly enriched in ^{15}N compared to the interior, indicating efficient escape of N_2 into space. In contrast, there is only a small difference in isotopic ratios between Earth's atmosphere and interior.

Titan's atmosphere contains mainly nitrogen and CH_4. Hydrogen and nitrogen are enriched in heavy isotopes, which suggests large atmospheric loss, but carbon isotopes are similar to those of Earth (Lammer et al., 2008). The details of how these isotopic fractionations occurred are hard to sort out because we do not know what Titan's original composition was or how it arose. Titan's orbit around Saturn means that it spends part of the year within the protection of Saturn's magnetosphere, where it is not affected by solar wind. But for the rest of its orbit, it is outside of Saturn's magnetosphere and exposed to the solar wind, and sputtering loss of light isotopes is expected. Several attempts to model Titan's atmospheric evolution have been made, but the results are not very satisfying. Overall, the current loss rates from Titan's atmosphere are too low to explain the isotopic compositions of its atmosphere.

8.2.3.2 *Fractionations in Extraterrestrial Organic Matter*

Organic matter in meteorites and interplanetary dust particles exhibits extreme enrichments in deuterium, reflected in D/H ratios hundreds to thousands of times higher than those seen on Earth. High $^{15}N/^{14}N$ ratios are also observed. Even more-extreme enrichments are observed in simple organic molecules in the interstellar medium, and it is inferred that the meteoritic enrichments reflect interstellar processes. The traditional explanation for these enrichments has been ion–molecule reactions in low-temperature gas in dense molecular clouds (e.g. Geiss and Reeves, 1981; Herbst, 2003), but the details of the relationship between deuterium-rich molecules in interstellar space and those in meteorites and interplanetary dust particles are not well understood.

At the low temperatures characteristic of dense molecular clouds ($T < 50$ K), chemical reactions between neutral molecules that are *endothermic* (require energy) or possess significant activation barriers will not proceed. However, *exothermic* ion–molecule reactions, which release energy, can proceed at very low temperatures. These reactions are expected to produce mass-dependent fractionation between hydrogen and deuterium at low temperatures. Bonds that involve deuterium have lower zero-point energies and thus are slightly stronger than those involving hydrogen, so deuterium enrichments are expected in molecules. Ions can be produced by interactions with high-energy cosmic rays or ultraviolet (UV) photons. An ion approaching a neutral molecule generates an attractive force. If the capture reaction is exothermic and the activation energy is not high, the ion can be captured. Consider this generic ion–molecule exchange reaction:

$$AH^+ + BD \leftrightarrow AD^+ + BH + \Delta E_0 \qquad 8.34$$

where A and B represent one or more atoms, and ΔE_0 is the exothermicity at 0 K. The forward reaction releases energy and thus can proceed at low temperatures. If BD is a significant reservoir for deuterium, and ΔE_0 is large enough, the forward reaction can convert most of the AH^+ to AD^+, and the AD^+/AH^+ ratio can greatly exceed D/H in elemental hydrogen. The reverse reaction requires energy from the environment and is very inefficient. It becomes increasingly efficient at higher temperatures, decreasing the overall isotopic fractionation.

There is good observational evidence that deuterium–hydrogen fractionation occurs in the interstellar medium. Several simple molecules have been shown to be enriched in deuterium relative to hydrogen by factors of up to 5000 compared to the cosmic ratio. Polycyclic aromatic hydrocarbons (PAHs) should also become enriched by gas-phase ion–molecule reactions in dense molecular clouds (Sandford et al., 2001). PAHs are considerably more stable and more abundant than smaller molecules that are enriched in deuterium, and are thus likely to represent the largest reservoir of materials enriched in deuterium by ion–molecule reactions. However, ion–molecule reactions only occur in the gas phase, and the majority of species are condensed onto grain surfaces at low temperature. So gas-phase ion–molecule reactions may not be the main contributor to D/H fractionations in refractory organic materials or to the bulk D/H fractionation in dense clouds.

Gas-grain reactions at <50 K can potentially produce large deuterium enrichments in organic compounds on grain surfaces. Most of the molecular species in molecular clouds condense out on grain surfaces, including species enriched in deuterium by gas-phase reactions. D/H ratios as high at 0.1 are expected for species like CH_4, NH_3, and H_2O, and ratios this high have been observed in interstellar ices. However, these light, deuterium-rich compounds must be converted to larger molecules in order to contribute to the carriers of deuterium in meteorites. This probably happens through photolysis reactions driven by UV light. UV photons ionize the organic molecules adsorbed on grain surfaces, allowing them to combine into larger species. Photolysis reactions generate organic mantles on dust grains in interstellar space and can fix the deuterium-rich compounds into larger organic molecules.

Some of the processes described here may also produce isotopic fractionations in carbon, nitrogen, and oxygen, with the fractionation in nitrogen likely to be the most significant. The ^{15}N enrichments in meteorites, and their possible correlation with δD, may be explainable in terms of low-temperature ion–molecule reactions, possibly followed by UV-driven exchange reactions in ices. But a tight correlation between deuterium and ^{15}N enrichments is not expected (Sandford et al., 2001).

Essentially all of the processes that occur in cold molecular clouds could also occur in the outer accretion disk, where temperatures are very low. The formation of the accretion disk may have destroyed much of the organic material produced in the Sun's parent molecular cloud. Determining the relative roles of interstellar and accretion-disk processes will require a better understanding of these environments and a better understanding of the products of each of the processes that synthesize organic compounds. In Chapter 11, we return to the question of where the deuterium- and ^{15}N-rich organic material originated and how it ended up in the meteorites.

8.3 Mass-Independent Isotope Effects

Prior to 1973, essentially all observed isotopic variations appeared to be explainable by processes operating within the solar system on normal solar system material. This was one of the pillars of the hot solar nebula model. Isotopic variations were due to either (1) radioactive decay, (2) cosmic-ray interactions (spallation reactions, neutron-capture reactions), or (3) mass-dependent processes as described above. In 1973, Robert Clayton and coworkers discovered that the oxygen isotopic compositions of CAIs in carbonaceous chondrites varied dramatically in ways that could not be explained by any of those processes (see below). This discovery was one of the first examples of isotope effects that could not be explained within the standard model for the solar system. These isotope effects were (and often still are) discussed in the literature using the term "isotopic anomaly," highlighting the fact that they did not fit the standard hot solar nebula model. The causes of mass-independent isotope variations fall into two basic categories: 1) mass-independent isotope fractionations produced by chemical processes and 2) isotope variations inherited from the nucleosynthesis processes that produced the elements.

8.3.1 Oxygen Isotopes

The well-ordered world of mass-dependent isotope effects in light elements was turned upside down with the discovery by Clayton et al. (1973) that the oxygen isotopic compositions of mineral separates from CAIs do not fall on the slope ~0.52 terrestrial mass fractionation line (Fig. 8.10). In the early 1970s, people were measuring $\delta^{18}O$ for a variety of minerals from meteorites and lunar samples with the goal of estimating the formation temperatures of the minerals (e.g. Onuma et al., 1970). They were not measuring $\delta^{17}O$ because in an equilibrium scenario where the temperature can be derived from oxygen-isotope partitioning, $\delta^{17}O$ and $\delta^{18}O$ would follow the same trends. Because the mass difference between ^{17}O and ^{16}O is only half as large as that between

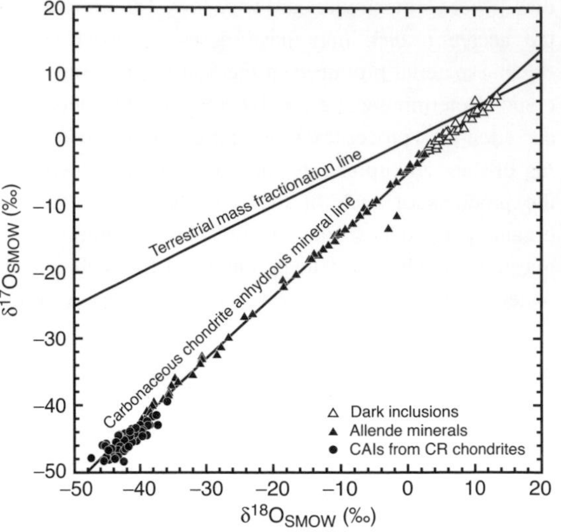

Figure 8.10 Oxygen isotopic compositions of various components of carbonaceous chondrites. Many meteoritic compositions fall along the carbonaceous chondrite anhydrous mineral line, which has a slope of ~1 on an oxygen three-isotope plot. Allende mineral data from Clayton et al. (1977), dark inclusion data from Clayton and Mayeda (1999), and CR CAI data from Makide et al. (2009).

^{18}O and ^{16}O, the isotope effect would be smaller by a factor of two in $\delta^{17}O$, and ^{17}O is five times less abundant than ^{18}O, making the measurement much harder. But the data that they were getting did not make sense. Clayton's group decided to measure $\delta^{17}O$ along with $\delta^{18}O$ and found out why. The data did not fall along the expected mass-dependent fractionation line and thus could not be explained by equilibrium partitioning of oxygen isotopes. Instead, they fell along a different line with a slope of ~1, which implied a variation in relative abundance of ^{16}O among the samples (Fig. 8.10).

A new notation was established to describe the mass-independent oxygen isotope effects. Building on the delta values used to describe mass-dependent variations, $\Delta^{17}O$ was established to describe mass-independent effects. $\Delta^{17}O$ describes the distance above or below the terrestrial mass fractionation line on the oxygen three-isotope plot and is given by:

$$\Delta^{17}O = \delta^{17}O - 0.52 \times \delta^{18}O \quad (8.35)$$

It is common to see the oxygen isotopic composition of meteorites or meteorite components described by this parameter. $\Delta^{17}O$ is the primary basis for the classification of meteorites by oxygen isotopes, discussed in Chapter 6 (see Figs. 6.4 and 6.15). Oxygen isotopes work well for classification, even though we do not yet know how the isotopic variations arose.

Clayton et al. (1973) proposed that the slope ~1 array was due to the incorporation of a nuclear component of nearly pure ^{16}O, and suggested that this component might be presolar in origin. This explanation has profound implications. Remember that at this time the hot solar nebula model was the framework for understanding all solar system materials. The survival of a distinct nucleosynthetic component posed a grave challenge to that model. Prior to this time, only one observation provided any evidence to contradict the hot solar nebula: the discovery of a neon component consisting of essentially pure ^{22}Ne (called Ne-E) in carbonaceous chondrites (Black and Pepin, 1969). In general, the discovery of Ne-E was largely ignored. The discovery of evidence for live ^{26}Al ($t_{1/2}$ = 717,000 years) in the early solar system (Lee et al., 1977) led to a new model that could potentially save the hot solar nebula. This was the supernova trigger model (Cameron and Truran, 1977), which proposed that the explosion of a supernova adjacent to or within the Sun's parent molecular cloud triggered the gravitational collapse of the cloud to form the solar system, and also injected both newly synthesized ^{26}Al and ^{16}O into the nebula.

The model proposed by Clayton et al. (1973) and the subsequent supernova trigger model based upon it assumed that no chemical mechanism could produce an isotopic effect that was not mass dependent. This assumption was demonstrated to be invalid by Thiemens and Heidenreich (1983), who showed that in the production of ozone from molecular oxygen by electric discharge, the product ozone is equally enriched in ^{17}O and ^{18}O, rather than exhibiting enrichments proportional to their masses. It is now well established that mass-independent fractionations are generated during the production of ozone in the terrestrial stratosphere by photochemical reactions (Thiemens, 2006). Although a complete understanding of the process for this mass-independent fractionation is not yet in hand, popular models suggest that it depends on molecular symmetry, with symmetrical molecules being more stable than other configurations. However, it is not clear that the processes operating in the stratosphere are relevant to the solar nebula.

8.3.1.1 *CO Self-Shielding*

Clayton (2002) and others have argued that the mass-independent oxygen isotopic variation observed in solar system materials arose from self-shielding during photodissociation of carbon monoxide (CO). CO is a major gas component in molecular clouds and accretion disks. It can be dissociated by UV radiation. This dissociation and recombination of carbon and oxygen into other molecules can produce isotope effects through a process called *self-shielding*. Each of the different isotopic combinations that make up CO is dissociated by a slightly different energy of UV radiation. The UV continuum has approximately the same number of photons that can dissociate each of the isotopic combinations. When UV radiation impinges on a molecular cloud, the UV photons that can dissociate $^{12}C^{16}O$ will be used up relatively quickly in the outermost cloud layers. $^{12}C^{18}O$ is ~500 times less abundant that $^{12}C^{16}O$ and so will be dissociated in a larger region at the edge of the cloud, and $^{12}C^{17}O$, which is 2500 times less abundant than $^{12}C^{16}O$, will be dissociated in a slightly larger region. Deeper in the cloud, CO will not be dissociated by UV because the molecules in the outer part of the cloud will have absorbed all of the relevant photons. The CO in the inner part of the cloud has thus been "shielded" from dissociation by the CO in the outer part of the cloud. Figure 8.11 is a schematic drawing of this process.

CO self-shielding becomes relevant to the oxygen isotopic variations in the early solar system if the isotopic differences in the dissociated carbon and oxygen in different parts of the cloud can somehow be preserved. Those atoms that reform CO do not leave any signature, but some of the dissociated oxygen may combine with hydrogen to form water. In the outer region of the cloud, the water that forms will have the same isotopic mixture as the cloud because all isotopologues of CO have been dissociated and all isotopes of oxygen are available to form water. But in the zone where only $^{12}C^{17}O$ and $^{12}C^{18}O$ are dissociated, the water that forms from the dissociated atoms will be enriched in ^{17}O and ^{18}O compared to the bulk cloud (Fig. 8.11). Water has a higher freezing point than CO and can be captured onto grains more easily. Consequently, in some regions, the grains will have a different bulk isotopic composition than the surrounding gas. If the water then reacts with the silicates in the dust, ^{17}O- and ^{18}O-rich dust can form.

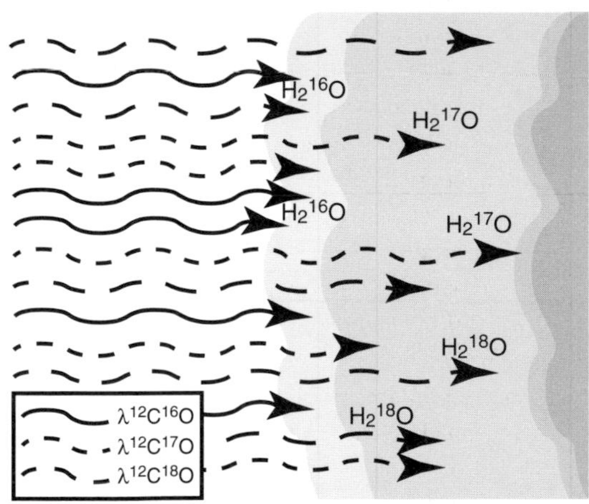

Figure 8.11 Schematic drawing of CO shelf-shielding showing equal numbers of UV photons that can dissociate $^{12}C^{16}O$, $^{12}C^{17}O$, and $^{12}C^{18}O$ impinging on a molecular cloud. The lengths of the arrows show how far a photon might penetrate before dissociating a molecule. Those that dissociate $^{12}C^{16}O$ are used up in the first layer of the cloud due to the high abundance of ^{16}O. Photons for $^{12}C^{17}O$ and $^{12}C^{18}O$ penetrate farther, but eventually they are exhausted as well. Some oxygen combines with hydrogen to form water. The gray area with $H_2{}^{17}O$ and $H_2{}^{18}O$ only has isotopically heavy water (no $H_2{}^{16}O$), whereas the first layer has isotopically normal water and the interior has no newly formed water.

CO self-shielding is currently a popular model for understanding the oxygen-isotope variation among solar system materials. A measurement of solar wind returned by the Genesis mission shows that the Sun is ^{16}O rich relative to the terrestrial composition (McKeegan et al., 2011; see Box 4.3). Isotopic measurements of a rare material known as "cosmic symplectite" found in a primitive carbonaceous chondrite gave oxygen compositions enriched by ~180‰ in both ^{17}O and ^{18}O relative to terrestrial oxygen (Sakamoto et al., 2007). It is postulated that cosmic symplectite formed by incorporating $^{17,18}O$-rich nebular water. These observations provide the basis for a relatively detailed model of oxygen isotopes in the solar system, with CAIs forming near the Sun from bulk solar system oxygen (^{16}O rich) and cosmic symplectite forming from $^{17,18}O$-rich nebular water (e.g. Krot et al., 2019). Other materials reflect mixing between the two end members. However, there are still gaps in the story, and the oxygen isotopic composition of solar system materials will be the topic of intense study for some time to come. A clean measurement of an oxygen-isotope gradient in a newly forming stellar system would go a long way toward validating or invalidating the self-shielding model.

8.3.1.2 *Combining Mass-Dependent and Mass-Independent Fractionations*

When working with oxygen-isotope data on meteorites, it is often necessary to consider the effects of mass-dependent and mass-independent fractionations together. The mass-dependent fractionations distribute related samples along slope ~0.5 lines on the oxygen three-isotope plot. These fractionations give information on processes that have affected the material. The $\Delta^{17}O$ value gives information on the source materials that went into making the sample. Most meteorite classes have unique $\Delta^{17}O$ values. The characteristic oxygen isotopic composition of a class of meteorites reflects the components that contributed to that class. Different rock and

Box 8.2 Are the Oxygen Isotopic Compositions of the Earth and the Moon the Same?

A cornerstone to our understanding of the Earth–Moon system is that the two bodies have the same oxygen isotopic composition. But this has been hard to reconcile with the giant-impact model for the formation of the Moon, in which the Moon formed when a Mars-sized body, named Theia, impacted the Earth (Chapter 14). Models indicate that it is very hard to mix material from two bodies with different oxygen isotopic compositions so completely that it would leave no isotopic trace. This leads to the suggestion that the Earth and the Moon accreted from the same "feeding zone" in the accretion disk and thus had the same isotopic compositions before the impact. Cano et al. (2020), using state-of-the-art laser-fluorination oxygen-isotope measurements, report evidence that the isotopic composition of the Moon is both heterogeneous and different from that of Earth. Measurements of partitioning between minerals in the lunar rocks also indicate disequilibrium, unlike similar data from Earth rocks. The isotopic compositions of the lunar rocks correlate with the chemical compositions of the rocks and appear to reflect the depth within the Moon where the rocks originated. The data are summarized in Figure 8.12.

Chakraborty et al. (2013) reported on an experimental study showing that when gas-phase SiO is oxidized to solid SiO_2, the SiO_2 exhibits mass-independent fractionation, with SiO_2 isotopically heavier than the SiO gas (Fig. 8.13). Cano et al. (2020) propose that this kind of oxidation occurred in the vapor produced by the giant impact that formed the Moon. The SiO_2 condensed into the Moon, leaving behind an increasingly light vapor. The condensed SiO_2 mirrored the vapor and became increasingly light as the Moon grew. The evolution of the lunar silicates is shown by deep-sourced very low-Ti (VLT) glass, intermediate-sourced low-Ti and high-Ti basalt, and lunar crust (Fig. 8.12). In this model, the Mars-sized impactor, Theia, was isotopically heavy compared to the Earth, and the deep-sourced rocks reflect Theia's composition. Cano et al. (2020) state that these data in the context of this model provide direct evidence that Theia formed farther from the Sun than Earth did.

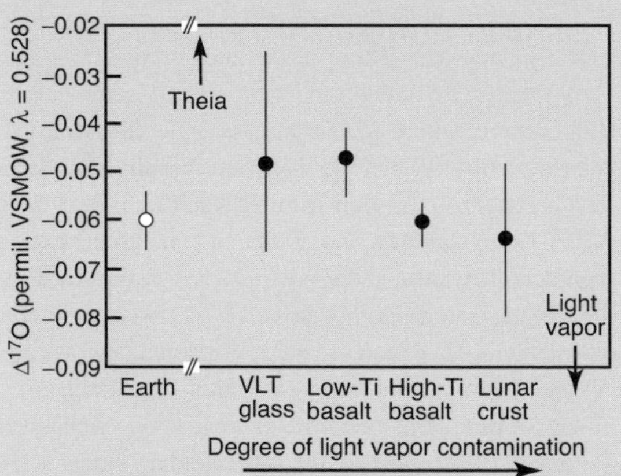

Figure 8.12 $\Delta^{17}O$ values for Earth rocks and for several types of lunar rocks. Compositions of the lunar rocks exhibit small differences from one another and from terrestrial rocks. The circles denote the mean values of the data and the vertical bars show the range of the data for each sample suite. The order of the data for the lunar samples shows increasing degrees of contamination by isotopically light vapor. Modified from Cano et al. (2020).

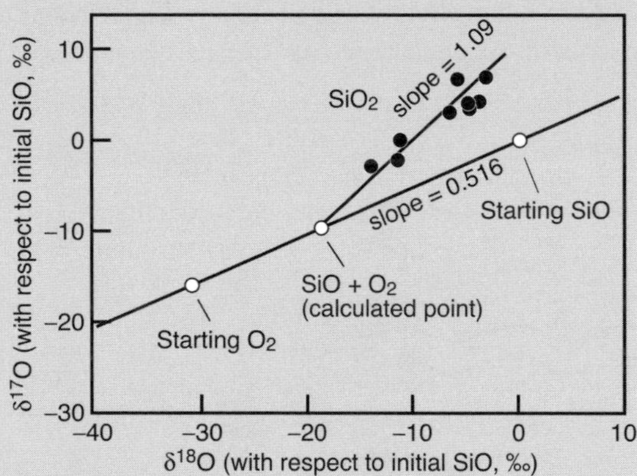

Figure 8.13 Results of an experimental study in which gaseous SiO was oxidized to solid SiO_2. The oxygen isotopes exhibit a clear mass-independent fractionation, with SiO_2 enriched in ^{17}O and ^{18}O relative to the starting SiO. After Chakraborty et al. (2013).

water reservoirs were typically the result of independent nebular processes.

Water and rock interact during aqueous alteration, generating arrays on the oxygen three-isotope diagram with slopes other than ~0.5. Such arrays reflect both mixing of two isotopically distinct reservoirs and mass-dependent fractionation as the water reacts with the rock and produces new minerals. Schrader et al. (2011) discussed how reactions between water and rock of different compositions under different conditions produced an array of bulk chondrite compositions with a slope of ~0.7 on the oxygen three-isotope plot. By considering both mass-dependent and mass-independent fractionations, one can extract information on the temperature and the water-to-rock ratio during aqueous alteration from the chemical and oxygen-isotope data. For example, Clayton and Mayeda (1984, 1999) inferred that aqueous alteration in CM chondrites took place at $<\sim 20°C$ and water-to-rock ratios of 0.35–0.6, while CI chondrites were altered at higher temperatures (100–150°C) and with higher water-to-rock ratios of ~2.5.

8.3.2 Sulfur Isotopes

Mass-independent sulfur isotope effects have been reported in terrestrial sediments formed at $>\sim 2.4$ Ga but not in younger rocks (Figure 8.14). The effects are reported as deviations from the Canyon Diablo troilite mass fractionation line, analogous to $\Delta^{17}O$ for oxygen isotopes:

$$\Delta^{33}S = \delta^{33}S - 0.515 \times \delta^{34}S \quad 8.36$$

and

$$\Delta^{36}S = \delta^{36}S - 1.89 \times \delta^{34}S \quad 8.37$$

These effects are attributed to photolysis of SO_2 by short-wave UV radiation in the upper stratosphere (Farquhar et al., 2000a). The sulfur-isotope ratios are of similar magnitude to those of oxygen (see Table 4.2). In order for photolysis to be effective, the column density of ozone (O_3) must be low enough to make the atmosphere transparent to UV. But production of ozone from O_2 by photolysis is efficient in the stratosphere, so the presence of mass-independent isotope effects is taken as evidence of a very low O_2 content before ~2.4 Ga. In addition, in order for mass-independent isotope effects to be transferred to other molecules and preserved, and then transferred to the surface to be incorporated into sediments, the O_2 concentration in the atmosphere would have been $<10^{-5}$ times the present atmospheric level (Pavlov and Kasting, 2002). The mass-independent isotopic signature in sulfur (Fig. 8.14) provides strong evidence that the Earth's atmosphere did not contain free oxygen until after about 2.4 Ga.

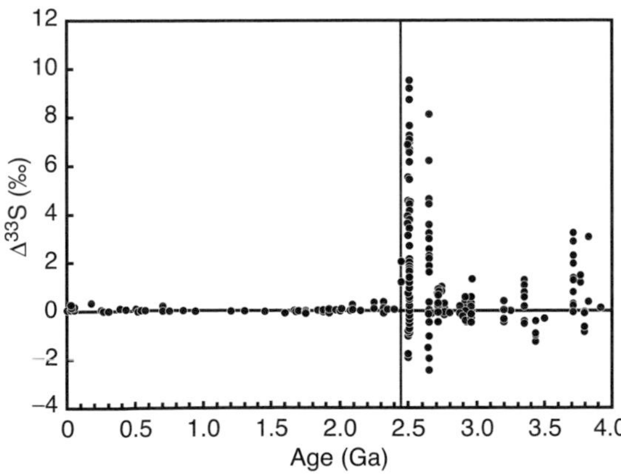

Figure 8.14 Compilation of $\Delta^{33}S$ versus age for terrestrial rock samples. There is a clear division between samples older than ~2.45 Ga (indicated by the vertical line), which have highly variable $\Delta^{33}S$, and those younger than 2.45 Ga, which have much smaller but still measurable $\Delta^{33}S$ values. The abrupt change is due to the presence of free oxygen in the atmosphere starting at ~2.45 Ga. Adapted from Farquhar et al. (2007).

Mass-independent variations in sulfur are also observed in martian meteorites (Farquhar et al., 2000b). These isotope anomalies have been interpreted to result from volcanic injections of SO_2 and H_2S into the atmosphere, followed by photolysis, which fractionated the sulfur isotopes. The anomalous sulfur was likely transferred to the martian crust by interaction between the anomalous aerosols and dust lofted into the atmosphere by dust storms.

8.4 Isotopic Anomalies Inherited from the Sun's Parent Molecular Cloud

At almost the same time that Clayton et al. (1973) discovered mass-independent isotope anomalies in oxygen, G. J. Wasserburg and coworkers discovered mass-independent isotopic variations in calcium, titanium, strontium, barium, samarium, and neodymium in two CAIs from the Allende meteorite (e.g. Fig. 8.15). These isotopic anomalies could not be explained by processes operating within the solar system. They were measured in parts per thousand (permil) or parts per ten thousand (epsilon units) and were interpreted as relict isotopic signatures of the nucleosynthesis that produced the elements that made up the solar system. These CAIs also exhibited large mass-dependent fractionations in oxygen, silicon, and magnesium and were given the name "FUN inclusions," for **f**ractionation and **u**nidentified **n**uclear effects (Wasserburg et al., 1977; Lee et al., 1979). The isotope anomalies, which were revealed by a new generation of mass spectrometers designed to analyze the Apollo samples, were orders of magnitude smaller than the anomalies carried by presolar grains that would be discovered ~15 years later (Chapter 5). Surviving presolar nucleosynthetic components were a direct challenge to the hot solar nebula model, but at first the anomalies could be ignored because they were only seen in a few, highly unusual FUN inclusions.

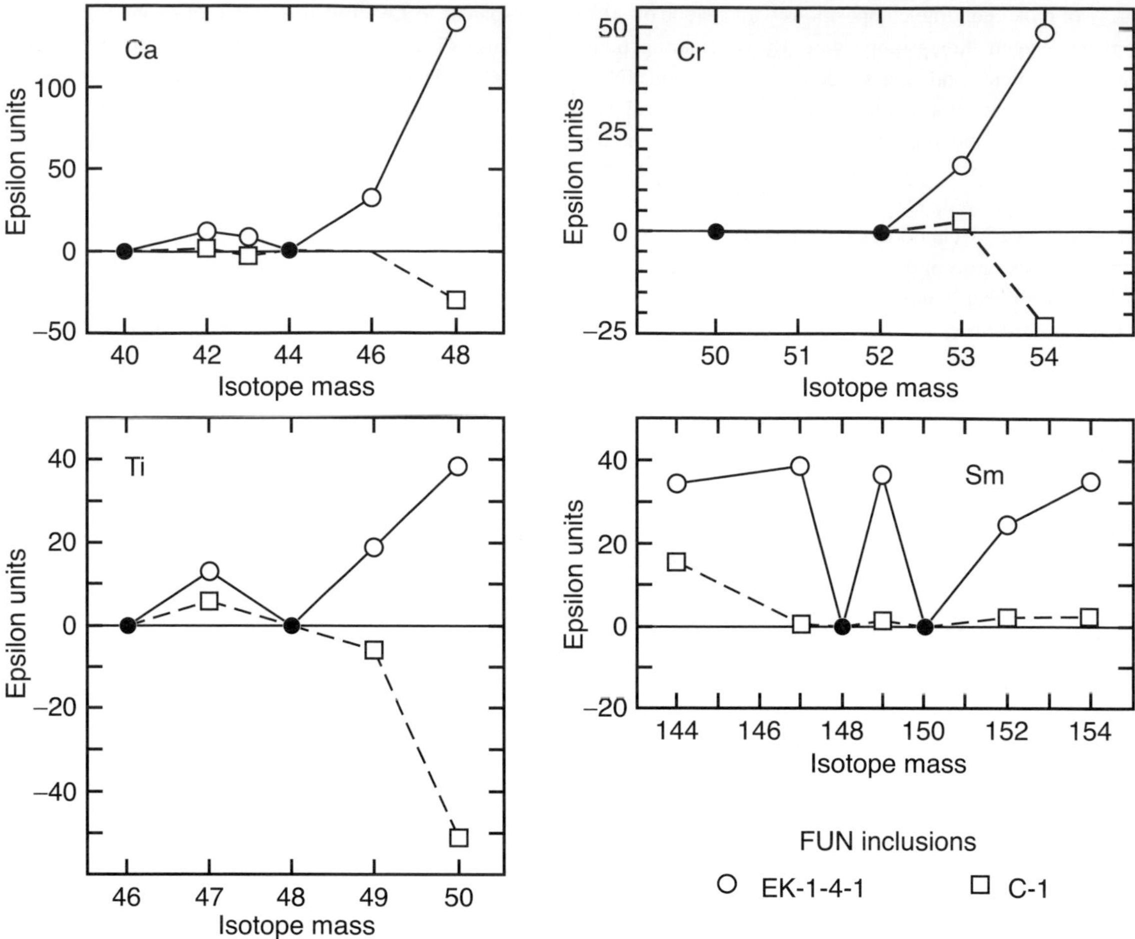

Figure 8.15 Mass-independent isotope effects for two FUN inclusions: EK-1-4-1 and C-1. EK-1-4-1 shows excesses in the neutron-rich isotopes of Ca, Cr, and Ti, while C-1 shows a deficit of ^{48}Ca, ^{50}Ti, and ^{54}Cr and only small effects in ^{46}Ca, ^{49}Ti, and ^{53}Cr. Modified from Wasserburg (1987).

However, within a few years, it became clear that titanium isotopic anomalies could be found in most CAIs and in some bulk meteorites, and that smaller anomalies in calcium, magnesium, and other elements of a similar character to those found in FUN inclusions were being found in "normal" CAIs as well (e.g. Jungck et al., 1984; Niemeyer and Lugmair, 1984; Brigham et al., 1985; Niederer et al., 1985). By the time presolar grains were discovered in 1987, isotopic anomalies could no longer be treated as rare special cases. The hot solar nebula model was under serious threat.

Today the majority of elements that have more than one isotope have been shown to exhibit isotopic anomalies in some samples. Figure 8.16 is a Periodic Table with elements coded to show the prevalence of isotopic anomalies in solar system materials. Elements that have only one isotope, and therefore cannot exhibit anomalous isotopic ratios, are shown in gray boxes. Most radioactive elements, listed in light gray text, do not exist in significant quantities in the solar system. Elements in bold text have been shown to have isotopic anomalies that were inherited from the Sun's parent molecular cloud. Those in plain text have not (yet) been demonstrated to exhibit isotopic anomalies. Elements marked with "R" exhibit anomalies from the decay of radioactive isotopes; Mg and Os also exhibit nucleosynthetic isotopic anomalies. An important point is that, of the 55 elements that could exhibit isotopic anomalies, 34 (62%) actually do. So does uranium, which is radioactive. Many of those that do not are relatively volatile elements (50% condensation temperatures below ~650°C) whose carriers are easily modified by solar system processes, thus erasing the anomalies

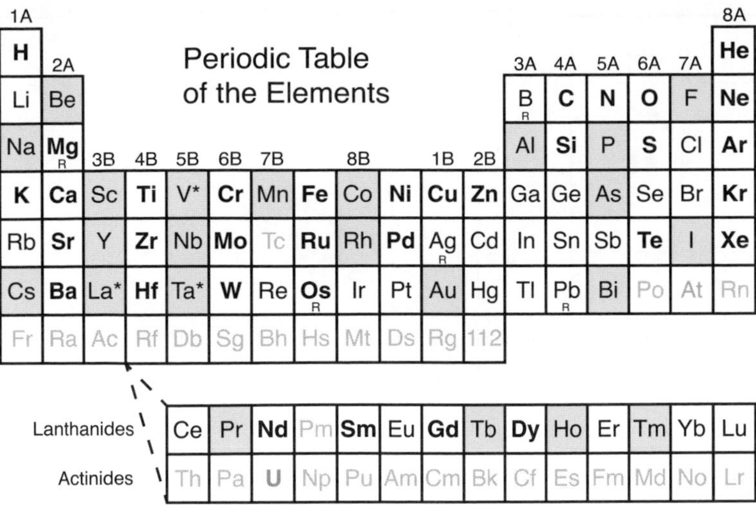

Figure 8.16 Periodic Table showing elements with isotopic anomalies. Elements in light gray text are radioactive, and for many of them, their original abundances in the solar system have decayed away. Elements in gray boxes are those that have only one isotope. Elements in plain text have multiple isotopes and have not (yet) exhibited isotopic anomalies. Elements with "R" in the box exhibit anomalies from the decay of radioactive isotopes; Mg and Os also exhibit nucleosynthetic isotopic anomalies. Elements in bold text have more than one isotope and have exhibited isotopic anomalies in natural materials. Of the 55 stable elements with multiple isotopes, 34 (62%) have been shown to exhibit isotopic anomalies in natural materials.

(e.g. chlorine, gallium, germanium, selenium, bromine, cadmium, indium, tin, antimony, mercury, thallium, and lead). Some of the rest are highly siderophile elements, which are likely to have been incorporated into metal during solar system processing, mixing out any anomalous isotopic signatures. There is really no way to argue that the solar system passed through a hot gaseous state that homogenized the isotopic compositions of the elements.

Warren (2011) first pointed out that the different classes of meteorites fall into two distinct isotopic groups. The carbonaceous chondrites define one isotopic group (the carbonaceous chondrite or CC group), and the terrestrial planets and noncarbonaceous stony meteorites form another group (the noncarbonaceous or NC group). Since that time, these two isotopic groups have been shown to exist in a variety of elements, and both groups have been shown to include achondritic and iron meteorites in addition to chondrites (Fig. 8.17). The two groups have been described as representing an isotopic dichotomy among solar system bodies.

A currently very popular model to explain the isotopic dichotomy proposes that the CC and NC meteorites come from two isotopically distinct regions of the solar system that were kept isolated from one another for millions of years. It is proposed that the reservoirs were kept isolated by the planet Jupiter, with the CC meteorites forming outside Jupiter's orbit and the NC meteorites forming inside Jupiter's orbit (Kleine et al., 2020). This model, and other possible explanations for the dichotomy in nucleosynthetic isotopes, will be discussed further in Chapter 15.

Summary

In this chapter, we examined how variations among stable isotopes can be used in cosmochemistry. Traditional stable-isotope cosmochemistry takes advantage of fractionations between isotopes that are a function of their masses. Bond strengths vary as a function of isotope mass, as do velocities of isotopes in gases, liquids, and solids. Small equilibrium, mass-dependent isotopic fractionations are widespread and provide the basis for isotope thermometry. Kinetic mass-dependent fractionations, which typically are caused by differences in the velocities of the atoms, can generate large isotopic effects. The extreme isotope fractionations observed in interstellar hydrogen in interstellar space typically reflect kinetic barriers due to very low temperatures.

More recently, processes that produce mass-independent isotope fractionations have been identified. Mass-independent oxygen isotope effects are

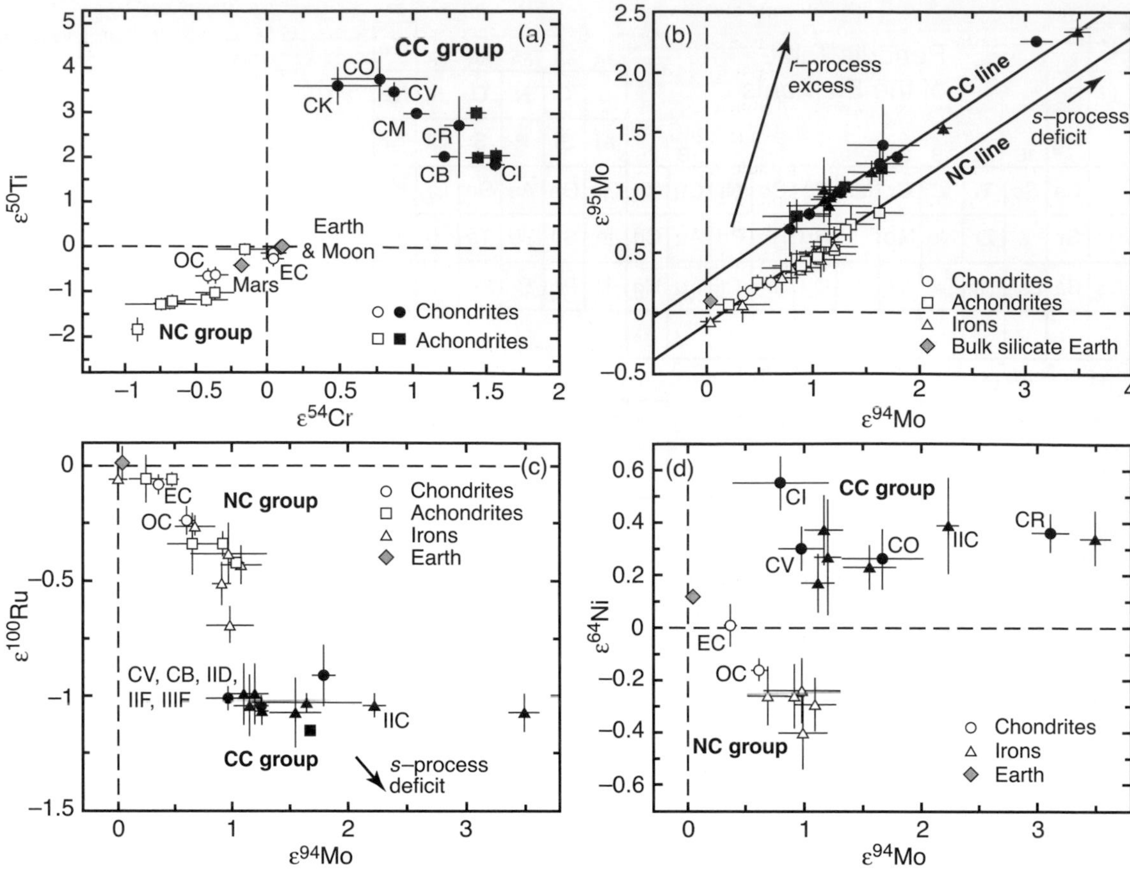

Figure 8.17 Four examples of the isotopic dichotomy between carbonaceous chondrites and related achondrites and iron meteorites (CC group, solid symbols) and noncarbonaceous meteorites and related objects (NC group, open symbols). Data are from Budde G. et al. (2019) and Kleine et al. (2020).

ubiquitous in solar system materials and are even used to classify meteorites. But a variety of other elements also show mass-independent isotopic fractionation. Isotopic anomalies that are not due to processes known to operate in the solar system were first identified in the 1970s. In most cases, the observed isotopic compositions record the nucleosynthetic history of the elements. A majority of all elements that have more than one stable isotope show evidence of these nucleosynthetic isotope anomalies. Interpretation of these isotope effects often uses end-member components derived from nucleosynthesis theory.

In the next chapter, we will examine radioactive-isotope systems. Radioactive isotopes provide one of the most powerful methods for determining the chronology of solar system events.

Questions

1. What are isotopes? Why do the isotopic compositions of most elements only vary by a small amount among solar system materials? What are some different causes of isotope variation in natural materials?
2. Calcite, magnetite, and fayalite precipitated together from the same fluid. How do you use their oxygen-isotope compositions to determine the temperature at which they precipitated? Can you determine the isotopic composition of the fluid? How?
3. What are the main differences between equilibrium and kinetically controlled evaporation? What conditions are necessary to generate large isotopic fractionations during evaporation? Why does evaporation from a solid typically not produce isotopic fractionation?

4. A meteorite has a very high D/H ratio and its oxygen is ^{17}O and ^{18}O rich. What can you say about where the meteorite or its constituents may have formed?
5. We now know that the majority of elements with more than one isotope show mass-independent isotope variations that can be attributed to nucleosynthetic processes. Describe how such "isotopic anomalies" could have survived to be incorporated into solar system objects. In spite of these anomalies, why are the isotopic compositions of most objects so similar?

Suggestions for Further Reading

Robert F., Gautier D., and Dubrulle B. (2000) The solar system D/H ratio: Observations and theories. *Space Science Reviews*, **92**, 201–224. This paper reviews what is known about hydrogen isotopes and what they can tell us about the history of the solar system.

Sharp Z. (2017) *Principles of Stable Isotope Geochemistry, 2nd Edition*. Open Textbooks. https://doi.org/10.25844/h9q1–0p82. A good recent textbook covering the basics of isotope fractionation and its application to geochemistry and cosmochemistry.

Other References

Ali A., Jabeen I., Nasir S. J., and Banerjee J. R. (2018) Oxygen isotope thermometry of DaG 476 and SaU 008 martian meteorites: Implications for their origin. *Geosciences*, **8**, article 15.

Barnes J. J., McCubbin F. M., Santos A. R. et al. (2020) Multiple early-formed water reservoirs in the interior of Mars. *Nature Geoscience*, **13**, 260–264.

Bindeman I. (2008) Oxygen isotopes in mantle and crustal magmas revealed by single crystal analysis. *Reviews in Mineralogy & Geochemistry*, **69**, 445–478.

Black D. C., and Pepin R. O. (1969) Tapped neon in meteorites II. *Earth & Planetary Science Letters*, **6**, 395–405.

Brigham C. A., Papanastassiou D. A., and Wasserburg G. J. (1985) Mg isotopic heterogeneities in fine-grained Ca-Al-rich inclusions (abstract). *Lunar & Planetary Science*, **XVI**, 93–94, Lunar & Planetary Institute, Houston.

Budde G., Burkhardt C., and Kleine T. (2019) Molybdenum isotopic evidence for the late accretion of outer solar system material to Earth. *Nature Astronomy*, **3**, 736–741.

Cameron A. G. W., and Truran J. W. (1977) The supernova trigger for formation of the solar system. *Icarus*, **30**, 447–461.

Cano E. J., Sharp Z. D., and Shearer C. K. (2020) Distinct oxygen isotope compositions of the Earth and Moon, *Nature Geoscience*, **13**, 270–274.

Carr M. H., and Head J. W. III (2003) Oceans on Mars: An assessment of the observational evidence and possible fate. *Journal of Geophysical Research*, **108**, E5, 5042.

Chacko T., Cole D. R., and Horita J. (2001) Equilibrium oxygen, hydrogen and carbon isotope fractionation factors applicable to geologic systems. *Stable Isotope Geochemistry*, Reviews in Mineralogy, 43, pp. 1–81.

Chakraborty S., Yanchulova P., and Thiemens M. H. (2013) Mass-independent oxygen isotopic partitioning during gas-phase SiO_2 formation. *Science*, **342**, 463–466.

Clayton R. N. (2002) Self-shielding in the solar nebula. *Nature*, **415**, 860–861.

Clayton R. N., and Mayeda T. K. (1984) The oxygen isotope record in Murchison and other carbonaceous chondrites. *Earth & Planetary Science Letters*, **67**, 151–166.

Clayton R. N., and Mayeda T. K. (1999) Oxygen isotope studies of carbonaceous chondrites. *Geochimica et Cosmochimica Acta*, **63**, 2089–2104.

Clayton R. N., Grossman L., and Mayeda T. K. (1973) A component of primitive nuclear composition in carbonaceous chondrites. *Science*, **182**, 485–488.

Clayton R. N., Onuma N., Grossman L., and Mayeda T. K. (1977) Distribution of the pre-solar component in Allende and other carbonaceous chondrites. *Earth & Planetary Science Letters*, **34**, 209–224.

Davis A. M., and Richter F. M. (2014) Condensation and evaporation of solar system materials. In *Treatise on Geochemistry, 2nd edition, Vol. 1: Meteorites and Cosmochemical Processes*, Davis A. M., editor, Elsevier, Oxford, pp. 335–360.

Davis A. M., Hashimoto A., Clayton R. N., and Mayeda T. K. (1990) Isotope mass fractionation during evaporation of forsterite (Mg_2SiO_4). *Nature*, **347**, 655–658.

Davis A. M., Richter F. M., Mendybaev R. A., et al. (2015) Isotopic mass fractionation laws for magnesium and their effects on ^{26}Al-^{26}Mg systematics in solar system materials. *Geochimica et Cosmochimica Acta*, **158**, 245–261.

Eiler J. M. (2007) "Clumped-isotope" geochemistry—The study of naturally-occurring, multiply-substituted isotopologues. *Earth & Planetary Science Letters*, **262**, 309–327.

Eiler J. M., Valley J. W., Graham C. M., and Fournelle J. (2002) Two populations of carbonate in ALH 84001: Geochemical evidence for discrimination and genesis. *Geochimica et Cosmochimica Acta*, **66**, 1285–1303.

Farquhar J., Bao H., and Thiemens M. (2000a) Atmospheric influence of Earth's earliest sulfur cycle. *Science*, **289**, 756–759.

Farquhar J., Savarino J., Jackson T. L., and Thiemens M. H. (2000b) Evidence of atmospheric sulphur in the martian regolith from sulphur isotopes in meteorites. *Nature*, **404**, 50–52.

Farquhar J., Peters M., Johnston D. T., et al. (2007) Isotopic evidence for Mesoarchaean anoxia and changing atmospheric sulphur chemistry. *Nature*, **449**, 706–709.

Geiss J., and Reeves H. (1981) Deuterium in the solar system. *Astronomy & Astrophysics*, **93**, 189–199.

Guo W., and Eiler J. M (2007) Temperatures of aqueous alteration and evidence for methane generation on parent bodies of the CM chondrites. *Geochimica et Cosmochimica Acta*, **71**, 5565–5575.

Halevy I., Fischer W. W., and Eiler J. M. (2011) Carbonates in the martian meteorite Allan Hills 84001 formed at 18 ± 4 °C in a near-surface aqueous environment. *Proceedings of the National Academy of Sciences USA*, **108**, 16,895–16,899.

Herbst E. (2003) Isotopic fractionation by ion-molecule reactions. *Space Science Reviews*, **106**, 293–304.

Hoefs J. (2009) *Stable Isotope Geochemistry, 6th edition*. Springer, Göttingen, Germany, 285 pp.

Hulston J. R., and Thode H. G. (1965) Variations in the ^{33}S, ^{34}S, and ^{36}S contents of meteorites and their relation to chemical and nuclear effects. *Journal of Geophysical Research*, **70**, 3475–3484.

Jilly-Rehak C. E., Huss G. R., Nagashima K., and Schrader D. L. (2018) Low-temperature aqueous alteration on the CR chondrite parent body: Implications from in situ oxygen-isotope analyses. *Geochimica et Cosmochimica Acta*, **222**, 230–252.

Jull A. J. T., Eastoe C. J., Xue S., and Herzog G. F. (1995) Isotopic composition of carbonates in the SNC meteorites and Allan Hills 84001 and Nakhla. *Meteoritics*, **30**, 311–318.

Jungck M. H. A., Shimamura T., and Lugmair G. W. (1984) Ca isotope variations in Allende. *Geochimica et Cosmochimica Acta*, **48**, 2651–2658.

Kasting J. F. (1988) Runaway and moist greenhouse atmospheres and the evolution of Earth and Venus. *Icarus*, **74**, 472–494.

Kleine T., Budde G., Burkhardt C., et al. (2020) The non-carbonaceous-carbonaceous meteorite dichotomy. *Space Science Reviews*, **216**, #55.

Krot A. N., Nagashima K., Fintnor K., and Pál-Molnár E. (2019) Evidence for oxygen-isotope exchange in refractory inclusions from Kaba (CV3.1) carbonaceous chondrite during fluid-rich interaction on the CV parent asteroid. *Geochimica et Cosmochimica Acta*, **246**, 419–435.

Lammer H., Kasting J. F., Chassefière E., et al. (2008) Atmospheric escape and evolution of terrestrial planets and satellites. *Space Science Reviews*, **139**, 399–436.

Lee T., Papanastassioui D. A., and Wasserburg G. J. (1977) Aluminum-26 in the early solar system. Fossil or Fuel? *Astrophysical Journal*, **211**, L107–L110.

Lee T., Russel W. A., and Wasserburg G. J. (1979) Calcium isotopic anomalies and the lack of aluminum-26 in an unusual Allende inclusion. *Astrophysical Journal*, **228**, L93–L98.

Leshin L. A. (2000) Insights into Martian water reservoirs from analyses of martian meteorite QUE 94201. *Geophysical Research Letters*, **27**, 2017–2020.

Leshin L. A., McKeegan K. D., Carpenter P. K., and Harvey R. P. (1998) Oxygen isotopic constraints on the genesis of carbonates from martian meteorite ALH 84001. *Geochimica et Cosmochimica Acta*, **62**, 3–13.

Makide K., Nagashima K., Krot A. N., and Huss G. R. (2009) Oxygen- and magnesium-isotope compositions of calcium-aluminum-rich inclusions from CR2 carbonaceous chondrites. *Geochimica et. Cosmochimica Acta,* **73**, 5018–5050.

Mathew K. J., and Marti K. (2001) Early evolution of martian volatiles: Nitrogen and noble gas components in ALH84001 and Chassigny. *Journal of Geophysical. Research*, **106**, E1, 1401–1422.

McKay D. S., Gibson E. K. Jr., Thomas-Keprta K. L., et al. (1996) Search for past life on Mars: Possible relic biogenic activity in martian meteorite ALHA84001. *Science*, **273**, 924–930.

McKeegan K. D., Kallio A. P. A., Heber V. S., et al. (2011) The oxygen isotopic composition of the Sun inferred from captured solar wind. *Science*, **332**, 1528–1532.

Niederer F. R., Papanastassiou D. A., and Wasserburg G. J. (1985) Absolute isotopic abundances of Ti in meteorites. *Geochimica et. Cosmochimica Acta*, **49**, 835–851.

Niemeyer S., and Lugmair G. W. (1984) Titanium isotopic anomalies in meteorites. *Geochimica et Cosmochimica Acta*, **48**, 1401–1416.

Onuma N., Clayton R. N., and Mayeda T. K. (1970) Oxygen isotope fractionation between minerals and an estimate of the temperature of formation. *Science*, **167**, 536–538.

Pavlov A. A., and Kasting J. F. (2002) Mass-independent fractionation of sulfur isotopes in Archean sediments: Strong evidence for an anoxic Archean atmosphere. *Astrobiology*, **2**, 27–41.

Rayleigh J. W. S. (1896) Theoretical considerations respecting the separation of gases by diffusion and similar processes. *Philosophy Magazine*, **42**, 493.

Richter F. M., Janney P. E., Mendybaev R. A., et al. (2007) Elemental and isotopic fractionation of Type B CAI-like liquids by evaporation. *Geochimica et Cosmochimica Acta*, **71**, 5544–5564.

Richter F. M., Watson E. B., Mendybaev R. A., et al. (2009) Isotopic fractionation of the major elements of molten basalt by chemical and thermal diffusion. *Geochimica et Cosmochimica Acta*, **73**, 4250–4263.

Russell W. A., Papanastassiou D. A., and Tombrello T. A. (1978) Ca isotope fractionation on Earth and other solar system materials. *Geochimica et Cosmochimica Acta*, **42**, 1075–1090.

Sakamoto N., Seto U., Itoh S., et al. (2007) Remnants of the early solar system water enriched in heavy oxygen isotopes. *Science*, **317**, 231–233.

Sandford S. A., Bernstein M. P., and Dworkin J. P. (2001) Assessment of the interstellar processes leading to deuterium enrichments in meteoritic organics. *Meteoritics & Planetary Science*, **36**, 1117–1133.

Schrader D. L., Franchi I. A., Connolly Jr., H. C., et al. (2011) The formation and alteration of the Renazzo-like carbonaceous chondrites I: Implications of bulk-oxygen isotopic composition. *Geochimica et Cosmochimica Acta*, **75**, 308–325.

Stone S. W., Yelle R. V., Benna M., et al. (2020) Hydrogen escape from Mars is driven by seasonal and dust storm transport of water. *Science*, **370**, 824–831.

Tachibana S., and Huss G. R. (2005) Sulfur isotope composition of putative primary troilite in chondrules from Bishunpur and Semarkona. *Geochimica et Cosmochimica Acta*, **69**, 3075–3097.

Thiemens M. H. (2006) History and applications of mass-independent isotope effects. *Annual Reviews of Earth and Planetary Sciences*, **34**, 217–262.

Thiemens M. H., and Heidenreich J. E. III (1983) The mass-independent fractionation of oxygen: A novel isotope effect and its possible cosmochemical implications. *Science*, **219**, 1073–1075.

Valley J. W., Eiler J. M., Graham C. M., et al. (1997) Low-temperature carbonate concretions in the martian meteorite ALH 84001: Evidence from stable isotopes and mineralogy. *Science*, **275**, 1633–1638.

Valley J. W., Peck W. H., King E. M., and Wilde S. A. (2002) A cool early Earth. *Geology*, **30**, 351–354.

Warren P. H. (2011) Stable-isotope anomalies and the accretionary assemblage of the Earth and Mars: A subordinate role for carbonaceous chondrites. *Earth & Planetary Science Letters*, **311**, 93–100.

Wasserburg G. J. (1987) Isotopic abundances: Inferences on solar system and planetary evolution. *Earth & Planetary Science Letters*, **86**, 129–173.

Wasserburg G. J., Lee T., and Papanastassiou D. A. (1977) Correlated O and Mg isotopic anomalies in Allende inclusions: II. Magnesium. *Geophysical Research Letters*, **4**, 299–302.

Zheng Y.-F. (1991) Calculation of oxygen isotope fractionation in metal oxides. *Geochimica et Cosmochimica Acta*, **55**, 2299–2307.

Zheng Y.-F. (2011) On the theoretical calculations of O isotope fractionation factors for carbonate-water systems. *Geochemical Journal*, **45**, 341–354.

9 Radioisotopes as Chronometers

Overview

So far we have discussed the materials that make up the solar system and the processes that caused those materials to be in their current state. We will now investigate the chronology of the events that led to the current state of the solar system. There are several different approaches to determine the timing of events. The sequence of events can often be established from spatial relationships among objects (e.g. younger things rest on older things). Absolute ages are provided by long-lived radioactive nuclides. Time intervals can be determined using short-lived radionuclides. Production of nuclides through irradiation by cosmic rays can also be used for age determinations. For a complete chronological picture, it is often necessary to use more than one method of age determination. In this chapter, we focus on the basic principles of radiometric dating. We review individual isotopic clocks, the types of materials that each can date, and the measurements that are made to determine the ages of different objects. In Chapter 10, we discuss the chronology of the solar system derived from these clocks.

9.1 Methods of Age Determination

Placing events in chronological order and attaching an absolute timescale to that order constitute major areas of research in cosmochemistry. There is no single clock that works for everything, so the chronology of the solar system has been built on a variety of observations and measurements. The methods of age determination can be divided into two main types. Some methods give *relative ages*. In other words, they tell us that one object is older than another or that one event occurred before another, but they cannot tell us exactly when the object formed or the event occurred. Relative age determination has a long history and is particularly important in geology. For example, consider a sandstone. The sand grains are older than the sandstone itself, because the sand had to exist before it was incorporated into the rock. In a geologic setting consisting of several layers of rock, the layer at the bottom of the pile must be the oldest because it had to be there before the others could be laid down on top of it. This type of relative age dating can be applied to extraterrestrial materials as well. For example, a chondritic meteorite consists of chondrules, refractory inclusions, metal, sulfide, and matrix, each of which formed independently. Like the sandstone, the chondrite is younger than the constituents that comprise it. We cannot tell how much younger, but the chondrite could not have formed before the constituents came to exist. Occasionally, we find a chondrule that contains an embedded refractory inclusion. There are currently no clear cases of a chondrule inside a refractory inclusion. This suggests that refractory inclusions are older than chondrules, and as we will see, radiometric dating shows this to be true. Short-lived radionuclides, whose primordial abundances have completely decayed away, provide a quantitative relative chronology. But to get ages in years before present, the relative ages must be tied to a chronometer that gives *absolute ages*. Absolute ages are provided by long-lived radionuclides, which are still present today in the solar system. Although typically the absolute ages provided by long-lived radionuclides are less precise than the relative ages provided by short-lived radionuclides, the combination of the two can provide a detailed chronology of the early solar system.

Cosmic-ray interactions provide the basis for another type of chronometer. Cosmic rays are high-energy protons and other atomic nuclei that originate either in the Sun or in interstellar space. They are ubiquitous in the interstellar medium and in the solar system. When a cosmic ray encounters an atomic nucleus, it may break the nucleus into two or more pieces. The resulting stable and radioactive nuclei accumulate over time in objects exposed to cosmic rays. The abundances of these *cosmogenic nuclides* provide information about how long an object was exposed to cosmic rays. Cosmic-ray exposure ages will be considered in Chapter 10.

9.2 Discussing Radiometric Ages and Time

You will see different ways to describe units of time in the literature and in this book. There are subtle differences in the meanings of these units that you should be aware of. For example, if we want to say that the formation of a river valley took one million years, we could write that it took one Myr. On the other hand, if we want to say that the meteorite that killed the dinosaurs struck the Earth 65 million years ago, we could write that it struck at 65 Ma. In this book, when we discuss the period of time over which something happened, we will use the symbols Kyr, Myr, and Gyr (kiloyears, megayears, and gigayears). When we discuss the timing of an event before the present day, we will use Ka, Ma, and Ga. This is a convenient convention, although it is not always used correctly in the literature.

9.3 Basic Principles of Radiometric Age Dating

Radiometric age dating is a powerful chronological tool. It is based on the accumulation of the daughter nuclide produced by the decay of the parent isotope. If the decay rate is known, and if one can measure the amounts of parent element and daughter isotope, then it is straightforward to calculate the time necessary for the daughter element to have accumulated to its measured abundance. In this section, we present the basic principles of radiometric dating. Details of the pathways of radioactive decay were discussed in Chapter 2.

A good way to think about radioactive decay is to envision a bag of coins. You want to take the coins to someone in a city far away. To get there you must travel through a series of toll gates, one every 10 miles. At each gate, a gatekeeper demands that you pay 1% of your coins in order to pass. At the first gate, when you have 1000 coins, you must pay 10 coins to pass. At the next gate, when you have 990 coins, you must pay 9.9 coins (the gatekeeper gives change). At the next gate, the toll is 9.8 coins. At each gate, the number of coins you have decreases by 1%, and as the number of coins gets smaller, the 1% that you pay gets smaller too. The toll is proportional to the number of coins you have, and your bag of coins gets depleted by the same percentage every 10 miles. Radioactive decay works the same way. The number of decays in a given period of time is proportional to the number of atoms present:

$$-\frac{dN}{dt} \propto N \tag{9.1}$$

This relationship is transformed into an equality by the introduction of a proportionality constant, λ, which represents the probability that an atom will decay within a stated period of time. The numerical value of λ is unique for each radionuclide and is expressed in units of reciprocal time. Thus, the equation describing the rate of decay of a radionuclide is

$$-\frac{dN}{dt} = \lambda N \tag{9.2}$$

where λN gives the rate of decay at time t. To determine the total number of atoms of a radioactive isotope, N, after a particular time t has elapsed, we integrate Equation 9.2:

$$-\int_{N_0}^{N} \frac{dN}{N} = \lambda \int_0^t dt$$

$$\ln N - \ln N_0 = -\lambda t - \lambda 0$$

Rearrange the terms (remembering that $\lambda 0 = 0$):

$$\ln \frac{N}{N_0} = -\lambda t$$

$$\frac{N}{N_0} = e^{-\lambda t}$$

$$N = N_0 e^{-\lambda t} \tag{9.3}$$

Equation 9.3 is the basic equation that describes all radioactive decay processes. It gives the number of atoms, N, of a radioactive parent isotope remaining at any time t from a starting number N_0 at time $t = 0$.

What about the daughter isotope? If the original number of atoms of the stable daughter isotope is taken to be zero at $t = 0$, and no daughter atoms are added to or lost from the system except by radioactive decay, then the number of daughter atoms (D^*) produced by the decay of the parent isotope at any given time is

$$D^* = N_0 - N \tag{9.4}$$

$$D^* = N_0 - N_0 e^{-\lambda t}$$

$$D^* = N_0\left(1 - e^{-\lambda t}\right) \tag{9.5}$$

Equation 9.5 gives the number of stable radiogenic daughter atoms (D^*) at any time t formed by the decay of a radioactive parent whose initial abundance at $t = 0$ was N_O.

When dating rocks and minerals, it is often more convenient to relate the number of daughter nuclei (D^*) to the number of parent atoms remaining (N), the quantity that can be measured, rather than to N_O. Starting from Equation 9.4

$$D^* = N_0 - N$$

and replacing N_O with $Ne^{\lambda t}$ (from Equation 9.3) gives

$$D^* = Ne^{\lambda t} - N = N\left(e^{\lambda t} - 1\right) \quad 9.6$$

In most natural systems, the number of atoms of the daughter nuclide (D) consists of initial atoms already present in the system (D_O) plus those resulting from *in situ* radioactive decay (D^*):

$$D = D_0 + D^* \quad 9.7$$

Substituting Equation 9.6 into 9.7 gives the basic equation that is used to calculate a date for a rock or mineral from the decay of a radioactive parent to a stable daughter:

$$D = D_0 + N\left(e^{\lambda t} - 1\right) \quad 9.8$$

Both D and N are measurable quantities and D_O is a constant whose value can either be assumed or calculated from the data (see below). Solving this equation for t gives the date:

$$\frac{D - D_0}{N} = e^{\lambda t} - 1$$

$$t = \frac{1}{\lambda} \ln\left[\frac{D - D_0}{N} + 1\right] \quad 9.9$$

The rate of decay of a radionuclide is often discussed in terms of its *half-life*. The half-life ($t_{1/2}$) is the time required for one-half of a given number of atoms of a radionuclide to decay. Mathematically, when $t = t_{1/2}$, $N = 0.5 \times N_O$. Substituting these values into Equation 9.3, we find

$$\frac{1}{2}N_0 = N_0 e^{-\lambda t_{1/2}}$$

$$\ln\left(\frac{1}{2}\right) = -\lambda t_{1/2}$$

$$\ln 2 = \lambda t_{1/2}$$

$$t_{1/2} = \frac{\ln 2}{\lambda} = \frac{0.693}{\lambda} \quad 9.10$$

Equation 9.10 gives the relationship between the half-life and the decay constant, λ.

Another parameter that is used to describe the decay of a radioactive species is the *mean life* (τ), which is the average life expectancy of a radioactive atom. The mean life is defined as

$$\tau = -\frac{1}{N_0}\int_{t=0}^{t=\infty} t \cdot dN \quad 9.11$$

From Equation 9.2,

$$-dN = \lambda N \cdot dt$$

Therefore,

$$\tau = \frac{1}{N_0}\int_{t=0}^{t=\infty} \lambda N t \cdot dt$$

Because

$$N = N_0 e^{-\lambda t}$$

$$\tau = \lambda \int_{t=0}^{t=\infty} t e^{-\lambda t} \cdot dt = -\left[\frac{\lambda t + 1}{\lambda} e^{-\lambda t}\right]_0^{\infty}$$

$$\tau = \frac{1}{\lambda} \quad 9.12$$

Thus, the mean life, τ, is equal to the reciprocal of the decay constant and is longer than the half-life by a factor of 1/0.693. The activity of a radionuclide is reduced by a factor of $1/e$ during each mean life. The decay of a radioactive nuclide can be discussed in terms of half-life or mean life, and you will see both in the cosmochemistry literature. We will use half-life in this chapter because this formulation is used more often in chronology applications. Discussions of galactic chemical evolution and the age of the elements (see Chapter 10) are often done in terms of mean life.

Figure 9.1 shows graphically the decay of the radionuclide ^{87}Rb to its stable radiogenic daughter ^{87}Sr as a function of time given in units of half-life. In this example, there are initially many atoms of ^{87}Rb and no atoms of the daughter, ^{87}Sr. In natural systems there is always some ^{87}Sr present initially; these atoms would shift the zero point on the y-axis of Figure 9.1 to a finite value. After one half-life, half of the atoms of ^{87}Rb have decayed to ^{87}Sr. After two half-lives, half of the atoms of ^{87}Rb remaining at the end of the first half-life have decayed and three-quarters of the original ^{87}Rb atoms now exist as ^{87}Sr. In general, the initial number of radionuclide atoms is reduced over time by 2^{-n}, where n is the number of half-lives, and the number of daughter nuclides approaches the initial abundance of ^{87}Rb asymptotically

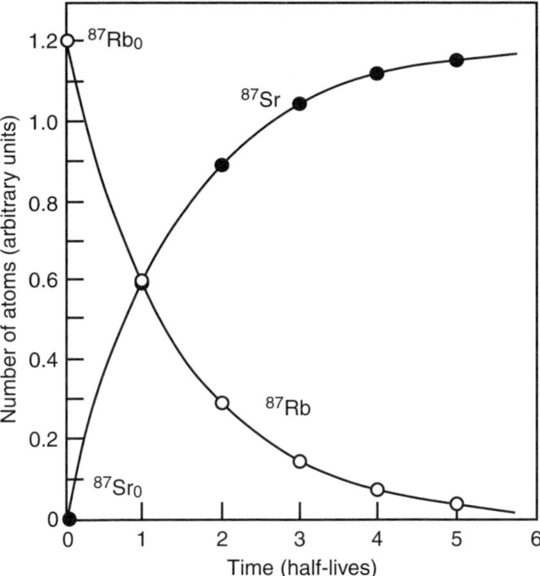

Figure 9.1 Number of atoms of radioactive ^{87}Rb and daughter ^{87}Sr as a function of time in units of half-life. For each half-life, the number of radioactive ^{87}Rb atoms drops by a factor of two as the ^{87}Rb atoms decay to ^{87}Sr. After 10 half-lives, the parent isotope is effectively gone.

as $t \rightarrow \infty$. For most purposes, a radionuclide can be considered to have decayed away after ~10 half-lives, when N_0 has been reduced by 2^{-10} (1/1024). Note that the rate of decay of a radionuclide is not constant, but decreases exponentially with time. What is constant is the proportion of the radionuclides that decay during one half-life. This is counter to our normal experience because the decay of an atom does not become more likely as the atom becomes older. Radioactive decay is a quantum-mechanical phenomenon, and the probability of decay of a radioactive nuclide can be accurately described from its basic quantum-mechanical properties.

In order for a radiometric date to provide a valid age, the following criteria must be met:

1. There must have been a specific event that homogenized the isotopic compositions of parent and daughter element. The homogenous composition of the parent element assures that additions to the amount of daughter isotope reflect only the amount of parent element and the time that has passed. The homogeneous composition of the daughter element is the background above which the accumulation of a radiogenic daughter isotope can be recognized and measured.
2. The isotopic system (e.g. a rock or mineral) must have remained closed, with no loss or gain of parent or daughter element since the event being dated.
3. The value of the decay constant (λ) must have remained constant over the age of the solar system and the galaxy, and it must be accurately known. As we discussed in Chapter 2, this assumption is well founded for conditions relevant to cosmochemistry. The concordance of dates given by systems using a variety of decay paths and astronomical observations of decay rates of newly synthesized elements over billions of years provide strong evidence that the decay rates have remained constant. In addition, detailed experiments and theoretical models have identified the extreme conditions (e.g. centers of stars) under which this assumption breaks down for certain isotopes, thereby identifying the "exceptions that prove the rule."
4. It must be possible to assign a realistic value to the initial abundance of the daughter element, D_0.

If these criteria are met, an isotope system can provide an age for the dated material. In practice, an additional criterion is also desirable for many isotope systems. The event being dated should somehow generate minerals or other components with different ratios of parent/daughter element. This helps us to recognize the portion of the daughter isotope that resulted from radioactive decay. The importance of this criterion will become clear below.

One can apply the formalism discussed above to a wide variety of systems to produce a radiometric date. In this book, we will use the word "date" to mean the time calculated from the ratio of a radioactive isotope and its daughter isotope using the equation for radioactive decay. An "age" is the time between a natural event and the present. A "date" becomes a valid "age" when the conditions described above are met. This terminology, suggested by Faure (1986), is not always used in the literature, where "age" and "date" are often used interchangeably. But there is value to the distinction because it helps a reader understand which numbers are significant.

Radiometric age dating is a powerful tool, but it cannot date everything. The clock must be started by a discrete event, the duration of which is short compared to the time elapsed since the event. Crystallization of a melt is an ideal event for dating. Melting will homogenize the parent and daughter elements, and crystallization will typically produce minerals with a variety of parent/daughter element ratios. Crystallization usually occurs relatively quickly. Precipitation of minerals from a fluid phase may also be datable if the dissolution and reprecipitation of the minerals is complete (i.e. the precipitates do not nucleate on pre-existing mineral phases). In contrast, metamorphic episodes are hard to date by radiometric methods. This is because the timescale for metamorphism, both on Earth and on extraterrestrial bodies, is typically long, and cooling from peak temperatures takes place slowly. Diffusion occurring at high temperatures could potentially homogenize the isotopic compositions of parent and daughter elements, but because elements stop diffusing in different minerals at different temperatures during cooling (*blocking temperatures*), the clock in one mineral may start at a different time than that in another mineral. Under special conditions, even these mineral clocks can be useful, but for the most part metamorphic events cannot be directly dated. Cold accumulation of objects, such as the accumulation of sediments to form sedimentary rocks or the accretion of chondrules and calcium-aluminum-rich inclusions (CAIs) to form a chondrite, cannot be dated by radiometric means because there is no discrete event that homogenizes the system and starts the clock. However, in the case where one body impacts another at hypervelocity, the resulting explosion may melt and homogenize material, providing an event that can be dated.

One of the criteria for obtaining a valid age is that the decay constant must be known. It turns out that decay constants can be very hard to measure accurately, and different measurements may not agree. If different laboratories use different decay constants, the dates they produce for the same object will be different. To get around this problem, the International Union of Geological Sciences Subcommission on Geochronology met and established recommended values for decay constants of radionuclides used in chronology. The results of the Subcommission's work were published by Steiger and Jäger (1977). This article does not claim to establish the true value of the decay constants of the various nuclides, but it does provide a single number for each nuclide that all laboratories can use to calculate ages. With the passage of time, better values for some decay constants have been determined, and some decay schemes now use values of the decay constants different than those proposed by Steiger and Jäger (1977). But in general, everyone uses the same decay constants and thus the published ages for a single isotope system can be directly compared. Comparing older ages with newer ones can be problematic, however. It should be noted that an agreement on which decay constants to use does not guarantee that ages determined by different chronometers will agree. This is because there are still significant uncertainties in the absolute values of the decay constants, and these uncertainties can result in systematic offsets between chronometers. This issue is seldom discussed in cosmochemistry papers, but users of these ages should be aware of it.

9.4 Long-Lived Radionuclides

Long-lived radionuclides are those with half-lives long enough that a significant fraction of the original atoms incorporated into the solar system when it formed are still present today. Long-lived radionuclides can give absolute ages for events all the way back to the formation of the solar system. If we can figure out how to apply them, they also have the potential to date events prior to solar system formation. Table 9.1 summarizes the properties of several long-lived radionuclides. Those given in bold type are commonly used as chronometers. In this section, we will briefly review each of these chronometers. We will not be able to cover all of the subtleties of each system, but the suggestions for further reading and the cited references provide a wealth of additional details.

9.4.1 The ^{40}K–^{40}Ar System

Potassium-40 decays to ^{40}Ar by electron capture (11%) and to ^{40}Ca by β^- decay (89%) with a combined half-life of 1.250×10^9 years. Although ^{40}Ca is the more abundant daughter isotope, ^{40}Ca is the major isotope of calcium, and because calcium is equally or more abundant than potassium in most materials, the fractional addition of radiogenic ^{40}Ca is typically very hard to detect. Thus, the ^{40}K–^{40}Ca system is only useful in systems with very high K/Ca ratios. In contrast, because argon is a gas and is effectively absent from solids when they crystallize from a melt or condense from a gas phase, the addition of radiogenic ^{40}Ar to the sample is easy to recognize. Thus, the ^{40}K–^{40}Ar system is

Table 9.1 Long-lived radionuclides used in cosmochemistry

Nuclide	Half-Life (years)	Daughter Isotope	Decay Mode
^{40}K	**1.25×10^9**	**^{40}Ar (11%)**	**E.C.**
		^{40}Ca (89%)	β^-
^{87}Rb	**4.88×10^{10}**	**^{87}Sr**	**β^-**
^{138}La	1.05×10^{11}	^{138}Ba (34%)	E.C.
		^{138}Ce (66%)	β^-
^{147}Sm	**1.06×10^{11}**	**^{143}Nd**	**α**
^{176}Lu	**3.72×10^{10}**	**^{176}Hf**	**β^-**
^{187}Re	**4.16×10^{10}**	**^{187}Os**	**β^-**
^{190}Pt	4.50×10^{11}	^{186}Os	α
^{238}U	**4.47×10^9**	**^{206}Pb**	**α (S.F.)**
^{235}U	**7.04×10^8**	**^{207}Pb**	**α (S.F.)**
^{232}Th	**1.40×10^{10}**	**^{208}Pb**	**α (S.F.)**

α: alpha decay, β^-: beta decay, E.C.: electron capture, S.F.: spontaneous fission.

the one primarily used for age dating. One problem with the $^{40}K–^{40}Ar$ system is that argon is relatively easily released from the crystal structures of minerals. If this happens, then the measured date will be too young. Argon-40 from the atmosphere can also be adsorbed onto or trapped in the structure of the rock or mineral to be dated. This ^{40}Ar can typically be recognized because atmospheric argon has two other isotopes, ^{36}Ar and ^{38}Ar.

9.4.1.1 *History*

The minor isotope ^{40}K was discovered by Nier (1935) and was shown to be radioactive by Smythe and Hemmendinger (1937). Aldrich and Nier (1948) found that the isotopic ratio $^{40}Ar/^{36}Ar$ was several times higher in argon extracted from potassium-bearing minerals than in the atmosphere – clear evidence that the ^{40}Ar was a radiogenic daughter of ^{40}K. Determining the decay constant for $^{40}K–^{40}Ar$ decay was a significant challenge. Direct experimental measurements gave a relatively high decay rate and a branching ratio between the $^{40}K–^{40}Ar$ and $^{40}K–^{40}Ca$ channels of 0.13, but measurements on minerals with known ages suggested a lower value. Wasserburg and Hayden (1955) used the newly developed isotope dilution technique to make precise measurements of potassium feldspars and concluded that the branching ratio should be 0.085 ± 0.005. But Wetherill et al. (1955) noted that feldspar gave lower $^{40}Ar/^{40}K$ ratios than coexisting micas, which implied that some of the ^{40}Ar had diffused out of the feldspars, making the branching ratio suggested by Wasserburg and Hayden (1955) too low. Further work confirmed this observation. The currently accepted values of the decay constants for the two channels shown in Table 9.2 give a branching ratio of 0.117.

9.4.1.2 *Technical Details*

As with all radiometric dating systems, the $^{40}K–^{40}Ar$ system will only produce a valid age if the basic assumptions of radiometric dating are satisfied. In particular, a valid age can only be obtained if the samples remained closed to diffusion of argon either into or out of the system. Two different approaches can be taken to date an object with the $^{40}K–^{40}Ar$ system.

$^{40}K–^{40}Ar$ Dating

In conventional $^{40}K–^{40}Ar$ dating, the abundance of potassium is measured on one aliquot of the sample and the argon abundance and isotopic composition are measured on another. The current abundance of ^{40}K is calculated using the measured content of potassium in the sample and the isotopic abundance of ^{40}K given in Table 9.2. The $^{40}Ar^*$ is calculated from the argon data by first correcting

Table 9.2 Decay constants for ^{40}K[1] and isotopic abundances of potassium[2] and argon[3]

Pathway	Half-Life (years)	Decay Constant (yr^{-1})
β decay	1.397×10^9	$\lambda_\beta = 4.962 \times 10^{-10}$
Electron capture	1.193×10^{10}	$\lambda_\varepsilon = 0.581 \times 10^{-10}$
Combined	1.250×10^9	$\lambda_C = 5.543 \times 10^{-10}$

Isotopic Abundances	(%)		(% in air)
^{39}K	93.2582	^{36}Ar	0.337
^{40}K	0.01167	^{38}Ar	0.063
^{41}K	6.7301	^{40}Ar	99.600

[1] Steiger and Jäger (1977).
[2] Garner et al. (1975).
[3] Nier (1950).

the raw data for instrumental fractionation effects through comparison with standards, and then subtracting the non-radiogenic ^{40}Ar from atmospheric argon or other components. The radiometric age can be calculated as follows (Equations 9.6 and 9.9):

$$^{40}Ar^* = \frac{\lambda_{EC}}{\lambda_{total}} {}^{40}K\left(e^{\lambda_{total}t} - 1\right) \quad 9.13$$

Solving for t gives

$$t = \frac{1}{\lambda_{total}} \ln\left(\frac{\lambda_{total}}{\lambda_{EC}} \frac{^{40}Ar^*}{^{40}K} + 1\right) \quad 9.14$$

Inserting the values for the constants from Table 9.2 gives

$$t = 1.804 \times 10^9 \ln\left(9.540 \frac{^{40}Ar^*}{^{40}K} + 1\right) \quad 9.15$$

$^{40}Ar–^{39}Ar$ Dating

The $^{40}Ar–^{39}Ar$ method, first developed by Merrihue and Turner (1966), has several advantages over the conventional $^{40}K–^{40}Ar$. In this method, the sample is irradiated with fast neutrons and a portion of stable ^{39}K is converted to ^{39}Ar. The argon in the sample is then measured by stepped heating and the ^{39}Ar is released with the radiogenic ^{40}Ar. Provided that the proportion of ^{39}K that was converted to ^{39}Ar is known, dates can be calculated from the argon released in each temperature step. The $^{40}Ar–^{39}Ar$ method eliminates the need for a separate measurement of the potassium content. But more importantly, it provides a way to directly associate $^{40}Ar^*$ with the potassium from which it decayed in specific sites in a rock or mineral. Because argon is first released in the experiments from the less retentive mineral sites, the effect of argon loss can be detected.

Each temperature step of the stepped heating procedure of the ^{40}Ar–^{39}Ar method yields a date. To calculate this date, one must determine how many ^{39}K atoms in the sample are represented by the measured number of ^{39}Ar atoms. To determine this number, a sample of known age is included as a monitor of the neutron flux. From measurements of this monitor, a parameter J can be calculated (see Appendix). The date for each step of the ^{40}Ar–^{39}Ar measurement can then be calculated from

$$t = \frac{1}{\lambda} \ln \left(\frac{^{40}\mathrm{Ar}^*}{^{39}\mathrm{Ar}} J + 1 \right) \qquad 9.16$$

In a sample that conforms perfectly to the assumptions listed above for the ^{40}K–^{40}Ar dating method, the dates for the temperature steps will be equivalent and will correspond to the age of the sample. However, in natural samples, this is not typically found. In order to make sense of the measurements, the "age-spectrum diagram" was developed, in which the calculated date of each temperature step is plotted against the cumulative percentage of the ^{39}Ar (= ^{39}K) released. Examples of such diagrams are shown for three lunar samples in Figures 9.2 and 9.3.

Typically, the low-temperature steps give younger dates than the high-temperature steps. This is due to loss of $^{40}Ar^*$ from the least retentive sites over time. Interpretation of an age-spectrum diagram involves identifying the gas release from sites that have quantitatively retained their $^{40}Ar^*$ and calculating the age from these steps. Results are often discussed in terms of a "plateau" on the age-spectrum diagram. For example, in Figure 9.2,

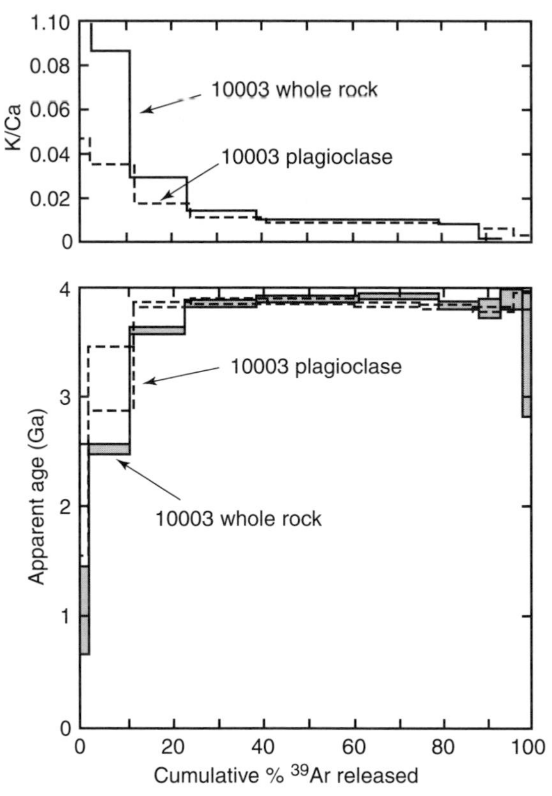

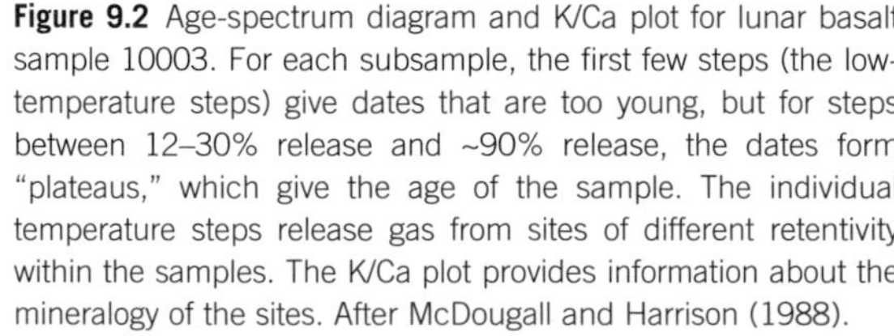
Figure 9.2 Age-spectrum diagram and K/Ca plot for lunar basalt sample 10003. For each subsample, the first few steps (the low-temperature steps) give dates that are too young, but for steps between 12–30% release and ~90% release, the dates form "plateaus," which give the age of the sample. The individual temperature steps release gas from sites of different retentivity within the samples. The K/Ca plot provides information about the mineralogy of the sites. After McDougall and Harrison (1988).

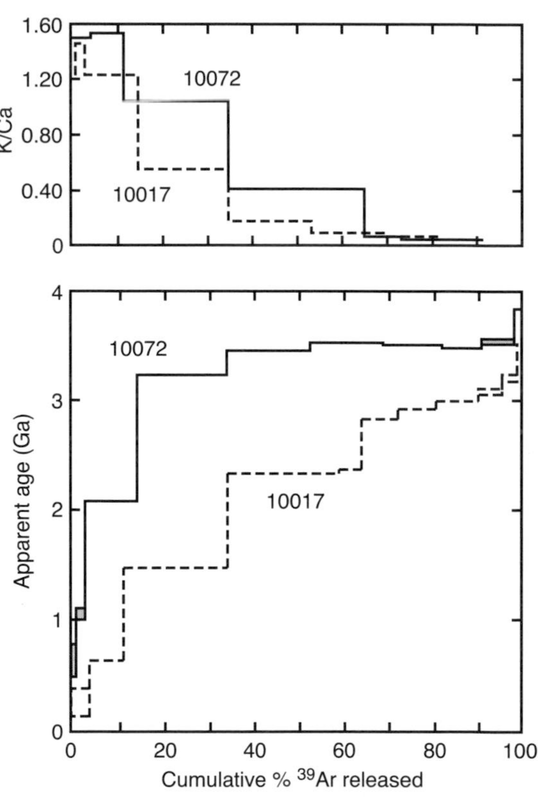

Figure 9.3 Age-spectrum diagram and K/Ca plot for two lunar basalt samples showing disturbance of the ^{40}Ar–^{39}Ar system. Basalt 10072 has experienced modest argon loss of ~14%, but shows a plateau for steps between 60% and 90% release and gives an age of ~3.5 Ga. In contrast, sample 10017 has experienced ~45% argon loss. Its age spectrum has no plateau and gives only a minimum age of ~3.2 Ga. After McDougall and Harrison (1988).

the plateau for plagioclase consists of the steps from ~12% to ~90% of the cumulative ^{39}Ar release. For the whole rock, and for ilmenite and clinopyroxene, the plateau starts at a higher cumulative release. The definition of what constitutes a plateau is somewhat arbitrary, but generally a plateau must consist of at least three contiguous steps that give the same age within errors, and these steps must contain a substantial fraction of the ^{39}Ar released. In unusual cases, it may be possible to identify two plateaus, representing either the formation age and a major metamorphic event or two metamorphic events, but more often a multistage history results in age-spectrum diagrams with no plateau (sample 10017 in Fig. 9.3).

The irradiation of the sample that is done to convert ^{39}K to ^{39}Ar also converts calcium to isotopes of argon. A small correction for this effect must be made to the calculated potassium abundance. This correction is done using ^{37}Ar, which is produced exclusively from calcium. In addition to providing a correction to better determine the potassium concentration, the ^{37}Ar also provides a means of estimating the calcium concentration. Plotting the K/Ca ratio versus cumulative ^{37}Ar release (Figs. 9.2 and 9.3) can provide information on the mineralogy of the site(s) that release the ^{40}Ar* at each temperature step.

9.4.1.3 *Applications*

Because potassium is relatively abundant in many materials, the ^{40}K–^{40}Ar chronometer has been widely used in geochemistry and cosmochemistry. It can often be used where other dating methods cannot. The half-life of ^{40}K is sufficiently long that the ^{40}K–^{40}Ar system can date objects that formed during the earliest epochs of solar system history, and is short enough to permit dating of objects that formed as recently as a few million years ago. The ^{40}K–^{40}Ar system is ideal for dating objects that crystallized from a melt or condensed from a gas phase because ^{40}Ar* from the decay of ^{40}K in the precursor material is typically completely degassed from the sample. The ^{40}K–^{40}Ar system has been used to date meteorites and samples from the Moon and Mars (SNC meteorites). Today, however, most ^{40}K–^{40}Ar dating in cosmochemistry is done with the ^{40}Ar–^{39}Ar method.

The ^{40}Ar–^{39}Ar method has been particularly useful in unraveling the history of the Moon. For example, Figure 9.2 shows an age-spectrum diagram for a whole-rock sample and a plagioclase separate of Apollo 11 basalt 10003. This is a well-behaved rock that gives a clear plateau consisting of >50% of the cumulative ^{39}Ar release for both subsamples. The age of eruption and crystallization on the lunar surface inferred for this basalt is 3.86 ± 0.02 (2σ) Ga ($\lambda = 5.543 \times 10^{-10}$ yr^{-1}). Other basalts are not so well behaved. For Apollo 11 basalt 10072, the plateau for 50–98% cumulative release gives an age of 3.5 Ga (Fig. 9.3), which is consistent with the ^{87}Rb–^{87}Sr age (3.57 ± 0.05 Ga) and the ^{147}Sm–^{143}Nd age (3.57 ± 0.03 Ga). The close agreement between the dates obtained by three independent techniques shows that ^{40}Ar–^{39}Ar ages are reliable when the system has remained closed and a good plateau is obtained on the age-spectrum diagram. On the other hand, Apollo 11 basalt 10017 does not give a plateau (Fig. 9.3) and only a minimum age of 3.2 Ga can be inferred.

9.4.2 The ^{87}Rb–^{87}Sr System

Rubidium-87 β^- decays to ^{87}Sr with a half-life of ~4.88 × 10^{10} years ($\lambda_\beta = 1.42 \times 10^{-11}$ yr^{-1}). Rubidium has two naturally occurring isotopes, ^{85}Rb and ^{87}Rb (Table 4.2). Rubidium is an alkali metal and belongs to Group 1A of the Periodic Table (Fig. 2.6). The ionic radius of rubidium (1.48 Å) is similar enough to that of potassium (1.33 Å) to permit it to substitute for potassium in all potassium-bearing minerals. As a consequence, rubidium does not form minerals of its own, but instead occurs in easily detectable amounts in many minerals.

Strontium has four naturally occurring isotopes (Table 4.2). It is a member of the alkaline earths (Group 2A) (Fig. 2.6). Strontium substitutes for calcium and is abundant in minerals such as plagioclase, apatite, and calcium carbonate.

9.4.2.1 *History*

The natural radioactivity of rubidium was demonstrated by Campbell and Wood (1906), and ^{87}Rb was identified as the radioactive isotope by Hemmendinger and Smythe (1937). Mattauch (1937) determined the isotopic composition of strontium separated from a rubidium-rich mica and found that it was nearly pure ^{87}Sr. This led Hahn and Walling (1938) to discuss using the ^{87}Rb–^{87}Sr system to determine ages of rocks. They envisioned determining abundances of rubidium and strontium by chemical means, which meant that their method was limited to minerals such as mica and potassium feldspar that form with high rubidium concentrations and essentially no common strontium. Although the chemical strontium method was used for several years, it was clear that isotopic analyses were desirable to permit detection of and correction for common strontium. With improved mass spectrometers and the newly developed isotope dilution technique for measuring very low elemental abundances (Inghram, 1954), the isotopic ^{87}Rb–^{87}Sr method replaced the chemical strontium method.

The ^{87}Rb–^{87}Sr technique was first applied to meteorites in 1956. However, the decay constant for rubidium was then poorly known. Measurements of one chondrite and one achondrite by Schumacher (1956) gave incorrectly old ages, 5.46 and 5.80 Ga, using the then-recommended half-life of 58 Gyr. Schumacher proposed a "geological value" for the half-life that gave ^{87}Rb–^{87}Sr ages for the meteorites of 4.7 ± 0.4 Ga, in reasonable agreement with the Pb–Pb ages determined by Patterson (1956). ^{87}Rb–^{87}Sr chronology was plagued by the uncertainty in the half-life of ^{87}Rb throughout the 1950s and 1960s. Values ranged from 47 Ga, as determined by direct counting, to a geological value of 50 Ga. The current uncertainty in the half-life is still on the order of 2%. Steiger and Jäger (1977) recommended a value of 48.8 billion years, but cosmochemists now often use values of 49.4–50 Ga. One can easily recalculate ages to a different decay constant using the following equation:

$$t_{1.42} = \frac{date \times 1.39 \times 10^{-11}}{1.42 \times 10^{-11}} \tag{9.17}$$

In this equation, $t_{1.42}$ is the date based on $\lambda = 1.42 \times 10^{-11}$ yr^{-1}, and the original date was calculated using $\lambda = 1.39 \times 10^{-11}$ yr^{-1}.

The ^{87}Rb–^{87}Sr method reached its modern level of maturity during the mid to late 1960s as scientists prepared for the return of lunar samples. The technology that enabled this work was a new generation of automated mass spectrometers with digital data acquisition. Leaders in developing this new technology were the Australian National University group headed by W. Compston and the Caltech group headed by G. J. Wasserburg. The other development was the application of the mineral or "internal" isochron technique to meteorites (see below). The ^{87}Rb–^{87}Sr system played an important role in unraveling the history of the Moon and in establishing the antiquity of meteorites. It is often used in conjunction with other techniques to evaluate the evolution of differentiated bodies.

9.4.2.2 *Technical Details*

There are two basic ways to apply the ^{87}Rb–^{87}Sr technique to natural samples. The original method is simply to measure the isotopic composition of strontium and the abundance of rubidium in a rock and then calculate a date. If the rock contains no common strontium, a date can be calculated from

$$t = \frac{1}{\lambda} \ln \left(\frac{^{87}Sr^*}{^{87}Rb} + 1 \right) \tag{9.18}$$

Here, $^{87}Sr^*$ denotes radiogenic ^{87}Sr from *in situ* decay. However, most rocks do contain common strontium, which must be subtracted from the measured ^{87}Sr to determine $^{87}Sr^*$:

$$t = \frac{1}{\lambda} \ln \left(\frac{^{87}Sr_{total} - {}^{87}Sr_0}{^{87}Rb} + 1 \right) \tag{9.19}$$

It is difficult from a single measurement to determine the amount of common Sr ($^{87}Sr_0$), so in practice, measurements are generally made on several cogenetic rocks or minerals, and the results are plotted on a ^{87}Rb–^{87}Sr evolution diagram.

Let us assume that a geologic event caused the strontium isotopes to be homogenized within a rock. This means that at time zero, all of the minerals in the rock had the same initial $^{87}Sr/^{86}Sr$ ratio, $(^{87}Sr/^{86}Sr)_0$. The Rb/Sr ratio differs among the minerals of a rock, so with the passage of time, the $^{87}Sr/^{86}Sr$ ratio in each mineral changes at a rate that is proportional to the Rb/Sr ratio (Fig. 9.4). Although Figure 9.4 illustrates what happens to the $^{87}Sr/^{86}Sr$ ratio with time in the various minerals in a rock, it is not particularly useful in dating the rock.

Returning to Equation 9.8 bovc, we can write

$$^{87}Sr = {}^{87}Sr_0 + {}^{87}Rb(e^{\lambda t} - 1) \tag{9.20}$$

Dividing through by ^{86}Sr, which is stable and is not involved in radioactive decay, the following equation is obtained:

$$\frac{^{87}Sr}{^{86}Sr} = \left(\frac{^{87}Sr}{^{86}Sr} \right)_0 + \frac{^{87}Rb}{^{86}Sr} (e^{\lambda t} - 1) \tag{9.21}$$

This is the equation for a straight line on a diagram of $^{87}Sr/^{86}Sr$ versus $^{87}Rb/^{86}Sr$, also known as a ^{87}Rb–^{87}Sr

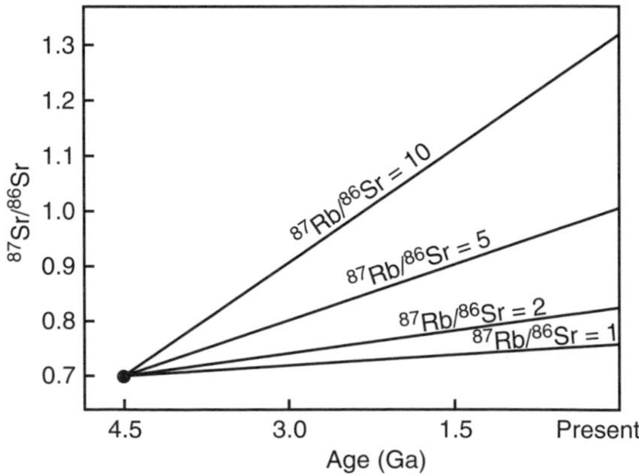

Figure 9.4 Compston–Jeffery diagram showing the growth of radiogenic ^{87}Sr as a function of time in minerals with different $^{87}Rb/^{86}Sr$ ratios.

evolution diagram. The line is an "isochron," with the slope and intercept given by

$$\text{slope} = e^{\lambda t} - 1$$

$$\text{intercept} = \left(\frac{^{87}\text{Sr}}{^{86}\text{Sr}}\right)_0$$

Figure 9.5 shows schematically the data from Figure 9.4 plotted on a ^{87}Rb–^{87}Sr evolution diagram, also known as an isochron diagram. At time zero, the event that we hope to date homogenized the strontium isotopes throughout the system, so each of the minerals started with the same ^{87}Sr/^{86}Sr ratio, (^{87}Sr/^{86}Sr)$_0$. With the passage of time, some of the ^{87}Rb decayed to ^{87}Sr, and the minerals each moved up and to the left on the diagram. Because the position of each mineral at any given time is described by Equation 9.21, the minerals always fall on a linear array, the slope of which is proportional to the amount of time that has passed. The date is thus given by solving the equation for the slope for time t:

$$t = \frac{1}{\lambda} \ln[\text{slope} + 1] \qquad 9.22$$

To apply this method, one measures the ^{87}Sr/^{86}Sr and ^{87}Rb/^{86}Sr ratios in several samples and plots the results on a ^{87}Rb–^{87}Sr evolution diagram. Each data point has a measurement uncertainty associated with it, so the points will not fall exactly on a straight line even in the best-behaved system. An error-weighted linear regression of the data provides the equation for the isochron. Regression methods that take into account the uncertainties in both ^{87}Sr/^{86}Sr and ^{87}Rb/^{86}Sr and provide as output the slope and intercept of the isochron along with their uncertainties and a measure of the goodness of fit to the data are available in the literature (York, 1966, 1969; Williamson, 1968; Mahon, 1996; Ludwig, 2003) and are used by most laboratories.

An isochron that is built from measurements of several mineral phases provides an internal check on whether or not the system has remained closed – one of the requirements for a valid age. If the system has not remained closed and has been affected by subsequent events, the isochron will be "disturbed." On the ^{87}Rb–^{87}Sr evolution diagram, the data will plot farther from the isochron than expected based on measurement errors. Such a line is known as an "errorchron." Figure 9.6 shows schematically what might happen to the isochron in Figure 9.5 if the rubidium and strontium are partially or totally redistributed. Disturbed isochrons may also result if the minerals being measured are not cogenetic (i.e. they did not form together from the same isotopic reservoir). If the isochron is highly disturbed, the problem is easy to

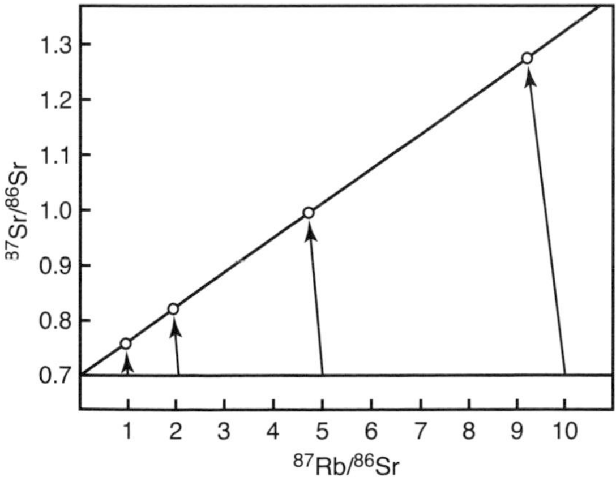

Figure 9.5 Rubidium–strontium evolution diagram showing the isochron generated by the minerals in Figure 9.4. Note that the ^{87}Rb/^{87}Sr ratio decreases slightly as ^{87}Rb decays to ^{87}Sr.

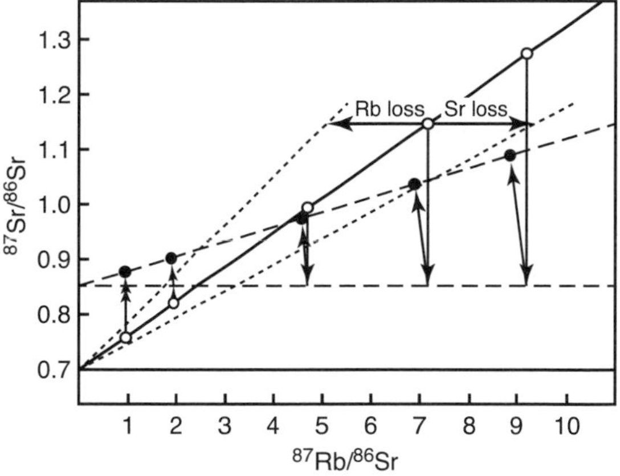

Figure 9.6 Rubidium–strontium evolution diagram illustrating what happens when an isochron is disturbed, perhaps by a later thermal event. If the thermal event happens in a closed system, the strontium isotopes simply re-equilibrate. Minerals with high ^{87}Sr/^{86}Sr ratios exchange with minerals with low ^{87}Sr/^{86}Sr ratios and, if the process goes to completion, a new equilibrium composition shown by the dashed horizontal line results. When the system cools, the decay products begin to accumulate again and a new, sloped isochron is generated (dashed line with solid points). If rubidium is lost or gained while the strontium isotopes remain undisturbed, the points on the original isochron (open symbols) move either left or right on the diagram and the correlation line becomes steeper or shallower. Unless the gain or loss is directly proportional to the amount of rubidium present, the correlation will be destroyed. Strontium gain or loss has a similar effect. Both kinds of disturbance can happen simultaneously. The result, if the system is not completely reset, is a general correlation that cannot be interpreted as an isochron.

recognize, but if it is only slightly disturbed, the fit to the data may look fairly good, but the errorchron will not give an accurate age.

Ideally, one would like to construct the isochron based on distinct mineral phases. In practice, it is not always possible to get clean mineral separates for analysis. However, if the rock meets the conditions necessary for a valid age, an isochron can be generated from impure mineral separates or even from different fragments of the rock that have different proportions of the constituent minerals. The different mixtures of minerals produce different $^{87}Rb/^{86}Sr$ ratios and provide the spread needed to generate the isochron. There is a danger in using this type of sample. Suppose a rock consists of two phases that did not form in a single event, such as plagioclase rich in strontium cut by a vein of a rubidium-rich orthoclase. Several samples of this rock containing different mixtures of the two phases could be measured, resulting in a series of $^{87}Rb/^{86}Sr$ and $^{87}Sr/^{86}Sr$ ratios that form a linear array on a $^{87}Rb–^{87}Sr$ evolution diagram. However, in this case, the array is not an isochron but a mixing line between the two phases. Figure 9.7 shows two examples of such arrays, along with the hypothetical mineral phases (end members of the mixing lines) from which they were generated. The data fall solidly on the linear arrays, so there is no way to determine from the fit of the data to the line that these are not isochrons. If the slope or intercept of the array were unreasonable, such as in the case of array C–D on Figure 9.7, one could tell, but array A–B could easily be mistaken for an isochron. Although such cases are rare, one must be aware of this possibility.

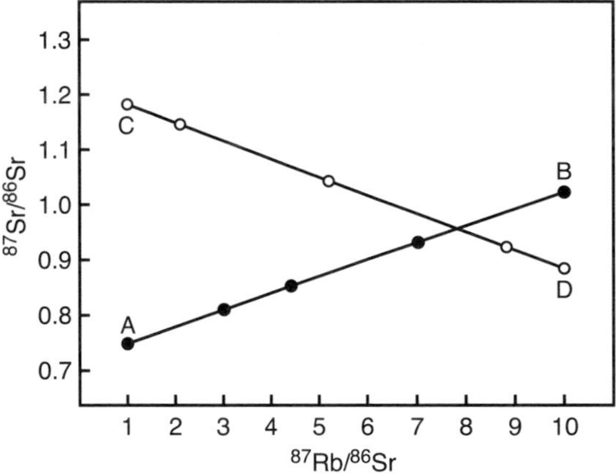

Figure 9.7 Two mixing lines on a rubidium–strontium evolution diagram. On any three-isotope diagram with a common denominator (or numerator), mixtures of two components will fall on straight lines connecting the two components. Different mixtures of components A and B produce a line with a positive slope that could be mistaken for an isochron. Different mixtures of C and D produce a line with a negative slope that is easy to recognize as a mixing line. If supposed internal isochrons are based only on different whole-rock samples, one must be aware of the possibility that a correlation on the evolution diagram could be a mixing line.

9.4.2.3 *Applications*

A wide variety of rock types can be dated by the $^{87}Rb–^{87}Sr$ system, provided that the samples satisfy the assumptions that the system was initially isotopically homogeneous (i.e. had a uniform $^{87}Sr/^{86}Sr$ ratio) and did not gain or lose rubidium or strontium after it formed. Both the slope and the intercept of the $^{87}Rb–^{87}Sr$ isochron give chronological information. If the conditions for radiometric dating are met, the slope of the array gives the age before present when an object formed. Differences in the intercept values among objects that formed from the same material reservoir also provide information about relative time differences.

Chronology Using the Slope of an Isochron

Two types of isochrons can be constructed. An *internal isochron* is one constructed from measurements of different cogenetic minerals in the same rock. The internal isochron gives information about the formation time of that rock. An isochron can also be constructed from bulk samples of different rocks thought to have formed from the same source material at the same time. Such isochrons are known as *whole-rock isochrons*. Early attempts to date chondrites by the $^{87}Rb–^{87}Sr$ method were carried out by constructing whole-rock isochrons, first for meteorites of all classes, and later for meteorites from single classes. These isochrons demonstrated that chondrites formed at ~4.5 Ga (e.g. Minster and Allègre, 1981).

Internal $^{87}Rb–^{87}Sr$ isochrons were particularly useful in unraveling the history of the Moon and Mars. As an example, consider the mare basalts collected by Apollo 11. Lunar basalts consist primarily of calcium-rich (anorthitic) plagioclase and low-calcium pyroxene, along with ilmenite, crystobalite, and other minor phases. Strontium is concentrated in the calcium site in plagioclase and is excluded from most other phases, resulting in a variation of a factor of several hundred in the strontium abundance among minerals. On the other hand, rubidium is present in most minerals and the abundance varies by only a factor of about ten. As a result, the $^{87}Rb/^{86}Sr$ ratio in the samples can vary by up to a factor of several hundred among the mineral phases. This large spread and the high-precision measurements of the $^{87}Sr/^{86}Sr$ ratio using thermal-ionization mass spectrometry (TIMS) permit precise age determinations. Figure 9.8 shows internal isochrons for two different Apollo 11 rocks, one

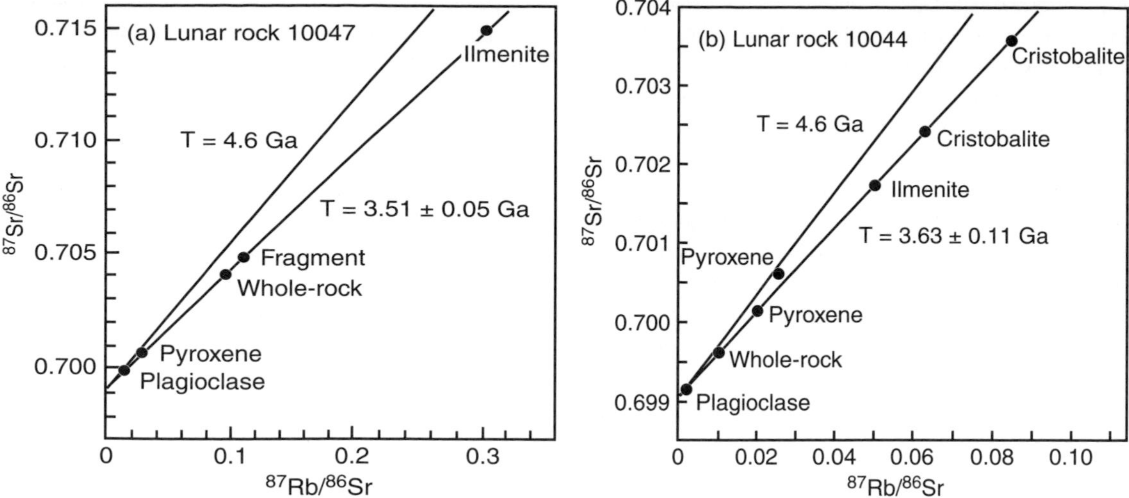

Figure 9.8 Rubidium–strontium evolution diagrams for two different Apollo 11 rocks. (a) Rock 10047 has high potassium and rubidium and the measured $^{87}Sr/^{86}Sr$ ratios of the rubidium-rich phases, pyroxene and ilmenite, are quite high, with the result that the isochron, produced from plagioclase, pyroxene, ilmenite, and two whole-rock fragments, has a precisely determined slope. (b) Rock 10044 is a low-potassium rock with lower rubidium content. The $^{87}Rb/^{86}Sr$ ratios are lower, as are the measured $^{87}Sr/^{86}Sr$ ratios. However, because the data are very precise, the isochron is still well determined. After Papanastassiou et al. (1970).

with high total potassium content and one with low total potassium. Mineral separates were obtained by a combination of handpicking, density separation, and magnetic separation. Note how small the measurement errors are compared to the spread in the isotope ratios among the samples and the precision of the resulting dates.

Chronology Using the Initial $^{87}Sr/^{86}Sr$ Ratio

The intercept of a ^{87}Rb–^{87}Sr isochron gives the initial $^{87}Sr/^{86}Sr$ ratio of the sample when it crystallized. The intercepts of isochrons will be different in objects that formed from the same reservoir at different times due to the continuing decay of ^{87}Rb to ^{87}Sr within the reservoir. As time passes, each new object that forms will have a slightly higher initial $^{87}Sr/^{86}Sr$ ratio, $(^{87}Sr/^{86}Sr)_0$. In order to use the initial value for chronology, it is necessary to determine it very precisely. The best samples for this work are those with very low Rb/Sr ratios, so that the $^{87}Sr/^{86}Sr$ ratio will not have evolved very much since the object formed. As an example, let us consider the $(^{87}Sr/^{86}Sr)_0$ ratio for eucrites, basaltic achondrites commonly thought to be samples from the asteroid Vesta. Eucrites have low Rb/Sr ratios and the resulting $^{87}Sr/^{86}Sr$ ratios have a total spread of only ~0.2%. Papanastassiou and Wasserburg (1969) measured the $^{87}Rb/^{86}Sr$ ratios and made high-precision measurements of the $^{87}Sr/^{86}Sr$ ratios in several of these meteorites. Their

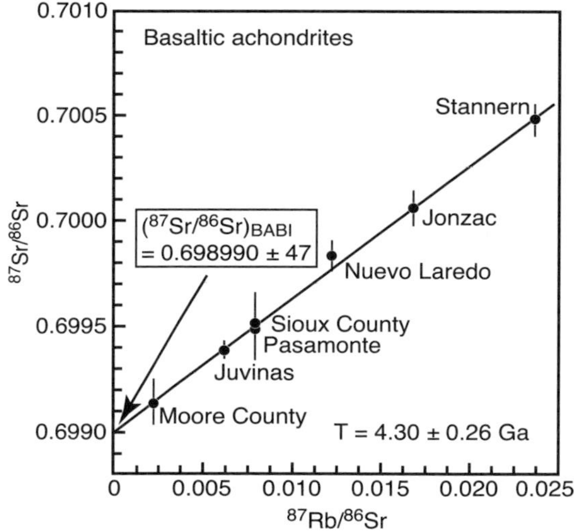

Figure 9.9 Rubidium–strontium evolution diagram for basaltic achondrites. A least-squared regression through the data gives an isochron age. The intercept of the isochron gives the $(^{87}Sr/^{86}Sr)_0$ ratio at the time the parent body of the basaltic achondrites differentiated. After Papanastassiou and Wasserburg (1969).

data are plotted on a ^{87}Rb–^{87}Sr evolution diagram (Fig. 9.9). The best-fit line through the data has a slope corresponding to an age of $(4.30 \pm 0.26) \times 10^9$ years (recalculated to $\lambda = 1.42 \times 10^{-11}$ yr^{-1}). The initial ratio indicated by the isochron is 0.698990 ± 0.000047.

The isochron age has an uncertainty of ~6%, but the initial ratio is determined to better than 0.01%. Papanastassiou and Wasserburg (1969) also calculated the initial ratios for each individual eucrite from the measured $^{87}Rb/^{86}Sr$ and $^{87}Sr/^{86}Sr$ ratios by assuming that the formation age for all meteorites was 4.5 Ga. If the $^{87}Rb/^{86}Sr$ ratios are low enough, the $^{87}Sr/^{86}Sr$ ratio evolved so little over 4.5 Gyr that the estimated initial ratio is not sensitive to the exact choice of the formation age of the meteorites. The two precisely measured meteorites with the lowest $^{87}Rb/^{86}Sr$ ratios (Juvinas and Sioux County), and therefore the ones with the least radiogenic ^{87}Sr, had the smallest uncertainties. The average of the initial ratios for these two meteorites gave $(^{87}Sr/^{86}Sr)_0 = 0.698976 \pm 0.000055$, which is essentially identical to the intercept of the isochron. Papanastassiou and Wasserburg (1969) defined the Basaltic Achondrite Best Initial value (BABI, Table 9.3) as an estimate of the $(^{87}Sr/^{86}Sr)_0$ ratio in the early solar system.

It was soon discovered that other solar system objects are older and have more primitive strontium than the basaltic achondrites. Table 9.3 compares the $(^{87}Sr/^{86}Sr)_0$ value determined for BABI with those from some other important samples from the early solar system. If Allende, Angra dos Reis, and the basaltic achondrites formed directly from the solar nebula, then the time intervals shown in the right-hand column of Table 9.3 are valid. Note that the uncertainties on these time intervals – a few million years – are much smaller than the uncertainty in the eucrite isochron shown in Figure 9.9 (~260 Myr). However, the validity of the time intervals determined from the initial ratios depends completely on the validity of the idea that they all formed directly from the bulk material of the solar nebula.

The $(^{87}Sr/^{86}Sr)_0$ ratio of a system depends not only on time but also on the details of the history of the system from which the final object formed. Figure 9.10 illustrates this point. Suppose that a batch of material evolves as a closed system for some period of time, but then experiences a fractionation event that produces a new material. An example might be differentiation of an asteroid or planet into a core, mantle, and crust. In our example, we will assume that the mantle has a lower $^{87}Rb/^{86}Sr$ ratio than the material from which it formed, so the $^{87}Sr/^{86}Sr$ ratio evolves more slowly, along the solid line with a shallower slope from F_1 to F_2 in Figure 9.10. The $(^{87}Sr/^{86}Sr)_{T1}$ ratio of the mantle reflects the time (T_1) that it separated from the original material. Now suppose that at a later time (T_2), the mantle generates a magma that crystallizes into several minerals. Each of the minerals has a different $^{87}Rb/^{86}Sr$ ratio and thus evolves along a line with its own slope. If the system remains closed until today, an isochron can be generated from these several minerals (inset in Fig. 9.10), and the time of crystallization can be determined from the isochron. The initial ratio obtained from this isochron reflects the evolution of the $^{87}Rb–^{87}Sr$ system in the mantle of the original body, but not the original body as a whole. The original body as a whole continued to evolve along the original trajectory, shown in Figure 9.10 as a dotted line. With appropriate knowledge of the $^{87}Rb/^{86}Sr$ ratio of the original body, a two-stage model can be constructed that accurately reflects the history of the system. However, if interpreted with a single-stage model, the history inferred from the initial ratio of the isochron will not be correct. A more complicated history that produces a magma with the $(^{87}Sr/^{86}Sr)_0$ shown schematically by $(^{87}Sr/^{86}Sr)_{T2}$ in Figure 9.10 could also be constructed. Reservoir evolution models typically are not unique and additional

Table 9.3 Initial strontium ($(^{87}Sr/^{86}Sr)_0$) ratios for solar system samples

Sample	$(^{87}Sr/^{86}Sr)_0$	$(^{87}Sr/^{86}Sr)_0$ Adjusted to NBS 987[1]	ΔT If Formed from Chondritic Reservoir (years)
Allende (ALL)[2]	0.69877 ± 0.00002	0.69888 ± 0.00002	0
Angra dos Reis (ADOR)[3]	0.69884 ± 0.00004	0.69895 ± 0.00004	$(4.5 \pm 2.8) \times 10^6$
Basaltic achondrites (BABI)[4]	0.69898 ± 0.00003	0.69909 ± 0.00003	$(13.5 \pm 2.3) \times 10^6$

[1] The numerical value of the NBS 987 standard was increased, so all numbers reported relative to that standard must also be increased to compare with modern values.

[2] Gray et al. (1973).

[3] Papanastassiou et al. (1970).

[4] Papanastassiou and Wasserburg (1969).

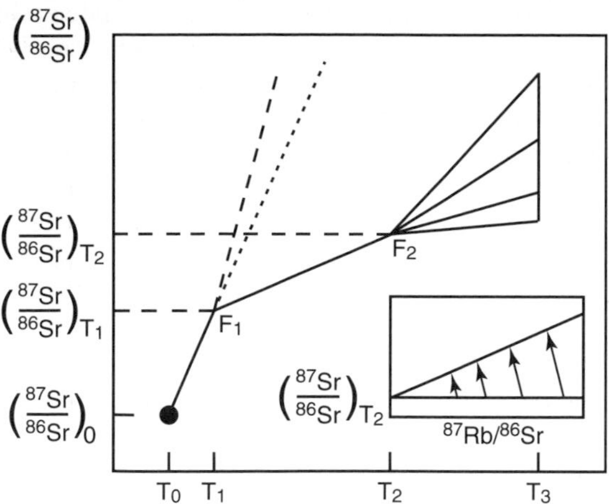

Figure 9.10 Schematic diagram showing a hypothetical multistage fractionation of rubidium and strontium and the subsequent isotopic evolution of strontium. At T_O, a system with fixed $^{87}Rb/^{86}Sr$ and $^{87}Sr/^{86}Sr$ ratios (black circle) begins to evolve. At time T_1, when the system has reached F_1, a fractionation event occurs which extracts material with a lower $^{87}Rb/^{86}Sr$ ratio. This new material evolves toward point F_2, while the original reservoir, which now has a slightly higher $^{87}Rb/^{87}Sr$ ratio evolves along the dashed line. The dotted line shows the evolution of the body as a whole. When the material reaches F2 at time T_2, a second fractionation event occurs that generates four new materials (minerals perhaps) with different $^{87}Rb/^{86}Sr$ ratios. Each mineral evolves along its own trend, and if the system remains closed, the four new minerals will generate an isochron (inset).

information is often required to choose the correct model. We will discuss reservoir evolution in more detail below.

9.4.3 The ^{147}Sm–^{143}Nd System

Samarium has seven naturally occurring isotopes, two of which, ^{147}Sm and ^{148}Sm, are radioactive (Table 4.2). The α decay of long-lived ^{147}Sm to ^{143}Nd ($t_{1/2} = 1.06 \times 10^{11}$ years; $\lambda_\alpha = 6.54 \times 10^{-12}$ yr^{-1}) is a widely used chronometer. In contrast, the half-life of ^{148}Sm, which α decays to ^{144}Nd with a half-life of ~7×10^{15} years, is too long for it to be useful for chronology. The now-extinct nuclide ^{146}Sm, which has a half-life of 1.03×10^8 years ($\lambda_\alpha = 6.71 \times 10^{-9}$ yr^{-1}), can be useful as a chronometer for the first few hundred million years of solar system history (see below). It also is the daughter of short-lived ^{150}Gd, which α decays to ^{146}Sm with a half-life of 1.8×10^6 years.

All three of the radioactive samarium isotopes decay to isotopes of neodymium. Neodymium has seven isotopes, one of which, ^{144}Nd, is slightly radioactive, with a half-life of 2.4×10^{15} years – too long to be of use for chronology. Both samarium and neodymium are rare earth elements (REEs) and are not fractionated from each other very much by melting and crystallization. They are both also refractory lithophile elements and so are not fractionated much by evaporation and condensation processes or by metal–silicate fractionation in the solar system. In igneous systems, neodymium is preferentially concentrated in the melt during partial melting because the slightly larger ionic radius of Nd^{3+} makes it more incompatible than Sm^{3+}. Thus, basaltic magmas have lower Sm/Nd ratios than the source rocks from which they form, and on Earth, crustal rocks such as granite have lower Sm/Nd ratios than basalts. But overall, the range in Sm/Nd ratios in igneous systems is only a factor of a few.

The ^{147}Sm–^{143}Nd decay scheme has been useful in both cosmochemistry and geochemistry in two different ways. The ^{147}Sm–^{143}Nd system can give an isochron, from which both a date and the initial $^{143}Nd/^{144}Nd$ ratio can be obtained. Because most of the materials of interest to cosmochemistry are old, the slow decay rate and relatively minimal fractionation of samarium from neodymium are overcome by long decay times. Very precise isotopic measurements also facilitate dating by the ^{147}Sm–^{143}Nd method. The ^{147}Sm–^{143}Nd system is also good for dating major fractionation events, such as the formation of crust and mantle. In this application, the $^{143}Nd/^{144}Nd$ ratio of the bulk system can be used to extract the time of differentiation. The geochemical behavior of the ^{147}Sm–^{143}Nd system is opposite to that of the ^{87}Rb–^{87}Sr system during partial melting. In the ^{147}Sm–^{143}Nd system, the radioactive isotope is left behind in the residual solids, whereas in the ^{87}Rb–^{87}Sr system, the radioactive isotope is concentrated in the melt phase. This difference means that the combination of the ^{147}Sm–^{143}Nd system and the ^{87}Rb–^{87}Sr systems is very powerful in investigating the petrogenesis of igneous rocks. Because it is difficult to fractionate samarium from neodymium under a variety of conditions, the ^{147}Sm–^{143}Nd system can provide chronological information in situations where the ^{87}Rb–^{87}Sr or U–Th–Pb systems have been compromised by later events.

9.4.3.1 *History*

The ^{147}Sm–^{143}Nd system was first used in cosmochemistry by Lugmair et al. (1975a), who published a precise isochron age for an Apollo 17 basalt (Fig. 9.11). The system developed rapidly thereafter. Note that the spread in $^{147}Sm/^{144}Nd$ in the mineral phases of this basalt is less than 50%, not factors of ten, several hundred, or even a few thousand, as can be seen in other systems. The fact that precise isochron ages can still be obtained is a testament to the care taken in chemical separation of samarium

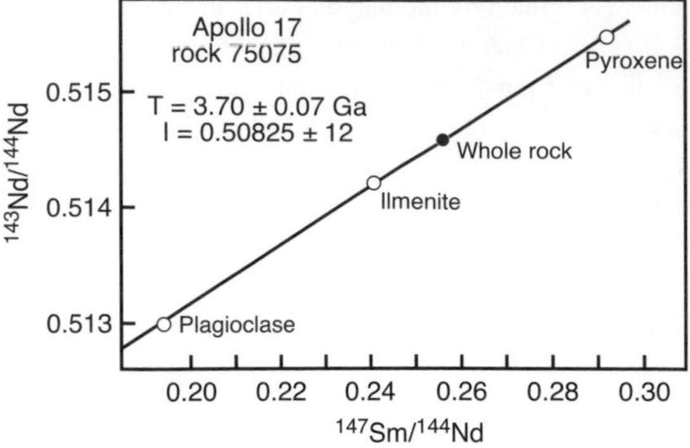

Figure 9.11 Samarium–neodymium evolution diagram for lunar mare basalt 75075. Data points for the total rock, plagioclase, ilmenite, and pyroxene form a precise linear array, the slope of which gives a crystallization age of 3.70 ± 0.07 Ga for this rock ($\lambda = 6.54 \times 10^{-12}$ yr^{-1}). After Lugmair et al. (1975a).

and neodymium and the precision of modern mass-spectrometric analysis.

Lugmair et al. (1975a) also showed that the initial ^{143}Nd/^{144}Nd ratio inferred from their isochron was measurably different from that expected from the closed-system evolution of a "chondritic" composition. DePaolo and Wasserburg (1976a, 1976b) expanded on this observation to create a methodology and reporting notation in which the timing of events that fractionated samarium from neodymium can be estimated from differences between the measured ^{143}Nd/^{144}Nd ratios and the ratio expected from the evolution of a "chondritic uniform reservoir" (CHUR). Using this system, the ratios measured at the present time are compared to the inferred composition of the CHUR today, in contrast to the ^{87}Rb–^{87}Sr system, where the initial ratio from the isochron is compared with an initial ratio inferred for the early solar system (e.g. BABI or ALL).

9.4.3.2 *Technical Details*

Several samarium and neodymium isotopes are isobars (Table 4.2) and cannot be separated by mass spectrometry. Thus, samarium and neodymium must be completely separated by chemical procedures prior to measuring them in a mass spectrometer. The ion-exchange chemistry necessary to separate these elements is now well developed (see Appendix). Isotope measurements can be done by either TIMS or ICPMS. The elemental ratios can be determined by isotope dilution using either TIMS or ICPMS (see Appendix).

Isochrons

In suitable systems, it is possible to obtain relatively precise internal isochrons from the ^{147}Sm–^{143}Nd system. In an igneous system, the REEs partition differently among the crystallizing minerals. In olivine and orthopyroxene, the light REEs are depleted relative to the heavy REEs, resulting in an enrichment of the parent samarium over the daughter neodymium. In plagioclase, the light REEs are enriched. This partitioning provides the fractionation of parent and daughter necessary to produce an isochron (Fig. 9.11). As with other isochron systems, the date is calculated from the slope of the isochron.

Reservoir Evolution

Figure 9.12 is a schematic diagram of the evolution of the ^{143}Nd/^{144}Nd ratio in the bulk solar system and in two planetary objects. The ^{143}Nd/^{144}Nd ratio of the solar system increases with time due to the decay of ^{147}Sm, and the increase can be numerically modeled if one knows the ^{147}Sm/^{144}Nd ratio in the solar system. In the original theoretical framework, the first solids began to condense from the hot solar nebula at ~4.57 Ga. But the starting time can also represent melting or local evaporation and condensation of the mixed gas and dust inherited from the molecular cloud. The exact time of formation of the first solids in the solar system is assumed because it cannot be determined from the ^{147}Sm–^{143}Nd system. Due to variations in the relative abundances of samarium and neodymium caused by thermal processing of nebular material, planetary bodies could accrete with different Sm/Nd ratios. The evolution of two such bodies is illustrated in Figure 9.12a. Planet A formed at time T_A with a lower Sm/Nd ratio than the bulk solar system, so its bulk ^{143}Nd/^{144}Nd ratio evolved more slowly with time than the bulk solar system. Planet B formed at time T_B with a higher Sm/Nd ratio, and thus its ^{143}Nd/^{144}Nd ratio evolved more quickly.

In the early solar system, most planetary bodies may have formed in only a few million years. The ^{147}Sm–^{143}Nd system is not particularly sensitive to short time intervals at 4.57 Ga because of the slow decay ratio of ^{147}Sm. In addition, because the fractionation of samarium from neodymium tends to be relatively small, the slopes of the lines on plots like Figure 9.12 tend to be similar. As a result, the precision required to resolve differences in formation time based on the intercepts of the isochrons is beyond current experimental capabilities. Also, in contrast to the situations for the ^{87}Rb–^{87}Sr and U–Th–Pb systems, for which large fractionations of parent and daughter elements have provided samples of strontium and lead with essentially primordial compositions, there are no examples of objects without significant samarium that can give a "direct" measurement of the (^{143}Nd/^{144}Nd)$_0$.

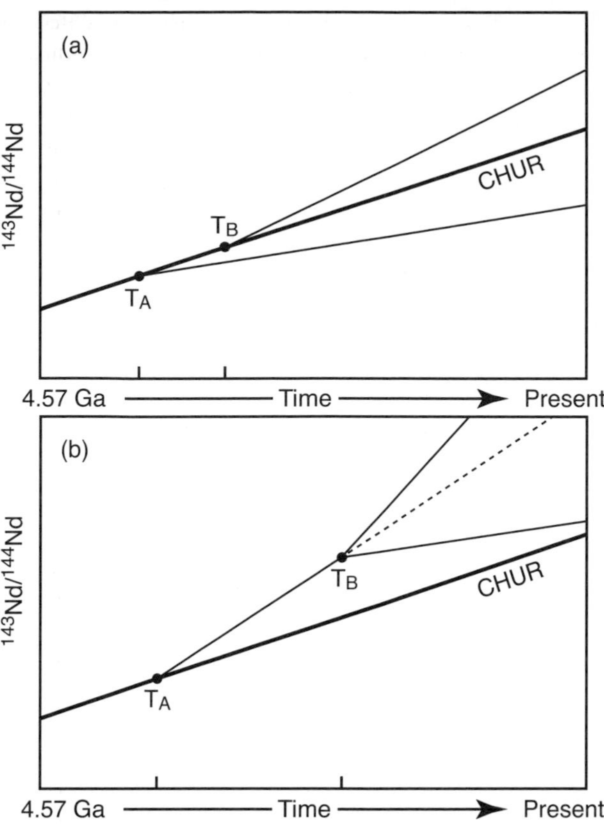

Figure 9.12 (a) Schematic diagram of $^{143}Nd/^{144}Nd$ versus time for the bulk solar system (CHUR: dark heavy line) and for two hypothetical planets formed from solar system material. When planet A formed at time T_A, it acquired a lower $^{147}Sm/^{144}Nd$ ratio than the parent reservoir, so from that point onward, the $^{143}Nd/^{144}Nd$ ratio evolved more slowly than the bulk solar system. Planet B formed slightly later with a higher $^{147}Sm/^{144}Nd$ ratio, and so its $^{143}Nd/^{144}Nd$ ratio evolved faster than the bulk solar system. (b) Schematic diagram of $^{143}Nd/^{144}Nd$ versus time for two-stage evolution of material separated from the CHUR reservoir. At time T_A, a planet forms with a higher $^{147}Sm/^{144}Nd$ ratio than the bulk solar system and subsequently evolves along a steeper trend. Later, at time T_B, the planet differentiates into two layers with different $^{147}Sm/^{144}Nd$ ratios. Each layer evolves from that point on with a different trend. It is generally not possible to determine from the samarium–neodymium data alone whether a sample is the result of a single- or multistage history. But in the example in panel (b), the planetary layer with the lowest $^{147}Sm/^{144}Nd$ ratio cannot be the result of a single-stage evolution because the evolution line intersects the CHUR line in the future.

DePaulo and Wasserburg (1976a, 1976b) showed how to get around these problems. Rather than attempting to determine the intercept of an isochron at extremely high precision and to compare it to a poorly known initial value for the solar system, they instead chose to compare the *present* bulk $^{143}Nd/^{144}Nd$ ratio with that expected for a uniform reservoir of chondritic composition (CHUR). In other words, they worked with the values that plot on the right axis of Figure 9.12, where the differences are the greatest. With sufficiently precise measurements of $^{143}Nd/^{144}Nd$ and $^{147}Sm/^{144}Nd$ for the sample, and well-known parameters for the CHUR, one can calculate the evolution of the $^{143}Nd/^{144}Nd$ ratio as a function of time for both and find out when the sample and the CHUR had the same composition. With current levels of analytical precision, the $^{143}Nd/^{144}Nd$ ratio can be measured with an accuracy of ~50 ppm and the $^{147}Sm/^{144}Nd$ ratio can be measured to ~0.05%. This translates into an uncertainty of ~10 ppm in the $^{143}Nd/^{144}Nd$ ratio at 4.57 Ga. In contrast, the uncertainty in the intercept of an isochron obtained with data of similar precision is on the order of 200 ppm.

Table 9.4 Accepted values for CHUR

	Present	4.56 Ga
$^{143}Nd/^{144}Nd$	0.512638	0.5066866
$^{147}Sm/^{144}Nd$	0.1966	0.20255

Jacobsen and Wasserburg (1980), Bouvier et al. (2008).

A critical input value necessary to use the bulk $^{143}Nd/^{144}Nd$ ratio to determine the time of Sm/Nd fractionation is the present-day composition of the CHUR. Samarium and neodymium are both refractory and are not expected to have fractionated very much in the accretion disk, so bulk chondrites are expected to give similar $^{147}Sm/^{144}Nd$ and $^{143}Nd/^{144}Nd$ ratios. Jacobsen and Wasserburg (1980) found that the $^{147}Sm/^{144}Nd$ ratios had a range of ~4% and the $^{143}Nd/^{144}Nd$ ratios exhibited a range of 5.3 parts per 10,000 among chondrites. The CHUR values (Table 9.4) have been refined only slightly since the original set of values proposed by Jacobsen and Wasserburg (1980).

The $^{143}Nd/^{144}Nd$ ratio of a sample is typically reported as

$$\varepsilon_j^0 \left[\frac{\left(^{143}\text{Nd}/_{144}\text{Nd}\right)^0_{meas}}{\left(^{143}\text{Nd}/_{144}\text{Nd}\right)^0_{\text{CHUR}}} - 1 \right] \times 10^4 \qquad 9.23$$

This "epsilon notation" gives the difference between the measured ratio in the sample and the CHUR value for the present time in parts per 10,000. The difference in the $^{147}Sm/^{144}Nd$ ratio relative to the CHUR (the elemental fractionation) is given by

$$f_j^{\text{Sm/Nd}} \frac{\left(^{147}\text{Sm}/_{144}\text{Nd}\right)_j}{\left(^{147}\text{Sm}/_{144}\text{Nd}\right)^0_{\text{CHUR}}} - 1 \qquad 9.24$$

To interpret these numbers, consider again the line for the bulk solar system (CHUR) in Figure 9.12. The $^{143}Nd/^{144}Nd$ ratio at any time, T, in the past can be calculated as follows:

$$\left(\frac{^{143}\text{Nd}}{^{144}\text{Nd}}\right)^T_{\text{CHUR}} = \left(\frac{^{143}\text{Nd}}{^{144}\text{Nd}}\right)^0_{\text{CHUR}} - \left(\frac{^{147}\text{Sm}}{^{144}\text{Nd}}\right)^0_{\text{CHUR}} \left[e^{\lambda T} - 1\right] \qquad 9.25$$

where the time, 0, refers to the present time. The ratio for 4.56 Ga is shown in Table 9.4. The same equation is used to calculate the evolution of a sample, with the values $^{143}Nd/^{144}Nd$ and $^{147}Sm/^{144}Nd$ obtained from the sample inserted in place of the CHUR values. The value of epsilon can be calculated for any specified time from Equation 9.23 by inserting the ratios for the sample and the CHUR calculated for that time using Equation 9.25, although care must be taken to specify the time if it is other than the present.

One can determine the time at which the parent reservoir for a sample was separated from the CHUR by setting Equation 9.25 for the CHUR equal to Equation 9.25 for the sample and solving for T:

$$T_{model} = \frac{1}{\lambda} \ln \left[1 + \frac{\left(\frac{^{143}\text{Nd}}{^{144}\text{Nd}}\right)^0_{Sample} - \left(\frac{^{143}\text{Nd}}{^{144}\text{Nd}}\right)^0_{\text{CHUR}}}{\left(\frac{^{147}\text{Sm}}{^{144}\text{Nd}}\right)^0_{Sample} - \left(\frac{^{147}Sm}{^{144}\text{Nd}}\right)^0_{\text{CHUR}}} \right] \qquad 9.26$$

The time, T_{model}, is a model age that assumes a single-stage evolution of the sample once it was separated from the CHUR. On Figure 9.12a, the time T_{model} is the time T_A for planet A, and the time T_B for planet B.

More complicated histories can be envisioned and modeled. Figure 9.12b shows a two-stage history. In this example, a planet separates from the CHUR at time T_A and its composition evolves along a steeper trend. At some later time, this body differentiates into a mantle and a crust, with the crust having a lower Sm/Nd ratio and the mantle a higher Sm/Nd ratio than the bulk planet. The crust and the mantle then each evolve along their own trends. Using the present-day $^{147}Sm/^{144}Nd$ and $^{143}Nd/^{144}Nd$ ratios for crust and mantle, one can calculate the time when all the layers had the same $^{143}Nd/^{144}Nd$ ratio. This is the time of the differentiation. If the bulk $^{147}Sm/^{144}Nd$ ratio of the planet can be reconstructed from the data on crust and mantle, the evolution of the bulk planet (the first stage of the two-stage evolution) can be calculated. But if a single-stage model is used to evaluate measurements of either the crust or the mantle, the model time of separation from the CHUR will be incorrect. Multistage models are not unique and are often difficult to constrain with $^{147}Sm–^{143}Nd$ data alone. However, in combination with other information, the $^{147}Sm–^{143}Nd$ system is a powerful tool for unraveling the differentiation history of a planetary body.

9.4.3.3 *Applications*

The $^{147}Sm–^{143}Nd$ system is particularly well suited for dating basaltic and ultramafic igneous rocks, which cannot be dated with the $^{87}Rb–^{87}Sr$ or U–Th–Pb systems. In general, the more mafic the rock, the lower the REE concentration and the higher the $^{147}Sm/^{144}Nd$ ratio. The REEs are not easily affected by weathering or by metamorphism, so crystallization ages can be obtained even from rocks that have been metamorphosed. These features make the $^{147}Sm–^{143}Nd$ system particularly well suited to dating terrestrial Archean rocks.

Chronology with $^{147}Sm–^{143}Nd$ Isochrons

The $^{147}Sm–^{143}Nd$ system has been widely used to date samples of differentiated bodies. The initial application of the $^{147}Sm–^{143}Nd$ system in cosmochemistry was to produce an isochron age for a lunar mare basalt (Fig. 9.11). The $^{147}Sm–^{143}Nd$ system is still used for lunar samples, although care must be taken to avoid the effects of secondary, impact-produced melting. Precision has improved over the years, as shown in Figure 9.13 for a lunar norite. The internal isochron gives an age of 4.334 ± 0.037 Ga (Edmunson et al., 2009). The authors also determined a concordant $^{207}Pb–^{206}Pb$ age of 4.333 ± 0.059 Ga for this rock,

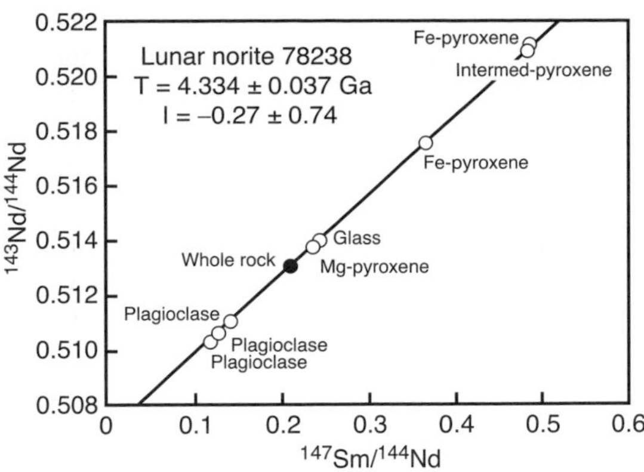

Figure 9.13 ^{147}Sm–^{143}Nd evolution diagram for nine mineral fractions and a whole-rock sample of lunar norite 78238. In spite of a small amount of scatter attributed to the inclusion of tiny fragments of other rock types in the measured sample, the isochron is very well defined. After Edmunson et al. (2009).

confirming that the ^{147}Sm–^{143}Nd system can produce precise and accurate ages for undisturbed rocks. This system has also played a major role in the dating of martian meteorites. Reliable ^{147}Sm–^{143}Nd ages of ~1.3 Ga have been obtained for the nakhlites and chassignites. These dates are concordant with dates obtained from the ^{40}Ar–^{39}Ar, ^{87}Rb–^{87}Sr, and U–Th–Pb systems (Nyquist et al., 2001b; Misawa et al., 2006). Dates for the martian shergottites are more controversial (see Chapter 10).

Reservoir Evolution

The ^{147}Sm–^{143}Nd system is actively used to investigate the differentiation and magmatic history of the Earth, the Moon, Mars, and Vesta. This system is used together with the ^{87}Rb–^{87}Sr, ^{176}Lu–^{176}Hf, U–Th–Pb, and ^{187}Re–^{187}Os systems to extract information on the timing and nature of melting and differentiation in planetary bodies. These results will be discussed in Chapters 10 and 14.

9.4.4 The U–Th–Pb System

Uranium has two long-lived isotopes, ^{235}U and ^{238}U, that decay to ^{207}Pb and ^{206}Pb, respectively. Thorium has one long-lived isotope, ^{232}Th, that decays to ^{208}Pb. The isotopic abundances of uranium and thorium are summarized in Table. 9.5. The isotopic abundances of terrestrial lead are given in Table 4.2.

The principal decay mode of ^{238}U is α decay, but a small fraction of the decays is by spontaneous fission. The emission of an α-particle initiates a series of decays known as the *uranium series* (Fig. 9.14). The uranium series can be summarized as

$$^{238}\text{U} \rightarrow {}^{206}\text{Pb} + 8\,{}^{4}\text{He} + 6\beta^- \qquad 9.27$$

The decay path splits at several points where decay can be by either α or β^- decay, but all paths end at ^{206}Pb (Fig. 9.14).

The decay of ^{235}U is also primarily by α decay through the *actinium series* (Fig. 9.15). The actinium series can be summarized as

$$^{235}\text{U} \rightarrow {}^{207}\text{Pb} + 7\,{}^{4}\text{He} + 4\beta^- \qquad 9.28$$

Similarly, the decay series initiated by the α decay of ^{232}Th (Fig. 9.16) can be summarized as

$$^{232}\text{Th} \rightarrow {}^{208}\text{Pb} + 6\,{}^{4}\text{He} + 4\beta^- \qquad 9.29$$

Although these decay series consist of 43 isotopes of 12 elements, none is a member of more than one series, so each decay chain always leads to a specific isotope of lead:

$$^{238}\text{U} \rightarrow {}^{206}\text{Pb} \quad {}^{235}\text{U} \rightarrow {}^{207}\text{Pb} \quad {}^{232}\text{Th} \rightarrow {}^{208}\text{Pb}$$

The half-lives of ^{238}U, ^{235}U, and ^{232}Th are all very much longer than those of the radioactive daughter isotopes in their decay chains. Therefore, a condition known as "secular equilibrium" is quickly established in which the decay rates of the daughter isotopes in the decay chain equal that of the parent isotope. In a closed system, once secular equilibrium is established, the rate of production of the stable daughter at the end of the chain equals the rate of decay of the parent isotope at the head of the chain.

Uranium and thorium are actinide elements. Their chemical behavior is similar under most conditions. Both are refractory elements, both occur in nature in the 4+ oxidation state, and their ionic radii are very similar (U^{4+} = 1.05 Å, Th^{4+} = 1.10 Å). However, uranium can also exist in the 6+ state as the uranyl ion (UO_2^{2+}), which forms compounds that are soluble in water. Thus, under oxidizing conditions, uranium can be separated from thorium through the action of water.

In contrast to refractory uranium and thorium, lead is a moderately volatile element. Uranium and thorium are lithophile, while lead can exhibit lithophile, siderophile, or chalcophile behavior. This means that in many cosmochemical situations, it is possible to strongly fractionate the daughter lead from parent uranium and thorium, a favorable situation for radiochronology. On the other hand, lead tends to be mobile at relatively low temperatures and can be either lost from a system or introduced at a later time. As already mentioned, uranium can also become mobile under oxidizing conditions. This means that the U–Th–Pb system is

Table 9.5 Decay constants and isotopic abundances of uranium and thorium

Isotope	Abundance	Decay Mode	Half-Life	Decay Constant
^{232}Th	100	Alpha	14.010×10^9 years	$\lambda_\alpha = 4.9475 \times 10^{-11}\ \text{yr}^{-1}$
^{235}U	0.7200	Alpha	0.7038×10^9 years	$\lambda_\alpha = 9.8485 \times 10^{-10}\ \text{yr}^{-1}$
^{238}U	99.2743	Alpha	4.468×10^9 years	$\lambda_\alpha = 1.55125 \times 10^{-10}\ \text{yr}^{-1}$

Steiger and Jäger (1977).

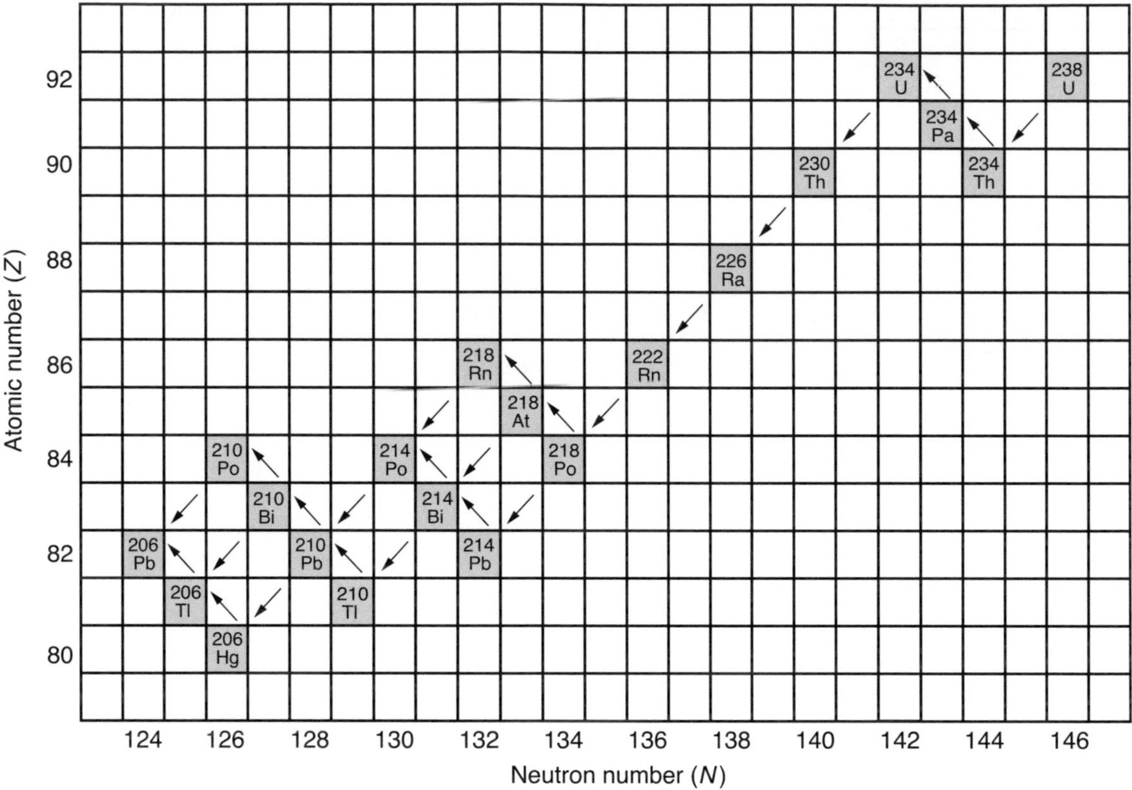

Figure 9.14 Portion of the chart of the nuclides illustrating the decay series of ^{238}U to ^{206}Pb. The decay occurs by a series of α decays, which cause the nuclide to move down and to the left, and β^- decays, which cause the nuclide to move up and to the left.

more susceptible to open-system behavior than several other commonly used dating techniques. However, as we discuss below, there are ways to recognize and account for the open-system behavior in many cases.

9.4.4.1 *History*

The first attempts to determine the ages of natural samples using uranium were chemical methods, because the nature of isotopes was not yet understood. Two methods based on uranium were initially used: the "chemical lead method" and the "helium method." Both methods assumed that the daughter element was initially absent from the rock or mineral analyzed and that the contribution of thorium to the daughter elements could be neglected. Thus, initially, these methods were applied only to uranium-rich, nearly lead-free materials. It was soon determined that most rocks and minerals slowly leak helium, so that dates determined by measuring the uranium content and the helium content of a rock were typically too young. It also turned out that the contribution from thorium was not negligible in many cases.

By the late 1920s, isotopic measurements had shown that uranium and thorium decayed to different isotopes of

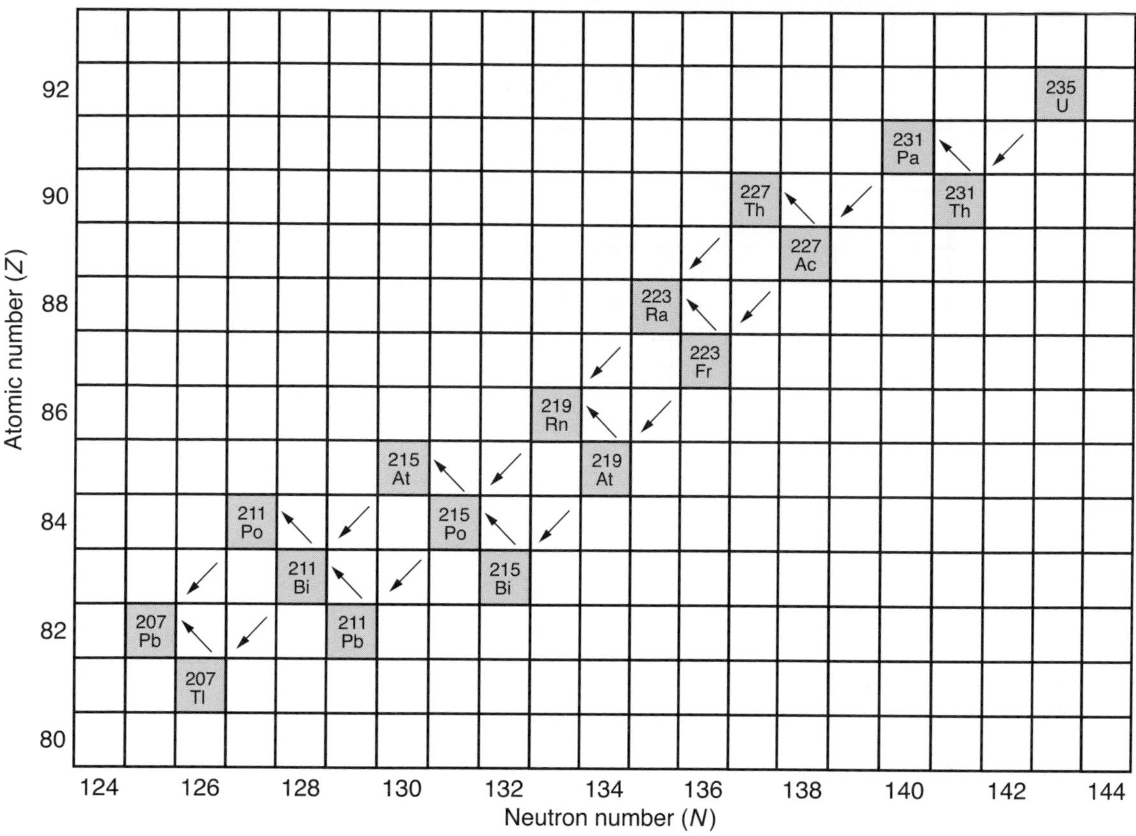

Figure 9.15 Portion of the chart of the nuclides illustrating the decay series of ^{235}U to ^{207}Pb. This decay series is often called the actinium series.

lead. A third lead isotope, ^{207}Pb, was also identified and was shown to be radiogenic. It appeared to be the product of the decay of actinium, but it was postulated that the ultimate source might be another isotope of uranium, ^{235}U. In a pair of classic papers, Nier (1939a, 1939b) placed the uranium–lead system on a much more solid footing. He determined that ^{235}U was indeed the parent of ^{207}Pb, measured the isotopic composition of lead extracted from uranium-rich minerals, and determined the decay constants for ^{235}U and ^{238}U to within 2% of their currently accepted values.

In 1946, Arthur Holmes and Fredrich Houtermans built on previous work to independently develop a general model for the isotopic evolution of lead in the Earth. The Holmes–Houtermans "common-lead" method enabled determination of the ages of common leads that have had single-stage histories, and it was used by Clair Patterson in the mid-1950s to determine the age of the Earth.

A significant problem in using the U–Pb system for age dating is that the system is relatively easily disturbed. However, George Wetherill showed that by using both the ^{238}U–^{206}Pb and ^{235}U–^{207}Pb systems together, isotopic disturbance can be recognized (Wetherill, 1956). Wetherill's concordia curve permits one to determine the formation age and the timing of an episode of lead loss in a rock (see below).

9.4.4.2 *Uranium–Thorium–Lead Methods*

Uranium, thorium, and lead comprise three independent chronometers that are described by the following equations (cf. Equation 9.21):

$$\frac{^{206}\mathrm{Pb}}{^{204}\mathrm{Pb}} = \left(\frac{^{206}\mathrm{Pb}}{^{204}\mathrm{Pb}}\right)_0 + \frac{^{238}\mathrm{U}}{^{204}\mathrm{Pb}}\left(e^{\lambda_{238}t} - 1\right) \qquad 9.30$$

$$\frac{^{207}\mathrm{Pb}}{^{204}\mathrm{Pb}} = \left(\frac{^{207}\mathrm{Pb}}{^{204}\mathrm{Pb}}\right)_0 + \frac{^{235}\mathrm{U}}{^{204}\mathrm{Pb}}\left(e^{\lambda_{235}t} - 1\right) \qquad 9.31$$

$$\frac{^{208}\mathrm{Pb}}{^{204}\mathrm{Pb}} = \left(\frac{^{208}\mathrm{Pb}}{^{204}\mathrm{Pb}}\right)_0 + \frac{^{232}\mathrm{Th}}{^{204}\mathrm{Pb}}\left(e^{\lambda_{232}t} - 1\right) \qquad 9.32$$

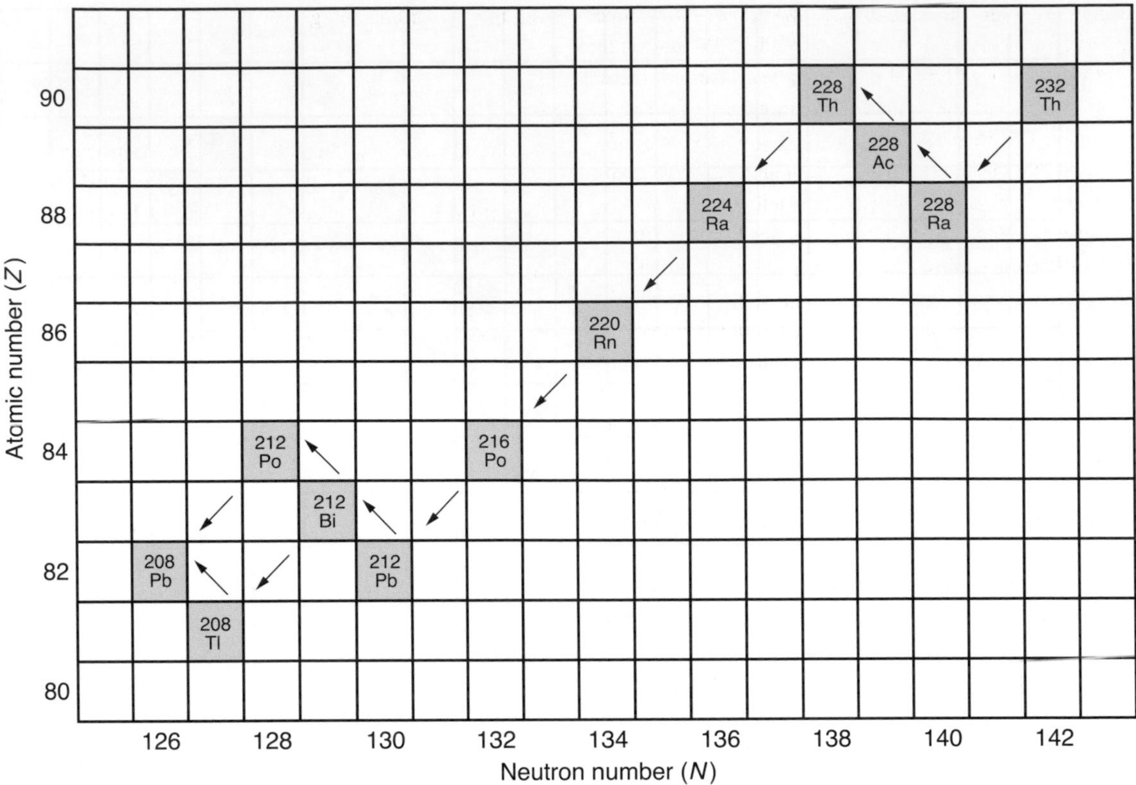

Figure 9.16 Portion of the chart of the nuclides illustrating the decay series of ^{232}Th to ^{208}Pb.

Under ideal conditions, all three chronometers should give the same age. However, lead loss is common in natural samples and can be difficult or impossible to recognize using just one of the above chronometers.

The Uranium–Lead Concordia Diagram

Wetherill (1956) showed how lead loss can be detected and even used to obtain additional information. Equations 9.30 and 9.31 can be rearranged to give the ratios of radiogenic $^{206}Pb^*$ to ^{238}U and $^{207}Pb^*$ to ^{235}U:

$$\frac{^{206}\text{Pb}^*}{^{238}\text{U}} = e^{\lambda_{238}t} - 1 \tag{9.33}$$

$$\frac{^{207}\text{Pb}^*}{^{235}\text{U}} = e^{\lambda_{235}t} - 1 \tag{9.34}$$

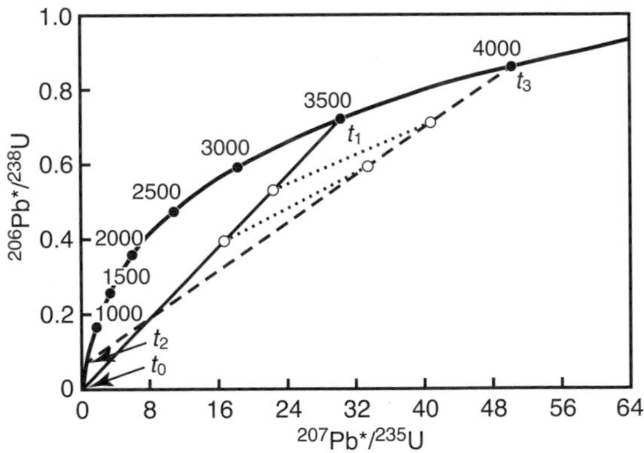

Figure 9.17 Wetherill's concordia diagram for interpreting discordant U–Pb dates caused by episodic loss of radiogenic lead.

These equations can be solved simultaneously for various times, t, generating a set of points that give concordant U–Pb dates. When these points are plotted (Fig. 9.17), they generate a curve describing all concordant dates that is known as *concordia* (Wetherill, 1956).

The concordia curve can provide a lot of information about a sample. When a rock or mineral first forms (at time t_0), it contains no radiogenic lead and will plot at the origin of Figure 9.17. As time passes, the rock will move up and to the right along the concordia curve. After 3.5 Gyr, the data will plot at the point labeled t_1, and this point will give the formation age of the rock or mineral. Now suppose that at time t_1 the system suffers the loss of some of its lead. The loss will lower both $^{206}Pb^*/^{238}U$ and $^{207}Pb^*/^{235}U$.

Because it is almost impossible to fractionate the isotopes of an element as heavy as lead, the $^{206}Pb/^{207}Pb$ ratio of the lead that is lost will equal that of the lead that is retained. Lead loss moves the system in a straight line toward the origin on Figure 9.17. This line is known as *discordia.* A group of rocks that formed together and suffered varying degrees of lead loss at the same time will spread out along the discordia line. If, following the lead-loss event, the system becomes closed again, each rock or mineral will evolve along a new line with a similar shape to concordia, but displaced on the diagram. The two points that mark the intersections of discordia with concordia will evolve along the concordia line (dashed line on Fig. 9.17). The point at the origin (labeled t_0) will move along concordia to point t_2, while the point labeled t_1 will move to point t_3. The original formation age of the rock or mineral is given by t_3. The interpretation of point t_2 depends on the rock's history. If the lead loss occurred during a single episode, then t_2 gives the time that the system closed again after that episode. But more complicated scenarios can be envisioned. Lead loss might occur in several episodes, by continuous diffusion over a long period of time, or by loss due to the radiation damage that crystals incur through the α decay of uranium, thorium, and their daughter isotopes. Models have been generated that can describe these latter two types of behavior. The geologic history of the sample may involve different formation episodes that did not fully reset the system. Zircons are a prime example: they often consist of cores that have one age, with one or more overgrowths of different ages. The ion microprobe (also known as secondary ion mass spectrometry, or SIMS) provides a means of dating each of the overgrowths independently, helping to sort out such complex systems.

Tera and Wasserburg (1972) introduced a second kind of concordia diagram to handle another complication. The U–Pb ages determined for many lunar rocks by standard methods appear to be significantly older than the $^{87}Rb–^{87}Sr$ and $^{40}Ar–^{39}Ar$ ages. The reason for the discrepancy is that many lunar rocks contain excess radiogenic lead that was not produced *in situ*. The Tera–Wasserburg concordia diagram does not require prior knowledge of the initial $^{206}Pb/^{204}Pb$ and $^{207}Pb/^{204}Pb$ ratios.

The number of atoms of ^{206}Pb and ^{207}Pb can be described as the number of atoms of each isotope present initially plus the number created *in situ* by radioactive decay. Tera and Wasserburg (1972, 1974) defined a concordia in parametric form where the x-coordinate is given by

$$\left(\frac{^{238}U}{^{206}Pb^*}\right) = \frac{1}{e^{\lambda_{238}t} - 1} \quad 9.35$$

and the y-coordinate is given by

$$\left(\frac{^{207}Pb}{^{206}Pb}\right)^* = \frac{1}{137.88}\left(\frac{e^{\lambda_{235}t} - 1}{e^{\lambda_{238}t} - 1}\right) \quad 9.36$$

The Tera–Wasserburg concordia is constructed by solving Equations 9.35 and 9.36 for various values of t. As with the Wetherill concordia (Fig. 9.17), the Tera–Wasserburg concordia curve (Fig. 9.18) is the locus of all concordant U–Pb ages, based on the decay constants in Table 9.5 and assuming that $^{235}U/^{238}U = 1/137.88$. Figure 9.18 shows a discordia line intersecting concordia at $t_1 = 3.2$ Ga and $t_2 = 0.2$ Ga. The system described by this discordia line formed originally at 3.2 Ga and contained no radiogenic lead. It then accumulated radiogenic lead in a closed system until 0.2 Ga, when the system was reset. The radiogenic lead that had accumulated in the system between time t_1 and t_2 would have an isotopic composition given by the intersection of the discordia line with the y-axis, and the numerical value is given by

$$I_0 = \frac{1}{137.88}\left(\frac{e^{\lambda_{235}t_1} - e^{\lambda_{235}t_2}}{e^{\lambda_{238}t_1} - e^{\lambda_{238}t_2}}\right) \quad 9.37$$

If a uranium-free mineral formed during the resetting event at t_2, then it would contain lead with a $^{207}Pb/^{206}Pb$ ratio given by Equation 9.37, and that lead would represent the history of the system from t_1 to t_2. If a lead-free mineral formed during the resetting event represented by t_2, then all of the lead measured in this

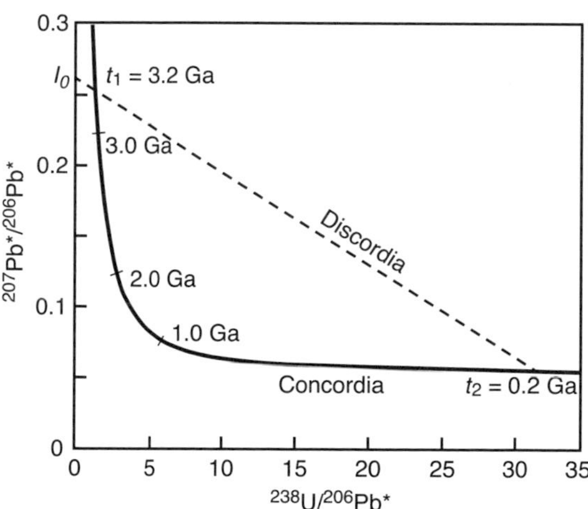

Figure 9.18 Tera–Wasserburg concordia diagram. This diagram does not require prior knowledge of the initial $^{206}Pb/^{204}Pb$ and $^{207}Pb/^{204}Pb$ ratios and permits determination of ages for samples that have inherited radiogenic lead. After Faure and Mensing (2005).

mineral would have accumulated by uranium decay after time t_2. In this case, both t_1 and t_2 give valid dates for events in the history of the system. Alternatively, the radiogenic $^{207}Pb/^{206}Pb$ given by I_0 could be the result of processes unrelated to the event at time t_2. In that case, the discordia line represents a mixing line between the inherited lead with $^{207}Pb/^{206}Pb = I_0$ and the U–Pb system given by point t_2. In this case, point t_1 has no chronological significance.

As an example of the application of the Tera–Wasserburg methodology, Figure 9.19 shows a Tera–Wasserburg concordia diagram for lunar basalt 14053. This rock gave a $^{87}Rb–^{87}Sr$ age of 3.88 ± 0.04 Ga ($\lambda = 1.42 \times 10^{-11}$ yr^{-1}) (Papanastassiou and Wasserburg, 1971) and $^{40}Ar–^{39}Ar$ ages of 3.81–3.83 Ga ($\lambda = 4.962 \times 10^{-10}$ yr^{-1}, $^{40}K/K = 1.167 \times 10^{-4}$) (Turner et al., 1971). The slope, m, of the discordia is related to the initial ($^{207}Pb^*/^{206}Pb^*$) ratio and to the age of the U–Pb system by

$$m = \frac{e^{\lambda_{235}t} - 1}{137.88} - \left(\frac{^{207}Pb}{^{206}Pb}\right)_i \left(e^{\lambda_{238}t} - 1\right) \qquad 9.38$$

The equation is solved by iteration to find the value that gives the same t in Equations 9.36 and 9.38. The time t can also be determined graphically from a plot of the slope of the concordia curve versus time. As shown in Figure 9.19, the intersection of discordia and the

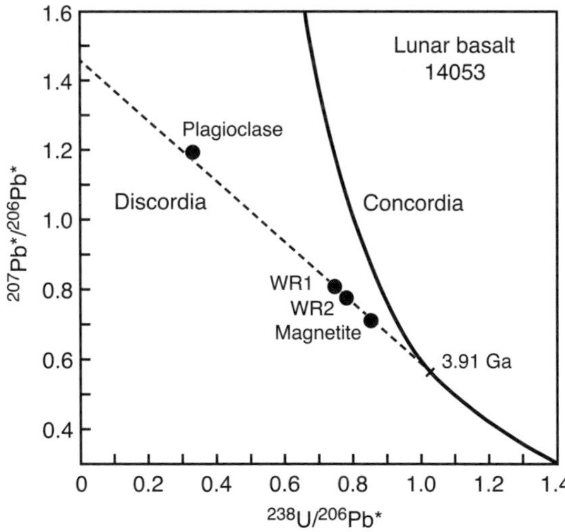

Figure 9.19 Tera–Wasserburg diagram for lunar basalt 14053. The measured data define a discordia line that intersects the y-axis at $(^{207}Pb/^{206}Pb)_i = 1.46$. The slope of the line, based on two whole-rock samples and a magnetite separate, is –0.88366. The intersection of the discordia line with the concordia curve gives the age of the sample. Data from Tera and Wasserburg (1972).

Tera–Wasserburg concordia gives an age of 3.91 Ga, quite consistent with the $^{87}Rb–^{87}Sr$ and $^{40}Ar–^{39}Ar$ ages. Using standard methods, this sample gave highly discordant U–Th–Pb dates in excess of 5 Ga.

The Wetherill concordia and the Tera–Wasserburg concordia can both be extended to include ^{232}Th. The ^{232}Th system simply takes the place of either ^{235}U or ^{238}U in the calculation of concordia. In general, this just introduces a potential additional complication because uranium can be fractionated from thorium. Fractionation of uranium from thorium would mean that the ratio of the two parent isotopes has changed and must be properly accounted for. Thus U–Th concordia treatments are rarely employed in cosmochemistry.

9.4.4.3 $^{207}Pb–^{206}Pb$ Dating

An alternative way to utilize the dual decay scheme, $^{235}U–^{207}Pb$ and $^{238}U–^{206}Pb$, is to calculate the date from the ratio of radiogenic isotopes $^{207}Pb^*$ and $^{206}Pb^*$ alone, without using the U/Pb ratio. The $^{207}Pb^*/^{206}Pb^*$ ratio is insensitive to recent fractionation between parent and daughter elements caused either by natural processes or by laboratory treatments such as chemical etching. Uranium-235 decays six times faster than ^{238}U, and only a little over 1% of the original abundance of ^{235}U in the solar system is still present. This means that the $^{207}Pb^*/^{206}Pb^*$ ratio evolved rapidly during the early history of the solar system. Thus, the precision of the date obtained for a given uncertainty in the measured $^{207}Pb^*/^{206}Pb^*$ ratio increases with the age of the sample. The current experimental uncertainties result in a precision in an absolute age determination of an early solar system object approaching those of short-lived nuclides.

The $^{207}Pb^*–^{206}Pb^*$ system can be applied in two ways. If the sample is sufficiently large and diverse that mineral separates with different U/Pb ratios can be produced, an internal isochron can be generated. The equation for the $^{207}Pb^*–^{206}Pb^*$ isochron can be obtained by combining Equations 9.30 and 9.31:

$$\frac{\frac{^{207}Pb}{^{204}Pb} - \left(\frac{^{207}Pb}{^{204}Pb}\right)_0}{\frac{^{206}Pb}{^{204}Pb} - \left(\frac{^{206}Pb}{^{204}Pb}\right)_0} = \frac{^{235}U}{^{238}U}\left[\frac{e^{\lambda_{235}t} - 1}{e^{\lambda_{238}t} - 1}\right] \qquad 9.39$$

where $^{235}U/^{238}U = 1/137.88$ (Table 9.5). If the $^{238}U–^{206}Pb$ and $^{235}U–^{207}Pb$ dates are concordant, this equation represents a family of straight lines (isochrons) with slopes of

$$m = \frac{1}{137.88}\left[\frac{e^{\lambda_{235}t} - 1}{e^{\lambda_{238}t} - 1}\right] = \left(\frac{^{207}Pb}{^{206}Pb}\right)^* \qquad 9.40$$

that pass through the point

$$\left(\frac{^{207}\text{Pb}}{^{204}\text{Pb}}\right)_0, \left(\frac{^{206}\text{Pb}}{^{204}\text{Pb}}\right)_0$$

The determination of an age using the ^{207}Pb*–^{206}Pb* isochron method relies on the assumption that all samples that define the isochron had the same initial lead isotopic composition, formed at the same time, and remained closed to uranium and lead until the very recent past.

If an internal isochron cannot be generated, a model age can be determined from the measured ^{207}Pb*/^{206}Pb* of the sample and the assumed initial lead isotopic ratios. For studies of the early solar system, this initial lead composition is assumed to be that measured in troilite (FeS) from the Canyon Diablo meteorite. Troilite is a uranium-free mineral and its host meteorite formed very early in the history of the solar system. Because the U/Pb ratio of the solar system is low, the lead incorporated into the troilite should not have evolved significantly from the initial composition in the solar system.

The most accurate ^{207}Pb*–^{206}Pb* dates on early solar system materials come from objects with very low abundances of common lead (= high U/Pb ratios). The smaller the correction for common lead, the less susceptible the calculated age is to variations in the composition of the common lead. In addition to the primordial lead, lead may also be introduced by later events in the history of the rock, including, in the case of meteorites, terrestrial weathering. Because the ^{207}Pb*–^{206}Pb* system is not affected by recent lead loss, samples are typically leached with acids to remove common lead. This leaching can also remove a significant fraction of the radiogenic lead, but this does not affect the measurement so long as the leaching does not fractionate the radiogenic lead isotopes and there is enough lead remaining to get an accurate measurement. In most cases, etching also disturbs the U/Pb ratio, eliminating the possibility of using U–Pb systematics to recognize a disturbed or multistage system.

One drawback of the ^{207}Pb*–^{206}Pb* system is that two-stage or three-stage evolution of the U–Pb isotopic system can produce linear arrays on a ^{207}Pb*–^{206}Pb* isochron plot (Gale and Mussett, 1973). If these arrays are interpreted as single-stage isochrons, incorrect dates may be obtained. In some cases, measurements of the U/Pb ratio can help identify multistage samples. But if several events occurred sufficiently early in solar system history, even a concordia approach may not be able to identify a multistage system (Tera and Carlson, 1999). Thus, to the extent possible, workers attempt to identify objects with simple histories, so that a single-stage evolution can be assumed.

There are several important assumptions that are made when using any of the methods based on the U–Th–Pb system. As the precision of the ^{207}Pb*–^{206}Pb* measurement improves, these assumptions should be re-evaluated. One assumption is that the ^{235}U/^{238}U ratio is constant in solar system materials. One place where this has been demonstrated not to be true is at Oklo, Gabon, in West Africa. At Oklo, the uranium is severely depleted in ^{235}U, apparently because two billion years ago, when the ^{235}U abundance was much higher, conditions were right to produce a natural fission reactor that burned out the ^{235}U. Fortunately, such situations are extremely rare.

Recently, it has been demonstrated that the ^{235}U/^{238}U ratio is not constant among CAIs (Brennecka et al., 2010; Connelly et al., 2012; Bollard et al., 2017). This means that ^{207}Pb*–^{206}Pb* ages measured assuming a ^{235}U/^{238}U ratio of 1/137.88 are incorrect. The variations in ^{235}U/^{238}U ratio were originally attributed to the presence of live ^{247}Cm, which decays to ^{235}U, in the early solar system (Brennecka et al., 2010). Subsequent work indicated that mass-dependent fractionation of uranium was the likely cause of the variations in ^{235}U/^{238}U ratio (Connelly et al., 2012). Another possibility, that the variations in ^{235}U/^{238}U ratio are nucleosynthetic in origin, is not supported by the observations. Although it is much more work, the ^{235}U/^{238}U ratios are measured when trying to get high-precision ^{207}Pb*–^{206}Pb* ages of CAIs and chondrules (Bollard et al., 2017). So far, work on differentiated meteorites has not shown any variability in ^{235}U/^{238}U ratio among samples (e.g. Connelly et al., 2012).

A second assumption implicit in most U–Th–Pb chronology is that the composition of primordial lead is the same as that currently measured in Canyon Diablo troilite. As measurements become more precise, more accurate corrections for the lead inherited from precursor materials must be made. The Canyon Diablo meteorite formed several million years after the earliest solar system solids. During that time, the lead isotopic composition evolved to some degree, even though the U/Pb ratio in the solar system is low. Tera and Carlson (1999) reanalyzed data for CAIs, chondrules, and matrix from the Allende meteorite and concluded that there is evidence for a more primitive composition than that found in Canyon Diablo troilite. If so, then high-precision ^{207}Pb*/^{206}Pb* dates that use Canyon Diablo troilite to correct for nonradiogenic lead would be slightly in error.

It is also possible that some samples may have inherited lead of a different composition than that inferred for the bulk solar system. Currently, the CAIs being dated have sufficiently small corrections that this effect is not significant, but at some point it may limit the ultimate accuracy of $^{207}Pb^*$–$^{206}Pb^*$ dates. A bigger problem today is that the nonradiogenic lead in most samples is a mixture of primordial lead and terrestrial lead acquired either from weathering on the Earth's surface or in the laboratory. The main method for dealing with this issue in $^{207}Pb^*$–$^{206}Pb^*$ dating is to measure samples with very radiogenic lead and to leach the sample to remove the majority of the nonradiogenic lead prior to measurement.

The absolute accuracy of $^{207}Pb^*$–$^{206}Pb^*$ dates also depends on knowing the decay constants of ^{235}U and ^{238}U very precisely. The current level of uncertainty in the decay constants puts a systematic error of ~9.3 million years on a date for an object formed 4.57 Ga ago. This is important primarily when comparing with dates determined by other long-lived chronometers. When comparing $^{207}Pb^*$–$^{206}Pb^*$ ages for early solar system objects, this uncertainty effectively cancels out.

9.4.4.4 *The Common-Lead Method*

The common-lead method looks at the isotopic evolution of lead in systems with U/Pb and Th/Pb ratios similar to or less than the ratios in bulk solar system materials. The original formulation, by Holmes and Houtermans, is a single-stage model that accounts for the isotopic composition of any sample of common lead in terms of primordial lead plus radiogenic lead produced in the source up to the time that lead was separated from uranium and thorium. Multistage models that more accurately describe the evolution of natural systems have been developed. The common-lead method is used in cosmochemistry primarily to study the time of differentiation and reservoir evolution in differentiated bodies such as the Earth, the Moon, Mars, and some asteroids. This method complements similar approaches used for the ^{147}Sm–^{143}Nd and ^{176}Lu–^{176}Hf systems.

The $^{206}Pb/^{204}Pb$ ratio of a uranium-bearing system of age T that has remained closed is (c.f. Equation 9.30)

$$\frac{^{206}\mathrm{Pb}}{^{204}\mathrm{Pb}} = \left(\frac{^{206}\mathrm{Pb}}{^{204}\mathrm{Pb}}\right)_0 + \frac{^{238}\mathrm{U}}{^{204}\mathrm{Pb}}\left(e^{\lambda_{238}T} - 1\right) \qquad 9.41$$

If lead was removed from such a system without isotope fractionation t years ago, the $^{206}Pb/^{204}Pb$ ratio of that lead is

$$\left(\frac{^{206}\mathrm{Pb}}{^{204}\mathrm{Pb}}\right)_t = \left(\frac{^{206}\mathrm{Pb}}{^{204}\mathrm{Pb}}\right)_0 + \frac{^{238}\mathrm{U}}{^{204}\mathrm{Pb}}\left(e^{\lambda_{238}T} - 1\right) - \frac{^{238}\mathrm{U}}{^{204}\mathrm{Pb}}\left(e^{\lambda_{238}t} - 1\right) \qquad 9.42$$

which can be simplified to

$$\left(\frac{^{206}\mathrm{Pb}}{^{204}\mathrm{Pb}}\right)_t = \left(\frac{^{206}\mathrm{Pb}}{^{204}\mathrm{Pb}}\right)_0 + \frac{^{238}\mathrm{U}}{^{204}\mathrm{Pb}}\left(e^{\lambda_{238}T} - e^{\lambda_{238}t}\right) \qquad 9.43$$

Analogous equations can be written for $(^{207}Pb/^{204}Pb)_t$ and $(^{208}Pb/^{204}Pb)_t$. It is useful at this point to introduce some new symbols to simplify the equations:

$$\frac{^{238}\mathrm{U}}{^{204}\mathrm{Pb}} = \mu \qquad \frac{^{235}\mathrm{U}}{^{204}\mathrm{Pb}} = \frac{\mu}{137.88} \qquad \frac{^{232}Th}{^{204}\mathrm{Pb}} = \mu\kappa$$

where $\kappa = {}^{232}Th/^{238}U$.

The resulting set of equations is

$$\left(\frac{^{206}\mathrm{Pb}}{^{204}\mathrm{Pb}}\right)_t = \left(\frac{^{206}\mathrm{Pb}}{^{204}\mathrm{Pb}}\right)_0 + \mu\left(e^{\lambda_{238}T} - e^{\lambda_{238}t}\right) \qquad 9.44$$

$$\left(\frac{^{207}\mathrm{Pb}}{^{204}\mathrm{Pb}}\right)_t = \left(\frac{^{207}\mathrm{Pb}}{^{204}\mathrm{Pb}}\right)_0 + \frac{\mu}{137.88}\left(e^{\lambda_{235}T} - e^{\lambda_{235}t}\right) \qquad 9.45$$

$$\left(\frac{^{208}\mathrm{Pb}}{^{204}\mathrm{Pb}}\right)_t = \left(\frac{^{208}\mathrm{Pb}}{^{204}\mathrm{Pb}}\right)_0 + \mu\kappa\left(e^{\lambda_{232}T} - e^{\lambda_{232}t}\right) \qquad 9.46$$

As can be seen, the isotopic shift observed in lead removed from a system at time t depends on the value of μ. Discussions of reservoir evolution are often cast in terms of the μ of the system.

By combining Equations 9.44 and 9.45, μ can be eliminated to give

$$\frac{\frac{^{207}\mathrm{Pb}}{^{204}\mathrm{Pb}} - \left(\frac{^{207}\mathrm{Pb}}{^{204}\mathrm{Pb}}\right)_0}{\frac{^{206}\mathrm{Pb}}{^{204}\mathrm{Pb}} - \left(\frac{^{206}\mathrm{Pb}}{^{204}\mathrm{Pb}}\right)_0} = \frac{1}{137.88}\left[\frac{e^{\lambda_{235}T} - e^{\lambda_{235}t}}{e^{\lambda_{238}T} - e^{\lambda_{238}t}}\right] \qquad 9.47$$

This equation gives the time, for a single-stage model, when lead was removed from the uranium-rich source.

Equation 9.47, with $t = 0$ and the composition of lead from meteoritic troilite used for the initial isotopic ratio of lead, was used by Clair Patterson (1955, 1956) to determine the age of the Earth. In the 1950s, the largest uncertainty in determining the age of the Earth was the composition of primordial lead. In 1953, Patterson solved this problem by using state-of-the-art analytical techniques to measure the composition of lead from troilite (FeS) in iron meteorites. Troilite has an extremely low U/Pb ratio because uranium was separated from the lead in troilite at near the time of solar system formation.

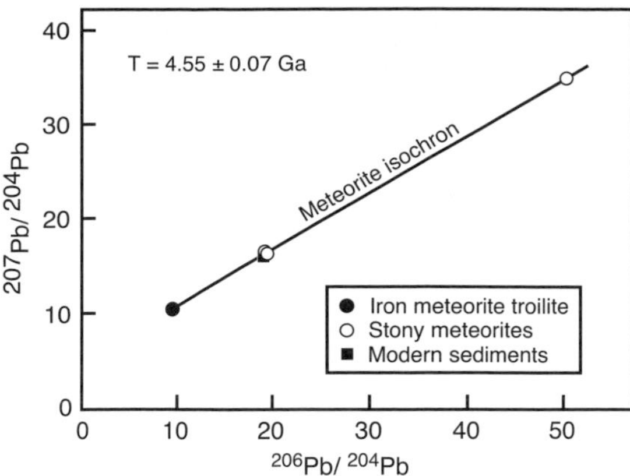

Figure 9.20 The Pb–Pb isochron used by Patterson (1956) to determine the age of the Earth. The isochron was constructed from troilite (FeS) from two iron meteorites and three bulk chondrites. Because troilite contains essentially no uranium, the lead in troilite is almost unchanged from the time the meteorite formed – it is "primordial" lead. Modern terrestrial sediments fall on the same isochron, indicating that the Earth and the meteorites are essentially the same age. The slope of the isochron gives an age of T = 4.55 ± 0.07 Ga using the decay constants used by Patterson (1956).

Patterson (1955) then measured the composition of lead from stony meteorites. In 1956, he demonstrated that the data from stony meteorites, iron meteorites, and terrestrial oceanic sediments all fell on the same isochron (Fig. 9.20). He interpreted the isochron age (4.55 ± 0.07 Ga) as the age of the Earth and of the meteorites. The value for the age of the Earth has remained essentially unchanged since Patterson's determination, although the age of the solar system has been pushed back by ~20 Myr.

9.4.5 The ^{187}Re–^{187}Os System

Rhenium has two naturally occurring isotopes (Table 4.2), one of which, ^{187}Re, β^- decays to ^{187}Os with a half-life of 4.16×10^{10} years ($\lambda_\beta = 1.666 \times 10^{-11}$ yr^{-1}). Osmium has seven naturally occurring isotopes (Table 4.2), one of which, ^{186}Os, is slightly radioactive, decaying by α decay to ^{182}W with a half-life of 2×10^{15} years. Two isotopes are daughters of radioactive isotopes. Osmium-187 is the daughter of ^{187}Re, and ^{186}Os is produced by α decay of ^{190}Pt with a half-life of 4.5×10^{11} years. The ^{190}Pt–^{186}Os system is also potentially useful in geochronology and cosmochronology. Rhenium is a member of Group 7B of the Periodic Table and has chemical properties similar to molybdenum. Osmium is a platinum group element. Rhenium and osmium are both highly siderophile and refractory, and in chondrites and iron meteorites, their abundances are highly correlated. They are extracted almost completely from planetary mantles into the metallic cores.

A direct experimental determination of the half-life of ^{187}Re was carried out by Lindner et al. (1989), giving a half-life of $(4.23 \pm 0.13) \times 10^{10}$ years ($\lambda_\beta = 1.64 \times 10^{-11}$ yr^{-1}). The half-life and decay constant have been refined based on high-precision whole-rock ^{187}Re–^{187}Os isochrons for iron meteorite groups. The current best estimate of the decay constant is $(1.666 \pm 0.017) \times 10^{-11}$ yr^{-1}, which has an uncertainty of about ± 1% (Shirey and Walker, 1998).

9.4.5.1 *History and Technical Details*

The ^{187}Re–^{187}Os method was first applied to extraterrestrial samples in the early 1960s when Hirt et al. (1963) reported a whole-rock isochron for 14 iron meteorites that gave an age of ~4 Ga. Further development of this system was hindered by several technical difficulties. Rhenium and osmium each exist in multiple oxidation states and can form a variety of chemical species, so complete digestion of the samples, which is required to chemically separate rhenium and osmium for mass spectrometry, is difficult. In addition, accurate determination of rhenium abundance and osmium isotopic composition requires spiking the samples with isotopically labeled rhenium and osmium, and equilibration of spikes and samples is challenging. A third problem is that osmium and particularly rhenium are very difficult to ionize as positive ions for mass spectrometry. These problems were only gradually overcome.

Following the work by Hirt et al. (1963), the next significant step in applying the ^{187}Re–^{187}Os method came when Shimizu et al. (1978) and Luck et al. (1980) used SIMS to measure rhenium and osmium separated from iron meteorites and metal grains from chondrites. Luck and Allègre (1983) showed that iron meteorites and metal grains fit the same isochron and therefore must have been separated from a homogeneous solar nebula within a short interval of ~90 Myr. But the poor ionization of rhenium and osmium was still a major problem. This problem was solved first when the Caltech group demonstrated a negative thermal-ionization technique (NTIMS) for rhenium and osmium, for which they obtained ionization efficiencies of 2–6% for osmium and >20% for rhenium (Creaser et al., 1991). Shortly thereafter, high-precision ICPMS also removed ionization efficiency as an issue. Reliable procedures for chemical dissolution and spike equilibration were first developed for iron meteorites and other metal phases (Shirey and Walker, 1995; Shen et al., 1996). Chondrites, which are mixtures of silicates, metal, and sulfides, proved to be more difficult, but by the late 1990s, a reliable chemical procedure had been developed for them as well (Chen et al., 1998).

9.4.5.2 *Applications*

The ^{187}Re–^{187}Os system in cosmochemistry has been primarily used to investigate the crystallization paths and ages of asteroidal cores through measurements of magmatic iron meteorites (e.g. Shen et al., 1996; Smoliar et al., 1996) and to characterize the budget of highly siderophile elements added to the terrestrial planets via late accretion (e.g. Meisel et al., 1996). This system also has some potential for investigating age relationships among different classes of metal-bearing chondrites.

Chronology with Re–Os Isochrons

The ^{187}Re–^{187}Os system does not lend itself to the determination of internal isochrons for most meteorites. Chen et al. (1998) produced an internal isochron for the St. Severin (LL6) chondrite using metal separates and whole-rock samples (Fig. 9.21). The slope of the isochron gives a date of 4.60 ± 0.15 Ga ($\lambda = 1.666\times10^{-11}$ yr^{-1}), consistent with expectations based on other chronometers, but not precise enough to improve upon other techniques.

Most of the work with the ^{187}Re–^{187}Os system in cosmochemistry has been done using "whole-rock" isochrons. For example, Shen et al. (1996) measured 16 iron meteorites from groups IIAB, IVA, IVB, IIIAB, and IAB. They found that all measured samples fell on a single isochron. Smoliar et al. (1996) were able to resolve small differences in the isochrons of various iron meteorite groups. When only the genetically related magmatic IIA and IIB irons are considered, both groups found the same isochron, within errors (Fig. 9.22).

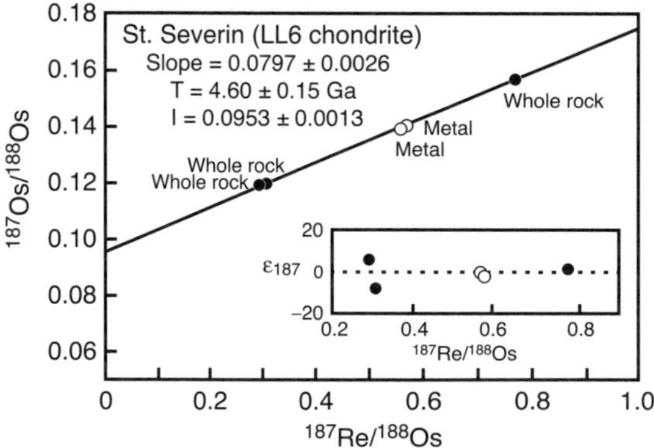

Figure 9.21 ^{187}Re–^{187}Os evolution diagram for the St. Severin LL6 chondrite. The isochron is generated from three whole-rock fragments and two metal separates. Deviations from the best-fit line are shown in part in 10^3 in the inset. Modified from Chen et al. (1998) using $\lambda = 1.666 \times 10^{-11}$ yr^{-1}.

Attempts to generate whole-rock isochrons for chondrites have proved less successful, primarily because of difficulties with the chemistry. As chemical procedures improved, the spread around isochrons determined for H chondrites decreased significantly (Meisel et al., 1996; Chen et al., 1998). The most recent data exhibit relatively little variation in ^{187}Re/^{188}Os ratios and the data cluster closely around the IIAB isochron. An attempt to use the ^{187}Re–^{187}Os system on a suite of angrite meteorites was unable to produce a chronologically useful isochron because of low-temperature terrestrial alteration among the samples (Riches et al., 2012).

Reservoir Evolution

As with the ^{87}Rb–^{87}Sr, ^{147}Sm/^{143}Nd, U–Th–Pb, and ^{176}Lu–^{176}Hf systems, the ^{187}Re–^{187}Os system can be used to investigate reservoir evolution in differentiated bodies. Most of this work has been carried out to investigate differentiation of the Earth and the degree to which a late addition of chondritic material may have been involved in Earth's accretion. Parameters for a CHUR have been established for the ^{187}Re–^{187}Os system (Table 9.6).

Differences between the current ^{187}Os/^{188}Os of a sample and the CHUR are expressed as

$$\gamma(\mathrm{Os})_j^0 = \left[\frac{\left({}^{187}\mathrm{Os}/{}^{188}\mathrm{Os}\right)^0_{meas}}{\left({}^{187}\mathrm{Os}/{}^{188}\mathrm{Os}\right)^0_{\mathrm{CHUR}}} - 1\right] \times 10^2 \qquad 9.48$$

Note that γ(Os) is expressed in percent, not parts per 10,000, as in the case of epsilon neodymium, primarily because the variations can be large and the measurement

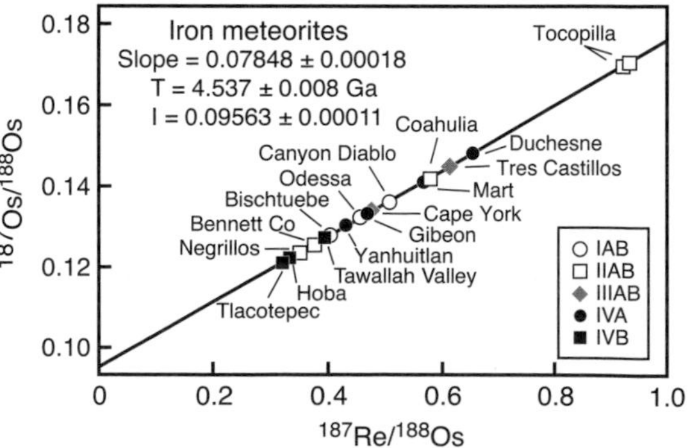

Figure 9.22 ^{187}Re–^{187}Os evolution diagram for 16 iron meteorites from diverse chemical groups. The line is the best-fit correlation for the IIA and IIB irons. Modified from Shen et al. (1996) using $\lambda = 1.666 \times 10^{-11}$ yr^{-1}.

Table 9.6 Accepted values for Re–Os CHUR

	Present	4.56 Ga
$^{187}Re/^{188}Os$	0.423 ± 0.007	0.4564
$^{187}Os/^{188}Os$	0.12863 ± 0.00046	0.095243

Chen et al. (1998).

Table 9.7 Proposed values for Lu–Hf CHUR

	Present	4.56 Ga
$^{176}Hf/^{177}Hf$	0.282785 (11)	0.279810
$^{176}Lu/^{177}Hf$	0.0336 (1)	0.03657

Bouvier et al. (2008).

precision is typically somewhat lower than for the $^{147}Sm/^{143}Nd$ system. Occasionally you will see osmium data presented as ε(Os) in units of parts per 10,000. Although most work has been done on terrestrial systems, there is potential for applications to the Moon and Mars.

9.4.6 The ^{176}Lu–^{176}Hf System

Lutetium has two naturally occurring isotopes (Table 4.2), one of which, ^{176}Lu, is radioactive. Lutecium-176 β^- decays to ^{176}Hf with a half-life of ~3.72×10^{10} years ($\lambda = 1.86 \times 10^{-11}$ yr^{-1}). Hafnium has six naturally occurring isotopes (Table 4.2). Of these, ^{174}Hf is radioactive and α decays to ^{170}Yb with a half-life of ~2×10^{15} years – too long to be useful for chronology. Both lutetium and hafnium are refractory lithophile elements that should not have been fractionated by either condensation/evaporation processes or metal–silicate fractionation.

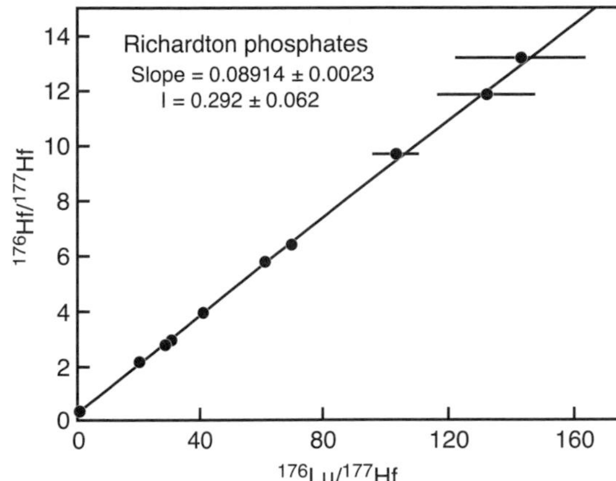

Figure 9.23 ^{176}Lu–^{176}Hf evolution diagram for phosphates from the Richardton (H5) chondrite. The data form a tight isochron. After Amelin (2005).

9.4.6.1 *History*

The ^{176}Lu–^{176}Hf system attracted the attention of chronologists around 1980 due to the work of Patchet and Tatsumoto (1980). However, it turned out to be a difficult system to use because of the very poor ionization efficiency of hafnium by TIMS. In addition, the $^{176}Hf/^{177}Hf$ and $^{143}Nd/^{144}Nd$ ratios in most terrestrial basalts are strongly correlated, which implies that in these rocks ^{176}Lu–^{176}Hf is redundant compared with the easier-to-use and more precise ^{147}Sm–^{143}Nd system. The technical limitations were largely removed with the introduction of high-precision ICPMS. Recent work has shown that phosphates in meteorites strongly fractionate lutetium and hafnium, which suggests applications for the ^{176}Lu–^{176}Hf system in situations where the ^{147}Sm–^{143}Nd system is not suitable (Fig. 9.23). However, for the most part, ^{176}Lu–^{176}Hf is used in conjunction with ^{147}Sm–^{143}Nd to evaluate reservoir evolution in differentiated bodies. Much of the effort in recent years has gone into establishing the parameters for the CHUR in this system (Table 9.7).

Another area of significant effort has been to determine the decay constant for ^{176}Lu. As with other long-lived radionuclides, direct laboratory measurements are difficult, so researchers often use decay constants derived from terrestrial and extraterrestrial samples. A recent high-precision laboratory measurement gave $\lambda = (1.8619 \pm 0.0145) \times 10^{-11}$, half-life = $(3.722 \pm 0.029) \times 10^{10}$ years (Hult et al., 2014). Attempts to use sample-based decay constants have produced variable results. Phosphates are a major carrier of REEs, including Lu, in meteorites, and it turns out that Lu can be lost from phosphates during metamorphism, while Hf is quantitatively retained (Debaille et al., 2017). Lutetium loss would result in a steepening of the ^{176}Lu–^{176}Hf isochron, resulting in either a higher decay constant or an older date. Many terrestrial rocks and extraterrestrial samples that have not been metamorphosed show less scatter in the inferred decay constant. The best estimates from these samples (e.g. $\lambda = (1.8640 \pm 0.0146) \times 10^{-11}$, half-life = $(3.718 \pm 0.029) \times 10^{10}$ years; Debaille et al., 2017) seem to be approaching the experimentally determined value.

9.4.6.2 *Technical Details*

As with the ^{147}Sm–^{143}Nd and ^{187}Re–^{187}Os systems, careful chemistry is required to separate cleanly the parent and daughter elements, because mass spectrometry cannot resolve ^{176}Lu from ^{176}Hf. The ion-exchange chemistry is similar to that for samarium–neodymium. In fact, fractions of samarium, neodymium, lutetium, and hafnium are often produced in a single procedure. Mass spectrometry is done by ICPMS because this is the only method that effectively ionizes hafnium.

As with ^{147}Sm/^{144}Nd, the ^{176}Lu/^{177}Hf ratio does not vary much in chondrites, making it very difficult to extract information about the initial ^{176}Hf/^{177}Hf of the solar system from isochron diagrams. Thus, the methodology developed for ^{147}Sm–^{143}Nd of comparing the current ^{143}Nd/^{144}Nd with the value for the CHUR has been adapted to the ^{176}Lu–^{176}Hf system. Table 9.7 shows recently proposed values for the ^{176}Lu–^{176}Hf CHUR. The evolution of the CHUR is calculated in the same way as for the ^{147}Sm/^{144}Nd system, and differences between the current ^{176}Hf/^{177}Hf of a sample and the CHUR are expressed as ε(Hf), which is defined in an analogous manner to ε(Nd):

$$\varepsilon(\mathrm{Hf})_j^0 = \left[\frac{\left({}^{176}\mathrm{Hf}/{}^{177}\mathrm{Hf}\right)^0_{meas}}{\left({}^{176}\mathrm{Hf}/{}^{177}\mathrm{Hf}\right)^0_{\mathrm{CHUR}}} - 1\right] \times 10^4 \qquad 9.49$$

As with ε(Nd), ε(Hf) values are reported as parts per 10,000.

9.4.6.3 *Applications*

As with ^{147}Sm/^{144}Nd, the ^{176}Lu/^{177}Hf system can potentially be used to date an object using the isochron method or to study differentiation processes. Figure 9.23 shows an isochron generated from phosphates in the Richardton chondrite. The isochron is precisely determined, but in this case, the isochron was used to estimate the decay constant of ^{176}Lu by applying an age determined from the U–Th–Pb system. Until the decay constant is known, using ^{176}Lu–^{176}Hf to directly date planetary materials will be problematic.

The ^{176}Lu–^{176}Hf system is often used together with the ^{147}Sm–^{143}Nd system to investigate planetary differentiation. The two systems can be highly correlated. All four elements are incompatible, so their relative abundances do not change much until the degree of fractional crystallization becomes extreme. A few minerals, such as phosphate, ilmenite, and garnet, can separate hafnium from the REEs. Garnet has been implicated to explain the isotopic relationships between the ^{147}Sm–^{143}Nd and ^{176}Lu–^{176}Hf systems in the source regions of terrestrial, lunar, and martian lavas. The eucrites require a different explanation because pressures on the eucrite parent body are not high enough for the source regions to be in the garnet stability field. This suggests that ilmenite is the cause of lutetium–samarium fractionation in eucrites and raises the possibility that ilmenite might also play a role in the evolution of larger bodies (Blichert-Toft et al., 2002). As with the ^{147}Sm–^{143}Nd system, multistage evolutionary models can be created for the ^{176}Lu–^{176}Hf system.

9.4.7 Other Long-Lived Nuclides of Potential Cosmochemical Significance

There are several other potential chronometers available. Most have experimental difficulties that make them hard to use, and they have not been developed to mainstream status. Examples include ^{40}K–^{40}Ca, U–Th–He, ^{190}Pt–^{190}Os, ^{138}La–^{138}Ba, and ^{138}La–^{138}Ce. We will not discuss them here.

9.5 Short-Lived Radionuclides

The short-lived radionuclides that were originally present in the early solar system have effectively decayed away. They are sometimes called *extinct radionuclides.* Short-lived radionuclides potentially provide high-resolution chronometers for early solar system events. The faster a nuclide decays, the shorter the timescale over which it provides time information, but the higher the time resolution that it can provide. Table 9.8 lists some short-lived radionuclides of interest in cosmochemistry. Those shown in bold have been developed as early solar system chronometers to varying degrees. These and others, such as ^{41}Ca, ^{60}Fe, and ^{107}Pd, are also of interest in determining the environment in which the solar system formed (see Chapter 10). Isotopes such as ^{10}Be, ^{14}C, ^{26}Al, and ^{36}Cl are produced continuously by cosmic-ray interactions and are useful in determining cosmic-ray exposure ages (Chapter 10). They were also present in the early solar system. The nuclides ^{22}Na and ^{44}Ti have half-lives that are much too short to have been incorporated live into the solar system after their synthesis in stars. However, the decay products of these nuclides have been valuable in establishing the stellar sources of some presolar grains (see Chapter 5).

Because the short-lived nuclides are extinct, a different approach must be taken to use them as chronometers. Equation 9.9 cannot be used to calculate a date because the number of parent nuclides, N, is zero and the equation is undefined. However, if a short-lived nuclide was homogeneously distributed throughout a system, then one could determine the order in which objects formed within that system based on the amount of radionuclide that was present when the object formed. The oldest object would form with the highest amount of the radionuclide relative to a stable isotope of the same element, and the youngest would have the lowest amount. Obviously, no chronological information could be obtained about objects that

Table 9.8 Short-lived radionuclides used in cosmochemistry

Nuclide N_R	Half-Life (years)	Daughter Isotope(s)	Decay Mode*	Stable Nuclide N_S	$(N_R/N_S)_0$	Ref
^{7}Be	53 days	^{7}Li	E.C.	^{9}Be		
^{10}Be	$1.39 \times 10^{6\dagger}$	^{10}B	β^-	^{9}Be	1×10^{-3}	[1]
^{14}C	5730	^{14}N	β^-	^{12}C		
^{22}Na	2.604	^{22}Ne	β^+	^{23}Na		
^{26}Al	**7.17×10^5**	**^{26}Mg**	**β^+, E.C.**	**^{27}Al**	**5.2×10^{-5}**	**[2]**
^{36}Cl	3.01×10^5	^{36}Ar (98.1%)	β^-	^{35}Cl	$1–3 \times 10^{-5}$	[3]
		^{36}S (1.9%)	E.C., β^+			
^{41}Ca	9.94×10^4	^{41}K	E.C.	^{40}Ca	4.2×10^{-9}	[4]
^{44}Ti	59.9	^{44}Ca (via ^{44}Sc)	E.C., E.C.			
^{53}Mn	**3.74×10^6**	**^{53}Cr**	**E.C.**	**^{55}Mn**	**6.7×10^{-6}**	**[5]**
^{60}Fe	$2.62 \times 10^{6\S}$	^{60}Ni (via ^{60}Co)	β^-, β^+	^{56}Fe	$(1–50) \times 10^{-8}$	[6,7]
^{92}Nb	36×10^6	^{92}Zr	E.C.			
^{107}Pd	6.5×10^6	^{107}Ag	β^-	^{108}Pd	5.9×10^{-5}	[8]
^{129}I	**1.7×10^7**	**^{129}Xe**	**β^-**	**^{127}I**	**1×10^{-4}**	**[9]**
^{146}Sm	**1.03×10^8**	**^{142}Nd**	**α**	**^{144}Sm**	**8×10^{-3}**	**[10]**
^{182}Hf	**8.9×10^6**	**^{182}W**	**β^-**	**^{180}Hf**	**1×10^{-4}**	**[11]**
^{244}Pu	8.2×10^7	^{208}Pb, Xe isotopes	α, S.F.	^{232}Th	3×10^{-3}	[9]

*E.C.: electron capture, β^-: beta decay, β^+: positron decay, α: alpha decay, S.F.: spontaneous fission.

† New half-life determination by Korschinek et al. (2010) and by Chmeleff et al. (2010).

§ New half-life determination by Rugel et al. (2009).

[1] McKeegan et al., 2000. [2] Jacobsen et al., 2008. [3] Tang et al., 2017. [4] Liu et al., 2012. [5] Davis and McKeegan, 2014. [6] Tang and Dauphas, 2012. [7] Telus et al., 2018. [8] Schönbächler et al., 2008. [9] Wasserburg et al., 2006. [10] Marks et al., 2014. [11] Kruijer et al., 2014.

formed after the radionuclide had reached a level too small to detect the radiogenic daughter isotope.

Consider a hypothetical system that has an initial abundance of a short-lived radionuclide, N_R. This nuclide is present as a fixed fraction of the parent element, and its abundance can be written as the ratio of the radionuclide to a stable reference isotope of the same element, N_S. When the short-lived nuclide has completely decayed to its daughter nuclide, D^*, we have:

$$\left(\frac{N_R}{N_S}\right)_0 = \frac{D^*}{N_S} \qquad 9.50$$

The initial ratio, $(N_R/N_S)_0$, is what we hope to determine. We can measure two quantities to extract this information. One is the ratio of the daughter isotope to a stable isotope of the same element. The abundance of the daughter isotope is the sum of the initial abundance, D_0, and the radiogenic component, D^*:

$$\frac{D_0 + D^*}{D_{ref}} \qquad 9.51$$

The other is the ratio of the abundance of the reference isotope of the parent element to the abundance of the reference isotope of the daughter element:

$$\frac{N_S}{D_{ref}} \qquad 9.52$$

If we know the initial ratio of the daughter isotope (D_0/D_{ref}), we can calculate the ratio of the radiogenic daughter isotope to the reference isotope:

$$\frac{D_{meas}}{D_{ref}} - \frac{D_0}{D_{ref}} = \frac{D_0 + D^*}{D_{ref}} - \frac{D_0}{D_{ref}} = \frac{D^*}{D_{ref}} \qquad 9.53$$

The initial ratio can then be calculated by dividing Equation 9.53 by Equation 9.52:

$$\frac{\frac{D^*}{D_{ref}}}{\frac{N_S}{D_{ref}}} = \frac{D^*}{N_S} = \left(\frac{N_R}{N_s}\right)_0 \qquad 9.54$$

If the ratio (N_s/D_{ref}) is very high, the initial ratio of the parent isotope to a stable isotope of the same element can be calculated very precisely by this method because the correction for the initial abundance of the daughter isotope is small.

In many cases, however, the elemental ratio is not very high and the contribution of the radiogenic component to the overall abundance of an isotope is small. An isochron diagram, analogous to those used for long-lived radionuclides, provides a powerful way to determine the initial abundance of the parent isotope. But in this case, the x-axis plots the ratio of a stable isotope of the parent element to a stable isotope of the daughter element. Figure 9.24 shows a schematic isochron diagram for the ^{26}Al–^{26}Mg system. As in any other isochron diagram, the ratio of the daughter isotope to a stable isotope of the same element (^{26}Mg/^{24}Mg in Fig. 9.24) increases as the radioactive isotope decays. The extinct radionuclide has now completely decayed away, so the isochron is frozen. The slope of the isochron gives the ratio of the radioactive isotope to a stable isotope of the same element at the time the object formed ((^{26}Al/^{27}Al)$_0$):

$$\text{Slope} = \frac{\Delta Y}{\Delta X} = \frac{\left(\frac{^{26}\text{Mg}}{^{24}\text{Mg}}\right)_{meas} - \left(\frac{^{26}\text{Mg}}{^{24}\text{Mg}}\right)_0}{\left(\frac{^{27}\text{Al}}{^{24}\text{Mg}}\right)_{meas} - 0}$$

The ^{24}Mg in this equation represents the indigenous magnesium in the sample and so is the same in both numerator and denominator and can be eliminated. The radiogenic ^{26}Mg* equals the measured ^{26}Mg minus the original ^{26}Mg in the sample, so the above equation becomes:

$$\text{Slope} = \frac{^{26}\text{Mg}_{meas} - {}^{26}\text{Mg}_0}{^{27}\text{Al}_{meas}} = \frac{^{26}\text{Mg}^*}{^{27}\text{Al}} = \left(\frac{^{26}\text{Al}}{^{27}\text{Al}}\right)_0 \quad 9.55$$

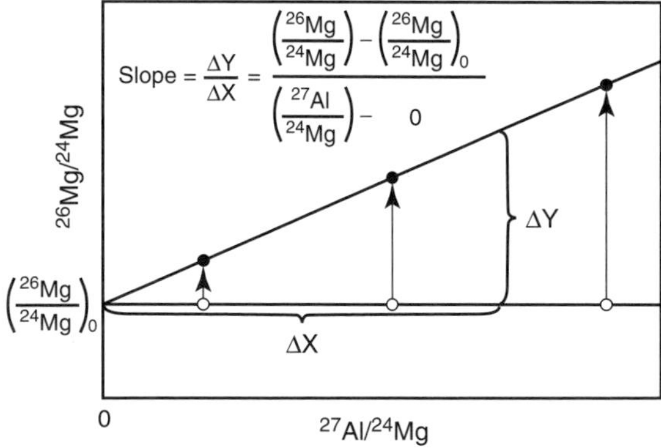

Figure 9.24 Schematic drawing of an ^{26}Al–^{26}Mg evolution diagram (isochron diagram). The slope of the isochron on such a diagram gives the (^{26}Al/^{27}Al)$_0$ ratio and the intercept gives the (^{26}Mg/^{24}Mg)$_0$ ratio for the system.

Three types of isochron diagrams are used in discussions of short-lived radionuclides. An *internal isochron* is one based entirely on measurements from the object being dated. By measuring several minerals with different parent/daughter elemental ratios, one can obtain an array of data that gives both the initial ratio ((N_R/N_S)$_0$) and the initial isotopic ratio of the daughter element (D_0/D_{ref}). Figure 9.25 shows an example of an internal isochron for an Allende CAI. In this figure, the initial ^{26}Mg/^{24}Mg ratio is given by the intersection of the isochron and the left-side y-axis. It is within uncertainties of normal terrestrial magnesium. The δ^{26}Mg values on the right axis give the ^{26}Mg/^{24}Mg ratios calculated relative to the terrestrial standard in parts per thousand.

Often, however, the sample does not lend itself to measurements of multiple phases. In these cases, one can create a *model isochron* in which the initial isotopic ratio of the daughter element (D_0/D_{ref}) is assumed and is used to anchor the y-intercept. With this assumption, a single measurement can provide the initial ratio ((N_R/N_S)$_0$) of an object. A third type of isochron diagram assumes that a group of bodies all formed from the same homogeneous isotopic reservoir at the same time. This *whole-rock isochron* is constructed from bulk measurements of each sample, and the resulting slope and initial ratio describe the reservoir from which all of the samples formed.

As with long-lived radiochronometers, some basic conditions must be satisfied for a short-lived nuclide to provide chronological information:

1. The object to be dated must have been isotopically homogeneous at the time that it formed in order for the

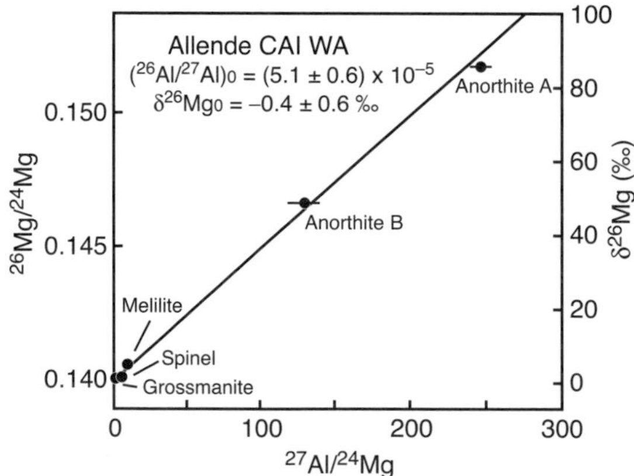

Figure 9.25 ^{26}Al–^{26}Mg evolution diagram of CAI WA from Allende, the first sample to show clear evidence of live ^{26}Al in the solar system. ^{26}Mg/^{24}Mg ratios are given on the left vertical axis and delta values calculated relative to terrestrial magnesium are given on the right. After Lee et al. (1977).

contribution of the radiogenic isotope to be recognized in the daughter element.

2. In order to generate an internal isochron, a chemical fractionation creating phases with different parent–daughter ratios must have occurred.

3. As with all radiometric dating systems, the system must have remained closed and undisturbed since the event being dated. A precise internal isochron provides a test of whether or not the system has remained closed. Because a model isochron can be generated with a single measurement, a chemically homogeneous object can be dated. In addition, even if an object has had its isotopes redistributed after formation, if the system has remained closed, a bulk measurement can still provide a valid model isochron. But an open system, in which the parent or daughter element has exchanged with the outside, cannot provide valid chronological information. Unfortunately, a model isochron provides no test of whether or not the bulk system has remained closed.

If the above conditions are satisfied and differences in $(N_R/N_S)_0$ are due to time of formation only, the time interval between the formation of two objects, 1 and 2, can be calculated as follows:

$$\Delta t = \frac{1}{\lambda}\ln\left[\frac{\left(\frac{N_{R1}}{N_{S1}}\right)_0}{\left(\frac{N_{R2}}{N_{S2}}\right)_0}\right] \qquad 9.56$$

However, to determine the absolute time of formation of the objects, one or both of the objects must also be dated with a long-lived chronometer.

9.5.1 The ^{129}I–^{129}Xe System

Iodine-129 decays to ^{129}Xe by β^- decay. The literature contains several values for the half-life of ^{129}I, ranging from 15.6 Myr to 19.7 Myr. The latest recommended value is 17 Myr ($\lambda = 4.077 \times 10^{-8}$ yr^{-1}) (Holden, 1990), which seems to give relative ages consistent with other techniques. The half-life is long enough that the ^{129}I–^{129}Xe system can potentially date events during the first 100 Myr of solar system history. Because the excesses of ^{129}Xe are often very large, the system has a high intrinsic precision and can resolve time differences as small as 100,000 years. Both iodine and xenon are volatile and mobile elements. These properties mean that the ^{129}I–^{129}Xe system is well suited for dating parent-body events using minerals with low isotopic closure temperatures. An excellent review of the ^{129}I–^{129}Xe system as a chronometer is given by Hohenberg and Pravdivtseva (2008).

9.5.1.1 *History*

Iodine-129 was the first short-lived nuclide shown to have been present in the early solar system. Brown (1947) proposed that the age of the elements might be determined if the decay product of an extinct radionuclide could be found in meteorites. Although the proposal itself was flawed, it led to a very important discovery. Initial searchers for ^{129}Xe from the decay of ^{129}I in the Beardsley H5 chondrite and in the Nuevo Laredo eucrite came up empty, but Reynolds (1960a, 1960b) found a large excess of ^{129}Xe in the Richardton H5 chondrite and an even larger excess in the Indarch EH4 chondrite. He argued that the excess was due to the decay of ^{129}I primarily on the basis of the improbability of the other options, but proof of an origin from ^{129}I came by irradiating samples with neutrons, which converts some ^{127}I to ^{128}Xe, and showing that the excess ^{129}Xe was associated with the neutron-produced ^{128}Xe (Jeffery and Reynolds, 1961).

Over the next several decades, much effort was devoted to using the ^{129}I–^{129}Xe system as an early solar system chronometer. However, because many of the results did not agree with the relative chronology indicated by the petrography of the samples and with ages determined by long-lived chronometers, and because people did not understand how the iodine was sited in the meteorites and the extent to which metamorphic heating and aqueous alteration could remobilize iodine and xenon, the ^{129}I–^{129}Xe chronometer fell into disrepute, with most workers concluding that it could not be used for age dating. Research since the mid-1990s has begun to redeem the reputation of the ^{129}I–^{129}Xe chronometer. Detailed work on phosphates, first from Acapulco and then from ordinary chondrites, showed that ^{129}I–^{129}Xe ages and ^{207}Pb*–^{206}Pb* ages give the same time differences relative to the Shallowater aubrite (Nichols et al., 1994; Brazzle et al., 1999). This work also resulted in the determination of a precise ^{207}Pb*–^{206}Pb* age for Shallowater, which put the ^{129}I–^{129}Xe ages on an absolute scale. When care is taken to produce clean mineral separates and to interpret the results in the context of the petrography of the materials, results that are consistent with other chronometers can be obtained.

9.5.1.2 *Technical Details*

If the iodine in a system is isotopically homogeneous, then objects forming at the same time from that system will have the same $(^{129}\text{I}/^{127}\text{I})_0$ ratio. Objects forming at later times will have lower $(^{129}\text{I}/^{127}\text{I})_0$ ratios due to decay of ^{129}I. If the system remains closed, then ^{129}Xe* from the decay of

^{129}I will remain associated with ^{127}I. Because it is difficult to measure the abundance of iodine in a sample by chemical means, the ^{129}I–^{129}Xe system is typically applied by irradiating a sample with thermal neutrons, turning a portion of the ^{127}I to ^{128}Xe*. The xenon isotopes are then measured and the ratio of radiogenic ^{129}Xe* from the decay of ^{129}I to the ^{128}Xe* produced in the irradiation is determined. The method is directly analogous to the ^{40}Ar–^{39}Ar method discussed above for the ^{40}K–^{40}Ar system.

The treatment of the data is different than in the ^{40}Ar–^{39}Ar system, however. In an iodine-bearing phase, the measured ^{129}Xe consists of radiogenic ^{129}Xe* from the decay of ^{129}I and the ^{129}Xe that was originally trapped in the sample (^{129}Xe$_t$). In the same way, the measured ^{128}Xe consists of ^{128}Xe* produced by irradiation from the ^{127}I and the ^{128}Xe$_t$ originally trapped in the sample. Thus, both ^{129}Xe and ^{128}Xe are two-component mixtures between a trapped component and an iodine-derived component, and, if the assumptions of radiometric dating have been met, measured samples of the same age will plot on a single line. In the case where the xenon isotopes are normalized to ^{132}Xe, the equation of the line, the isochron, on a ^{129}Xe/^{132}Xe versus ^{128}Xe/^{132}Xe plot such as Figure 9.26 is

$$\frac{^{129}\mathrm{Xe}}{^{132}\mathrm{Xe}} = M \times \frac{^{128}\mathrm{Xe}}{^{132}\mathrm{Xe}} + \left[\left(\frac{^{129}\mathrm{Xe}}{^{132}\mathrm{Xe}}\right)_t - M \times \left(\frac{^{128}\mathrm{Xe}}{^{132}\mathrm{Xe}}\right)_t\right] \tag{9.57}$$

where M is the slope (^{129}Xe*/^{128}Xe*) and the term in brackets is the trapped component represented by the end point of the line at the lower left on Figure 9.26 (analogous plots and equations can be created if ^{130}Xe is used for normalization). The trapped component is typically

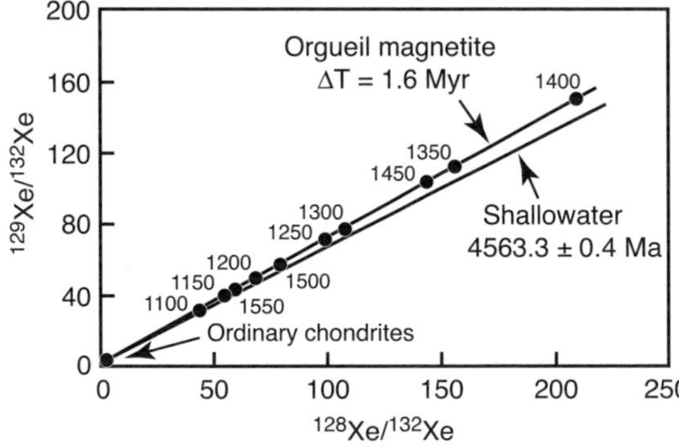

Figure 9.26 ^{129}I–^{129}Xe isochrons for magnetite separated from the Orgueil CI chondrites and for the Shallowater meteorite (the irradiation standard). The difference in the slopes of the two correlation lines corresponds to the difference in closure times of the ^{129}I–^{129}Xe systems. After Hohenberg and Pravdivtseva (2008).

the dominant noble gas component in chondrites, known as Xe-P1 or Q-Xe (see Chapter 11). Although the measurements must be corrected for the blank in the mass spectrometer and for adsorbed air, these corrections are usually very small and the composition of atmospheric xenon has similar isotopic ratios to the trapped component, so uncertainties in the corrections do not affect the slope of the isochron.

A piece of the Shallowater aubrite is typically included in every irradiation as a flux monitor, and the inferred (^{129}I/^{127}I)$_0$ ratios of the samples are determined relative to the ratio for Shallowater (Fig. 9.26). Over the last few years, the absolute age of the Shallowater meteorite (4563.3 ± 0.4 Ma) has been determined through use of several isotope chronometers. This has placed the ^{129}I–^{129}Xe ages determined relative to Shallowater on an absolute scale.

9.5.1.3 *Applications*

The greatest asset of the ^{129}I–^{129}Xe technique is its ability to investigate the postformational evolution of early solar system materials. The ^{129}I–^{129}Xe chronometer has been successfully used to investigate the secondary alteration of carbonaceous chondrites (e.g. Pravdivtseva et al., 2018). It has also been applied to chondrules, which often show two isochrons: a primary one reflecting formation and a secondary one reflecting later alteration. An attempt has also been made to use the ^{129}I–^{129}Xe system to evaluate the cooling history of IAB iron meteorites. The ^{129}I–^{129}Xe system is probably the best tool for investigating secondary processes.

9.5.2 The ^{26}Al–^{26}Mg System

Aluminum-26 decays to ^{26}Mg by positron emission and electron capture with a half-life of ~717,000 years ($\lambda = 9.67 \times 10^{-7}$ yr^{-1}). Magnesium is the sixth most abundant element in the solar system; aluminum is a factor ~13 less abundant. Thus, relatively large elemental fractionations favoring aluminum are required for the detection of ^{26}Al effects. Aluminum is a refractory element with a condensation temperature of ~1650 K, while magnesium condenses at ~1340 K (Table 7.1). Thus, both elements resist vaporization, and thermal processing tends to increase the Al/Mg ratio. In minerals where magnesium is a major structural element, magnesium tends to be resistant to thermal metamorphism. But in minerals such as feldspar, where magnesium does not fit into the structure, mild heating will cause any magnesium present to migrate out of the crystal. Overall, the ^{26}Al–^{26}Mg system is relatively sensitive to thermal resetting. The half-life of 717,000 years means that the ^{26}Al–^{26}Mg system is suitable for dating events during the first 5 Myr of solar system history.

Aluminum-26 is produced by stellar nucleosynthesis in a wide variety of stellar sites. Its abundance relative to other short-lived nuclides provides information about the stellar source(s) for short-lived nuclides and the environment in which the Sun formed. Aluminum-26 is also produced by interactions between heavier nuclei such as silicon atoms and cosmic rays. Aluminum-26 is one of several nuclides used to estimate the cosmic-ray exposure ages of meteorites as they traveled from their parent asteroids to Earth.

9.5.2.1 *History*

The idea that ^{26}Al might have been present in the early solar system and might have provided a heat source for melting and differentiation in early solar system bodies was first presented by Urey (1955). With the development of high-precision TIMS in the late 1960s, it became possible to search for evidence of ^{26}Al in early solar system materials. The initial searches were unsuccessful, in part because the wrong samples were being measured. However, as scientists prepared to study the lunar samples due to arrive in 1969, the newly fallen Allende meteorite provided ideal samples to renew the search. CAIs from Allende, then interpreted as primitive nebular condensates, were ideal candidates to measure for ^{26}Al. In 1974, isotopic anomalies were found in magnesium (Gray and Compston, 1974; Lee and Papanastassiou, 1974), but they could not be directly attributed to decay of ^{26}Al. Clear proof that excesses of ^{26}Mg were due to decay of ^{26}Al came when Lee et al. (1977) measured large, clean plagioclase crystals and mineral separates of spinel, melilite, and grossmanite from an Allende CAI. The plagioclase crystals showed large excesses of ^{26}Mg and the excesses correlated with the Al/Mg ratio (Fig. 9.25). The initial ^{26}Al/^{27}Al ratio inferred from this inclusion was $(5.1 \pm 0.6) \times 10^{-5}$. A good summary of the early development of the ^{26}Al–^{26}Mg system as a chronometer is given by Wasserburg and Papanastassiou (1982).

A significant technical advance came when Hutcheon (1982) obtained reliable ^{26}Al–^{26}Mg data using SIMS. This technique permitted internal isochrons to be measured on individual CAIs and chondrules while keeping the petrologic context intact. Since that time, considerable ^{26}Al–^{26}Mg data have been collected on CAIs, which are believed to be the oldest objects formed in the solar system (see Chapter 10). Many pristine inclusions from CV, CO, CR, and other types of carbonaceous chondrites exhibit $(^{26}\text{Al}/^{27}\text{Al})_0$ ratios of $\sim 5 \times 10^{-5}$, and this has come to be called the "canonical" ratio for solar system objects. Other objects give low initial ratios, in many cases due to later resetting or isotopic disturbance caused by parent-body metamorphism. Some apparently unaltered CAIs exhibit $(^{26}\text{Al}/^{27}\text{Al})_0$ ratios of $< 1 \times 10^{-5}$. Among these are the FUN inclusions, which also show large isotopic anomalies in many constituent elements. ^{182}Hf/^{182}W dating indicates that FUN inclusions formed at the same time as normal CAIs (Holst et al., 2013). This result and identification of ^{16}O-rich corundum grains with low ^{26}Al initial ratios (Makide et al., 2011) suggest that ^{26}Al was delivered to the solar system as the first solids were forming. This raises the possibility that ^{26}Al cannot be used for chronology, a point that is still under debate.

As measurement capabilities improved, inferences based on the ^{26}Al chronometer became more detailed. Two research groups have produced bulk-CAI isochrons from CV chondrites that give precise and effectively identical values for the solar system: $(^{26}\text{Al}/^{27}\text{Al})_0$ ratio of $(5.23 \pm 0.12) \times 10^{-5}$ (Jacobsen et al., 2008) and $(5.25 \pm 0.02) \times 10^{-5}$ (Larsen et al., 2011). SIMS measurements now show slight but resolvable differences in formation times for chondrules from L and LL chondrites ($(6–9) \times 10^{-6}$; Siron et al., 2020), CV chondrites ($(2.6–6.3) \times 10^{-6}$; Nagashima et al., 2017), CO chondrites ($(4.6–6.0) \times 10^{-6}$; Kita et al., 2020), and CR chondrites ($< 6 \times 10^{-6}$ to undetectable; Nagashima et al., 2014). High-precision multicollector ICPMS (MC-ICPMS) measurements now seem to be showing resolvable differences in formation times for CAIs from CV and CR chondrites (Larsen et al., 2020). Again, CR CAIs seem to be slightly younger.

9.5.2.2 *Technical Details*

The ^{26}Al–^{26}Mg system can be studied by TIMS and ICPMS measurements of bulk samples and mineral separates. For these studies, aluminum and magnesium are chemically separated and purified before isotopic measurement. These measurements can be very precise, allowing studies into the differences in the $(^{26}\text{Mg}/^{24}\text{Mg})_0$ ratios before ^{26}Al decayed. The ^{26}Al–^{26}Mg system can also be studied by laser-ablation ICPMS and SIMS measurements of individual phases in thin section, keeping the petrologic context intact. In all cases, it is necessary to find samples that meet the requirements for valid radiometric dates.

9.5.2.3 *Applications*

Because its half-life is ideally suited to investigations of the first 5 Myr of solar system history, and because aluminum-rich minerals are available to measure, the ^{26}Al–^{26}Mg system has been used extensively to study the formation and history of CAIs and chondrules from carbonaceous, ordinary, and enstatite chondrites. The ^{26}Al–^{26}Mg system has also given information about the metamorphic history of chondritic bodies, although the timescale for metamorphism is a bit too long. There is evidence that ^{26}Al was present in the eucrites when they crystallized.

There is clear evidence that live ^{26}Al was incorporated into presolar silicon carbide and aluminum oxide grains (see Chapter 5). These grains acquired their ^{26}Al in the atmospheres of the dying stars in which they formed. It had almost certainly decayed away long before the grains reached the solar system. In presolar grains, ^{26}Al serves as a probe of nucleosynthesis in the parent stars.

Aluminum-26 is an important nuclide for investigating the cosmic-ray exposure history of meteorites on their way to Earth from the asteroid belt. It can also be used to estimate the terrestrial age of a meteorite. In both of these applications, the ^{26}Al is alive in the samples, having been produced by cosmic-ray interactions with elements heavier than aluminum, primarily silicon. Cosmic-ray-exposure dating will be discussed in Chapter 10.

9.5.3 The ^{41}Ca–^{41}K System

Calcium-41 decays by electron capture to ^{41}K with a half-life of 9.94×10^4 yrs ($\lambda = 6.97 \times 10^{-6}$ yr^{-1}). Calcium is a refractory alkaline earth element, while potassium is a moderately volatile alkali metal. Thus, calcium and potassium can be separated by both chemical and thermal processes. The short half-life of ^{41}Ca means that the abundance in the early solar system was very low. Therefore, minerals with very high Ca/K ratios are required in order for the radiogenic ^{41}K to be detected.

9.5.3.1 *History*

Hutcheon et al. (1984) found strong hints of radiogenic $^{41}K^*$ in pyroxene from two Allende CAIs, but they were unable to completely exclude the possibility that the effects were due to an interference. The first clear evidence of ^{41}Ca was found in pyroxene and perovskite from two CAIs from the Efremovka CV3 chondrite (Srinivasan et al., 1994). The initial $^{41}Ca/^{40}Ca$ ratio inferred for these inclusions was $(1.5 \pm 0.3) \times 10^{-8}$. These results were confirmed by subsequent measurements of CAIs and refractory mineral grains from several carbonaceous chondrites, although the inferred initial ratio is now somewhat lower (Srinivasan et al., 1996; Liu et al., 2012; Liu, 2017).

9.5.3.2 *Technical Details*

In general, it is not possible to obtain enough suitable material to do a chemical separation of calcium and potassium. This limits measurements of the ^{41}Ca–^{41}K system to SIMS. But the ^{41}Ca–^{41}K system presents an additional, very tough experimental problem because $^{40}Ca^{42}Ca^{++}$ has a mass/charge ratio that is so close to that of $^{41}K^+$ that it cannot be separated by mass spectrometry. This means that there is an unresolvable interference that correlates with Ca/K ratio in exactly the same way that radiogenic $^{41}K^*$ should correlate with Ca/K ratio. Even worse, the interference tends to be large compared to the amount of radiogenic $^{41}K^*$. To get around this problem, careful monitoring of the $^{40}Ca^{43}Ca^{++}/^{43}Ca^+$ ratio is required to provide a means to accurately correct for the $^{40}Ca^{42}Ca^{++}$ interference on $^{41}K^+$.

9.5.3.3 *Applications*

To date, the ^{41}Ca–^{41}K system has not been used for chronology of the early solar system. The short half-life and the experimental challenges mean that measurements are confined to demonstrating that ^{41}Ca was present in an object. While this constrains the object to have formed well within the first million years of solar system history, the presence of ^{41}Ca is more useful as a means of constraining the time of nucleosynthesis of the short-lived nuclides and the formation environment of the solar system (see Chapter 10).

9.5.4 The ^{53}Mn–^{53}Cr System

Manganese-53 decays by electron capture to ^{53}Cr with a half-life of 3.7 Myr ($\lambda = 1.87 \times 10^{-7}$ yr^{-1}). Manganese and chromium are both relatively abundant and are thus present in most materials. Both are moderately volatile elements, with manganese being somewhat more volatile than chromium. They undergo fractionation in evaporation/condensation processes, in magmatic processes, and during aqueous alteration. The half-life of ^{53}Mn is long enough that the ^{53}Mn–^{53}Cr system can be used to date events over the first ~20 Myr of solar system history. Its half-life is also long enough that a significant fraction of the initial solar system inventory could have been inherited from the ambient interstellar medium.

9.5.4.1 *History*

The first suggestion that ^{53}Mn might have been present in the early solar system came from ^{53}Cr deficits in the low Mn/Cr fractions of some Allende CAIs. These deficits were interpreted to reflect formation of the manganese-poor CAIs before ^{53}Mn had decayed (Birck and Allègre, 1985). Birck and Allègre (1988) found the first clear evidence that ^{53}Mn was alive in the early solar system in the form of excesses of ^{53}Cr that correlated with Mn/Cr ratio in mineral separates of an Allende inclusion.

Lugmair and Shukolyukov (1998) presented an extensive dataset on a wide variety of meteorites – mostly differentiated meteorites – and constructed a relative chronology for their formation. To get absolute ages, they tied this chronology to the angrite LEW 86010, a meteorite with a simple igneous history and a precisely determined $^{207}Pb^{*}$–$^{206}Pb^{*}$ age of 4557.8 ± 0.5 Ma (Lugmair and Galer, 1992). The ^{53}Mn–^{53}Cr system has also been applied to bulk chondrites, chondrules from a variety of chondrite classes (e.g. Nyquist et al., 2001c), sulfides from enstatite chondrites (e.g. Wadhwa et al., 1997), and secondary carbonates and fayalite from carbonaceous chondrites (e.g. Endress et al., 1996; Hutcheon et al., 1998). Much of the extensive SIMS work on secondary carbonates and fayalite used San Carlos olivine to standardize the measurements. However, San Carlos olivine is not a good matrix match for either carbonates or fayalite. Recently, new measurements with appropriate standards have significantly decreased the scatter in the data, constraining the aqueous alteration in carbonaceous chondrites to 4–5 million years after CAI formation (Fujiya et al., 2013; Doyle et al., 2015; Jilly-Rehak et al., 2017).

There has been some inconsistency between the ^{53}Mn–^{53}Cr system and other chronometers, particularly with respect to the time interval between CAI formation and the formation of other solar system objects. Nyquist et al. (2009) suggested that the CAIs may contain excess ^{53}Mn due to irradiation with solar cosmic rays and proposed a lower $(^{53}Mn/^{55}Mn)_0$ for the solar system of 9.1×10^{-6}. Others have proposed even lower initial ratios (e.g. $(6.7 \pm 0.6) \times 10^{-6}$; Davis and McKeegan, 2014). These lower ratios may have resolved this problem (see Chapter 10).

Lugmair and Shukolyukov (1998) also presented data on initial $^{53}Cr/^{52}Cr$ ratios that they interpreted to show a radial gradient in initial $^{53}Mn/^{55}Mn$ in the early solar system. The proposed gradient was too subtle to have affected the ^{53}Mn–^{53}Cr chronometer in the asteroid belt. Later authors have argued that the data can be explained by fractionation of manganese from chromium due to their relative volatilities and that the isotopic composition of manganese was homogeneous (e.g. Birck et al., 1999).

9.5.4.2 *Technical Details*

The ^{53}Mn–^{53}Cr system can be studied by TIMS, ICPMS, and SIMS techniques. For TIMS and ICPMS work, bulk samples or mineral separates are dissolved and the solutions are passed through ion-exchange columns to produce clean solutions of manganese and chromium. For minerals with high Mn/Cr ratios, SIMS can obtain isotopic data while retaining the petrographic context of the measurements. Appropriate standards are required. The chromium isotopic compositions may have to be corrected for small additions of chromium from spallation reactions induced by cosmic rays. This is particularly important in iron-rich meteorites.

Direct measurements of the initial abundance of ^{53}Mn in the early solar system have proven unreliable. CAIs, the first solids to form in the early solar system, are highly depleted in both manganese and chromium, both of which are moderately volatile. Operationally, the relative ^{53}Mn–^{53}Cr timescale has been anchored to an angrite, LEW 86010 (see above) or D'Orbigny. Age differences can be calculated relative to, for example, D'Orbigny from

$$\Delta t = \frac{1}{\lambda_{53}} \ln \left[\frac{\left(\frac{^{53}Mn}{^{55}Mn}\right)_{D'Orb}}{\left(\frac{^{53}Mn}{^{55}Mn}\right)_{Sample}} \right] \qquad 9.58$$

where λ_{53} is the decay constant of ^{53}Mn and the absolute age is determined from the uranium-corrected $^{207}Pb^{*}$–$^{206}Pb^{*}$ age of 4563.37 ± 0.25 Ma.

9.5.4.3 *Applications*

The ^{53}Mn–^{53}Cr system is widely used to investigate early solar system chronology. Because of its relatively long half-life, the ^{53}Mn–^{53}Cr chronometer is well suited to provide precise relative ages for differentiated meteorites. It is also well suited for dating aqueous alteration. Fayalite, hedenbergite, and carbonates that precipitated from aqueous fluids typically have very high Mn/Cr ratios. The ^{53}Mn–^{53}Cr system can also be used to date chondrule formation, but because the half-life is relatively long, because manganese is more volatile than chromium leading to relatively low Mn/Cr ratios, and because manganese and chromium are easily mobilized in thermal metamorphism, the results have not been as precise as those from other chronometers.

9.5.5 The ^{60}Fe–^{60}Ni System

Iron-60 β^{-} decays to ^{60}Ni by way of ^{60}Co with a half-life of 2.62 Myr ($\lambda = 2.65 \times 10^{-7}$ yr^{-1}). This half-life is suitable for dating events in the first ~10 Myr of solar system history. Iron is the ninth most abundant element in the solar system (by number); its abundance is similar to those of magnesium and silicon. Nickel is the fifteenth most abundant element. Both exist as oxides, sulfides, carbides, and metal, raising the possibility that the ^{60}Fe–^{60}Ni system might be widely applicable to dating early solar system materials. However, unfavorable nickel isotopic abundances (Table 4.2), the mobility of iron and nickel during mild metamorphism, and limited elemental fractionations in many materials significantly degrade its usefulness. Iron and nickel are both moderately volatile elements, so little fractionation occurs by evaporation/condensation. Iron can be fractionated from nickel by melting and crystallization and by aqueous processes.

9.5.5.1 *History*

The first hints that ^{60}Fe was present in the early solar system came from ^{60}Ni excesses in Allende CAIs (Birck and Lugmair, 1988). However, these excesses were accompanied by excesses at ^{62}Ni and could not be shown to correlate with the Fe/Ni ratio. The first clear evidence of live ^{60}Fe in early solar system materials was found through TIMS measurements of three eucrites (Shukolyukov and Lugmair, 1993a, 1993b). The $(^{60}\text{Fe}/^{56}\text{Fe})_0$ ratios inferred for these meteorites differed by an order of magnitude (from ~4 $\times$ 10^{-10} to ~4 $\times$ 10^{-9}), but the $(^{53}\text{Mn}/^{55}\text{Mn})_0$ ratios in the same meteorites are quite similar. This indicates that the Fe–Ni system was disturbed in at least some of these eucrites.

The first evidence for ^{60}Fe in chondritic material came from SIMS measurements of troilite in type 3.0–3.1 ordinary chondrites (Tachibana and Huss, 2003; Mostefaoui et al., 2005). The $(^{60}\text{Fe}/^{56}\text{Fe})_0$ ratio inferred for the sulfides in Bishunpur and Krymka were in the range of (1–2) $\times$ 10^{-7} (Tachibana and Huss, 2003), but Semarkona sulfides gave a higher value (~1.6 $\times$ 10^{-6}) (Mostefaoui et al., 2005). Studies of type 3 enstatite chondrites showed that the ^{60}Fe–^{60}Ni system in sulfides is easily disturbed (Guan et al., 2007). Work shifted to silicate minerals in chondrules, particularly Fe-rich pyroxene, because the Fe–Ni system in silicates is less susceptible to secondary thermal disturbance. Inferred $(^{60}\text{Fe}/^{56}\text{Fe})_0$ ratios obtained for pyroxene-rich chondrules were in the range of (2–4) $\times$ 10^{-7}, somewhat higher than in the Bishunpur and Krymka sulfides (Tachibana et al., 2006) and implying $(^{60}\text{Fe}/^{56}\text{Fe})_0$ for the solar system of (0.5–1.0) $\times$ 10^{-6}. Such high ratios imply that a supernova supplied the ^{60}Fe and other short-lived nuclides to the solar system.

However, Ogliore et al. (2011) pointed out a statistical artifact that shows up in counting experiments that involve small numbers of counts. If the number of counts in the denominator of a ratio is too low, the ratio will be biased high by an amount that is inversely proportional to the number of denominator counts. This is an insidious effect because the bias can correlate with the Fe/Ni ratio in the same way as excess ^{60}Ni from the decay of ^{60}Fe. It turned out that most of the SIMS data for the ^{60}Fe–^{60}Ni system taken up to that time were inaccurate and the inferred initial ratios were too high.

At about this time, other workers were addressing the ^{60}Fe problem by measuring bulk meteorite samples and bulk chondrules with ICPMS, a technique capable of much higher precision in the isotope ratios, allowing measurements of samples with lower Fe/Ni ratios, but at the cost of losing petrographic context. Tang and Dauphas (2012) presented data for a large number of such samples and found that HED meteorites and the D'Orbigny angrite gave isochrons implying $(^{60}\text{Fe}/^{56}\text{Fe})_0$ ratios of ~3.4 $\times$ 10^{-9}. Projecting this ratio back in time gave an $(^{60}\text{Fe}/^{56}\text{Fe})_0$ ratio for the solar system of ~1.17 $\times$ 10^{-8}. Tang and Dauphas also measured a group of individual chondrules from ordinary chondrites. Although the initial ratio from their isochron was not resolved from zero, it was consistent with the ~1 $\times$ 10^{-8} derived from achondrites.

Work also continued using SIMS with a revised measurement protocol to minimize ratio bias. SIMS continued to show excesses of ^{60}Ni generally correlated with Fe/Ni ratio, but the excesses were smaller and not as well correlated (e.g. Telus et al., 2018). This suggested that the ^{60}Fe–^{60}Ni system might be disturbed. Using synchrotron X-ray fluorescence, Telus et al. (2016) found that essentially all chondrules in the most primitive chondrites show evidence that iron and nickel have been mobile. This explains the isotopic disturbance observed in the chondrules. Telus et al. (2018) used synchrotron mapping of iron and nickel to guide new measurements of chondrules from unequilibrated ordinary chondrites. They still did not get clean isochrons, but they did get clear excesses of ^{60}Ni that could only be explained by decay of ^{60}Fe. Modeling these results, they inferred $(^{60}\text{Fe}/^{56}\text{Fe})_0$ ratios for the early solar system between ~8.5 $\times$ 10^{-8} and ~5 $\times$ 10^{-7}. Another group continued to get higher initial ratios by SIMS (~7 $\times$ 10^{-7}; e.g. Mishra et al., 2016). Trappitsch et al. (2018) used the newly developed technique of resonance ionization mass spectrometry (RIMS) to measure a chondrule previously measured by SIMS. RIMS measures a much higher percentage of the atoms in the sample and, when fully developed, will get much better precision for ^{60}Fe–^{60}Ni measurements. The RIMS data were consistent with previous SIMS data, indicating a solar system $(^{60}\text{Fe}/^{56}\text{Fe})_0$ ratio of ~(7 $\pm$ 12) $\times$ 10^{-8}. The range of initial ratios obtained by ICPMS, SIMS, and RIMS is currently too large to constrain the origin of short-lived radionuclides in the solar system.

9.5.5.2 *Technical Details*

Nickel-60 is the second most abundant nickel isotope (26.2%), so either very high Fe/Ni ratios or very high-precision measurements are required to detect radiogenic

^{60}Ni*. For SIMS, Fe/Ni ratios approaching 10^4 are required to detect ^{60}Fe at the abundance it has been found in chondritic materials. For multicollector ICPMS and TIMS, both of which can achieve precisions on the order of 10 ppm, samples with elemental ratios of less than 100 can give resolvable excesses of ^{60}Ni. For these studies, the iron and nickel must be separated from bulk samples and purified in ion-exchange columns.

9.5.5.3 *Applications*

Because ^{60}Fe can only be synthesized in stars, it will likely play an important role in understanding the source of the latest contribution of newly synthesized elements to the early solar system and the environment in which the solar system formed (see Chapters 10 and 15). But the solar system $(^{60}Fe/^{56}Fe)_0$ ratio will have to be refined considerably in order to distinguish between stellar environments.

The ^{60}Fe–^{60}Ni system has the potential to be a useful chronometer for early solar system events, but it is far from mature. Early work showed that iron-rich olivine and pyroxene exclude nickel when they crystallize, and both are stable against the degrees of metamorphism found in type 3.0–3.1 chondrites. However, iron and nickel are mobile in the presence of even small amounts of water, such as affected most primitive chondrites. A way around this problem might be to measure minerals that formed by precipitation from the aqueous fluids (e.g. fayalite and hedenbergite) and were not subsequently disturbed.

If abundant enough, ^{60}Fe could also have been a significant heat source in the early solar system.

9.5.6 The ^{107}Pd–^{107}Ag System

Palladium-107 β^- decays to ^{107}Ag with a half-life of 6.5 Myr ($\lambda = 1.07 \times 10^{-7}$ yr^{-1}) and is suitable for dating events during the first ~40 Myr of solar system history. Both elements are siderophile, but silver will also partition into sulfide. Melting and segregation of metal and troilite produces segregation of palladium from silver, but major chemical fractionation of palladium from silver primarily reflects volatility control. Palladium condenses at ~1320 K, while silver condenses at ~993 K (Table 7.1), so palladium is expected to be highly enriched relative to silver in metal that is depleted in volatile elements.

9.5.6.1 *History*

Following the discovery that ^{129}I was present in the early solar system, searches were carried out for evidence of other short-lived nuclides. In the early 1960s, there were reports of excess ^{107}Ag in iron meteorites, but these reports turned out to be in error and the search for evidence of ^{107}Pd was suspended for over a decade. The discovery that ^{26}Al was present in the early solar system renewed the interest in searching for evidence of ^{107}Pd. Kelly and Larimer (1977) suggested that IVB irons, which are strongly depleted in volatile siderophile elements and have high Pd/Ag ratios, might be a good place to look for evidence of ^{107}Pd. Based on this suggestion, Kelly and Wasserburg (1978), using new chemical procedures and careful mass spectrometry, found clear excesses of ^{107}Ag that correlated with Pd/Ag in the Santa Clara (IVB) meteorite. Kaiser and Wasserburg (1983) confirmed this initial work and extended the observations to include several other IVB meteorites, as well as meteorites from other classes. However, this work also showed that silver can be mobile in iron meteorites, making age determinations problematic (Chen and Wasserburg, 1990). Development of MC-ICPMS improved measurement sensitivity for these elements by more than an order of magnitude (Carlson and Hauri, 2001). The higher sensitivity has permitted isotopic measurements of chondrites and has led to a new determination for the initial ^{107}Pd/^{108}Pd ratio in the early solar system that appears to produce a chronology for iron meteorites that is consistent with those of other isotopic systems (Schönbächler et al., 2008).

9.5.6.2 *Technical Details*

Measurement of the ^{107}Pd–^{107}Ag system requires careful chemistry to remove terrestrial silver and palladium and to eliminate interferences in the mass spectrometer. Because silver has only two isotopes, great care must be taken to account for mass-dependent fractionation, both in the mass spectrometer and in the samples. Interpretation of the ^{107}Pd–^{107}Ag system in iron meteorites is not straightforward. In many iron meteorites, fractionation of palladium from silver took place early, followed by an extended period of slow cooling during which time the radiogenic silver could migrate throughout a body. Shock metamorphism may affect the ^{107}Pd–^{107}Ag systematics, and cosmic-ray spallation can produce palladium and silver isotopes both during and after the decay of the original inventory of ^{107}Pd.

9.5.6.3 *Applications*

The ^{107}Pd–^{107}Ag system has been applied most effectively to dating iron meteorites and pallasites (Chen and Wasserburg, 1990; Carlson and Hauri, 2001; Theis et al., 2013). Measurements of carbonaceous chondrites (Schönbächler et al., 2008) have provided a reasonably precise estimate of $(^{107}Pd/^{108}Pd)_0$ ratio in the early solar system (~5.9×10^{-5}), which is a key parameter for using the ^{107}Pd–^{107}Ag system as a chronometer.

9.5.7 The ^{146}Sm–^{142}Nd System

Samarium-146 α decays to ^{142}Nd with a half-life of 103 Myr ($\lambda_\alpha = 6.7 \times 10^{-9}$ yr^{-1}). Its half-life is too long to provide a high-resolution chronometer of the first few million years of solar system history, but it can provide time information about the first several hundred million years. As with the long-lived ^{147}Sm–^{143}Nd system, both parent and daughter are REEs that are not fractionated very much by melting and crystallization or evaporation and condensation. When used together with the ^{147}Sm–^{143}Nd system, the ^{146}Sm–^{142}Nd system can help identify objects that are isotopically disturbed in spite of apparently precise isochrons.

9.5.7.1 *History*

The possible importance of ^{146}Sm in the early solar system was first suggested by Kohman (1954), although he cautioned that it might have only limited usefulness due to the small degree of chemical fractionation between parent and daughter in natural systems. Audouze and Schramm (1972) suggested that it might be used as a chronometer for the synthesis of *p*-process elements. The first search for evidence of ^{146}Sm in early solar system materials was carried out by Lugmair et al. (1975b) in the Juvinas eucrite. Although they only found a hint of its presence, their work prompted Jacobsen and Wasserburg (1984) to extend the search to several other achondrites. But again, only hints of ^{146}Sm were found. Prinzhofer et al. (1992) and Stewart et al. (1994) found clear evidence of live ^{146}Sm in several meteorites and were able to establish an initial ^{146}Sm/^{144}Sm ratio for the early solar system of 0.008–0.009. Subsequent work has concentrated on using the ^{146}Sm–^{142}Nd system along with the long-lived ^{147}Sm–^{143}Nd system to investigate planetary accretion and differentiation.

9.5.7.2 *Technical Details*

Chemistry and mass spectrometry for the ^{146}Sm–^{142}Nd system are basically the same as for the long-lived ^{147}Sm–^{143}Nd system. The low abundance of the *p*-process isotopes and the small degree of elemental fractionation of parent and daughter in natural systems result in relatively large uncertainties on the isochrons and initial ratios determined from meteorites. In studies of planetary accretion and differentiation, the isotopic shifts in ^{142}Nd/^{144}Nd are reported in parts per 10,000 relative to the terrestrial standard:

$$\varepsilon^{142}\mathrm{Nd} = \left[\frac{\left(\frac{^{142}\mathrm{Nd}}{^{144}\mathrm{Nd}}\right)_{sample}}{\left(\frac{^{142}\mathrm{Nd}}{^{144}\mathrm{Nd}}\right)_{std}} - 1\right] \times 10^4 \qquad 9.59$$

9.5.7.3 *Applications*

The ^{146}Sm–^{142}Nd system is used primarily to investigate the nature and timing of the accretion of planetary bodies, such as the Earth, the Moon, Mars, and Vesta. Current work is aimed at unraveling the roles of nebular and planetary processes in the isotopic systematics of these bodies.

9.5.8 The ^{182}Hf–^{182}W System

Hafnium-182 β^- decays to ^{182}W with a half-life of 8.9 Myr ($\lambda = 7.8 \times 10^{-8}$ yr^{-1}). (Osmium-186 also α decays to ^{182}W, but its half-life is so long [2×10^{15} yr] that additions to ^{182}W from this source are insignificant.) The ^{182}Hf–^{182}W system is thus useful for events occurring in the first ~60 Myr of solar system history. Both hafnium and tungsten are highly refractory elements and therefore are expected to occur in chondritic proportions in most solar system objects. Hafnium is strongly lithophile and tungsten is moderately siderophile. Thus, core formation and metal–silicate fractionation are events well suited to ^{182}Hf–^{182}W dating. Within silicates, fractionation of hafnium from tungsten can occur because tungsten is more incompatible than hafnium. Tungsten is one of the most incompatible elements, while hafnium can be incorporated into high-calcium pyroxene and ilmenite.

9.5.8.1 *History*

The ^{182}Hf–^{182}W system was slow to develop because it is extremely difficult to measure tungsten by TIMS. The development of NTIMS allowed some early measurements to be made (e.g. Völkening et al., 1991; Harper and Jacobsen, 1996), but it took the development of MC-ICPMS to make this system truly accessible to study (e.g. Halliday et al., 1996). Initial interest in the ^{182}Hf–^{182}W system was to date core formation (e.g. Harper and Jacobsen, 1996), but Hf–W fractionation also occurred during mantle differentiation. This system has been used to determine the formation times and early history of both magmatic and nonmagmatic iron meteorites. Hafnium is also partitioned among silicate phases, permitting investigation of the chronology of many types of stony meteorites. These studies showed that the parent bodies of differentiated meteorites accreted earlier than those of chondritic meteorites, consistent with the ^{26}Al heat source being more powerful earlier in solar system history (Kleine et al., 2005). High-precision measurements of CAIs have generated a well-defined isochron (Fig. 9.27) that defines the initial ε^{182}W and ^{182}Hf/^{180}Hf at the time of solar system formation (Kruijer et al., 2014). A good review of the ^{182}Hf–^{182}W system is given by Kleine and Walker (2017).

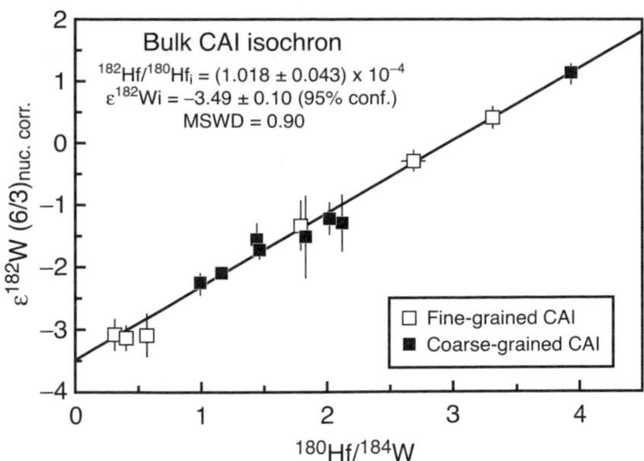

Figure 9.27 ^{182}Hf–^{182}W data for coarse-grained and fine-grained CAIs, mostly from Allende. The slope of the isochron gives the best estimate of the $(^{182}Hf/^{180}Hf)_0$ of the solar system, and the intercept gives the best estimate for the initial $\varepsilon^{182}Hf$. The data were normalized to $^{186}W/^{183}W$ to correct for instrument effects and were corrected for nucleosynthetic effects in tungsten based on a correlation between $\varepsilon^{182}W$ and $\varepsilon^{183}W$. After Kruijer et al. (2014); see this paper for detailed discussion.

9.5.8.2 *Technical Details*

The most reliable measurements of the ^{182}Hf–^{182}W system are made by MC-ICPMS, although NTIMS is sometimes used as well. The elements must be isolated from the sample and purified using ion-exchange chromatography. Excesses of ^{182}W are typically reported relative to ^{184}W, the most abundant isotope, and are expressed as deviations in parts per 10,000 from the ratio in the bulk silicate Earth (BSE):

$$\varepsilon^{182}W = \left[\frac{\left(\frac{^{182}W}{^{184}W}\right)_{sample}}{\left(\frac{^{182}W}{^{184}W}\right)_{BSE}} - 1\right] \times 10^4 \qquad 9.60$$

Chronological interpretations may be affected by both nucleosynthetic tungsten isotope variations (e.g. in CAIs, Kruijer et al., 2014) and by contributions of cosmogenic ^{182}W (e.g. in lunar samples) or cosmogenic burnout of ^{182}W (in iron meteorites) (Kleine and Walker, 2017).

9.5.8.3 *Applications*

The ^{182}Hf–^{182}W system is particularly powerful for constraining the timescales for accretion and metal–silicate fractionation (e.g. core formation). Because no fractionation of hafnium and tungsten is expected based on volatility, most solar system objects are expected to have a chondritic hafnium/tungsten ratio. If a body remains unfractionated while ^{182}Hf is alive, all parts of the body will have the same tungsten isotopic composition. Carbonaceous chondrites fit this criterion. But Hf–W fractionations within a body provide chronological information. Other types of chondrites and their constituents and differentiated bodies exhibit such internal fractionations. The 8.9 Myr half-life means that the ^{182}Hf–^{182}W system can investigate the first ~50 Myr of solar system history. The system has been applied to CAIs, to establish the $(^{182}W/^{184}W)_0$ for the solar system. It has been applied to all types of meteorites with particular emphasis on metal–silicate fractionation.

9.5.9 The ^{10}Be–^{10}B System

Beryllium-10 β^- decays to ^{10}B with a half-life of 1.39 Myr ($\lambda = 5.0 \times 10^{-7}$ yr^{-1}). Beryllium and boron (along with lithium) are several orders of magnitude less abundant than the other light elements in the solar system because, except for ^7Li, they are not produced in stars. They are produced when high-energy cosmic rays, mostly protons, fragment atomic nuclei into smaller pieces in a process called *spallation*. Beryllium-10 is constantly being produced at low levels by spallation in the solar system, and its abundance in bulk meteorites is used to estimate their cosmic-ray exposure ages (see Chapter 10). Beryllium is the lightest alkaline earth element. It has the same electronic structure as, and commonly substitutes for, magnesium. Boron is a Group 3A element and has the same electronic structure as aluminum. Beryllium is a refractory element that condenses at ~1490 K, whereas boron, a moderately volatile element, condenses at ~964 K (Table 7.1). This strong contrast in volatility leads to high Be/B ratios in refractory materials.

9.5.9.1 *History*

The first clear evidence that ^{10}Be was alive in the early solar system was provided by McKeegan et al. (2000), who found large excesses of ^{10}B that correlated with $^9Be/^{11}B$ ratio in an Allende type B CAI. These results were quickly confirmed in other type B and in type A CAIs, and an initial ratio $((^{10}Be/^9Be)_0)$ for the early solar system of ~1 × 10^{-3} was inferred. An irradiation origin was proposed, and the high $(^{10}Be/^9Be)_0$ implied production by solar cosmic rays. At this time, the solar system abundance of ^{60}Fe, which is only produced in stars, was thought to be very low, so a model that produced all short-lived radionuclides by irradiation with solar cosmic rays was proposed (Gounelle et al., 2001). The model predicted that the ^{10}Be and ^{26}Al abundances should be correlated, but this turned out not to be the case (Marhas et al., 2002; MacPherson et al., 2003; Srinivasan et al., 2013) and this universal-irradiation model lost favor. Desch et al. (2004) proposed that the ^{10}Be might have been inherited from the Sun's parent molecular cloud, where it accumulated as trapped cosmic rays. This model did not gain much traction, and it is now generally thought that ^{10}Be in early solar system material originated through spallation driven by solar cosmic rays.

Another short-lived isotope of beryllium, ^{7}Be, decays to ^{7}Li by electron capture with a half-life of ~53 days. This half-life is so short that any atoms present in chondrite components must have been produced in the solar system. A hint of the presence of ^{7}Be in the form of large excesses of ^{7}Li in an Allende CAI was presented by Chaussidon et al. (2006). However, these authors were not able to demonstrate a tight correlation with the ^{9}Be/^{6}Li ratio.

9.5.9.2 *Applications*

To date, the ^{10}Be–^{10}B system has not been shown to be useful as a chronometer for the early solar system. A prerequisite for being able to use a short-lived nuclide as a chronometer is that it was homogeneously distributed, so that differences in initial ^{10}Be/^{9}Be ratios among objects can be interpreted in terms of time. It is unlikely that the ^{10}Be–^{10}B system will ever be very useful as a chronometer because most of the ^{10}Be was probably locally produced and heterogeneously distributed in time and space. However, ^{10}Be may be extremely useful as a measure of the irradiation environment experienced by various objects in the early solar system.

Beryllium-10, like ^{14}C, ^{26}Al, and ^{36}Cl, is used to infer the cosmic-ray exposure history of a meteorite or a planetary surface.

Summary

In this chapter, we have discussed the basic principles of using radioactive isotopes as chronometers. The requirements that a sample must meet in order to provide a valid age are:

1. The decay constants must have remained constant over the age of the solar system. We know this to be true from the concordance of chronologies based on radioactive isotopes that decay by very different mechanisms and by the identical time evolution of radioactivity-driven light curves of supernovae that exploded 10 billion years ago and a few million years ago.
2. The sample must have experienced an event that homogenized the isotopic compositions of parent and daughter elements. This starts the clock. Ideally, the homogenizing event will also produce a fractionation resulting in two or more phases with different parent-to-daughter elemental ratios, permitting construction of an isochron.
3. The system must have remained closed to exchange with the outside since the event being dated, and in order for an internal isochron to be produced, the system must also have been closed to internal redistribution of the isotopes.

One of the most difficult aspects of radiometric dating is finding samples that clearly meet these last two requirements.

We discussed two different types of radiochronometers. Those based on long-lived radionuclides for which a portion of the primordial abundance is still present provide absolute ages relative to the present time on suitable samples. Examples of how these chronometers are used to date individual objects (chondrules, CAIs, achondrites) and fractionation events (planetary differentiation, magma generation) were discussed.

Short-lived radionuclides can provide high-resolution relative chronometers. The shorter the half-life, the higher the time resolution. Short-lived systems currently in use can resolve time differences of a few hundred thousand years at 4.57 Ga. But in order to get absolute time, the short-lived systems must be tied to a long-lived radionuclide.

The individual isotope systems differ in the objects that they can date, the details of how they are measured, and the specifics of how they are applied. Some systems are best suited to dating fractionation events based on volatility. Others are best suited to dating igneous processes. Still others are most useful for secondary events. Together they provide the keys to unlocking the time history of the solar system. In Chapter 10, we will summarize the current state of knowledge of the chronology of the solar system obtained from these radiochronometers.

Questions

1. Radiometric dating is one way to get age information. What are some of the other ways of measuring time? How do radiochronometers complement these methods?
2. Why is it necessary for an object to be dated by a radiochronometer to have been isotopically homogeneous when it formed?
3. Which radiochronometer(s) would be best suited to determine the crystallization age of a basaltic meteorite? What are some of the complications that might result in an incorrect age determination?

4. What are the best radiochronometers to use to determine the formation times of CAIs and chondrules? Why?
5. What radiochronometers are best suited for dating aqueous alteration in chondritic meteorites? Why?
6. Which radiochronometers use nuclear reactions as part of the analytical procedure? What are the advantages of this approach?

Suggestions for Further Reading

Faure G., and Mensing T. M. (2005) *Isotopes: Principles and Applications, 3rd Edition.* John Wiley and Sons, Hoboken, 897 pp. A tremendous resource for details about various techniques of radiometric dating.

Nyquist L. E., Bogard D. D., and Shif C.-Y. (2001a) Radiometric chronology of the Moon and Mars. In *The Century of Space Science*, Bleeker A. M., Geiss J., and Huber M. C., editors, pp. 1325–1376, Klewer Academic Publishers, Amsterdam. An easy-to-read review of the history of radiometric chronology as it relates to the Moon and Mars.

Wasserburg G. J., and Papanastassiou D. A. (1982) Some short-lived nuclides in the early solar system – a connection with the placental ISM. In *Essays in Nuclear Astrophysics*, Barnes C. A., Clayton D. D., and Schramm D. N., editors, pp. 77–140, Cambridge University Press, Cambridge. A good review of the early history of chronology using short-lived radionuclides.

Other References

Aldrich L. T., and Nier A. O. (1948) Argon-40 in potassium minerals. *Physical Review*, **74**, 876–877.

Amelin Y. (2005) Meteorite phosphates show constant ^{176}Lu decay rate since 4557 million years ago. *Science*, **310**, 839–841.

Audouze J., and Schramm D. M. (1972) ^{146}Sm: A chronometer for *p*-process nucleosynthesis. *Nature*, **237**, 447–449.

Birck J. L., and Allègre C. J. (1985) Evidence for the presence of ^{53}Mn in the early solar system. *Geophysical Research Letters*, **12**, 745–748.

Birck J. L., and Allègre C. J. (1988) Manganese chromium isotope systematics and the development of the early solar system. *Nature*, **331**, 579–584.

Birck J. L., and Lugmair G. W. (1988) Nickel and chromium isotopes in Allende inclusions. *Earth & Planetary Science Letters*, **90**, 131–143.

Birck J. L., Rotaru M., and Allègre C. J. (1999) ^{53}Mn-^{53}Cr evolution of the early solar system. *Geochimica et Cosmochimica Acta*, **63**, 4111–4117.

Blichert-Toft J., Boyet M., Télouk P., and Albarède F. (2002) ^{147}Sm-^{143}Nd and ^{176}Lu-^{176}Hf in eucrites and the differentiation of the HED parent body. *Earth & Planetary Science Letters*, **204**, 167–181.

Bollard J., Connelly J. N., Whitehouse M. J., et al. (2017) Early formation of planetary building blocks inferred from Pb isotopic ages of chondrules. *Science Advances*, **3**, e1700407.

Bouvier A., Vervoort J. D., and Patchett P. J. (2008) The Lu-Hf and Sm-Nd isotopic composition of CHUR: Constraints from unequilibrated chondrites and implications for the bulk compositions of the terrestrial planets. *Earth & Planetary Science Letters*, **273**, 48–57.

Brazzle R. H., Pravdivtseva O. V., Meshik A. P., and Hohenberg C. M. (1999) Verification and interpretation of the I-Xe chronometer. *Geochimica et Cosmochimica Acta*, **63**, 739–760.

Brennecka G. A., Weyer S., Wadhwa M., et al. (2010) ^{238}U/^{235}U variations in meteorites: Extant ^{247}Cm and implications for Pb-Pb dating. *Science*, **327**, 449–451.

Brown H. (1947) An experimental method for the estimation of the age of the elements. *Physical Review*, **72**, 348.

Campbell N. R., and Wood A. (1906) The radioactivity of the alkali metals. *Proceedings of the Cambridge Philosophical Society*, **14**, 15–21.

Carlson R. W., and Hauri E. H. (2001) Extending the ^{107}Pd-^{107}Ag chronometer to low Pd/Ag meteorites with multicollector plasma-ionization mass spectrometry. *Geochimica et Cosmochimica Acta*, **65**, 1839–1848.

Chaussidon M., Robert F., and McKeegan K. D. (2006) Li and B isotopic variations in an Allende CAI: Evidence for the *in situ* decay of short-lived B-10 and for the possible presence of the short-lived nuclide Be-7 in the early solar system. *Geochimica et Cosmochimica Acta*, **70**, 224–245.

Chen J., and Wasserburg G. J. (1990) The isotopic composition of Ag in meteorites and the presence of ^{107}Pd in protoplanets. *Geochimica et Cosmochimica Acta*, **54**, 1729–1743.

Chen J. H., Papanastassiou D. A., and Wasserburg G. J. (1998) Re-Os systematics in chondrites and the fractionation of the platinum group elements in the early solar system. *Geochimica et Cosmochimica Acta*, **62**, 3379–3392.

Chmeleff J., von Blanckenburg R., Kossert K., and Jakob D. (2010) Determination of the ^{10}Be half-life by multicollection ICP-MS and liquid scintillation counting. *Nuclear Instruments & Methods in Physics Research Section B: Beam Interactions with Materials & Atoms*, **268**, 192–199.

Connelly J. N., Bizzarro M., Krot A. N., et al. (2012) The absolute chronology and thermal processing of solids in the solar protoplanetary disk. *Science*, **338**, 651–655.

Creaser R. A., Papanastassiou D. A., and Wasserburg G. J. (1991) Negative thermal ion mass spectrometry of osmium, rhenium, and iridium. *Geochimica et Cosmochimica Acta*, **55**, 397–401.

Davis A. M., and McKeegan K. D. (2014) Short-lived radionuclides and early solar system chronology. In *Treatise on Geochemistry, 2nd Edition, Vol. 1: Meteorites and Cosmochemical Processes*, Davis A. M., editor, pp. 361–395, Elsevier, Oxford.

Debaille V., Van Orman J., Yin Q.-Z., and Amelin Y. (2017) The role of phosphates for the Lu-Hf chronology of meteorites. *Earth & Planetary Science Letters*, **473**, 52–61.

DePaolo D. J., and Wasserburg G. J. (1976a) Nd isotopic variations and petrogenetic models. *Geophysical Research Letters*, **3**, 249–252.

DePaolo D. J., and Wasserburg G. J. (1976b) Inferences about magma sources and mantle structure from variations of ^{143}Nd/^{144}Nd. *Geophysical Research Letters*, **3**, 743–746.

Desch S. J., Connolly H. C., Jr., and Srinivasan G. (2004) An interstellar origin for the beryllium-10 in calcium-rich, aluminum-rich inclusions. *Astrophysical Journal*, **602**, 528–542.

Doyle P. M., Jogo K., Nagashima K., et al. (2015) Early aqueous activity on the ordinary and carbonaceous chondrite parent bodies recorded by fayalite. *Nature Communications*, **6**, 7444.

Edmunson J., Borg L. E., Nyquist L. E., and Asmerom Y. (2009) A combined Sm-Nd, Rb-Sr, and U-Pb isotopic study of Mg-suite norite 78238: Further evidence for early differentiation of the Moon. *Geochimica et Cosmochimica Acta*, **73**, 514–527.

Endress M., Zinner E., and Bischoff A. (1996) Early aqueous activity on primitive meteorite parent bodies. *Nature*, **379**, 701–703.

Faure G. (1986) *Principles of Isotope Geology, 2nd Edition*. John Wiley and Sons, New York.

Fujiya W., Sugiura N., Sano Y., and Hiyagon H. (2013) Mn-Cr ages of dolomites in CI chondrites and the Tagish Lake ungrouped carbonaceous chondrite. *Earth & Planetary Science Letters*, **362**, 130–142.

Gale N. H., and Mussett A. E. (1973) Episodic uranium-lead models and interpretation of variations in isotopic composition of lead in rocks. *Reviews in Geophysics*, **11**, 37–86.

Garner E. L., Murphy T. J., Bramlich J. W., et al. (1975) Absolute isotopic abundance ratios and the atomic weight of a reference sample of potassium. *Journal of Research of the National Bureau of Standards – A. Physics and Chemistry*, **79A**, 713–725.

Gounelle M., Shu F. H., Shang H., et al. (2001) Extinct radioactivities and protosolar cosmic rays: Self-shielding and light elements. *Astrophysical Journal*, **548**, 1051–1070.

Gray C. M., Papanastassiou D. A., and Wasserburg G. J. (1973) The identification of early condensates from the solar nebula. *Icarus*, **20**, 213–239.

Gray C. M., and Compston W. (1974) Excess ^{26}Mg in the Allende meteorite. *Nature*, **251**, 495–497.

Guan Y., Huss G. R., and Leshin L. A. (2007) ^{60}Fe-^{60}Ni and ^{53}Mn-^{53}Cr isotope systems in sulfides from unequilibrated ordinary chondrites. *Geochimica et Cosmochimica Acta*, **71**, 4082–4091.

Hahn O., and Walling E. (1938) Über die Möglichkeit geologischer Altersbestimmungen rubidiumhaltiger Mineralen and Gesteine. *Zeitschrift Anorganishen Allgemeine Chemie*, **236**, 78–82.

Halliday A. N., Rehkämper M., Lee D.-C., and Yi W. (1996) Early evolution of the Earth and Moon: New constraints from Hf-W isotope geochemistry. *Earth & Planetary Science Letters*, **142**, 75–89.

Harper C. L., and Jacobsen S. B. (1996) Evidence for ^{182}Hf in the early solar system and constraints on the timescale of terrestrial core formation. *Geochimica et Cosmochimica Acta*, **60**, 1131–1153.

Hemmendinger A., and Smythe W. R. (1937) The radioactive isotope of rubidium. *Physical Review*, **51**, 1052–1053.

Hirt B., Tilton G. R., and Hoffmeister W. (1963) The half-life of ^{187}Re. In *Earth Science and Meteorites*, Geiss J., and Goldberg E. D., editors, pp. 273–280, North Holland, Amsterdam.

Hohenberg C. M., and Pravdivtseva O. V. (2008) I-Xe dating: From adolescence to maturity. *Chemie der Erde*, **68**, 339–351.

Holden N. E. (1990) Total half-lives for selected nuclides. *Pure & Applied Chemistry*, **62**, 941–958.

Holmes A. (1946) An estimate of the age of the Earth. *Nature*, **157**, 680–684.

Holst J. C., Olsen M. B., Paton C., et al. (2013) ^{182}Hf–^{182}W age dating of a ^{26}Al-poor inclusion and implications for the origin of the short-lived radioisotopes in the early solar system. *Proceedings of the National Academy of Sciences, USA*, **110**, 8819–8823.

Houtermans F. G. (1946) Die Isotopenhäufigkeiten im natürlichen Blei und das Alter den Urans. *Naturwissenschaften*, **33**, 186–219.

Hult M., Vidmar R., Rosengard U., et al. (2014) Half-life measurements of lutetium-176 using underground HPGe detectors. *Applied Radiation & Isotopes*, **87**, 112–117.

Hutcheon I. D. (1982) Ion probe magnesium isotope measurements of Allende inclusions. *ACS Symposium Series*, **176**, 95–128.

Hutcheon I. D., Armstrong J. T., and Wasserburg G. J. (1984) Excess ^{41}K in Allende CAI: A hint re-examined (abstract). *Meteoritics*, **19**, 243–244.

Hutcheon I. D., Krot A. N., Keil K., et al. (1998) ^{53}Mn/^{53}Cr dating of fayalite formation in the CV3 chondrite Mokoia: Evidence for asteroidal alteration. *Science*, **282**, 1865–1867.

Inghram M. G. (1954) Stable isotope dilution as an analytical tool. *Annual Review of Nuclear Science*, **4**, 81–92.

Jacobsen B., Yin Q.-Z., Moynier F., et al. (2008) ^{26}Al-^{26}Mg and ^{207}Pb-^{206}Pb systematics of Allende CAIs: Canonical solar initial ^{26}Al/^{27}Al ratio reinstated. *Earth & Planetary Science Letters*, **272**, 353–364.

Jacobsen S. B., and Wasserburg G. J. (1980) Sm-Nd isotopic evolution of chondrites. *Earth & Planetary Science Letters*, **50**, 139–155.

Jacobsen S. B., and Wasserburg G. J. (1984) Sm-Nd isotopic evolution of chondrites and achondrites. 2. *Earth & Planetary Science Letters*, **67**, 137–150.

Jeffery P. M., and Reynolds J. H. (1961) Origin of excess Xe^{129} in stone meteorites. *Journal of Geophysical Research*, **66**, 3582–3583.

Jilly-Rehak C. E., Huss G. R., and Nagashima K. (2017) ^{53}Mn-^{53}Cr radiometric dating of secondary carbonates in CR chondrites: Timescales for parent body aqueous alteration. *Geochimica et Cosmochimica Acta*, **201**, 224–244.

Kaiser T., and Wasserburg G. J. (1983) The isotopic composition and concentration of Ag in iron meteorites. *Geochimica et Cosmochimica Acta*, **47**, 43–58.

Kelly W. R., and Larimer J. W. (1977) Chemical fractionations in meteorites, VIII. Iron meteorites and the cosmochemical history of the metal phase. *Geochimica et Cosmochimica Acta*, **41**, 93–111.

Kelly W. R., and Wasserburg G. J. (1978) Evidence for the existence of ^{107}Pd in the early solar system. *Geophysical Research Letters*, **5**, 1079–1082.

Kohman T. P. (1954) Geochronological significance of extinct natural radioactivity. *Science*, **119**, 851–852.

Kita N. T., Fukuda K., Siron G., and Kimura M. (2020) Younger Al-Mg ages of chondrules in CO chondrites than L/LL chondrites. *Goldschmidt Abstracts*, **2020**, 1329.

Kleine T., and Walker R. J. (2017) Tungsten isotopes in planets. *Annual Review of Earth & Planetary Sciences*, **45**, 389–417.

Kleine T., Mezger K., Palme H., et al. (2005) Early core formation in asteroids and late accretion of chondrite parent bodies: Evidence from ^{182}Hf–^{182}W in CAIs, metal-rich chondrites, and iron meteorites. *Geochimica et Cosmochimica Acta*, **69**, 5805–5818.

Korschinek G., Bergmaier A., Faestermann T., et al. (2010) A new value for the half-life of ^{10}Be by heavy-ion elastic recoil detection and liquid scintillation counting. *Nuclear Instruments & Methods in Physics Research Section B: Beam Interactions with Materials & Atoms*, **268**, 187–191.

Kruijer T. S., Kleine T., Fischer-Godde M., Burkhardt C., and Wieler R. (2014) Nucleosynthetic W isotope anomalies and the Hf-W chronometry of Ca-Al-rich inclusions. *Earth & Planetary Science Letters*, **403**, 317–327.

Larsen K. K., Trinquier A., Paton C., et al. (2011) Evidence for magnesium isotope heterogeneity in the solar protoplanetary disk. *Astrophysical Journal Letters*, **735**, L37.

Larsen K. K., Wielandt D., Schiller M., et al. (2020) Episodic formation of refractory inclusions in the solar system and their presolar heritage. *Earth & Planetary Science Letters*, **535**, 116088.

Lee T., and Papanastassiou D. A. (1974) Mg isotopic anomalies in the Allende meteorite and correlation with O and Sr effects. *Geophysical Research Letters*, **1**, 225–228.

Lee T., Papanastassiou D. A., and Wasserburg G. J. (1977) Aluminum-26 in the early solar system: Fossil or fuel? *Astrophysical Journal Letters*, **211**, L107–L110.

Lindner M., Leich D. A., Russ G. P., et al. (1989) Direct determination of the half-life of ^{187}Re. *Geochimica et Cosmochimica Acta*, **53**, 1597–1606.

Liu M.-C. (2017) The initial ^{41}Ca/^{40}Ca ratios in two type A Ca-Al-rich inclusions: Implications for the origin of short-lived ^{41}Ca. *Geochimica et Cosmochimica Acta*, **201**, 123–135.

Liu M.-C., Chaussidon M., Srinivasan G., and McKeegan K. D. (2012) A lower initial abundance of short-lived ^{41}Ca in the early solar system and its implications for solar system formation. *Astrophysical Journal*, **761**, 137.

Luck J.-M., and Allègre C. J. (1983) ^{187}Re-^{187}O systematics in meteorites and cosmochemical consequences. *Nature*, **302**, 130–132.

Luck J.-M., Birck J. L., and Allègre C. J. (1980) ^{187}Re-^{187}O systematics in meteorites: Early chronology of the solar system and age of the galaxy. *Nature*, **283**, 256–259.

Ludwig K. R. (2003) Isoplot-3.00, a Geochronological Toolkit for Microsoft Excel. *Berkeley Geochronology Center Special Publication No. 4*, 70 pp.

Lugmair G. W., and Galer S. J. G. (1992) Age and isotopic relationships among the angrites Lewis Cliff 86010 and Angra dos Reis. *Geochimica et Cosmochimica Acta*, **56**, 1673–1694.

Lugmair G. W., and Shukolyukov A. (1998) Early solar system timescales according to the ^{53}Mn-^{53}Cr system. *Geochimica et Cosmochimica Acta*, **62**, 2863–2886.

Lugmair G. W., Scheinin N. B., and Marti K. (1975a) Sm-Nd age and history of Apollo 17 basalt 75075: Evidence for early differentiation of the lunar exterior. *Proceedings of the 6th Lunar Science Conference, Geochimica et Cosmochimica Acta Supplement*, **6**, 1419–1429.

Lugmair G. W., Scheinin N. B., and Marti K. (1975b) Search for extinct ^{146}Sm, 1. The isotopic abundance of ^{142}Nd in the Juvinas meteorite. *Earth & Planetary Science Letters*, **27**, 79–84.

MacPherson G. J., Huss G. R., and Davis A. M. (2003) Extinct ^{10}Be in type A calcium-aluminum-rich inclusions from CV chondrites. *Geochimica et Cosmochimica Acta*, **67**, 3165–3179.

Mahon K. I. (1996) The new "York" regression: Application of an improved statistical method to geochemistry. *International Geology Review*, **38**, 293–303.

Makide K., Nagashima K., Krot A. N., et al. (2011) Heterogeneous distribution of ^{26}Al at the birth of the solar system. *Astrophysical Journal Letters*, **733**, L31.

Marhas K. K., Goswami J. N., and Davis A. M. (2002) Short-lived nuclides in hibonite grains from Murchison: Evidence for solar system evolution. *Science*, **298**, 2182–2185.

Marks N. E., Borg L. E., Hutcheon I. D., et al. (2014) Samarium-neodymium chronology and rubidium-strontium systematics of an Allende calcium-aluminum-rich inclusion with implications for ^{146}Sm half life. *Earth & Planetary Science Letters*, **405**, 15–24.

Mattauch J. (1937) Das Paar Rb^{87}-Sr^{87} and die Isobarenregel. *Naturwissenschaften*, **25**, 189–191.

McDougall I., and Harrison M. T. (1988) Geochronology and Thermochronology by the ^{40}Ar/^{39}Ar method. In *Oxford Monographs on Geology and Geophysics No. 9*, Oxford University Press, Oxford, 212 pp.

McKeegan K. D., Chaussidon M., and Robert F. (2000) Incorporation of short-lived ^{10}Be in a calcium-aluminum-rich inclusion from the Allende meteorite. *Science*, **289**, 1334–1337.

Meisel T., Walker R. J., and Morgan J. W. (1996) The osmium isotopic composition of the Earth's primitive upper mantle. *Nature*, **383**, 517–520.

Merrihue C. M., and Turner G. (1966) Potassium-argon dating by activation with fast neutrons. *Journal of Geophysical Research*, **71**, 2852–2857.

Minster J. F., and Allègre C. J. (1981) ^{87}Rb-^{87}Sr dating of LL chondrites. *Earth & Planetary Science Letters*, **5**, 361–376.

Misawa K., Shih C.-Y., Reese Y., et al. (2006) Rb-Sr, Sm-Nd and Ar-Ar isotopic systematics of Martian dunite Chassigny. *Earth & Planetary Science Letters*, **246**, 90–101.

Mishra R. K., Marhas K. K., and Sameer L. (2016) Abundance of ^{60}Fe inferred from nanoSIMS study of QUE 97008 (L3.05) chondrules. *Earth & Planetary Science Letters*, **436**, 71–81.

Mostefaoui S., Lugmair G. W., and Hoppe P. (2005) ^{60}Fe: A heat source for planetary differentiation from a nearby supernova explosion. *Astrophysical Journal*, **625**, 271–277.

Nagashima K., Krot A. N., and Huss G. R. (2014) ^{26}Al in chondrules from CR2 chondrites. *Geochemical Journal*, **48**, 561–570.

Nagashima K., Krot A. N., and Komatsu M. (2017) ^{26}Al-^{26}Mg systematics in chondrules from Kaba and Yamato 980145 CV3 carbonaceous chondrites. *Geochimica et Cosmochimica Acta*, **201**, 303–319.

Nichols R. H., Hohenberg C. M., Kehm K., et al. (1994) I-Xe studies of the Acapulco meteorite: Absolute ages of individual phosphate grains and the Bjurböle standard. *Geochimica et Cosmochimica Acta*, **58**, 2523–2561.

Nier A. O. (1935) Evidence for the existence of an isotope of potassium of mass 40. *Physical Review*, **48**, 283–284.

Nier A. O. (1939a) The isotopic composition of uranium and the half-lives of uranium isotopes. *Physical Review*, **55**, 150–153.

Nier A. O. (1939b) The isotopic constitution of radiogenic leads and the measurement of geological time. II. *Physical Review*, **55**, 153–163.

Nier A. O. (1950) A redetermination of the relative abundances of the isotopes of carbon, nitrogen, oxygen, argon, and potassium. *Physical Review*, **77**, 789–793.

Nyquist L. E., Bogard D. D., Shih C.-Y., et al. (2001b) Ages and geologic histories of martian meteorites. *Space Science Reviews*, **96**, 105–164.

Nyquist L., Lindstrom D., Mittlefehldt D., et al. (2001c) Manganese-chromium formation intervals for chondrules from the Bushunpur and Chainpur meteorites. *Meteoritics & Planetary Science*, **36**, 911–938.

Nyquist L. E., Kleine T., Shih C.-Y., and Reese Y. D. (2009) The distribution of short-lived radioisotopes in the early solar system and the chronology of asteroid accretion, differentiation, and secondary mineralization. *Geochimica et Cosmochimica Acta*, **73**, 5115–5136.

Ogliore R. C., Huss G. R., and Nagashima K. (2011) Ratio estimation in SIMS analysis. *Nuclear Instruments & Methods in Physics Research, Section B: Beam interactions with Materials & Atoms*, **269**, 1910–1918.

Papanastassiou D. A., and Wasserburg G. J. (1969) Initial strontium isotopic abundances and the resolution of small time differences in the formation of planetary objects. *Earth & Planetary Science Letters*, **5**, 361–376.

Papanastassiou D. A., and G. J. Wasserburg (1971) Rb-Sr ages of igneous rocks from the Apollo 14 mission and the age of the Fra Mauro Formation. *Earth & Planetary Science Letters*, **12**, 36–48.

Papanastassiou D. A., Wasserburg G. J., and Burnett D. S. (1970) Rb-Sr ages of lunar rocks from the Sea of Tranquility. *Earth & Planetary Science Letters*, **8**, 1–19.

Patchett P. J., and Tatsumoto M. (1980) Lu-Hf total-rock isochron for eucrite meteorites. *Nature*, **288**, 571–574.

Patterson C. C. (1955) The Pb^{207}/Pb^{206} ages of some stone meteorites. *Geochimica et Cosmochimica Acta*, **7**, 151–153.

Patterson C. C. (1956) Age of meteorites and the earth. *Geochimica et Cosmochimica Acta*, **10**, 230–237.

Pravdivtseva O., Krot A. N., and Hohenberg C. M. (2018) I-Xe dating of aqueous alteration in the CI chondrite Orgueil: I. Magnetite and ferromagnetic separates. *Geochimica et. Cosmochimica Acta*, **227**, 38–47.

Prinzhofer A., Papanastassiou D. A., and Wasserburg G. J. (1992) Samarium-neodymium evolution of meteorites. *Geochimica et Cosmochimica Acta*, **56**, 797–815.

Reynolds J. H. (1960a) Determination of the age of the elements. *Physical Reviews Letters*, **4**, 8–10.

Reynolds J. H. (1960b) Isotopic composition of xenon from enstatite chondrites. *Zeitschrift für Naturforschung*, **15a**, 1112–1114.

Riches A. J. V., Day J. M. D., Walker R. J., et al. (2012) Rhenium-osmium isotope and highly-siderophile-element abundance systematics of angrite meteorites. *Earth & Planetary Science Letters*, **353–354**, 208–218.

Rugel G., Faestermann T., Knie K., et al. (2009) New measurement of the ^{60}Fe half life. *Physical Review Letters*, **103**, 072502.

Schönbächler M., Carlson R. W., Horan M. F., et al. (2008) Silver isotope variations in chondrites: Volatile depletion and the initial ^{107}Pd abundance of the solar system. *Geochimica et Cosmochimica Acta*, **72**, 5330–5341.

Schumacher E. (1956) Alterbestimmung von Steinmeteoriten mit der Rubidium-Strontium-Methode. *Zeitschrift für Naturforschung*, **11a**, 206.

Shen J. J., Papanastassiou D. A., and Wasserburg G. J. (1996) Precise Re-Os determinations and systematics of iron meteorites. *Geochimica et Cosmochimica Acta*, **60**, 2887–2900.

Shimizu N., Semet M. P., and Allègre C. J. (1978) Geochemical applications of quantitative ion microprobe analysis. *Geochimica et Cosmochimica Acta*, **42**, 1321–1334.

Shirey S. B., and Walker R. J. (1995) Carius tube digestions for low-blank rhenium-osmium analysis. *Analytical Chemistry*, **67**, 2136–2141.

Shirey S. B., and Walker R. J. (1998) The Re-Os isotope system in cosmochemistry and high-temperature geochemistry. *Annual Reviews of Earth & Planetary Science*, **26**, 423–500.

Shukolyukov A., and Lugmair G. W. (1993a) Live iron-60 in the early solar system. *Science*, **259**, 1138–1142.

Shukolyukov A., and Lugmair G. W. (1993b) ^{60}Fe in eucrites. *Earth & Planetary Science Letters*, **119**, 159–166.

Siron G., Fukuda K., Kimura M., and N. T. Kita (2020) Al-Mg chronology of anorthite-bearing chondrules from unequilibrated ordinary chondrites: Clues on short duration of chondrules formation. *Lunar and Planetary Science*, **51**, 1574.

Smoliar M. I., Walker R. J., and Morgan J. W. (1996) Re-Os ages of Groups IIA, IIIA, IVA, and IVB iron meteorites. *Science*, **271**, 1099–1102.

Smythe W. R., and Hemmendinger A. (1937) The radioactive isotope of potassium. *Physical Review*, **51**, 178–182.

Srinivasan G., and Chaussidon M. (2013) Constraints on ^{10}B and ^{41}Ca distribution in the early solar system from ^{26}Al and ^{10}Be studies of Efremovka CAIs. *Earth & Planetary Science Letters*, **374**, 11–23.

Srinivasan G., Ulyanov A. A., and Goswami J. N. (1994) ^{41}Ca in the early solar system. *Astrophysical Journal*, **431**, L67–L70.

Srinivasan G., Sahijpal S., Ulyanov A. A., and Goswami J. N. (1996) Ion microprobe studies of Efremovka CAIs: II. Potassium isotope compositions and ^{41}Ca in the early solar system. *Geochimica et Cosmochimica Acta*, **60**, 1823–1835.

Steiger R. H., and Jäger E. (1977) Subcommission on geochronology: Convention on the use of decay constants in geo- and cosmochronology. *Earth & Planetary Science Letters*, **36**, 359–362.

Stewart B. W., Papanastassiou D. A., and Wasserburg G. J. (1994) Sm-Nd chronology and petrogenesis of mesosiderites. *Geochimica et Cosmochimica Acta*, **58**, 3487–3509.

Tachibana S., and Huss G. R. (2003) The initial abundance of ^{60}Fe in the solar system. *Astrophysical Journal Letters*, **588**, L41–L44.

Tachibana S., Huss G. R., Kita N. T., et al. (2006) ^{60}Fe in chondrites: Debris from a nearby supernova in the early solar system? *Astrophysical Journal Letters*, **639**, L87–L90.

Tang H., and Dauphas N. (2012) Abundance, distribution, and origin of ^{60}Fe in the solar protoplanetary disk. *Earth & Planetary Science Letters*, **359–360**, 248–263.

Tang H., Liu M-C., McKeegan K. D., et al. (2017) ^{36}Cl-^{36}S in Allende CAIs: Implications for the origins of ^{36}Cl in the early solar system. Lunar and Planetary Science, **48**, #2618.

Telus M., Huss G. R., Ogliore R. C., et al. (2016) Mobility of iron and nickel at low temperatures: Implications for ^{60}Fe-^{60}Ni systematics of chondrules from unequilibrated ordinary chondrites. *Geochimica et Cosmochimica Acta*, **178**, 87–105.

Telus M., Huss G. R., Nagashima K., et al. (2018) In situ ^{60}Fe-^{60}Ni systematics of chondrules from unequilibrated ordinary chondrites. *Geochimica et Cosmochimica Acta*, **221**, 342–357.

Tera F., and Wasserburg G. J. (1972) U-Th-Pb systematics in three Apollo 14 basalts and the problem of initial Pb in lunar rocks. *Earth & Planetary Science Letters*, **14**, 281–304.

Tera F., and Wasserburg G. J. (1974) U-Th-Pb systematics on lunar rocks and inferences about lunar evolution and the age of the Moon. *Proceedings of the 5th Lunar Science Conference, Geochimica et Cosmochimica Acta Supplement*, **5**, 1571–1599.

Tera F., and Carlson R. W. (1999) Assessment of the Pb-Pb and U-Pb chronometry of the early solar system. *Geochimica et Cosmochimica Acta*, **63**, 1877–1889.

Theis K. J., Schönbächler M., Benedix G. K., et al. (2013) Palladium-silver chronology of IAB iron meteorites. *Earth & Planetary Science Letters*, **361**, 402–411.

Trappitsch R., Boehnke P., Stephan T., et al. (2018) New constraints on the abundance of ^{60}Fe in the early solar system. *Astrophysical Journal Letters*, **857**, L15.

Turner G., Huneke J. C., Podosek F. A., and Wasserburg G. J. (1971) ^{40}Ar-^{39}Ar ages and cosmic-ray exposure ages of Apollo 14 samples. *Earth & Planetary Science Letters*, **12**, 19–35.

Urey H. C. (1955) The cosmic abundances of potassium, uranium and thorium and the heat balances of the Earth, the Moon and Mars. *Proceedings of the National Academy of Sciences, USA*, **41**, 127–144.

Völkening J., Köppe M., and Heumann K. G. (1991) Tungsten isotope ratio determinations by negative thermal ionization mass spectrometry. *International Journal of Mass Spectrometry & Ion Processes*, **107**, 361–368.

Wadhwa M., Zinner E., and Crozaz G. (1997) Mn-Cr systematics in sulfides of unequilibrated enstatite chondrites. *Meteoritics & Planetary Science*, **32**, 281–292.

Wasserburg G. J., and Hayden R. J. (1955) Age of meteorites by the A^{40}-K^{40} method. *Physical Review*, **97**, 86–87.

Wasserburg G. J., Busso M., Gallino R., and Nollett K. M. (2006) Short-lived nuclei in the early solar system: Possible AGB sources. *Nuclear Physics A*, **777**, 5–69.

Wetherill G. W., Aldrich L. T., and Davis G. L. (1955) ^{40}Ar/^{40}K ratios of feldspars and micas from the same rock. *Geochimica et Cosmochimica Acta*, **8**, 171–172.

Wetherill G. W. (1956) Discordant uranium-lead ages. *Transactions – American Geophysical Union*, **37**, 320–326.

Williamson J. H. (1968) Least-squares fitting of a straight line. *Canadian Journal of Physics*, **46**, 1845–1847.

York D. (1966) Least-squares fitting of a straight line. *Canadian Journal of Physics*, **44**, 1079–1086.

York D. (1969) Least squares fitting of a straight line with correlated errors. *Earth & Planetary Science Letters*, **5**, 320–324.

10 Chronology of the Solar System from Radioactive Isotopes

Overview

In this chapter, we review what is known about the chronology of the solar system, based on radioisotope systems described in Chapter 9. We start by discussing the ages of materials that formed the solar system. Short-lived radionuclides also provide information about the galactic environment in which the solar system formed. We then consider how the age of the solar system is estimated from its oldest surviving materials – the refractory inclusions in chondrites. We discuss time constraints on the accretion of chondritic planetesimals and their subsequent metamorphism and alteration. Next, we discuss the chronology of differentiated asteroids, and of the Earth, the Moon, and Mars. Finally, we consider how to date the impact histories of solar system bodies, and how to use cosmogenic nuclides to assess the timescales for the transport of meteorites from their parent bodies to the Earth, as well as the residence time of meteorites on the Earth's surface.

10.1 Age of the Elements and the Sun's Formation Environment

The presence of radioactive isotopes in meteorites and planets demonstrates that the atoms that make up our solar system are not infinitely old. On the other hand, the abundance ratios for radioactive species, such as $^{235}U/^{238}U$, in the solar system when it formed were much lower than the ratios in which they were produced in the stellar sources, as determined from detailed nucleosynthesis models. This indicates that these isotopes were not produced at the time of solar system formation. For example, the production ratio for $^{235}U/^{238}U$ is ~1.79, but the ratio inherited by the solar system was ~0.31. Clearly much of the ^{235}U ($t_{1/2} = 7.04 \times 10^8$ years) originally produced with the ^{238}U ($t_{1/2} = 4.47 \times 10^9$ years) had decayed away before the solar system formed. It is not possible to estimate the age of the elements from ratios like this alone, because nucleosynthesis was not a discrete event in time. We must be able to model stellar nucleosynthesis as a function of time, and this in turn requires knowledge of the physical evolution of the galaxy. Evaluation of the relative abundances of various long-lived radiochronometers indicates that the nucleosynthesis began between 12 and 18 Ga. This is broadly consistent with the age of the galaxy as determined by independent means. By fixing the age of the galaxy in the models, useful information can be extracted about the radioactive isotopes inherited by the early solar system. Let's consider a standard model for nucleosynthesis in the galaxy and its implications for the early solar system.

Our galaxy, the Milky Way, formed at ~12 Ga. The elements that make up our solar system were synthesized in stars over ~7.5 Gyr of galactic history prior to the formation of the solar system. Star formation apparently started almost as soon as the galaxy began to form, and once the galactic disk formed, the rate of star formation closely tracked the mass of the gas. The initial mass function (Chapter 3) shows that in a single episode of star formation, many more low-mass stars form than high-mass stars. Because the lifetime of a star is a strong function of its mass, nucleosynthesis early in galactic history was dominated by massive stars, and the contributions of less massive stars have increased with time. This, along with the increase in overall metallicity with time, resulted in shifts in the isotopic compositions of the elements in the galaxy. Because the rate of nucleosynthesis tracks the mass of the gas, and because the mass of gas is greater in the center of the galaxy, there are gradients in metallicity and in isotopic compositions with galactic radius. We will therefore construct our model considering an annulus of the galaxy centered on the solar radius.

Many discussions of the galaxy assume that it started out with essentially its current mass. However, a more realistic view is that the galactic mass built up over time through the infall of metal-free gas and the assimilation of smaller galaxies. Much of the buildup appears to have occurred relatively early, but the galaxy took some time to reach its current mass. We infer this because there are too few low-mass, metal-poor stars for the galaxy to have had its present mass at 12 Ga.

The mass of gas at any time in the galaxy is a function of the rate of infall of new gas from the outside, the sequestration of gas into low-mass stars and stellar remnants, and the input of newly synthesized material from dying stars. Figure 10.1 shows a "closed-box" model in which infall is set to zero and the galaxy starts with its current mass. The mass of gas declines with time due to sequestration in stars until it reaches the currently observed value of ~10% of the total mass after 12 Gyr. Figure 10.2 shows a more plausible model in which the galaxy builds up from infalling metal-poor gas. The initial infall rate is high and decreases slowly with time. The mass of gas peaks after a few Gyr and then declines because star formation sequesters gas faster than new gas is added from the outside.

When were the atoms that became our solar system synthesized? The star-formation rate in the galaxy was highest early in galactic history and a lot of heavy elements were produced. However, most of this material was subsequently incorporated into stars and much of that has been "permanently" sequestered. On the other hand, much of the recently synthesized material has not yet been incorporated into a new generation of stars. This balance between synthesis and sequestration means that the birth dates of the elements that became the solar system are roughly evenly distributed over the 7.5 Gyr of galactic history prior to the solar system's birth (Clayton, 1988).

The abundances of radioactive isotopes over time in the galaxy can be modeled based on the above considerations. With an approximately constant production rate, the abundance of a stable nuclide will grow and will be proportional to the time over which it has been produced. In contrast, the abundance of a radionuclide will reach a steady state between production and decay in about eight mean lifetimes. (We will use mean life [τ] instead of half-life in this discussion because the mathematics is cleaner. Remember that $\tau = t_{1/2}/\ln 2$.) This is because the abundance of the radionuclide is controlled by the amount produced during the last mean life. Figure 10.3 shows

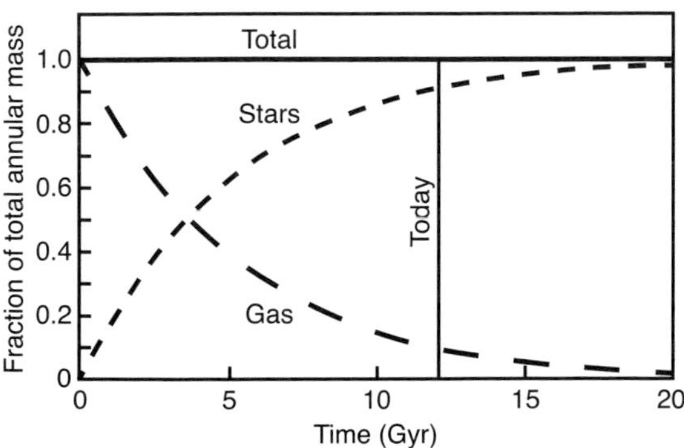

Figure 10.1 "Closed-box" model for the mass of the galaxy and the distribution of mass between stars and gas as a function of time. The mass of gas is constrained to match the currently observed value of ~10% at 12 Gyr.

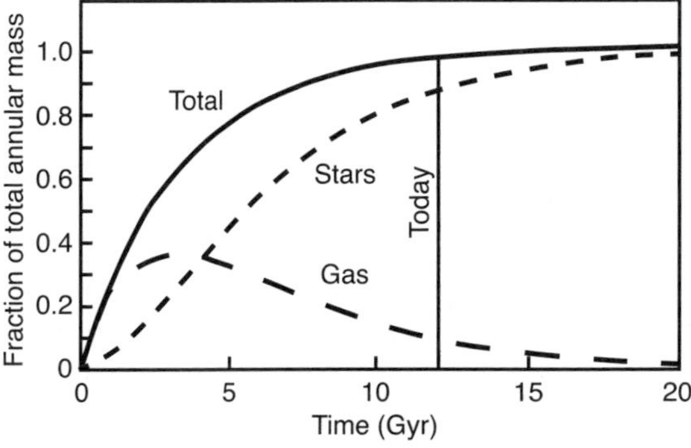

Figure 10.2 More general model for the mass of the galaxy and the distribution of mass between stars and gas as a function of time. This model includes infall of metal-poor gas onto the galactic disk, sequestration of gas in stellar remnants, and return of newly synthesized elements from stars. The mass of gas is constrained to match the currently observed value of ~10% at 12 Gyr.

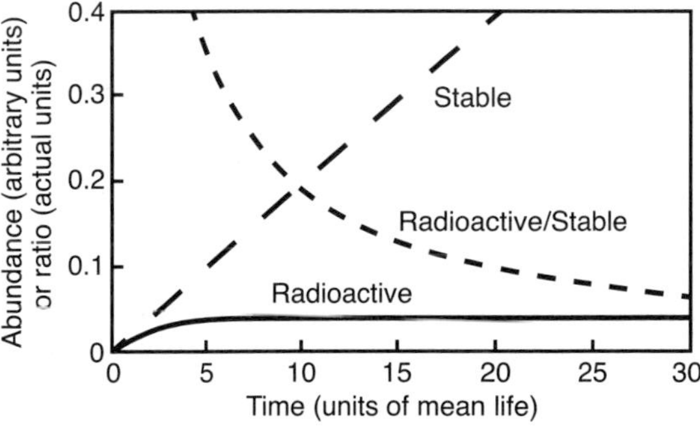

Figure 10.3 The abundances of hypothetical stable and radioactive nuclides as a function of time when both are produced at the same rate. The abundance of the stable isotope increases continuously, but the abundance of the radioactive isotope reaches a steady state. Thus, the ratio of the radioactive to the stable isotope decreases with time from the production ratio (in this case, 1).

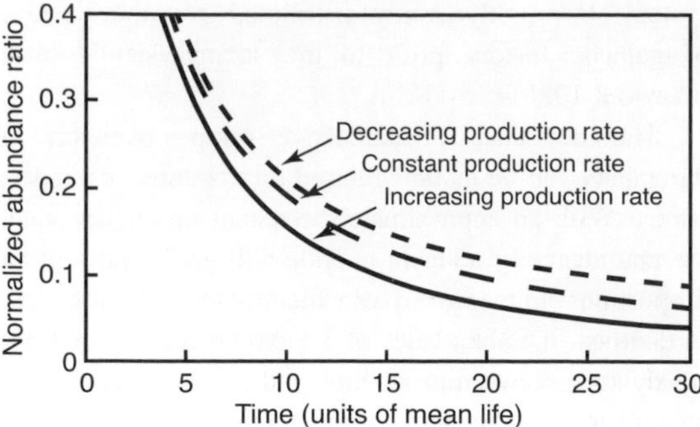

Figure 10.4 When the abundance ratio in Figure 10.3 is normalized to production rate, the second-order effect of a changing production rate is seen. A constant production rate gives the same curve as in Figure 10.3. But because the abundance of the radioactive species only reflects production during the last mean life, an increasing production rate results in a lower integrated ratio because more stable nuclide has been produced overall than in the constant-production-rate case. On the other hand, a decreasing production rate results in a higher integrated ratio because less stable nuclide has been produced overall than in the constant-production-rate case.

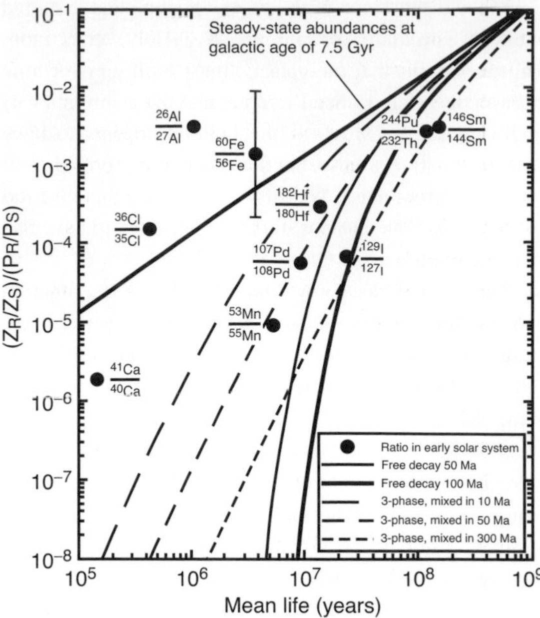

Figure 10.5 Abundances of short-lived nuclides in the early solar system compared to abundances predicted by different models for the interstellar medium. The y-axis of this diagram is the observed abundance ratio for a short-lived nuclide divided by its effective production ratio. For continuous production and times that are long with respect to the mean life (τ), the abundance ratios normalized to their production ratios plot on a straight line with slope τ/t. This line can be considered to represent the average interstellar medium. For nuclides with longer mean lives, the straight line begins to curve because there has not been enough time to establish a steady state. But these steady-state ratios are not the ratios that would be inherited by a newly forming star. Two models for those ratios are shown by the solid and dashed curves (see text for explanation). The model curves show that nuclides with mean lives of >5 Myr can be explained by the galactic background abundances, but those with shorter mean lives (^{41}Ca, ^{36}Cl, ^{26}Al, and ^{60}Fe) require a late addition of newly synthesized material. Adapted from Huss et al. (2009).

these two individual trends with time (in units of mean life), along with the trend for the ratio of the radionuclide to the stable nuclide, which decreases with time. A changing production rate has only a second-order effect on the abundance ratio. Figure 10.4 shows a hypothetical abundance ratio normalized to the production rate for the cases of constant production rate (same as Fig. 10.3), increasing production rate, and decreasing production rate. Note that except for the special case of a rapidly increasing production rate, the curves all decrease with time.

We can now construct a model that describes the evolution of the abundances of stable and radioactive isotopes in the gas phase of the galaxy as a function of time. Such a model provides an estimate of the abundances of the radioactive nuclides that should have been available in the average interstellar medium at the time the solar system formed. By comparing these predictions with the abundances inferred for the early solar system from meteorites, we can investigate the environment in which the solar system formed.

Figure 10.5 compares the abundance ratios of short-lived radionuclides in the early solar system to the abundance ratios predicted by a standard model for galactic chemical evolution. The x-axis of the plot is the mean life (τ) of the radionuclide. Note that the axis does not extend beyond 10^9 years. The y-axis is the abundance ratio of the radionuclide to a stable nuclide of the same or a closely related element (Table 9.8), normalized to the production ratios of the same isotopes. Let's first consider the lines on the diagram – the predictions of the model.

The straight solid line in Figure 10.5 represents the average, steady-state abundances of the radionuclides at a galactic age of 7.5 Ga, when the solar system formed. For times that are much longer than the mean lifetimes of the short-lived radionuclides, the slope of this line is given by the mean life, τ, divided by the galactic age, *t*. As the mean life approaches *t*, the steady-state abundance falls below the τ/*t* line and the trend asymptotically approaches the effective production ratio (y = 1 on this plot). The vertical position of this line and the curves derived from it depend on our choice of *t*.

When Reynolds (1960) first showed that ^{129}I was present in the early solar system, he quickly realized that the inferred initial ratio for the solar system (~1 × 10^{-4}) was well below that expected from stellar nucleosynthesis (~1). He proposed a period of free decay between the last input of newly synthesized material and the formation of the solar system to explain the low observed ratio and estimated the free-decay time at ~3 × 10^8 years. As more-realistic models of the expected abundances of radionuclides in interstellar space were developed, this interval decreased to ~10^8 years. The curved solid lines in Figure 10.5 show the effect on the steady-state abundances of short-lived radionuclides of free-decay intervals of 5 × 10^7 and 10^8 years. Nuclides with mean lives of less than ~5 Myr would have completely decayed away. Thus, when Lee et al. (1977) discovered ^{26}Al and inferred an initial $^{26}Al/^{27}Al$ abundance ratio of 5.1 × 10^{-5} (see Chapter 9), they were forced to conclude that ^{26}Al had been synthesized immediately before or during the formation of the solar system. Models of stellar nucleosynthesis and of local irradiation in the solar system were investigated to understand how the ^{26}Al was produced, and very quickly a standard model of solar system formation emerged. In this model, the gravitational collapse of the solar system was triggered by a supernova explosion, which also supplied ^{26}Al (and other isotopes) to the early solar system (Cameron and Truran, 1977).

Subsequent to that time, new generations of telescopes have revealed a tremendous amount about the structure of the interstellar medium (Fig. 10.6). In a classic paper, McKee and Ostriker (1977) described a three-phase interstellar medium regulated by supernova explosions. An adaptation of this model (Clayton, 1988) considers these phases to be: (1) molecular clouds from which stellar systems form, (2) large HI (neutral H) clouds that are too large to be evaporated by supernova shocks, and (3) smaller HI clouds that can be evaporated by supernova shocks along with the surrounding hot medium. Figure 10.6 shows how molecular clouds interact with the hot interstellar medium. Newly synthesized material is injected primarily into the hot medium and mixes into the large neutral clouds and then into the dense molecular clouds on timescales of 10^7–10^8 years. In this picture, the so-called free-decay interval is actually the time it takes on average for newly synthesized matter to be distributed through the three phases in the interstellar medium and to become available for incorporation into a stellar system. The dashed lines in Figure 10.5 show the

Figure 10.6 Spitzer space telescope image of the W5 star-forming region in the constellation Cassiopeia. Wispy molecular clouds surround cavities carved out by radiation from the region's most massive stars. The winds push the molecular cloud gas together, potentially triggering star formation. White knotty areas are where the youngest stars are forming. Figure from NASA/JPL Caltech.

abundances of short-lived radionuclides that would be present in molecular clouds based on a three-phase model. The three lines assume different times for mixing between the phases. The hot medium would plot on a nearly horizontal line near the top of the diagram, and the large neutral clouds would plot near, but perhaps slightly above, the line for steady-state abundances.

The model abundances for molecular clouds shown in Figure 10.5 are broadly consistent with the early solar system abundances of short-lived radionuclides with mean lives $\geq$5 Myr. For these isotopes, the choice of model plays a critical role in determining whether or how much of a late injection is required. But no version of this model can explain the abundances of ^{41}Ca, ^{36}Cl, ^{26}Al, and ^{60}Fe in the early solar system. A source for these isotopes just before or during solar system formation is required. The initial abundance of ^{60}Fe in the solar system is still quite uncertain (Chapter 9).

The three-phase model for the abundances of short-lived radionuclides in the interstellar medium illustrates the stochastic nature of nucleosynthesis. Stars inject newly synthesized matter into the interstellar medium at discrete times. Between injections, the radionuclides in the interstellar medium decay, and if a longer than average time passes, the abundances can locally drop well below the steady-state line for the galaxy as a whole (curved solid lines in Fig. 10.5). When a supernova injects newly synthesized material into the interstellar medium, the local abundances are dramatically higher than the average for the interstellar medium as a whole and plot above the straight solid line on Figure 10.5. But this material is not easily available for incorporation into a newly forming star because of its high temperature ($>10^6$ K). This hot gas interacts with cold molecular clouds across a shock front. A shock front is characterized by an abrupt, nearly discontinuous change in properties, such as temperature, pressure, and density. Efficient mixing only occurs when the gas cools and the shock front dissipates. This takes a long time for very hot gas because the gas has no efficient way of radiating thermal energy. On timescales of a few million years, the hot gas can cool and mix with molecular cloud gas, producing material enriched in newly synthesized radionuclides compared to the matter in the cloud prior to mixing. If several supernovae explode relatively close together in time and space, molecular clouds in that region of the galaxy may be enriched relative to the average steady-state abundances in molecular clouds (dashed lines in Figure 10.5). Such "self-enrichment" may be a feature of molecular clouds undergoing star formation. However, the timescale for the exchange of matter between phases to form a molecular cloud and for that cloud to generate a new generation of stars means that self-enrichment is unlikely to be the explanation for solar system abundances.

Comparing the relative abundances of short-lived nuclides in the solar system with those predicted by stellar models can provide a constraint on the nature of the stellar source. Iron-60 can only be produced efficiently by stellar nucleosynthesis, and the upper part of the abundance range implicates a supernova source (e.g. Meyer, 2005; Huss et al., 2009). A problem with a supernova source is that supernovae seem to overproduce ^{53}Mn. This problem may be avoided if the deeper layers of the star, where ^{53}Mn is produced, fall back onto the stellar remnant rather than being ejected by the supernova, sequestering much of the ^{53}Mn in the remnant (Cameron et al., 1995; Takigawa et al., 2008). Supernovae are typically found in star-forming regions because their progenitors are massive stars whose lifetimes are on the order of a few million years, too short for them to drift away from the region where they formed before they explode. In fact, star formation in molecular clouds is typically terminated when the intense stellar winds and radiation of massive stars, and their subsequent supernova explosions, disperse the cloud.

If the ^{60}Fe abundance was at the low end of the range shown in Figure 10.5, other stellar sources are possible. Stars with initial masses greater than ~33 $M_{\odot}$ experience mass winds that can eject the entire envelope, exposing newly synthesized material at the surface. The winds of these Wolf–Rayet stars carry the products of hydrogen and helium burning and *s*-process synthesis, including ^{26}Al, ^{41}Ca, and ^{107}Pd, but not much ^{60}Fe. If the solar system incorporated large amounts of Wolf–Rayet wind during collapse and became a compact object that could withstand the inevitable supernova explosion without either evaporating or acquiring too much material from the interior layers of the supernova, a Wolf–Rayet star could have seeded the solar system with short-lived nuclides (Arnould et al., 2006). An asymptotic giant branch (AGB) star of 5–8 $M_{\odot}$ is also a potential source of short-lived radionuclides in the newly forming solar system if the ^{60}Fe abundance was relatively low. Such stars can produce the right amounts of ^{26}Al, ^{41}Ca, ^{107}Pd, and relatively small amounts of ^{60}Fe (Wasserburg et al., 2006). But the probability of the newly forming solar system encountering an AGB star of the right mass in the right stage of its evolution is extremely small. At the present time, the jury is still out on what kind of star contributed the short-lived radionuclides to the early solar system. More-reliable data on the abundance of ^{60}Fe in the early solar system will be necessary to answer this question.

What about the environment where the solar system formed? Two types of environments can be envisioned: regions like Taurus, where stars appear to be forming in near isolation, and regions like Orion, where a large number of stars are forming almost simultaneously in what is known as "cluster star formation." In such an environment, a burst of star formation can locally disperse a region of a giant molecular cloud, while at the same time compressing a nearby region and triggering a new burst of star formation. Lada and Lada (2003) estimated that 70–90% of all stars form in regions of cluster star formation. Thus, based on simple probabilities, it is likely that the Sun formed in a region of cluster star formation. A supernova or Wolf–Rayet source for the short-lived radionuclides in the early solar system would be quite consistent this conclusion.

Before closing this section, it is worth mentioning that γ-rays from the decay of ^{26}Al and ^{60}Fe have been observed in interstellar space using the Compton orbiting observatory and INTEGRAL satellite. These observations can provide ground truth for models of nucleosynthesis in the galaxy. The measurements indicate that there are $2.8 \pm 0.8\ M_{\odot}$ of ^{26}Al in the galaxy. Cosmochemists do not know what to do with solar masses of ^{26}Al; we much prefer to discuss the ^{26}Al/^{27}Al ratio. Converting solar masses of ^{26}Al to a ratio is potentially a problem because we need to know the abundance of ^{27}Al, which is not directly measurable. Published estimates of the ^{26}Al/^{27}Al ratio in the galaxy assume that the galaxy has the metallicity and composition of the solar system. Scaling to the mass of the galaxy gives a mass of ^{27}Al and produces a ^{26}Al/^{27}Al ratio of ~8.4×10^{-6} (Diehl et al., 2006). A similar calculation for ^{60}Fe based on the observed ^{60}Fe/^{26}Al γ-ray flux ratio of 0.148 (Wang et al., 2007), equivalent to an ^{60}Fe/^{26}Al abundance ratio of 0.31, gives an ^{60}Fe/^{56}Fe ratio for the galaxy of 2.7×10^{-7}. However, there are two problems with these calculations. First, the galaxy has evolved for ~4.57 Gyr since the solar system formed, and as we have already discussed, the abundance ratios of short-lived nuclides to stable isotopes of the same element decrease with time due to production of the stable nuclides. Thus, the current ^{26}Al/^{27}Al and ^{60}Fe/^{56}Fe ratios must be lower than the published numbers. Huss et al. (2009) estimated the effects of 4.57 Gyr of evolution on the ^{27}Al and ^{56}Fe abundances using a galactic chemical evolution model. Using the higher abundances of ^{27}Al and ^{56}Fe, they calculated that the observed abundances imply that the present ratios should be ^{26}Al/^{27}Al = ~4×10^{-6} and ^{60}Fe/^{56}Fe = ~4.4×10^{-8}, significantly lower than the published numbers. Clearly, galactic chemical evolution must be considered when modeling the abundance ratios for radioactive nuclides.

10.2 Age of the Solar System

In order to determine the age of the solar system, we must first decide what its age really means. Astronomers and theoreticians tell us that the process of forming a star from a dense core of gas and dust in a molecular cloud takes on the order of a few million years. The process can be divided into four stages that are defined by the characteristics of four classes of observed stellar objects (Fig. 10.7). *Class 0* objects are young protostars in the early main accretion phase. A hydrostatic core has formed, surrounded by a dense accretion disk, but the core has not yet accreted the majority of its final mass. At this stage the stars are embedded in optically thick dust clouds and are observed in reradiated energy at submillimeter wavelengths. These objects have bipolar outflows and suffer periodic outbursts during which the mass-ejection rate in the outflows increases by several orders of magnitude. Each outburst may last only a few years, and they have been given the name "FU-Orionis outbursts" after the star in which they were first observed. This stage lasts 10,000–30,000 years. *Class I* objects are protostars in the late accretion phase. These objects are embedded in less-dense gas and dust and are observed by reradiated emission in the infrared. Class I objects also have bipolar outflows, but they are less powerful and less well collimated than those of Class 0 objects. This stage lasts 100,000–200,000 years. *Class II* objects, also known as classical T Tauri stars, are pre-main sequence stars with optically thick protoplanetary disks. They are no longer embedded in their parent cloud, and they are observed in optical and infrared wavelengths. They still exhibit bipolar outflows and strong stellar winds. This stage lasts 1–10 million years. *Class III* objects are the so-called "weak line" or "naked" T Tauri stars. They have optically thin disks, perhaps debris disks in some cases, and there are no outflows or other evidence of accretion. They are observed in the visible and near-infrared and have strong X-ray emission. These stars may have planets around them.

Cosmochemists have access to a variety of samples from the earliest history of our solar system, including CAIs, chondrules, and matrix from primitive chondrites and interplanetary dust from comets. Using a combination of long-lived and short-lived chronometers, we can determine which of these is the oldest and estimate its absolute age. But at the moment, we have no way of connecting the astronomers' stellar stages with the cosmochemical

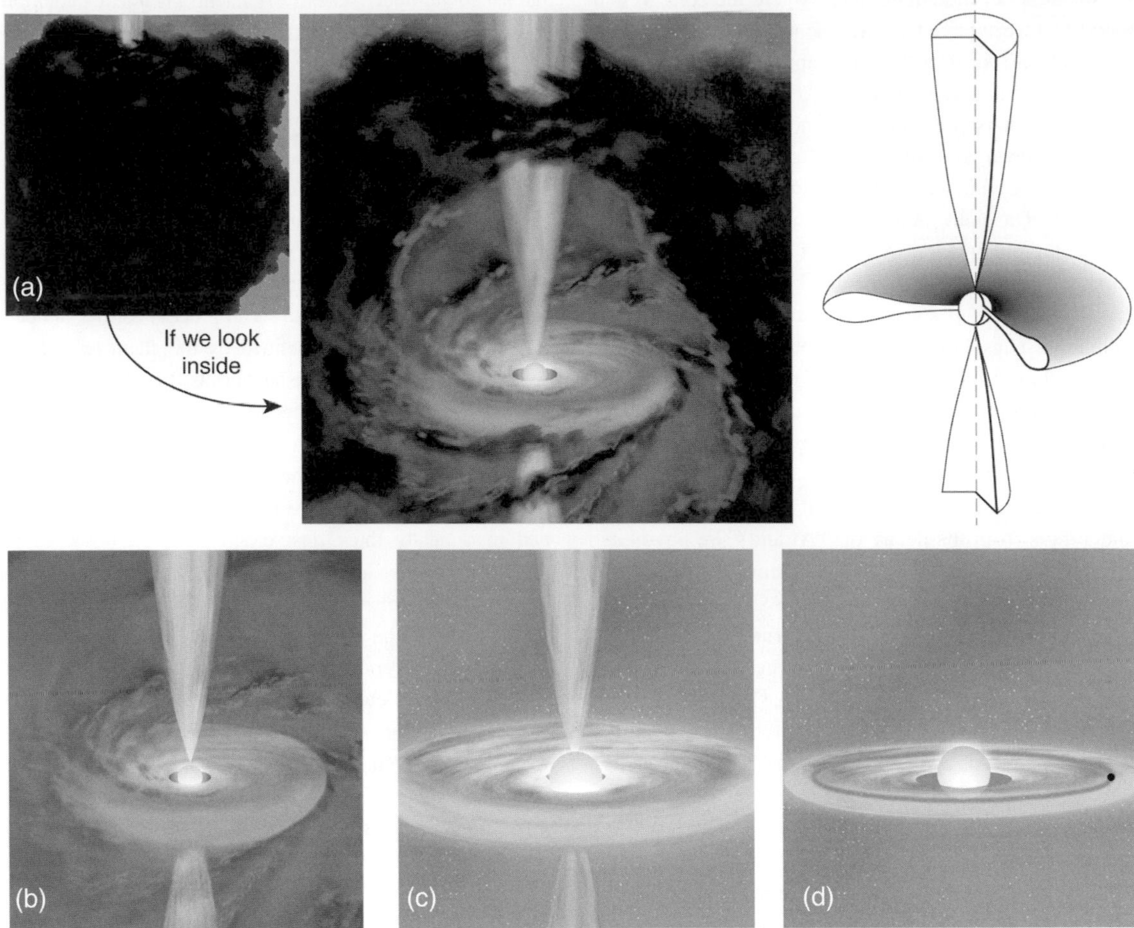

Figure 10.7 Artistic rendering of four observed stages of star formation. (a) Class 0 object: a deeply embedded hydrostatic core surrounded by a dense accretion disk. Strong bipolar jets remove angular momentum. (b) Class I object: protostar in the later part of the main accretion phase. (c) Class II object or T Tauri star: pre-main sequence star with optically thick protoplanetary disk. (d) Class III object: naked T Tauri star which has an optically thin disk and thus can be directly observed; some may have planets.

timescale. Gravitational collapse of a portion of a molecular cloud is not a datable event. Did the objects whose ages we measure form when the Sun was a Class 0, Class I, or Class II object? All of these seem plausible. In the Class 0 stage, the Sun was very energetic and disk temperatures near the Sun were high, perhaps suitable for making CAIs and chondrules, but the material in the disk at this stage was rapidly accreting onto the star and may not have survived to become incorporated into meteorites. During the Class I stage, the accretion disk was more stable, but the temperatures in the disk, particularly farther out in the disk where the meteorites formed, were probably relatively low. This stage would seem to require transient high-temperature events that have not been directly observed, or a mechanism to transport materials from near the Sun outward, or both. Stage II is also a plausible time for the formation of the components of chondrites. It is likely that the meteorites represent the end stages of accretion disk evolution. The records of earlier events may have been wiped out by accretion of their products onto the Sun. It is probable that by the time the Sun was a Class III star, planetesimals and planets had already formed. For the time being, we must acknowledge that we cannot connect the astronomical and cosmochemical timescales and simply do the best we can to accurately date the materials we have.

The tools that we have to determine the age of the solar system and the chronology of early solar system events are

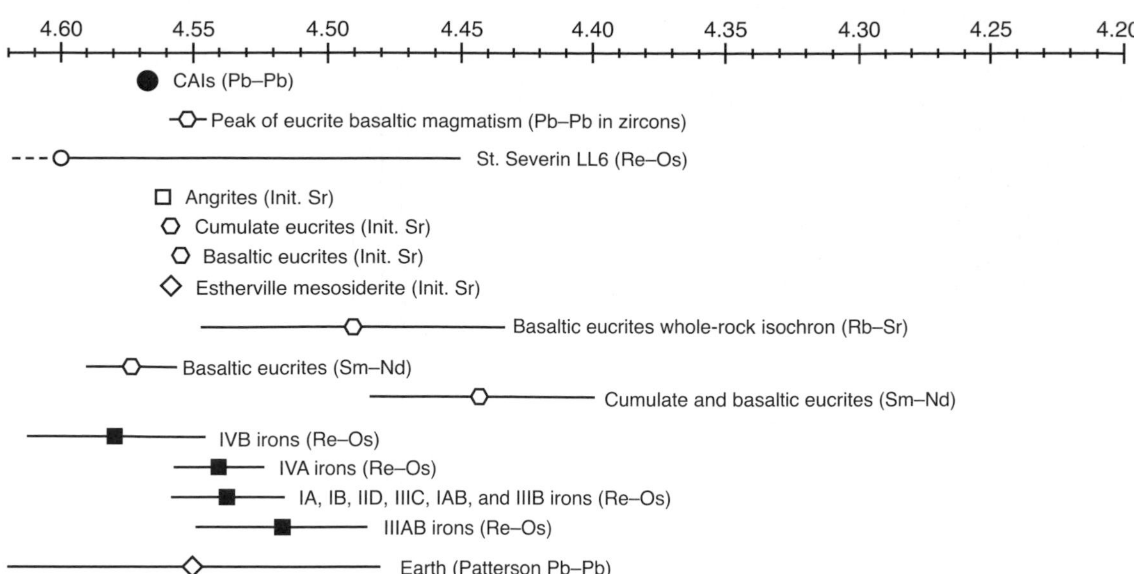

Figure 10.8 Formation ages and crystallization ages (given in Ga, at the top of the figure) of various objects as determined by long-lived radiochronometers. The initial strontium isotopic ratio is inherently precise, but a large systematic uncertainty in the half-life gives absolute age determinations relatively large uncertainties. Initial strontium data for eucrites and angrites support the inference from short-lived nuclides that the parent bodies accreted very early. ^{87}Rb–^{87}Sr and ^{147}Sm–^{143}Nd isochron ages indicate that magmatism on the eucrite parent body continued for more than 100 million years. In spite of relatively low precision, the ^{187}Re–^{187}Os data for iron meteorites suggest that their parent bodies differentiated early in solar system history. Data from Smoliar (1993), Wadhwa et al. (2006), and Day et al. (2016) and references therein.

the long-lived and short-lived radiochronometers described in Chapter 9. The long-lived radionuclides tell us that the oldest objects in the solar system formed at ~4.5–4.6 Ga (Fig. 10.8). But the precision of most of these measurements is not sufficient to investigate the details of those early times. In addition, there are uncertainties in the half-lives of the different radionuclides that translate into absolute uncertainties of several million to tens of millions of years. Initial strontium can provide a time resolution approaching a million years in minerals with low Rb/Sr, but it is difficult to integrate the information from this system with ^{207}Pb–^{206}Pb results because of uncertainties in the half-lives. In addition, realization of the inherent precision of initial strontium requires a clear understanding of the strontium evolution of the system from which all objects formed, and at present that understanding is lacking. Short-lived radionuclides provide the time resolution necessary to unravel the details of early solar system history.

10.2.1 Early Solar System Chronology

Detailed information about the chronology of the early solar system comes from the ^{26}Al–^{26}Mg, ^{53}Mn–^{53}Cr, ^{182}Hf–^{182}W, and ^{129}I–^{129}Xe short-lived systems (Table 9.8) and the ^{207}Pb–^{206}Pb system. The time resolution of these clocks ranges from less than 100,000 to about a million years. In order for the short-lived radionuclides to give time information, they must have been homogeneously mixed throughout the region where the objects being dated formed, so that differences in abundances of the daughter isotopes are attributable only to differences in time of formation. For most systems, the distribution of radionuclides appears to have been homogenous, but recent work on the ^{26}Al–^{26}Mg system suggests that ^{26}Al may not have been homogeneously distributed in the CAI-forming region (see below).

Constructing a self-consistent timeline from the various radiochronometers is harder than it may first appear. Each sample providing a date for the timeline must meet the criteria for a valid age (see Chapter 9). This is not always easy to demonstrate, in part because we may not know the closure temperatures for diffusion for each mineral in the system. The dates must be reliable from an experimental point of view. As each new technique is brought on line, there is a period of development during which scientists learn how to make reliable

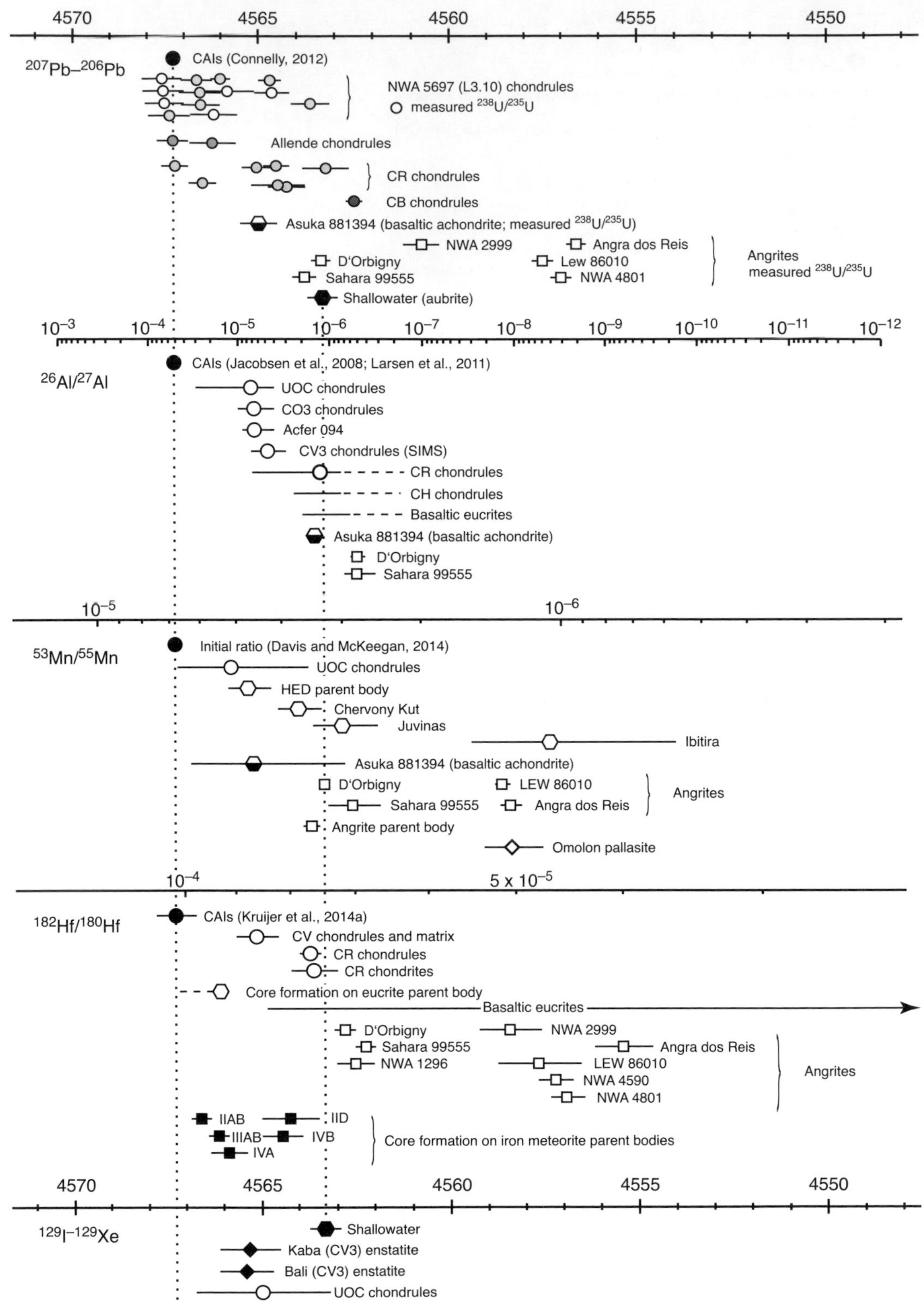

Figure 10.9 Chronology of the primitive components in chondrites and of differentiated meteorites in the early solar system as given by five radiochronometers. The ^{207}Pb–^{206}Pb system provides the time

Figure 10.9 (*cont.*) anchor for the short-lived nuclides (given in Ma at the top of the figure). The anchor points – CAIs for ^{26}Al–^{26}Mg and ^{182}Hf–^{182}W, D'Orbigny for ^{53}Mn–^{53}Cr, and Shallowater for ^{129}I–^{129}Xe – are shown by dotted vertical lines. The short-lived nuclides are plotted on scales of their initial ratios, except for ^{129}I–^{129}Xe (see text). The arrow for ^{182}Hf–^{182}W data for basaltic eucrites describes a range of dates that extend beyond the range of the chronometer. Data for the ^{207}Pb–^{206}Pb system from Connelly et al. (2012), Bollard et al. (2017), Wimpenny et al. (2019), Kleine et al. (2012), Hohenberg and Pravdivtseva (2008); for the ^{26}Al–^{26}Mg system from Jacobsen et al. (2008), Larsen et al. (2011), Nagashima et al. (2018), Wimpenny et al. (2019), Spivak-Birndorf et al. (2009); for the ^{53}Mn–^{53}Cr system from Davis and McKeegan (2014), Nyquist et al. (2009) and references therein, Lugmair and Shukolyukov (1998), Shukolyukov and Lugmair (2007); for the ^{182}Hf–^{182}W system from Kruijer et al. (2014a, 2014b), Kleine et al. (2018), Touboul et al. (2015), Kleine et al. (2012); and for ^{129}I–^{129}Xe from Hohenberg and Pravdivtseva (2008).

measurements. Eventually, bad measurements become obvious as more measurements are made, but in the early stages, it is not always obvious which measurements are reliable. Even if all of the ages are reliable, different systems may not reflect precisely the same time due to differences in the closure temperatures for each system. In fast-cooled systems, this is not an issue, but in slowly cooled systems, such as differentiated asteroids or metamorphosed chondrites, differences in closure temperature can produce large differences in the dates given by different systems. But sometimes this is an advantage. In spite of these difficulties, we now have a basic outline of the sequence of events in the early solar system. The details of the chronology we present below are still the subject of intense study, and the picture will change as new measurements are made.

10.2.1.1 *Primitive Components in Chondrites*

Figure 10.9 shows the chronological results from five different isotope systems: ^{207}Pb–^{206}Pb, ^{26}Al–^{26}Mg, ^{53}Mn–^{53}Cr, ^{182}Hf–^{182}W, and ^{129}I–^{129}Xe. The ^{207}Pb–^{206}Pb system provides the absolute timescale for all systems. Careful measurements have shown that the ^{238}U/^{235}U ratio in the CAI-forming region was not homogeneous (Brennecka et al., 2010), so for a reliable age, the uranium isotope ratio must be measured for each object. To date, four CAIs have had uranium-corrected ^{207}Pb–^{206}Pb measurements. These four CAIs give the best estimate for the age of the CAIs – the oldest objects that we know of that formed in the solar system – of 4567.30 $\pm$ 0.16 Ma (Connelly et al., 2017). For the ^{26}Al–^{26}Mg, ^{53}Mn–^{53}Cr, and ^{182}Hf–^{182}W systems, Figure 10.9 shows the initial ratios for each of the objects, plotted so that the scales are tied to the absolute timeline by the half-life of the nuclide and by an "anchor," an object for which reliable measurements exist both for the ^{207}Pb–^{206}Pb system and for the short-lived nuclide. For the ^{26}Al–^{26}Mg system, the anchor is the CAIs, for which we assign a $(^{26}\mathrm{Al}/^{27}\mathrm{Al})_0$ ratio of 5.2×10^{-5} (Jacobsen et al., 2008; Larsen et al., 2011). For the ^{53}Mn–^{53}Cr system, CAIs do not seem to work as an anchor. Their inferred initial ratios are too high, so the system is anchored to angrite meteorites. Angrites have a simple igneous history: they apparently cooled rapidly and did not suffer subsequent alteration. Originally, LEW 86010 was used as the anchor, but more recently, D'Orbigny has been preferred. Using the D'Orbigny anchor and an inferred initial ratio for the solar system of $(6.7 \pm 0.56) \times 10^{-6}$ (Davis and McKeegan, 2014), the ^{53}Mn–^{53}Cr system gives results concordant with the other systems in Figure 10.9. For the ^{182}Hf–^{182}W system, we will use CAIs and the initial ratio determined by Kruijer et al. (2014a) of $(^{182}\mathrm{Hf}/^{180}\mathrm{Hf})_0 = (1.018 \pm 0.043) \times 10^{-4}$ as the anchor. For the ^{129}I–^{129}Xe system, we adopt the Shallowater aubrite, which has a ^{207}Pb–^{206}Pb age of 4563.3 $\pm$ 0.4 Ma as the anchor. The $(^{129}\mathrm{I}/^{127}\mathrm{I})_0$ ratios for objects being dated are determined relative to the Shallowater aubrite standard, which is included in every irradiation (see Appendix). The $(^{129}\mathrm{I}/^{127}\mathrm{I})_0$ ratio is not directly determined. Instead, the age is calculated as a difference relative to Shallowater and is reported as an absolute age. Thus, the ^{129}I–^{129}Xe system is referenced to the absolute timeline.

CAIs

Several isotope systems indicate that CAIs are the oldest objects to have formed in the solar system. Uranium-corrected ^{207}Pb–^{206}Pb ages for four CAIs are within errors of each other and give a precise CAI formation time of 4567.30 $\pm$ 0.16 Ma (Connelly et al., 2012). This is the current best estimate of the age of the solar system. When considering how well we know the age of the solar system, we must add an absolute uncertainty of ~0.2% (9 Myr at the age of the solar system) to the ^{207}Pb–^{206}Pb age of CAIs, because of uncertainties in the decay rates of the uranium isotopes. This uncertainty is not included in the discussions of early solar system chronology because it applies to the system as a whole, not to the differences in ages between individual objects.

CAIs also give the highest $(^{26}Al/^{27}Al)_0$ ratios among solar system materials. The first inclusion shown conclusively to have formed with live ^{26}Al gave $(^{26}Al/^{27}Al)_0$ of $(5.1 \pm 0.6) \times 10^{-5}$ (Fig. 9.25). Two types of isochrons have been produced: whole-rock isochrons, in which bulk samples of several different CAIs are measured and plotted on a single diagram, and internal isochrons, which are based on minerals from a single CAI. Two whole-rock isochrons give nearly identical $(^{26}Al/^{27}Al)_0$ ratios of $(5.23 \pm 0.15) \times 10^{-5}$ (Jacobsen et al., 2008) and $(5.25 \pm 0.02) \times 10^{-5}$ (Larsen et al., 2011). The tight fits to both of these isochrons indicate that the CAIs formed within ~30,000 years of each other. Whole-rock isochrons need not give the crystallization age of the CAIs, but may instead represent the time the CAIs or their precursors were isolated from the bulk solar nebula. But for the purposes of Figure 10.9, an initial ratio of 5.2×10^{-5} will be used.

Most work on the ^{26}Al–^{26}Mg system has been done using internal isochrons. The $(^{26}Al/^{27}Al)_0$ ratio of ~5×10^{-5} has been found repeatedly in CAIs of different types and from different classes of meteorites, and it has become known as the "canonical" ratio. Small differences in the $(^{26}Al/^{27}Al)_0$ ratio between different CAIs appear to reflect real differences in formation time. But in many cases with lower inferred $(^{26}Al/^{27}Al)_0$ ratios, the CAIs have been reheated, either disturbing or completely resetting the ^{26}Al–^{26}Mg clock. Some CAIs were reprocessed in the nebula, and sometimes this extended history is recorded in a single CAI. For example, one Allende inclusion gave different Al–Mg isochrons for the main inclusion and for three igneous objects incorporated within it, suggesting that this object formed over a period of ~300,000 years (Hsu et al., 2000). Some CAIs were even incorporated into chondrules (e.g. Krot et al., 2006). Many CAIs have experienced secondary processing on the meteorite parent body that has disturbed or reset the ^{26}Al–^{26}Mg system. Some inclusions give lower initial ratios due to poor standardization. CAIs with corundum or calcium dialuminate often exhibit low initial ratios because the standards used in ion probe measurements are not appropriately matrix matched.

A rare group of inclusions give low $(^{26}Al/^{27}Al)_0$ ratios, show larger than normal isotopic anomalies in many elements, and are highly mass-fractionated. These so-called FUN inclusions (for **f**ractionation and **u**nidentified **n**uclear effects, Chapter 8) and platy hibonite crystals (known as PLACs) from CM2 chondrites have $(^{26}Al/^{27}Al)_0$ ratios of much less than 5×10^{-5} (Park et al., 2017; Kööp et al., 2016). The low ratios do not appear to be due to parent-body metamorphism. There are three possible interpretations for their low $(^{26}Al/^{27}Al)_0$ ratios: 1) the inclusions formed or were last remelted late, after the ^{26}Al had decayed; 2) they formed early, before ^{26}Al was introduced into the solar system; or 3) they formed at the same time as other CAIs, but from material that did not contain ^{26}Al. Late formation seems unlikely. The distribution of these inclusions in chondrites and the presence of isotopic anomalies in many elements argue for early formation before ^{26}Al was fully integrated into the solar system (Park et al., 2017). Micron-sized corundum grains from ordinary and carbonaceous chondrites show a bimodal distribution, with just over half of the grains showing $(^{26}Al/^{27}Al)_0$ ratios of ~5×10^{-5} and the rest showing no evidence of ^{26}Al (Makide et al., 2011). Both types of corundum are ^{16}O rich and are interpreted as nebular condensates. The difference in ^{26}Al content is interpreted as evidence of formation before ^{26}Al was integrated into solar system material (Makide et al., 2011). But others argue that normal CAIs formed from ^{26}Al-rich dust that was produced less than 5 Myr before solar system formation, and FUN CAIs formed from older dust that had lost its ^{26}Al to radioactive decay (e.g. Larsen et al., 2020). At this time, there is no consensus as to the origin of these unusual inclusions.

If ^{26}Al was not homogeneously distributed during the earliest epoch of solar system history, then chronology based on ^{26}Al is called into question. We cannot explore all the implications of this possibility here. In the discussion to follow, we will present a standard model of ^{26}Al chronology that interprets the data in terms of a solar system where ^{26}Al was homogeneously distributed and had an initial ratio at the time the CAIs formed of 5.2×10^{-5}. Most of the data are consistent with homogeneous distribution of ^{26}Al by the time the majority of CAIs formed, but alternative models are being discussed and will be mentioned briefly.

The ^{182}Hf–^{182}W system does not exhibit evidence of a late addition of ^{182}Hf. Its half-life (8.9 Myr) is long enough for it to have come into the solar system as part of the bulk molecular cloud material. Hafnium and tungsten are partitioned relatively strongly among CAI minerals, allowing internal isochrons to be generated. Grossmanite and perovskite are the main carriers of hafnium, and tiny metal nuggets are the primary carriers of tungsten. Because bulk CAIs also show significant variation in $^{180}Hf/^{184}W$, whole-rock CAI isochrons can also be generated (Fig. 9.27). Tungsten in CAIs also exhibits nucleosynthetic anomalies caused by variable abundances of an *s*-process tungsten carrier. In order to generate an isochron, the data must first be corrected for these nucleosynthetic anomalies

(Kruijer et al., 2014a). After making this correction, a $(^{182}Hf/^{180}Hf)_0$ ratio of $(1.018 \pm 0.043) \times 10^{-4}$ is obtained for bulk CAIs. When anchored to the D'Orbigny angrite, the ^{182}Hf–^{182}W system gives an age for the CAIs of 4567.9 ± 0.7 Ma (Kruijer et al., 2014a), within error of the age of 4567.3 ± 0.16 Ma given by ^{207}Pb–^{206}Pb.

Chondrules

Chronological information on chondrule formation comes from the ^{207}Pb–^{206}Pb, ^{26}Al–^{26}Mg, ^{53}Mn–^{53}Cr, ^{182}Hf–^{182}W, and ^{129}I–^{129}Xe systems (Fig. 10.9). The ^{207}Pb–^{206}Pb system gives chondrule ages that start at the same time as CAI formation and continue for ~4 Myr. Chondrules for which $^{238}U/^{235}U$ was measured are shown with open symbols in Figure 10.9. For chondrules, the uranium measurement is less critical than for CAIs because a study of many chondrules and solar system materials other than CAIs indicate that, outside the CAI-forming region, the solar system is homogeneous and has a present-day $^{238}U/^{235}U$ ratio of 1/137.786 (Connelly et al., 2017).

The ^{26}Al–^{26}Mg system has been used extensively to date chondrule formation. The ^{26}Al–^{26}Mg system dates the melting and crystallization of chondrules. Internal isochrons can be constructed from plagioclase and/or mesostasis glass (high Al/Mg phases) and olivine and pyroxene (low Al/Mg phases). Only the least metamorphosed chondrites can be used for this work because magnesium begins to move around at temperatures of 300–400°C (Nagashima et al., 2018). It took a number of years to figure out that only some type 3 chondrites were sufficiently unmetamorphosed to give reliable ^{26}Al–^{26}Mg data. The scale of petrologic types for chondrites described in Chapter 6 has been modified so that meteorites suitable for ^{26}Al–^{26}Mg dating (and other studies) could be reliably identified (type 3 chondrites were divided into 20 subtypes). Chondrules from several different meteorite classes have been measured for ^{26}Al–^{26}Mg systematics (Fig. 10.10). Data for Semarkona (LL3.00), Yamato 81020 (CO3.05), Acfer 094 (3.00), and Kaba (CV3.1) give $(^{26}Al/^{27}Al)_0$ distributions that are resolved from zero. The spreads of the plotted data are slightly larger than they should be due to simple counting statistics. This makes it hard to determine if the chondrules in each group all formed in a single event or if they represent formation over an extended period. Distributions from individual analysts typically exhibit somewhat less scatter and seem to show a systematic decrease in $(^{26}Al/^{27}Al)_0$ from Semarkona ($(7.1 \pm 1.7) \times 10^{-6}$) to Yamato 81020 and Acfer 094 ($(6.7 \pm 2.0) \times 10^{-6}$ and $(6.6 \pm 1.5) \times 10^{-6}$, respectively) to Kaba ($(4.8 \pm 1.1) \times 10^{-6}$). These initial ratios correspond to time intervals of millions of years (2–2.5 Myr) relative to CAIs (Fig. 10.10).

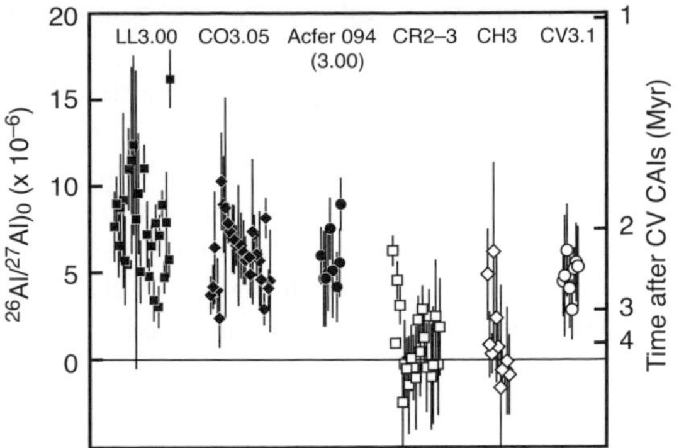

Figure 10.10 Initial $(^{26}Al/^{27}Al)_0$ ratios obtained from internal isochrons by SIMS analysis of chondrules from the least metamorphosed ordinary (LL3.00) and carbonaceous (CO3.05, Acfer 094 [ungrouped type 3.00], CR2–3, CH3, and CV3.1) chondrites. The right axis shows the formation times relative to CAIs, the oldest solids formed in the solar system. These data indicate that chondrule formation started at least 1 Myr after CAIs formed and continued for at least 3 Myr. Many of the late-forming chondrules show no evidence of live ^{26}Al when they formed. Diagram adapted from and references given in Nagashima et al. (2018).

Chondrules from CR chondrites seem to have formed later than chondrules from other classes (Fig. 10.10). The ^{26}Al–^{27}Al chronometer gives ages for CR2 chondrules ranging from 2.2 ± 0.2 to >4 Myr after CAIs, with many measurements not resolved from zero (Nagashima et al., 2014). ^{182}Hf–^{182}W data for a group of CR chondrules give an age of 3.7 ± 0.8 Myr after CAIs (Kleine et al., 2018). The ^{207}Pb–^{206}Pb data for CR2 chondrules are more ambiguous, with ages ranging from almost as old as CAIs to 4563.2 ± 0.6 Ma, ~ 4 Myr after CAIs (Bollard et al., 2017). These data imply that most CR2 chondrules formed 1–2 Ma after those in ordinary and CO chondrites (Figs. 10.9, 10.10).

Metal-rich chondrites (CH, CB) contain two types of chondrules: 1) ordinary type 1 and type 2 porphyritic chondrules that apparently formed in the same way as most other chondrules, and 2) nonporphyritic, magnesian chondrules that may have formed from the vapor plume generated in a collision of asteroid-sized bodies (Krot et al., 2005). The nonporphyritic magnesian chondrules formed later than most chondrules, giving a tight cluster of ^{207}Pb–^{206}Pb ages of 4562.5 Ma, ~4.8 Myr after CAIs (Bollard et al., 2015). Similar ages for chondrules and metal from CB chondrites have been determined using ^{53}Mn–^{53}Cr, ^{129}I–^{129}Xe, and ^{182}Hf–^{182}W (Bollard et al., 2015 and references therein). The more-normal chondrules also give young ages, with a few giving ^{26}Al–^{26}Mg ages similar to CR chondrules and most being unresolved from zero, implying low ^{26}Al contents and young formation ages.

Chondrule age determinations made using ^{53}Mn–^{53}Cr, ^{182}Hf–^{182}W, and ^{129}I–^{129}Xe support this ^{26}Al–^{26}Mg chronology (Fig. 10.9). The precision of the ^{53}Mn–^{53}Cr and ^{129}I–^{129}Xe dates are not as high as those for ^{26}Al–^{26}Mg, but the range of formation times is essentially the same. There is too little tungsten in chondrules to measure individual chondrules, but measurements of groups of hundreds to thousands of chondrules from CV and CR chondrites give tight isochrons that support the ^{26}Al–^{26}Mg and ^{53}Mn–^{53}Cr chronologies (Budde et al., 2016, 2018). They do not support the ^{207}Pb–^{206}Pb data that indicate chondrule formation concurrent with CAI formation.

Larsen et al. (2016) suggested a possible way to reconcile the difference in ages between ^{207}Pb–^{206}Pb and ^{26}Al–^{26}Mg for chondrules. They propose that the CAI-forming region had a much higher ^{26}Al/^{27}Al ratio than the rest of the nebula. Based on the initial ^{26}Mg/^{24}Mg ratios, reported as μ^{26}Mg*, they estimated the (^{26}Al–^{26}Mg)$_0$ ratio for most of the solar system to be (~1–2) × 10^{-5}. With this initial ratio, chondrule formation could have been contemporaneous with CAI formation. Larsen et al. (2016) present a detailed model to reconcile the ages of other bodies as well as degrees of asteroidal heating by ^{26}Al with this revised chronology. But, while this model may reconcile the ^{207}Pb–^{206}Pb and ^{26}Al–^{26}Mg ages of chondrules, it does not address the resulting disagreement between the ^{182}Hf–^{182}W and ^{26}Al–^{26}Mg chronologies. Much work remains to sort out this aspect of early solar system chronology.

In this book, we accept the general picture that CAIs were the first solids to form in the solar system, that they formed relatively quickly, and that many were subsequently altered in the nebula and in the host meteorites. Most chondrules probably formed 1–3 Myr after CAIs, although those in metal-rich chondrites formed or were reprocessed considerably later. Direct evidence that CAIs are older that chondrules comes from the discovery of several CAIs inside chondrules (e.g. Krot et al., 2006). To date, no chondrules have been discovered inside CAIs.

10.2.1.2 *Accretion and Thermal History of Chondritic Parent Bodies*

The chondrite parent bodies obviously could not have accreted before their constituent chondrules formed. Based on the formation times of chondrules, accretion of the ordinary chondrite parent bodies began ~3 Myr after CAIs (therefore at ~4564 Ma). The end of accretion can be inferred from the metamorphic history of the chondrite parent bodies. Isotopic data from metamorphic assemblages, coupled with thermal modeling of the chondrite parent bodies, suggest that the time of peak metamorphism for the H chondrite parent body was at ~4563 Ma. As will be discussed in Chapter 12, it is likely that the source of heat for metamorphism on chondrite parent bodies was the decay of ^{26}Al, perhaps with a contribution from ^{60}Fe. Thermal evolution models indicate that accretion of chondritic asteroids could not have occurred earlier than ~2 Myr after CAI formation, or they would have melted.

Once formed, the chondrite parent bodies experienced a variety of processes, including thermal metamorphism, aqueous alteration, shock metamorphism due to impacts, and even disruption from large impacts. Several radiochronometers can provide information on the timing of metamorphism and aqueous alteration. The chronology of this processing is summarized in Figure 10.11.

Ordinary Chondrites

Information about the timing of metamorphism in ordinary chondrites is derived from the ^{207}Pb–^{206}Pb, ^{40}Ar–^{39}Ar, ^{53}Mn–^{53}Cr, ^{26}Al–^{26}Mg, ^{129}I–^{129}Xe, and ^{182}Hf–^{182}W systems and from fission-track dating. The dates provided by these chronometers reflect the times when the meteorites cooled sufficiently to become closed to isotopic diffusion. Each isotopic system has a different closure temperature in each mineral. The most-detailed information about metamorphic history comes from the H chondrites. Figure 10.12 shows dates from three different isotopic systems for H4, H5, and H6 chondrites. Among these systems, the U–Pb system in pyroxene has the highest closure temperature (~780°C), followed by the U–Pb system in phosphate (closure at ~475°C), and followed in turn by the K–Ar system (closure at ~275°C). For each of the three classes of H chondrite, the dates become younger as the closure temperature decreases. Cooling rates calculated from these data show that H4 chondrites cooled most quickly, followed by H5 chondrites, and then H6 chondrites. This same pattern is observed for ^{182}Hf–^{182}W dating, which has a closure temperature of 800–850°C (Kleine et al., 2008), and for fission-track dating, which has a closure temperature similar to the K–Ar system (Trieloff et al., 2003). A detailed study of cooling rates and closure temperatures for metal in H chondrites showed that H6 chondrites generally have both slower cooling rates and later closure times than H5 chondrites, which in turn have slower cooling rates and later closure times than H4 chondrites (Hellmann et al., 2019). These observations are consistent with an "onion-shell" model for the H chondrite parent

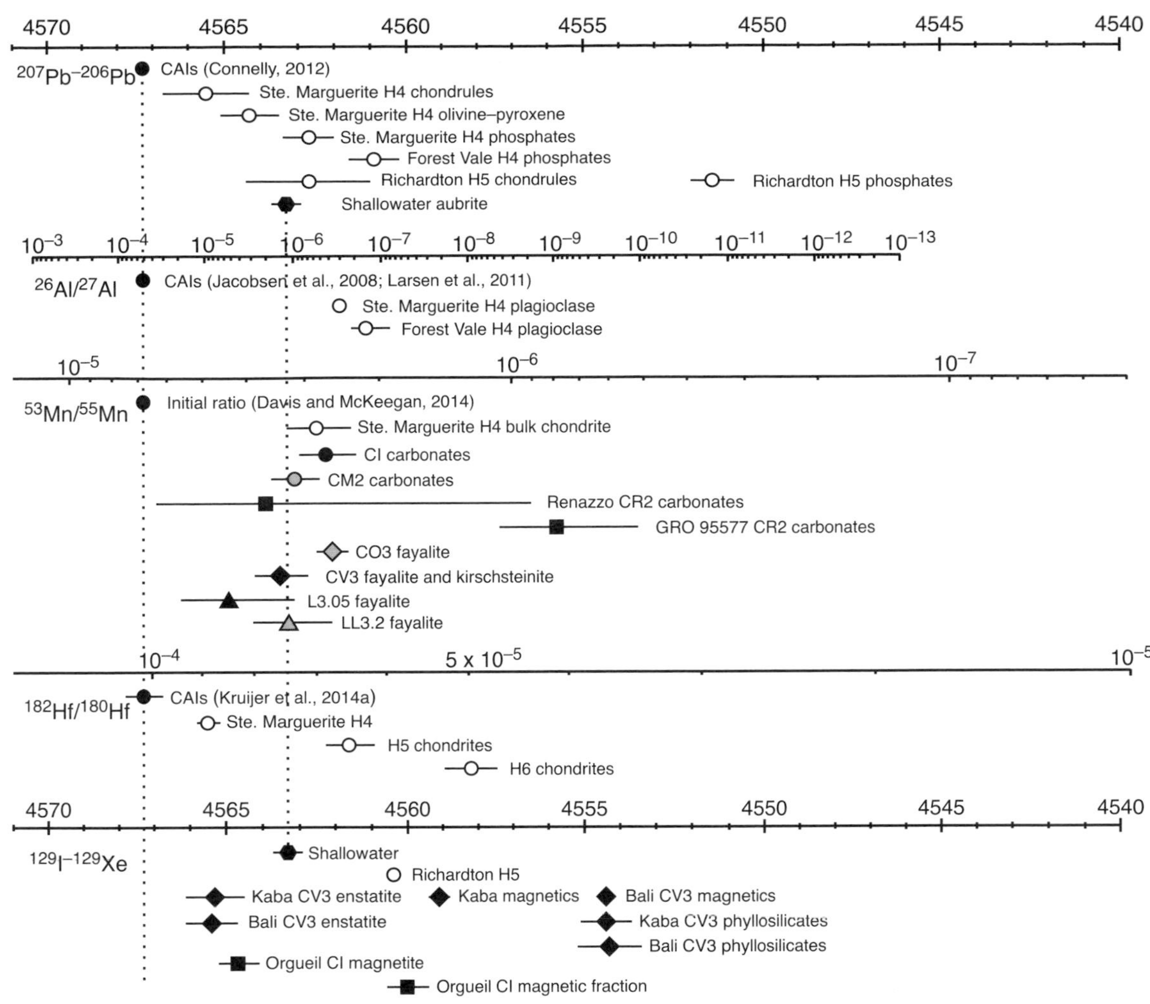

Figure 10.11 Chronology of secondary processes in the early solar system. Ages in Ma are given at the top and bottom of the figure. Plot format and anchor points are the same as in Figure 10.9. Dates related to thermal metamorphism are shown as open symbols, and dates related to aqueous alteration are shown as filled symbols. Both thermal metamorphism and aqueous alteration continued for tens to as much as 100 Myr after CAIs. Data from Bouvier et al. (2007), Göpel et al. (1994), Telus et al., (2014), Fujiya et al. (2012), Jilly-Rehak et al. (2017), Doyle et al. (2015), MacPherson et al. (2017), Kleine et al. (2008), and Hohenberg and Pravdivtseva (2008).

body, with the H6 chondrites in the center, surrounded in turn by H5 chondrites, H4 chondrites, and H3 chondrites near the surface (discussed more fully in Chapter 12).

A few H chondrites do not follow the expected cooling rate trends. Ste. Marguerite (H4), Forest Vale (H4), and Beaver Creek (H4) have faster cooling rates than other H chondrites of the same type. These meteorites were probably excavated from the interior of the parent body by impacts while the parent asteroid was still hot (Telus et al., 2014). For L and LL chondrites, the isotopic data do not correlate as well with petrologic type as they do for H chondrites. This could reflect earlier disruption of the parent asteroids or the subsequent shock history or both. In general, the ages of L chondrites tend to be younger than those of H chondrites, but the overall timescales are similar for H, L, and LL chondrites.

CV and CO Chondrites

As with the ordinary chondrites, the onset of accretion for CV and CO chondrite parent bodies can be estimated from the times of chondrule formation, which extended to ~3–3.5 Myr after CAIs (Fig. 10.10). This is slightly later than, but within uncertainties of, the time indicated for accretion of ordinary chondrite parent bodies. The end

(2013) measured U–Pb and ^{207}Pb–^{206}Pb ages of zircons in five basaltic eucrites and found a weighted average ^{207}Pb–^{206}Pb age of 4541 ± 11 Ma and a concordia age of 4525 ± 2 Ma. Based on these and other data, they infer core–mantle differentiation at 4564 ± 2 Ma (3.3 Myr after CAIs) and a peak of basaltic magmatism at 4552 ± 7 Ma, which slowly diminished over the next 50 Myr. Touboul et al. (2015) found that basaltic eucrites fall into three distinct groups with ^{182}Hf–^{182}W ages of ~4 Myr, 11 Myr, and 22 Myr after CAIs. ^{182}Hf–^{182}W measurements of three cumulate eucrites imply formation times of 38 ± 21 Myr after CAIs. Lugmair and Shukolyukov (1998) obtained ^{53}Mn–^{53}Cr isochrons for three basaltic eucrites and found ages of ~4564 to ~4557 Ma, corresponding to ~3.3 to ~10 Myr after CAIs (Fig. 10.9). The ^{26}Al–^{26}Mg system is almost too short-lived to date eucrites, but data for several basaltic eucrites give an age of 4563.5 ± 0.9 Ma, ~3.8 Myr after CAIs (Wadhwa et al., 2006 and references therein).

Taken together, these data imply that the eucrite parent body accreted and differentiated during the first 1–2 Myr after CAIs, and that eruption of eucritic magmas was actively occurring by ~7 Myr after CAIs. The basaltic eucrites were produced over at least the next ~20 Myr, and the cumulate eucrites, which did not erupt on the surface, continued to be emplaced for another ~20 million years. Long-lived radionuclides, including ^{86}Rb–^{87}Sr and ^{147}Sm–^{143}Nd, suggest that magmatism continued for at least another 100 million years (Fig. 10.8).

Asuka 881394

This basaltic achondrite appears to be unrelated to the eucrites, implying formation on a different asteroid. High-precision ^{207}Pb–^{206}Pb data for Asuka 881394 yield a precise and ancient age of 4564.95 ± 0.53 Ma (Wimpenny et al., 2019), just over 2 Myr after CAIs. The ^{26}Al–^{26}Mg and ^{53}Mn–^{53}Cr systems give similar ages (4563.5 and 4565.2, respectively). These ages imply that the parent body accreted, melted and differentiated, and erupted the Asuka 881394 basalt all within the first 2–3 Myr of solar system history.

Angrites

The angrites seem to have well-behaved isotope systematics. The ^{207}Pb–^{206}Pb, ^{26}Al–^{26}Mg, ^{53}Mn–^{53}Cr, and ^{182}Hf–^{182}W systems are generally concordant in angrites (Fig. 10.9). Based on a whole-rock isochron, the differentiation of the angrite parent body had occurred by ~3.6 Myr after CAIs (Kleine et al., 2012). Individual angrites started crystallizing within 0.1–0.2 Myr after the differentiation of the angrite parent body, and they continued to form over about 7 Myr (Fig. 10.9). The lack of isotopic disturbance among the angrites implies that magmatic activity ceased on this asteroid shortly after the youngest known angrite formed.

Mesosiderites

These meteorites consist of silicate and metal that were brought together by impact events on the meteorite parent bodies. The silicate portion of the Vaca Muerta mesosiderite gives a $(^{26}Al/^{27}Al)_0$ ratio slightly higher than the silicates in eucrites (Bizzarro et al., 2005), suggesting that the parent body from which they are derived also formed and began to differentiate during the first million years of solar system history.

Pallasites

There is not much chronological information about pallasites in the literature because the simple mineralogy of these meteorites does not provide many phases that can be dated. A ^{53}Mn–^{53}Cr measurement of the Omolon pallasite gives a formation time of 4558.3 Ma (Fig. 10.9), while a similar but less well constrained measurement of Springwater suggests that it is ~1.3 Myr younger than Omolon (Lugmair and Shukolyukov, 1998).

Iron Meteorites

In Chapter 6, we learned that iron meteorites are divided into magmatic irons, which apparently formed by fractional crystallization during the slow cooling of asteroidal cores, and nonmagmatic irons, which formed by a variety of different mechanisms. Dating of iron meteorites by long-lived radionuclides such as the ^{187}Re–^{187}Os system shows that iron meteorites formed near the beginning of the solar system, but the dates do not provide much precision (Fig. 10.9). The most reliable constraints on the timing of core formation in the parent asteroids of the magmatic irons come from ^{182}Hf–^{182}W measurements. To obtain reliable data, one must correct for cosmogenic effects (or avoid these effects by measuring meteorites with low cosmic-ray exposure ages) and for nucleosynthetic effects in tungsten. Careful, high-precision measurements that take these effects into account have been made on IIAB, IIIAB, IID, IVA, and IVB irons (Kruijer et al., 2014b). The resulting ages range from 4566.6 Ma for IIAB irons to 4564.2 Ma for IID irons (Fig. 10.9). These ages date core formation, which means that the parent bodies of these irons accreted, melted, and differentiated within 1–3 Myr after CAIs. This is clear evidence for the accretion of parent bodies of differentiated meteorites before the accretion of chondrite parent bodies.

10.3 Accretion, Differentiation, and Igneous History of Planets and the Moon

Given our focus on cosmochemistry, we will consider only chronology based on radiogenic isotopes. Most planetary chronology is presently based on crater counting (discussed briefly in Box 10.1), and only the Earth, the Moon, and Mars have provided samples that can be analyzed in the laboratory for radioisotopes. However, these three bodies illustrate what could be learned from samples of other terrestrial planets.

10.3.1 Age of the Earth

The origin of the Earth is difficult to date because the geologic record of its first 500 Myr is missing. Most long- and short-lived radionuclides do not provide useful constraints on this event. Only the U–Pb and ^{182}Hf–^{182}W isotopic systems allow valid constraints (reviewed by Halliday, 2014). Both of these systems basically define the timing of the Earth's differentiation, because core formation fractionates the parent/daughter ratios (uranium and hafnium are lithophile, while lead is both siderophile and chalcophile and tungsten is moderately siderophile). In this regard, we caution that these chronometers are not able to distinguish between early accretion with late core formation, and late accretion with concurrent late core formation. Also, if core formation was protracted, then the "age" derived does not define any particular event but instead a kind of weighted average. That said, there is now a consensus that the Earth's core formation was basically concurrent with accretion, because the conversion of kinetic energy into heat would have allowed silicate and metal melting temperatures to be achieved and because many of the accreting planetesimals would have already been differentiated.

The U–Pb system has been a chronometer of choice for the Earth's age since the pioneering study of Clair Patterson in 1956, as discussed in Chapter 9. No samples exist from the earliest epoch of Earth history. However, it is now clear that most U–Pb model ages for the Earth (reviewed by Allègre et al., 1995) are significantly younger than the ages of chondrites, achondrites, and their constituents (Halliday, 2014). Tera (1981) used the lead isotopic compositions of four of the oldest conformable galena deposits to infer a model age for their source region of 4.54 Ga, only ~30 Myr after CAIs. Various ^{207}Pb–^{206}Pb ages for mid-ocean-ridge source regions give ~4.4 Ga (Allègre et al., 1995). The oldest surviving material is a detrital zircon with a U–Pb age of 4.4 Ga (Wilde et al., 2001). Models with an exponentially decreasing growth rate and continuous core formation give an accretionary mean lifetime for the Earth of between 15 and 50 Myr (Halliday, 2014). The timing of the Moon-forming impact, which probably added the last ~15% to the mass of the Earth, is a little better determined. Various estimates based on the U–Pb system converge on 69 $\pm$ 10 Myr after CAIs (Maltese and Mezger, 2020).

The ^{182}Hf–^{182}W system can provide some constraints on the Earth's core formation. The bulk silicate Earth has a ^{182}W excess of ~190 ppm relative to chondrites, showing that live ^{182}Hf was present in the bulk silicate Earth after much of the tungsten had been extracted into the core. If the core was extracted in a single, instantaneous event, the ^{182}Hf–^{182}W system gives a core-formation time of 34 $\pm$ 3 Myr after CAIs (Kleine and Walker, 2017). Episodic or continuous core formation is more likely. Although this system cannot tell how long core formation persisted, it does constrain that a significant fraction of the Earth's core must have formed within the first 10 Myr of solar system history. This system can also indirectly constrain accretion. Yin et al. (2002) estimated the mean time for Earth's accretion at ~11 Myr. Halliday (2014), taking into account the uncertainty in the initial ^{182}W abundance, estimated the mean age for the Earth's accumulation (with an exponentially decreasing accretion rate) of 12–15 Myr. But the 8.9 Myr half-life of ^{182}Hf means that this system cannot distinguish between a model with a final contribution to the core at ~30 Myr after CAIs and one where core formation continued for $>$100 Myr (Kleine and Walker, 2017).

10.3.2 Age and Igneous History of the Moon

Like for the Earth, the time of formation of the Moon must be determined by indirect means. The Moon must obviously be older than its oldest rocks, which are thought to have formed in a magma ocean. Ferroan anorthosites are interpreted to be early cumulates from the lunar magma ocean (see Chapter 14). An assessment of their ^{147}Sm–^{143}Nd isochron ages, which are most resistant to thermal resetting, gives dates ranging from 4.57 to 4.29 Ga (Fig. 10.14; Borg et al., 2014 and references therein). Mg-suite rocks were originally thought to represent plutons intruded after solidification of the magma ocean, but their ^{147}Sm–^{143}Nd isochron ages overlap those of the anorthosites (Fig. 10.14). However, isochrons can only date the lunar magma ocean if the isotopic systems have not been disturbed, which is unlikely given that these ancient rocks are mostly impact breccias. Borg et al. (2014) considered that the oldest ages for lunar crustal rocks are unreliable, and preferred an age

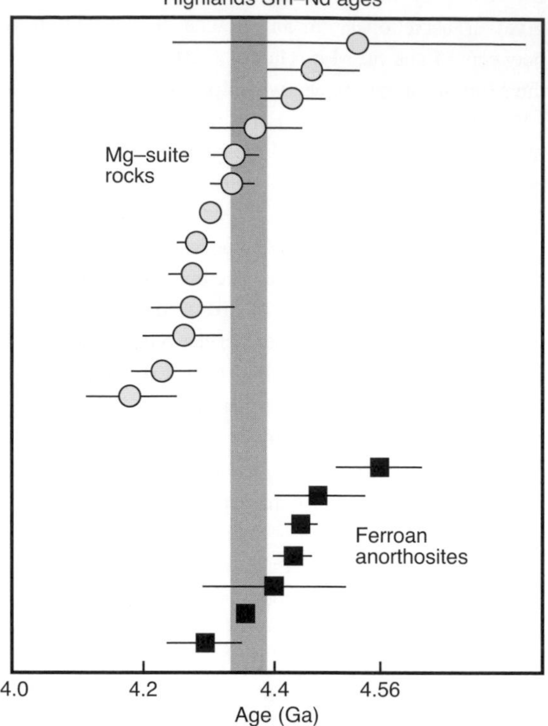

Figure 10.14 Sm–Nd ages for lunar highlands rock samples, thought to best represent the chronology of the most ancient rocks on the Moon (ferroan anorthosites and Mg-suite rocks). The shaded vertical bar represents model ages for KREEP and mare basalt source regions. After Borg et al. (2014).

of 4353 ± 45 Ma for solidification of the magma ocean or some other widespread magmatic event.

Another approach involves model ages for isotopic systems. Model ages for the ^{176}Lu–^{176}Hf system (4353 ± 37 Ma) and for the ^{147}Sm–^{143}Nd system (4389 ± 45 Ma) for KREEP, the end stage of lunar magma ocean crystallization, were derived by Gaffney and Borg (2014) and Borg et al. (2014), respectively. Snape et al. (2016) derived lead isotope model ages of 4376 ± 18 Ma for the source regions of lunar mare basalts. All these model ages, shown by the shaded vertical bar in Figure 10.14, are virtually identical to the preferred measured ages of anorthosites and Mg-suite rocks (Borg et al., 2014). Lead model ages have also been used in an attempt to determine the time of the catastrophic impact thought to have formed the Moon. A two-stage lead isotope evolution model for lunar crustal rocks (Connolly and Bizzarro, 2016) interpreted an inflection in the U/Pb ratio at 4426–4417 Ma to have been caused by lead loss due to the giant impact. Maltese and Mezger (2020) reviewed a variety of estimates based on lead isotope evolution and concluded that the Moon-forming impact occurred 69 ± 10 Myr after CAIs. However, lead model ages are somewhat uncertain because U/Pb fractionation on the Moon is poorly constrained. In any case, this model age is consistent with the measured ages of ancient crustal rocks already described.

With the exception of Maltese and Mezger (2020), all these measured and model ages are ~100 Myr younger than that predicted by dynamical models of the Moon-forming impact or solidification models for the lunar magma ocean. An older age is provided by some U–Pb analyses of zircons that crystallized from the lunar magma ocean. Zircons are not generally susceptible to isotopic disturbance and are amenable to high-precision chronology. Although the zircon ages peak at 4335 Ma, the oldest ages are >4.4 Ga (Borg et al., 2014 and references therein). By combining U–Pb and ^{176}Lu–^{176}Hf isotope systematics in individual zircon grains, Barboni et al. (2017) constructed a minimum model age of 4.51 ± 0.01 Ga for the differentiation of the Moon. This age is 120–150 Myr older than the isochron ages for ancient lunar crustal rocks.

Models of the ^{129}I–^{129}Xe and I–Pu–Xe systematics suggest that the Moon-forming impact occurred at ~40 Myr (range 30–60 Myr) after CAIs (Avice and Marty, 2014). Taken together with constraints from the short-lived ^{182}Hf–^{182}W isotope system that indicate the Moon formed ~50 Myr after CAIs (Thiemens et al., 2019) and the zircon model age from zircons (Barboni et al., 2017), these data suggest that the Moon's formation occurred within the first ~50–60 Myr of solar system history.

Lunar mare basalts have usually been dated by the ^{147}Sm–^{143}Nd and ^{87}Rb–^{87}Sr techniques. Their crystallization ages generally range from 4.0 Ga to 3.1 Ga, with a peak in eruption rates between 3.6 and 3.8 Ga (Heisinger et al., 2011; Snape et al., 2016 and references therein). Analyzed mare basalts do not bound the full age range for mare volcanism. Crater counting (see Box 10.1) of other maria suggest volcanism may have extended to ~1.2 Ga before surface igneous activity ceased (Heisinger et al., 2011).

10.3.3 Age and Igneous History of Mars

Because accretion is not directly datable by any of the radiometric clocks, we must rely on dates from early differentiation events to put constraints on the formation time of Mars. Core formation probably began very early, perhaps during the last part of the planet's accretion. The timing of core formation can be investigated using the ^{182}Hf–^{182}W system, and crust formation can be studied using the ^{146}Sm–^{142}Nd system. Two-stage models for the

Box 10.1 Relative Timescales: Crater Counting and Stratigraphy

Absolute time measurements using radioactive isotopes are presently limited to samples that can be carefully processed and analyzed in the laboratory. Estimates of the *relative times* of formation of various geologic units on other planets can be gained from spacecraft imagery, by crater counting. The older a surface is, the longer it has been exposed to meteorite impacts and thus the higher the density of craters on that surface. However, very old surfaces can become saturated with craters, when every new crater effectively obliterates an older one.

In practice, crater diameters on a particular geologic unit are measured and assigned to various size bins. Then a log–log plot of the cumulative number of craters above some size versus crater diameter is constructed, giving a diagram like that shown in Figure 10.15. The crater distribution has a negative slope, because small craters are always more abundant than large craters. As time passes, the crater distribution curve shifts to the right, so comparison of the crater distributions for several geologic units allows us to determine their relative ages. In Figure 10.15, the high density of craters within the Orientale basin on the Moon indicates its antiquity, relative to craters like Eratosthenes and even younger craters with bright rays.

It is beyond the scope of this book to discuss the methodology of crater counting in detail (there are lots of details and caveats). However, we should note that measuring the radiometric age of a rock sample obtained from a mapped geologic unit provides a calibration point for crater counting. If absolute ages are determined for multiple units with different crater densities, a crater production curve for that planet can be constructed; then, the absolute ages of other,

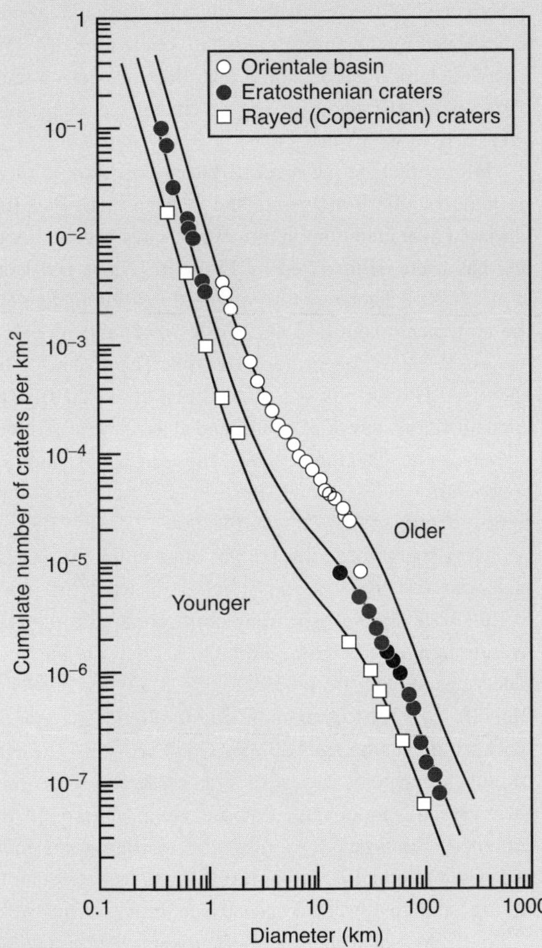

Figure 10.15 Plot of the cumulative frequency of craters versus crater diameter for three geologic units on the Moon. Crater density measurements provide a means of ordering units in relative time. Modified from Neukum et al. (2001).

Box 10.1 (cont.)

unsampled units on the same planet can be estimated from their crater densities. Such a crater production curve has been determined for the Moon, based on the radiometric ages of rocks returned by Apollo astronauts.

The geologic timescale for the Earth is based mostly on stratigraphy (the depositional succession of layers from bottom to top). Most strata are sedimentary rocks. The relative ages of the sedimentary layers are determined by the sequence of deposition and by the fossils that they contain. It is not possible to use radioactive isotopes to measure the time that a sediment was deposited, because deposition does not reset radiometric clocks. However, the absolute ages of sediments have been determined by measuring the radiometric ages of volcanic ash layers in the sequence or lavas that crosscut sedimentary strata (e.g. Bowring et al., 2006).

The geologic timescale for the Moon is also based on stratigraphy, but in this case the layers are blankets of impact ejecta. Huge craters, occurring mostly before about 3.9 Ga, spread ejecta over much of the lunar surface, producing global-scale stratigraphic units. This serviceable system was first suggested by geologist Eugene Shoemaker, who used it to devise the lunar timescale and to map the Moon's surface. Some limits on the absolute ages of lunar ejecta blankets come from impact ages, determined from isotopic systems like ^{40}Ar–^{39}Ar that are reset by shock. The geologic timescale for Mars presently has only three gross subdivisions, and its only constraints are based on crater counting and an assumed crater production rate. The relative and absolute timescales for geologic units on the Moon and Mars are discussed further in Chapter 14.

evolution of these systems using the data for shergottites indicate that crystallization of a magma ocean occurred within 20–25 Myr of CAIs, followed by crust formation ~15 Myr later (Kruijer et al., (2017). This scenario is supported by ^{176}Lu–^{176}Hf dating, which indicates that accretion, differentiation of the core, and magma ocean crystallization were completed in the first 20 Myr of solar system history (Bouvier et al., 2018).

The only ancient martian meteorites are NWA 7034 and NWA 7533, paired samples on which considerable work has been done. Mineral separates from these meteorites have ^{147}Sm–^{143}Nd and ^{146}Sm–^{142}Nd ages of ~4.44 Ga (Nyquist et al., 2016); the Rb–Sr system is too disturbed for age dating. These breccias contain igneous crustal clasts with zircons, which have been dated by U–Pb methods (Humayun et al., 2013; Bellucci et al., 2018). Their ages, as ancient as 4.476 Ga (Bouvier et al., 2018), indicate that a stable crust formed within the first ~100 Myr.

There are significant differences in the ε^{142}Nd and ε^{182}W values obtained for the different types of martian meteorites (e.g. Lee and Halliday, 1997). These variations indicate that the meteorites were derived from distinct mantle reservoirs that were established early in the history of Mars. Modeling based on ^{129}I–^{129}Xe and ^{244}Pu–^{136}Xe systematics indicates that Pu–I fractionation in the martian mantle occurred $\leq$ 35 Myr after Mars formed, although degassing of xenon and other volatiles from the mantle may have continued for ~300 Myr (Marty and Marti, 2002). The preservation of distinct mantle reservoirs over time in Mars and the limited degassing experienced by the martian mantle after the initial period, as indicated by the almost quantitative retention of ^{244}Pu-produced fission xenon, show that Mars has been a static planet with no mantle mixing or crustal recycling since very early in its history.

Information on the igneous history of Mars is given by the crystallization ages of the martian meteorites (the chronological literature on these meteorites is voluminous and has been summarized by Filiberto [2017] and Udry et al. [2020]). There are only two old martian meteorites: the aforementioned 4.44 Ga NWA 7034 (and its pairs), and ALH 84001, an orthopyroxenite. The ^{147}Sm–^{143}Nd age of ALH 84001 is ~4.1 Ga (Lapen et al., 2010). The crystallization ages of nakhlites and chassignites are much younger, ~1.3 Ga (Fig. 10.16). These ages are based on concordant dates from the ^{40}Ar–^{39}Ar, ^{87}Rb–^{87}Sr, ^{147}Sm–^{143}Nd, and U–Pb systems.

Most shergottites (by far the most common martian meteorites) give ^{147}Sm–^{143}Nd, ^{87}Rb–^{87}Sr, and ^{176}Lu–^{176}Hf mineral isochrons corresponding to crystallization ages of ~175 Ma, although a small number give older ages ranging from ~290 to 575 Ma (vertical axis of Fig. 10.16). The uncommon augite-phyric shergottites have crystallization ages of ~2.4 Ga. The ^{40}Ar–^{39}Ar ages of most shergottites appear to be considerably older than their crystallization ages, but are compromised by the presence of trapped ^{40}Ar from the martian atmosphere and possibly inherited radiogenic ^{40}Ar as well. In contrast to the mineral (internal) isochrons, whole-rock ^{207}Pb–^{206}Pb isochrons for shergottites suggest a much older age of ~4.1 Ga, and Bouvier et al. (2008) interpreted

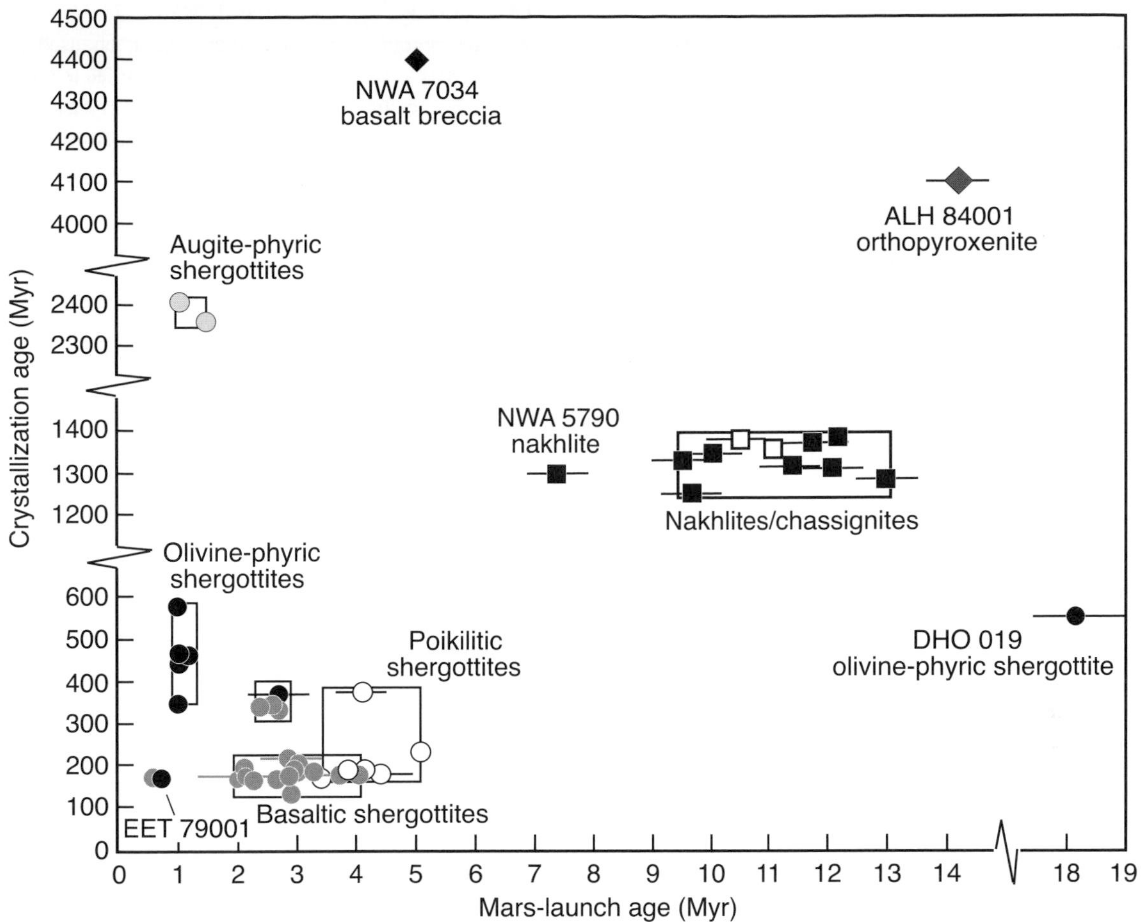

Figure 10.16 Comparison of igneous crystallization ages and Mars-launch ages for martian meteorites. Only meteorites for which both kinds of ages are available are plotted. Crystallization ages are generally Rb–Sr and/or Sm/Nd ages. Launch ages are cosmic-ray exposure ages plus terrestrial ages, although the latter are usually too short to be significant. The meteorites interpreted to have come from specific launch sites are enclosed by boxes; the meteorites from a specific site generally share the same launch age and a limited range of crystallization ages, as well as the same petrologic classification and geochemical characteristics (described later in Chapter 14). Data from Udry et al. (2020), Herzog and Caffee (2014), and Wieler et al. (2016).

this age as indicating the time of crystallization of shergottite magmas. However, concordant ages based on other isotopic systems are younger, and radiogenic crustal lead may have contaminated these meteorites.

10.4 Shock Ages and Impact Histories

10.4.1 Shock Ages of Meteorites

As discussed in Chapter 9, isotopic chronometers can give the time of formation of an object provided that specific assumptions are met. One of those conditions is that the system must remain undisturbed since its formation. However, a large impact can partially or completely reset an isotopic clock. Susceptibility to impact resetting varies widely among the various chronometers. Chronometers that involve noble gases are particularly sensitive because a noble gas atom is not incorporated into the structure of a mineral; rather, it is interstitial. The U–He and ^{40}K–^{40}Ar systems are often used to evaluate shock history. Other systems that can be reset include, in order of decreasing susceptibility to shock, the ^{87}Rb–^{87}Sr, ^{207}Pb–^{206}Pb, and ^{147}Sm–^{143}Nd systems. Because the isotopic clocks have generally not been completely reset, the exact timing of an event is often impossible to determine.

The timing is usually inferred by comparing more than one isotopic system or by the clustering of shock ages from many meteorites of the same type. Relatively severe shocks are required to affect these radiometric clocks. In general, the final collision that released a meteorite from its parent body cannot be dated by these clocks, in part because it was so recent, but also because the collision was usually not sufficiently violent. The timing of this last event is inferred from the cosmic-ray exposure ages (see below).

Many chondrites show evidence of having experienced shock from impacts, and a significant fraction show isotopic disturbances that can be interpreted to give shock ages. A histogram of $^{40}Ar–^{39}Ar$ shock ages for ordinary chondrites is shown in Figure 10.17. There is a small cluster of ages in the range of 3.4–4.1 Ga, but most shock ages are <1 Ga. The inset in Figure 10.17 shows these younger shock ages broken down by chondrite type. The large group of L chondrites with shock ages of ~500 Ma has been associated with a major breakup of the L chondrite parent body. Because of the nature of shock ages, the precise timing of this breakup is not known. However, a recent discovery that a large number of L chondrites fell to Earth in the early Ordovician period ~480 Myr ago seems to support this conclusion (see Box 10.2).

The majority of HED meteorites show evidence of isotopic resetting over a limited period of time between ~3.4 and ~4.1 Ga ago. Most of the evidence comes from the $^{40}Ar–^{39}Ar$ system, but the $^{87}Rb–^{87}Sr$ system was also disturbed at this time in some meteorites.

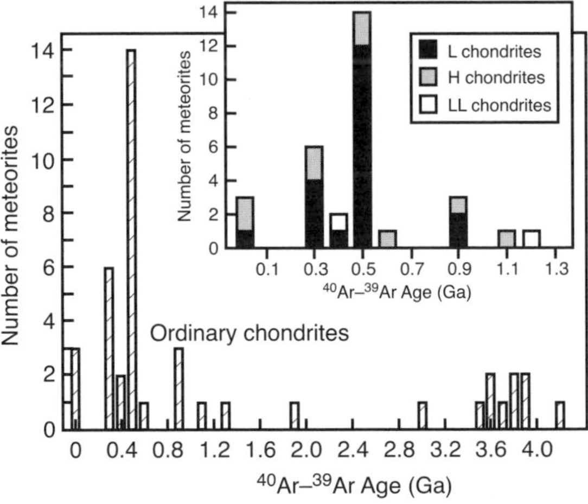

Figure 10.17 Shock ages of ordinary chondrites as determined from $^{40}Ar–^{39}Ar$ measurements. The inset gives an expanded version of the youngest ages separated by composition group. After Bogard (1995).

10.4.2 Shock Ages of Lunar Rocks

The same isotopic techniques that have been used to determine shock ages in meteorites can be applied to lunar samples. Logically, these studies focus on impact-melt breccias, which are most likely to have had their isotopic ages fully reset by impact events. $^{87}Rb–^{87}Sr$ data, and to a lesser extent $^{147}Sm–^{143}Nd$ data, for lunar impact breccias give ages ranging from 4.45 to 3.89 Ga (reviewed by Warren, 2004). Because these breccias are composed of rock clasts, crystal fragments, and impact-melt glasses that were likely derived from multiple shock events, their ages represent an average of impact events. Differences in breccia shock ages indicate the scale of sampling of these materials.

The most useful shock ages are based on $^{40}Ar–^{39}Ar$ data for impact-melt breccias. Most lunar breccias yield $^{40}Ar–^{39}Ar$ plateau ages younger than ~3.85 Ga. By carefully selecting breccias from different Apollo landing sites that are thought to have formed during specific large impact events (that is, they are associated with specific impact basins), the absolute ages of these basins have been defined. These basins form the basis for the lunar timescale, as discussed in Chapter 14. Most lunar basins formed during the period between ~3.9 and 3.75 Ga. For example, Nectaris, Imbrium, and Orientale are inferred to have formed at 3.92, 3.85, and 3.75 Ga, respectively. Serenitatis and Crisium are also thought to have formed during this interval. The youngest lunar basins are the rayed craters Copernicus and Tycho, with inferred ages of ~1.0 and ~0.3 Ga, respectively.

10.4.3 A Late Heavy Bombardment?

The pronounced peak in large impact events seen in the ages of lunar impact-melt breccias has led to the concept of a lunar cataclysm. Although this is commonly attributed to an enhanced flux, or spike, in large impactors at ~3.85 Ga, some workers have suggested that it represents an inflection in a basically monotonic decline of impacts during the waning stages of planetary accretion. Skeptics of the spike idea suggest that the clustering of shock ages reflects a "stonewall" effect, where the lunar crust became so saturated with craters that earlier radiometric clocks were reset in a heated mega-regolith. The term *late heavy bombardment* could apply to either model. A peak in the shock ages for HED meteorites also corresponds to the period of bombardment on the Moon. This is unlikely to be a coincidence and may provide further support for this important event in solar system history. In Chapter 15, we will explore the possibility of a late heavy bombardment in more detail and consider a dynamical model that might account for a spike in the impact history of the Moon.

Box 10.2 An Ordovician Meteorite Rain

In a limestone deposit at Kinnekulle, Sweden, more than 100 highly altered meteorites (Fig. 10.18) totaling ~8 kg have been collected during routine quarrying operations (Schmitz et al., 2001, 2016; Schmitz, 2013). The ~3.2 meter-thick limestone layer in which they were found was deposited over ~1.75 Myr in the mid-Ordovician. Despite being almost completely replaced by calcite, barite, and phyllosilicates, the meteorites are easily identified by their chondritic texture. Their identification as meteorites is confirmed by measurements of platinum-group elements. The chemical characteristics of relict spinels indicate that, with only one exception, all are either L or LL chondrites. The odd specimen is apparently a type of meteorite not falling on Earth today (Schmitz et al., 2016).

How did so many meteorites become incorporated into this limestone? An obvious possibility is that they represent a single large meteorite that broke in the atmosphere and scattered across the landscape. But this possibility can be ruled out by their positions in the stratigraphic column. The meteorites typically lie on hardgrounds, where they accumulated together with nautiloid shells (Fig. 10.18) and other large objects. In many cases, the specific horizon at which individual meteorites were deposited can be identified. Based on their stratigraphic positions, at least 12 separate meteorite falls are represented. This is a tremendous number of meteorites to have fallen in a ~2700 m^2 area over $\leq$1.75 Myr. The data imply that the infall rate was significantly higher than today.

For many years, we have known that many of the L chondrites experienced a severe shock event at ~500 Ma. One interpretation of these data was that the L chondrite parent body was disrupted by an impact at that time. Could this catastrophic event have produced the rain of meteorites in the Ordovician? Detailed stratigraphic evaluation places the age of the meteorite-bearing limestone at 467.3 ± 1.6 Ma. Recent ^{40}Ar–^{39}Ar dating of one of the meteorites that experienced the major shock event gave an age for that event of 470 ± 6 Ma, the same age or slightly older than the

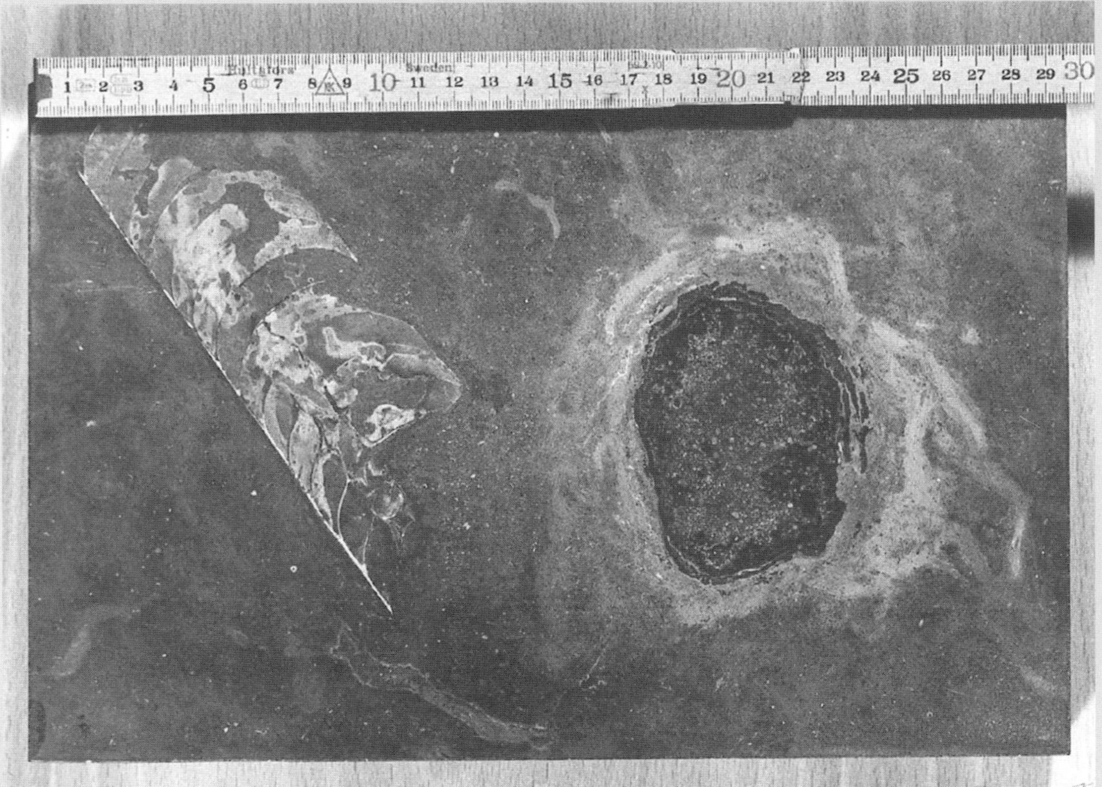

Figure 10.18 Fossil meteorite together with a nautiloid shell in a limestone plate sawed parallel to the seafloor surface. Image from Schmitz et al. (2001).

Box 10.2 (cont.)

limestone layer. Scientists who study orbital dynamics then went to work to identify an asteroid of the correct spectral type that might be in a position to have transferred large numbers of fragments into Earth-crossing orbits quickly after a disrupting impact. Bottke et al. (2009) found that the Gefion asteroid family, which consists of ~2200 individual asteroids, is in just the right place, near the 5:2 mean motion resonance with Jupiter. The resonance provides a means of transferring fragments of Gefion-family asteroids into the inner solar system. A dynamical analysis of this family suggests that the parent asteroid was disrupted ~485 Myr ago (Bottke et al., 2009), making it just the right age as well.

So the data indicate that a catastrophic collision 470–480 Myr ago created the Gefion family of asteroids. Fragments from this catastrophic breakup were transferred through a resonance into Earth-crossing orbits. These fragments rained down on Earth at a rate perhaps 25 times the current meteorite influx rate for several million years. Additional support for this model comes from an apparent spike in the cratering rate on Earth at about this time and by a high abundance of dispersed meteoritic chromite of L chondrite affinity in condensed Ordovician limestone in other locations in Sweden, in western Russia, and in central China, confirming that this was a worldwide event (Schmitz, 2013).

10.5 Cosmogenic Nuclides in Meteorites

10.5.1 Cosmic-Ray Exposure Ages

Interplanetary space is permeated by *cosmic rays*. These particles, mostly protons and α-particles, originate both in the Sun (solar cosmic rays) and in interstellar space (galactic cosmic rays). Solar cosmic rays have a power-law distribution of energies from ~10 to a few hundred MeV. Galactic cosmic rays are more energetic, with energies of 10^2–10^4 MeV, again with a power-law distribution. However, because they are charged particles, they are affected by the Sun's magnetic field. Solar activity thus governs how far the galactic cosmic rays, particularly the low-energy ones, penetrate into the inner solar system.

Energetic particles react with solid matter in a variety of ways. Low-energy particles in the solar wind (~1 KeV/nucleon) are implanted into solids to depths of ~50 nm. Energetic heavy particles penetrate more deeply and disrupt the crystal lattice, leaving behind tracks that can be imaged by transmission electron microscope or chemically etched and observed in an optical microscope. Particles with energies of several MeV or more may induce a nuclear reaction. The two main modes of production of cosmogenic nuclides are spallation reactions and neutron capture.

Spallation occurs when a high-energy cosmic ray breaks a target nucleus into two or more pieces. These interactions commonly eject neutrons. The secondary neutrons slow down to thermal energies and eventually react with other nuclei in the target material to generate heavier species. Production of cosmogenic nuclides by secondary neutrons increases with depth to a peak at 0.5–1 meter below the surface. Therefore, in order to get an accurate exposure age, it is necessary to estimate the depth in the pre-atmospheric body where the sample was located. This can be done by looking at isotope ratios, which vary with depth.

The fraction of incoming cosmic rays that generate nuclear reactions is quite low. In a meteorite traveling in space, about 1 in 10^8 of the target atoms undergoes a nuclear reaction in a 10 Myr period. However, the *cosmogenic nuclides* that they produce can be measured to estimate the time that an object has been exposed to cosmic rays. Table 10.1 shows some of the nuclides that are used to estimate cosmic-ray exposure ages in meteorites and in materials from planetary surfaces.

Cosmogenic nuclides are measured by a number of techniques. Helium-3, ^{21}Ne, and ^{38}Ar are measured by conventional noble gas mass spectrometry. In many samples, concentrations are high enough, especially for

Table 10.1 Nuclides used for cosmic-ray exposure ages

Nuclide	Half-Life (years)	Produced From
^{3}H	12.32	O, Mg, Si, Fe
^{3}He	Stable	O, Mg, Si, Fe
^{10}Be	1.39×10^6	O, Mg, Si, Fe
^{14}C	5730	O, Mg, Si, Fe
^{21}Ne	Stable	Mg, Al, Si, Fe
^{22}Na	2.6	Mg, Al, Si, Fe
^{26}Al	7.17×10^5	Si, Al, Fe
^{38}Ar	Stable	Fe, Ca, K
^{36}Cl	3.01×10^5	Fe, Ca, K
^{81}Kr	2.29×10^5	Rb, Sr, Y, Zr

helium and neon, that only small corrections for trapped gases are required. They are stable isotopes so their abundances increase proportionally with their cosmic-ray exposure. Krypton-81 is present in lower abundances and is radioactive, and thus requires a more complicated measurement procedure to extract age information. Radioactive isotopes such as ^{3}H, ^{14}C, ^{10}Be, ^{22}Na, ^{26}Al, and ^{36}Cl can be measured by counting the γ-rays emitted by their decay. These isotopes can also be counted directly in an accelerator mass spectrometer.

Cosmogenic nuclides can provide information about the travel time between a meteorite parent body and the Earth (the meteorite's *cosmic-ray exposure age*), the irradiation environment in interplanetary space, the number of collisions suffered by an object on its way to Earth, whether the object got too close to the Sun, and even the residence time of the meteorite on Earth. In order for a sample to give a cosmic-ray exposure age, several assumptions must be satisfied: 1) the flux of primary cosmic rays was constant in time, 2) the flux of primary cosmic rays was constant in space, 3) the shape of the sample did not change, 4) the chemical composition of the sample did not change, 5) any cosmogenic contributions from previous irradiations are known, 6) all noncosmogenic contributions to the inventory of the nuclide of interest are known, and 7) the sample did not lose nuclides of interest except by radioactive decay. Many meteorites appear to fit these requirements. To calculate the age from measured data, the cosmic-ray flux must be known and a model of the shape and size of the preatmospheric object and where within it the sample resided is highly desirable (Herzog, 2004; Eugster et al., 2006).

Most cosmic-ray exposure ages are calculated from helium, neon, and argon isotopic abundances. Although all isotopes of these noble gases are produced by cosmic rays, the low natural abundance of ^{38}Ar and the extremely low natural abundances of ^{3}He and ^{21}Ne make the cosmogenic component easy to see. In addition, because all three isotopes are stable, the age is easy to determine by dividing the amount of excess ^{3}He, ^{21}Ne, or ^{38}Ar by the rate at which they are produced by cosmic rays. In some cases, the ^{3}He abundance is unusually low relative to ^{21}Ne or ^{38}Ar. This may mean that the sample was heated for an extended period, as would be the case if the orbit passed near the Sun. For meteorites with cosmic-ray exposure ages of <4 Myr, radionuclides such as ^{10}Be, ^{26}Al, and ^{36}Cl are also used, particularly if the meteorites have high abundances of trapped noble gases, as in the cases of CM and CI chondrites. When radionuclides are used, one must take into account the terrestrial age of the meteorite, during which time the nuclides decay according to their half-life, to determine the actual time of exposure to cosmic rays.

Figures 10.19–10.21 summarize the cosmic-ray exposure ages for various classes of meteorites. The CI and CM chondrites have the shortest exposure ages, ranging from ~10^5 to a few × 10^6 years (Fig. 10.19). Most other chondrites have exposure ages ranging from ~1 Myr to nearly 100 Myr (Fig. 10.20). The distributions of exposure ages among chondrites are inconsistent with continuous delivery of meteorites to Earth. The distributions more likely reflect superposition of debris from discrete collisional events in the asteroid belt. Note the strong peak in exposure ages for the H chondrites at 6–7 Ma. This most likely represents a major collision on the H chondrite parent body at ~7 Ma. An additional large collision at 33 Ma is also indicated. The L chondrites show a different distribution, with collisions on their parent body occurring at ~28 Ma and ~40 Ma (these ages reflect impact sampling of their immediate parent bodies, rather than the collision that disrupted the original L chondrite asteroid, as noted in Box 10.2). Like the majority of chondrites, the achondrites have exposure ages of a few Myr to nearly 100 Myr. The HED meteorites (Fig. 10.21) exhibit a strong peak at ~22 Ma and a smaller one at ~36 Ma. The eucrites show an additional peak at ~12 Ma.

The iron meteorites generally exhibit much longer cosmic-ray exposure ages than do stony meteorites, with most having exposure ages of several hundred Myr (Fig. 10.21). The progression of cosmic-ray exposure ages from ~0.1 to <10 Ma for CI and CM chondrites, to 1–100 Ma for other types of chondrites, to several hundred Ma for the irons parallels the sequence of their

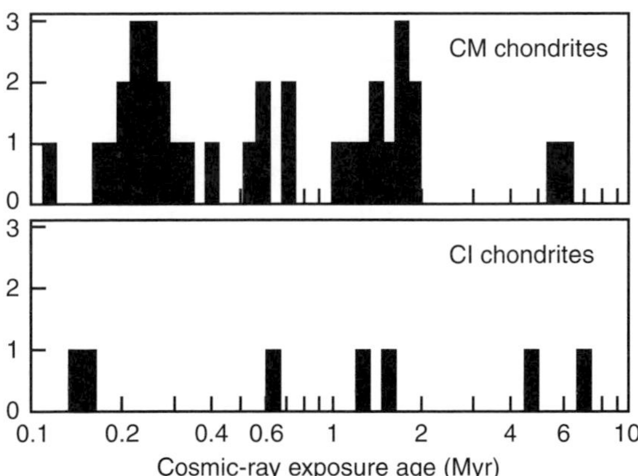

Figure 10.19 Cosmic-ray exposure ages of CI and CM chondrites. Two CM chondrites with exposure ages of <1 million years are not shown. Modified from Eugster et al. (2006).

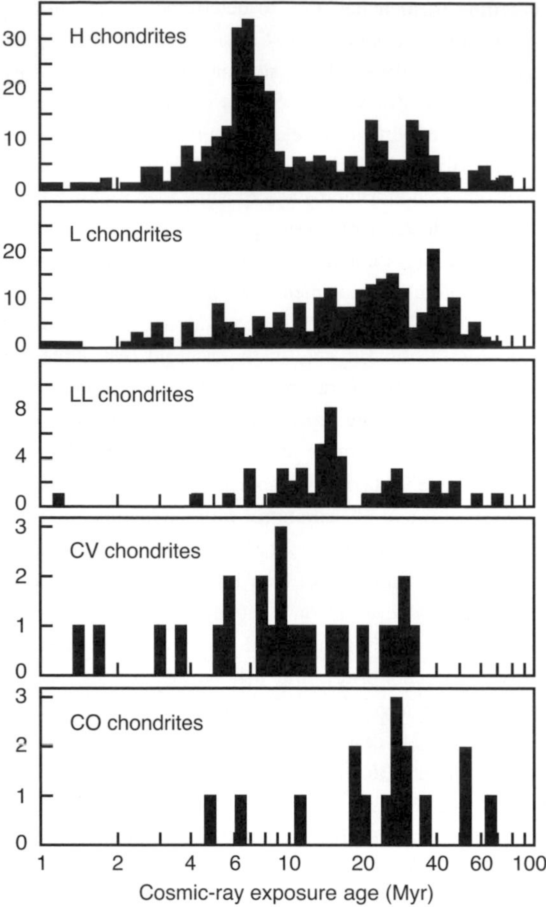

Figure 10.20 Cosmic-ray exposure ages of ordinary, CV, and CO chondrites. Modified from Eugster et al. (2006).

physical strength. This indicates that collisions and physical abrasion in space are major factors, along with orbital dynamics, in controlling if and when a meteorite reaches the Earth.

More than 250 meteorites from Mars, comprising ~146 separate falls, have been recovered. Several distinct groups of martian meteorite have been identified by their petrologic and geochemical relationships and crystallization ages. The Mars-ejection age is the sum of the cosmic-ray exposure age and the terrestrial age (the latter are usually inconsequential). The Mars-ejection ages for martian meteorites tend to be clustered and to correlate with petrologic classification and crystallization ages (Fig. 10.16). From such data, it is possible to identify different launch sites, although the specific location on Mars where any one of these groups of meteorites originated has not yet been determined.

We also have over 150 distinct lunar meteorites in our collections, consisting of ~415 specimens. Because the

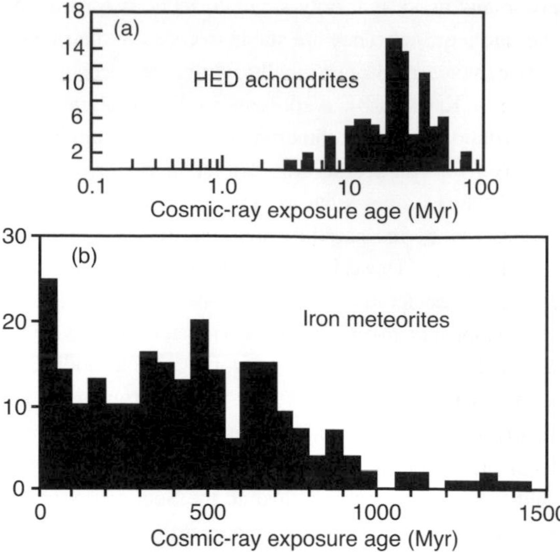

Figure 10.21 Cosmic-ray exposure ages of differentiated meteorites: (a) HED achondrites and (b) iron meteorites. Modified from Eugster et al. (2006).

Moon has no atmosphere, the irradiation history of these meteorites can include an extended period in the lunar regolith. The transit times from the Moon to the Earth range from a few × 10^4 years to nearly 10 Myr.

10.5.2 Terrestrial Ages

Meteorites are not in chemical equilibrium with the Earth's surface and thus are quickly degraded and destroyed after they fall. The most striking example of this lack of equilibrium is the presence of FeNi metal in many meteorites, which quickly rusts on Earth. The rate at which meteorites weather and disintegrate can be measured. The time since a meteorite fell is known as its *terrestrial age*.

When a meteorite falls onto the Earth's surface, it becomes for the most part shielded from the effects of cosmic rays. The cosmic-ray-produced nuclides cease to form so the exposure clock stops. However, because many of the cosmic-ray-produced nuclides are radioactive, several new clocks start that can be used to estimate the length of time that a meteorite has been on the Earth's surface.

The cosmogenic clocks work as follows (e.g. Jull, 2006): in space, while the meteorite is exposed to cosmic rays, the abundances of both radioactive and stable cosmogenic isotopes increase. However, the radioactive ones also decay, and eventually, after about 10 half-lives, the abundance of a radioactive species reaches a steady state between production and decay. For a typical stony meteorite, exposure ages are long enough that the abundances of cosmogenic ^{14}C, ^{26}Al, ^{36}Cl, and ^{10}Be have reached a steady state. The abundance level for each nuclide in a meteorite can be inferred from the bulk composition of the meteorite and its exposure history, as measured by stable cosmogenic nuclides. When the meteorite falls to the Earth, production of the radioactive isotopes ceases and their abundances begin to drop. By comparing the abundances of one or more of these nuclides in the meteorite now with the expected steady-state abundances, the terrestrial age can be calculated. The ^{14}C clock is useful for terrestrial ages younger than ~30,000 years, while ^{26}Al, ^{10}Be, and ^{36}Cl can provide terrestrial ages for objects that have been on Earth in excess of a million years. The development of the accelerator mass spectrometer, which permits quantitative measurements of isotopes with extremely low abundances (detection limit of a few parts in 10^{15}), has greatly facilitated the work on terrestrial ages of meteorites.

A large number of meteorites have been collected from hot desert areas. In these areas, rainfall is low, which helps to preserve the meteorites. In addition, the dry conditions make meteorites easier to find because vegetation is scarce and the wind blows away the fine dust, leaving rocks and meteorites as a lag deposit. Numerous meteorites have been collected in Western Australia and terrestrial ages have been determined on ~50 of them. Ages range from very young to around 40,000 years. There is a rough exponential decline in the number of meteorites as a function of age. The distribution of ages gives a mean residence time of ~10,000 years in this region. Thousands of meteorites have also been recovered from the deserts of North Africa. Terrestrial ages of these meteorites extend out to ~50,000 years, and again the number of meteorites shows a rough exponential decrease with age. The mean residence time inferred for these meteorites is ~12,000 years. A third desert site is Eastern New Mexico, where nearly 200 meteorites have been found. Terrestrial ages for 32 of these meteorites fall in the range of 10,000–50,000 years. They do not show an exponential decrease in abundance with age. Most of these meteorites were found in blowouts – areas where the topsoil had blown away, exposing an older surface. The distribution of ages suggests that the meteorites were somehow protected from weathering in this area. A study of 19 larger meteorites from the Great Plains of the United States indicated a shorter mean residence time of ~5200 years. Meteorites that fall in wetter climates typically do not last as long as in drier climates.

Antarctica has turned out to be a wonderful place to collect meteorites. The meteorites are found in areas of blue ice. These areas form when ice is ablated by the wind. The blue ice formed at depths below ~100 meters where the pressure of overlying snow forced the air bubbles out of it. Thus, if blue ice is on the surface, then at least 100 meters of overlying snow and ice have been removed, and all of the meteorites in that ice that fell over the thousands of years when the snow fell and the ice formed are now concentrated on the surface. Horizontal compression of ice against a barrier, such as a mountain range, can further increase the number of meteorites on the ice. Terrestrial ages of Antarctic meteorites are older than in other parts of the world, with ages ranging from recent to >480,000 years. Different ice fields have meteorites of different ages. Antarctic ice is apparently a very good storage box for meteorites. However, when the meteorites are exposed on the surface of the ice, they begin to weather. There is currently no good way to measure the survival age for a meteorite on the surface of Antarctic ice, but it is probably not longer than a few thousand years.

10.6 Up Next: Flight Instruments for In Situ Dating

A long-standing desire of cosmochemists has been directly to date surfaces of planetary bodies. Radiometric dating would be the gold standard. Several approaches are being undertaken to develop *in situ* techniques for the dating of rocks by spacecraft (Cohen et al., 2019). The Chemistry and Dating Experiment (Anderson et al., 2015) uses laser-ablation resonance-ionization mass spectrometry to obtain ^{87}Rb–^{87}Sr isochrons on geologic samples. Laser ablation coupled to a time-of-flight mass spectrometer can be used to measure chemistry. Resonance ionization lasers are shown through the ablation plume to get rubidium and strontium isotopic data. The prototype system was used to date the Zagami martian meteorite, giving an age of 360 ± 90 Myr (Anderson et al., 2015), consistent with ages determined in dedicated terrestrial labs, though much less precise. Considerable work is still required to convert the prototype into a flight instrument.

Cohen et al. (2014) have developed a system called KArLE (Potassium–Argon Laser Experiment) for *in situ* ^{40}K–^{40}Ar dating. The potassium content of the sample is determined by laser-induced breakdown spectroscopy (LIBS). The argon released by the LIBS ablation laser is admitted into a quadrupole mass spectrometer and measured. The potassium content and argon abundance can be related by sample mass, which is determined by optical measurements of the ablated volume (Cohen et al., 2014). Although not specifically designed for the purpose, instruments on the Curiosity rover have been employed to obtain ^{40}K–^{40}Ar dates and cosmic-ray exposure ages on Mars surface rocks. In this case, the potassium content was determined by α-particle X-ray spectrometry, and ^{40}Ar and cosmogenic ^{3}He, ^{21}Ne, and ^{36}Ar were measured in a quadrupole mass spectrometer. A mudstone in an exposed bedrock outcrop gave the best result, yielding a date of 4.21 ± 0.35 Ga (Farley et al., 2014). All of the ^{40}Ar was assumed to result from *in situ* decay of ^{40}K. The mineralogy of the rock suggests that the potassium was cited in minerals derived from volcanic rocks. If these components came from a single volcanic formation, the date may be the age of that volcanic formation (Cohen et al., 2019). The ^{3}He, ^{21}Ne, and ^{36}Ar abundances gave concordant cosmic-ray exposure ages and indicate that the mudstone has been exposed at the surface for 78 ± 30 Myr (Farley et al., 2014).

Summary

We have seen that the age of the elements that now comprise the solar system can be estimated from models of stellar nucleosynthesis in the galaxy. These models are constrained by the abundances of short-lived radionuclides in meteorites. The age of the solar system is estimated to be ~4.567 Ga, determined from CAIs, the oldest materials dated so far. Chondrules, the other important component of chondrites, appear to be several million years younger. The ages of chondrules and the timing of planetesimal accretion and differentiation, relative to CAI formation, can be determined using short-lived radionuclides like ^{26}Al and ^{182}Hf. The absolute ages of these events require that they be anchored to U–Pb ages. Accretion of differentiated asteroids occurred within 2 Myr of CAI formation, as appropriate for their heating by ^{26}Al decay. Accretion of chondritic asteroids occurred slightly later, so that they only experienced thermal metamorphism or aqueous alteration.

The ages for planet formation are more difficult to assess and must be constrained from the times of differentiation. The age of the Earth, estimated from U–Pb data, is 59–79 Myr after CAIs. ^{182}Hf–^{182}W data, which date formation of the core, suggest an average accretion time of 12–15 Myr after CAIs, although core formation could have been more protracted. The age of the Moon, based on ^{182}Hf–^{182}W, ^{129}I–^{129}Xe, Pu–Xe, and a zircon model age from U–Pb and ^{176}Lu–^{176}Hf, is ~50–60 Myr after formation of the solar system. Mare basalts were erupted between 4.0 and 3.1 Ga, although the dated sample collection does not include the full range of basalts, which may have extended until ~1.2 Ga. Based on ^{182}Hf–^{182}W and ^{176}Lu–^{176}Hf isotopes in martian meteorites, core formation on Mars occurred within 20–25 Myr of CAIs. U–Pb ages of the oldest zircons suggest a stable crust formed within ~100 Myr.

Impact ages of meteorites and lunar samples can be determined from radioisotope chronometers that are reset by shock. These involve radiogenic nuclides that are gases (helium, argon) and are thus easily lost. Impact ages for lunar impact-melt breccias and for HED meteorites reveal a late heavy bombardment at 3.9–3.75 Ga.

Cosmic-ray exposure ages are determined from spallation-produced radioactive nuclides. Cosmic irradiation normally occurs while a meteoroid is in space, but surface rocks unshielded by an atmosphere may also have cosmogenic nuclides. These measurements provide information on orbital lifetimes of meteoroids and constrain orbital calculations. Terrestrial ages of meteorites can be estimated by comparing the decay of cosmogenic nuclides having different half-lives, once they fall to Earth and become shielded from cosmic rays. These ages provide information on meteorite survival relative to weathering.

Flight instruments capable of *in situ* radiometric dating of samples on the surfaces of extraterrestrial bodies are under development, although technical challenges remain.

Now that we have an understanding of radioactive isotopes and their utility in constraining the timing and rates of nebular, planetesimal, and planetary processes, we will turn our attention to the most volatile elements and compounds. Stable isotopes in these provide additional constraints on cosmochemical processes.

Questions

1. How accurately do we know the age of the solar system? What is the largest uncertainty in the age? Why can we determine the sequence and timing of events in the early solar system with higher precision than we know the absolute age?
2. Summarize the evidence that chondrules are younger than CAIs. Hint: several lines of evidence do not involve radionuclides.
3. Give two lines of reasoning to support the idea that differentiated asteroids accreted before chondritic asteroids.
4. Describe the cosmic-ray exposure ages of meteorites and summarize what they tell us about the history of meteorites.
5. What is the evidence that martian meteorites came to Earth as a result of a limited number of discrete impact events on the martian surface?

Suggestions for Further Reading

Huss G. R., Meyer B. S., Srinivasan G., et al. (2009) Stellar sources of the short-lived radionuclides in the early solar system. *Geochimica et Cosmochimica Acta*, **73**, 4922–4945. This paper gives an overview of galactic chemical evolution as it relates to short-lived nuclides and discusses the nucleosynthesis of these nuclides. It provides a good entry point into the literature.

Nyquist L. E., Kleine T., Shih C.-Y., and Reese Y. D. (2009) The distribution of short-lived radioisotopes in the early solar system and the chronology of asteroid accretion, differentiation, and secondary mineralization. *Geochimica et Cosmochimica Acta*, **73**, 5115–5136. Presents an interesting discussion of using multiple radionuclides to extract chronological information about the solar system.

Kleine T., and Walker R. J. (2017) Tungsten isotopes in planets. *Annual Reviews of Earth and Planetary Science*, **45**, 389–417. A nice review of the ^{182}Hf–^{182}W data on meteorites and planets.

Bogard D. D. (1995) Impact ages of meteorites: A synthesis. *Meteoritics*, **30**, 244–268. A nice summary of the impact ages of meteorites.

Eugster O., Herzog G. F., Marti K., and Caffee M. W. (2006) Irradiation records, cosmic-ray exposure ages, and transfer time of meteorites. In *Meteorites and the Early Solar System II*, Lauretta D. S., and McSween H. Y., editors, pp. 829–851, University of Arizona Press, Tucson. A good summary of what is known about cosmic-ray exposure ages and the transfer of meteorites from the asteroid belt to Earth.

Jull A. J. T. (2006) Terrestrial ages of meteorites. In *Meteorites and the Early Solar System II*, Lauretta D. S., and McSween, H. Y., editors, pp. 889–905, University of Arizona Press, Tucson. This chapter reviews what is known about the terrestrial ages of meteorites.

Other References

Allègre C. J., Manhès G., and Göpel C. (1995) The age of the Earth. *Geochimica et Cosmochimica Acta*, **59**, 1445–1456.

Anderson F. S., Levine J., and Whitacker T. J. (2015) Dating the martian meteorite Zagami by the ^{87}Rb-^{87}Sr isochron method with a prototype in situ resonance ionization mass spectrometer. *Rapid Communications in Mass Spectrometry*, **29**, 191–204.

Arnould M., Goriely S., and Meynet G. (2006) The production of short-lived radionuclides by new non-rotating and rotating Wolf-Rayet model stars. *Astronomy & Astrophysics*, **453**, 653–659.

Avice G., and Marty B. (2014) The iodine-plutonium-xenon age of the Moon-Earth system revisited. *Philosophical Transactions of the Royal Society A*, **372**, 20130260.

Barboni M., Boehnke P., Keller B., et al. (2017) Early formation of the Moon 4.51 billion years ago. *Science Advances*, **3**, e1602365.

Bellucci J. J., Nemchin A. A., Whitehouse M. J., et al. (2018) Pb evolution in the martian mantle. *Earth & Planetary Science Letters*, **485**, 79–87.

Bizzarro M., Baker J. A., Haack H., and Lundgaard K. L. (2005) Rapid timescales for accretion and melting of differentiated planetesimals inferred from ^{26}Al-^{26}Mg chronometry. *Astrophysical Journal*, **632**, L41–L44.

Bollard J., Connelly J. N., and Bizzarro M. (2015) Pb-Pb dating of individual chondrules from the CBa chondrite Gujba: Assessment of the impact plume formation model. *Meteoritics & Planetary Science*, **50**, 1197–1216.

Bollard J., Connelly J. N., Whitehouse M. J., et al. (2017) Early formation of planetary building blocks inferred from Pb isotopic ages of chondrules. *Science Advances*, **3**, e1700407.

Borg L. E., Gaffney A. M., and Shearer C. K. (2014) A review of lunar chronology revealing a preponderance of 4.34-4.37 Ga ages. *Meteoritics & Planetary Science*, **50**, 715–732.

Bottke W. F., Nesvorny D., Vokrouhlicky D., and Borbidelli A. (2009) The Gefion family as the probably source of the L chondrite meteorites. *Lunar and Planetary Science*, **60**, abstract #1445, CDROM.

Bouvier A., Blichert-Toft J., Moynier F., et al. (2007) Pb-Pb dating constraints on the accretion and cooling history of chondrites. *Geochimica et Cosmochimica Acta*, **71**, 1583–1604.

Bouvier A., Blichert-Toft J., Vervoort J. D., et al. (2008) The case for old basaltic shergottites. *Earth & Planetary Science Letters*, **266**, 105–124.

Bouvier L. C., Costa M. M., Connelly J. N., et al. (2018) Evidence for extremely rapid magma ocean crystallization and crust formation on Mars. *Nature*, **558**, 586–589.

Bowring S. A., Schoene B., Crowley J. L., et al. (2006) High-precision U-Pb zircon geochronology and the stratigraphic record: Progress and promise. *Paleontological Society Papers*, **12**, 25–45.

Brennecka G. A., Weyer S., Wadhwa M., et al. (2010) ^{238}U/^{235}U variations in meteorites: Extant ^{247}Cm and implications for Pb-Pb dating. *Science*, **237**, 449–451.

Budde G., Kleine T., Kruijer T. S., et al. (2016) Tungsten isotopic constraints on the age and origin of chondrules. *Proceedings of the National Academy of Sciences, USA*, **113**, 2686–2691.

Budde G., Kruijer T. S., and Kleine T. (2018) Hf-W chronology of CR chondrites: Implications for the timescales of chondrule formation and the distribution of ^{26}Al in the solar nebula. *Geochimica et Cosmochimica Acta*, **222**, 284–304.

Cameron A. G. W., and Truran J. W. (1977) The supernova trigger for formation of the solar system. *Icarus*, **30**, 447–461.

Cameron A. G. W., Hoeflich P., Myers P. C., and Clayton D. D. (1995) Massive supernovae, Orion gamma rays, and the formation of the solar system. *Astrophysical Journal*, **447**, L53–L57.

Clayton D. D. (1988) Nuclear cosmochronology with analytic models of the chemical evolution of the solar neighborhood. *Monthly Notices of the Royal Astronomical Society*, **234**, 1–36.

Cohen B. A., Miller J. S., Li Z.-H., et al. (2014) The Potassium-Argon Laser Experiment (KArLE): In situ geochronology for planetary robotic missions. *Geostandards and Geoanalytical Research*, **38**, 421–439.

Cohen B. A., Malespin C. A., Farley K. A., et al. (2019) In situ geochronology on Mars and the development of future instrumentation. *Astrobiology*, **19**, 1303–1314.

Connelly J. N., and Bizzarro M. (2016) Lead isotope evidence for a young formation age of the Earth-Moon system. *Earth & Planetary Science Letters*, **452**, 36–43.

Connelly J. N., Bizzarro M., Krot A. N., et al. (2012) The absolute chronology and thermal processing of solids in the solar protoplanetary disk. *Science*, **338**, 651–655.

Connelly J. N., Bollar J., and Bizzarro M. (2017) Pb-Pb chronometry and the early solar system. *Geochimica et Cosmochimica Acta*, **201**, 345–363.

Davis A. M., and McKeegan K. D. (2014) Short-lived radionuclides and early solar system chronology. In *Treatise on Geochemistry, 2nd Edition, Vol. 1: Meteorites and Cosmochemical Processes*, Davis, A. M., editor, pp. 361–395, Elsevier, Oxford.

Day J. M. D., Brandon A. D., and Walker R. J. (2016) Highly siderophile elements in Earth, Mars, the Moon, and asteroids. *Reviews in Mineralogy & Geochemistry*, **81**, 161–238.

Diehl R., Halloin H., Kretschmer K., et al. (2006) ^{26}Al in the inner galaxy. *Astronomy & Astrophysics*, **449**, 1025–1031.

Doyle P. M., Jogo K., Nagashima K., et al. (2015) Early aqueous activity on the ordinary and carbonaceous chondrite parent bodies recorded by fayalite. *Nature Communications*, **6**, 7444.

Eugster O., Herzog G. F., Marti K, and Caffee M. W. (2006) Irradiation records, cosmic-ray exposure ages, and transfer time of meteorites. In *Meteorites and the Early Solar System II*, Lauretta D. S., and McSween, H. Y., editors, pp. 829–851, University of Arizona Press, Tucson.

Farley K. A., Malespin C., Mahaffy P., et al. (2014) In situ radiometric and exposure age dating of the martian surface. *Science*, **343**, doi:10.1126/science.1247166.

Filiberto J. (2017) Geochemistry of martian basalts with constraints on magma genesis. *Chemical Geology*, **466**, 1–14.

Fujiya W., Sugiura N., Hotta H., et al. (2012) Evidence for the late formation of hydrous asteroids from young meteoritic carbonates. *Nature Communications*, **3**, 627.

Gaffney A. M., and Borg L. E. (2014) A young solidification age for the lunar magma ocean. *Geochimica et Cosmochimica Acta*, **140**, 227–240.

Göpel C., Manhès G., and Allègre C.-J. (1994) U-Pb systematics of phosphates from unequilibrated ordinary chondrites. *Earth & Planetary Science Letters*, **121**, 153–171,

Halliday A. N. (2014) The origin and earliest history of the Earth. In *Treatise on Geochemistry, 2nd Edition, Vol. 2: Planets, Asteroids, Comets and the Solar System*, Davis A. M., editor, pp. 149–211, Elsevier, Oxford.

Heisinger H., Head J. W., Wolf U., et al. (2011) Ages and stratigraphy of lunar mare basalts: A synthesis. *Geological Society of America Special Paper*, **447**, doi:10.1130/SPE447.

Hellmann J. L., Kruijer T. S., Van Orman J. A., et al. (2019) Hf-W chronology of ordinary chondrites. *Geochimica et Cosmochimica Acta*, **258**, 290–309.

Herzog G. F. (2004) Cosmic-ray exposure ages of meteorites. In *Treatise on Geochemistry, Vol. 1: Meteorites, Comets, and Planets*, Davis A. M., editor, pp. 347–380, Elsevier, Oxford.

Herzog G. F., and Caffee M. W. (2014) Cosmic-ray exposure ages of meteorites. In *Treatise in Geochemistry, 2nd Edition, Vol. 1: Meteorites and Cosmochemical Processes*, Davis A. M., editor, pp. 419–454, Elsevier, Oxford.

Hohenberg C. M., and Pravdivtseva O. V. (2008) I-Xe dating: From adolescence to maturity. *Chemie der Erde*, **68**, 339–351.

Hsu W., Wasserburg G. J., and Huss G. R. (2000) High time resolution using ^{26}Al in the multistage formation of a CAI. *Earth & Planetary Science Letters*, **182**, 15–29.

Humayun M., Nemchin A., Zanda B., et al. (2013) Origin and age of the earliest martian crust from meteorite NWA 7533. *Nature*, **503**, 513–516.

Jacobsen B., Yin Q.-Z, Moynier F., et al. (2008) ^{26}Al-^{26}Mg and ^{207}Pb-^{206}Pb systematics of Allende CAIs: Canonical solar initial ^{26}Al/^{27}Al ratio reinstated. *Earth & Planetary Science Letters*, **272**, 353–364.

Jilly-Rehak C. E., Huss G. R., and Nagashima K. (2017) ^{53}Mn-^{53}Cr radiometric dating of secondary carbonates in CR chondrites: Timescales for parent body aqueous alteration. *Geochimica et Cosmochimica Acta*, **201**, 224–244.

Kleine T., Touboul M., Van Orman J. A., et al. (2008) Hf-W thermochronometry: Closure temperature and constraints on the accretion and cooling history of the H chondrite parent body. *Earth & Planetary Science Letters*, **270**, 106–118.

Kleine T., Hans U., Irving A. J., and Bourdon B. (2012) Chronology of the angrite parent body and implications for core formation in protoplanets. *Geochimica et Cosmochimica Acta*, **84**, 186–203.

Kleine T., Budde G., Hellmann J. L., et al. (2018) Tungsten isotopes and the origin of chondrules and chondrites. In *Chondrules, Records of Protoplanetary Disk Processes*, Russell S. S., Connolly H. C., Jr, and Krot A. N., editors, pp. 276–299, Cambridge University Press, Cambridge.

Kööp L., Davis A. M., Nakashima D., et al. (2016) A link between oxygen, calcium and titanium isotopes in ^{26}Al-poor hibonite-rich CAIs from Murchison and implications for the heterogeneity of dust reservoirs in the solar nebula. *Geochimica et Cosmochimica Acta*, **189**, 70–95.

Krot A. N., Amelin Y., Cassen P., and Meibom A. (2005) Young chondrules in CB chondrites from a giant impact in the early solar system. *Nature*, **436**, 989–992.

Krot A. M., McKeegan K. D., Huss G. R., et al. (2006) Aluminum-magnesium and oxygen isotope study of relict Ca-Al-rich inclusions in chondrules. *Astrophysical Journal*, **639**, 1227–1237.

Kruijer T. S., Kleine T., Fischer-Godde M., et al. (2014a) Nucleosynthetic W isotope anomalies and the Hf-W chronometry of Ca-Al-rich inclusions. *Earth & Planetary Science Letters*, **403**, 317–327.

Kruijer T. S., Touboul M., Fischer-Godde M., et al. (2014b). Protracted core formation and rapid accretion of protoplanets. *Science*, **344**, 1150–1154.

Kruijer T. S., Kleine T., Borg L. E., et al. (2017) The early differentiation of Mars inferred from Hf-W chronometry. *Earth & Planetary Science Letters*, **474**, 345–354.

Lada C. J., and Lada E. A. (2003) Embedded clusters in molecular clouds. *Annual Reviews of Astronomy & Astrophysics*, **41**, 57–115.

Lapen T. J., Righter M., Brandon A. D., et al. (2010) A younger age for ALH 84001 and its geochemical link to shergottite sources in Mars. *Science*, **328**, 347–351.

Larsen K. K., Trinquier A., Paton C., et al. (2011) Evidence for magnesium isotope heterogeneity in the solar protoplanetary disk. *Astrophysical Journal Letters*, **735**, L37.

Larsen K. K., Schiller M., and Bizzarro M. (2016) Accretion timescales and style of asteroidal differentiation in an ^{26}Al-poor protoplanetary disk. *Geochimica et Cosmochimica Acta*, **176**, 295–315.

Larsen K. K., Wielandt D., Schiller M., et al. (2020) Episodic formation of refractory inclusions in the Solar System and their presolar heritage. *Earth & Planetary Science Letters*, **535**, 116088.

Lee D. C., and Halliday A. N. (1997) Core formation on Mars and differentiated asteroids. *Nature*, **388**, 854–857.

Lee T., Papanastassiou D. A., and Wasserburg G. J. (1977) Aluminum-26 in the early solar system: Fossil or fuel? *Astrophysical Journal Letters*, **211**, L107–L110.

Lugmair G. W., and Shukolyukov A. (1998) Early solar system timescales according to the ^{53}Mn-^{53}Cr system. *Geochimica et Cosmochimica Acta*, **62**, 2863–2886.

MacPherson G. J., Nagashima K., Krot A. N., et al. (2017) ^{52}Mn-^{53}Cr chronology of Ca-Fe silicates in CV3 chondrites. *Geochimica et Cosmochimica Acta*, **201**, 260–274.

Makide K., Nagashima K., Krot A. N., et al. (2011) Heterogeneous distribution of ^{26}Al at the birth of the solar system. *Astrophysical Journal Letters,* **733**, L31.

Maltese A., and Mezger K. (2020) The Pb isotope evolution of bulk silicate Earth: Constraints from its accretion and early differentiation history. *Geochimica et Cosmochimica Acta*, **271**, 179–193.

Marty B., and Marti K. (2002) Signatures of early differentiation of Mars. *Earth & Planetary Science Letters*, **196**, 251–263.

McKee C. F., and Ostriker J. P. (1977) A theory of the interstellar medium – three components regulated by supernova explosions in an inhomogeneous substrate. *Astrophysical Journal*, **218**, 148–169.

Meyer B. S. (2005) Synthesis of short-lived radioactivities in a massive star. In *Chondrites and the Protoplanetary Disk*, Krot A. N., Scott E. R. D., and Reipurth B., editors, pp. 515–526, Astronomical Society of the Pacific, San Francisco.

Nagashima K, Kita N. T., and Luu T.-H. (2018) ^{26}Al–^{27}Al systematics of chondrules. In *Chondrules, Records of Protoplanetary Disk Processes*, Russell S. S., Connolly H. C., and Krot A. N., editors, pp. 247–275, Cambridge University Press, Cambridge.

Nagashima K., Krot A. N., and Huss G. R. (2014) ^{26}Al in chondrules from CR2 chondrites. *Geochemical Journal*, 48, 561–570.

Neukum G., Ivanov B. A., and Hartmann W. K. (2001) Cratering records in the inner solar system in relation to the lunar reference system. *Space Science Reviews*, **96**, 55–86.

Nyquist L. E., Shih C.-Y., McCubbin F. M., et al. (2016) Rb-Sr and Sm-Nd isotopic and REE studies of igneous components in the bulk matrix domain of martian breccia Northwest Africa 7034. *Meteoritics & Planetary Science*, **51**, 483–498.

Park C., Nagashima K., Krot A. N., et al. (2017) Calcium-aluminum-rich inclusion with fractionation and unidentified nuclear effects (FUN CAIs): II. Heterogeneities of magnesium isotopes and ^{26}Al in the early solar system inferred from *in situ* high-precision magnesium-isotope measurements. *Geochimica et Cosmochimica Acta*, **201**, 6–24.

Reynolds J. H. (1960) Determination of the age of the elements. *Physical Reviews Letters*, **4**, 8–10.

Schiller M., Baker J., Creech J., et al. (2011) Rapid timescales for magma ocean crystallization on the howardite-eucrite-diogenite parent body. *Astrophysical Journal Letters*, **740**, L22.

Schmitz, B. (2013). Extraterrestrial spinels and the astronomical perspective on Earth's geological record and evolution of life. *Chemie der Erde*, *73*, 117–145.

Schmitz B., Tassinari M., and Peucker-Ehrenbrink B. (2001) A rain of ordinary chondritic meteorites in the early Ordovician. *Earth & Planetary Science Letters*, **194**, 1–15.

Schmitz B., Yin Q.-Z., Sanborn M. E., et al. (2016) A new type of solar-system material from Ordovician marine limestone. *Nature Communications*, **7**, 11851.

Shukolyukov A., and Lugmair G. W. (2007) The Mn-Cr isotope systematics of bulk angrites. *Lunar and Planetary Science*, **38**, abstract #1423.

Smoliar M. I. (1993) A survey of Rb-Sr systematics of eucrites. *Meteoritics*, **28**, 105–113.

Snape J. F., Nemchin A. A., Bellucci J. J., et al. (2016) Lunar basalt chronology, mantle differentiation and implications for determining the age of the Moon. *Earth & Planetary Science Letters*, **451**, 149–158.

Spivak-Birndorf L., Wadhwa M., and Janney P. (2009) ^{26}Al/^{26}Mg systematics in D'Orbigny and Sahara 99555 angrites: Implications for high-resolution chronology using extinct chronometers. *Geochimica et Cosmochimica Acta*, **73**, 5202–5211.

Takigawa A., Miki J., Tachibana S., et al. (2008) Injection of short-lived radionuclides into the early solar system from a faint supernova with mixing fallback. *Astrophysical Journal*, **688**, 1382–1387.

Tera F. (1981) Aspects of isochronism in Pb isotope systematics – Application to planetary evolution. *Geochimica et Cosmochimica Acta*, **45**, 1439–1448.

Telus M., Huss G. R., Nagashima K., and Ogliore R. C. (2014) Revisiting ^{26}Al-^{26}Mg systematics of plagioclase in H4 chondrites. *Meteoritics & Planetary Science*, **49**, 929–945.

Thiemens M. M., Sprung P., Ronseca R. O. C., et al. (2019) Early Moon formation inferred from hafnium-tungsten systematics. *Nature Geoscience*, **12**, 696–700.

Touboul M., Sprung P., Aciego S. M., et al. (2015) Hf-W chronology of the eucrite parent body. *Geochimica et Cosmochimica Acta*, **156**, 106–121.

Trieloff M., Jessberger E. K., Herrwerth I., et al. (2003) Structure and thermal history of the H-chondrite parent asteroid revealed by thermochronometry. *Nature*, **422**, 502–506.

Udry, A., Howarth, G. H., Herd, C. D. K., Day, J. M. D., Lapen, T. J. D., & Filiberto, J. (2020) What martian meteorites reveal about the interior and surface of Mars. *Journal of Geophysical Research: Planets*, **125**, e2020JE006523

Wadhwa M., Srinivasan G., and Carlson R. W. (2006) Timescales of planetesimal differentiation in the early solar system. In *Meteorites and the Early Solar System II*, Lauretta D. S., and McSween H. Y., editors, pp. 715–731, University of Arizona Press, Tucson.

Wang W., Harris M. J., Diehl R., et al. (2007) SPI observations of the diffuse ^{60}Fe emission in the galaxy. *Astronomy & Astrophysics*, **469**, 1005–1012.

Warren P. H. (2004) The Moon. In *Treatise on Geochemistry, Vol. 1: Meteorites, Comets, and Planets*, Davis, A. M., editor, pp. 559–599, Elsevier, Oxford.

Wasserburg G. J., Busso M., Gallino R., and Nollett K. M. (2006) Short-lived nuclei in the early solar system: Possible AGB source. *Nuclear Physics A*, **777**, 5–69.

Wieler R., Huber L., Busemann H., et al. (2016) Noble gases in 18 martian meteorites and angrite Northwest Africa 7812 – Exposure ages, trapped gases, and a re-evaluation of evidence for solar cosmic ray-produced neon in shergottites and other achondrites. *Meteoritics & Planetary Science*, **51**, 407–428.

Wilde S. A., Valley J. A., Peck W. H., and Graham C. M. (2001) Evidence from detrital zircons for the existence of continental crust and oceans on Earth 4.4 Gyr ago. *Nature*, **409**, 175–178.

Wimpenny J., Sanborn M. E., Koefoed P., et al. (2019) Reassessing the origin and chronology of the unique achondrite Asuka 881394: Implications for distribution of ^{26}Al in the early solar system. *Geochimica et Cosmochimica Acta*, **244**, 478–501.

Yin Q. Z., Jacobsen S. B., Yamashita K., et al. (2002) A short timescale for terrestrial planet formation from Hf-W chronometry of meteorites. *Nature*, **418**, 949–952.

Zhou Q., Yin Q.-Z., Young E. D., et al. (2013) SIMS Pb-Pb and U-Pb age determination of eucrite zircons at <5 μm scale and the first 50 Ma of the thermal history of Vesta. *Geochimica et Cosmochimica Acta*, **110**, 152–175.

11 The Most Volatile Elements and Compounds

Ices, Noble Gases, and Organic Matter

Overview

Highly volatile elements and compounds are depleted in rocky asteroids and the terrestrial planets, but are concentrated in comets, the giant planets, and icy moons. We begin this chapter by considering ices, which are surprisingly abundant phases in the outer solar system. Ices condensed in the frigid outer reaches of the solar nebula and in interstellar space. Besides H_2O, ices include compounds containing carbon (CO, CO_2, CH_4) and nitrogen (N_2, NH_3). Next, we look at the noble gases, which do not condense as solids and thus are strongly depleted in meteorites and planets relative to solar system abundances. Noble gas isotopes allow the identification of nuclear, solar, and planetary components with different origins. The concentrations and varied isotopic compositions of noble gases in meteorites and planetary atmospheres provide unique perspectives on processes occurring in the early solar system and during planetary differentiation. Finally, we consider the sometimes bewilderingly complex molecules composed of carbon and hydrogen, often combined with other elements like oxygen, nitrogen, sulfur, and phosphorus (organic compounds). Following an introduction to terminology and structures, we focus on the organic matter in carbonaceous chondritic meteorites. This material appears to be a mixture of compounds inherited from the interstellar medium (ISM) and synthesized within the nebula and solar system bodies. Finally, we review the occurrence of ices and noble gases in comets, planets, and moons, as well as discoveries of organic matter in extraterrestrial bodies. All these volatile substances are related, in that ices trapped noble gases and provided sites for the formation of simple organic compounds in space.

11.1 Volatility

In Chapter 7, we discussed element *volatility* and its role in cosmochemical fractionations. Here, we focus on some of the most volatile constituents in planets and meteorites: ices, noble gases, and organic compounds. Each of these actually constitutes a voluminous subject of its own in cosmochemistry, and we can only provide overviews of these interesting substances.

This may seem like an eclectic choice of materials to share space in a chapter of this book. However, these components have a number of related characteristics. At the frigid temperatures of the ISM, ices condensed onto tiny dust motes. They also condensed in the outer reaches of the solar nebula and were then incorporated into planetesimals and planets. Noble gases did not condense at all, at least not directly; instead, they were adsorbed onto or trapped within ices. The simplest organic compounds are thought to have formed by low-temperature reactions on ice-coated grains in molecular clouds. Although some large organic compounds are decidedly not volatile, these molecules were synthesized from simpler organic precursors that were highly volatile. Organic polymers also served as hosts for trapped noble gases. Thus, the components described in this chapter can be considered fellow travelers in cosmochemical environments.

Ice mantles are important constituents of interstellar grains in molecular clouds, and large icy bodies dominate the outer reaches of the solar system in the Kuiper belt and the Oort cloud. The region of the solar system where ices were stable increased with time as the solar system formed, as accretion rates of materials onto the disk waned and the disk cooled. The giant planets and their satellites formed, in part, from these ices, and also from the nebular gas itself.

Noble gases, occupying the rightmost column of the Periodic Table, have completely filled electron shells and thus do not easily form chemical compounds. More so than any other elements, they partition into a gas phase. Because the Earth and the rest of the bodies in the inner solar system were made from solids, excluding nebular gases for the most part, noble gases are very scarce in these objects, accounting for the synonym "rare gases." As we shall see, the important roles that helium, neon, argon, krypton, and xenon play in cosmochemistry are actually related to their miniscule abundances in meteorites and planets.

Organic molecules are composed of carbon and hydrogen, usually combined with other atoms such as oxygen, nitrogen, sulfur, and phosphorus. In modern usage, the term "organic" does not necessarily imply formation by or occurrence in organisms, although that was its original connotation. Abiotic organic compounds are important constituents of chondrites and other solar system objects, and they have been telescopically observed in significant quantities in interstellar space. These molecules are of obvious interest as precursors for life, but organic matter also provides a window into processes in the ISM, the nebula disk, and planetesimals.

11.2 Condensation of Ices

Our understanding of ices in the solar system is rudimentary because, unlike solids in the form of meteorites, extraterrestrial ices have been studied only by remote sensing. We also gain information about ices indirectly, from the study of materials altered by reaction with fluids produced by melting ices.

The ISM is predominantly gas, with approximately 1% of its mass in the form of micron-sized dust particles. In cold molecular clouds, some of the gases condense out onto the grains as ices. Water is a major component of the ices. In the frigid environment of molecular clouds, water ice may have an amorphous structure. Other compounds that may condense as ices on grains include CO, CO_2, N_2, CH_4, NH_4, and H_2CO, with CO being volumetrically most important. All these other ices are more volatile than H_2O. The icy mantles on interstellar grains provide ideal sites for the synthesis of organic compounds and the trapping of noble gases in the ISM.

Interstellar grains with ice mantles probably comprised a significant amount of the material that collapsed to form the solar nebula. Heating of this material caused the icy mantles to sublimate, producing a vapor that subsequently condensed as crystalline ices as the nebula cooled. By mass, water ice rivals rock in terms of potentially condensable matter from a gas of cosmic composition (see Box 11.1). The amount of water ice depends, of course, on the extent to which oxygen is otherwise tied up with carbon as CO and/or CO_2. Ice condensation in the solar nebula occurred outboard of the *snowline*, defined as the radial distance along the nebular midplane where water ice was first stable. Calculation of the snowline radius is complicated and its exact position likely varied with time (Lunine, 2005). Also, each different ice composition had its own snowline.

11.3 Accretion of Ices and the Snowline

In the region beyond the snowline, condensed ices ultimately accreted along with rocky materials to form planetesimals. There is some controversy about exactly where the nebular water snowline was situated. The occurrence of some classes of asteroids that formerly contained ice (that melted, causing aqueous alteration like that seen in carbonaceous chondrites) or that presently contain ice (observed in remote-sensing spectra) has prompted the idea that the snowline was located within the Main asteroid belt between Mars and Jupiter. A newer notion is that the snowline was farther out, and that ice-bearing planetesimals formed in the giant planet region and subsequently drifted or were perturbed inward. Comets and bodies now in the Kuiper belt (KBOs) and the distant Oort cloud clearly formed at distances beyond the Main belt, although models posit that they too were perturbed (in this case, outward) to their present orbital distances.

The occurrences of ices of various compositions in asteroids, comets, planets, and icy moons are described at the end of this chapter. The melting of ices and the consequent alteration of chondrites caused by these aqueous fluids will be considered in Chapter 12.

11.4 Noble Gases and How They Are Analyzed

Noble gases – helium, neon, argon, krypton, and xenon – are found in Group 8 of the Periodic Table. They are most abundant in planetary atmospheres, although even there they are only minor components, and they are extremely depleted in solid phases (Cartwright, 2015). We will consider their utility in understanding planetary differentiation and atmospheric evolution shortly, but first we will focus on their rather miniscule abundances in meteorites and other extraterrestrial materials.

Because the chemistry of noble gases is virtually an oxymoron, what is it that makes them so useful in

Box 11.1 Relative Amounts of Ices and Rock in the Solar System

The left-hand pie diagram in Figure 11.1 shows the relative proportions of hydrogen, helium, and metals (everything else) in the solar nebula (abundances in wt. %). To illustrate how abundant ices are versus rock in the solar system (i.e. in the tiny metals wedge in this figure), let's perform a back-of-the-envelope estimate of their proportions. We can safely assume that there is more than enough hydrogen to bond with other elements in ices. In a nebula dominated by hydrogen, the most abundant elements that make ices are oxygen (H_2O), carbon (CH_4), and nitrogen (NH_3). Of course, these elements could potentially combine into other forms, such as CO_2 and CO, but ices using anions other than hydrogen will produce less total ice. We will assume that rock is a mixture of olivine (Mg_2SiO_4) plus Fe metal (probably the most common minerals in the solar system; other cations will certainly combine to make other minerals, but olivine plus Fe metal is a reasonable estimate for rock).

From the solar system elemental abundance table (Table 4.1), we obtain the following abundances (atoms/10^6 Si):

$H = 3.090 \times 10^{10}$, but who's counting, since there is more than enough
$C = 8.318 \times 10^6$
$N = 2.089 \times 10^6$
$O = 15.14 \times 10^6$
$Fe = 0.873 \times 10^6$

Let's first make rock: 10^6 Si combines with 4×10^6 O to make 10^6 molecules of Mg_2SiO_4. This leaves 11.14×10^6 O atoms remaining. To this we add 0.873×10^6 atoms of Fe as metal.

Combining the remaining 11.14×10^6 O atoms with $2(11.14 \times 10^6)$ H produces 11.14×10^6 molecules of H_2O.

Combining 8.318×10^6 C atoms with $4(8.318 \times 10^6)$ H produces 8.318×10^6 molecules of CH_4.

Finally, combining 2.089×10^6 N atoms with $3(2.089 \times 10^6)$ H produces 2.089×10^6 molecules of NH_3.

Converting these atomic proportions (molecules) to weight proportions by multiplying by their respective molecular weights (g/molecule):

Olivine $= (1 \times 10^6)140.7 = 140.7 \times 10^6$ g
Fe metal $= (0.873 \times 10^6)55.85 = 48.8 \times 10^6$ g
Rock = olivine + Fe metal $= 189.5 \times 10^6$ g

H_2O ice $= (11.14 \times 10^6)18.02 = 200.7 \times 10^6$ g
CH_4 ice $= (8.318 \times 10^6)16.05 = 133.5 \times 10^6$ g
NH_3 ice $= (2.089 \times 10^6)17.04 = 35.6 \times 10^6$ g

Summing all of these, the total weight of rock and ice is 559.3×10^6 g. The weight % proportions of the components are thus:

Rock = 34%
H_2O ice = 36%
CH_4 ice = 24%
NH_3 ice = 6%

These approximate proportions are shown in the right-hand pie diagram in Figure 11.1. From this simple calculation, we can see that the mass proportion of ices greatly exceeds that of rock, if the solar nebular gas were fully condensed.

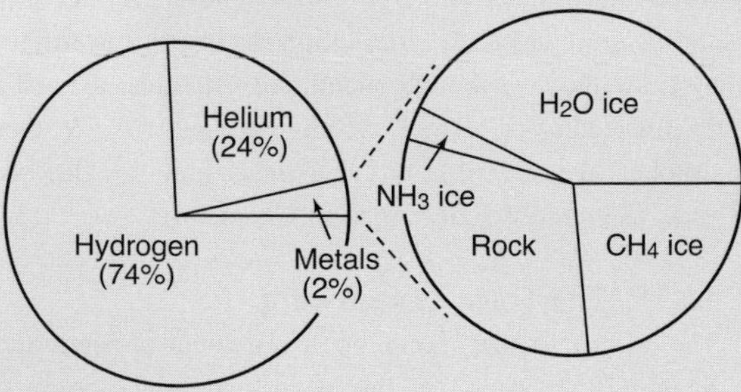

Figure 11.1 Weight proportions of hydrogen, helium, and metals in the solar system composition, and of ices and rock in the metals fraction.

cosmochemistry (and geochemistry)? Noble gas nuclides are produced by a variety of nuclear transformation processes. Although the quantity of newly produced nuclei is typically very small, the even-smaller background abundances of noble gases allow these nuclear transformations to be detected, unlike the situation for more common elements. It is understandable, then, that noble gas cosmochemistry mostly deals with isotopes. When applied to measuring noble gas isotopes, mass spectrometry is also much more sensitive than it is for other elements.

Different nuclear processes produce distinct isotopes or suites of isotopes in specific proportions, usually called "components." Multiple noble gas components commonly occur within a sample, so the game is to recognize and unmix these components. This can be done by analyzing samples that may contain the components in different proportions, or by stepwise heating or oxidation to release components occurring in various phases (techniques also used in the study of organic compounds). The isotopic components can be visualized by use of a three-isotope diagram, in which two isotope ratios, each with the same reference isotope as its denominator, are plotted against each other. An example using xenon isotopes is illustrated in Figure 11.2. Isotopically distinct components plot at different points on this diagram, and two-component mixtures will define a straight line joining the end members. Mixtures of three components will plot within a triangle defined by the three end members, as in Figure 11.2.

Some noble gas components are referred to as *nuclear*, meaning that they were produced *in situ* within a given meteorite sample by nuclear transmutations.

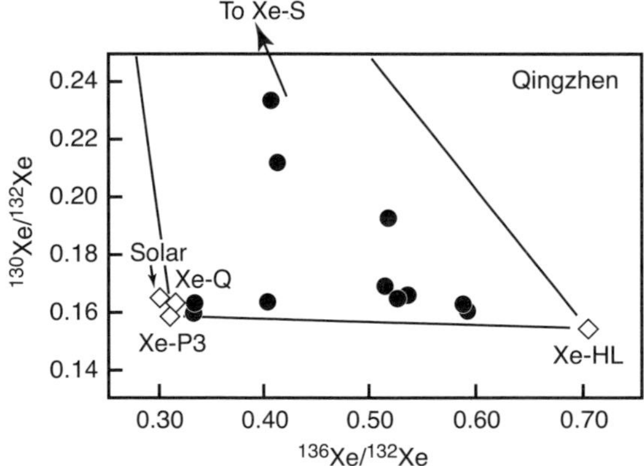

Figure 11.2 Three-isotope plot of xenon isotopes in the Qingzhen (EH3) chondrite, released at different temperatures. The measurements are mixtures of various trapped components (solar, Xe-Q, Xe-P3, Xe-S, and Xe-HL), explained in the text. Modified from Huss and Lewis (1995).

These are described in the immediately following section. Other components were *trapped* by the meteorite or its precursor materials. In 1963, cosmochemists Peter Signer and Hans Suess distinguished two kinds of trapped noble gas components in meteorites, which they referred to as *solar* and *planetary*. After discussing nuclear components, we will consider solar and planetary noble gases in turn.

11.5 Noble Gas Components in Extraterrestrial Samples

11.5.1 Nuclear Components

The decay of naturally occurring radionuclides provides one mechanism for producing noble gas isotopes. The best-known examples are ^{4}He, produced by alpha decay of ^{232}Th, ^{238}U, and ^{235}U, and ^{40}Ar, a beta decay product of ^{40}K. The use of ^{40}Ar to determine radiometric ages was discussed in Chapter 9. A few short-lived (now extinct) radionuclides have also yielded noble gas isotopes. Meteorites are so old that they incorporated these short-lived nuclides when they were still alive, and their noble gas decay products can be measured against the low-abundance background. The former presence of ^{129}I (with a half-life of only 17 Myr) was documented by measuring its ^{129}Xe decay product in meteorites.

Spontaneous fission of certain radionuclides (especially ^{238}U) can also produce noble gas isotopes. Rather than produce one specific daughter isotope, the fission of an actinide nucleus produces several isotopes of xenon and krypton (and other elements, though these are not of interest here) in well-defined proportions. For example, the relative proportions of ^{136}Xe, ^{134}Xe, ^{132}Xe, ^{131}Xe, and ^{129}Xe can be used to distinguish between components produced by fission of different isotopes. The prior presence of ^{244}Pu, a short-lived radioisotope with a half-life of 82 Myr, in the early solar system has been inferred from its fission-produced xenon and krypton isotopes.

Noble gas isotopes are also produced through irradiation by cosmic rays. These rays are mostly high-energy protons that produce a cascade of secondary particles when they bombard other target nuclei in a process called *spallation*. Cosmic-ray irradiation occurs on the surfaces of airless bodies like the Moon and asteroids, as well as on small chunks of rock orbiting in space. Using these isotopes, it is possible to calculate cosmic-ray exposure ages, as previously described in Chapter 10.

11.5.2 The Solar Component

The original "solar" noble gas component of Signer and Suess (1963) was recognized in meteorites based on elemental ratios. The five noble gases occur in relative abundances close to the solar system ratios. Subsequently, the solar noble gas component was determined to be implanted gases from the solar wind. Solar wind has been directly implanted into regoliths on the surfaces of airless bodies like the Moon and asteroids. The depths to which solar wind gases are implanted are very shallow, typically just a few hundred nanometers. Because noble gases are not depleted in the Sun, as they are in solids in the solar system, solar wind should be less affected by the addition of tiny amounts of nuclear components (described in the previous section). The Sun contains most of the mass of the solar system, and thus is most representative of the original bulk composition of the solar system. The one exception is its ^{3}He abundance. This is because the deuterium in the proto-Sun was fused into ^{3}He during the initial nuclear burning, increasing the ^{3}He abundance substantially. Minor production from incomplete hydrogen burning also contributes to the ^{3}He in the solar photosphere. The best estimate for the ^{3}He/^{4}He ratio in the solar nebula comes from the atmosphere of Jupiter.

The solar noble gas component in meteorites is most abundant in regolith breccias, which consist of material that was on the surfaces of asteroids. Numerous studies of implanted solar gases in these so-called "gas-rich" meteorites and in lunar soils returned during the Apollo program have been carried out. A complication in interpreting the results from lunar and meteorite samples is that spallation by cosmic rays has also added another (nuclear) noble gas component to the soils. However, the spallation component is sited more deeply inside soil particles, owing to the higher energies of cosmic rays, so the solar and spallation components can be separated by stepwise heating or etching. The spallation component also has a very different isotopic composition. Some analyses of solar noble gases in lunar soils and in the Pesyanoe gas-rich meteorite are illustrated in Figure 11.3a (plotted relative to solar system abundances). Ideally, solar gas abundances should define a horizontal line on this diagram. The modest slants suggest that some portion of the lighter elements has been lost. The Genesis mission provided an independent measure of the elemental and isotopic composition of the solar wind (see Box 4.3). These new data minimize the fractionations due to implantation into natural samples (Fig. 11.3a). However, another potential problem in inferring the composition of the Sun from solar wind data is that some fractionation, both elemental and isotopic, may have occurred when

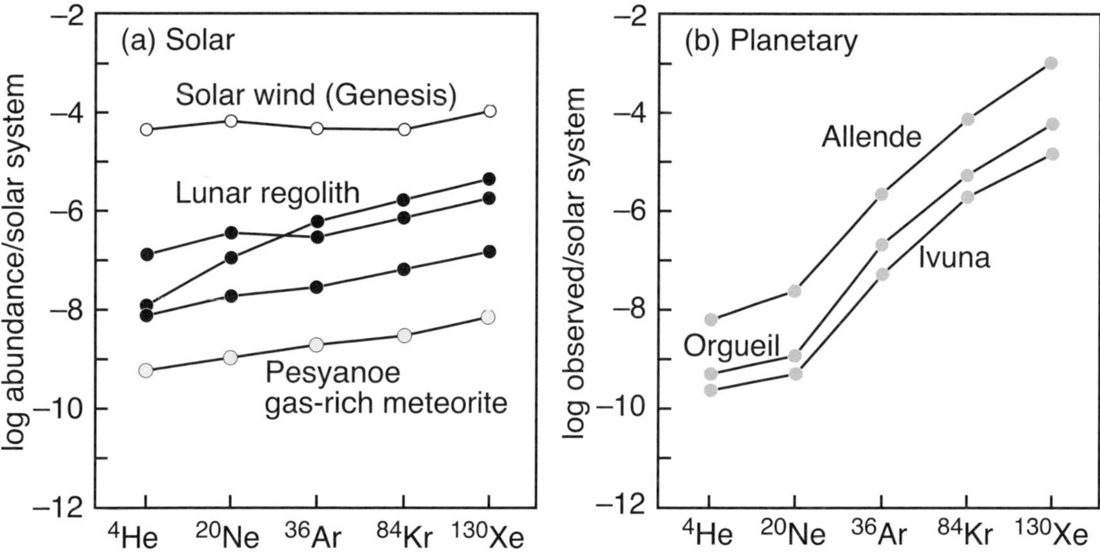

Figure 11.3 Abundances of trapped noble gas abundances in lunar soils and meteorites. (a) Elemental abundance patterns for trapped solar wind in lunar soils and the Pesyanoe gas-rich meteorite are compared to the pattern for solar wind obtained by the Genesis mission; all compositions are normalized to xenon and to solar system abundances, which are inferred from theory (Chapter 4). (b) Elemental abundance patterns for planetary noble gases, normalized to xenon and to solar system abundances, in the Orgueil (CI), Ivuna (CI), and Allende (CV) chondrites. This diagram is intended to illustrate patterns only; vertical positions are arbitrary. Modified from Ozima and Podosek (2002). Genesis data from Pepin et al. (2012) and Meshik et al. (2014).

atoms in the solar atmosphere were accelerated into the solar wind. Determining the magnitude of those fractionation effects is the focus of current work on noble gases in the Sun.

11.5.3 Planetary Components

The planetary noble gas component of Signer and Suess (1963) refers to a fractionated elemental abundance pattern highly depleted in the light elements (Fig. 11.3b). This component was designated "planetary" because the elemental ratios mimic those in the Earth's atmosphere. However, trapped planetary gases are much more complex than solar gases and, in fact, the planetary noble gas component is now recognized as being a mixture of several different components. To make matters worse, noble gases in the terrestrial planets do not have exactly the composition of the meteoritic planetary component, and the noble gases in planets probably originated from a completely different mechanism. Although the term "planetary noble gases" is thus a misnomer (Ott, 2014), in the absence of an alternate term we will use this term to describe an ensemble of meteorite components with strongly depleted light elements relative to heavy ones (Fig. 11.3b).

The terminology used to describe the various admixed components of planetary noble gases has a long and tortured history. Table 11.1 gives the names for the different components that we will introduce in this book and briefly summarizes their properties. Detailed discussions of these components were provided by Wieler et al. (2006), Ott (2014), and Cartwright (2015). Each exotic component in Table 11.1 is carried within a specific type of presolar grain: diamond, silicon carbide, or graphite. The exception is the Q gases, which likely reside in organic matter. Thus, carbon plays an extraordinary role as host phases for trapped noble gases. Each component consists of an assortment of noble gases with particular elemental and isotopic proportions (Fig. 11.4) that formed by specific nucleosynthetic processes. This should sound familiar – some of these noble gas components were the tracers used to isolate various presolar grains, as described in Chapter 5. Some of the components differ radically from solar composition, whereas others show only moderate deviations from solar isotopic abundances.

Diamonds are host to the HL component, so named because it is enriched in both heavy and light isotopes of xenon (Fig. 11.4b). The high abundances of heavy isotopes suggest *r*-process nucleosynthesis, whereas the abundant light isotopes suggest the *p*-process. Both the *r*- and *p*-processes occur in supernovae. However, it is not obvious why products of the two nucleosynthetic processes would be coupled, and this is a subject of current research. Diamonds also contain other, less anomalous noble gas components (P3 and P6). Rather than being products of specific nucleosynthetic processes in specific stars, these components may represent mixtures of many stellar nucleosynthetic contributions, as might be found in the ISM (and our own solar system). Recall from Chapter 5 that presolar diamonds are too small to be analyzed separately, so the assortment of diamonds in a chondrite probably samples multiple stellar sources.

Silicon carbide contains the G component (Fig. 11.4). The xenon and krypton isotope abundances in this component can be understood as arising from *s*-process nucleosynthesis in carbon-rich asymptotic giant branch (AGB) stars. In fact, the "G" stands for "giant," referring to these huge, red stars. The G component also includes neon that is especially enriched in ^{22}Ne, consistent with helium burning in AGB stars. The N planetary component in SiC (Fig. 11.4) is thought to represent the outer envelope of the same AGB stars in which helium burning generates the G component.

The origins of presolar graphite grains are diverse. Bulk samples of presolar graphite release large amounts

Table 11.1 Trapped planetary noble gas components in meteorites

Component	Synonyms	Carriers	Constituents	Characteristics
P1	Q gases	Phase Q (organic?)	All 5 gases	Dominates Ar, Kr, Xe
P3		Diamond	All 5 gases	Similar to but distinct from P1
P6		Diamond	All 5 gases	Composition poorly defined
HL	Ne-A2,* Xe-X, Xe-HL	Diamond	All 5 gases	Isotopically anomalous (*r*-process?)
G	Ne-E(H), Kr-S, Xe-S	SiC and graphite	All 5 gases	Isotopically anomalous (*s*-process)
N		SiC and graphite	All 5 gases	Isotopically normal
Ne-E(L)	R	Graphite	Pure ^{22}Ne	Comes from decay of ^{22}Na

*Ne-A2 is a mixture of Ne-HL and Ne-P6 that is inseparable experimentally.

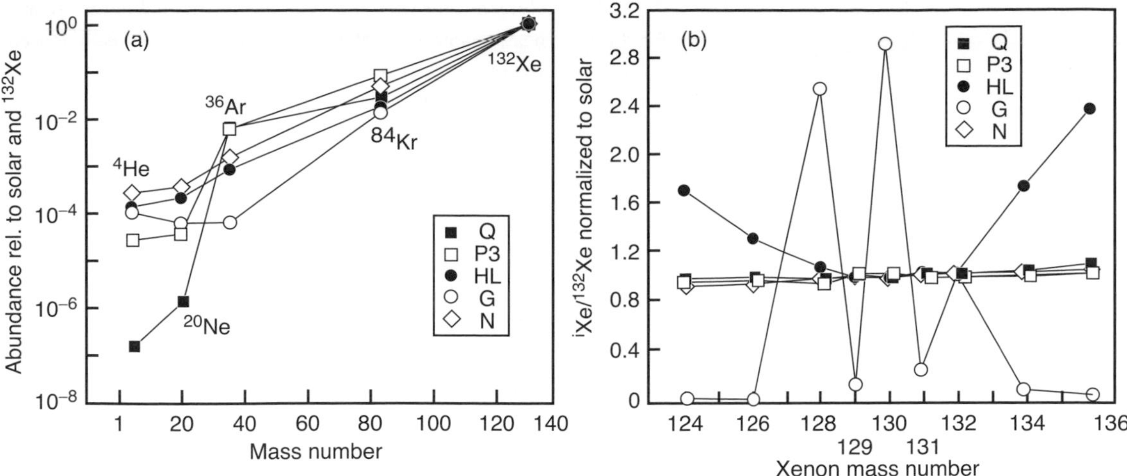

Figure 11.4 (a) Isotope abundances and (b) xenon isotopic abundances for some planetary noble gas components (defined in Table 11.1) in meteorites. Modified from Wieler et al. (2006).

of neon consisting almost entirely of ^{22}Ne at low temperatures. This Ne-E(L) is thought to come from the decay of radioactive ^{22}Na (half-life = 2.6 years), which is produced by explosive nucleosynthesis, probably in novae. The heavy noble gases, like those in SiC, reflect formation by the *s*-process in AGB stars. Recall that studies of individual graphite grains identified grains from AGB stars and grains from supernovae (Chapter 5). Physical properties and noble gas components thus indicate multiple production sites for presolar graphite.

Last but not least, the Q (or P1) component dominates the planetary argon, krypton, and xenon. It is carried in phase Q, named for "quintessence." This component is easily lost when acid residues of chondrites are oxidized, and its carrier is probably an organic phase. This component actually constitutes everything that remains after separation of all the other exotic noble gas components described in Table 11.1. The isotopic composition of Q is normal, perhaps suggesting that it is of local (nebular) origin, but the concentration of the Q component closely tracks those of the presolar components in meteorites, suggesting that it is also presolar. The extremely low helium and neon abundances relative to heavy noble gases in the Q component (Fig. 11.4a) suggest that the gases may have been trapped in their carrier at low temperatures, when argon, krypton, and xenon were frozen out into ices and helium and neon were still in the gas phase. It is not resolved whether the Q component was trapped into its carrier in the Sun's parent molecular cloud, making it a presolar component, or in the solar system.

11.6 Planetary Atmospheres

Atmospheres are a natural consequence of the origin and evolution of planets. If planets were sufficiently massive, they could have captured nebular gas while they formed. Accretionary and radioactive heating also released gases that were brought into the planet in solid carriers. The atmospheres of Venus, Earth, and Mars are composed of the same gases (CO_2, N_2, H_2O, O_2), but in markedly different amounts and proportions, reflecting the planets' different evolutionary histories. The relative proportions of CO_2 and N_2 in the atmospheres of Venus and Mars are similar, implying that atmospheres dominated by these gases should be the norm. Earth is the oddball here, because its CO_2 has been almost completely removed and sequestered in limestone, petroleum, and coal, while O_2 from photosynthesis has accumulated in the atmosphere. The masses of the atmospheres are also very different between the three planets, with surface pressures of 95.6, 1, and 0.0064 bars on Venus, Earth, and Mars, respectively. Mercury and the Moon have very tenuous atmospheres composed mostly of noble gases and moderately volatile elements like sodium that were volatilized by cosmic-ray bombardment. The noble gases are much less affected by many of the processes that occur in planetary atmospheres and have retained many of the characteristics established early in the histories of the planets.

Atmospheric noble gases are highly depleted in the terrestrial planets relative to their abundances in the Sun (Fig. 11.5). Plotted in this figure are nonradiogenic isotopes, the abundances of which do not change over time.

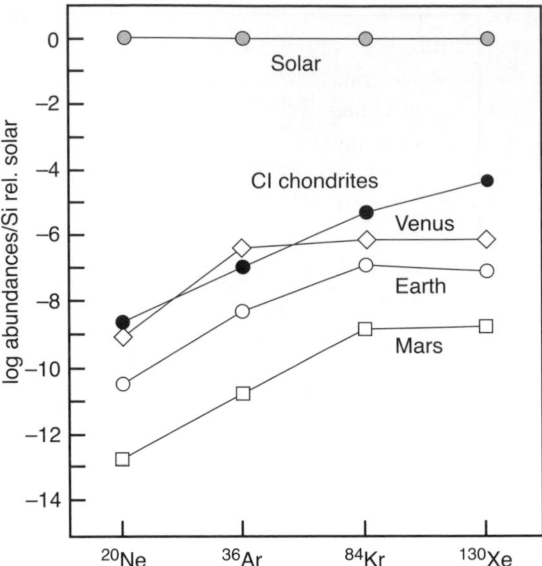

Figure 11.5 Noble gas abundances in planetary atmospheres and CI carbonaceous chondrites, relative to silicon and solar abundances. After Porcelli and Pepin (2004).

Depletions of the light elements are greater than those of the heavy elements, which we previously recognized as the planetary pattern. Helium is not plotted in Figure 11.5 because it is not gravitationally bound to any of the terrestrial planets. The absolute concentrations of noble gases vary by three orders of magnitude, from gas-rich Venus to gas-poor Mars. The low and varied abundances of noble gases in the terrestrial planets show that their atmospheres cannot be primary (that is, acquired directly from the gaseous nebula), but must have been outgassed from planetary interiors. Earth and Venus are massive enough to have captured atmospheres from the solar nebula if the gas persisted until they grew to approximately their present sizes. However, if these planets captured such an atmosphere, much of it was subsequently lost and the noble gas abundances were markedly changed. The noble gases in the present atmospheres of the terrestrial planets could have been originally acquired in several ways. They could have been implanted into the precursor material by the solar wind, or carried to the planets by comets or meteorites (Porcelli and Pepin, 2004). Perhaps multiple processes played a role, with the relative importance varying between planets. Earth and Mars have nearly parallel abundance patterns (Fig. 11.5) that are unlike the Sun but similar to CI chondrites, suggesting that noble gases were brought in by accreted planetesimals or comets. Earth's atmosphere is particularly depleted in heavy xenon isotopes (when corrected for fission xenon and for fractionation favoring heavy isotopes) relative to chondrites, as is a Rosetta spacecraft analysis of xenon in comet Churyumov–Gerasimenko. This depletion is consistent with a mixing model of ~78% chondrite and ~22% comet for Earth's atmospheric composition (Marty et al., 2017). Venus' gases are somewhat more solar-like, suggesting gravitational capture of nebular gases or accretion of particles into which solar wind was implanted.

Neon, argon, krypton, and xenon on Earth and Mars, and neon and argon on Venus, are all isotopically heavier than the gases in the Sun. In other words, the light isotopes are preferentially depleted. The depletions are smoothly mass dependent relative to the solar wind ratios, which suggests that the original inventory of noble gases was depleted by loss into space. The implication is that the current inventories of noble gases are remnants of the original inventories. As discussed in Chapter 8, D/H ratios indicate that Venus and Mars have also lost much of their original inventories of hydrogen, which was predominantly present as water. Warming of the martian atmosphere by dust storms increases water vapor, which is dragged along as dust swirls into the upper atmosphere, where it is lost (Stone et al., 2020). Loss of water and light gases by thermal escape or sputtering from the top of the exosphere has also affected Mars, Venus, and Earth. But these mechanisms do not work for heavy argon, krypton, and xenon. Hydrodynamic escape, whereby noble gases are entrained by a light gas such as hydrogen as it escapes from a hot atmosphere, is a possible mechanism for the loss and mass fractionation of the heavy gases. On Earth, hydrodynamic escape could have been driven by the Moon-forming impact (Pepin, 2006). Noble gases carried by planetesimals could also have been lost during accretion.

Unlike the terrestrial planets, the giant planets are massive enough to have captured and retained nebular gases directly. Thus, they are mostly hydrogen, with small but varying amounts of helium. However, concentrations of argon, krypton, and xenon measured in Jupiter's atmosphere by the Galileo spacecraft are ~3 times solar (Mahaffy et al., 2000), which may imply that its atmosphere preferentially lost hydrogen and helium over the age of the solar system.

Some noble gases are also retained in planetary interiors, so a planet's total inventory is the sum of retained and degassed volatiles. We have some limited data on noble gases in the mantles of the Earth (from basaltic lavas) and Mars (from martian meteorites). The compositions of the noble gases in planetary interiors differ from those in the atmosphere due to the decay of radioactive isotopes. For example, the ratios of $^{3}He/^{4}He$ in magmas derived from the Earth's mantle vary considerably from the atmospheric ratio, reflecting the mixing of primordial ^{3}He with radiogenic helium produced in the mantle. The ^{40}Ar in the atmospheres of the terrestrial planets results from decay of ^{40}K in the interior and partial outgassing of the daughter ^{40}Ar. The relative contributions of radiogenic and nonradiogenic argon in the atmosphere differ dramatically between Earth, Mars, and Venus. Radiogenic components produced in planetary interiors and then partially outgassed are also seen in the heavy krypton and xenon isotopes (from fission of U and Pu) and in ^{129}Xe (from decay of ^{129}I). Noble gases sequestered in the interior could have avoided the processing experienced by atmospheric gases. For example, some deep mantle sources on Earth have $^{3}He/^{4}He$ and $^{20}Ne/^{21}Ne$ ratios similar to those in solar wind. For a long time the low Xe/Kr ratios in the atmospheres of Earth and Mars relative to meteoritic gases were thought to reflect sequestration of xenon in the interior or in another reservoir. However, the "missing" xenon has not been found and the low Xe/Kr ratio is now believed to be a property of the planets.

11.7 Extraterrestrial Organic Matter: Occurrence and Complexity

Organic compounds have been observed in interstellar and circumstellar gases using telescopes fitted with infrared and microwave spectrometers. The bonds in organic molecules bend and stretch when absorbing certain wavelengths of energy, and this bending and stretching can be detected by spectroscopic measurements. These absorptions are diagnostic for certain compounds. An updated list of organic molecules telescopically detected in the ISM is maintained by the National Radio Astronomy Observatory website (www.nrao.edu).

Organic matter is nearly ubiquitous in asteroids, comets, and planets, but it is difficult to study by remote sensing. The most important sources of detailed information about extraterrestrial organic compounds are chondritic meteorites. Organic matter in meteorites can be extracted and analyzed using the full panoply of methods available in the laboratory. One major problem, though, is contamination by terrestrial organic matter. Simply holding a meteorite in one's fingers adds enough contaminants to complicate organic analyses, and there are numerous examples of meteorite studies that have been compromised by terrestrial contamination. In spite of this problem, developments in understanding the organic components in meteorites have proceeded rapidly, as improvements have been made in analytical capabilities. Organic compounds occurring in trace quantities can now be identified, and the isotopic composition of carbon, hydrogen, oxygen, nitrogen, and sulfur in specific compounds can be measured. It is also useful to determine *chirality* – the identification of compounds that can exist in mirror-image structures (see Box 11.2). All these properties place constraints on the formation and history of organic compounds.

In this chapter, we will focus primarily on what has been learned from analyses of carbonaceous chondrites, which contain more organic matter than most other chondrite types and consequently have been the focus of most research in this area. The organic matter in carbonaceous chondrites can be divided into two very different kinds of materials. The first consists of high-molecular-weight compounds ("macromolecules") that cannot be readily extracted from the meteorites with solvents and strong acids. The organic matter in the CO, CV, and CK chondrite classes is mostly macromolecules. The organic matter in the CI and CM chondrites also includes up to 30% low-molecular-weight compounds that can be extracted with solvents. These extractable compounds constitute almost all of the organic suite in CR chondrites, and because they can be readily separated, they can be more easily characterized.

Especially large amounts (100 kg) of the uncontaminated CM chondrite Murchison have been available since its fall in Australia in 1969, and the first hint that it contained organic matter came from eyewitness reports that commented on solvent smells from the meteorite samples. Much of the modern research on extraterrestrial organic matter has focused on this meteorite. The organic matter in Murchison has long been considered representative of carbonaceous chondrites, although studies of other CMs like Tagish Lake (Pizzarello et al., 2001) and Sutter's Mill (Pizarello et al., 2013) show considerable variations in the relative proportions of organic compounds present. Research has also been done on organic matter in the Orgueil CI chondrite, although this meteorite has been contaminated during its nearly 150 years on Earth. More recently, analyses of organic compounds in CR chondrites (finds preserved in clean conditions in Antarctic ice) have offered important new insights into abiotic chemical evolution (Pizzarello and Shock, 2010).

11.7.1 Extractable Organic Matter in Carbonaceous Chondrites

Soluble organic compounds in carbonaceous chondrites are highly varied in structure, composition, and size

Box 11.2 A Crash Course in Organic Nomenclature

The basic skeleton of bonded carbon atoms can be arranged in simple straight chains, branched chains, one or more rings, or combinations of these structures. The simplest organic molecules are *hydrocarbons*, which consist solely of carbon and hydrogen atoms. Carbon atoms have four valence electrons that must be satisfied either by single covalent bonds to four separate atoms or by some combination of single and double (or less commonly triple) bonds. A double bond is one that involves two valence electrons. Hydrocarbons that have only single bonds are said to be *saturated*, whereas those that contain double or triple bonds are *unsaturated*.

Carbon atoms can be linked together to form either straight-chain (*aliphatic*) or cyclical (*aromatic*) structures. A saturated aliphatic hydrocarbon is called an "alkane," where the suffix "–ane" refers to the lack of any double bonds in the molecule. An unsaturated aliphatic hydrocarbon is called an "alkene," where the suffix "–ene" denotes the presence of a double or triple bond. Alkanes (and alkenes) are named according to the number of carbon atoms in the chain, as in methane (one), ethane (two), propane (three), butane (four), pentane (five), and so on. Hydrocarbon structures can be represented by a variety of two-dimensional drawings (Fig. 11.6a). All the atoms and bonds can be illustrated (as in the top pentane diagram), all the bonds except those between the carbons in the chain can be omitted (middle pentane diagram), or the chain can be represented by a "sawtooth" drawing (bottom pentane diagram) in which the kink at each line segment represents a carbon atom. Leftover valence electrons (those not involved in bonding the carbon atoms together) attach hydrogen atoms to the molecule at each kink.

Stable configurations of carbon atoms also occur where double (C=C) bonds alternate with single (C–C) bonds to form a pattern like C–C=C–C=C–C. This configuration, referred to as "conjugated," is common in polyunsaturated compounds. If conjugation occurs in a ring structure with three double and three single bonds (Fig. 11.6a), the configuration is *aromatic*. Aromatic compounds are thus unsaturated and typically very stable. Two representations of the simplest aromatic ring (benzene) are shown in Figure 11.6a. In the simple line sketch, the alternating single and double bonds between carbon atoms are shown, carbon atoms are understood to occur at each corner, and the single hydrogen atoms attached to each carbon are not illustrated.

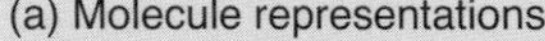

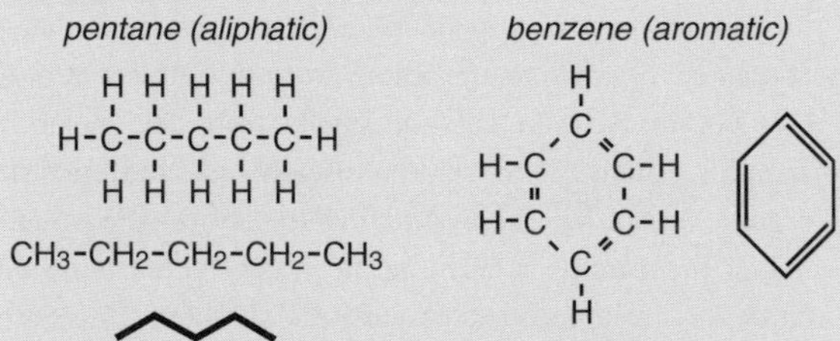

(b) Functional groups (R represents various groups)

		compound
R - OH *hydroxyl*		alcohol (R = aliphatic) phenol (R = aromatic)
- C = O, R *carbonyl*		aldehyde (R = H) ketone (R = aliphatic)
- C = O, OH *carboxyl*	(or - COOH)	carboxylic acid (sulfonic and phosphonic acids if S or P replace C)
- N(H)(H) *amino*	(or - NH_2)	amine

Figure 11.6 Molecular representations and functional groups in organic compounds. (a) Different ways to represent the structures of aliphatic (pentane) and aromatic (benzene) compounds. (b) Some important functional groups attached to organic skeletons ("R"), and the corresponding names of organic compounds.

Box 11.2 (cont.)

Groups of elements bonded to the carbon backbone (whether it's a chain or a ring, and generically represented by the letter "R") are called *functional groups*, because they are usually reactive and thus control the chemical behavior of the compound. Functional groups (some common ones are illustrated in Fig. 11.6b) involve combinations of hydrogen atoms with one or more atoms of another element, such as carbon, oxygen, nitrogen, sulfur, or phosphorus. The presence of nonbonding electrons in elements other than carbon in functional groups makes them likely sites for chemical reactions. Organic compounds are often grouped under the name of the functional group they contain. For example, alcohols and phenols are based on combinations of the hydroxyl functional group –OH with aliphatic or aromatic groups, respectively. Aldehydes and ketones feature the carbonyl group –COR, where R can be H, a chain, or a ring. Carboxylic acids include the carboxyl group –COOH, and the carbon atom in a carboxyl can be replaced by sulfur or phosphorus to produce sulfonic or phosphonic acids. Amines are based on the nitrogen-containing amino group $-NH_2$. Massive organic molecules include sugars, which are aliphatic compounds with many hydroxyl groups in place of hydrogen atoms. Polycyclic aromatic hydrocarbons (PAHs) involve multiple aromatic rings hooked together.

Although we typically represent organic compounds with two-dimensional drawings, these molecules are actually three-dimensional structures. Several different spatial configurations of atoms attached to a central carbon atom can exist; these *isomers* have the same chemical formula, but have different arrangements of atoms and different properties. Certain amino acids and sugars have identical compositions but differ by the placement of their functional groups on one side or the other of the central carbon atom. Isomers that are mirror images of each other are called *enantiomers*. Your hands provide an illustration of this phenomenon – each hand has the same structure of four fingers and a thumb, but you cannot superimpose them with both palms facing up (the *chirality* discussed in Section 11.7.1). Enantiomers rotate polarized light in opposite directions. Compounds that rotate light clockwise are dextrorotatory (denoted by the prefix D-), whereas counterclockwise rotation is referred to as levorotatory (denoted by L-). Abiotic organic compounds usually occur as *racemic mixtures*, which have D- and L-enantiomers in equal proportions. In contrast, biologically produced organic matter strongly favors one enantiomer. However, such enantiomers tend to racemize over relatively short geologic timescales.

(molecular weight). The relative abundances of soluble compounds also vary drastically in CM and CR chondrites (Fig. 11.7), and the compounds reflect the much greater abundance of nitrogen in CRs. Let's consider what is known about some specific compounds (Gilmour, 2004; Pizzarello et al., 2006; Pizzarello and Shock, 2010; Remusat, 2015).

Amino acids, which contain both the amino and carboxylic acid functional groups (Fig. 11.8), are essential components of proteins in the Earth's biosphere. Twenty common terrestrial amino acids are almost exclusively L-enantiomers, and this chirality is indispensable for the function of living organisms. Amino acids in Murchison show much greater structural diversity, with more than 100 species having up to eight carbon atoms (the limit of solubility). In CR chondrites, the shorter-chain-length amino acids are overabundant compared to longer-chain species. L- and D-enantiomers would be expected to occur in approximately equal amounts for abiotic amino acids; however, both Murchison and CR chondrites show an excess of L-forms in some amino acids. Terrestrial contamination (pure L-enantiomers) could cause such an excess, but careful analyses of amino acids not known to occur naturally on Earth also showed chirality (Glavin et al., 2020). One hypothesis proffered to explain the chirality in meteorite amino acids is photolysis by ultraviolet, circularly polarized starlight in interstellar space. Another possibility is that L-enantiomers have slightly higher thermodynamic stability than D-forms. The common association of L-amino acids with higher degrees of aqueous alteration in CM chondrites suggests yet another explanation – that formation of organic compounds on mineral substrates during alteration might induce chirality.

Ammonia (NH_3) is abundant in CR chondrites but not in Murchison, and amines are nitrogen-containing compounds found in all meteorites that contain amino acids. The abundance of amines is much greater in CR chondrites than in Murchison (Fig. 11.7), and abundances decrease with increasing numbers of carbon atoms. Similarities between amines and amino acids in the same meteorite have led to the suggestion that amines formed

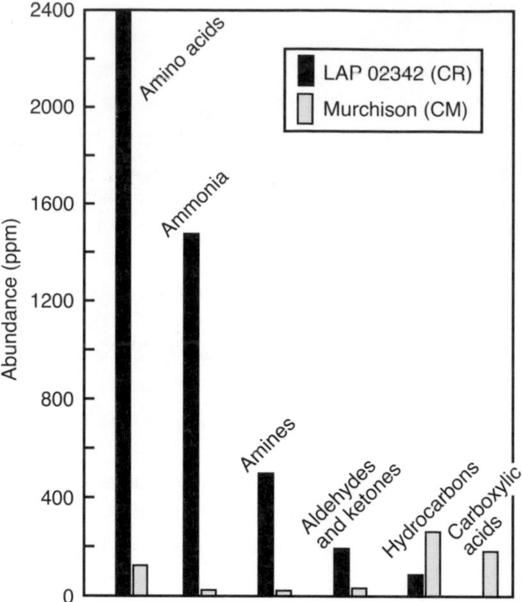

Figure 11.7 Comparison of significantly different abundances of the major soluble organic compounds in the Murchison (CM2) and LAP 02342 (CR2) chondrites. Modified from Pizzarello and Shock (2010).

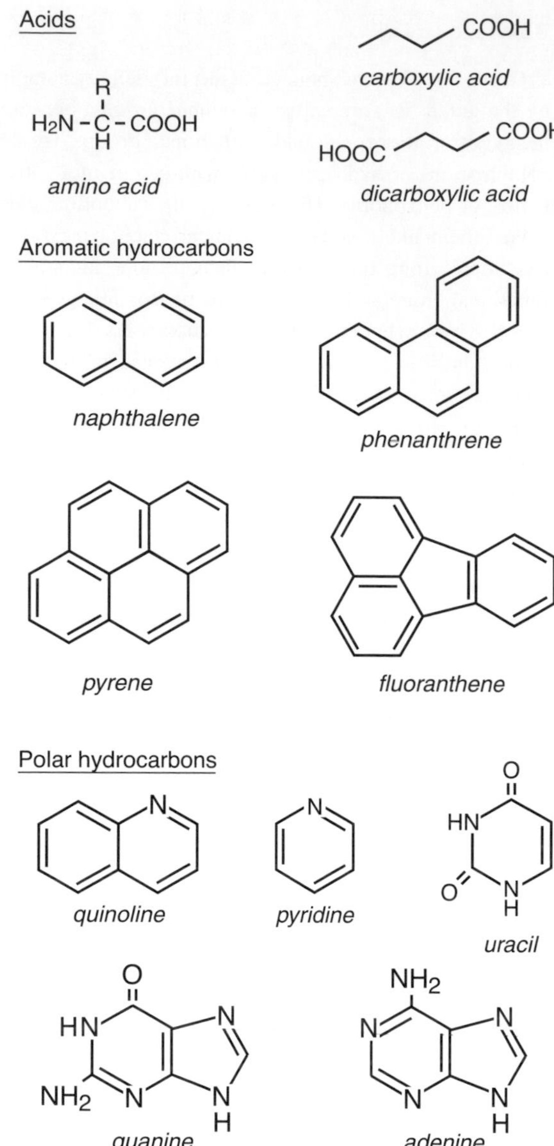

Figure 11.8 Structures of some soluble organic compounds found in carbonaceous chondrites.

from the amino acids by decarboxylation (loss of the carboxylic acid functional group). This process commonly occurs upon heating.

Carboxylic acids (Fig. 11.8) are structurally diverse, with equal concentrations of straight-chain and branched compounds. The abundances of carboxylic acids decrease as the numbers of carbon atoms in the structure increase. Dicarboxylic acids, containing two carboxyl groups (Fig. 11.8), also occur. These acids are abundant in Murchison but not in CR chondrites. Sulfonic and phosphonic acids contain sulfur and phosphorus in their structures, substituting for carbon in the carboxyl functional group. They are fairly abundant in Murchison but have not been studied extensively.

Aldehydes and ketones are abundant in CR chondrites but not in CMs (Fig. 11.7). Both are carbonyl compounds, meaning that a carbon double-bonded to oxygen is also bonded to one or two hydrogens (aldehyde) or to two carbons (ketone), as illustrated in Figure 11.8. These hydrocarbons have been proposed as the reactants (with HCN, in the presence of water and ammonia) that formed amino acids in chondrites (Pizzarello and Shock, 2010).

Other hydrocarbons are much more abundant in CM than in CR chondrites (Fig. 11.7). Aliphatic hydrocarbons in Murchison range in size from methane (which has one carbon and thus really isn't a chain) to long saturated chains (alkanes) with up to 30 carbon atoms. There has been a long debate about whether these hydrocarbons are indigenous meteorite components or terrestrial contaminants. In the 1960s, analyses of alkanes in the Orgueil CI chondrite spawned a heated controversy about extraterrestrial life, but later analyses revealed that these were contaminants. However, analyses of clean meteorite samples have now demonstrated that most alkanes are indigenous to chondrites.

The literature on soluble aromatic hydrocarbons (PAHs) reveals considerable differences in the relative abundances of different compounds. The most abundant aromatic compounds in Murchison are the four-ring PAHs pyrene and fluoranthene (Fig. 11.8). The extractable PAHs are quite complex and include compounds with up to 22 carbons. However, low-molecular-weight PAHs such as naphthalene and phenanthrene (Fig. 11.8) are also fairly abundant.

We did not mention "polar" hydrocarbons in the crash course on organic nomenclature (Box 11.2). This term refers to compounds containing one or more nitrogen, oxygen, or sulfur atoms replacing carbon atoms in ring structures or sometimes attached to them. These elements are more electronegative than carbon, so C–N, C–O, and C–S bonds are polar, hence the name. In Murchison, polar compounds are abundant, comprising nearly two-thirds of the extractable organic matter.

To summarize, the soluble organic compounds in CM and CR chondrites show stark contrasts. The CM compounds are highly diverse and structurally complex, whereas those in CR chondrites are simpler and dominated by nitrogen-bearing species.

11.7.2 Insoluble Macromolecules in Carbonaceous Chondrites

The great majority of organic carbon (70–99%) in all carbonaceous chondrites (except CRs) is present as complex, insoluble macromolecular material, often referred to as *insoluble organic matter* (IOM) or kerogen. It can be separated from meteorites only by demineralizing the samples with harsh acid treatments. Besides carbon, the insoluble organic fraction contains hydrogen, oxygen, nitrogen, and sulfur, and the proportions of these elements vary from chondrite to chondrite. The average elemental composition of this material, expressed relative to 100 carbon atoms, is $C_{100}H_{60}N_3O_{18}S_2$ for CM chondrite Murchison, and $C_{100}H_{80}N_3O_{16}S_7$ for CR chondrite Al Rais (Remusat, 2015). Identifying the compounds present in these macromolecules poses a daunting analytical challenge. Viewed by transmission electron microscope (TEM), this material is unstructured and heterogeneous and reveals little about its makeup. Because solvents are not useful in separating the various compounds in IOM, macromolecular organic material has mainly been analyzed using harsh heating or chemical degradation techniques that break the huge molecules into smaller fragments that can be studied.

Chemical degradation together with detailed analysis of the abundance and composition of the fragmented products have provided a general picture of IOM (Alexander et al., 2017). It consists primarily of aromatic ring clusters, bridged by aliphatic chains containing S, N, and O, with –OH and –COOH peripheral functional groups. One possible structural model for a constituent of IOM is illustrated in Figure 11.9 (Pizzarello and Shock, 2010). Pyrolysis experiments on IOM release aromatic compounds such as benzene and naphthalene (Fig. 11.8), leaving behind an essentially unmodified residue of PAHs with higher molecular weights. This may suggest that two aromatic components are actually present: one more labile and one refractory. The labile compounds tend to be more deuterium rich. It has been suggested that the labile component may have formed in the ISM, and the other during aqueous alteration on asteroid parent bodies. Other aromatic molecules such as benzene carboxylic acids and aromatic ketones are also present in pyrolysis products, as are nitrogen-containing polar compounds.

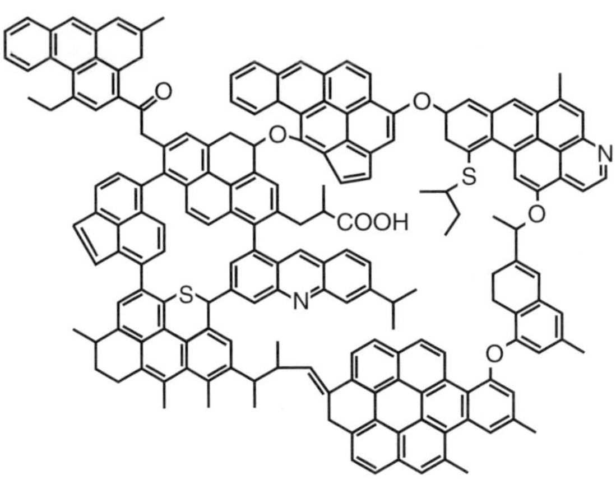

Figure 11.9 A suggested structural model for IOM in chondrites. Adapted from Pizzarello and Shock (2010).

Nuclear magnetic resonance (NMR) spectra of bulk macromolecular material provide information on bonding and the nature of functional groups. The major functional groups identified are illustrated in the NMR spectrum of Figure 11.10, which also shows the prominent peak corresponding to aromatic carbon. NMR spectra confirm the conclusion from decomposition studies that this material is mostly aromatic. The relative abundances of the aromatic and aliphatic compounds comprising IOM differ significantly among the carbonaceous chondrite classes (Fig. 11.11).

In addition to macromolecular organic compounds, IOM also contains graphitic flakes, nanotubes, and spheres (Garvie and Buseck, 2004), some of which are hollow (Fig. 11.12). These particles, which constitute up to 10% of the IOM in the Tagish Lake chondrite, have very low contents of hydrogen, oxygen, and nitrogen compared to the rest of the macromolecular material.

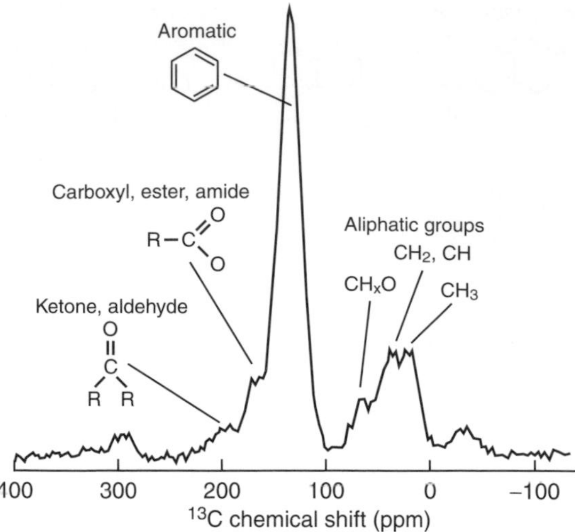

Figure 11.10 ^{13}C-NMR spectra of IOM in Murchison show a prominent peak for aromatic carbon, demonstrating its high abundance. Other, smaller peaks correspond to the various identified functional groups. The small peaks at each end of the spectrum are artifacts of the method. After Cody et al. (2002).

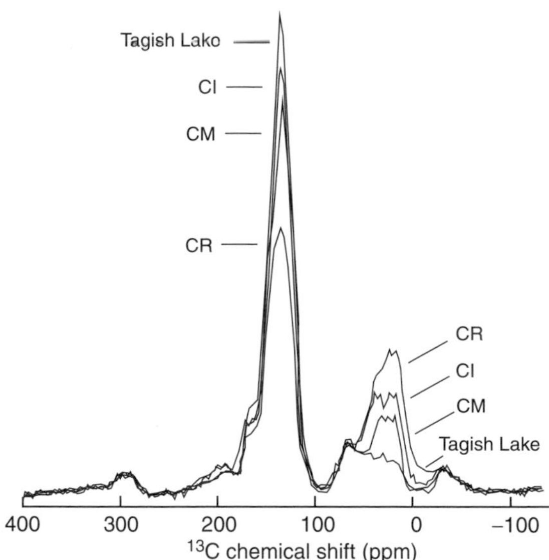

Figure 11.11 Comparison of ^{13}C-NMR spectra of CR, CI, CM, and Tagish Lake chondrites, illustrating differences in the relative abundances of aromatic and aliphatic compounds in their IOM. After Cody and Alexander (2005).

Although CM chondrites are dominated by IOM, it is in low abundance in CR chondrites. The water-soluble, relatively low-molecular-weight, nitrogen-containing organic matter in CRs has been characterized as "prebiotically desirable" as a feedstock for life (Pizzarello and Shock, 2010).

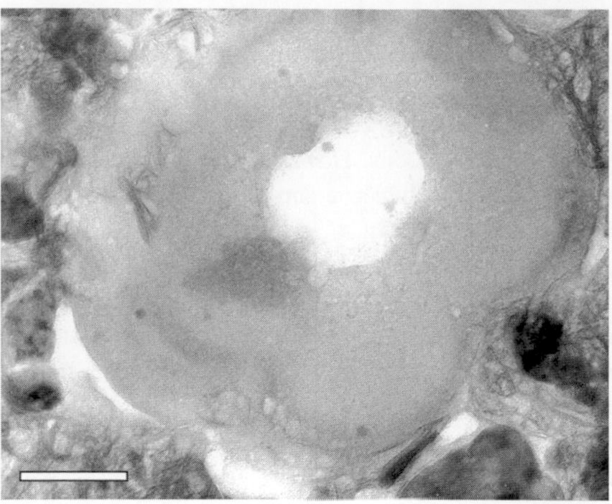

Figure 11.12 TEM image of a macromolecular particle found in the Tagish Lake CM-like chondrite. Up to 10% of the IOM in CM meteorites consists of hollow spheres and flakes like these. The scale bar is 100 nm. Image courtesy of P. Buseck and K. Nakamura.

11.7.3 Stable Isotopes in Organic Compounds

Traditionally, isotopic compositions of extraterrestrial organic matter have been measured on bulk samples. Isotopic measurements of bulk organic matter extracted from small samples of carbonaceous chondrite matrix reveal huge anomalies and heterogeneities, hinting at isotopic variations in individual compounds. Values of δD ranging from 1700 to 19,400 permil and δ^{15}N ranging from 400 to 3200 permil have been reported (Busemann et al., 2006).

In the 1990s, developments in instrumentation began to allow direct measurement of the isotopic compositions of individual compounds (*compound-specific isotope analysis*). The approach, relying on gas-chromatographic separation, works best for highly volatile compounds with relatively few functional groups. Compound-specific isotope measurements for hydrogen and carbon, and for nitrogen or sulfur (if present), have now been made on aliphatic hydrocarbons, acids (amino, carboxylic, and sulfonic), and aromatic compounds. Polar compounds are less volatile, and they must be chemically modified to increase their volatility and make them amenable to gas-chromatographic analysis.

All compounds analyzed have much higher δD values than terrestrial samples, confirming the enrichments seen in bulk measurements. Some individual compounds have very high δD values, in one case as high as 3600 permil (36 times the solar system ratio).

A major conclusion from analyses of carbon isotopes in organic compounds in chondrites is that δ^{13}C values decrease as the number of carbon atoms in the molecule increases. This consistent pattern is seen in aliphatic hydrocarbons, carboxylic acids, amino acids, and sulfonic acids in Murchison. This finding is interpreted to indicate that larger molecules formed from smaller ones (this has euphemistically been referred to as the "Lego principle"). The addition of a ring to an aromatic framework likewise increases the ^{12}C content of the resulting compound. Several models have been proposed to explain this trend. One idea is that the ^{13}C-rich compounds formed in interstellar space, and the final compounds resulted from transformation of the interstellar organic precursors by aqueous reactions on the meteorite parent bodies (Sephton and Gilmour, 2000). This idea would also explain high δD values. Alternatively, addition of atoms to the carbon skeleton or rings to the aromatic framework under kinetic control could favor ^{12}C. This kinetic isotope effect is not seen in terrestrial PAHs, which mostly form at high temperatures.

The $\delta^{15}N$ values of individual amino acids have been measured in an attempt to determine whether the reported excess in L-enantiomers is of extraterrestrial origin. The L- and D-isomers of amino acids have higher $\delta^{15}N$ values than their terrestrial counterparts, arguing against terrestrial contamination. Similarities in the nitrogen isotopic compositions of amino acids and amines in the same meteorite support the hypothesis (mentioned above) that heating may transform amino acids into amines.

Sulfur isotope compositions in sulfonic acids show a mass-independent enrichment in ^{33}S. Determining fractionations that are not controlled by mass is made possible because sulfur has so many stable isotopes. This feature has been attributed to ultraviolet irradiation of carbon disulfide in space, prior to its reaction to make sulfonic acid.

Determining the isotopic compositions of compounds that comprise the IOM in meteorites is more difficult. Stepped combustion heating releases fractions of different volatility at different times. Studies have determined that the relatively labile and refractory components of IOM differ in their carbon, hydrogen, and nitrogen isotopic compositions (we suggested earlier that these different components might have formed in different environments).

11.8 Are Organic Compounds Interstellar, Nebular, or Planetary?

Determining the origins of organic molecules is a major goal of cosmochemistry. Some hints are provided by the structures of organic molecules. Structures rich in rings and double or triple bonds linking carbon atoms (in other words, structures with lower ratios of hydrogen to carbon) suggest that reactive hydrogen was uncommon during the assembly of the molecule. Some carbon skeletons can accommodate strain, and the survival of strained molecules suggests that they have remained cool since the time they formed. The identity of functional groups and the occurrence of chirality also provide evidence about a molecule's origin. For example, alcohols, aldehydes, and carboxylic acids indicate progressively greater levels of oxidation. The occurrence of functional groups that would react with water would indicate that the environments of synthesis and storage were dry. Patterns within mixtures of organic molecules can also be informative about their origins. For example, a preference for molecules with even numbers of carbon atoms would indicate that C_2 reactive units were important in synthesis. Certain catalysts may also provide precise controls on the compositions of organic mixtures. The stable-isotope compositions of elements that are used to construct organic molecules are among the best indicators of origin.

Let's now turn to the specific question of where organic molecules in chondrites formed. Organics are omnipresent in the ISM and in the solar system. When the solar system formed, it inherited organic material from the parent molecular cloud. Thus, one possibility is that the organic materials were inherited from the ISM. However, we must keep in mind that the collapse of the presolar cloud to form a protostar and accretion disk may have destroyed or modified much of the inherited organic material, at least close to the Sun. Synthesis of new organic compounds then took place in the nebula. Thermal metamorphism and aqueous alteration on the meteorite parent bodies also provided pathways for production and destruction of organic molecules. The organic materials that we observe today in meteorites are the products of this multistage history, so it is unlikely that we will find simple direct connections to a single specific environment.

We can make some predictions about the structures and isotopic compositions of compounds that might have been inherited from the molecular cloud, based on what is known about organic species in interstellar space. The spectroscopically observed organic molecules in the ISM are not those expected for conditions of thermodynamic equilibrium. Why is that? Although we sometimes describe molecular clouds as dense, in reality they are highly rarified. The low particle density limits chemical reactions, restricting them to two-body processes, as opposed to three-body reactions that are more common in laboratory experiments. If one body is a positively charged ion and the other is a neutral molecule, reaction is enhanced because they are attracted to each other (that is to say, their reaction has negligible activation energy). Such ion–molecule reactions are fast and are important in ISM chemistry. These reactions occur on dust grains, which develop mantles of condensed ices at the low temperatures (~10 K) of interstellar space. Ultraviolet-light photons and cosmic-ray interactions dissociate molecules in the icy mantles into free radicals, which then react with other molecules. The end result of these ion–molecule reactions is the synthesis of a variety of organic molecules, commonly unsaturated. Ion–molecule reactions are also a route by which significant deuterium and ^{15}N enhancements occur in organic molecules in the ISM. Isotope fractionations are most efficient at low temperatures, and hydrogen has the potential for the most fractionation because of the high relative mass difference of the $^{2}H/^{1}H$ pair, relative to $^{15}N/^{14}N$ and $^{13}C/^{12}C$. Spectra of interstellar dust show dramatic D/H enrichment, which might serve as a fingerprint for ISM synthesis of organic molecules.

Some of the PAHs in the ISM may have a different origin: formation in carbon-rich stars. The spectra of interstellar dust reveal that PAHs are widespread but constitute only 5–19% of total carbon.

Another possibility is that organic matter in meteorites formed within the protoplanetary disk. Synthesis reactions like those in the ISM could occur on the surfaces of nebular grains in distant, cold parts of the disk. As in the ISM, ion–molecule reactions could generate fractionations, so high D/H and $\delta^{15}N$ in organic materials may not be due to interstellar processes. High-energy, ionizing radiation in parts of the disk could have caused dehydrogenation of molecules and breaking of aromatic rings. However, fractionations in the solar nebula might be more modest than those produced in frigid molecular clouds.

Chemical reactions on the surfaces of planetesimals are inherently different from those that occur in space. Small bodies with low gravity tend to lose hydrogen preferentially, so that their surfaces become oxidizing, and organic molecules synthesized there would be relatively oxidized. Laboratory studies demonstrate that irradiation can produce complex organic molecules (and also break them down). Heating by short-lived radionuclides causes ices incorporated into planetesimals to melt, producing aqueous fluids that could also promote synthesis and destruction of organics within the interiors of asteroids. At least some of the amino acids in Murchison are thought to have formed in the parent asteroid. Studies of CR chondrites with differing degrees of alteration (Changela et al., 2018) and of clasts altered to varying degrees in the Tagish Lake meteorite breccia (Quirico et al., 2018) show that soluble organic matter was polymerized to IOM by hydrothermal processing in the parent asteroid.

In summary, interstellar chemistry is dominated by ion–molecule reactions on grains at low temperatures, producing radicals and unsaturated organic molecules with high deuterium enrichment and potential enrichment in ^{15}N. Nebular chemistry could feature similar reactions, but the nebula was also the site of high-energy processes that can destroy molecules or cause loss of hydrogen. Cosmic-ray irradiation of planetesimal surfaces produces complex organic molecules, as do reactions with aqueous fluids circulating within planetesimal interiors. The organic matter in carbonaceous chondrites appears to have formed by random recombination of small free radicals, producing a great variety of structural isomers. Isotopic evidence suggests that ion–molecule reactions in the ISM played an important role. The resulting compounds, when exposed to liquid water in chondrite parent bodies, reacted to form compounds that resemble those in the meteorites. Thus, the organic matter in asteroid materials probably owes its properties to formation and processing in several cosmic environments (Pizzarello and Shock, 2010). The organic matter in comets (briefly discussed below) appears to be more primitive and may be more closely related to the molecules inherited from the ISM.

11.9 Ices, Noble Gases, and Organic Matter in Planetesimals and Planets

Ices in the solar system are much more common than liquids because temperatures and pressures are not generally consistent with liquid stability. Noble gases are trapped within ices in small bodies and are outgassed from bodies undergoing igneous activity. Organic matter is ubiquitous, but samples are required to detect and identify most compounds.

11.9.1 Asteroids

Ices likely occur in many asteroids located in the outermost asteroid belt, and telescopic spectroscopy has identified water ice in a few asteroids (e.g. Campins et al., 2010). The observation of comet-like effusions of gas in a few asteroids further indicates ices, and the former presence of water ice in carbonaceous chondrite parent bodies is inferred from the aqueous alteration they have experienced.

The best measurements of noble gases in asteroids come, of course, from meteorite samples derived from them, as already discussed. A few small grains of the regolith of asteroid Itokawa, returned by the Hayabusa spacecraft, have been analyzed for He, Ne, and Ar, which represent implanted solar wind (Nagao et al., 2013).

Organic matter has been identified in a few asteroids by visible/near-infrared spectra and searched for unsuccessfully in Itokawa samples (Naraoka et al., 2012). An absorption band at 3.4 μm in reflectance spectra of IOM deepens with increasing H/C for H/C >0.4 (Kaplan et al., 2019). This band has been detected in only a few large asteroids: 24 Themis (Campins et al., 2010), 65 Cybele (Licandro et al., 2011), and dwarf planet Ceres (De Sanctis et al., 2017). This aliphatic absorption is also seen in bulk carbonaceous chondrite spectra, though aromatic bands are not seen. Analyses of meteorites still provide the best – although indirect – characterizations of organic matter in asteroids. We have already discussed the well-studied carbonaceous chondrites, but lesser amounts of organic matter have also been described in other chondrite classes (e.g. Chan et al., 2012).

Future analyses of samples returned by Hayabusa2 and OSIRIS-REx should improve the database of noble gases and organic matter in asteroids.

11.9.2 Comets

Depending on conditions, frozen volatiles in comet nuclei can be crystalline ices, amorphous ices, and *clathrate hydrates* (compounds in which cages in the ice lattice can host guest molecules). Water is clearly the most important ice in comets. H_2O that condenses below 120 K is amorphous, but upon heating it crystallizes into cubic (above ~150 K) or hexagonal (above ~200 K) crystals. Amorphous ice has a remarkable ability to accommodate other volatiles such as CO and noble gases; as much as 5% of its mass can be guest molecules. The transformation of amorphous to crystalline ice is exothermic, providing heat for sublimation and the release of volatiles. Although models of cometary activity often assume that amorphous ice is present, there is some doubt that amorphous ice could actually have formed in the nebula. Even if amorphous ice was originally incorporated into comets, various heat sources (especially the decay of short-lived radionuclides like ^{26}Al) may have caused the ice to crystallize.

Compositions of the ices in comets have been determined from telescopic observations. Spectral line measurements of gases in a comet's coma allow the identification of molecules and radicals. An inherent difficulty in spectral measurements is that volatiles in the coma are commonly broken apart by photolysis, so it is sometimes difficult to discern the parent species. Spectroscopically determined volatiles in cometary comae include H_2O, CO, CH_4, C_3H_6, HCN, NH_3, H_2CO, CH_3OH, and CH_4CN. Some of these compounds may have formed by degrading more complex molecules. Cometary outbursts occurring at great distances from the Sun (corresponding to temperatures <100 K) are probably controlled by ices more volatile than H_2O. For example, comet Hale–Bopp exhibited emission of highly volatile CO at great solar distances. Trapped CO was presumably released by crystallization of amorphous ice or sublimation of CO ice crystals at very low temperatures. Spectroscopy of solid comet nuclei is more difficult owing to their faintness. Ices and organic compounds commonly have diagnostic spectral features in the near-infrared, but models of these spectra are complex and nonunique.

Other volatile species occur in lesser abundances. The most comprehensive analysis of cometary volatiles to date is for the long-period (Oort cloud) comet Hale–Bopp (Bockelée-Morvan et al., 2003). Hale–Bopp was the largest comet to pass through the inner solar system in the last 500 years. Its relative abundances of volatiles, expressed as production rates of molecules compared to H_2O, are given in Table 11.2. The largest molecules identified were

Table 11.2 Chemistry of comet Hale-Bopp, based on production rates of molecules, relative to water.

Molecule	X/H_2O
H_2O	100
HDO	0.06
CO	23
CO_2	20
CH_4	0.6
C_2H_2	0.2
CH_3OH	2.4
H_2CO	1.1
HCOOH	0.09
NH_3	0.7
HCN	0.25
DCN	6×10^{-4}
HNCO	0.10
HNC	0.04
CH_3CN	0.02
HC_3N	0.02
NH_2CHO	0.015
H_2S	1.5
OCS	0.4
SO	0.3
CS	0.2
SO_2	0.2
H_2CS	0.02
NS	0.02

After Bockelee-Morvan and Crovisier (2002).

aromatic compounds with three and four rings (phenanthrene and pyrene, Fig. 11.8) (Bertaux and Lallement, 2017). These compounds and their abundances are similar to those inferred for interstellar ices and molecular clouds. Such similarity suggests that the molecules comprising comets may have formed in extrasolar environments and were transported into the solar nebula, where they were trapped in condensing ices.

Several spacecraft missions that have flown by comets carried mass spectrometers to identify gas molecules. However, identification was sometimes hampered by overlaps in the masses of certain species; for example, CO, N_2, and C_2H_4 all have masses of 28 atomic mass units. The collision of the impactor from the Deep Impact spacecraft with comet Tempel 1 excavated material from the interior that contained higher concentrations of organic matter than the surface materials. The Rosetta spacecraft orbited comet Churyumov–Gerasimenko (Fig. 11.13) and detected water ice on the surface, as well as H_2O, CO_2, CO, and O_2 in the coma (Le Roy et al., 2015).

The major isotopes of noble gases (argon, krypton, xenon) were detected in the coma of comet Churyumov–Gerasimenko by a mass spectrometer on board Rosetta (Rubin et al., 2018). Argon and krypton were approximately solar in composition, but xenon was isotopically distinct (Marty et al., 2017). Attempts have been made to analyze He and Ne in Stardust particle tracks, and the isotopic compositions resemble the Q component and solar wind (Marty et al., 2008). Interplanetary dust particles that are thought to be comet grains have been analyzed for He, Ne, Ar, and Xe; they are generally isotopically similar to Q, although most of the efforts have focused on helium to determine cosmic-ray exposure and heating during atmospheric entry to confirm their cometary origin (Nier and Schlutter, 1993).

Direct analysis of organic matter in particles derived from comet nuclei has been accomplished by studying IDPs, samples returned by the Stardust spacecraft from comet Wild 2, and *in situ* surface dust from comet Churyumov–Gerasimenko by Rosetta's Philae lander. Chondritic porous IDPs are usually coated with organic matter consisting of aromatic and aliphatic compounds (Flynn et al., 2013). Organic compounds in Stardust samples (Sandford et al., 2006) are similar to the organics in IDPs, but are relatively enriched in oxygen- and nitrogen-containing compounds and poorer in aromatics. In some grains, PAHs consisting of one (benzene) and two (naphthalene) rings occur without accompanying PAHs with greater numbers of rings, which distinguishes them from the organic matter in CM chondrites. However, these PAHs might have been synthesized by heating during collection. Other grains contain PAHs consisting of two-, three-, and four-ring aromatics, many of which contain alkane functional groups. These may be indigenous to the comet. Aliphatic compounds include amines and one amino acid. At least 16 organic compounds, including many nitrogen-bearing but no sulfur-bearing molecules, were detected on a comet surface by the Philae lander (Goesmann et al., 2015).

Extreme enrichments in D/H (Fig. 11.14) and ^{15}N (McKeegan et al., 2006; Bertaux and Lallement, 2017) suggest that the organic molecules, or at least their precursors, in comet spectra and in comet samples may have originated in cold molecular clouds.

Figure 11.13 Rosetta image of the nucleus of comet Churyumov–Gerasimenko, showing loss of volatiles from its surface. ESA image.

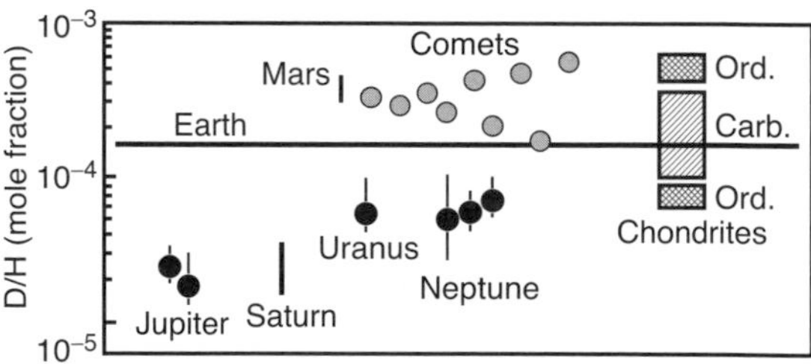

Figure 11.14 Hydrogen isotopic compositions, expressed as molar D/H ratios, of solar system bodies and chondrites. Ratios of D/H in comets show deuterium enrichments relative to Earth's oceans (horizontal line). The relatively low D/H values in the atmospheres of Jupiter and Saturn are similar to those in the early Sun, whereas D/H ratios for Uranus and Neptune are intermediate between Jupiter and Saturn and those of comets. Mars values are from martian meteorites. Adapted from McKeegan et al. (2006) and Bertaux and Lallement (2017).

11.9.3 The Moon, Mercury, and Venus

Although the Moon and Mercury have long been viewed as dry, permanently shadowed regions in craters near the poles on both bodies are very cold (<120 K) and act as cold traps for volatiles. Neutron spectrometers and radar measurements on a number of lunar orbiters and on MESSENGER have detected enhanced hydrogen in these shaded regions (Lawrence, 2017). The Lunar Reconnaissance Orbiter Centaur upper stage flew to and impacted the Moon, closely followed by the LCROSS experiment, which measured the composition of the impact plume. LCROSS detected water and other volatiles, including CO_2, sulfur species, and hydrocarbons (Schultz et al., 2010). The Moon Mineralogy Mapper spectrometer aboard the Chandrayaan-1 spacecraft also found water ice in permanently shadowed craters (Shuai et al., 2018). The water may have been delivered to the Moon and Mercury by comets, or perhaps produced when solar wind hydrogen ions reacted with oxygen in minerals. In either case, the resulting water vapor could migrate to the poles and condense there. Volatile abundances on Mercury are much higher than on the Moon.

We have already discussed the noble gases implanted in lunar soils, which have been instrumental in understanding the solar wind. An indigenous noble gas component in lunar rocks has eluded investigators, although Bekaert et al. (2017) suggested that an isotopically unusual Xe component in anorthosites might qualify. The LADEE spacecraft measured He, Ar, and Ne in the Moon's exosphere. Owing to the fact that Mercury has almost no atmosphere, the only noble gas that has been detected there is helium. The abundances of noble gases, as well as CO_2, N_2, and H_2O, in Venus' atmosphere were discussed earlier.

Very low concentrations of amino acids found in lunar soils returned by the Apollo missions are dominated by L-enantiomers, suggesting that they represent terrestrial contamination. We know nothing of organic matter on Mercury. Although it has been postulated that hydrocarbons exist on Venus, there is as yet no supporting spectroscopic evidence other than a reported discovery of phosphene.

11.9.4 Mars

Mars has two polar caps composed of H_2O and CO_2 ices, and carbon dioxide cycles seasonally between the poles. The total volume of ices in both caps is $>3 \times 10^6$ km^3. Bright radar reflections discovered by the Mars Express orbiter might suggest that a lake of liquid water may exist below the south polar cap (Orosei et al., 2018). Glaciers of water ice occur on the upper slopes of the giant Tharsis volcanoes. Orbital measurements of hydrogen by Mars Odyssey (Fig. 11.15) (Feldman et al., 2004) revealed extensive subsurface water ice at mid- to high latitudes. Water in the parent magmas of martian meteorites constrain H_2O contents for bulk Mars (137 ppm), the mantle (14–250 ppm), and crust (1410 ppm) (McCubbin et al., 2016).

Noble gas abundances and isotopic compositions have been measured in the present martian atmosphere by the Viking landers and more recently by the Sample Analysis at Mars (SAM) instrument on the Curiosity rover (Ott et al., 2019). Noble gas isotopes analyzed by Viking provided the necessary comparison to confirm the identity of martian meteorites, but SAM provided more

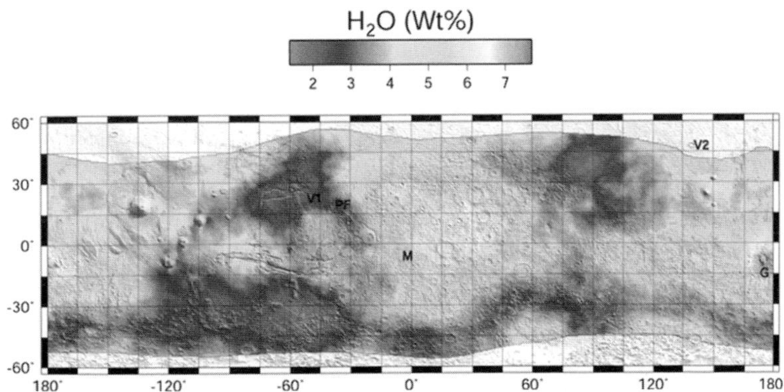

Figure 11.15 Global distribution map of water, calculated from measured hydrogen, on Mars. Hydrogen was mapped in the upper few meters of the surface by the Neutron and Gamma Ray Spectrometer on Mars Global Surveyor. Near the equator, the water occurs as permafrost and bound in minerals, but it only occurs as permafrost at higher latitudes. From Boynton et al. (2008), with permission. *For the colour version, refer to the plate section.*

complete measurements of atmospheric Ar, Kr, and Xe (Atreya et al., 2013; Conrad et al., 2016). These measurements of noble gases in the modern atmosphere can be compared with ancient atmosphere by analyzing gases trapped in martian meteorites (Pepin, 2006). The isotopic difference between the atmosphere and the interior on Mars is unlike the difference observed on Earth and points to distinctive volatile outgassing and loss histories.

Organic compounds were not found by life-detection experiments on the Viking landers on Mars, which led to the idea that any organic matter in the soil may have been destroyed by ultraviolet radiation or highly reactive compounds. However, organic compounds have been discovered by the SAM instrument on the Curiosity rover (Freissinet et al., 2015; Eigenbrode et al., 2018). The identified compounds included simple aromatic molecules and sulfur-bearing compounds, consistent with the breakdown of either abiotic organic matter that occurs in carbonaceous chondrites or decomposed organisms. Periodic seasonal releases of methane (CH_4) in the martian atmosphere have also been found (Webster et al., 2018), sourced from surface or subsurface reservoirs; however, to date, the Trace Gas Orbiter has found no sign of methane higher in the atmosphere. Organic matter has been analyzed in martian meteorites, and was one of the lines of evidence cited to support the hypothesis for extraterrestrial life in the ALH 84001 meteorite (see Box 11.3). A comprehensive review of the organic carbon in shergottites, nakhlites, and other martian meteorites (Steele et al., 2016) found amorphous carbon grading to graphite, amino acids and other aliphatic compounds, and PAHs, and concluded that its origin is best described as a combination of abiotic organic synthesis and infall of exogenic carbonaceous material.

11.9.5 The Giant Planets

As noted earlier, the atmospheres of the giant planets consist mostly of hydrogen and helium. Jupiter and Saturn are "gas giants" in which molecular hydrogen in the interior becomes metallic (ultradense and electrically conducting) hydrogen at high pressures. Conversely, Uranus and Neptune are "ice giants" in which volatiles occur mostly as ices. Helium variations in the giant planet atmospheres may reflect depletions relative to hydrogen,

Box 11.3 Cosmochemical Controversies about Life in Meteorites

At a meeting of the New York Academy of Sciences in 1961, Nagy et al. presented a paper comparing the chemistry of organic matter in the Orgueil CI chondrite to that of living tissue. They concluded that life must have existed on the parent body from which the meteorite came. Eight months later, Claus and Nagy (1961) reported the discovery of tiny clumps of organic matter, which they called "organized elements," in Orgueil. Moreover, they noted that this organic material exhibited chirality, as does organic matter from terrestrial organisms. They suggested that these microscopic clumps were the fossil remains of extraterrestrial organisms. A firestorm of media attention followed, prompting other scientists to search for organized elements in Orgueil. Detailed studies, however, showed that most of these tiny particles were actually oddly shaped mineral grains, not organic matter. A few were demonstrably from living things: pollen grains of ragweed and a few starch particles. The hypothesis crashed on the realization that the organized elements were either abiotic or products of the current denizens of Earth. The meteorite had apparently been contaminated before it was collected or while displayed in less-than-sterile museum cases.

In 1996, a NASA team led by David McKay published a fascinating but controversial report (McKay et al., 1996) describing putative biochemical markers, biogenic minerals, and possible microfossils in the Mars meteorite ALH 84001. This meteorite, an ancient (~4.09 Ga) achondrite, contains chemically zoned carbonate globules (Fig. 11.16a) that McKay and coworkers suggested were precipitated by organisms. The carbonates contain very small quantities of PAHs, which they interpreted as the decayed products of such organisms. However, PAHs commonly form by abiotic processes, as noted earlier, and the PAHs in ALH 84001 comprise all the structural forms ranging from C_{14} to C_{22}, unlike the selected hydrocarbon molecules formed by the decay of terrestrial organisms. Steele et al. (2012) also found amorphous to poorly ordered graphitic carbon in ALH 84001; its enclosure in high-temperature igneous crystals demonstrates its magmatic (abiotic) origin. To further confound the interpretation of organic matter, terrestrial organisms have contaminated the meteorite, and the amino acids in it can be explained as terrestrial contamination; moreover, the measured carbon isotopic composition of organic matter ($\delta^{13}C = -22$ to -25 permil) is mostly indistinguishable from terrestrial values. Other evidence cited by McKay and coworkers included very tiny (nanophase) magnetite crystals with unusual morphologies. They proposed that these crystals were produced by bacteria. On Earth, magnetotactic bacteria form chains of tiny magnetite crystals in their cells, used to orient the

Box 11.3 (cont.)

organisms relative to the magnetic field. The magnetites in ALH 84001 do not occur in chains. Moreover, TEM images of them (Fig. 11.16b) actually show a variety of morphologies, including twisted whiskers and platelets – morphologies seen in crystals formed by vapor condensation. The most plausible origin for these magnetites appears to be thermal decomposition of siderite (iron carbonate) in the globules during an impact. Siderite is more readily affected by shock than magnesium or calcium carbonates, and an impact could have produced a vapor from which magnetite condensed. Finally, some intriguing elongated, apparently segmented forms (Fig. 11.16c), described as possible microbial fossils by McKay and coworkers, have been subsequently identified by other workers as either magnetite whiskers or man-made features resulting from gold coating of the sample prior to SEM examination. Whatever their origin, the putative fossils appear to be far too small to represent living forms and could at best be appendages or fragments of organisms. The controversy over life in this martian meteorite has now largely subsided, with the consensus that abiotic processes provide more compelling explanations for all the observed evidence. However, the interest in ALH 84001 has generated new research into the authenticity of various chemical and physical biomarkers and spawned a reinvigorated Mars exploration program focused on the search for life. McSween (2019) provided a complete summary of findings from the voluminous, and often conflicting, literature on ALH 84001.

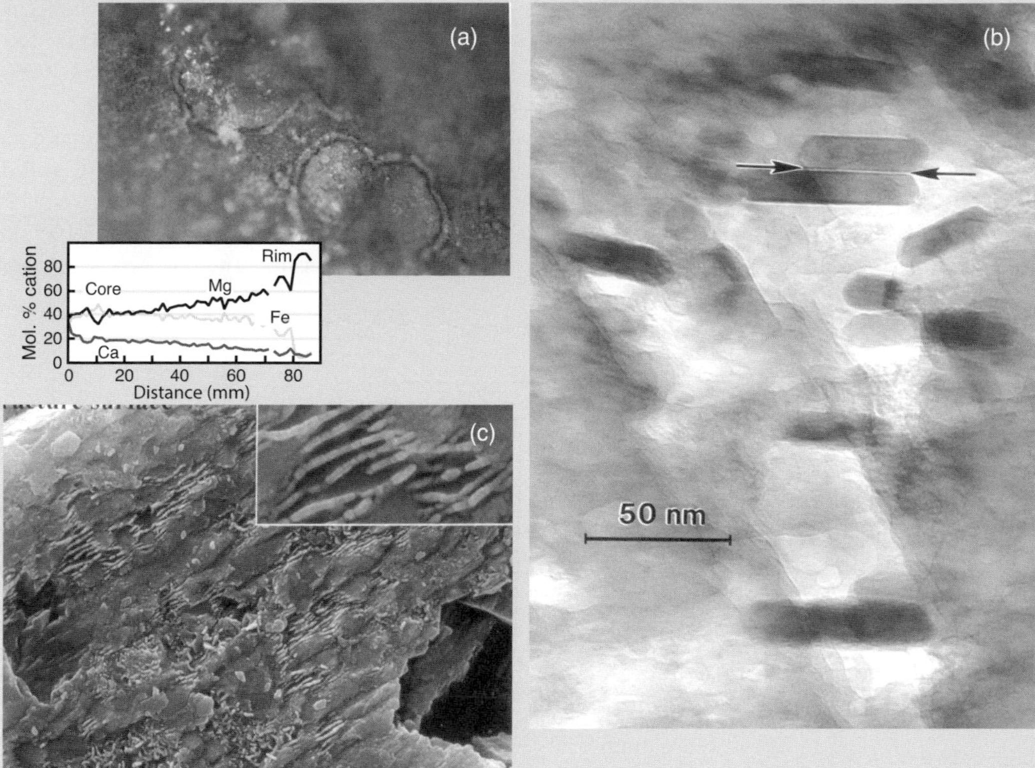

Figure 11.16 Images of the ALH 84001 Martian meteorite. (a) Backscattered electron image of a zoned carbonate globule. Compositional profiles for magnesium, iron, and calcium in one carbonate grain, measured by electron microprobe, are shown in the inset. (b) TEM image of elongated magnetite grain in ALH 84001 carbonate. (c) Electron microscope image of objects in ALH 84001 carbonate previously misinterpreted as microfossils. Images in (b) and (c) courtesy of J. Bradley.

compared to solar system abundances; other noble gases have only been measured for Jupiter. Methane and several photochemically produced hydrocarbons such as acetylene (C_2H_2) and ethane (C_2H_6) derived from methane have been found in the atmospheres of Jupiter, Saturn, Uranus, and Neptune. The D/H ratios of the giant planets are lower than that of the Earth (Fig. 11.14), more like the solar nebula. The D/H ratios for Uranus and Neptune are higher than for Jupiter and Saturn, thought to reflect their accretion in ices as opposed to nebular gas. Some forms of ices in the interiors of ice giants may be unlike those with which we are familiar. For example, "superionic" water ice, in which hydrogen is free to move around within the lattice of oxygen, has been documented in high-pressure experiments and suggested to occur in Neptune and Uranus (Millot et al., 2018).

11.9.6 Ocean Worlds: Icy Moons and Dwarf Planets

The dwarf planets Ceres and Pluto, as well as most of the moons of the giant planets, contain considerable quantities of ices. In some cases, they have subsurface liquid water (leading to the designation "ocean worlds"). With the exception of Enceladus, discussed below, the compositions of subsurface oceans are unknown. Induction signals detected for the Galilean satellites indicated that dissolved ions must be present in their oceans, because pure water has very low conductivity.

For small icy bodies, any subsurface ocean will rest directly atop a silicate rock core and be covered by a solid ice shell. The ocean can thus react with the core if it is permeable, perhaps accounting for the dissolved ions. For larger icy satellites, internal pressures are high enough to transform ices into high-pressure forms, as shown in the phase diagram in Figure 11.17 (Nimmo and Pappalardo, 2016). These phases are denser than ice and have melting temperatures that increase with pressure, so they will occur in contact with the silicate core, and the liquid ocean will be "sandwiched" between the high-pressure solid phases and the outer ice shell. If the outer shell is breached, ocean worlds may even exhibit active or recent eruptions of liquids and gases (a phenomenon called "cryovolcanism").

Triton, Neptune's only large (2700 km diameter) satellite, was first imaged by Voyager 2 (Fig. 11.18a) as the spacecraft rushed past in 1989 on its way out of the solar system. Triton's retrograde orbit and 29 degree tilt suggest that it may be a captured KBO, an inference bolstered by its spectral similarity to Pluto. Triton has a youthful surface, estimated at only ~100 Myr old. Orbital capture would likely have caused intense tidal heating, which might account for its tectonically deformed surface. Triton's frigid crust contains a veneer of methane and nitrogen ices over a substrate of water ice, and it displays active geyser-like eruptions of volatiles and entrained organic molecules. These geysers are possibly

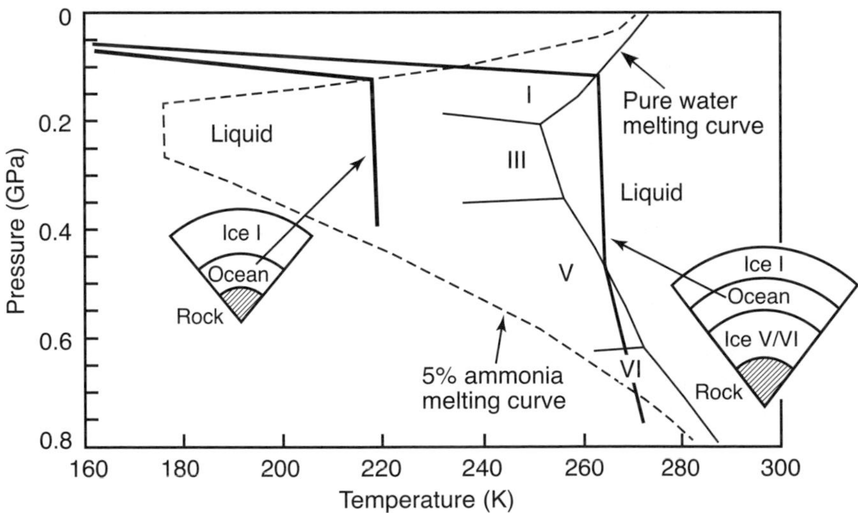

Figure 11.17 Phase diagram for water ice (I) and its high-pressure forms (III, V, VI), showing melting conditions and the effect of adding 5% ammonia on the melting curve (dashed line). Heavy solid lines are geothermal gradients for small and large icy bodies, with accompanying sketches of their interior structures. Modified from Nimmo and Pappalardo (2016).

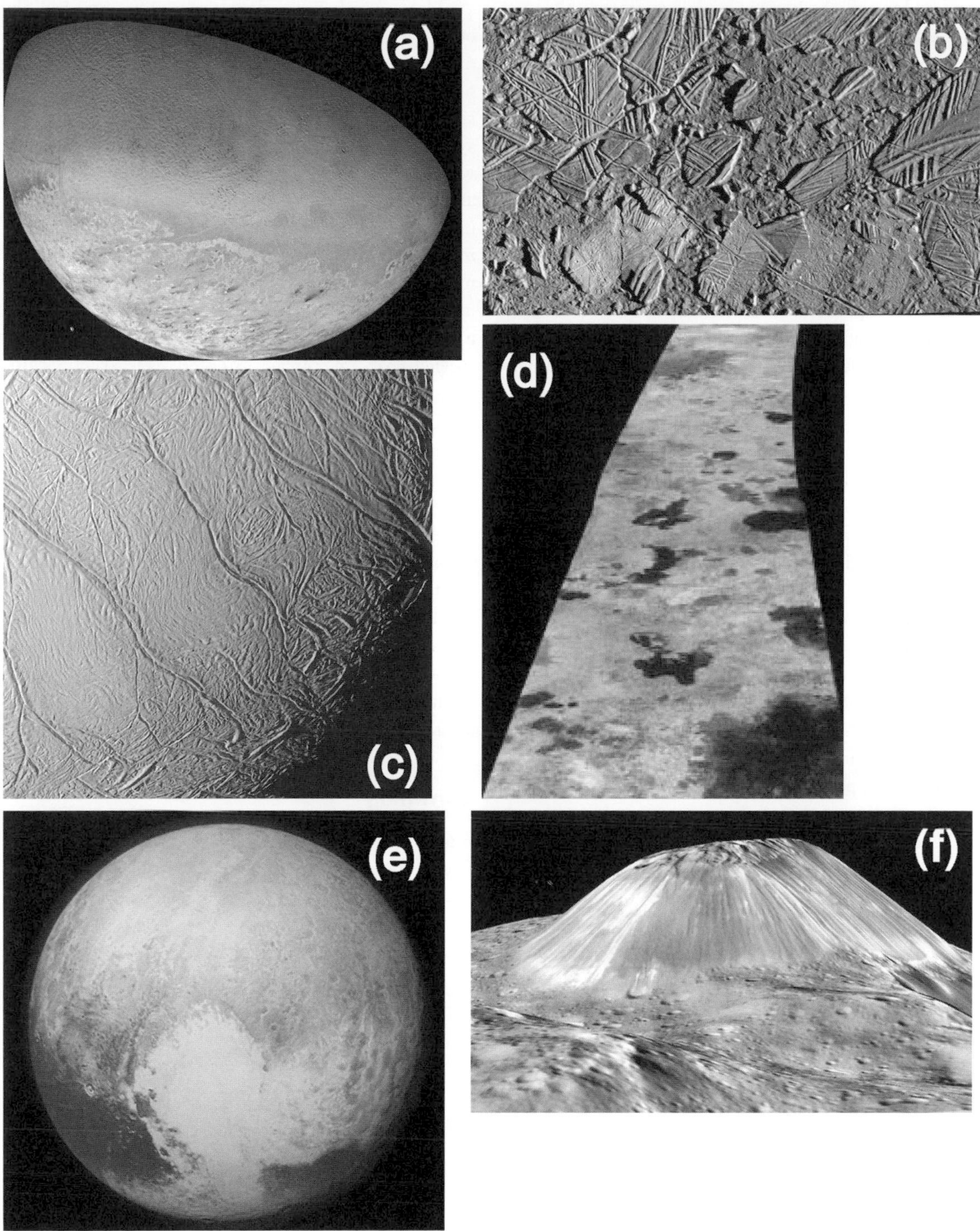

Figure 11.18 Spacecraft images of icy moons and dwarf planets. (a) Voyager 2 image of Triton, showing complex "cantaloupe" terrain and a polar cap of nitrogen ice; smooth areas near the equator may be frozen lakes. (b) Jumbled kilometer-sized blocks of water ice on the surface of Europa, revealed by Galileo imagery. Rafting of these plates suggests a substrate of water or soft ice. (c) Cassini image of Enceladus, showing warm fractures from which volatile jets are actively erupting. (d) Cassini radar image of Titan showing hydrocarbon lakes (dark). (e) New Horizons image of Pluto, showing terrains of nitrogen, methane, and carbon monoxide ices. (f) Cryovolcano on Ceres, imaged by the Dawn orbiter, that erupted cold brine which precipitated salts. NASA images.

powered by sunlight, because eruptions appear only near the latitude that receives maximum solar illumination.

Europa, a moon of Jupiter with a diameter of 3210 km, has received considerable attention because of evidence that its icy crust hides a global ocean. Infrared spectra indicate abundant water ice and some organic compounds on its surface. Spacecraft images reveal blocks of ice jumbled by swarms of rifts (Fig. 11.18b). The surface has low relief, and visible bands may be loci of cryovolcanic water eruptions. Pits and spots may represent diapirs of warmer ice rising through the colder ice of the crust. Spectra from the Galileo spacecraft suggest that reddish features on Europa may be rich in hydrated silicates and salts such as magnesium sulfate, deposited from evaporating water.

On Saturn's moon Enceladus (500 km diameter), jets of gas erupt to heights of 500 km from warm, active fractures (Fig. 11.18c) in its south polar region. This activity is probably the result of tidal heating. The Cassini spacecraft determined that water was the dominant plume constituent, although trace amounts of several organic compounds (acetylene and propane) were also measured (Postberg et al., 2009). The jets arise from a subsurface sea beneath the moon's icy crust.

Titan, Saturn's largest satellite (5150 km diameter, approximately the size of Mercury), has a Europa-sized rock core enclosed within a mantle of water ice. It has a dense (1.5 bar) nitrogen-methane atmosphere that might be compositionally similar to Triton's surface ices; Triton is so cold that much of its nitrogen and methane freezes onto the surface. Titan's orange-colored haze results from organic polymers derived from methane. Oxygen-containing molecules (CO, CO_2, and H_2O) may result from oxygen ions flowing into Titan's atmosphere from Enceladus. High-mass ions in the upper atmosphere, impossible to identify, are thought to be aromatic organic molecules like naphthalene, anthracene, and other PAHs (Horst, 2017). The Cassini–Huygens probe peered through Titan's obscuring atmosphere and revealed a young surface apparently sculpted by flowing hydrocarbon liquids and winds. Hundreds of radar-dark areas in the northern hemisphere are lakes (Fig. 11.18d), some composed of methane and others of ethane. Methane rain may feed rivers, and springs of liquid hydrocarbons are predicted to bubble up from the crust. Benzene snow is thought to fall, dissolve in the lakes, and ultimately precipitate as sludge on the bottom. Dark materials (perhaps organic powders) have been blown across the surface to collect in dunes, and rounded organic rocks suggest erosion and transport.

Pluto, which has a diameter of 2380 km, has now been demoted from being the smallest planet to being one of the largest KBOs. Pluto and its satellite Charon could be considered a binary system because they are closer in size than any other known celestial pair in the solar system and the barycenter of their orbits does not lie within either body. There are also four smaller moons, and all six bodies are likely KBOs with similar compositions. Pluto has a thin atmosphere containing N_2, with minor CH_4, CO, and Ar. Its surface is predominantly nitrogen ice, with traces of methane and carbon monoxide. Pluto (Fig. 11.18e) was studied during the flyby of the New Horizons spacecraft (Stern et al., 2018), which imaged peaks interpreted to be cryovolcanoes that may have erupted flows of water and ammonia.

Ceres, formerly the largest asteroid (930 km diameter) and now the asteroid belt's only dwarf planet, was studied extensively by the Dawn orbiter (Russell et al., 2016). It is partly differentiated, with a rocky core, a mantle of highly altered rock, and a crust composed mostly of hydrated and ammoniated phyllosilicates, carbonates, and ice. Its composition is similar to the mostly highly altered carbonaceous chondrites (McSween et al., 2018). Although its surface is mostly ancient, recent eruptions of brines formed a cryovolcano (Fig. 11.18f) and seeped from fractures in the floors of recent craters. Sublimation of these brines caused precipitation of sodium carbonate.

Summary

Admittedly, this chapter is difficult reading. We have considered some of the most complex constituents of meteorites and planets – ices which are unsampled and about which little is really known, isotopically diverse noble gases, and structurally complicated organic molecules.

Ices formed as mantles on silicate grains in interstellar space, trapping noble gases and providing sites for the synthesis of organic compounds. As the solar system formed, infalling ices were vaporized, particularly in the warmer regions near the Sun. Ices composed of H_2O, CO_2, CO, CH_4, and NH_3 condensed beyond their respective snowlines and combined with rocky material and surviving interstellar material to form planetesimals.

Noble gases in meteorites comprise an array of components with distinct origins, recognizable by their elemental and isotopic patterns. Nuclear components formed *in situ* by radioactive decay, fission, or though spallation by cosmic rays. The solar wind, which has distinctive noble gas isotope abundances, is continuously implanted into meteorites in space and into regoliths exposed on the surfaces of airless bodies. Planetary noble gases are strongly depleted in light elements, and include an assortment of exotic components found in interstellar grains and produced in other stars. Atmospheric noble gas abundances and isotopic compositions vary considerably from planet to planet, and provide constraints on the origins of planetary volatile inventories and outgassing histories.

The organic matter in carbonaceous chondrites consists of compounds that can be extracted using acids and other solutes and thus can be readily characterized, and abundant, high-molecular-weight material that is insoluble. Moderately abundant extractable compounds include aliphatic and aromatic hydrocarbons, amino acids and other kinds of organic acids, and amines. All these compounds are structurally diverse, with all possible forms present. IOM consists of many cross-linked aromatic rings. The organic material in chondrites has had a complicated history. Isotopic compositions of hydrogen, nitrogen, and carbon indicate that some of the compounds or their precursors originated in interstellar space. Nebular heating and irradiation destroyed some of the interstellar molecules and new ones were synthesized. Aqueous processing on meteorite parent bodies further modified the accreted organic compounds.

We have now set the stage for the next chapter on planetary building blocks. These objects contain the ices, noble gases, and organic matter that we have just learned about, as well as the rocky materials considered in earlier chapters, and they provide us with the best record of primitive materials in the solar system.

Questions

1. How much water ice, relative to rocky minerals, is potentially condensable from a solar gas?
2. What makes noble gases so useful in cosmochemistry?
3. What noble gas components have been identified in meteorites, and what are their origins?
4. Explain the difference between extractable and insoluble organic matter in chondrites, and note why this distinction is significant.
5. What elements are found in organic molecules and how can their isotopes help constrain their origins?
6. What observations or properties of organic molecules suggest that they formed in the interstellar medium, in the solar nebula, or in planetesimals?

Suggestions for Further Reading

Lunine J. I. (2005) Origin of water ice in the solar system. In *Meteorites and the Early Solar System II*, Lauretta D. S., and McSween H. Y., editors, pp. 309–319, University of Arizona Press, Tucson. A thoughtful review of the condensation of ices in the nebula and the delivery of ices to the terrestrial planets.

Several comprehensive reviews provide excellent summaries of the complicated cosmochemistry of noble gases; here are two:

Ott U. (2014) Planetary and pre-solar noble gases in meteorites. *Chemie der Erde*, **74**, 519–544.

Podosek F. A. (2004) Noble gases. In *Treatise on Geochemistry, Vol. 1: Meteorites, Comets, and Planets*, Davis A. M., editor, pp. 381–405, Elsevier, Oxford.

Pizzarello S., and Shock E. (2010) The organic composition of carbonaceous meteorites: The evolutionary story ahead of biochemistry. In *Cold Spring Harbor Perspectives in Biology*, Deamer D., and Szostak J. W., editors, Cold Spring Harbor Laboratory Press, Long Island, NY, doi:10.1101/cshperspect.a002105. A comprehensive review of organic matter in carbonaceous chondrites.

Other References

Alexander C. M. O'D., Cody G. D., De Gregorio B. T., et al. (2017) The nature, origin and modification of insoluble organic matter in chondrites, the major source of Earth's C and N. *Chemie der Erde-Geochemistry*, **77**, 227–256.

Atreya S. K., Trainer M. G., Franz H. B., et al. (2013) Primordial argon isotope fractionation in the atmosphere of Mars as measured by the SAM instrument on Curiosity, and implications for atmospheric loss. *Geophysical Research Letters*, **40**, 5605–5609.

Bekaert D. V., Avice G., Marty B., et al. (2017) Stepwise heating of lunar anorthosites 60025, 60215, 65315 possibly reveals an indigenous noble gas component of the Moon. *Geochimica et Cosmochimica Acta*, **218**, 114–131.

Bertaux J.-L., and Lallement R. (2017) Diffuse interstellar bands carriers and cometary organic material. *Monthly Notices of the Royal Astronomical Society*, **469**, S646–S660.

Bockelée-Morvan D., and Crovisier J. (2002) Lessons of comet Hale-Bopp for coma chemistry. *Earth, Moon, & Planets*, **89**, 53–71.

Bockelée-Morvan D., Crovisier J., Mumma M. J., and Weaver H. A. (2003) The composition of cometary volatiles. In *Comets II*, Festou M., Keller, H. U., Weaver, H. A., editors, pp. 391–423, University of Arizona Press, Tucson.

Boynton W. V., Taylor G. J., Karunatillake S., et al. (2008) Elemental abundances determined via the Mars Odyssey GRS. In *The Martian Surface: Composition, Mineralogy, and Physical Properties*, Bell J. F., editor, pp. 105–124, Cambridge University Press, Cambridge.

Busemann H., Young A. F., Alexander C. M. O'D., et al. (2006) Interstellar chemistry recorded in organic matter from primitive meteorites. *Science*, **312**, 727–730.

Campins H., Hargrove K., Pinilla-Alonso N., et al. (2010) Water ice and organics on the surface of the asteroid 24 Themis. *Nature*, **464**, 1320–1321.

Cartwright J. A. (2015) Noble gas chemistry of planetary materials. *Planetary Mineralogy, EMU Notes in Mineralogy*, **15**, 165–212.

Chan H.-S., Martins Z., and Sephton M. A. (2012) Amino acid analyses of type 3 chondrites Colony, Ornans, Chainpur, and Bishunpur. *Meteoritics & Planetary Science*, **47**, 1502–1516.

Changela H. G., Le Guillou C., Bernard S., and Brearley A. J. (2018) Hydrothermal evolution of the morphology, molecular composition, and distribution of organic matter in CR (Renazzo-type) chondrites. *Meteoritics & Planetary Science*, **53**, 1006–1029.

Claus G., and Nagy B. (1961) A microbiological examination of some carbonaceous chondrites. *Nature*, **192**, 594–596.

Cody G., and Alexander C. M. O'D. (2005) NMR studies of chemical structural variation of insoluble organic matter from different carbonaceous chondrite groups. *Geochimica et Cosmochimica Acta*, **69**, 1085–1097.

Cody G., Alexander C. M. O'D., and Tera F. (2002) Solid state (^{1}H and ^{13}C) NMR spectroscopy of the insoluble organic residue in the Murchison meteorite: A self-consistent quantitative analysis. *Geochimica et Cosmochimica Acta,* **66**, 1851–1865.

Conrad P. G., Malespin C. A., Franz H. B., et al. (2016) In situ measurement of atmospheric krypton and xenon on Mars with Mars Science Laboratory. *Earth & Planetary Science Letters*, **454**, 1–9.

De Sanctis M. C., Ammannito E., McSween H. Y., et al. (2017) Localized aliphatic organic material on the surface of Ceres. *Science*, **355**, 719–722.

Eigenbrode J. L., Summons R. E., Steele A., et al. (2018) Organic matter preserved in 3-billion-year-old mudstones at Gale crater, Mars. *Science*, **360**, 1096–1101.

Feldman W. C., Prettyman T. H., Maurice S., et al. (2004) The global distribution of near-surface hydrogen on Mars. *Journal of Geophysical Research*, **109**, E09006.

Flynn G. J., Wirick S., and Keller L. P. (2013) Organic grain coatings in primitive interplanetary dust particles: Implications for grain sticking in the solar nebula. *Earth, Planets, & Space*, **65**, 1159–1166.

Freissinet C., Glavin D. P., Mahaffy P. R., et al. (2015) Organic molecules in the Sheepbed mudstone, Gale Crater, Mars. *Journal of Geophysical Research, Planets*, **120**, 495–514.

Garvie L. A. J., and Buseck P. R. (2004). Nanosized carbon-rich grains in carbonaceous chondrite meteorites. *Earth & Planetary Science Letters*, **184**, 9–21.

Gilmour I. (2004) Structural and isotopic analysis of organic matter in carbonaceous chondrites. In *Treatise on Geochemistry, Vol. 1: Meteorites, Comets, and Planets*, Davis A. M., editor, pp. 269–290, Elsevier, Oxford.

Glavin D. P., McLain H. L., Dworkin J. P., et al. (2020) Abundant extraterrestrial amino acids in the primitive CM carbonaceous chondrite Asuka 12236. *Meteoritics & Planetary Science*, **55**, 1979–2006.

Goessmann F., Rosenbauer H., Bredehoft J. H., et al. (2015) Organic compounds on comet 67P/Churyumov-Gerasimenko revealed by COSAC mass spectrometry. *Science*, **349**, doi:10.1126/science.aab0689.

Horst S. M. (2017) Titan's atmosphere and climate. *Journal of Geophysical Research, Planets*, **122**, 432–482.

Huss G. R., and Lewis R. S. (1995) Presolar diamond, SiC, and graphite in primitive chondrites: Abundances as a function of meteorite class and petrologic type. *Geochimica et Cosmochimica Acta*, **59**, 115–160.

Kaplan H. H., Milliken R. E., Alexander C. M. O'D., and Herd C. D. K. (2019) Reflectance spectroscopy of insoluble organic matter (IOM) and carbonaceous chondrites. *Meteoritics & Planetary Science*, **54**, 1051–1068.

Lawrence D. J. (2017) A tale of two poles: Toward understanding the presence, distribution, and origin of volatiles at the polar regions of the Moon and Mercury. *Journal of Geophysical Research, Planets*, **122**, 21–52.

Le Roy L., Atwegg K., Balsiger H., et al. (2015) Inventory of the volatiles on comet 67P/Churyumov-Gerasimenko from Rosetta/ROSINA. *Astronomy & Astrophysics*, **583**, A1.

Licandro J., Campins H., Kelly M., Hargrove K., et al. (2011) 65 Cybele: Detection of small silicate grains, water-ice, and organics. *Astronomy & Astrophysics*, **525**, A34.

Mahaffy P. R., Niemann H. B., Alpert A., et al. (2000) Noble gas abundance and isotope ratios in the atmosphere of Jupiter from the Galileo Probe Mass Spectrometer. *Journal of Geophysical Research, Planets*, **105**, 15061–15071.

Marty B., Palma R. L., Pepin R. O., et al. (2008) Helium and neon abundances and compositions in cometary matter. *Science*, **319**, 75–78.

Marty B., Altwegg K., Balsinger H., et al. (2017) Xenon isotopes in 67P/Churyumov-Gerasimenko show that comets contributed to Earth's atmosphere. *Science*, **354**, 1069–1072.

McCubbin F. M., Boyce J. W., Srinivasan P., et al. (2016) Heterogeneous distribution of H_2O in the martian interior: Implications for the abundance of H_2O in depleted and enriched mantle sources. *Meteoritics & Planetary Science*, **51**, 2036–2060.

McKay D. S., Gibson E. K., Thomas-Keprta K. L., et al. (1996) Search for past life on Mars: Possible relic biogenic activity in martian meteorite ALH84001. *Science*, **273**, 924–930.

McKeegan K. D., Aléon J., Bradley J., et al. (2006) Isotopic compositions of cometary matter returned by Stardust. *Science*, **314**, 1724–1728.

McSween H. Y. (2019) The search for biosignatures in martian meteorite Allan Hills 84001. In *Biosignatures for Astrobiology*, Cavalazzi B., and Westall F., editors, pp. 167–182, Springer Nature, Switzerland.

McSween H. Y., Emery J. P., Rivkin A. S., et al. (2018) Carbonaceous chondrites as analogs for the composition and alteration of Ceres. *Meteoritics & Planetary Science*, **53**, 1793–1804.

Meshik A., Hohenberg C., Pravdivtseva O., and Burnett D. (2014) Heavy noble gases in solar wind delivered by Genesis mission. *Geochimica et Cosmochimica Acta*, **127**, 326–347.

Millot M., Hamel S., Rugg J. R., et al. (2018) Experimental evidence for superionic water ice using shock compression. *Nature Physics*, **14**, 297–302.

Nagao K., Okazaki R., Miura Y. N., et al. (2013) Noble gas analysis of two Hayabusa samples as the first international A/OP investigation: a progress report. *Lunar and Planetary Science*, **44**, abstract #1976.

Nagy B., Meinschein W. G., and Henessy D. J. (1961) Mass spectroscopic analysis of the Orgueil meteorite: Evidence for biogenic hydrocarbons. *Annals of the New York Academy of Sciences*, **93**, 27–35.

Naraoka H., Mita H., Hamase M., et al. (2012) Preliminary organic compound analysis of microparticles returned from asteroid 25143 Itokawa by the Hayabusa mission. *Geochemical Journal*, **46**, 61–72.

Nier A. O., and Schlutter D. J. (1993) The thermal history of interplanetary dust particles collected in the Earth's stratosphere. *Meteoritics*, **28**, 675–681.

Nimmo F., and Pappalardo R. T. (2016) Ocean worlds in the outer solar system. *Journal of Geophysical Research, Planets*, **121**, 1378–1399.

Orosei R., Lauro S. E., Pettinelli E., et al. (2018) Radar evidence of subglacial liquid water on Mars. *Science*, **361**, 490–493.

Ott U. (2014) Planetary and pre-solar noble gases in meteorites. *Chemie der Erde*, **74**, 519–544.

Ott U., Swindle T. D., and Schwenzer S. P. (2019) Noble gases in martian meteorites: Budget, sources, sinks, and processes. In *Volatiles in the Martian Crust*, Filiberto J., and Schwenzer S. P., editors, pp. 35–70, Elsevier, Oxford.

Ozima M., and Podosek F. (2002) Noble Gas Geochemistry, 2nd edition, Cambridge University Press, Cambridge, 286 pp.

Pepin R. O. (2006) Atmospheres on the terrestrial planets: Clues to origin and evolution. *Earth & Planetary Science Letters*, **252**, 1–14.

Pepin R. O., Schlutter D. J., Becker R. H., and Reisenfeld D. B. (2012) Helium, neon, and argon composition of the solar wind as recorded in gold and other Genesis collector materials. *Geochimica et Cosmochimica Acta*, **89**, 62–80.

Pizzarello S., Huang Y., Becker L., et al. (2001) The organic content of the Tagish Lake meteorite. *Science*, **293**, 2236–2239.

Pizzarello S., Cooper G. W., and Flynn G. J. (2006) The nature and distribution of the organic material in carbonaceous chondrites and interplanetary dust particles. In *Meteorites and the Early Solar System II*, Lauretta D. S., and McSween H. Y., editors, pp. 625–651,University of Arizona Press, Tucson.

Pizzarello S., Davidowski S. K., Holland G. P., and Williams L. B. (2013) Processing of meteoritic organic materials as a possible analog of early molecular evolution in planetary environments. *Proceedings of the National Academy of Sciences, USA*, **110**, 15614–15619.

Porcelli D., and Pepin R. O. (2004) The origin of noble gases and major volatiles in the terrestrial planets. In *Treatise on Geochemistry, Vol. 4: The Atmosphere*, Keeling R. F., editor, pp. 319–344, Elsevier, Oxford.

Postberg F. S., Kempf J., Schmidt N. et al. (2009) Sodium salts in E-ring ice grains from an ocean below the surface of Enceladus. *Nature*, **459**, 1098–1101.

Quirico E., Bonal L., Beck P., Alexander C. M. O'D., et al. (2018) Prevalence and nature of heating processes in CM and C2-ungrouped chondrites as revealed by insoluble organic matter. *Geochimica et Cosmochimica Acta*, **241**, 17–37.

Remusat L. (2015) Organics in primitive meteorites. *Planetary Mineralogy, EMU Notes in Mineralogy*, **15**, 33–65.

Rubin M., Altwegg K., Balsiger H., et al. (2018) Krypton isotopes and noble gas abundances in the coma of comet 67P/Churyumov-Gerasimenko. *Science Advances*, **4**, doi:10.1126/sciadv.aar6297.

Russell C. T., Raymond C. A., Ammannito E., et al. (2016) Dawn arrives at Ceres: Exploration of a small, volatile-rich world. *Science*, **353**, 1008–1010.

Sandford S. A., and 56 coauthors (2006) Organics captured from Comet 81P/Wild 2 by the Stardust spacecraft. *Science*, **314**, 1720–1724.

Schultz P. H., Hermalyn B., Colaprete A., et al. (2010) The LCROSS cratering experiment. *Science*, **330**, 468–472.

Sephton M., and Gilmour I. (2000) Aromatic moieties in meteorites: Relicts of interstellar grain processes? *Astrophysical Journal*, **540**, 588–591.

Shuai L., Lucey P. G., Milliken R. E., et al. (2018) Direct evidence of surface exposed water ice in the lunar polar regions. *Proceedings of the National Academy of Sciences, USA*, **115**, 8907–8912.

Signer P., and Suess H. (1963) Rare gases in the Sun, in the atmosphere, and in meteorites. In *Earth Science and Meteorites*, Geiss J., and Goldberg E. D., editors, pp. 241–272, North Holland, Amsterdam.

Steele A., McCubbin F. M., Fries M., et al. (2012) A reduced organic carbon component in martian basalts. *Science*, **337**, 212–215.

Steele A., McCubbin F. M., and Fries M. D. (2016) The provenance, formation, and implications of reduced carbon phases in martian meteorites. *Meteoritics & Planetary Science*, **51**, 2203–2225.

Stern S. A., Grundy W. M., McKinnon W. G., et al. (2018) The Pluto system after New Horizons. *Annual Review of Astronomy and Astrophysics*, **56**, 357–392.

Stone S. W., Yelle R. V., Benna M., et al. (2020) Hydrogen escape from Mars is driven by seasonal and dust storm transport of water. *Science*, **370**, 824–831.

Webster C. R., Mahaffy P. R., Atreya S. K., et al. (2018) Background levels of methane in Mars' atmosphere show strong seasonal variations. *Science*, **360**, 1093–1096.

Wieler R., Busemann H., and Franchi I. A. (2006) Trapping and modification processes of noble gases and nitrogen in meteorites and their parent bodies. In *Meteorites and the Early Solar System II*, Lauretta D. S., and McSween H. Y., editors, pp. 499–517, University of Arizona Press, Tucson.

12 Planetesimals

Leftover Planetary Building Blocks

Overview

Planetesimals are small bodies that can be either dry or hydrated, reflecting nebular accretion of rock with or without ices. Samples of asteroids in hand (meteorites) include various classes of chondrites, as well as chunks of the crusts, mantles, and cores of differentiated planetesimals. Ice-bearing bodies in the Kuiper belt and the Oort cloud enter the inner solar system as comets and are sampled, on a fine scale, as CP IDPs. The taxonomy of asteroids is based on spectroscopy, and their connections to samples in hand can be made by spectral comparisons. The orbital distributions of the various classes of planetesimals provide insights into their formation, although models involving the migration of giant planets suggest that the present distributions of planetesimals are not primordial. Asteroids have evolved by thermal metamorphism or aqueous alteration (depending on whether or not they contained ices), or by melting and differentiation (depending on how rapidly they accreted relative to the decay of short-lived radionuclides). Collisions have disrupted most planetesimals, producing asteroid families or reaccreted porous bodies. Glancing collisions have, in some cases, stripped off silicate mantles to expose iron cores. The millions of small planetesimals in the Main asteroid belt and the Kuiper belt are mostly impact debris. This chapter provides the context for understanding the compositions of planetesimals described in the next chapter.

12.1 Millions and Millions

The vast space between Mars and Jupiter – the location of the *Main asteroid belt* – contains millions of small to modest-sized planetesimals. Although science-fiction movies make careening through the Main belt in a spacecraft seem hazardous, the reality is that spacecraft traversing the belt can safely navigate because the asteroids are usually several million kilometers apart. Like the planets, asteroids generally orbit close to the ecliptic plane, although orbital inclinations from the ecliptic can be significant. But the Main belt is not the only place where asteroids are found. The *Trojan asteroids* cluster in stable orbits at Lagrange points 60 degrees ahead of and behind Jupiter in its orbit. The *Centaur asteroids* orbit between the giant planets, generally in unstable orbits. *Near-Earth objects* (NEOs) are just that – asteroids in highly elliptical orbits that cross or come close to the path of our planet.

Extending beyond Neptune to approximately 50 AU is the Edgeworth–Kuiper belt (usually shortened to just the *Kuiper belt*), the home of more than 1000 known *Kuiper belt objects* (KBOs) and likely millions more. The total mass of KBOs has been estimated to be perhaps 200 times that of the asteroids in the Main belt. Pluto and its five moons represent the innermost KBOs, although Neptune's moon Triton is also thought to be a captured KBO. Pluto's orbital inclination is far larger than that of any planet, and other KBOs are commonly inclined to the ecliptic. The discovery of a handful of Pluto-sized KBOs, with the prospect of many more still to find, prompted the demotion of Pluto itself from planet to *dwarf planet*, so as not to require constant changes to the planetary architecture. At even greater heliocentric distance is the *Oort cloud*, a spherical volume of bodies that marks the outmost part of the solar system. This hypothetical cloud extends from 2000 to 200,000 AU. Samples of the planetesimals in the Kuiper belt and Oort cloud that are perturbed into the inner solar system become comets.

With so many millions of asteroids and cometary bodies, it is worth recalling that there once were many more. Uncounted numbers of planetesimals were assembled to form the planets, so the ones that remain can rightly be viewed as unused planetary building blocks. Moreover, some dynamical models of the early solar system suggest that inward and then outward migration of the giant planets drastically perturbed the orbits of planetesimals, largely depopulating the space around them.

Planetesimals are of special importance to cosmochemistry: these small bodies are primordial objects dating from the earliest era of solar system history. As such, they provide the raw materials (meteorites, IDPs, and a few samples returned by spacecraft missions) that are the focus of cosmochemistry. Laboratory studies of them reveal physical and chemical processes that occurred within the nebula, the chronology of the early solar system, and the compositions of the solids that formed the terrestrial planets. Along with ices and nebula gas, which are as yet unsampled directly, they also accreted to form the giant planets.

In this chapter we describe planetesimals to provide context for understanding and interpreting their compositions, which will be the subject of the following chapter.

12.2 Physical Properties of Planetesimals

12.2.1 Asteroids

Spacecraft encounters with asteroids offer the most detailed information on their sizes, shapes, masses, surface features, and internal structures. Beginning in the 1990s, a number of spacecraft missions have flown by asteroids. In 2000, NASA'S NEAR Shoemaker orbited and ultimately touched down on Eros. Five years later, JAXA's Hayabusa returned a tiny sample from the surface of Itokawa, which is really only an asteroid fragment, just ~300 m in diameter. The Hayabusa2 spacecraft reached the asteroid Ryugu in 2018, photographed and studied it frm orbit, and collected a 5.4 g sample that was returned to Earth in late 2020. NASA's OSIRIS-REx orbited Bennu in 2019–2020 and collected a large sample of the asteroid's surface. The OSIRIS-REx spacecraft is on its way back to Earth at time of writing. Images obtained by missions to small asteroids are given in Figure 12.1.

All of these small asteroids are irregular in shape and are marked by craters in various states of degradation. One crater on Mathilde is so large that it is difficult to envision how the asteroid survived the impact. The surfaces of asteroids are covered by fine-grained regoliths and by boulders of various sizes.

Itokawa and Eros have dumbbell shapes that suggest they may actually be two connected asteroidal fragments held together by their mutual gravity. Asteroid shapes can

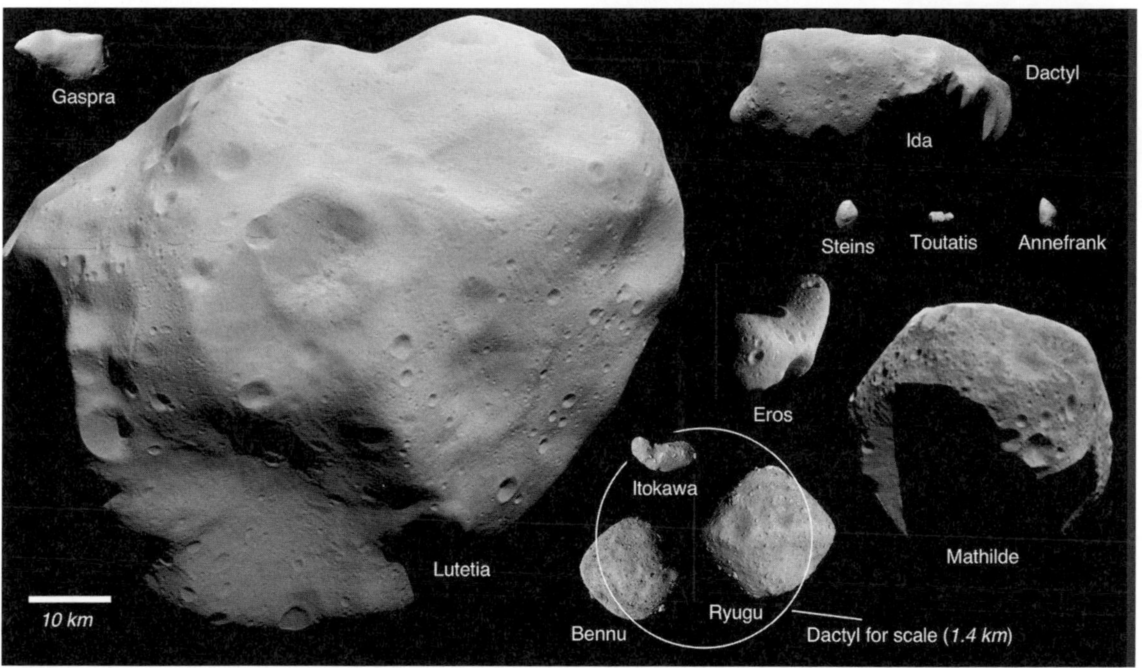

Figure 12.1 Images of small asteroids visited by spacecraft, all to scale except tiny Itokawa, Bennu, and Ryugu, which are scaled to Dactyl (white circle). Spacecraft missions for specific asteroids are: Annefrank (Stardust, 2002), Bennu (OSIRIS-REx, 2019), Eros (NEAR Shoemaker, 2000), Gaspra (Galileo, 1991), Ida and Dactyl (Galileo, 1993), Itokawa (Hyabusa, 2005), Lutetia (Rosetta, 2010), Mathilde (NEAR Shoemaker, 1997), Ryugu (Hayabusa2, 2018), Steins (Rosetta, 2008), and Toutatis (Chang'E 2, 2012). NASA, ESA, and JAXA images.

also be determined remotely using radar observations, and other dumbbell-shaped (*contact-binary*) objects have been found. Ida has a tiny moon, Dactyl. Small asteroids like those in Figure 12.1, comprising most of the Main belt, are fractured collisional remnants of once-larger bodies.

Although small asteroids number in the millions, more than half of the mass of the Main belt is contained in just the four largest bodies: Ceres, Vesta, Pallas, and Hygiea. These large bodies are rounded (or nearly so) and apparently intact, unlike the collisional rubble that comprises much of the belt. Bottke et al. (2015) suggested that most asteroids with diameters >100 km may be intact. Vesta (525 km diameter) and Ceres (930 km diameter) were explored by NASA's Dawn spacecraft. Rocky Vesta has a lunar-like appearance, whereas icy Ceres is dark with bright splotches of salts and cryovolcanic domes, as described in Chapter 11. Images of Vesta and Ceres are compared in Figure 12.2.

The masses of asteroids have been estimated from their gravitational effects on the trajectories of passing or orbiting spacecraft, and, together with volume measurements, are used to calculate mean densities. The densities of some asteroids not visited by spacecraft can be estimated from perturbations by other large asteroids, but they have large uncertainties. The generally low calculated mean densities of many rocky asteroids (as low as 1.3 g/cm^3; Carry, 2012) suggest high porosities. Densities correlate roughly with size, indicating that small asteroids are more likely to be porous. Laboratory measurements of the porosities of meteorites (Flynn et al., 2018) vary from 0% to 40%, similar to the calculated porosities of asteroids, so high porosities imply that many asteroids are fractured internally. Vesta is an exception, with a mean density of 3.4 g/cm^3, reflecting the presence of a massive iron core at its center. Asteroids inferred to be metallic have densities averaging ~5.3 g/cm^3 (Carry, 2012) and may be cores that have been largely stripped of their silicate mantles.

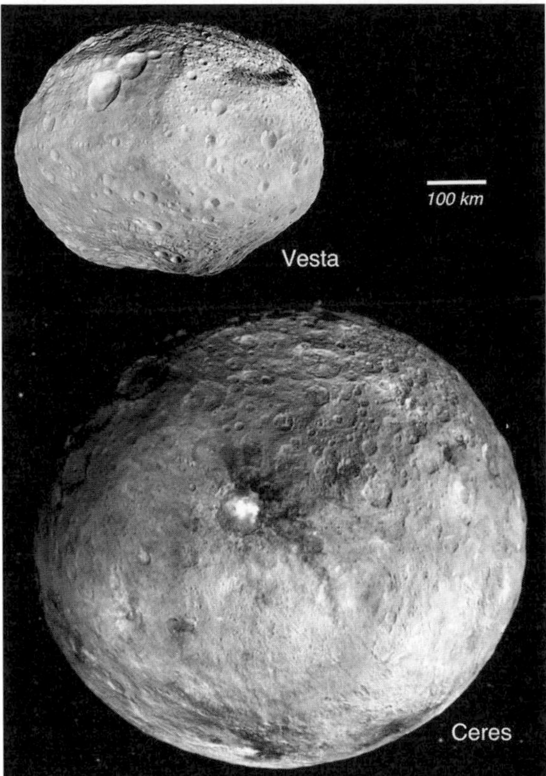

Figure 12.2 Images of asteroid Vesta and dwarf planet Ceres, obtained by the Dawn spacecraft. NASA images.

12.2.2 Comets and KBOs

A popular term for describing comets is "dirty snowballs," although that description probably understates the proportion of rock and dust relative to ices. The activity that characterizes a comet is driven by solar heating. As the comet approaches the Sun, jets of dust and gas erupt from active areas, as they periodically rotate into the sunlight. The nucleus becomes surrounded by a spherical "coma" formed by the emitted gas and dust. Emitted gas becomes ionized due to interaction with solar ultraviolet radiation, and the ions are swept outward by the solar wind to form the comet's ion tail. A separate dust tail commonly has a different orientation, reflecting differences in the velocities of solid particles and ions.

Dust, representing less-volatile materials, is liberated from comets and propelled outward by the expanding jets of gas. Appreciable quantities of dust are sometimes released; for example, the dust production rate from comet Hale–Bopp exceeded its gas production rate by a factor of five. Particle impacts onto spacecraft suggest that solid grains in the coma can be of millimeter size.

Comets are considered to be weakly consolidated, and active comets are commonly observed to split into fragments. This is sometimes due to the tidal forces of a close planetary encounter, such as affected comet Shoemaker–Levy when it passed close to Jupiter in 1992 and broke into 21 pieces. More commonly, a comet spontaneously fragments multiple times over its orbital period, without any obvious cause. Disintegrating comets leave trails of small particles in their wakes. These trails are known as *meteor streams*, and when the Earth passes through a meteor stream, as it does several times a year, a *meteor shower* occurs. Meter-sized rocks are known to occur within cometary meteor streams.

Comet nuclei that have been visited by spacecraft are generally a few kilometers in size. The densities of Halley and Tempel2 were estimated to be ~0.3 g/cm^3, consistent with highly porous objects. Wild2 appears to be a more coherent object, implying that it may have a higher density, estimated at ~0.6 g/cm^3, and the well-determined volume and mass of comet Churyumov–Gerasimenko give a density of 0.53 g/cm^3.

Images of comet nuclei taken during various spacecraft encounters illustrate both similarities and differences. Figure 12.3 compares images of five comets, taken at very different spatial resolutions. Despite being roughly comparable in size, these objects have different shapes and varying proportions of smooth and rough areas, sometimes with visible layering. Comet Wild2 has pinnacles more than 100 m high and numerous pockmarks, some of which may be impact craters. ESA's Rosetta spacecraft orbited comet Churyumov–Gerasimenko in 2014, conducting the most extensive set of comet observations to date. Its Philae lander unfortunately was put down in shadow and keeled over, so it was only able to make a few of the planned surface observations and measurements. This comet and Hartley2 are interpreted to be contact binaries.

For all comets visited by spacecraft, the measured reflectivity (*albedo*) is $<5\%$, sometimes described as being like "black velvet." This dark coloration is due to black crusts that cover most of the nuclei. The materials now comprising the crusts were presumably left behind when ice sublimated from the surface or spewed out of the interior at speeds below the escape velocity, so they can be considered "lag" deposits. The portions of the surfaces covered by such deposits are inert, and active areas releasing gas and dust constitute only a small percentage of a comet's total surface area.

Compared to comets, we have less information about the physical properties of KBOs, and we can only guess about Oort cloud objects. KBOs have not experienced the repeated solar heating that alters and eventually destroys active comets. The discovery of some rather large KBOs (the diameter of the largest, Eris, is >2400 km, slightly larger than Pluto), coupled with their expected dynamical histories, suggest that many smaller KBOs may be collisional fragments of larger objects. The New Horizons spacecraft provided a resolved image of one small KBO, called Arrokoth, in 2019. It consists of two connected spheres (Fig. 12.4), like contact-binary asteroids and comets, and measures 31 km in length (Spencer et al., 2020). Its dual-lobe nature suggests it is a direct result of accretion rather than impact.

Spacecraft images of Neptune's moon Triton (2710 km diameter, density ~2.1 g/cm^3, and thought to be a captured KBO) and Pluto (2370 km diameter, density ~1.9 g/cm^3) reveal surprisingly complex surfaces composed of ices. Images of Triton and Pluto were shown previously in Figures 11.18a and 11.18e. In Box 11.4, we discussed their surface features and cryovolcanic activity.

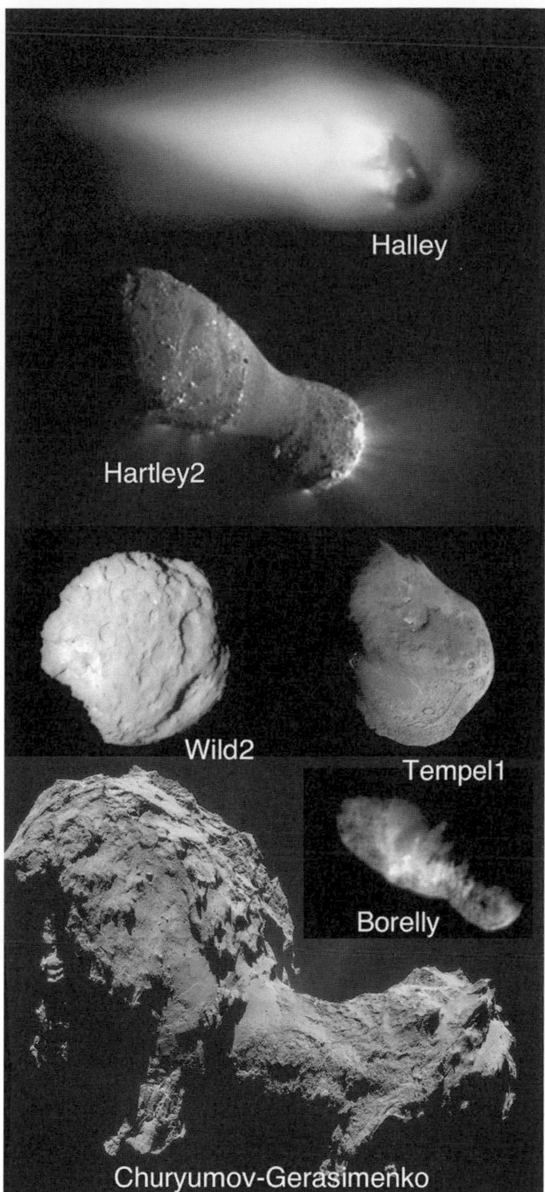

Figure 12.3 Images of comet nuclei taken during spacecraft encounters: Halley (Giotto, 1986), Hartley2 (Deep Impact, 2010), Wild2 (Stardust, 2004), Tempel1 (Deep Impact, 2005), Borelly (Deep Space 1, 2001), and Churyumov–Gerasimenko (Rosetta, 2014). Images are not to scale. NASA and ESA images.

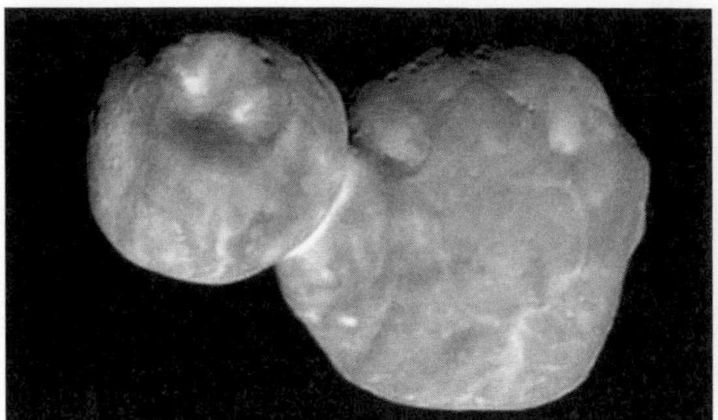

Figure 12.4 The resolved image of a KBO, Arrokoth, taken by the New Horizons spacecraft. NASA image.

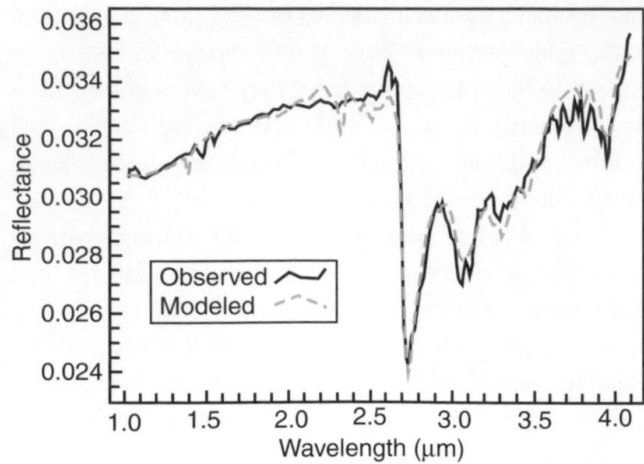

Figure 12.5 The spectrum of Ceres obtained by the Dawn spacecraft, compared with a modeled spectrum obtained by mixing serpentine, ammonium-bearing clay, magnetite, and carbonate. Adapted from De Sanctis et al. (2015).

12.3 Spectroscopy and Taxonomy of Planetesimals, and Relation to Samples

12.3.1 Linking Asteroids to Meteorites

Incident sunlight is partially absorbed by minerals on an asteroid's surface, and the fraction of light that is reflected can vary as a function of wavelength. Spectrophotometric studies of asteroids have been carried out by telescopes for decades. Most measurements are made at visible wavelengths ranging from ~0.4 to 1.0 μm, because the illuminating solar flux peaks in the visible region and the Earth's atmosphere is relatively transparent at these wavelengths. More-recent studies also commonly use near-infrared spectroscopy (~1.0–3.0 μm). Asteroid spectra may show ultraviolet absorptions due to Fe^{2+} charge transfer in minerals, absorption features centered near 1 and 2 μm due to Fe-bearing silicates like pyroxene and olivine, and an increase in slope longward of 0.55 μm (*spectral reddening*) due to iron–nickel metal or complex organic molecules.

Minerals containing bound water (OH) or water molecules (H_2O) give rise to prominent spectral absorptions near 3 μm that are observed in some asteroids (Rivkin et al., 2015). These absorption bands have different shapes: the hydroxyl feature is sharp and the H_2O feature is more subdued. Near-infrared spectra for asteroids with high albedos are generally characterized by the water feature, whereas spectra of low-albedo asteroids tend to have the sharp OH feature. The sharp ~3 μm feature indicates that hydrous phyllosilicates are present on their surfaces (Pieters and McFadden, 1994; Vilas et al., 1994), i.e. they are aqueously altered. An example, illustrated in Figure 12.5, is the near-infrared spectrum of Ceres. This spectrum is especially clear, since it was taken nearby using a spectrometer on the orbiting Dawn spacecraft, rather than using a telescope on the Earth's surface. The absorption feature at 2.75 μm is due to serpentine, and that at 3.1 μm indicates ammonium-bearing clay; bands at longer wavelengths are due to organics and carbonate.

Asteroids are classified based on their spectral shapes and albedo. A number of different taxonomies have been proposed, each using statistical clustering techniques on spectra. One somewhat dated but still widely used classification (Table 12.1), based on an eight-color (wavelength) asteroid survey, identifies 15 spectral classes (Tholen and Barucci, 1989). A more recent classification (Bus et al., 2002, subsequently expanded by DeMeo et al., 2009 to include near-infrared spectra), crafted from a much larger spectral dataset, places most asteroids within three large superclasses (called the C-, S-, and X-complexes), with a handful of smaller groupings for outlying objects. The correlation between the two classification schemes is shown in Table 12.1.

Asteroid reflectance spectra can be compared to the spectra of meteorites measured in the laboratory, providing a way to link some kinds of meteorites to possible parent bodies. The Tholen and Bus asteroid taxonomies have been interpreted as representing the types of meteorites indicated in Table 12.1 and Figure 12.6. In the Bus et al. (2002) taxonomy, the C-complex asteroids are probably hydrated carbonaceous chondrite (CI or CM) objects (Fornasier et al., 2014). Some S-complex asteroids are ordinary chondrite parent bodies, but this superclass is very diverse and includes many other meteorite types as well. The X-complex includes objects with spectra that resemble enstatite chondrites and aubrites, and some irons and stony irons, although other X-complex asteroids are

Table 12.1 Asteroid taxonomy based on spectra (Tholen and Barucci, 1989; Bus et al., 2002; DeMeo et al., 2009)

Tholen Class	Bus Class	Possible Phases	Possible Meteorite Analogs
C	C	Phyllosilicates, organics	CM, CI chondrites
S	S	Olivine, pyroxene, metal	Ordinary chondrites, achondrites*, primitive achondrites*
P	X	Olivine, organics	Carbonaceous chondrites
E	X	Enstatite, oldhamite	Aubrites, E chondrites
M	X	Metal?	Irons, CB chondrites, E chondrites
Q	Q	Olivine, pyroxene, metal	Ordinary chondrites
B, F, G	C	Phyllosilicates, carbon	Carbonaceous chondrites
T	T	Troilite?, metal?	
A, R	A, R	Olivine, pyroxene	Pallasites, brachinites, R chondrites
K	K	Olivine	CV and CO chondrites
D	D	Organics, opaques	Some IDPs
V	V	Pyroxene, plagioclase	HED achondrites

*Angrites, acapulcoites-lodranites, winonaites, ureilites.

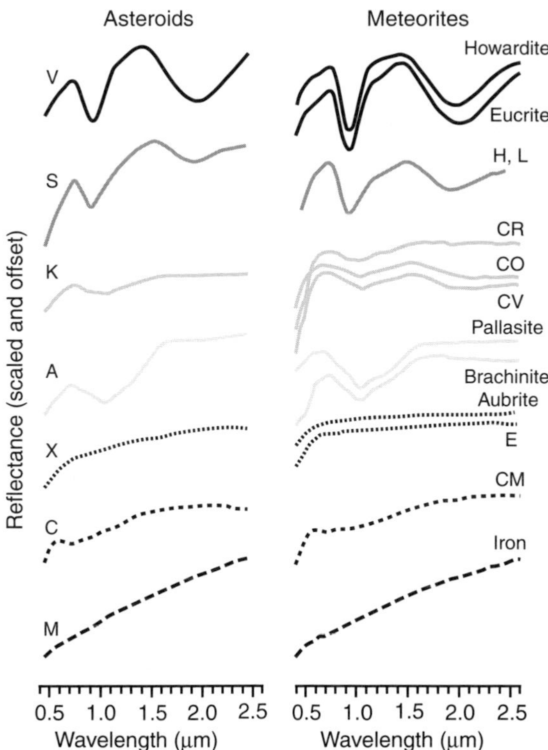

Figure 12.6 Comparison of telescopic spectra used to classify asteroids (asteroid taxonomy by DeMeo et al., 2009) with laboratory spectra of meteorites. Spectral similarities provide a means of identifying possible parent bodies for classes of meteorites.

unlike known meteorite types. The D-class asteroids have featureless spectra and are dark red in color (Jones et al., 1990), thought to indicate surfaces coated with complex organic compounds (tholins) formed by interactions with ultraviolet radiation and cosmic rays. These asteroids have no clear meteorite analogs, although some IDPs have similar spectra. A few asteroid spectra are unique and provide more definitive connections to meteorites, such as between Vesta (V-type) and HED meteorites. Determining the presence of minerals with specific compositions from asteroid spectra can provide more quantitative correlations with meteorites (Reddy et al., 2015).

In all, there are 14 well-defined groups of chondrites (refer back to Figure 6.3), as well as a number of chondrites that are unique or members of grouplets that are too small for adequate characterization. There are many more differentiated meteorite groups, including numerous groups or grouplets of achondrites and stony irons and ~60 irons. The meteorite collection has been estimated to represent as many as 110 asteroidal parent bodies (Greenwood et al., 2017). While this may seem like a diverse collection, it represents only a small portion of the approximately 10^6 asteroids with diameters >1 km in the Main belt.

We have only a few small samples of asteroids that have been returned by spacecraft. The Japanese Hayabusa mission returned a tiny sample from the NEO Itokawa in 2010. The composition of its olivine and pyroxene fragments indicate that Itokawa is an equilibrated LL chondrite (Nakamura et al., 2014). The Japanese Hayabusa2

Box 12.1 Space Weathering: Complicating Spectral Interpretation

Many asteroid spectra tend to be "redder" than the spectra of the corresponding meteorites. Reddened spectra have lower reflectivity, weaker absorptions, and flatter slopes. This phenomenon has been attributed to *space weathering*, a catchall term referring to any process that modifies the optical properties of the surfaces of airless bodies exposed to the space environment (Clark et al., 2002).

Space weathering is most prominently displayed in the spectra of lunar soils. As surface soils "mature" over millions of years of exposure to micrometeorites and cosmic rays, they darken noticeably and diagnostic absorption bands weaken. Electron microscope observations demonstrate that these optical changes result from submicroscopic metallic iron grains that occur in silicate glasses coating the surfaces of fine soil particles (Pieters and Noble, 2016). These tiny ("nanophase") iron grains are deposited from vapor produced by micrometeorite impacts or solar wind sputtering. SiO_2 and FeO volatilize before most other components, and the gaseous molecules thermally dissociate into elements. The vapor containing silicon and iron atoms then condenses to form silicate glass and iron particles. In some cases, other combinations of silicon and iron, such as the mineral hapkeite (Fe_2Si), also condense in small amounts. This assemblage represents kinetically controlled condensation, instead of equilibrium condensation as described in Chapter 7, and the formation of metallic iron may require solar-wind-implanted hydrogen.

Although no samples of the unconsolidated surface regoliths of asteroids are available for laboratory analysis, a similar process is thought to have affected the spectra of many asteroid surfaces. The spectra of some asteroids, particularly the S-types, are reddened relative to the spectra of the ordinary chondrites thought to be derived from them. Space weathering has also been seen in close-up images of asteroids. Ejecta blankets around relatively recent craters on Gaspra and Ida are less spectrally reddened than the surrounding surface materials. Only materials directly exposed to the space environment experience space weathering, and impact excavation exposes unweathered materials from deeper in the regolith. For decades, the inability to match precisely the spectra of asteroids and meteorites was a conundrum, but space weathering now provides an explanation.

Figure 12.7 Millbillillie regolith breccia in plane-polarized light, 2 cm across. Clasts of basaltic eucrite are visible in a crushed matrix, crosscut by a vein of shock-melted glass at the lower right. Image from Lauretta et al. (2014), with permission.

Box 12.1 (cont.)

Meteorite regolith breccias (Fig 12.7) are mixtures of fine-grained surface materials and coarser subsurface grains. Although some of them are perceptibly darkened and affected by shock, they do not show the spectral reddening that is characteristic of space weathering. Consequently, most regolith breccias launched from asteroids must come from deeper in the regolith.

Although space weathering specifically denotes processes that affect an asteroid's spectrum, there are other effects of meteorite bombardment that do not necessarily alter their spectra. For example, impact melting produces *agglutinates* – small particles of minerals or rocks bonded together by glass. Agglutinates are much more abundant in lunar soils than in meteorite regolith breccias, suggesting that impacts at lower velocities in the Main belt may produce less melting than on the Moon. Both lunar regolith samples and asteroid regolith breccias contain implanted solar wind gases (mostly hydrogen, but also helium and other noble gases, as discussed in Chapter 11).

mission returned a 5.4 g sample from the asteroid Ryugu, which has C-type and B-type spectral properties. The sample had not been characterized as of time of writing. NASA's OSIRIS-REx is scheduled to return a relatively large sample from the B-type asteroid Bennu in September 2023. These missions will be important in improving the interpretation of spectral data.

12.3.2 Linking Comets and KBOs to IDPs

Frozen ices in comet nuclei are inferred from spectra of gases in the coma, as previously discussed in Chapter 11. The mineralogical composition of coma dust has been inferred from telescopic spectra. Silicates were first detected from a 10 μm infrared emission feature resulting from the stretching of Si–O bonds. This silicate feature can only be observed in small (<2 μm) grains. The fine structure in the silicate feature provides more specific mineralogical information. A small bump at 11.2 μm on the 10 μm feature (Fig. 12.8a), observed for several comets, is generally interpreted as indicating crystalline olivine. Another shoulder at 11.9 μm and a slope change at 9.2 μm (Fig. 12.8a) are attributed to pyroxene or amorphous grains having pyroxene composition. The 11.2 μm feature is not seen in spectra of the interstellar medium and molecular clouds. A common astronomical interpretation is that amorphous ISM silicates were annealed to produce olivine crystals after they were incorporated into the solar nebula. A more likely possibility, strongly supported by new data (see below), is that olivine incorporated into comets formed within the solar nebula.

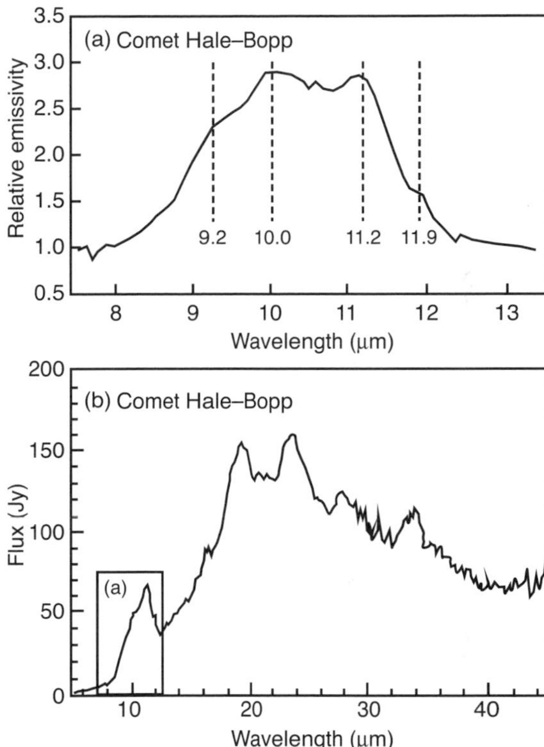

Figure 12.8 Spectra of comet Hale–Bopp, showing features attributable to silicate minerals. (a) Profile of the fine structure in the 10 μm silicate emission feature; a peak at 11.2 μm and a shoulder at 11.9 μm are due to olivine, and a slope change at 9.2 μm results from pyroxene. (b) Expanded infrared spectrum exhibiting a number of sharp peaks due to magnesian olivine and pyroxene. The region of (a) is bounded by a small box. Modified from Crovisier et al. (2000) and Hanner and Bradley (2003).

Figure 12.8b shows an expanded spectrum of comet Hale–Bopp (Crovisier et al., 2000) covering the range from 6 to 45 μm, a region that is highly sensitive to mineralogy. Sharp peaks in this spectrum correspond to forsterite (Mg_2SiO_4), and minor features are attributed to enstatite ($Mg_2Si_2O_6$). Forsterite and enstatite are among the earliest silicate phases to form in condensation calculations, and it is difficult to see how magnesian end members could have formed by the annealing of more iron-rich compositions. Condensation of these minerals is also suggested by the fact that forsterite and enstatite spectra are also measured around young stars.

Mass spectrometers were carried aboard the Vega and Giotto spacecraft that visited comet Halley and the Stardust spacecraft that encountered comet Wild2. Impacting dust particles generated ions whose masses were determined from their flight times in a drift tube. The particles were described as either "CHON" (composed mostly of carbon, hydrogen, oxygen, and nitrogen) or silicates (mainly of magnesium and iron in roughly chondritic proportions), or as mixtures of both. The continuum of compositions indicates that carbonaceous and silicate materials are mixed at the submicron scale, and that the bulk solid composition is close to chondritic except for higher abundances of carbon and nitrogen.

IDPs, specifically the chondritic porous (CP) IDPs described in Chapter 6, are thought to be comet dust. Spectral comparisons with comets are difficult, because of the small, unrepresentative sizes of IDPs, but the similarities in mineralogy are clear. Collecting IDPs at times when the Earth is passing through meteor streams associated with known comets offers an opportunity to associate particles with specific comets.

Using visual records from contemporary reports of the fall of the Orgueil CI chondrite, Gounelle et al. (2006) derived a possible orbit that is similar to comets. This orbit supports a suggestion that some high-strength rocks capable of surviving atmospheric passage as meteorites could come from comets.

The telescopic spectra of KBOs are limited. Initially, only colors were determined, defining two populations that are either grey-red or very red (Wong and Brown, 2017). The color variation is hypothesized to result from retention or loss of H_2S ice. Spectra of the largest KBOs, as well as Pluto and Triton, indicate surfaces dominated by ices of CH_4, N_2, and CO, and H_2O ice has also been detected (Brown, 2002; Schaller, 2010). The volatile ices are sustained by the very low temperatures (<50 K) in the Kuiper belt. Although nitrogen ice is dominant on Pluto and Triton, methane ice is most easily detected due to its strong features in the near-infrared. Identified hydrocarbons include ethane, ethylene, and acetylene, all probably derived from methane. The infrared spectra of the large KBOs Eris, Pluto, and Makemake are compared in Figure 12.9. KBOs >2000 km in size have enough gravity to prevent the thermal escape of the more highly volatile methane, nitrogen, and carbon monoxide. The spectra of smaller KBOs either show absorption features

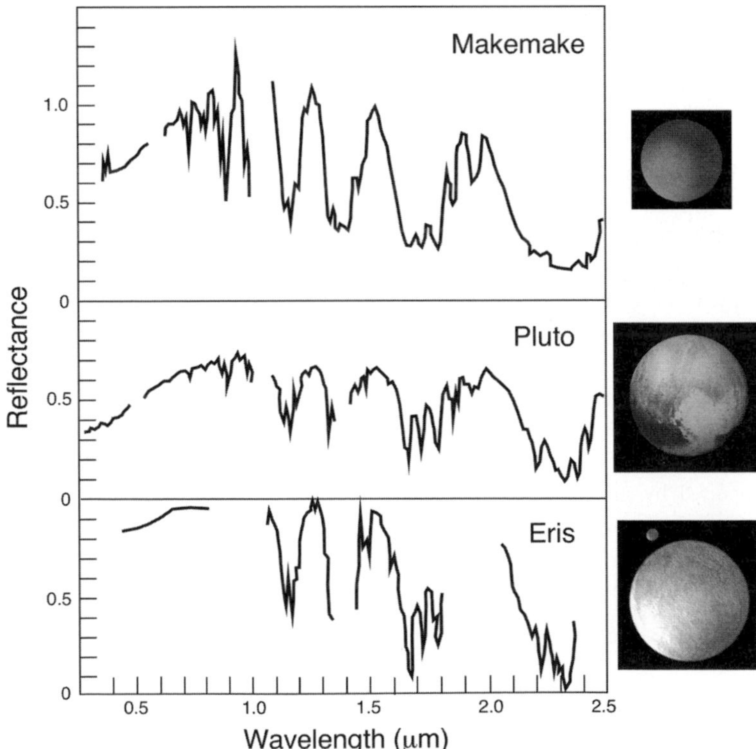

Figure 12.9 Infrared spectra of the KBOs Makemake, Pluto, and Eris. Modified from Schaller (2010). Images of Makemake and Eris are artists' renditions, and the Pluto image is from New Horizons (NASA). *For the colour version, refer to the plate section.*

only of water ice, or are spectrally featureless, possibly indicating a refractory residue of tholins. However, methanol ice and complex tholins dominate the spectrum of small KBO Arrokoth, with no clear spectral detection of water ice (Grundy et al., 2020).

Particles collected from comet Wild2 (Fig 12.10) were returned to Earth by the Stardust spacecraft (see Box 12.2). Most of the particles in the aerogel (the dust collection medium, described below) were disaggregated during capture. Examination of these grains suggests that they were originally fine-grained, loosely bound aggregates, often containing some larger crystals (Zolensky et al., 2006). Care must be exercised in interpreting analyses of these grains, because impact heating during their collection has

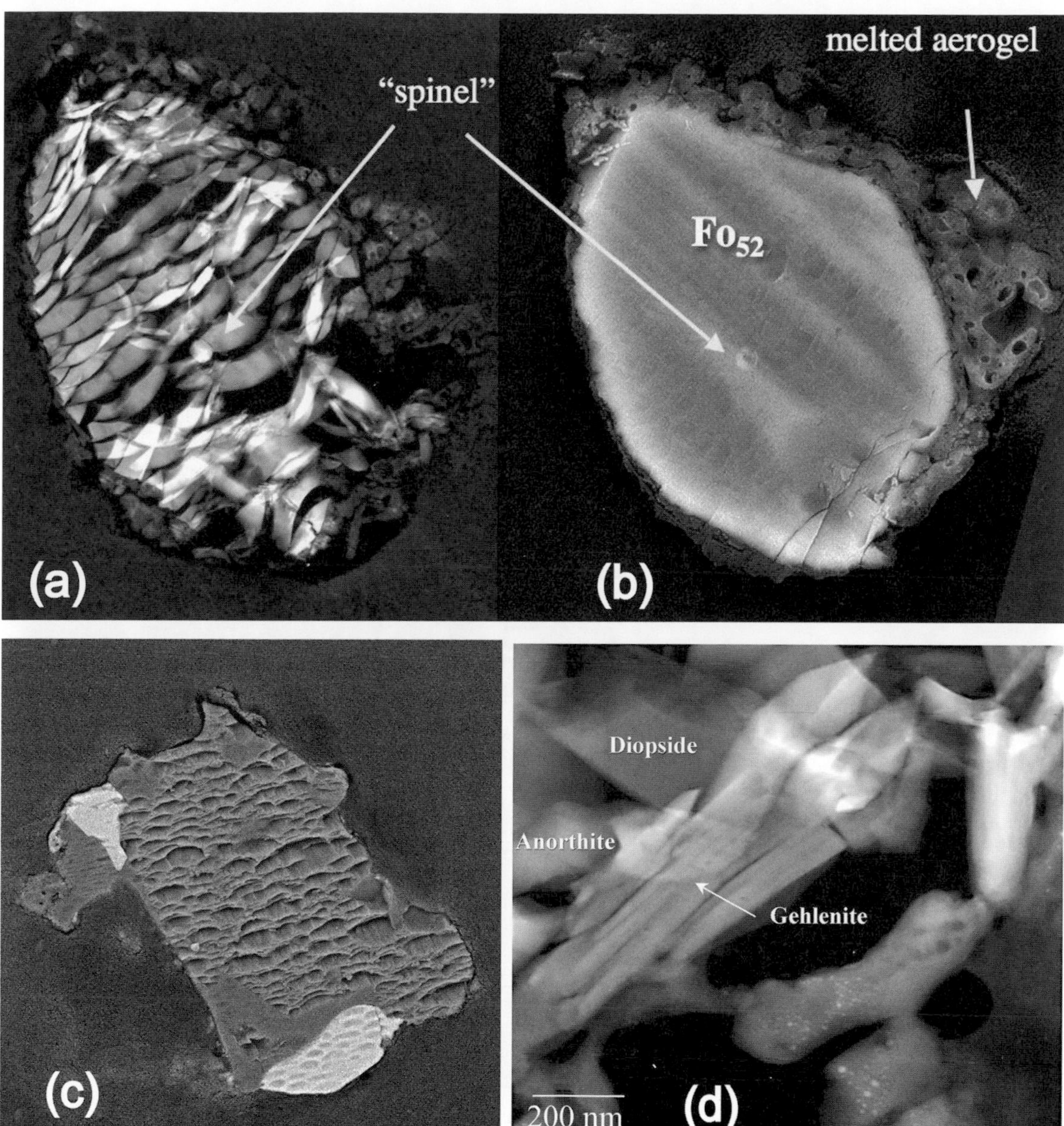

Figure 12.10 Examples of comet particles collected and returned to Earth by the Stardust spacecraft. (a, b) Two ways to image a particle made of iron-rich olivine, with minor spinel; the left view is a thin section viewed with plane-polarized light, and the right view is a scanning electron microscope image of the same grain, field of view (FOV) = 4 mm. (c) Particle composed of pyroxene and sulfides, FOV = ~1 μm. (d) Particle composed of minerals that form CAIs. Images courtesy of D. Brownlee.

Box 12.2 The Stardust Mission: Collecting Comet Dust

In 2004, the NASA Stardust spacecraft passed through the dust cloud surrounding the nucleus of comet Wild2 and captured more than 10,000 particles ranging from 1 to 300 μm in size (Brownlee et al., 2006). These particles were returned to Earth for study in terrestrial laboratories in 2006.

At the time of the encounter, Wild2 was at 1.86 AU and thus was very active. Comets release thousands of tons of dust during cometary activity – the dust is there for the taking, and the trick is to snatch it without destroying the spacecraft. Two technological achievements by scientists at the Jet Propulsion Laboratory made the collection of comet dust possible: the clever design of a trajectory that allowed the spacecraft to encounter the comet coma at a relatively modest speed (6.1 m/sec), and the development of a capture medium that slowed and trapped the particles without destroying them.

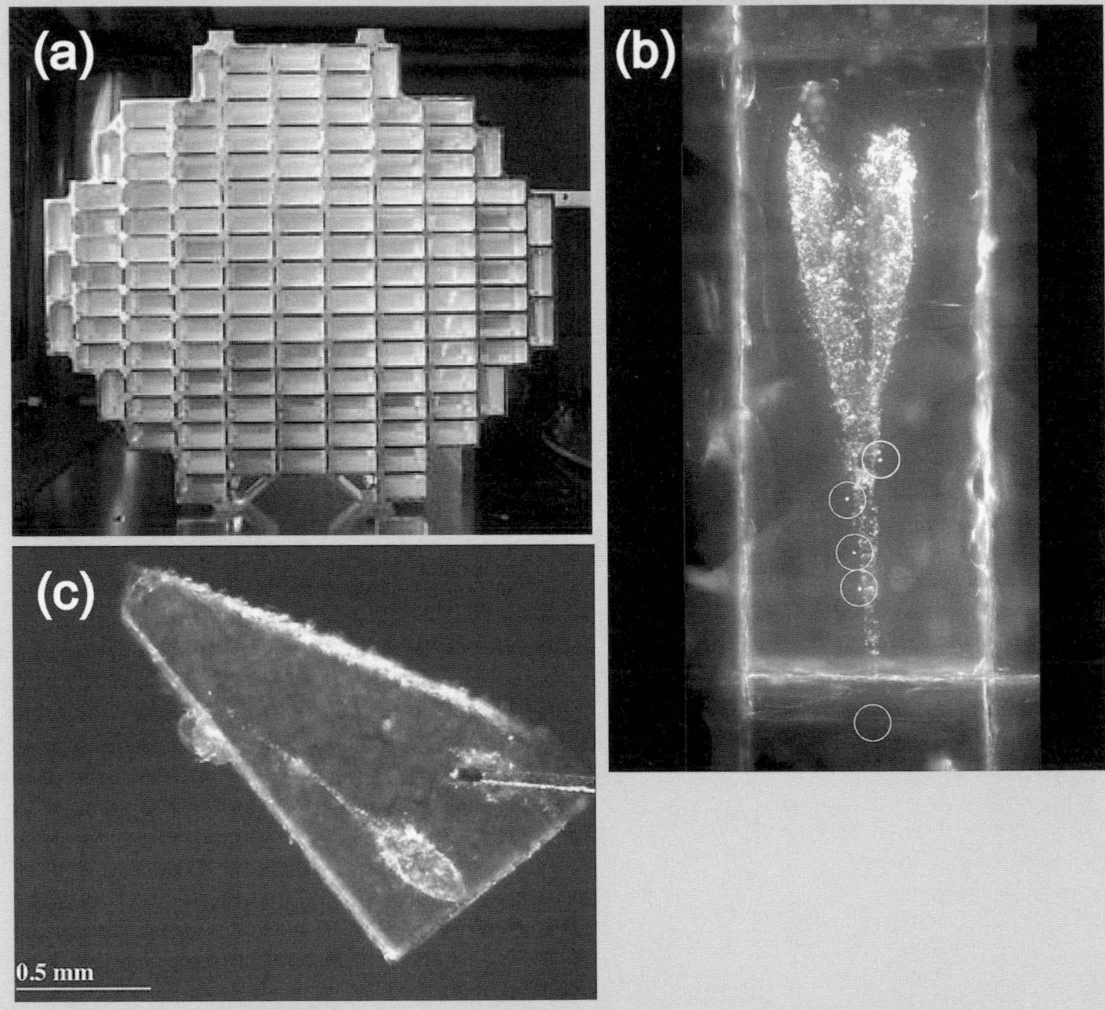

Figure 12.11 Images of aerogel used on the Stardust spacecraft. (a) A bank of aerogel collectors for cometary dust. (b) Tadpole-shaped track made by comet dust particles captured in aerogel. Impacting particles undergo devolatilization, producing the bulbous entryway. Large (2–15 μm) dust grains (circled) can be seen toward the end of a thin trail, and other finer material occurs along the track. (c) A keystone, containing a particle track, cut out of aerogel. Figures courtesy of D. Brownlee.

Box 12.2 (cont.)

The capture medium was aerogel (Tsou et al., 2004), a highly porous silica foam with a density comparable to air. In order to slow the particles gradually, the aerogel collectors (Fig. 12.11a) were crafted to have densities varying from <0.01 g/cm^3 at the surface to 0.05 g/cm^3 at 3 cm depth. Impacts into the aerogel produced tapered cavities (tracks), some shaped like carrots and others like turnips with single or multiple roots (Fig 12.11b). The thinner tracks were formed by particles that suffered little fragmentation on capture, and the bulbous tracks were formed from disaggregated particles. Although it is tempting to attribute the bulbous ends of tracks to impact vaporization of cometary ice, most particles are thought to have lost any icy component they may have had during their hours-long exposure to sunlight in space prior to collection. Terminal particles located at the ends of tracks represent intact cometary dust, whereas particles distributed along the tracks have usually been devolatilized, melted, or mixed with melted aerogel. All the particles were affected to some degree by capture, with the smallest effects in particles larger than a micron in size. Aluminum foil covering the frame used to hold the aerogel was also impacted by dust particles, which produced tiny impact craters lined with melted projectile residue.

For particle handling, wedges of aerogel (Fig 12.11c) containing entire tracks, called "keystones," were cut from the aerogel tiles (Westphal et al., 2004). From the keystones, individual particles and materials lining the tracks can be isolated and studied. The great advantage offered by these Stardust particles is the ability to bring to bear the analytical technology of Earth-based laboratories. Research on these cometary dust particles has drawn on the prior investments made in developing the cosmochemical techniques used for other microscopic samples, specifically IDPs and presolar grains.

modified them in ways that cannot always be ascertained (Brownlee et al., 2006). Laboratory simulations have been critical in unraveling the effects of capture in aerogel.

Most of the larger particles consist of olivine, pyroxene, or FeNi sulfides (Zolensky et al., 2006; Ishii et al., 2008). Olivine grains exhibit a surprisingly large compositional range, from Fo_{100} to Fo_4, although most grains are magnesium rich. Likewise, pyroxenes range from En_{100} to En_{52}, with most being magnesian, and they do not show the whisker morphologies seen in CP IDPs. Another unusual component, called "kool" grains, consists of Fe-rich olivine and Na- and Cr-rich augite, sometimes with albite (Joswiak et al., 2009); the only other occurrence of kool grains is in CP IDPs. Sulfides are either troilite or pyrrhotite (it is not possible to determine the exact stoichiometry of these tiny crystals) and pentlandite. Grains of FeNi metal also occur. This mineral assemblage resembles that of unaltered (type 3) carbonaceous chondrites. Among the comet grains are also particles that superficially resemble GEMS, consisting of metal and sulfide embedded in glass. However, in this case the glass is melted aerogel. One grain curiously contained a phase with the composition of K-feldspar. Another particle, called "Inti" and consisting of anorthite, pyroxene, melilite, spinel, corundum, osbornite, and perovskite, has the same mineralogy as CAIs from chondrites (see Chapter 6). Additional refractory particles were found in at least five separate tracks (Westphal et al., 2017).

Before the return of the Stardust samples, it was thought that comets were mostly presolar grains. However, the oxygen isotopic compositions of Stardust silicate grains resemble those of chondrites, suggesting that they formed within the solar nebula (McKeegan et al., 2006). Only one grain was found with anomalous oxygen isotopes, demonstrating its extrasolar origin. Oxygen isotopes in the CAI particle Inti are like those of other refractory inclusions.

The high abundance of crystalline minerals in comet dust was also unexpected, because most interstellar grains are amorphous. Some comet crystals appear to be fragments of microchondrules or refractory inclusions (Brownlee et al., 2012) that presumably formed in the inner solar system and were transported outward. At least in the Kuiper belt where Wild2 formed, a large fraction of the material had been cycled through the inner solar system before being incorporated into the comet (Westphal et al., 2009). Some Wild2 cometary grains appear to be loaded with noble gases that were probably implanted by energetic processes near the early Sun (Marty et al., 2008). Radial transport of material over large distances in the early solar system seems to have been remarkably efficient. Several models have been proposed to explain such large-scale mixing. One class calls on the bipolar outflows that are ubiquitous in young stellar objects to eject gas and dust outward, out of the plane of the accretion disk so that it can fall back onto the outer disk. The X-wind model (e.g. Shu et al., 1996) is an example. Turbulent diffusion is a natural consequence of disk evolution, and outward diffusion has been invoked as a possible explanation for the presence of high-temperature grains in comets (e.g. Ciesla, 2007; Hughes and Armitage, 2010). However, the formation location of the high-temperature particles in comets is not yet fully resolved.

12.4 Orbital Distributions of Planetesimals

12.4.1 Asteroids

It is instructive to compare the spectroscopic classifications of Main belt asteroids with their orbits. The distribution of the major asteroid classes (using the taxonomy of Tholen and Barucci, 1989, presented in Table 12.1) is illustrated in Figure 12.12a, which shows relative abundances as a function of orbital semimajor axis (heliocentric distance). This pattern was originally interpreted to indicate that asteroids were stratified according to their thermal histories, with melted (differentiated) bodies near the Sun, thermally metamorphosed or aqueously altered bodies farther out, and unaltered bodies at even larger heliocentric distance. A similar exercise, using the newer superclass taxonomy (Bus et al., 2002) and a large asteroid database (Mothe-Diniz et al., 2003), does not indicate such a clear picture (Fig. 12.12b). Nevertheless, S-complex asteroids dominate the inner and middle belt out to

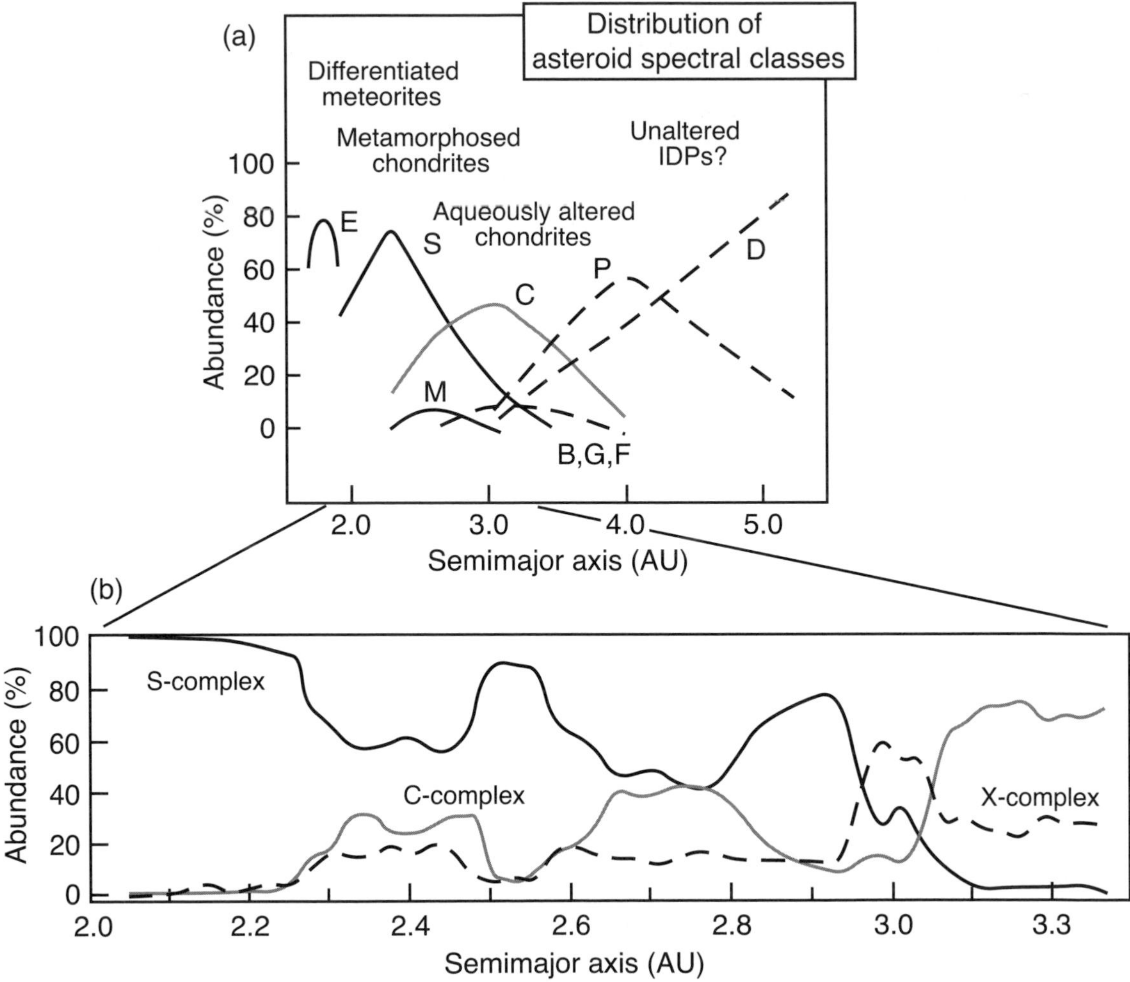

Figure 12.12 Heliocentric distribution of asteroid spectral types in the Main belt. (a) and (b) show distributions using two different asteroid taxonomies, from Tholen and Barucci (1989) and Bus et al. (2002), respectively. Possible petrologic interpretations of the various spectral classes are shown in (a).

2.95 AU, C-complex asteroids are most common in the outer belt, and X-complex bodies are most common beginning at about 3 AU. Note that the distribution in Figure 12.12b represents only a part of the Main asteroid belt shown in Figure 12.12a. Because of uncertainties in the interpretation of S-complex objects as either chondrites or achondrites, we can no longer say that the innermost asteroids are differentiated, but we can infer that S-complex bodies were at least heated (recall from Chapter 6 that chondrite classes other than carbonaceous are mostly metamorphosed). CI and CM2 chondrites have suffered extensive aqueous alteration, suggesting they formed beyond a "snowline" marking the condensation of ice that later melted; that snowline likely marks the transition to mostly C-complex objects at about 3 AU.

The Trojan asteroids are mostly D-class asteroids, although a few are classified as P and C objects. Centaurs have spectra that reveal water ice and CO, and a few exhibit cometary activity.

The orbital distribution of asteroids in the Main belt between Mars and Jupiter shows gaps (*Kirkwood gaps*) in both heliocentric distance and inclination. Several kinds of gravitational resonances act on asteroids at these locations, depopulating the gaps by perturbing the asteroids into different orbits. Several mean-motion resonances with Jupiter are illustrated in Figure 12.13.

12.4.2 Bodies in the Kuiper Belt and the Oort Cloud

The Kuiper belt stretches from roughly 30–55 AU. The greatest part of the belt, sometimes called the "classical belt," which contains most of the discovered KBOs, lies between two mean-motion resonances with Neptune: the 2:3 resonance at 39.5 AU and the 1:2 resonance at 48 AU. These resonances produce gaps in the Kuiper belt, analogous to the Kirkwood gaps in the Main asteroid belt. The Kuiper belt is quite thick, extending to 10 degrees outside the ecliptic, with a diffuse distribution several times beyond that; it resembles a doughnut that is inclined to the ecliptic by several degrees.

The classical belt consists of two populations: dynamically "cold" objects in nearly circular orbits with low inclinations, and dynamically "hot" objects with much higher inclinations. The names do not refer to temperatures, but are analogies to gas particles that increase velocities as they are heated. The cold KBOs are spectrally redder, occur more commonly as binary objects, and contain fewer large bodies. One hypothesis is that the cold population formed in its current position, whereas the hot KBOs formed near Neptune and were scattered outward during giant planet migration.

Orbits for objects in the Oort cloud have not been determined directly, except when they enter the inner solar

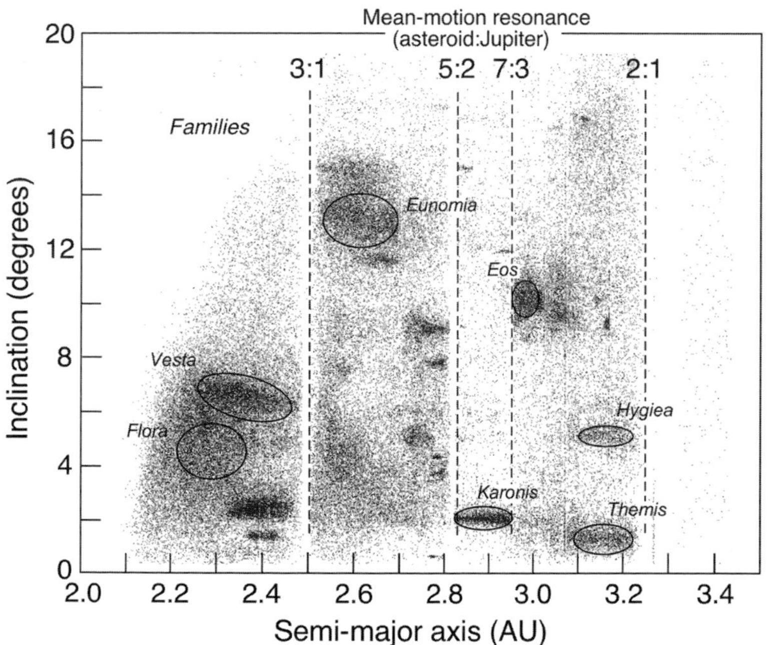

Figure 12.13 Distribution of orbital semi-major axes and inclinations of asteroids in the Main belt. Gaps occur where asteroid orbits are in resonance with Jupiter (dashed, labeled lines indicate orbit ratios for asteroids in that position with Jupiter). Some major asteroid families are circled and named. Modified from Nesvorny et al. (2015).

system as long-period comets. The Oort cloud is hypothesized to occupy a spherical volume surrounding the solar system, beginning at perhaps 2000 AU and extending almost a quarter of the distance to the nearest star.

12.4.3 Living in the Fast Lane

The Kirkwood gaps in the Main asteroid belt serve as "escape hatches" that provide a means of scattering materials knocked off asteroids by impacts into the inner solar system, where they become NEOs or meteorites. Other mechanisms that do not depend on gravity, such as the Yarkovsky effect (resulting from asymmetric solar heating of small bodies that causes their orbits to drift), can also cause asteroid fragments to enter the inner planet region. Gaps in the Kuiper belt, resulting from resonances with Neptune, also likely deliver short-period comets to the planetary region. Long-period comets come from the Oort cloud, possibly perturbed into highly elliptical or hyperbolic orbits by gravitational interactions with passing stars.

Thousands of NEOs are large enough to be tracked, and a few may eventually pose an impact hazard to our planet. Such impacts in the geologic past have caused biologic extinctions. Near-Earth objects and comets are the immediate source of most meteorites and IDPs, liberated by impacts onto NEOs or shed during cometary activity. Because of their proximity, NEOs are the targets for spacecraft missions that return samples to Earth.

12.5 Thermal Metamorphism, Aqueous Alteration, and Melting of Planetesimals

12.5.1 Heat Sources

Various heat sources for asteroid metamorphism and melting of rock or ice have been considered. First, let's consider what is probably implausible. The slow decay of long-lived radionuclides (isotopes of potassium, uranium, and thorium), a primary heat source for large planets, is largely ineffective for asteroids because the timescale for the release of energy is long compared to that for conductive heat loss from small bodies. Impacts have been invoked as a heat source for asteroids, and localized melting can certainly occur during cratering events. However, the global temperature increase from collisions of asteroids is only a few degrees, because small bodies do not produce large amounts of collisional energy. Impacts can be effective heat sources if they involve very large objects, but energy transfer to the interior of the target is inefficient (Keil et al., 1997). Electromagnetic induction, whereby heat is produced by resistance to the flow of electric currents induced by outflows from the young Sun, has also been promulgated as a cause of asteroid heating (Herbert and Sonnett, 1980). However, outflows from young stars are focused at the stellar poles, avoiding the nebular disk where planetesimals occur. Induction models are also highly dependent on the choice of parameters, few of which are constrained, so it is difficult to ascertain if these models are reasonable.

The most plausible heat source for heating planetesimals is decay of short-lived radionuclides, such as ^{26}Al and possibly ^{60}Fe. The decay products of these nuclides have been found in both chondrites and achondrites, and their wide distribution and high decay energy make them attractive sources of asteroid heating. There is some uncertainty in the abundance of ^{60}Fe (refer back to Fig. 10.5), which would affect its contribution to asteroid heating. In any case, recent evidence that the parent bodies of achondrites formed very early, within 1 Myr of the formation of CAIs (Chapter 10), means that abundances of any short-lived radionuclide would have been very high in these bodies, leading to melting. Chondrite parent bodies, which formed $>$2 Myr after CAIs, may not have inherited enough of the short-lived nuclides to melt silicates or metal.

12.5.2 Thermal Evolution of Asteroids without Ice

The peak temperatures that asteroids have experienced can be estimated from chemical exchange reactions between minerals – so-called *geothermometer*s. For example, the exchange of calcium between coexisting orthopyroxene and clinopyroxene in highly metamorphosed chondrites has been used to estimate their equilibration temperatures (e.g. Slater-Reynolds and McSween, 2005). For ordinary chondrites, these temperatures range up to ~1175 K. The experimental conditions at which achondrites melt provide minimum temperatures for their parent bodies. Partial melting of achondrites typically requires temperatures of $>$1200 K.

Metamorphic reactions in anhydrous chondrites (O, E, R classes) involved textural recrystallization, along with mineral changes. During progressive metamorphism, chemically zoned minerals (especially olivine) become homogenized, pyroxene structures change from monoclinic to orthorhombic, and chondrule glasses devitrify to form feldspars. These changes provide a means of classifying chondrites in terms of petrologic type, as catalogued earlier in Table 6.1. Another metamorphic reaction in ordinary chondrites involves oxidation:

$$\underset{\text{metal}}{Fe} + O_2 + \underset{\text{pyroxene}}{(Mg, Fe)SiO_3} = \underset{\text{olivine}}{(Mg, Fe)_2SiO_4}$$

which changes the relative proportions of pyroxene and olivine (McSween and Labotka, 1993). Several studies have observed progressive changes in the olivine/pyroxene ratio and in the proportion of metal with increasing degrees of metamorphism in ordinary chondrites (e.g. Dunn et al., 2010). Certain spectral features are sensitive to the olivine/pyroxene ratio, so this change can be seen in asteroid spectra.

The rates at which parent bodies cooled also provide constraints on thermal models. A method for determining the cooling rates for iron meteorites is described in Box 12.3. A similar method for estimating chondrite cooling rates is also based on the compositions of metal grains. Cooling rates can also be estimated from knowing the blocking temperatures of various radioisotope systems in meteorites.

The basis for asteroid thermal models is the heat transfer equation:

$$\frac{\partial T}{\partial t} = \frac{1}{R^2}\frac{\mathrm{d}}{\mathrm{d}R}\left(R^2\kappa\frac{\partial T}{\partial R}\right) + \frac{Q}{\rho C_v} \qquad 12.1$$

where T is temperature, R is the radius of the asteroid, t is time, κ is thermal diffusivity, Q is the heat generated by radioactive decay, ρ is density, and C_v is the specific heat at constant volume. Let's see how each term in the equation might affect the solution. The term on the left is the rate of change of temperature with time in a layer of infinitesimal thickness at any arbitrary depth in the asteroid. The first term on the right gives the amount of heat gained in (or lost by) the infinitesimal layer by conduction. The amount of heat transmitted by conduction is proportional to the rate at which the product of thermal gradient (∂T/∂R) and thermal diffusivity changes. The last term on the right gives the amount of heat generated in the asteroid by radioactive decay.

By necessity, asteroid thermal models must make assumptions about model parameters and initial conditions. However, models based on ^{26}Al heating have been remarkably successful in reproducing asteroid peak temperatures and cooling histories.

Thermal models for anhydrous asteroids that do not melt, such as the parent bodies for ordinary chondrites, are relatively straightforward, because heat migrates through these bodies by conduction. Decay of ^{26}Al results in rapid heating within the first few million years after accretion. These models (McSween et al., 2003) suggest that asteroid interiors experienced the highest temperatures, producing highly metamorphosed (type 6) chondrites, and that successively less-metamorphosed shells occur toward the surface. The resulting thermal structure, called an *onion shell*, is illustrated in Figure 12.14a. Cooling rates estimated for chondrites

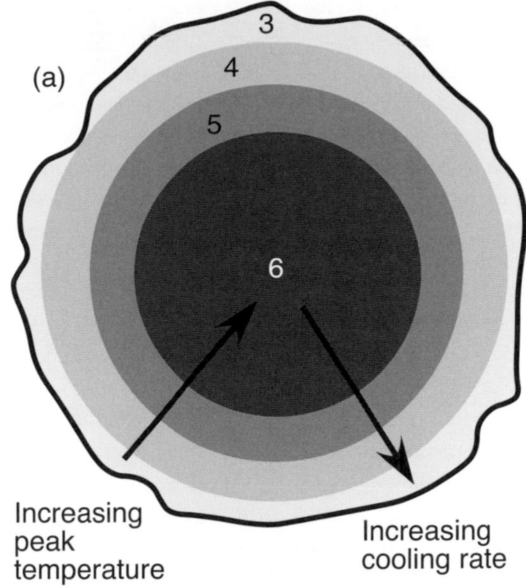

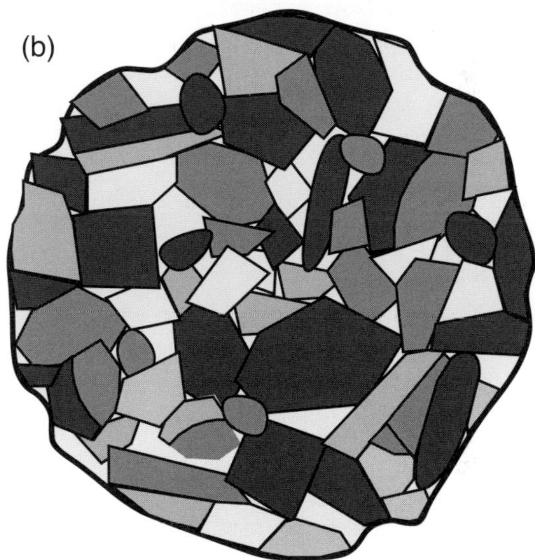

Figure 12.14 Sketches of the interiors of chondritic asteroids, illustrating (a) onion-shell and (b) rubble-pile structures. The numbers and shadings identify petrologic types.

of higher metamorphic grade are slower than those of lower grade (Fig. 12.15), because diffusion of heat from the deep interior is slower than from near the surface, supporting the onion-shell model. These cooling rates are based on various radiogenic isotope chronometers, which block at different temperatures (see also Fig. 10.12). For asteroids with regoliths, the outer fine-grained layer insulates the interior, preventing efficient heat loss by radiation to space. Consequently, small

Box 12.3 Iron Meteorite Cooling Rates: A Study of Nickel Diffusion

Nickel and iron form a solid solution at high temperatures, but an alloy on cooling will unmix to form oriented plates of low-Ni kamacite within high-Ni taenite. These plates constitute the familiar Widmanstätten pattern seen in many iron meteorites (see Fig. 6.10). Experiments indicate that both phases become more nickel rich with further cooling, which is accomplished as kamacite plates grow thicker at the expense of taenite. Kamacite growth and taenite shrinkage occur by diffusion of nickel, the rate of which varies with temperature. Nickel diffusion in kamacite is rapid, as these plates tend to be homogeneous. However, at some point during cooling, diffusion becomes sluggish enough that nickel atoms in taenite begin to pile up at the margins of the plates, producing chemically zoned taenite in which the nickel content decreases toward the center of the grains. By approximately 725 K, diffusion effectively grinds to a halt and the zoning pattern becomes frozen in, as illustrated in Figure 12.18. Electron microprobe analyses of spots in a traverse across a taenite grain define a nickel profile that is often described as "M-shaped" (only half of the M is illustrated in Fig. 12.18).

Nickel diffusion profiles in taenite depend on nickel diffusion rates, the bulk nickel content, the nucleation temperature of kamacite, and the dimensions and cooling history of the system. The last parameter is the most interesting. Other things being equal, the more slowly an iron meteorite cools, the greater the opportunity for nickel to diffuse to the interiors of the taenite crystals. By matching measured nickel profiles with calculated profiles, meteoriticists have estimated the cooling rates for iron meteorites. The most recent version of this method is given by Yang and Goldstein (2005). Iron meteorite cooling rates have been found to vary from less than 1 K to several thousand K per million years.

Cooling rates have important implications for the sizes of meteorite parent bodies. Because of its high thermal conductivity, a metallic core should have a uniform temperature, but its rate of cooling is controlled by outer silicate layers that act like insulation. Larger asteroids cool more slowly, and Haack et al. (1990) developed an approximate relationship between the radius of an asteroid, R, and the cooling rate, CR, of its metallic core:

$$R = 149(CR)^{-0.465}$$

Using this equation, the calculated radii of most iron meteorite parent bodies are found to have been ~10–100 km.

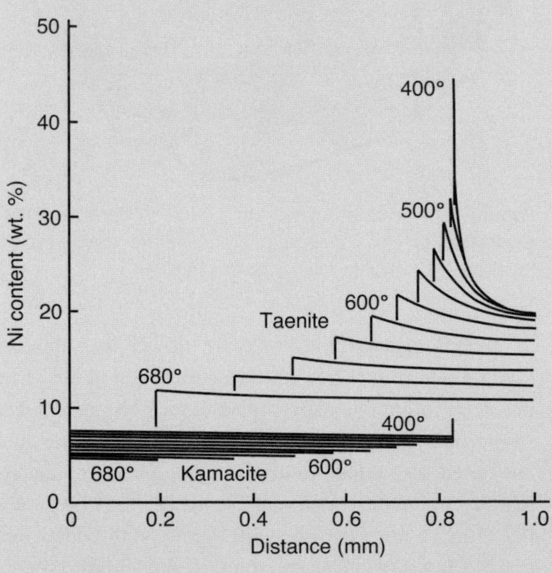

Figure 12.18 Computation of the development of nickel profiles in kamacite and taenite within an iron meteorite. The left edge of the figure represents the midpoint of a kamacite plate. The thickness of the plate expands and its nickel content increases with falling temperature. The right edge is the center of a residual taenite area. Its nickel content also increases with falling temperature, first throughout the crystal and later only at its edges, producing an M-shaped profile (only half of which appears in this sketch). Temperatures are given in °C.

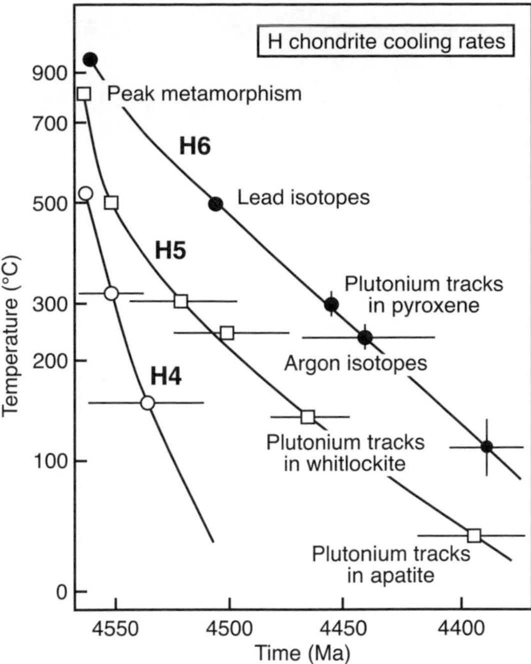

Figure 12.15 Cooling rates for H chondrites of different petrologic types can be estimated from the blocking temperatures for diffusion in various radioisotope systems. The age measured by each radioisotope indicates the time at which the meteorite passed through that blocking temperature. Slower cooling rates for more highly metamorphosed H5 and H6 chondrites are taken as evidence that their parent body had an onion-shell structure.

asteroids with thick regoliths can produce the same cooling rates as larger asteroids.

12.5.3 Thermal Evolution of Asteroids Containing Ice

The interiors of P- and D-class asteroids are thought have accreted water ice that never melted, based on their spectra. The C-class asteroids also originally accreted with ice, but the ice subsequently melted to form aqueous fluids. Such fluids reacted with anhydrous minerals like olivine, pyroxene, and metal to produce a host of secondary phases. The alteration minerals include serpentines and clays, as well as a variety of carbonates, sulfates, oxides, sulfides, halides, and oxyhydroxides, some of which are pictured in Figure 12.16. The identities of these phases in carbonaceous chondrites reveal the conditions of alteration, and thereby constrain thermal models for ice-bearing asteroids. The alteration of carbonaceous chondrites has been discussed extensively in the literature (reviewed by Brearley, 2006). In CI chondrites, the alteration was pervasive and almost no unaltered minerals remain (King et al., 2015). CM chondrites contain mixtures of heavily altered and partially altered materials. The degree of alteration among CM chondrites is quite variable, and several alteration indices based on their mineral assemblages and chemical compositions have been proposed (Browning et al., 1996; Rubin et al., 2007; Howard et al., 2015). In CR2 and $CV3_{oxB}$ chondrites, matrix minerals have been moderately altered and chondrules show some effects of aqueous alteration. For other carbonaceous chondrite groups, the alteration is subtle and secondary minerals are uncommon. In some CV chondrites, a later thermal metamorphic overprint has dehydrated serpentine to form iron-rich olivine.

These differences suggest that the compositions and amounts of fluids and the temperatures of reactions in asteroids were quite variable. Temperatures and water/rock ratios have been constrained from oxygen-isotope exchange between various minerals (e.g. Box 8.1). Mass-dependent isotopic fractionation between coexisting calcite and phyllosilicates indicates that aqueous alteration of CM chondrites took place at ~0–80°C, whereas alteration in CI chondrites occurred at temperatures between 100 and 150°C (Clayton and Mayeda, 1984, 1999). Mass-independent isotopic differences among phases permit estimates of the amount of water with which the rock reacted. Modeling of isotopic exchange during aqueous alteration between rock and fluid leads to estimated water/rock ratios for alteration of CM chondrites of 0.36–0.6, and similar modeling for CI chondrites gives water/rock ratios of ~2.5 (Clayton and Mayeda, 1999).

Thermal evolution models for hydrated carbonaceous chondrite asteroids (e.g. Grimm and McSween, 1989; Young, 2001) are similar to those for anhydrous chondritic asteroids, although there are a few important differences. The oxygen isotopic constraints suggest that alteration temperatures were within ~100°C of the melting temperature of water ice. The large heat of fusion of ice, the high heat capacity of water, and the ability of circulating water to enhance heat loss buffered temperature excursions when short-lived radionuclides decayed. Consequently, carbonaceous chondrites experienced aqueous alteration at low temperatures rather than thermal metamorphism. However, hydration reactions are exothermic and thus could have liberated large amounts of heat if reaction rates were rapid relative to ice melting. High vapor pressures associated with ice melting may also have caused fracturing and venting of gases.

Thermal models presume that aqueous alteration occurred in parent bodies. However, Metzler et al. (1992) argued that altered rims around chondrules in unbrecciated CM chondrites formed by hydration reactions in the nebula, rather than within hydrous asteroids. This proposal led to controversy about the environment for aqueous alteration. Reactions of anhydrous minerals either with water vapor in

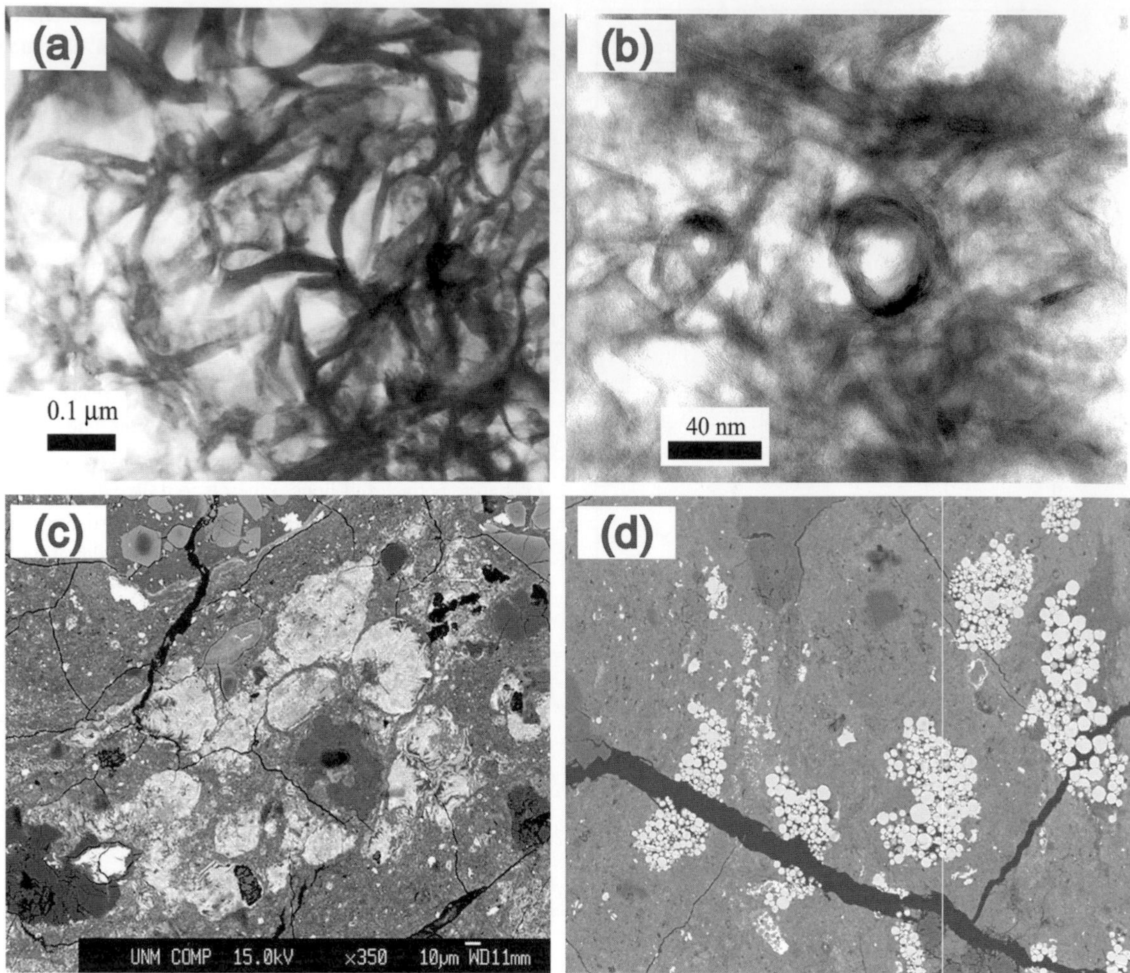

Figure 12.16 Some alteration minerals in carbonaceous chondrites. (a) Transmision electron microscope (TEM) image of twisted phyllosilicate plates in the Ivuna CI chondrite. (b) TEM image of hollow serpentine tubes in the Murchison CM2 chondrite. (c) Backscattered electron image of intergrowths of tochilinite and serpentine (light-colored clots) in the Y-791198 CM2 chondrite. (d) Reflected light image of clusters of spherical magnetite grains, called "framboids," in the EET 93042 CR2 chondrite. Figures courtesy of A. Brealey.

the solar nebula or during the passage of shock waves through regions with elevated ice/dust ratios might have produced aqueous alteration minerals. More traditional models envision alteration occurring within parent bodies after accretion and melting of ice and are generally favored. Experimental and theoretical constraints on mineral stability and slow reaction rates in the nebula appear to support parent-body alteration. Brearley (2006) has summarized the observations used in support of each model.

12.5.4 Planetesimal Melting and Differentiation

Thermal models for differentiated asteroids are more complex than for chondritic bodies, because radiogenic heat sources move around during the simulation. When partial melting occurs, aluminum (and thus ^{26}Al) partitions into silicate magma and iron (and thus ^{60}Fe) is concentrated into metallic melt. The dense metallic melt drains into the core and the less dense silicate melt ascends towards the surface to form a basaltic crust, as illustrated in Figure 12.17. Such separations are not instantaneous, because melts can "wet" solid residues, so some threshold value of partial melting is usually required to initiate phase separation. Because no individual samples of core, mantle, or crust retain the asteroid's bulk composition, there is some uncertainty in estimating the composition of the parent body prior to

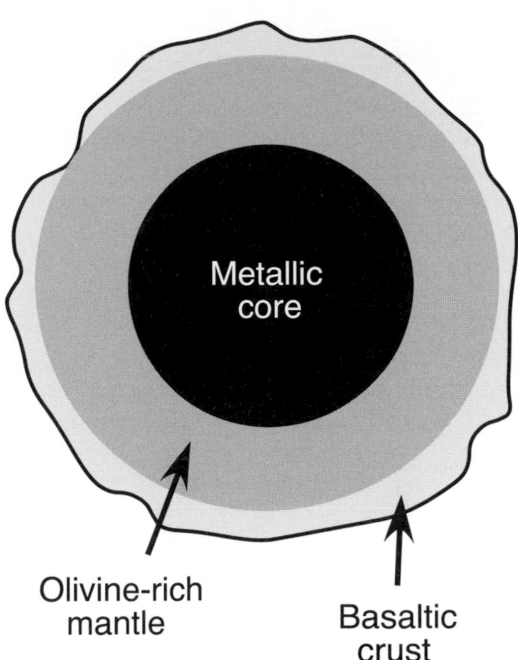

Figure 12.17 Sketch of the interior of a differentiated rocky asteroid.

differentiation. The bulk compositions of differentiated bodies are generally thought to be chondritic but, as we learned in Chapter 7, they are more depleted in volatile elements than are chondrites. Thermal models for asteroid Vesta (Mandler and Elkins-Tanton, 2013, is just one of many), the HED parent body, indicate rapid melting and differentiation, with high temperatures being maintained for tens of millions of years. Such models are broadly consistent with the radiometric ages for HED meteorites, as discussed in Chapter 10.

A variation on differentiation models is based on the Dawn spacecraft's exploration of dwarf planet Ceres. In this case, ice melted and the brine separated from the rock matrix and likely formed an icy crust on the surface. Some of that ice is preserved in Ceres' outer shell (Prettyman et al., 2017), although most of it was probably lost through impacts and sublimation. Thermal models that involve melting of ice, rather than rock, can result in fractionation of water and rock (Castillo-Rogez and McCord, 2010).

12.6 Compositional and Thermal Structure of the Asteroid Belt

A roughly heliocentric pattern of asteroid types, with S- and C-class objects closer to the Sun and P- and D-class objects farther out, persists for the largest asteroids (DeMeo et al., 2015). This gradation is not only compositional but also thermal, because the S- and C-class objects are interpreted to have suffered metamorphism or aqueous alteration, and the P- and D-class objects are thought to be primitive. The gradient is less obvious for smaller asteroids, probably because of subsequent dynamical stirring of orbits and ejection of bodies from the Main belt. Differentiated objects appear to have formed earlier than chondritic bodies, and dynamical modeling suggests they may have even accreted inside 2 AU.

This pattern may be a natural consequence of the fact that the time of accretion was earlier where the density of materials in the swarm was higher, and density decreased at greater solar distances. The planetesimals that accreted closer to the Sun and hence earlier would have had larger proportions of live ^{26}Al available to drive heating. A model summarizing this effect is illustrated in Figure 12.19 (Grimm and McSween, 1993). This diagram shows contours of peak temperature on a plot of asteroid size versus solar distance (the latter is assumed to be equivalent to accretion time, shown at the top of the diagram). The vertical bar at 2.7 AU represents an approximate boundary between bodies that melted or were thermally metamorphosed and those that experienced melting of ice and subsequent aqueous alteration. The vertical bar at 3.4 AU denotes the transition to unaltered asteroids that might still contain ice. The accretion times at the top of Figure 12.19 produce peak temperatures appropriate for silicate melting (1375 K) and ice melting (273 K) for 100 km asteroids at the distances of the bars.

Ghosh et al. (2006) formulated a more complex model that incorporated incremental rather than instantaneous accretion. Because the timescale of accretion is comparable to that of ^{26}Al decay, asteroids were actually heated as they grew. By coupling a thermal model with an accretion model, they were able to calculate the stratification of asteroids in a more realistic way.

The model in Figure 12.19, in which asteroids accreted at approximately their current locations, represents one end member of asteroid formation scenarios. Another, currently more widely accepted end member suggests that the asteroid belt of today may be very different from how it was when it formed. Bottke et al. (2006) proposed that differentiated asteroids formed closer to the Sun and were then scattered into the inner Main belt by interactions with growing planets. The Nice and Grand Tack models (summarized by Morbidelli et al., 2015) posit that the formation of C-, P-, and D-class asteroids (and the snowline) originally occurred in the

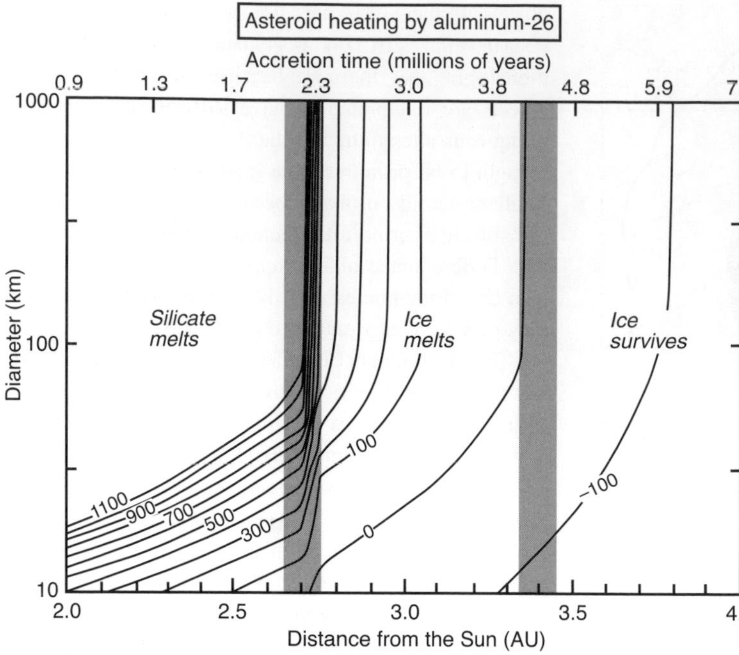

Figure 12.19 Contours of peak temperature (K) in asteroids as functions of body diameter and heliocentric distance. Assumed accretion times corresponding to various solar distances are given at the top of the figure (this model assumes accretion times increase with distance). Shaded vertical bars mark major divisions in the asteroid belt based on spectral interpretations of asteroid thermal histories. Modified from Grimm and McSween (1993).

outer solar system, and that these bodies were later implanted into the outer Main belt by stirring of their orbits as Jupiter and Saturn migrated inward. Such dynamic models, described in Chapter 15, are also used to explain the ejection of primitive planetesimals near Neptune into the Kuiper belt. In these models, the compositional and thermal gradient with heliocentric distance is not primordial.

12.7 Collisions among Planetesimals

Collisions among asteroids have been commonplace over solar system history. In general, impactors form craters roughly 10 times their own diameter. However, the physics of cratering changes as crater diameters approach the diameters of the target asteroids themselves. In that case, both the impactor and the target asteroid may be catastrophically disaggregated. If most of the fragments do not exceed the escape velocity and instead settle back, or reaccrete, a new asteroid is formed. The structure of such a body, in which interior and exterior fragments are mixed haphazardly, is called a *rubble pile* (Fig. 12.14b).

Evidence for rubble-pile asteroids comes from a variety of observations. The low densities of many asteroids imply that they have high porosities, which likely results from the assembly of loose fragments. Spectral variations seen in some S-class asteroids as they rotate also support rubble piles. The variations suggest that portions of the surface have experienced different degrees of thermal metamorphism. Catastrophic collision and reassembly has transformed bodies that formerly had onion-shell structures into rubble piles.

On a smaller scale, materials excavated by impact craters may become compacted and cemented together, forming breccias, which reveal further evidence of rubble-pile parent bodies. For example, the Cangas de Onis ordinary chondrite breccia contains metal grains that cooled at rates varying from 1 to 1000 K per million years (Fig. 12.20). (Metallographic cooling rates for chondrites are based on diffusion of nickel in iron, as discussed previously for iron meteorites in Box 12.3.) The slowest measured cooling rates could only have occurred in the center of the asteroid, so this one rock contains fragments that formed at various depths ranging from the center to the surface. Any impact that excavated materials from an asteroid center would certainly have disrupted it, so the formation of this breccia requires subsequent reassembly.

Other indications of cooling rate ("speedometers") have also been applied to chondrites. Besides metallography, $^{40}Ar/^{39}Ar$ ages and ^{244}Pu fission-track ages with relatively low closure temperatures ($\leq$500°C) have been used extensively (e.g. Treiloff et al., 2003), but they are not sensitive to collisional breakup near peak metamorphic temperatures. Thermometers based on diffusional exchange of REEs between high-Ca and low-Ca pyroxenes, or diffusion of Ca in olivine, close at high

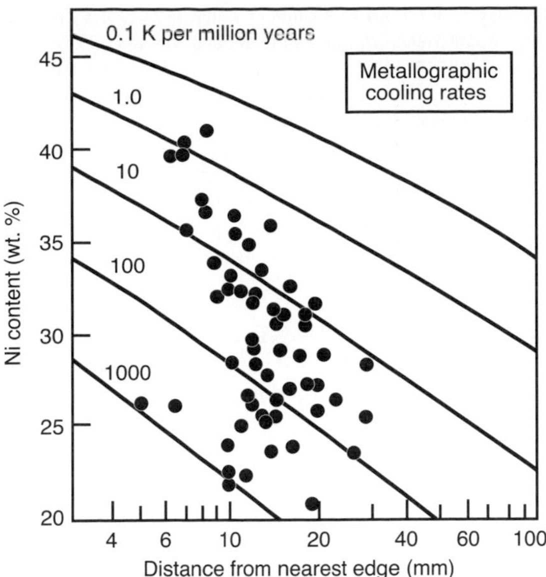

Figure 12.20 Metal grain compositions in the Cangas de Onis chondrite breccia, and calculated metallographic cooling rates (curves). The cooling rates are estimated from modeling nickel diffusion profiles in metal grains. The slowest cooling rates of 1 K/Myr require that those grains formed in the asteroid center, whereas grains with cooling rates of ~1000 K/Myr must have formed near the surface. This huge range of cooling rates in one meteorite is evidence that the parent body was disrupted by impact and then reassembled into a rubble pile. Modified from Williams et al. (1985).

temperatures and thus are better indications of fragmentation and reassembly. Such data support the contention that H, L, and LL chondrite parent bodies were rubble piles (Lucas et al., 2020).

When asteroid collisions are especially violent, sufficient kinetic energy may be imparted to launch large fragments at greater than escape velocities. In that case, separate asteroids are formed. Such fragments generally share similar orbital characteristics and are referred to as *asteroid families*. Several prominent families are indicated by concentrations of asteroids in Figure 12.13. The members of most asteroid families share the same spectral characteristics, further linking them together. Families composed of fragments of differentiated asteroids can potentially provide important information on their internal compositions.

Some asteroid collisions may be glancing blows rather than head-on collisions. In such a case, one body may strip the other of its outer layers, as discussed in Chapter 7. If the target asteroid was differentiated, this would provide a means of removing the silicate crust and mantle, leaving behind a naked metal core (Asphaug et al., 2006). M-class asteroids may be the result of such impact stripping. This process may also explain the unusual cooling rates of several classes of iron meteorites. If metallic cores cooled slowly inside thermally insulating silicate mantles, the metallographic cooling rates of all samples of that core should be the same. However, the cooling rates of IVA irons vary from 90 to 7000 K per million years, and IIIAB irons show smaller but still significant differences in cooling history. The exteriors of cores suddenly exposed to frigid space by glancing impacts would cool more rapidly than their interiors. Impact removal may also account for the fact that olivine-rich mantle materials are uncommon among meteorites, whereas metallic cores are abundant. Repeated impacts might have ground the mantle materials to tiny particles over time.

Summary

Planetesimals occupy a very important place in cosmochemistry, because samples of them provide the data used to understand nebular composition, processes, and chronology. Planetesimals occur as asteroids that accreted as rock, with or without ices. Asteroid taxonomy is based on spectroscopy, and spectral comparisons provide connections between asteroid classes and various kinds of meteorites. It is unclear whether asteroids located beyond the snowline accreted in place or accreted farther out and were emplaced later. In any case, most of the ice-bearing planetesimals were swept out of the planetary region by gravitational interactions with the giant planets, and their distribution now extends out to the limits of the Sun's grasp. Icy bodies in the distant Kuiper belt and Oort cloud are sometimes sampled as comets and the dust they shed (CP IDPs). The orbital distributions of the various classes of planetesimals help constrain their formation, although dynamic models suggest that some radial mixing has occurred. Gaps in the Main asteroid belt and Kuiper belt, produced by resonances with giant planets, provide a means of delivering objects into the inner solar system.

Thermal metamorphism or aqueous alteration in chondritic asteroids and melting in differentiated asteroids are driven by heat produced by the decay of short-lived radionuclides (especially ^{26}Al). Thermal models can reproduce the peak temperatures and cooling rates estimated for meteorite parent bodies. Such models for chondrites produce onion-shell bodies, with more highly metamorphosed materials at the center and materials of lower metamorphic grade near the surface. Models for ice-bearing asteroids result in aqueous alteration at low temperatures, because of thermal buffering by ice. Models for melted asteroids produce bodies with metallic cores, olivine-rich mantles, and basaltic crusts, or, in the case of ice-bearing asteroids, altered rocky interiors and icy crusts. KBOs have suffered less thermal processing, but smaller bodies have still lost the most volatile ices.

Impacts have disrupted onion-shell asteroids, producing asteroid families with similar orbital parameters, and reaccretion has produced porous rubble piles. Glancing impacts have stripped away silicate mantles, leaving naked iron cores. KBOs have also been disrupted by impacts, producing millions of smaller objects.

This chapter is intended to provide geologic context for understanding the compositions of planetesimals, which is the subject we consider next.

Questions

1. How are spectroscopic observations used to relate various kinds of meteorites to asteroid parent bodies?
2. What are plausible heat sources for planetesimals, and how does heating manifest itself in ice-free and ice-bearing bodies?
3. What are onion-shell and rubble-pile structures for asteroids, and how do they form?
4. How are the various classes of planetesimals distributed in the solar system, and what can we infer about their origin from their distributions?
5. What do we know about the compositions of Kuiper belt objects and the processes that have affected them?
6. It has sometimes been said that there is a compositional continuum between asteroids and comets. What does this mean, and what are the implications of this statement?
7. How have collisions affected planetesimals?

Suggestions for Further Reading

The University of Arizona Press has published a number of books on asteroids, and these two are the most current. Each is a wonderful resource on the properties, evolution, and exploration of asteroids.

Bottke W. F., Cellino A., Paolicchi P., and Binzel R. P., editors (2002) *Asteroids III*. University of Arizona Press, Tucson, 785 pp.

Michel P., Demeo F. E., and Bottke W. F., editors (2015) *Asteroids IV*. University of Arizona Press, Tucson, 895 pp.

Bradley J. P. (2014) Early solar nebula particles - interplanetary dust particles. In *Treatise on Geochemistry, 2nd edition, Vol. 1: Meteorites and Cosmochemical Processes*, Davis A. M., editor, pp. 287–308, Elsevier, Oxford. This review provides an excellent summary of the voluminous literature that describes and interprets IDPs.

Brearley A. J. (2006) The action of water. In *Meteorites and the Early Solar System II*, Lauretta D. S., and McSween H. Y., editors, pp. 587–624, University of Arizona Press, Tucson. The best available review of aqueous alteration processes and materials in chondritic meteorites.

Brownlee D. E. (2014) Comets. In *Treatise on Geochemistry, 2nd edition, Vol. 2. Planets, Asteroids, Comets and the Solar System*, Davis A. M., editor, pp. 335–363, Elsevier, Oxford. A comprehensive and thoughtful review of what is known about the composition and structure of comets and the bodies within the belts that supply comets.

Other References

Asphaug E., Agnor C. B., and Williams Q. (2006) Hit-and-run planetary collisions. *Nature*, **439**, 155–160.

Bottke W. F., Nesvorny D., Grimm R. E., et al. (2006) Iron meteorites as remnants of planetesimals formed in the terrestrial planet region. *Nature*, **439**, 821–824.

Bottke W. F., Broz M., O'Brien D. P., et al. (2015) The collisional evolution of the Main asteroid belt. In *Asteroids IV*, Michel P., Demeo F. E., and Bottke W. F., editors, pp. 701–724, University of Arizona Press, Tucson.

Brown M. E. (2002) The compositions of Kuiper Belt objects. *Annual Review of Earth and Planetary Sciences*, **40**, 467–494.

Browning L. B., McSween H. Y., and Zolensky M. E. (1996) Correlated alteration effects in CM carbonaceous chondrites. *Geochimica et Cosmochimica Acta,* **60**, 2621–2633.

Brownlee D., and 181 coauthors (2006) Comet 81P/Wild2 under a microscope. *Science*, **314**, 1711–1716.

Brownlee D., Joswiak D., and Matrajt G. (2012) Overview of the rocky component of Wild2 comet samples: Insights into the early solar system, relationship with meteoritic materials and the differences between comets and asteroids. *Meteoritics & Planetary Science*, **47**, 453–470.

Bus, S. J., Vilas, F., and Barrucci, M. A. (2002) Visible-wavelength spectroscopy of asteroids. In *Asteroids III*, Bottke, W. F., Cellino, A., Paolicchi, P., and Binzel, R. P., editors, pp. 169–182, University of Arizona Press, Tucson.

Carry B. (2012) Density of asteroids. *Planetary & Space Science*, **73**, 98–118.

Castillo-Rogez J. C., and McCord T. B. (2010) Ceres' evolution and present state constrained by shape data. *Icarus*, **205**, 443–459.

Ciesla F. J. (2007) Outward transport of high-temperature materials around the midplane of the solar nebula. *Science*, **318**, 613–615.

Clark B. E., Hapke B., Pieters C., and Britt D. (2002) Asteroid space weathering and regolith evolution. In *Asteroids III*, Bottke W. F., Cellino A., Paolicchi P. and Binzel R. P., editors, pp. 585–602, University of Arizona Press, Tucson.

Clayton R. N., and Mayeda T. K. (1984) The oxygen isotope record in Murchison and other carbonaceous chondrites. *Earth & Planetary Science Letters*, **67**, 151–166.

Clayton R. N., and Mayeda T. K. (1999) Oxygen isotope studies of carbonaceous chondrites. *Geochimica et Cosmochimica Acta*, **63**, 2089–2104.

Crovisier J., et al. (2000) The thermal infrared spectra of comets Hale-Bopp and 103P/Hartley 2 observed with the Infrared Space Observatory. In *Thermal Emission Spectroscopy and Analysis of Dust, Disks, and Regoliths, ASP Conference 196*, Sitko M. L., Sprague A. L., and Lynch D. K., editors, pp. 109–117, Astronomical Society of the Pacific, San Francisco.

DeMeo F. E., Binzel R. P., Silvan S. M., and Bus S. J. (2009) An extension of the Bus asteroid taxonomy into the near-infrared. *Icarus*, **202**, 160–180.

DeMeo F. E., Alexander C. M. O'D., Walsh K. J., et al. (2015) The compositional structure of the asteroid belt. In *Asteroids IV*, Michel P., Demeo F. E., and Bottke W. F., editors, pp. 13–42, University of Arizona Press, Tucson.

De Sanctis M. C., Ammannito E., Marchi S., et al. (2015) Ammoniated phyllosilicates with a likely outer solar system origin on (1) Ceres. *Nature*, **528**, 241–245.

Dunn T. L., Cressey G., McSween H. Y., and McCoy T. J. (2010) Analysis of ordinary chondrites using powder X-ray diffraction: 1. *Modal mineral abundances. Meteoritics & Planetary Science*, **45**, 123–134.

Flynn G. J., Consolmagno G. J., Brown P., and Macke R. J. (2018) Physical properties of the stone meteorites: Implications for the properties of their parent bodies. *Chemie der Erde*, **78**, 269–298.

Fornasier S., Lantz C., Barucci M. A., and Lazzarin M. (2014) Aqueous alteration on main belt primitive asteroids: Results from visible spectroscopy. *Icarus*, **233**, 163–178.

Ghosh A., Weidenschilling S. J., McSween H. Y., and Rubin A. (2006) Asteroid heating and thermal stratification of the asteroid belt. In *Meteorites and the Early Solar System II*, Lauretta D. S., and McSween H. Y., editors, pp. 555–566, University of Arizona Press, Tucson.

Gounelle M., Spurny P., and Bland P. A. (2006) The orbit and atmospheric trajectory of the Orgueil meteorite from historical records. *Meteoritics & Planetary Science*, **41**, 135–150.

Greenwood R. C., Burbine T. H., Miller M. F., and Franchi I. A. (2017) Melting and differentiation of early-formed asteroids: The perspective from high precision oxygen isotope studies. *Chemie der Erde*, **77**, 1–43.

Grimm R. E., and McSween H. Y. (1989) Water and the thermal evolution of carbonaceous chondrite parent bodies. *Icarus*, **82**, 244–280.

Grimm R. E., and McSween H. Y. (1993) Heliocentric zoning of the asteroid belt by aluminum-26 heating. *Science*, **259**, 653–655.

Grundy W. M., Bird M. K., Britt D. T., et al. (2020) Color, composition, and thermal environment of Kuiper Belt object (486958) Arrokoth. *Science*, **367**, 999.

Haack H., Rasmussen K. L., and Warren P. H. (1990) Effects of regolith/megaregolith insulation on the cooling histories of differentiated asteroids. *Journal of Geophysical Research*, 95, 5111–5124.

Hanner M. S., and Bradley J. P. (2003) Composition and mineralogy of comet dust. In *Comets II*, Festou M., Keller H. U., and Weaver H. A., editors, pp. 555–564, University of Arizona Press, Tucson.

Herbert F., and Sonnett C. P. (1980) Electromagnetic inductive heating of the asteroids and moon as evidence bearing on the primordial solar wind. In *The Ancient Sun*, Pepin R. O., Eddy J. A., and Merrill R. B., editors, pp. 563–576, Pergamon, New York.

Howard K. T., Alexander C. M. O'D., Schrader D. L., and Dyl K. A. (2015) Classification of hydrous meteorites (CR, CM and C2 ungrouped) by phyllosilicate fraction: PSD-SRD modal mineralogy and planetesimal environments. *Geochimica et Cosmochimica Acta*, **149**, 206–222.

Hughes A. L. H., and Armitage P. J. (2010) Particle transport in evolving protoplanetary disks: Implications for results from Stardust. *Astrophysical Journal*, **719**, 1633–1653.

Ishii H. A., Bradley J. P., Dai Z. R., et al. (2008) Comparison of comet 81P/Wild2 dust with interplanetary dust from comets. *Science*, **319**, 447–450.

Jones T. D., Lebofsky L. A., Lewis J. S., and Marley M. S. (1990) The composition and origin of the C, P, and D asteroids: Water as a tracer of thermal evolution in the outer belt. *Icarus*, **88**, 172–192.

Joswiak D. J., Brownlee D. E., Matrajt G., et al. (2009) Kosmochloric Ca-rich pyroxenes and FeO-rich olivines (kool grains) and associated phases in Stardust tracks and chondritic porous interplanetary dust particles: Possible precursors to FeO-rich type II chondrules in ordinary chondrites. *Meteoritics & Planetary Science*, **44**, 1561–1588.

Keil K., Stoffler D., Love S. G., and Scott E. R. D. (1997) Constraints on the role of impact heating and melting in asteroids. *Meteoritics and Planetary Science*, **32**, 349–363.

King A. J., Schofield P. E., Howard K. T., and Russell S. S. (2015) Modal mineralogy of CI and CI-like chondrites by X-ray diffraction. *Geochimica et Cosmochimica Acta*, **165**, 148–160.

Lauretta D. S., Bartels A. E., Barucci M. A., et al. (2014) The OSIRIS-REx target asteroid (101955) Bennu: Constraints on its physical, geological, and dynamical nature from astronomical observations. *Meteoritics & Planetary Science*, 50, 834–849.

Lucas M. P., Dygert N., Ren J., et al. (2020) Evidence for early fragmentation-reassembly of ordinary chondrite (H, L, and LL) parent bodies from REE-in-two-pyroxene thermometry. *Geochimica et Cosmochimica Acta*, **290**, 366–390.

Mandler B. E., and Elkins-Tanton L. T. (2013) The origin of eucrites, diogenites, and olivine diogenites: Magma ocean crystallization and shallow magma chamber processes on Vesta. *Meteoritics & Planetary Science*, **48**, 2333–2349.

Marty B., Palma R. L., Pepin R. O., et al. (2008) Helium and neon abundances and compositions in cometary matter. *Science*, **319**, 75–78.

McKeegan K. D., plus 45 coauthors (2006) Isotopic compositions of cometary matter returned by Stardust. *Science*, **314**, 1724–1728.

McSween H. Y., and Labotka T. C. (1993) Oxidation during metamorphism of the ordinary chondrites. *Geochimica et Cosmochimica Acta*, **57**, 1105–1114.

McSween H. Y., Ghosh A., Grimm R. E., et al. (2003) Thermal evolution models of asteroids. In *Asteroids III*, Bottke W. F., Cellino A., Paolicchi P. and Binzel R. P., editors, pp. 559–571, University of Arizona Press, Tucson.

Metzler K., Bischoff A., and Stoffler D. (1992) Accretionary dust mantles in CM chondrites: Evidence for solar nebula processes. *Geochimica et Cosmochimica Acta*, **56**, 2873–2897.

Morbidelli A., Walsh K. J., O'Brien D. P., Minton D. A., and Bottke W. F. (2015) The dynamical evolution of the asteroid belt.

In *Asteroids IV*, Michel P., Demeo F. E., and Bottke W. F., editors, pp. 493–507, University of Arizona Press, Tucson.

Mothe-Diniz T., Carvano J. M., and Lazzaro D. (2003) Distribution of taxonomic classes in the main belt of asteroids. *Icarus*, **162**, 10–21.

Nakamura T., Nakato A., Ishida H., et al. (2014) Mineral chemistry of MUSES-C Regio inferred from analysis of dust particles collected from the first- and second-touchdown sites on asteroid Itokawa. *Meteoritics & Planetary Science*, **49**, 215–227.

Nesvorny D., Broz M., and Carruba V. (2015) Identification and dynamical properties of asteroid families. In *Asteroids IV*, Michel P., Demeo F. E., and Bottke W. F., editors, pp. 297–322, University of Arizona Press, Tucson.

Pieters C. M., and McFadden L. A. (1994) Meteorite and asteroid reflectance spectroscopy: Clues to early solar system processes. *Annual Review of Earth & Planetary Sciences*, **22**, 457–497.

Pieters C. M., and Noble S. K. (2016) Space weathering on airless bodies. *Journal of Geophysical Research, Planets*, **121**, 1865–1884.

Prettyman T. H., Yamashita N., Toplis M. J., et al. (2017) Extensive water ice within Ceres' aqueously altered regolith: Evidence from nuclear spectroscopy. *Science*, **355**, aah6765, doi:10.1126/science.aah6765.

Reddy V., Dunne T. L., Thomas C. A., et al. (2015) Mineralogy and surface composition of asteroids. In *Asteroids IV*, Michel P., Demeo F. E., and Bottke W. F., editors, pp. 43–64, University of Arizona Press, Tucson.

Rivkin A. S., Campins H., Emery J. P., et al. (2015) Astronomical observations of volatiles on asteroids. In *Asteroids IV*, Michel P., Demeo F. E., and Bottke W. F., editors, pp. 65–88, University of Arizona Press, Tucson.

Rubin A. E., Trigo-Rodriguez J. M., Huber H., and Wasson J. T. (2007) Progressive aqueous alteration of CM carbonaceous chondrites. *Geochimica et Cosmochimica Acta*, **71**, 2361–2382.

Schaller E. L. (2010) Atmospheres and surfaces of small bodies and dwarf planets in the Kuiper Belt. *EPJ Web of Conferences*, **9**, 267–276, doi:10.105/epjconf/201009021.

Shu F. H., Shang H., and Lee T. (1996) Toward an astrophysical theory of chondrites. *Science*, **271**, 1545–1552.

Slater-Reynolds V., and McSween H. Y. (2005) Peak metamorphic temperatures in type 6 ordinary chondrites: An evaluation of pyroxene and plagioclase geothermometry. *Meteoritics & Planetary Science*, **40**, 745–754.

Spencer J. R., Stern S. A., Moore J. M., et al. (2020) The geology and geophysics of Kuiper Belt object (486958) Arrokoth. *Science*, **367**, eaay3999.

Tholen D. J., and Barucci M. A. (1989) Asteroid taxonomy. In *Asteroids II*, Binzel R. P., Gehrels T., and Matthews M. S.., editors, pp. 298–315, University of Arizona Press, Tucson.

Treiloff M., Jesberger E. K., Herrwerth I., et al. (2003) Structure and thermal history of the H-chondrite parent asteroid revealed by thermochronometry. *Nature*, **442**, 502–506.

Tsou P., plus 19 coauthors (2004) Stardust encounters comet 81P/Wind 2. *Journal of Geophysical Research*, **109**, E12S01, doi:10.1029/2004JE002317.

Vilas F., Jarvis K. S., and Gaffey M. J. (1994) Iron alteration minerals in the visible and near-infrared spectra of low-albedo asteroids. *Icarus*, **109**, 274–283.

Westphal A. J., Snead C., Butterworth A., et al. (2004) Aerogel keystones: Extraction of complete hypervelocity impact events from aerogel collectors. *Meteoritics & Planetary Science*, **39**, 1375–1386.

Westphal A. J., Fakra S. C., Gainsforth Z., et al. (2009) Mixing fraction of inner solar system material in come 81P/Wild2. *Astrophysical Journal*, **694**, 18–28.

Westphal A. J., Bridges J. C., Brownlee D. E., et al. (2017) The future of Stardust science. *Meteoritics & Planetary Science*, **52**, 1859–1898.

Williams C. V., Rubin A. E., Keil K., and San Miguel A. (1985) Petrology of the Cangas de Onis and Nulles regolith breccias: Implications for parent body history. *Meteoritics*, **20**, 331–345.

Wong I., and Brown M. E. (2017) The bimodal color distribution of small Kuiper belt objects. *Astronomical Journal*, **153**, 145.

Yang J., and Goldstein J. I. (2005) The formation of the Widmanstätten structure in meteorites. *Meteoritics & Planetary Science*, **40**, 239–253.

Young E. D. (2001) The hydrology of carbonaceous chondrite parent bodies and the evolution of planet progenitors. *Philosophical Transactions of the Royal Society of London*, ***A359***, 2095–2109.

Zolensky M. E., and 74 coauthors (2006) Mineralogy and petrology of comet 81P/Wild2 nucleus samples. *Science*, **314**, 1735–1739.

13 Chemistry of Planetesimals and Their Samples

Overview

The compositions of undifferentiated planetesimals are determined from chemical analyses of chondrites and interplanetary dust particles (IDPs). Differentiated planetesimals have chondritic bulk compositions, but their samples (achondrites and irons) have been modified by fractionations occurring during melting and crystallization. Because planetesimals are unused raw materials from formation of the terrestrial and giant planets, they hold keys to understanding planetary accretion, as well as a host of nebular processes that predated planet formation. A few planetesimals have been studied by spacecraft missions, and remote-sensing methods that produce limited chemical data provide context and support for interpretations made from analyses of samples.

13.1 The Value of Bulk Chemical Analyses

We described meteorites and IDPs (samples of planetesimals) in Chapter 6, and the planetesimals themselves (asteroids, comets, Kuiper belt objects) in Chapter 12. In this chapter we focus on their bulk chemical compositions. Bulk analyses of chondritic samples define the compositions of undifferentiated asteroids and even of the solar system itself. They also reveal nucleosynthetic processes that predate the solar system and nebular processes that affected the compositions of protoplanetary materials. Bulk analyses of achondrites and irons constrain fractionation processes in differentiated bodies. The classifications of meteorites and planetesimals are based largely on their bulk compositions (not only elements but also oxygen isotopes, as previously discussed in Chapter 6). Remote-sensing analyses of planetesimal surfaces by spacecraft instruments further support the conclusions from laboratory analyses of samples, and sample compositions provide a basis for rigorous interpretation of spacecraft data. Because planetesimals are leftover planetary building blocks, their bulk compositions are also used to model the compositions of planets (to be described in Chapter 15).

13.2 Compositions of Chondrites and Primitive Planetesimals

Despite the common usage of the term "chondritic" to refer to unfractionated compositions, the various classes of chondrites have different chemistry and mineralogy. "Chondritic" actually identifies samples that have relative abundances of refractory elements that are approximately solar. In Chapter 7, we discussed various types of fractionation processes that have affected chondrite bulk compositions: sorting and mixing of components formed under different conditions, depletion of volatile elements due to evaporation or condensation, and oxidation/reduction.

Chondritic meteorites are among the most thoroughly analyzed materials known (Brearley and Jones, 1998), and numerous compilations of their chemical compositions are available. Especially notable are analyses by Eugene Jarosewich, who laboriously separated metal and sulfide before performing wet chemical analyses, so that his data distinguish iron in its various oxidation states. Among modern analyses performed by neutron activation, those of Gregory Kallemeyn are especially complete, although they do not report silicon abundances or iron oxidation states.

Here we separate noncarbonaceous and carbonaceous chondrites, because of the important distinction in their nucleosynthetic isotopes described in Chapter 15.

13.2.1 Noncarbonaceous Chondrite Groups

The noncarbonaceous chondrite groups are anhydrous and thus represent samples of asteroids that did not

Box 13.1 Size Matters: The Perils of Analyzing Tiny Particles and Breccias

Extraterrestrial samples are precious, and curators are appropriately judicious in allocating samples for destructive chemical analyses. IDPs, as well as asteroid and comet samples returned by spacecraft, are extremely small and potentially not representative of their source. Meteorites are larger, but many – perhaps most – are breccias and therefore are heterogeneous. Determining whether a particular sample volume (whether analyzed in the laboratory or by spacecraft remote sensing) is representative can be challenging. This is illustrated in Figure 13.1, showing bulk Fe/Si and Mg/Si ratios inferred from comet Halley dust-grain spectra obtained by the Vega 1 and Giotto spacecraft and from laboratory analyses of comparably sized (~1 μm) spots of matrix in a CI chondrite (Lawler et al., 1989). The scatter in both datasets demonstrates the difficulty in accurately determining the bulk composition from unrepresentative samples.

Instrumental neutron activation analysis (INAA) is the technique of choice for determining the bulk compositions of meteorites. INAA can give precise data for most elements of samples with masses of only a few hundred milligrams. The problem is that samples of that size do not give a representative bulk composition for most meteorites.

Multiple analyses of chondrites provide a means of assessing the minimum mass required to obtain a representative sample analysis. Jarosewich (1990) estimated that a 10 g sample is sufficient to provide representative sampling of an ordinary chondrite. To clarify, that mass should be powdered and well mixed, although a smaller amount of a few grams of homogenized powder is actually analyzed.

Determining representative sampling of achondrites is even harder, because they tend to be comprised of coarser grains or clasts that are not uniformly distributed. Mittlefehldt et al. (2013) cautioned that the masses (0.35–5 g) of howardites, eucrites, and diogenites (HEDs) that are typically analyzed for geochemistry might introduce uncertainties in coarser-grained samples. Jarosewich (1990) indicated that at least 15 g were required to insure a representative sampling of mesosiderites.

Unfortunately, large databases of published meteorite analyses (e.g. Nittler et al., 2004; Meteoritical Bulletin Database) do not typically assess data quality, which may vary because of unrepresentative analyzed masses and interlaboratory biases. Let the user beware.

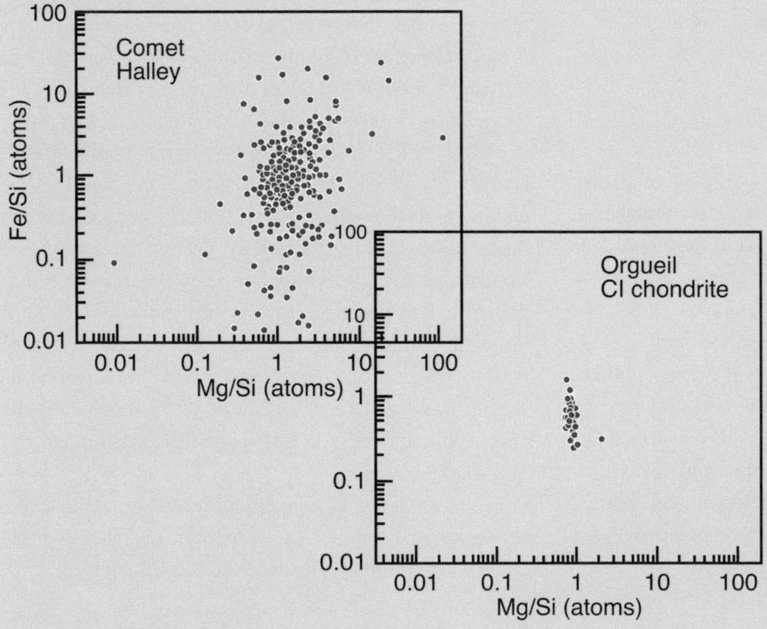

Figure 13.1 Fe/Si versus Mg/Si ratios inferred from comet Halley dust spectra are significantly more scattered than analyses of comparably sized spots in the Orgueil CI chondrite. After Lawler et al. (1989).

accrete ices. These include the ordinary (H, L, and LL) chondrites, as well as Rumuruti (R) and enstatite (EH and EL) chondrites. The average elemental abundances – encompassing most of the Periodic Table – for these chondrite groups are given in Table 13.1. Although most of these chondrites experienced thermal metamorphism, due to the decay of short-lived radionuclides, there are no significant chemical differences between the various petrologic types.

Figures 13.2a and 13.2b show variations in lithophile elements and in siderophile and chalcophile elements, respectively (all normalized to CI chondrite abundances and magnesium, and plotted in order of increasing volatility from left to right in each diagram) for the major classes of noncarbonaceous meteorites. As is apparent in these figures, the greatest disparities occur among volatile elements, which are depleted to varying degrees.

Another important difference between these chondrite groups is their oxidation states. Enstatite chondrites are highly reduced, ordinary chondrites are moderately reduced, and R chondrites are more oxidized (refer back to Figure 7.13). These differences are expressed in the relative proportions of FeNi metal, sulfides, and oxidized Fe in silicates and oxides. In Chapter 7, we learned that differences in chondrite compositions cannot be simply explained by fractionation of metallic iron, but must also involve oxidation or reduction of iron.

Table 13.1 Average element abundances for noncarbonaceous chondrite groups

Element	H	L	LL	R	EH	EL
Li (ppm)	1.7	1.85	1.8		1.9	0.7
Be (ppb)	30	40	45		21	
B (ppm)	0.4	0.4	0.7		1	
C (wt. %)	0.21	0.25	0.31	0.058	0.39	0.43
N (ppm)	48	43	70		420	240
O (wt. %)	35.7	37.7	40.0		28.0	31.0
F (ppm)	125	100	70		155	140
Na (wt. %)	0.611	0.690	0.684	0.663	0.688	0.5770
Mg (wt. %)	14.1	14.9	15.3	12.9	10.7	13.8
Al (wt. %)	1.06	1.16	1.18	1.06	0.82	1.00
Si (wt. %)	17.1	18.6	18.9	18.0	16.6	18.8
P (ppm)	1200	1030	910		2130	1250
S (wt. %)	2.0	2.2	2.1	4.07	5.6	3.1
Cl (ppm)	140	270	200	<100	570	230
K (ppm)	780	920	880	780	840	700
Ca (wt. %)	1.22	1.33	1.32	0.914	0.85	1.02
Sc (ppm)	7.8	8.1	8.0	7.75	6.1	7.7
Ti (ppm)	630	670	680	900	460	550
V (ppm)	73	75	76	70	56	64
Cr (ppm)	3500	3690	3680	3640	3300	3030
Mn (ppm)	2340	2590	2600	2960	2120	1580
Fe (wt. %)	27.2	21.8	19.8	24.4	30.5	24.8
Co (ppm)	830	580	480	610	870	720
Ni (wt. %)	1.71	1.24	1.06	1.44	1.84	1.47
Cu (ppm)	94	90	85		215	120
Zn (ppm)	47	57	56	150	290	18
Ga (ppm)	6.0	5.4	5.3	8.1	16.7	11
Ge (ppm)	10	10	10		38	30
As (ppm)	2.2	1.36	1.3	1.9	3.5	2.2
Se (ppm)	8.0	8.5	9.0	14.1	25	15
Br (ppm)	<1	<2	1.0	0.55	2.7	0.8
Rb (ppm)	2.3	2.8	2.2		3.1	2.3

Table 13.1 *(cont.)*

Element	H	L	LL	R	EH	EL
Sr (ppm)	8.8	11	13		7.0	9.4
Y (ppm)	2.0	1.8	2.0		1.2	
Zr (ppm)	7.3	6.4	7.4		6.6	7.2
Nb (ppm)	0.4	0.4				
Mo (ppm)	1.4	1.2	1.1	0.9		
Ru (ppb)	1100	750		960	930	770
Rh (ppb)	210	155				
Pd (ppb)	845	620	560		820	730
Ag (ppb)	45	50	75		280	85
Cd (ppb)	<10	30	40		705	35
In (ppb)	<1.5	<20	<20		85	4
Sn (ppb)	350	540			1360	
Sb (ppb)	66	78	75	72	190	90
Te (ppb)	520	460	380		2400	930
I (ppb)	60	70			210	80
Cs (ppb)	<200	<500	150		210	125
Ba (ppm)	4.4	4.1	4.0		2.4	2.8
La (ppb)	301	318	330	310	240	196
Ce (ppb)	763	970	880	830	650	580
Pr (ppb)	120	140	130		100	70
Nd (ppb)	581	700	650		440	370
Sm (ppb)	194	203	205	180	140	149
Eu (ppb)	74	80	78	72	52	54
Gd (ppb)	275	317	290		210	196
Tb (ppb)	49	59	54		34	32
Dy (ppb)	305	372	360	29	230	245
Ho (ppb)	74	89	82	59	50	51
Er (ppb)	213	252	240		160	160
Tm (ppb)	33	38	35		24	23
Yb (ppb)	203	226	230	216	154	157
Lu (ppb)	33	34	34	32	25	25
Hf (ppb)	150	170	170	150	140	210
Ta (ppb)	21	21				
W (ppb)	164	138	115	<180	140	140
Re (ppb)	78	47	32	43	55	57
Os (ppb)	835	530	410	690	660	670
Ir (ppb)	770	490	380	610	570	560
Pt (ppm)	1.58	1.09	0.88	<1.0	1.29	1.25
Au (ppb)	220	156	146	183	330	240
Hg (ppb)		30	22		60	
Tl (ppb)	<1	<5	<30		100	7
Pb (ppb)	240	40			1500	240
Bi (ppb)	<10	14	<30		90	13
Th (ppb)	38	42	47	<50	30	38
U (ppb)	13	15	15	<25	9.2	7.0

Data sources given by Lodders and Fegley (1998).

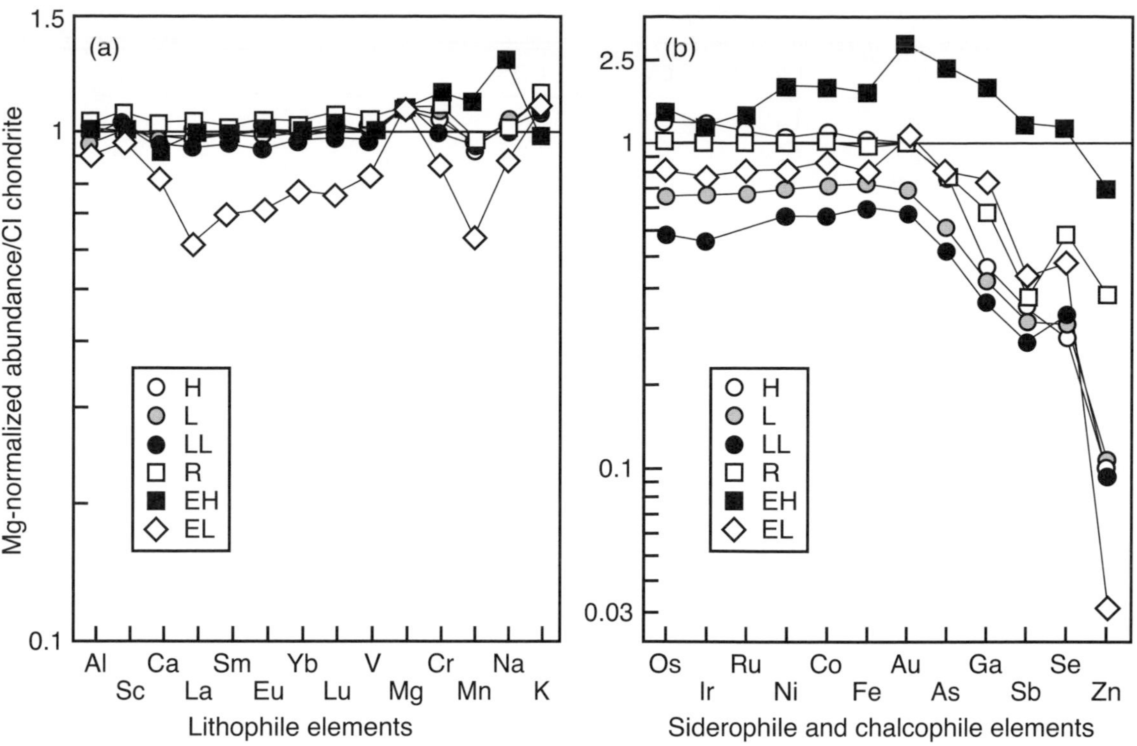

Figure 13.2 Compositional variations among noncarbonaceous chondrites. (a) Lithophile and (b) siderophile and chalcophile elements in ordinary (H, L, LL), enstatite (EH, EL), and Rumuruti (R) chondrites. Elements are plotted from left to right in order of increasing volatility. Lithophile elements are normalized to CI chondrites and Mg; siderophile and chalcophile elements are normalized to CI chondrites. Modified from Krot et al. (2014).

13.2.2 Carbonaceous Chondrite Groups

The carbonaceous chondrites include both hydrous and anhydrous groups, reflecting the presence or absence of accreted ices in their parent bodies. The CI and CM chondrites have experienced extensive aqueous alteration, and minor alteration occurs in some CR and CV chondrites. The CH, CO, CK, CB, and most CR and CV chondrites are relatively anhydrous. The average chemical compositions of the carbonaceous chondrite groups are given in Table 13.2.

Figure 13.3 illustrates some differences in lithophile, siderophile, and chalcophile element abundances among the carbonaceous groups using diagrams like those used to describe the noncarbonaceous chondrites. The elevated refractory lithophile element abundances reflect the incorporation of refractory inclusions in these meteorites. As in the noncarbonaceous chondrites, the greatest differences are in the depletion of volatile elements. For example, highly volatile and moderately volatile elements in CM chondrites are present at 50–60% of the abundances of the refractory elements. The volatile elements are primarily located in the matrix, and the matrix comprises 50–60% of CM chondrites. This implies that the matrix has essentially CI abundances of all elements, whereas the chondrules and refractory inclusions have CI-relative abundances of refractory elements but are highly depleted in volatile elements. The transition in the region of moderately volatile elements is sloping rather than abrupt, indicating that the chondrules retain some fraction of these elements. Other carbonaceous chondrite groups generally have both lower matrix abundances and lower volatile element contents in their matrices compared to CM chondrites. The CI and CM chondrites are also more oxidized than other chondrite types.

A perplexing observation is that aqueous alteration, caused by fluids produced by the melting of ice by ^{26}Al decay, appears to have been largely isochemical. The CI chondrites, which provide the closest match to solar abundances, show the most extensive mineralogical alteration. Likewise, the

Table 13.2 Average element abundances for carbonaceous chondrite groups

Element	CI	CM	CV	CO	CR	CH	CK
H (wt. %)	2.02	1.4	0.28	0.07			
Li (ppm)	1.5	1.5	1.7	1.8			1.4
Be (ppm)	0.025	0.04	0.05				
B (ppm)	0.87	0.48	0.3				
C (wt. %)	3.45	2.2	0.53	0.44	2.0	0.78	0.22
N (ppm)	3180	1520	80	90	620	190	
O (wt. %)	46.4	43.2	37.0	37.0			
F (ppm)	60	38	24	30			20
Na (wt. %)	0.50	0.39	0.34	0.420	0.330	0.180	0.310
Mg (wt. %)	9.70	11.5	14.3	14.5	13.7	11.3	14.7
Al (wt. %)	0.865	1.13	1.68	1.40	1.15	1.05	1.47
Si (wt. %)	10.64	12.7	15.7	15.8	15.0	13.5	15.8
P (ppm)	950	1030	1120	1210	1030		1100
S (wt. %)	5.41	2.7	2.2	2.2	1.9	0.35	1.7
Cl (ppm)	700	430	250	280			260
K (ppm)	550	370	360	360	315	200	290
Ca (wt. %)	0.926	1.29	1.84	1.58	1.29	1.30	1.7
Sc (ppm)	5.9	8.2	10.2	9.5	7.8	7.5	11
Ti (ppm)	440	550	870	730	540	650	940
V (ppm)	55	75	97	95	74	63	96
Cr (ppm)	2650	3050	3480	3520	3415	3100	3530
Mn (ppm)	1940	1650	1520	1620	1660	1020	1440
Fe (wt. %)	18.2	21.3	23.5	25.0	23.8	38.0	23.0
Co (ppm)	505	560	640	680	640	1100	620
Ni (wt. %)	1.10	1.23	1.32	1.42	1.31	2.57	1.31
Cu (ppm)	125	130	104	130	100	120	90
Zn (ppm)	315	180	110	110	100	40	80
Ga (ppm)	9.8	7.6	6.1	7.1	6.0	4.8	5.2
Ge (ppm)	33	26	16	20	18		14
As (ppm)	1.85	1.8	1.5	2.0	1.5	2.3	1.4
Se (ppm)	21	12	8.7	8.0	8.2	3.9	8.0
Br (ppm)	3.5	3.0	1.6	1.4	1.0	1.4	0.6
Rb (ppm)	2.3	1.6	1.2	1.3	1.1		
Sr (ppm)	7.3	10	14.8	13	10		15
Y (ppm)	1.56	2.0	2.6	2.4			2.7
Zr (ppm)	3.9	7.0	8.9	9.0	5.4		8
Nb (ppm)	0.25	0.4	0.5		0.5		0.4
Mo (ppm)	0.92	1.4	1.8	1.7	1.4	2	0.38
Ru (ppm)	0.71	0.87	1.2	1.08	0.97	1.6	1.1
Rh (ppm)	0.14	0.16	0.17				0.18
Pd (ppm)	0.56	0.63	0.71	0.71	0.69		0.58
Ag (ppb)	200	160	100	100	95		
Cd (ppb)	690	420	350	8	300		
In (ppb)	80	50	32	25	30		
Sn (ppm)	1.70	0.79	0.68	0.89	0.73		0.49
Sb (ppb)	135	130	85	110	80	90	60
Te (ppm)	2.3	1.3	1.0	0.95	1.0		0.8

Table 13.2 (*cont.*)

Element	CI	CM	CV	CO	CR	CH	CK
I (ppm)	0.43	0.27	0.16	0.2			0.02
Cs (ppm)	0.19	0.11	0.09	0.08	0.084		
Ba (ppm)	2.35	3.1	4.55	4.3	3.4	3.0	4.7
La (ppb)	235	320	469	380	310	290	460
Ce (ppb)	620	940	1190	1140	750	870	1270
Pr (ppb)	94	137	174	140			
Nd (ppb)	460	626	919	850	790		990
Sm (ppb)	150	204	294	250	230	185	290
Eu (ppb)	57	78	105	96	80	76	110
Gd (ppb)	200	290	405	390	320	290	440
Tb (ppb)	37	51	71	60	50	50	
Dy (ppb)	250	332	454	420	280	310	490
Ho (ppb)	56	77	97	96	100	70	100
Er (ppb)	160	221	277	305			350
Tm (ppb)	25	35	48	40		40	
Yb (ppb)	160	215	312	270	220	210	320
Lu (ppb)	25	33	46	39	32	30	46
Hf (ppb)	105	180	230	220	150	140	250
Ta (ppb)	14	19					
W (ppb)	93	160	160	150	110	150	180
Re (ppb)	38	50	57	58	50	73	60
Os (ppb)	490	670	800	805	710	1150	815
Ir (ppb)	465	580	730	740	670	1070	760
Pt (ppm)	1.0	1.1	1.25	1.24	0.98	1.7	1.3
Au (ppb)	145	150	153	190	160	250	120
Hg (ppb)	310						
Tl (ppb)	142	92	58	40	60		
Pb (ppm)	2.5	1.6	1.1	2.15			0.8
Bi (ppb)	110	71	54	35	40		20
Th (ppb)	29	41	58	80	42		58
U (ppb)	8	12	17	18	13		15

Data sources given by Lodders and Fegley (1998).

chemical compositions of CM chondrites are nearly uniform, despite significant mineralogical differences reflecting variations in their degree of aqueous alteration. Aqueous fluids can dissolve significant amounts of soluble materials during mineral reactions, but there is little evidence in bulk CI and CM chondrites for significant dissolution and loss. However, as noted earlier in Chapter 7, the abundance of organic matter in these meteorites was affected by aqueous alteration, and the stable-isotope compositions of hydrogen, oxygen, and nitrogen were modified by exchange with fluids (Alexander et al., 2015).

13.2.3 Chondritic Asteroids Analyzed by Spacecraft Remote Sensing and Sampling

In situ chemical analyses of asteroids have been made by several spacecraft, using methods described in Box 13.2. The abundances of the handful of elements that can be measured remotely are generally interpreted by comparison with the compositions of meteorite analogs (e.g. Nittler et al., 2004).

Two near-Earth chondritic asteroids have been analyzed using an X-ray fluorescence spectrometer (XRF). The NEAR Shoemaker orbiting spacecraft analyzed Eros (McCoy et al., 2001; Nittler et al., 2001) during a period

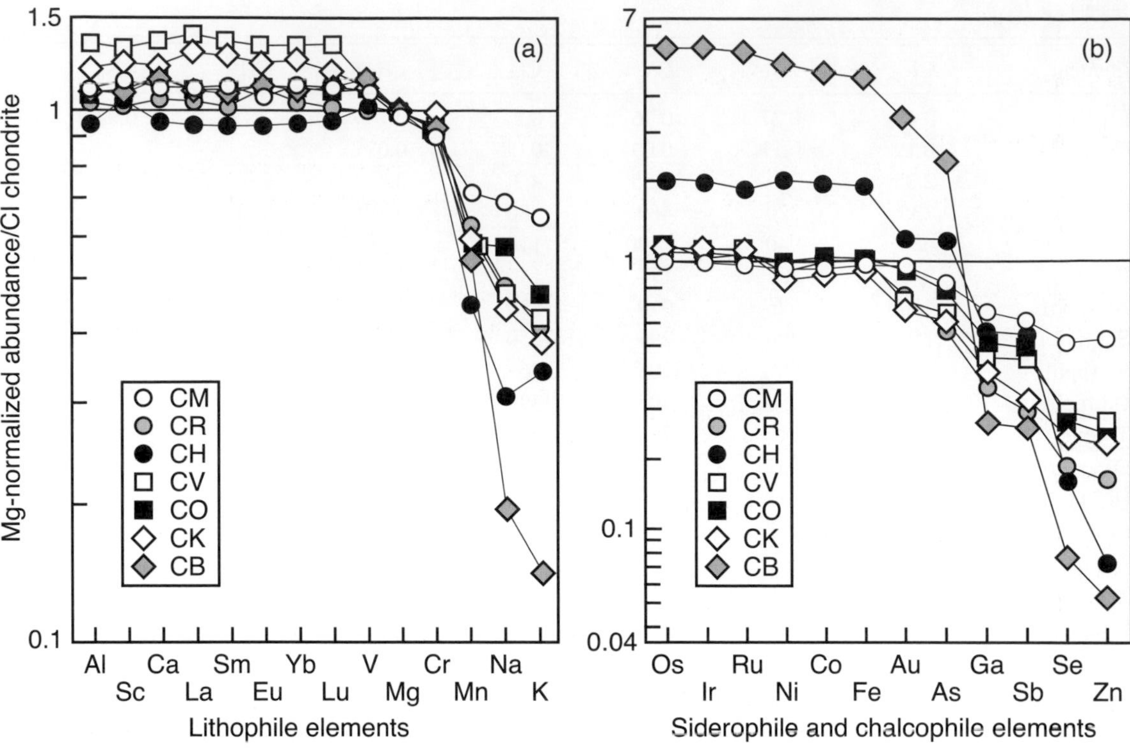

Figure 13.3 Differences in the abundances of (a) lithophile and (b) siderophile and chalcophile elements in carbonaceous chondrite groups. Elements are plotted from left to right in order of increasing volatility. Lithophile elements are normalized to CI chondrites and Mg; siderophile and chalcophile elements are normalized to CI chondrites. Modified from Krot et al. (2014).

when the Sun was near the maximum of its 11-year activity cycle. The chemical composition of this asteroid is expressed as element ratios (Table 13.3) because there was no internal calibration for absolute abundances. An analysis of Itokawa by the Hayabusa spacecraft (Okada et al., 2006) occurred during quiet solar conditions, so only magnesium, aluminum, and silicon data were obtained (Table 13.3). Both asteroids are classified as S-type, and the compositions of both are similar to ordinary chondrites, although anhydrous carbonaceous chondrites (CV, CO, CR) and primitive achondrites were not initially ruled out. Especially interesting is the low measured S/Si ratio for Eros, which might result from impact volatilization or sputtering of sulfur in surface materials or limited extraction of an FeS-rich partial melt.

NEAR Shoemaker also carried a gamma-ray spectrometer (GRS) (Evans et al., 2001). The instrument was designed to collect data from orbit, but a late arrival at Eros and operational considerations did not allow sufficient time at a low enough altitude to provide acceptable measurements. At the end of the mission, the spacecraft made a controlled descent to the asteroid's surface, and the GRS-determined element abundances (again expressed as ratios) obtained over a week-long integration on the surface are given in Table 13.3.

Hayabusa returned a small sample of Itokawa to Earth in 2010. Analyses of oxygen isotopes (Yurimoto et al., 2011) are consistent with its classification as an ordinary (LL) chondrite. Hayabusa2 returned a small sample of asteroid Ryugu, which has both B- and C-class spectral properties, in 2020. The OSIRIS-REx mission carries an XRF and has collected chemical analyses during its orbit of asteroid Bennu, a B-class asteroid linked to carbonaceous chondrites (Lauretta et al., 2019). It has collected a much larger sample and is expected to return it to Earth in 2023.

13.3 Geochemical Exploration of Dwarf Planet Ceres

Dwarf planet Ceres is the largest body in the Main asteroid belt. It is compositionally similar to carbonaceous

Box 13.2 Chemical Remote-Sensing Methods Used by Orbiting Spacecraft

The options for orbiting spacecraft instruments that conduct chemical analyses of planetesimals are limited.

X-Ray Fluorescence

Solar X-rays impinging on an asteroid's surface generate fluorescence X-rays that can be measured using an X-ray fluorescence spectrometer (XRF). During times when the Sun is quiet (i.e. not producing many solar flares), the X-ray flux is fairly low, so that Kα line spectra of only a few low-atomic-weight, rock-forming elements, such as magnesium, aluminum, and silicon, can be measured. An example of such a spectrum for asteroid Itokawa is shown in Figure 13.4. During solar flare periods, heavier elements such as calcium, iron, and sulfur may be detectable. Conversion of the observed photon fluxes to element abundances requires a concurrent measurement of the solar spectrum.

Gamma-Ray and Neutron Spectroscopy

Gamma-ray spectroscopy (GRS) can determine some element concentrations from a spatially resolved portion of an asteroid's surface. Characteristic γ-ray line emissions for specific elements are produced by nuclear spallation reactions initiated by high-energy galactic cosmic rays hitting the surface (Fig. 13.5). Spallation reactions are of two types. Inelastic scattering reactions occur when energetic (fast) neutrons produced by spallation lose energy to a target nucleus, leaving the nucleus in an excited state. The excited nucleus then produces γ-rays that can be used to identify that element. Neutron capture occurs when less-energetic neutrons produced by spallation are absorbed by a target nucleus, again leaving the nucleus in an excited state. As with inelastic scatter, the nucleus de-excites and produces characteristic γ-rays. Elements that can be measured from spallation reactions potentially include iron, titanium, magnesium, aluminum, calcium, silicon, and oxygen (the list of elements actually measured is typically shorter and depends on the abundance of each element in the target and on what elements are located near the instrument on the spacecraft itself). Finally, there are also a few naturally occurring radioactive elements – thorium, uranium, and potassium – that on decay produce enough γ-rays that their abundances can sometimes be measured from orbit (Fig. 13.5).

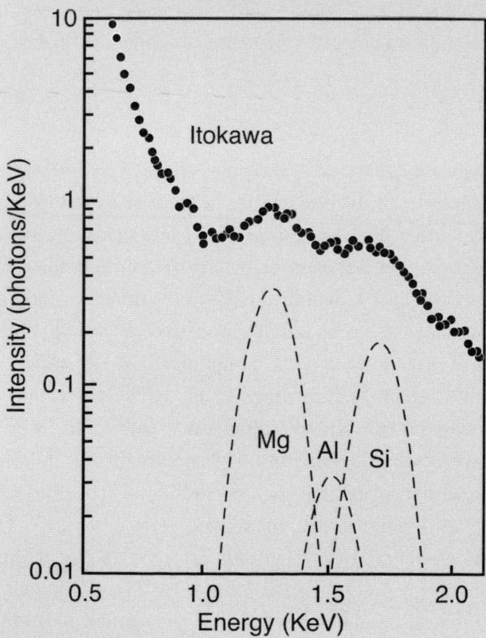

Figure 13.4 X-ray spectrum of asteroid Itokawa measured by the Hayabusa spacecraft. Locations of the Kα lines for magnesium, aluminum, and silicon are also shown by dashed lines. Modified from Okada et al. (2006).

Box 13.2 (cont.)

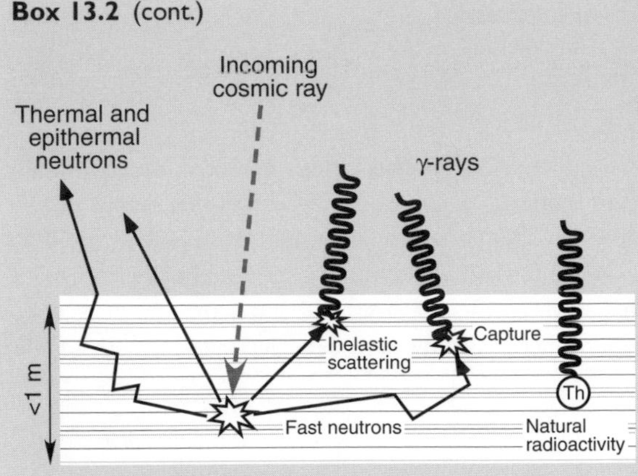

Figure 13.5 Production of γ-rays and neutrons used for *in situ* geochemical analysis. Galactic cosmic rays penetrate <1 m into the surface and collide with target atomic nuclei. Fast (energetic) neutrons are dislodged. These can be slowed by collisions with hydrogen atoms to produce thermal and epithermal neutrons, or can undergo other kinds of nuclear reactions to produce γ-rays. Decay of natural radioactive isotopes (Th, U, K) also generates γ-rays.

Gamma-ray spectrometers use scintillator detectors. These spectrometers measure γ-rays from all directions and hence have large "footprints," with sizes determined by orbital elevation above the surface. The γ-rays come from depths of less than a meter in the target material.

Neutrons generated by cosmic rays (Fig. 13.5) also provide a sensitive measurement of hydrogen in an asteroid's near-surface environment. The sensitivity stems from the fact that neutrons and protons (hydrogen nuclei) have essentially the same mass. When particles with the same mass collide, the energy transferred from the neutron to the proton is greater than in the case of a collision of a neutron with a heavier nucleus. (An analogy is colliding billiard balls, where significant energy is transferred to the target ball, in contrast to the collision of a billiard ball with a bowling ball, where the billiard ball bounces off and retains its energy.) Consequently, neutrons impacting hydrogen atoms are slowed dramatically – they become "epithermal" neutrons that are easily absorbed. Thus, the flux of epithermal neutrons that leak out of an asteroid's surface will be decreased in proportion to the hydrogen content. Measurements by orbiting neutron spectrometers of the epithermal neutrons that do escape can thus provide information on the abundance of hydrogen, commonly interpreted as H_2O (either ice or bound in hydrous minerals) and reported as *water-equivalent hydrogen*. In some cases, thermal neutrons can also be used to measure the absolute abundances of other elements. However, transforming the neutron spectrum into elemental abundances can be quite involved.

chondrites, which provide the best meteoritic analog (McSween et al., 2018). The Dawn spacecraft orbited Ceres for three years, carrying out extensive mineralogical and geochemical investigations. Dawn carried a Gamma Ray and Neutron Detector (GRaND) and a Visible and InfraRed Spectrometer (VIR) for the analysis and mapping of surface materials.

VIR spectra indicate a surface composed of serpentine, ammoniated clay, carbonate, and a darkening agent, possibly graphitized organics and/or magnetite (De Sanctis et al., 2018). Ceres has an icy crust and its rocky material is more highly altered than CM and CI chondrites (Castillo-Rogez et al., 2018). Measured global abundances of iron and potassium are given in Table 13.3, and spatial variations in iron are illustrated in Figure 13.6. Hydrogen measurements (expressed as water equivalents) by GRaND (Prettyman et al., 2018) vary from 16 wt. % at the equator to 29 wt. % at the poles (Fig. 13.6), indicating decreasing depth to the ice table with increasing latitude. Bulk water ice for Ceres, estimated from its bulk density, is 17–30 wt. %. Carbon contents are greater than the ~3.5 wt. % found in CI chondrites. Compositional variability on the surface of Ceres is also indicated by GRaND high-energy γ-rays, although the abundances of elements are difficult to quantify (Lawrence et al., 2018).

Localized bright spots on Ceres (see Fig. 12.2) correspond to recent effusions of brines in some craters. The sublimating brines precipitated salts (sodium carbonate and ammonium bicarbonate or chloride) (De Sanctis et al., 2018). A young cryovolcano on Ceres is also composed of these minerals. These materials reveal that aqueous fluids exist or are formed periodically in Ceres' interior, and reveal that the body is geologically active.

Table 13.3 Remote-sensing analyses of near-Earth and Main belt asteroids by spacecraft

Elements and wt. ratios	Eros XRF	Eros GRS	Itokawa XRF	Ceres GRNS	Vesta GRNS
Mg/Si	0.85 ± 0.11	0.75	0.78 ± 0.09		
Fe/Si	1.65 ± 0.27	0.8			0.54 ± 0.07
Al/Si	0.068 ± 0.022		0.07 ± 0.03		
S/Si	0.014 ± 0.017				
Ca/Si	0.077 ± 0.006				
Si/O		0.53–0.67			
Fe/O		0.16–0.44			0.30 ± 0.04
Fe				16 ± 1 wt. %	
K		0.07 wt. %		410 ± 40 μg/g	
K/Th					900 ± 400
H (water equivalent)				17 ± 2 wt. %	variable
C				>3.5 wt. %	

Eros XRF data from Nittler et al. (2001); Eros GRS data from Evans et al. (2001); Itokawa XRF data from Okada et al. (2006); Ceres gamma ray and neutron spectroscopy (GRNS) data from Prettyman et al. (2017, 2018); Vesta GRNS data from Prettyman et al. (2012, 2015).

The interior structure of Ceres (Fig. 13.7), constrained by gravity models based on orbital tracking of spacecraft, has a crust ~40 km thick composed of water ice, salts, silicates, and clathrates (the latter is inferred to explain its rigid rheology), giving it a density of ~1.3 g/cm^3 (Ermakov et al., 2017). The underlying mantle features an upper muddy layer with a density of ~2.4 g/cm^3, composed mostly of hydrous silicates, which grades into mantle rock with a density of ~2.9 g/cm^3 (Park et al., 2016). From available gravity data, it is not possible to determine the nature of the deeper mantle (thought to be CI/CM chondrite-like) or whether or not Ceres has a core. The separation of water to produce an icy crust (or perhaps even a former ice shell that has been lost to impacts and sublimation; Castillo-Rogez et al., 2018) indicates global differentiation of volatile species. Despite Ceres' compositional similarity to CI and CM carbonaceous chondrites, the smaller parent bodies of those aqueously altered meteorites apparently did not undergo such pervasive alteration.

13.4 Compositions of IDPs and Comet Samples

As described in Chapters 6 and 12, chondritic porous (CP) IDPs are thought to be cometary samples. Element ratios for hundreds of analyzed CP particles are roughly chondritic (Fig. 13.8) (Schramm et al., 1989). An exception, though, is carbon, which is significantly more abundant in IDPs. The mean carbon content of IDPs is ~13 wt. %, relative to ~3.5 wt. % for CI chondrites (Bradley, 2014). The abundances of trace elements in bulk IDPs scatter from ~0.3 to ~3 times CI, and volatile elements especially tend to be enriched (Flynn and Sutton, 1992). Higher abundances of carbon and volatile elements, relative to the most solar-like carbonaceous chondrites, support the contention that CP IDPs are among the most primitive materials known.

The bulk chemical composition of comet Wild2 dust returned by the Stardust spacecraft, obtained by averaging the compositions of particles in numerous aerogel tracks (Fig. 13.9a) and impact crater residues (Fig. 13.9b), is nearly chondritic for iron, silicon, titanium, chromium, manganese, nickel, germanium, and selenium, within the 2σ confidence level (Flynn et al., 2006). Copper, zinc, and gallium are enriched in these particles, relative to CI chondrite values. These same chemical characteristics have been previously reported for CP IDPs.

The oxygen-isotope compositions of Stardust silicate grains (Fig. 13.10) resemble those of chondrites, indicating formation in the solar nebula (McKeegan et al., 2006). Oxygen in a refractory inclusion particle in this sample is isotopically like CAIs.

The organic matter in these comet grains was described in Chapter 11. Hydrogen in these particles is correlated with carbon and is presumed to occur in organic matter. Analyses of the isotopic composition of hydrogen (McKeegan et al., 2006) show D/H enrichments (Fig. 13.11). The D/H ratios

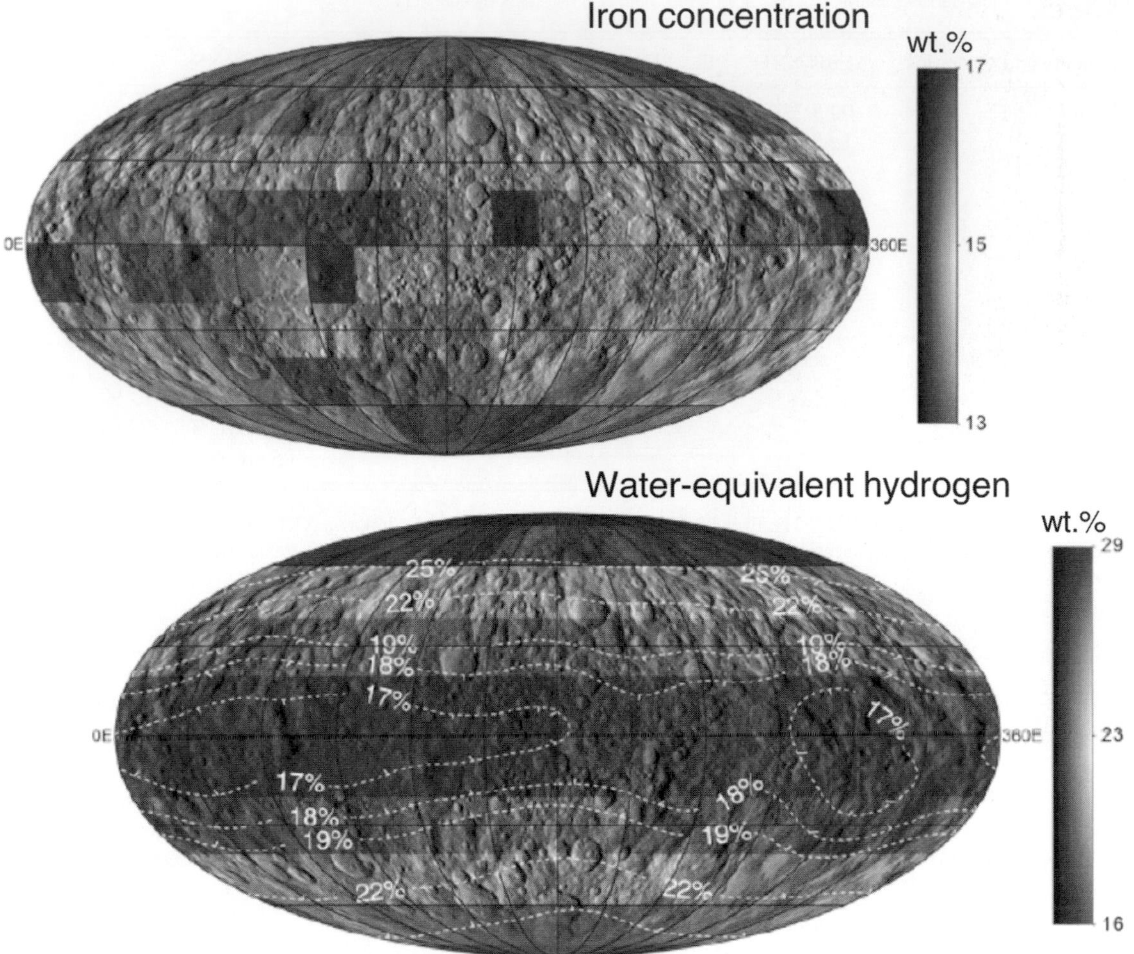

Figure 13.6 Variations in the concentrations of iron and water-equivalent hydrogen in the upper meter of Ceres' regolith, as measured by GRaND. Adapted from Prettyman et al. (2018). *For the colour version, refer to the plate section.*

overlap those of H_2O in comets but are well below those determined for organic matter in CP IDPs. Measurements of nitrogen isotopes show ^{15}N enhancements, suggesting an interstellar contribution.

13.5 Compositions of Differentiated Meteorites

Achondrites, irons and stony irons are fragments of asteroids that underwent melting and differentiation resulting from the decay of short-lived ^{26}Al. Differentiated meteorites were introduced in Chapter 6, and magmatic fractionations were described in Chapter 7. For the time being, we will ignore planetary materials (martian meteorites and Mars remote-sensing data) and lunar materials (samples and remote sensing) and focus on differentiated planetesimals and their samples.

13.5.1 Primitive Achondrites

Primitive achondrites are the solid residues left when partial melts (silicate and/or metal/sulfide) are extracted. Some representative chemical analyses of primitive achondrites (acapulcoites-lodranites, winonaites, brachinites, and ureilites) are given in Table 13.4. The near-chondritic compositions of acapulcoites, which experienced very limited partial melting, are illustrated in Figure 13.12, where they are compared with ordinary (H-group) chondrites. This similarity accounts for the prefix "primitive," although that name was affixed before their origin as residues

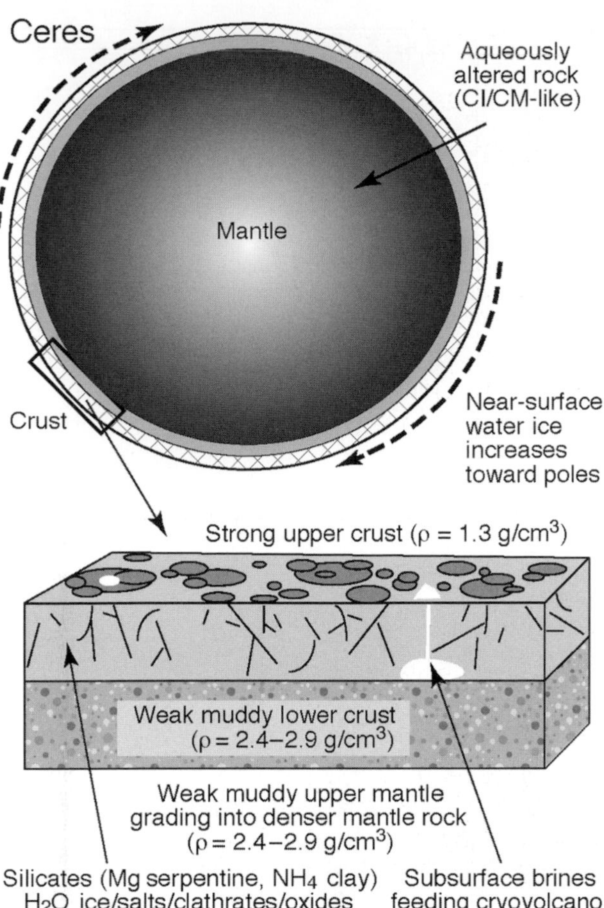

Figure 13.7 Compositional model of Ceres, based on data from the Dawn spacecraft. The surface regolith of the dwarf planet is composed of phyllosilicates, carbonates, and oxides (similar to highly altered CI or CM chondrites), with water ice increasing in abundance toward the poles. Ceres has a strong crust, ~40 km thick, over a mantle that grades downward from a weak, muddy upper layer into denser rock. The crust contains as much as 30–40 vol. % water ice, accounting for its low density, and clathrates may explain its rigidity. The mantle has a higher density and thus contains less ice. The weakness of the upper mantle explains the relaxation of large, but not small, craters on Ceres. A more rigid mantle (likely similar to aqueously altered carbonaceous chondrite) likely exists at greater depth, although available geophysical data do not provide much information on its properties.

was recognized. The compositions of lodranites, which are residues from the acapulcoite parent body and from which a greater amount of melt was extracted, are more depleted in those elements that partition into liquid. Winonaites are also roughly chondritic in composition (Fig. 13.12). Brachinites and ureilites apparently represent residues from more extensive melting, and their compositions are more highly fractionated from chondritic abundances.

Primitive achondrites have varying oxygen-isotope compositions, of the same magnitude as chondrites, indicating that partial melting did not diffusively homogenize their parent bodies (see Fig. 6.15). They also show varying Fe contents at constant (chondritic) Mn/Mg ratios (see Fig. 6.8), interpreted to result from the heterogeneous distribution of metal or variable redox states in their precursor materials.

The first melts of metal-bearing chondrites would have a metal/sulfide eutectic composition (which is mostly sulfide). It has been generally accepted that this melt would shortly be joined by a silicate melt of basaltic composition, by analogy with melts derived from the Earth's ultramafic mantle. Surprisingly, however, melting experiments on chondritic compositions first produce melts enriched in silica and alkali elements, more akin to andesite than basalt (Collinet and Grove, 2020). Magmatic achondrites with this composition are extremely rare, and evidence for them may have been erased by subsequent melting which produced basalts.

13.5.2 Magmatic Achondrites

Magmatic achondrites represent crystallized magmas or cumulates from such magmas. The compositional diversity among magmatic achondrites reflects, in part, the compositional differences among their chondritic precursors, but also the geochemical changes that accompany melting and crystallization. Environmental conditions also play a role, as melting of similar chondritic compositions under different oxidation states can produce distinct

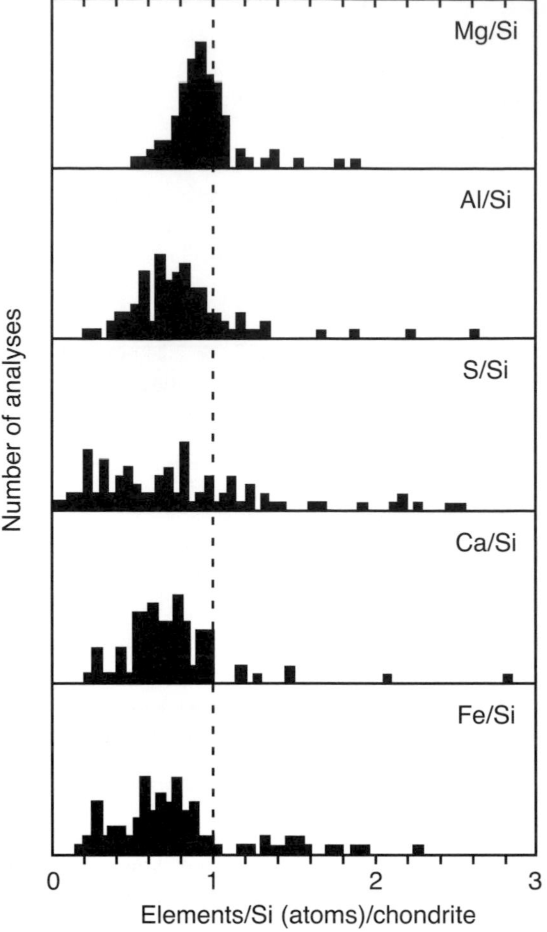

Figure 13.8 Chondrite-normalized element/silicon ratios measured in CP IDPs. CI chondrite ratios are shown by the vertical dashed line. Modified from Schramm et al. (1989).

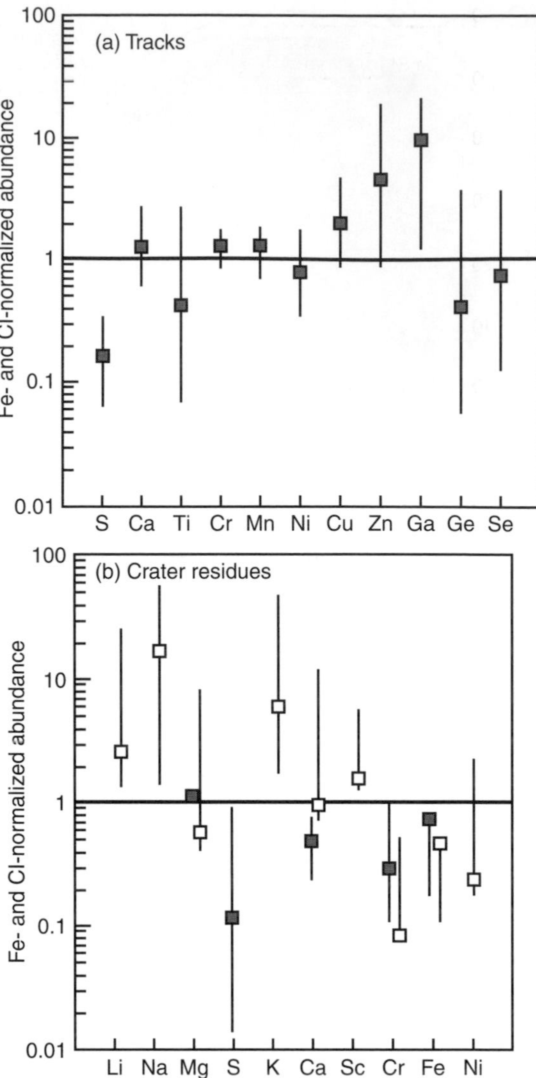

Figure 13.9 Mean compositions of tracks and crater residues produced by Stardust particles, compared to CI chondrite ratios (horizontal lines). (a) Fe- and CI chondrite-normalized composition determined by averaging 23 track analyses by SXRM (filled squares). (b) Si- and CI chondrite-normalized composition of seven crater residue analyses by SEM-EDX (filled squares), and TOF-SIMS analyses for five craters (open squares). Modified from Flynn et al. (2006).

types of achondrites. The reasons that elements are partitioned differently between melts and solids were explained in Chapter 7.

The compositions of some magmatic achondrites are given in Table 13.5. Especially useful tabulations are given by Lodders and Fegley (1998), Mittlefehldt et al. (1998), and Mittlefehldt (2014). Angrites are basalts that show extreme depletions in silica and in moderately volatile elements such as the alkalis. They formed by the melting of a chondritic source, apparently unlike known chondrite classes, under oxidizing conditions. In contrast, the aubrites are the most reduced achondrites, and they share many compositional and mineralogical characteristics with enstatite chondrites. A large body of literature on the compositions of HED achondrites exists, owing to their abundance (>1000 named meteorites). Eucrites are basalts with very limited variations in major-element composition but significant variations in trace elements. Diogenites and cumulate eucrites show somewhat greater compositional variations due to differences in the proportions of accumulated minerals. The HED suite is depleted in volatile and siderophile elements, reflecting the composition of the chondritic precursor (or prior extraction of a silica- and alkali-rich melt) and the extraction of metallic elements

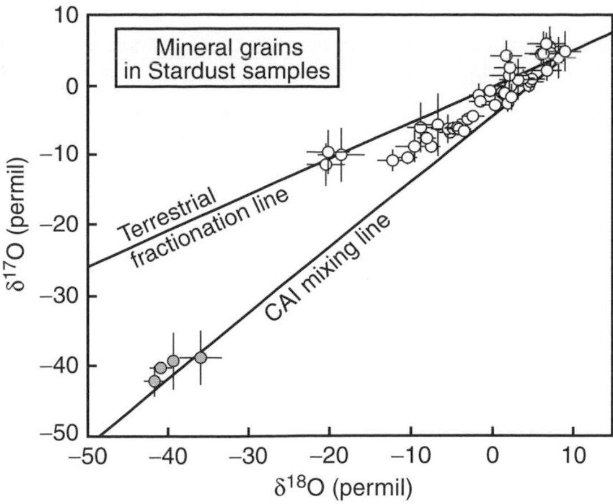

Figure 13.10 Oxygen isotopic compositions (relative to standard mean ocean water, SMOW) of a variety of grains from Stardust tracks. The grains include olivine, pyroxene, chromite, chondrule mesostasis, and amorphous material. Shaded circles are from a CAI comet particle. After Ogliore et al. (2015), with CAI data from McKeegan et al. (2006).

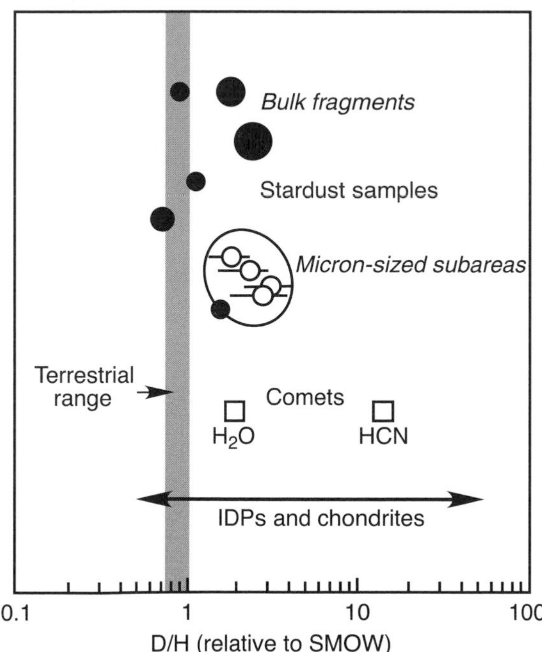

Figure 13.11 Measured D/H ratios in six bulk Stardust particles (black dots; sizes represent relative particle sizes), and in micron-sized subareas in one particle (open circles enclosed by the oval), measured by ion microprobe. The particle compositions overlap D/H ratios in comet H_2O and HCN, and in IDPs and chondrites. Modified from McKeegan et al. (2006).

into a core, respectively. Howardite breccias have compositions intermediate between eucrites and diogenites.

The oxygen-isotope compositions of magmatic achondrites from the same parent body lie on a common mass-fractionation line (see Fig. 6.15), as appropriate for melts. Fractional crystallization produces varying Fe/Mg ratios at nearly constant Fe/Mn (see Fig. 6.8).

A few rare magmatic achondrites, such as GRA 06128 and NWA 11575, have andesite-like compositions and are rich in sodic plagioclase. Similar rocks occur as clasts in ureilite breccias and in some iron meteorites. These may represent the silica- and alkali-rich magmas formed at the onset of chondrite melting (Collinet and Grove, 2020), as opposed to highly fractionated, evolved melts (Day et al., 2009).

13.5.3 Irons and Stony Irons

Irons are mostly samples of once-molten asteroid cores. The recognized groups sample at least 11 different metallic cores (Chabot and Haack, 2006), and compositionally anomalous irons sample many more. As with magmatic achondrites, the compositional variations of magmatic irons reflect differences in parent-body chemistry, as well as changes wrought by fractional crystallization and mixing of liquid and crystals in cores (see Fig. 6.12). The exceptions are "nonmagmatic" irons, which show distinct chemical variations from the fractionation trends. Pallasites may represent samples of core–mantle boundaries, and mesosiderites represent as-yet poorly understood mixing of crust and core materials, probably by impact.

The compositions of most iron meteorites have been reported by John Wasson and his colleagues in an extensive set of papers published over the past 40 years (reviewed by Benedix et al., 2014). The chemistry of these meteorites is highly varied; average compositions for nine siderophile elements in the major iron groups are given in Table 13.6. In part, these differences reflect nebular processes that established siderophile element abundances in chondritic precursors. Group IIAB irons have the lowest nickel concentrations, IIIAB irons have the highest abundance of phosphates, and IVA irons have low volatile element abundances and a distinctive Ir/Au ratio. However, within each group, some elements vary along fairly smooth trends that reflect fractional crystallization. Some examples of compositional variations within one core (IIIAB irons) are illustrated in Figure 13.13. Other factors that may have affected the composition of core samples are the mechanism of crystal growth (dendritic or concentric), the trapping of liquid in pockets between growing crystals, the concentration of

Table 13.4 Representative chemical analyses of primitive achondrites

Element	Acapulco *acapulcoite*	Gibson *lodranite*	Pontlyfni *winonaite*	Udei Station *IAB silicate*	Havero *ureilite*	ALH84025 *brachinite*
Na (mg/g)	6.42	2.8	6.42	5.25	0.242	0.51
Mg (mg/g)	153.8	188	119.2	197	224	184
Al (mg/g)	11.2	3.4	13.1	10	1.1	0.9
Si (mg/g)	176.5	221	144.8		188	170
P (mg/g)	1.6		0.48			
S (mg/g)	21.1		70.4			
K (μg/g)	530	190	700	614	35	
Ca (mg/g)	9.8	9.15	11.8	8.3		18
Sc (μg/g)	8.5	10.5	5.79	12.6	4.9	13.6
Ti (mg/g)	0.7	0.72	0.6		0.36	
V (μg/g)	92			68		108
Cr (mg/g)	7.00	2.8	2.09	3.23	4.09	3.94
Mn (mg/g)	3.26	4.2	2.02	2.37	2.66	2.53
Fe (mg/g)	204.5	159	296	95	150	250
Co (μg/g)	640	720	883	110	101	365
Ni (mg/g)	13.3	8.1	23.4	2.75	0.92	5.1
Zn (μg/g)	250	62	130	237	235	164
Ga (μg/g)	7.78	8.1	16.6	5.3	1.13	2.1
As (μg/g)	1.57	2.12	3.2	0.33		0.55
Se (μg/g)	10.8	0.51	16	10.4		12.4
Br (ng/g)		3200		450		330
Rb (ng/g)		2600				
Sb (ng/g)	61	150	390	36		56
La (ng/g)	742	170	260	158	70	65
Ce (ng/g)		700	700			
Nd (ng/g)		550				
Sm (ng/g)	250	210	103	92	14	34
Eu (ng/g)	100	20	100	60	4.1	33
Tb (ng/g)	62	50				
Yb (ng/g)	300	270	90	147	25	98
Lu (ng/g)	40	50	13	26	8.5	16
Hf (ng/g)	120	50	110			
Ta (ng/g)		14				
W (ng/g)		190			180	
Os (ng/g)	670		780	50		250
Ir (ng/g)	536	1220	586	37	240	123
Au (ng/g)	149	90	250	38	24	61
Th (ng/g)		34				
U (ng/g)	70	50			6	

Data sources given by Mittlefehldt (2014).

PLATES

Figure 1.10 A family portrait of Mars rovers. Sojourner (Mars Pathfinder rover) is in the foreground, Spirit (Mars Exploration Rover, MER) is on the back left, and Curiosity (Mars Science Laboratory, MSL) is on the right. Perseverance (Mars 2020 rover, not pictured) is approximately the size of Curiosity. Image from NASA and JPL.

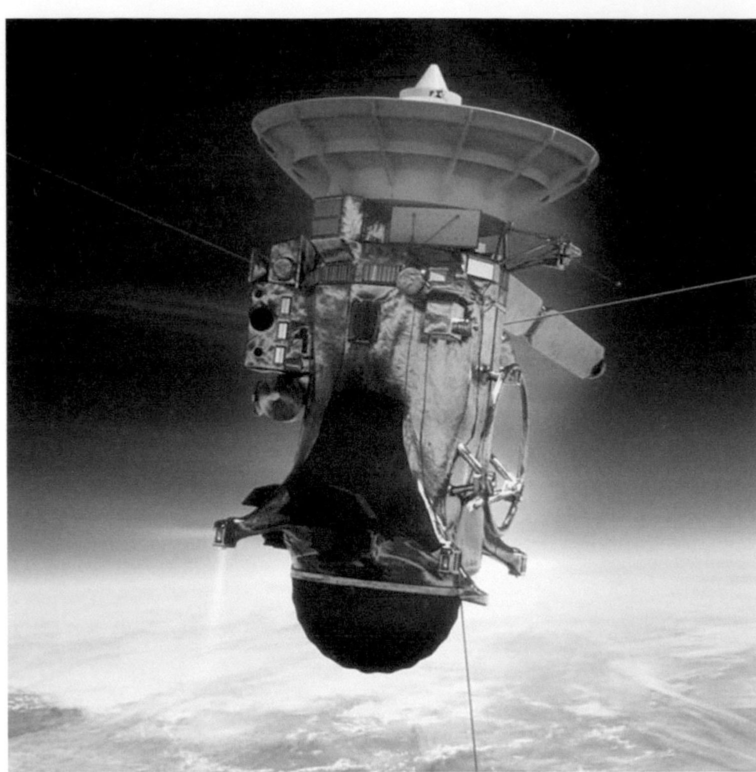

Figure 1.11 The nuclear-powered Cassini spacecraft, launched towards Saturn in 1997, was active for nearly 20 years. The ring on the right side held the Huygens probe, which parachuted onto Titan's surface. NASA image.

Figure 1.12 A frozen meteorite exposed on Antarctic ice. The numbers at the bottom of the identifying counter at the right are a centimeter scale. NASA image.

Figure 1.14 A piece of the ALH 84001 martian meteorite, ~8 cm across. Dark material on the surface is fusion crust, formed during passage through the Earth's atmosphere. This sample created a stir when it was proposed to contain evidence for extraterrestrial life. Image courtesy of the Smithsonian Institution.

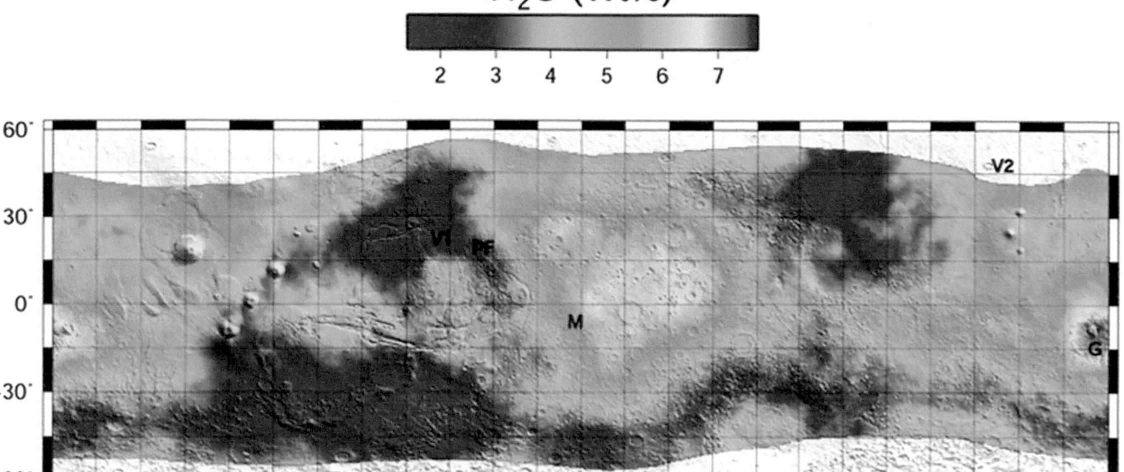

Figure 11.15 Global distribution map of water, calculated from measured hydrogen, on Mars. Hydrogen was mapped in the upper few meters of the surface by the Neutron and Gamma Ray Spectrometer on Mars Global Surveyor. Near the equator, the water occurs as permafrost and bound in minerals, but it only occurs as permafrost at higher latitudes. From Boynton et al. (2008), with permission.

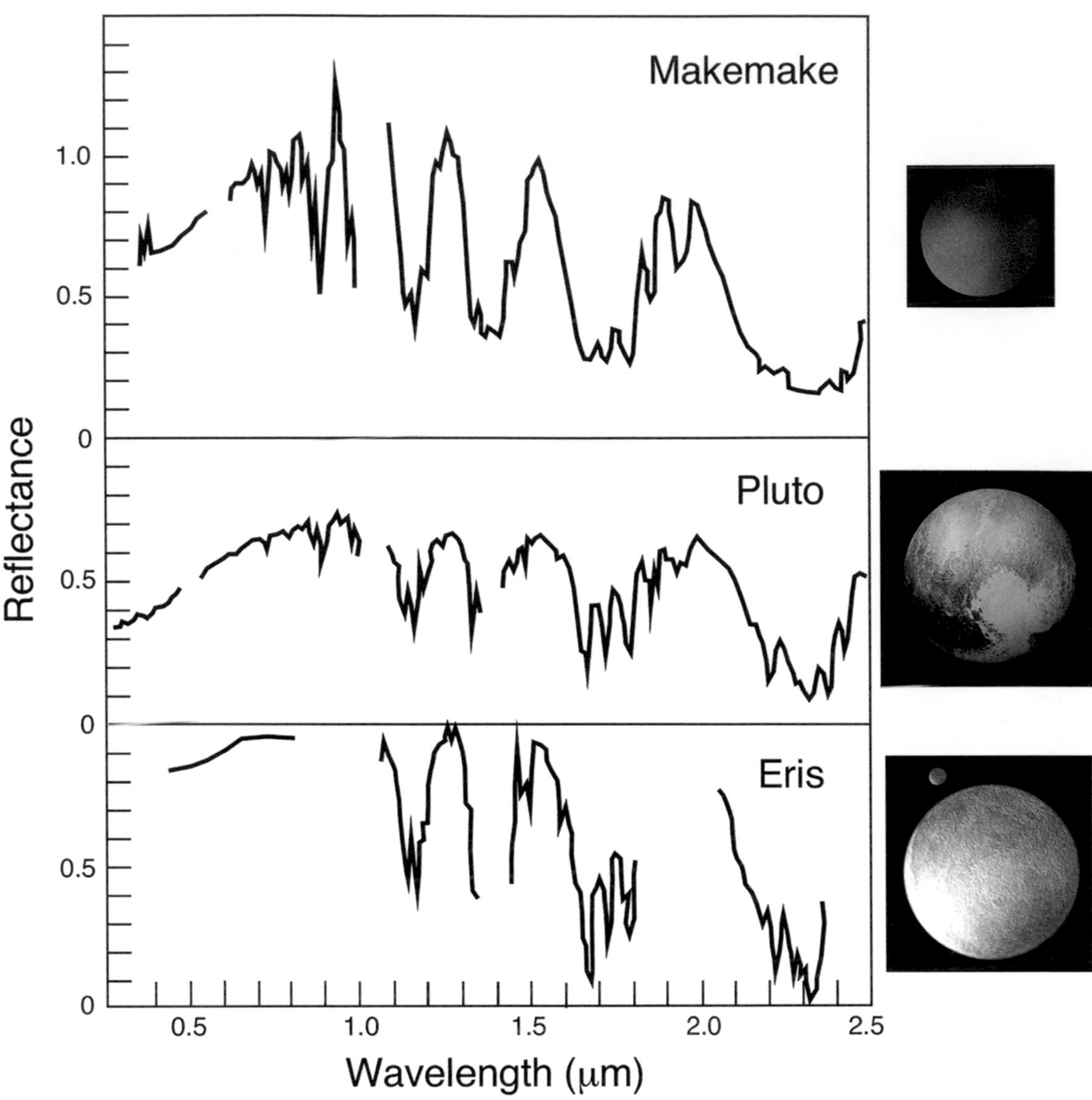

Figure 12.9 Infrared spectra of the KBOs Makemake, Pluto, and Eris. Modified from Schaller (2010). Images of Makemake and Eris are artists' renditions, and the Pluto image is from New Horizons (NASA).

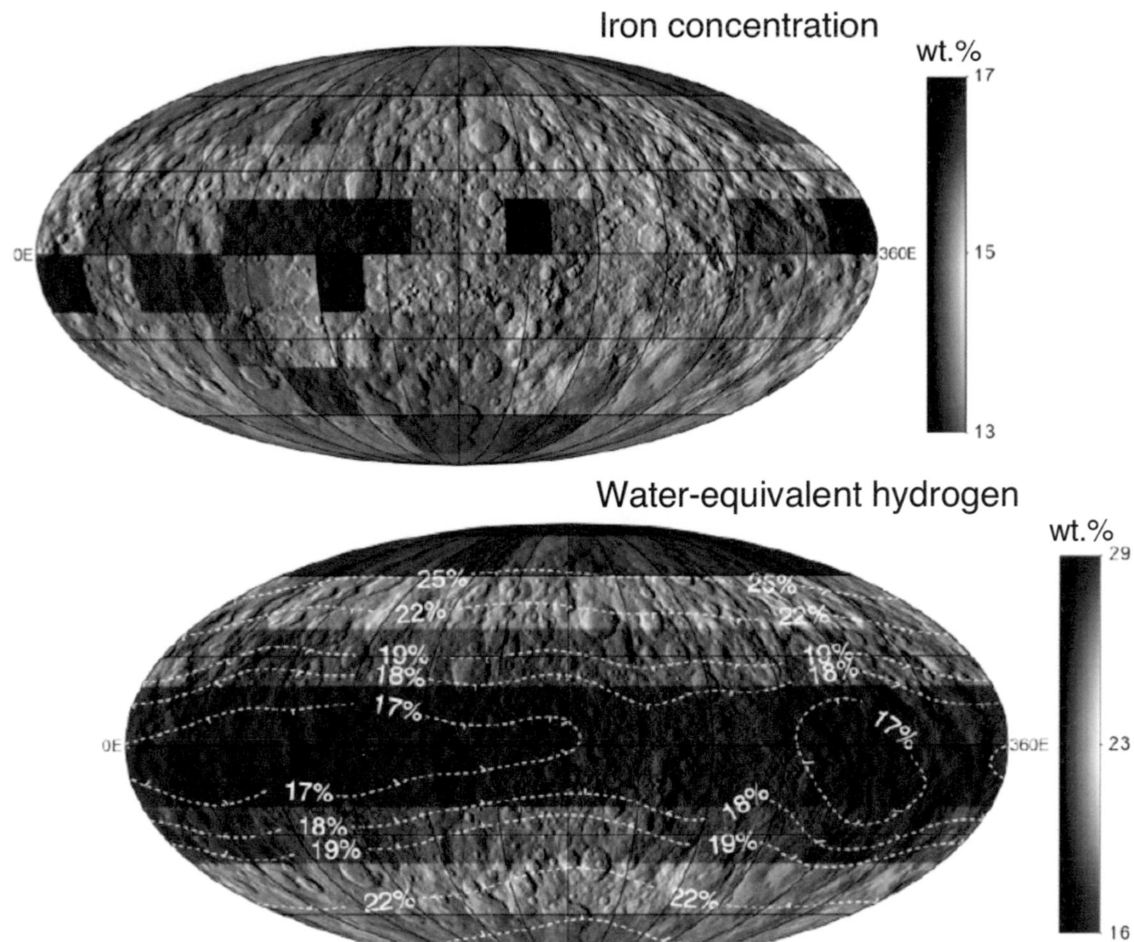

Figure 13.6 Variations in the concentrations of iron and water-equivalent hydrogen in the upper meter of Ceres' regolith, as measured by GRaND. Adapted from Prettyman et al. (2018).

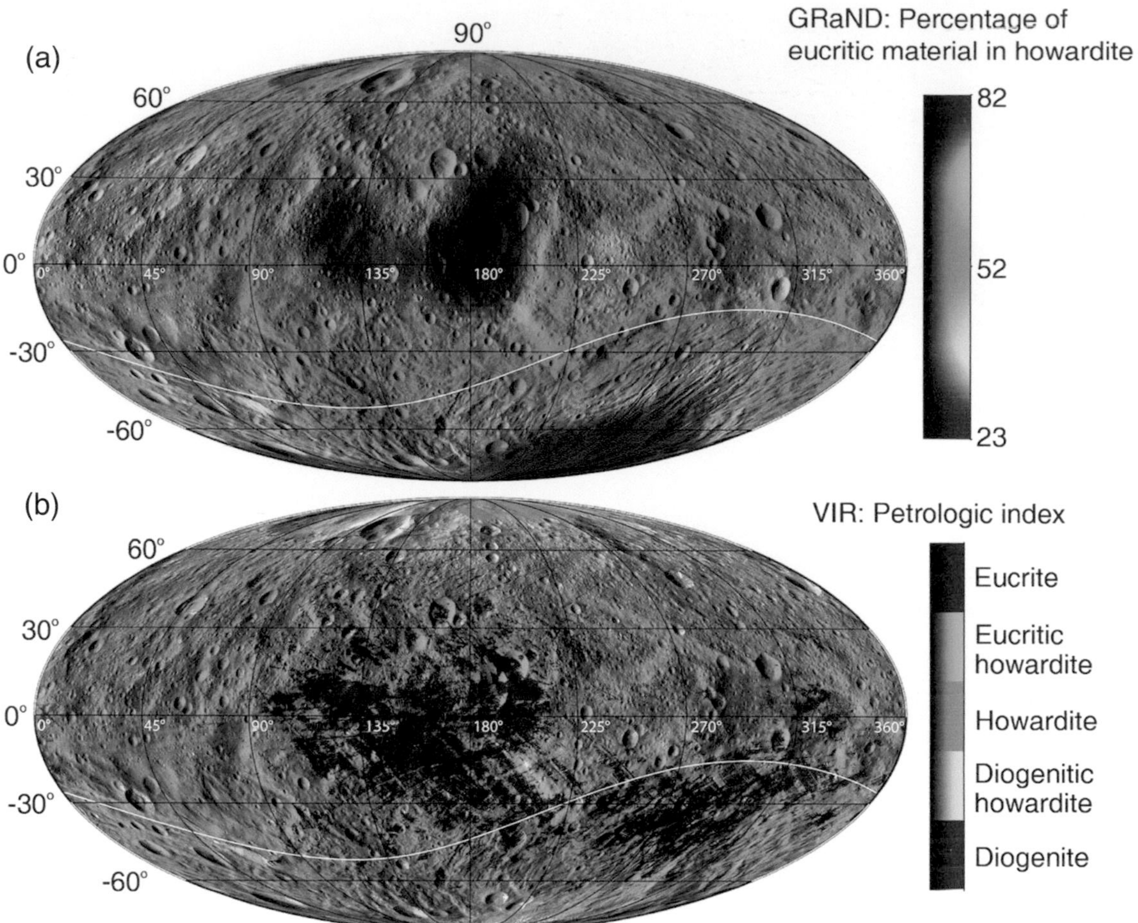

Figure 13.14 Compositional maps of Vesta in terms of proportion of eucrite (relative to diogenite and howardite), based on data from two Dawn instruments: (a) GRaND and (b) VIR. Methods used to construct the maps were described by Prettyman et al. (2013) and Ammannito et al. (2013), respectively. The white lines in (a) and (b) indicate the outline of the Rheasilvia basin at the south pole. From Raymond et al. (2017), with permission.

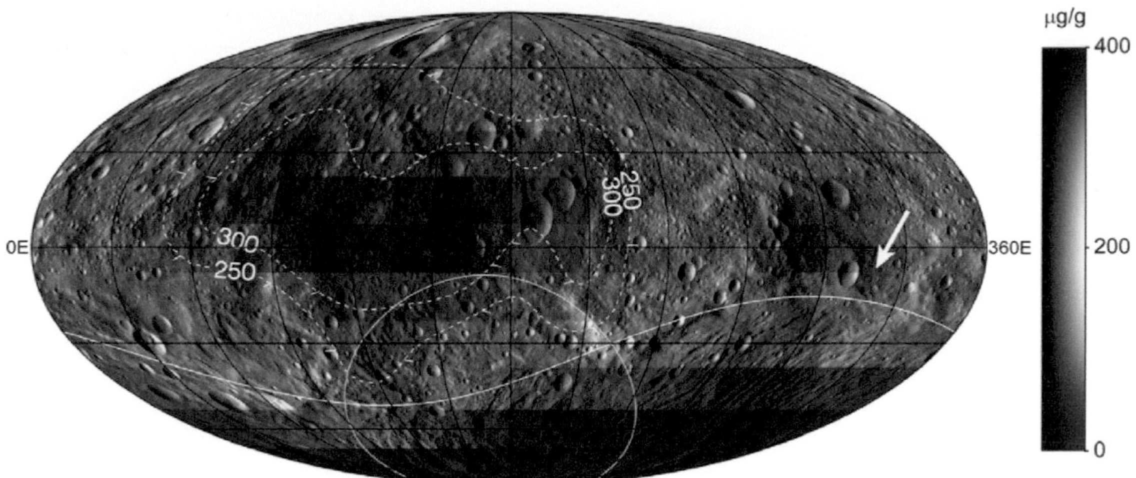

Figure 13.15 Map of hydrogen, expressed as water equivalents, on Vesta, based on GRaND epithermal neutron absorption data. The solid white lines indicate the outlines of Rheasilvia and a smaller, superimposed basin, and the dashed white lines are contours of high water abundance defining an area impacted by a carbonaceous chondrite body that contained hydrous minerals. Adapted from Prettyman et al. (2012).

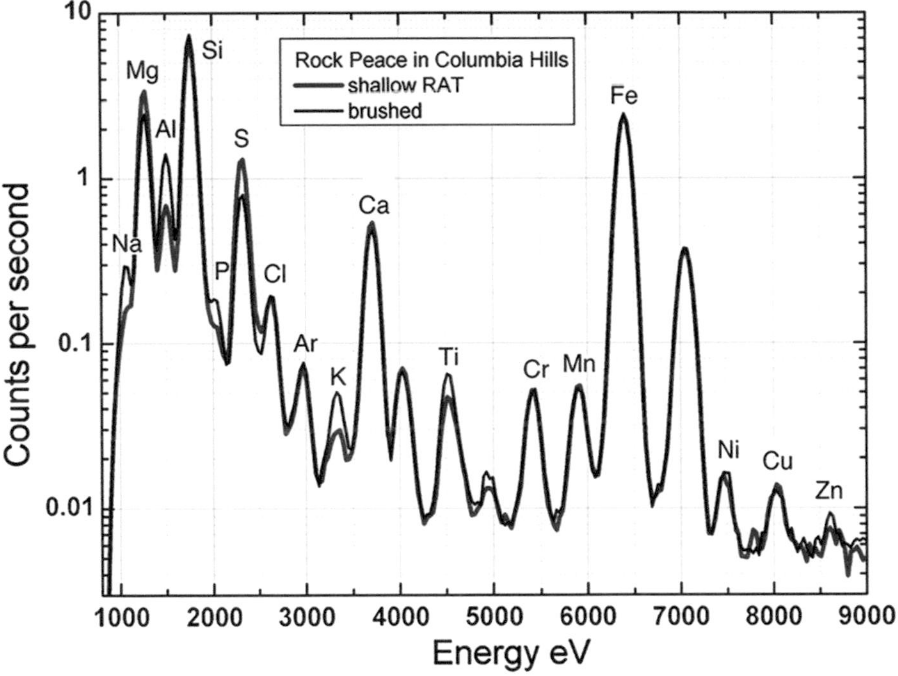

Figure 14.14 Image of an APXS showing a detector surrounded by an array of radioactive curium α-particle sources. The APXS spectra below are for a rock at the Spirit rover landing site that was brushed to remove surface dust and RAT-ground to remove any surficial weathering.

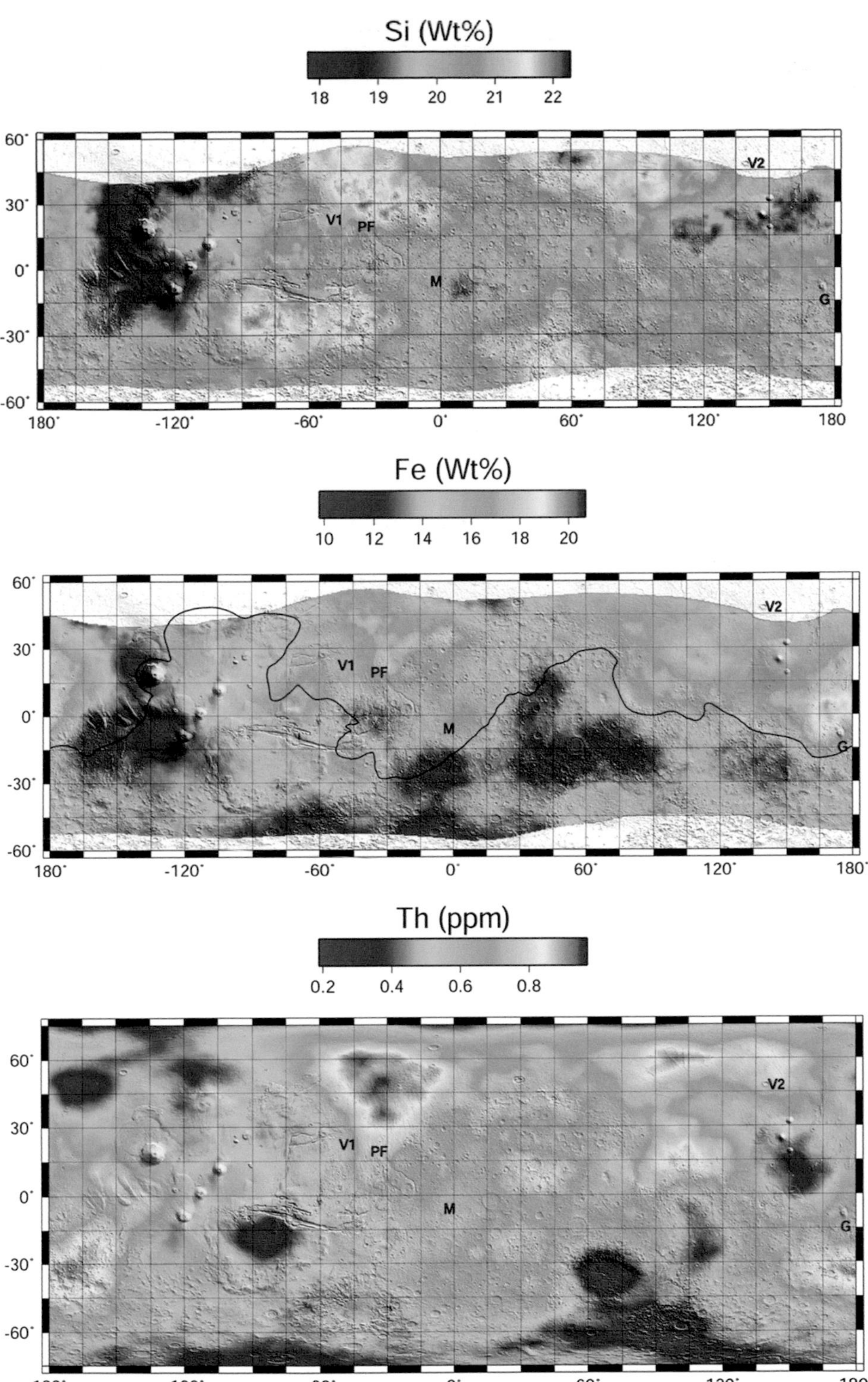

Figure 14.18 Global GRS compositional maps of Mars for (a) silicon, (b) iron, and (c) thorium. Letters represent spacecraft landing sites: V1 and V2 = Viking, PF = Mars Pathfinder, M = Merdiani (Opportunity), and G = Gusev (Spirit). The irregular line in (b) separates the southern highlands from the northern lowlands. From Boynton et al. (2008), with permission.

Figure 15.1 Hubble Space Telescope image of gas and dust in the star-forming region of the Carina Nebula. Two massive stars out of the field of view to the upper right and upper left are evaporating the molecular cloud, leaving pillars of gas pointing at the stars. Dense knots of cloud material are EGGs containing new stellar systems. Image courtesy of NASA and Hubble 20th Anniversary Team (STScI).

Figure 15.3 ALMA image of the accretion disk around HL Tauri, a Sun-like star, showing gaps interpreted to have been opened by unseen planets.

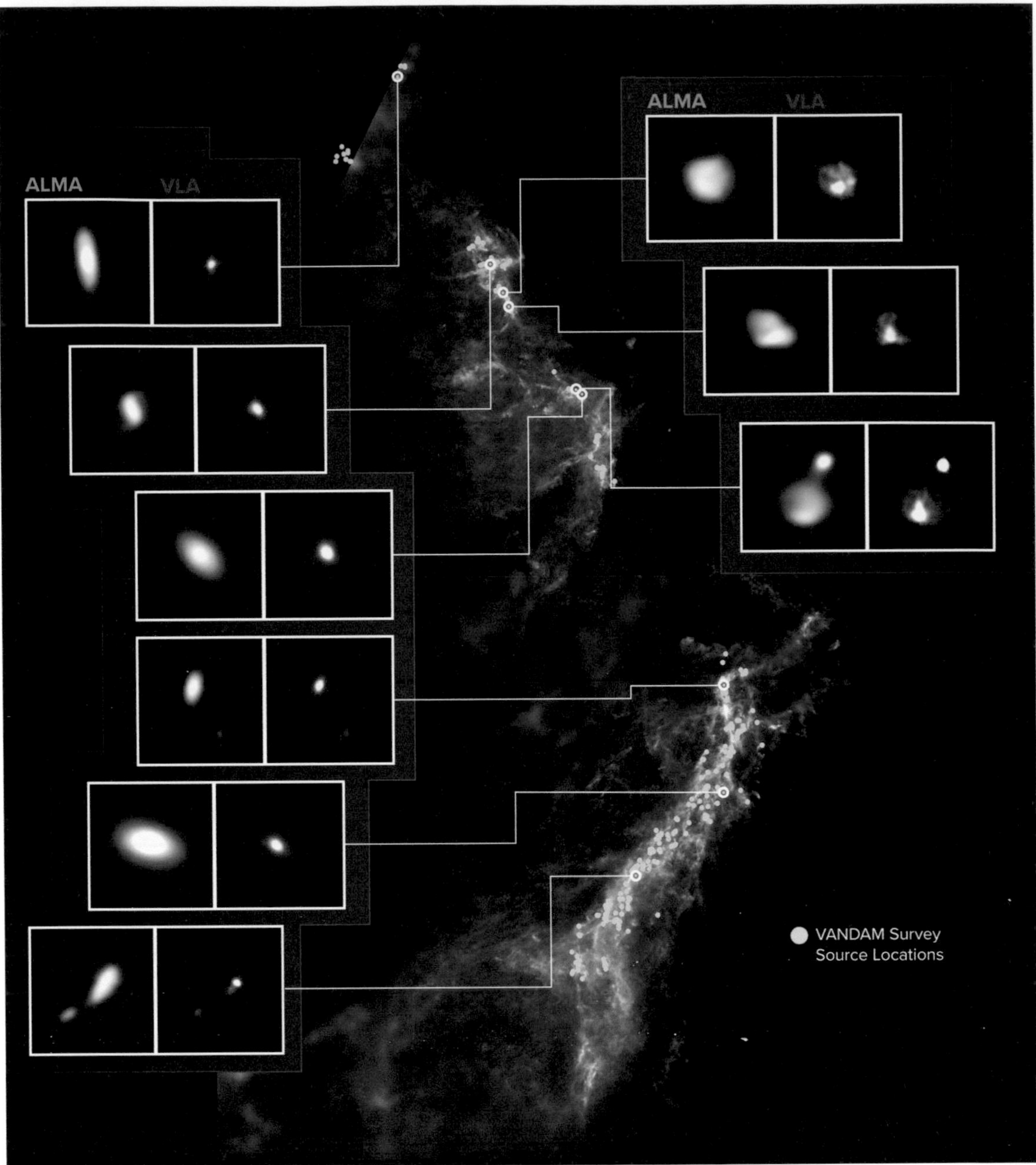

Figure 15.4 ALMA and VLA data complement each other. ALMA images the outer disk structure, visualized in blue, and VLA images the inner disk and star core (orange) in the Orion Molecular Clouds. Image courtesy of ALMA (ESO/NAOJ/NRAO), J. Tobin; VLA (NRAO/AUI/NSF), S. Dagnello; and Herschel/ESA.

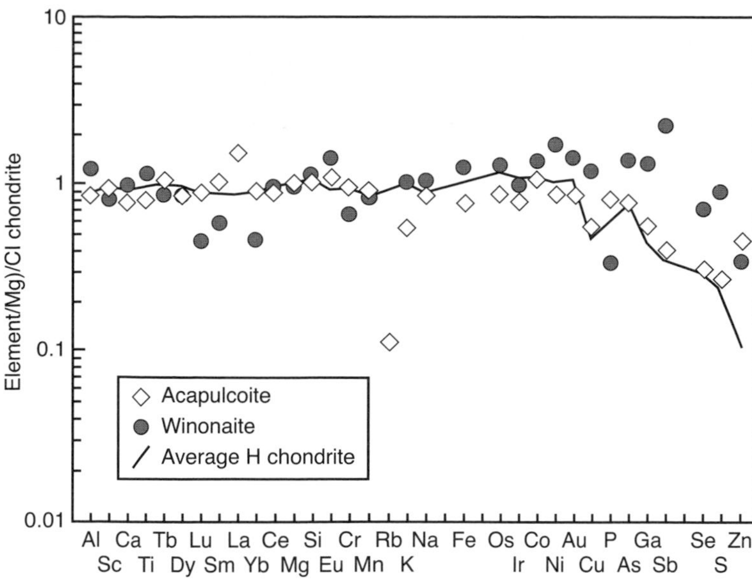

Figure 13.12 Elemental abundances, normalized to Mg and CI chondrites, for two groups of primitive achondrite (acapulcoites and winonaites) that experienced low degrees of partial melting. These abundances are similar to chondritic abundances (average H-chondrite composition is illustrated). Modified from Mittlefehldt (2014).

sulfur (which is insoluble in solid metal and thus becomes concentrated in fractionating liquid), and the late-stage immiscibility of phosphorus- and sulfur-rich melts (Chabot and Haack, 2006).

Because pallasites and mesosiderites are so coarse-grained, it is very difficult to obtain representative bulk chemical analyses, so we will not discuss these. The Psyche spacecraft mission, now under development, is intended to explore an asteroid thought to be a stony-iron asteroid.

13.6 Geochemical Exploration of Asteroid Vesta

Asteroid Vesta is particularly noteworthy by virtue of the fact that it is one of only a handful of intact, differentiated protoplanets, and because it is the parent body for the most voluminous class of achondrites (HEDs, as summarized by McSween et al., 2013). The fact that such a large number of meteorites come from this body is attributed to excavation by the impactor that created the huge, ~1 Ga-old Rheasilvia impact basin and launched materials into a nearby escape hatch (orbital resonance with Jupiter).

Vesta (see Fig. 12.2) was investigated by the Dawn spacecraft mission (Russell et al., 2012; Raymond et al., 2017). The GRaND and VIR instruments on the Dawn orbiter each produced a global compositional map of Vesta, presented in terms of the abundances of eucrite, diogenite, and howardite (Fig. 13.14). The GRaND map is based on neutron absorption as calibrated with HEDs (Prettyman et al., 2013), and the VIR map utilizes the positions of absorption bands for HEDs (Ammannito et al., 2013). These maps illustrate the preponderance of howardite in the surface regolith, at least at the coarse spatial scales of these instruments. Plutonic rocks (diogenites) predictably occur in the Rheasilvia basin and in its ejecta. However, olivine is deficient in the crust and upper mantle of Vesta, relative to chondrites; olivine is likely sequestered in the lower mantle, either left behind as a residue from partial melting or concentrated by crystal settling in a solidifying magma ocean.

GRaND measured globally averaged iron, silicon, and oxygen abundances for Vesta, as well as a K/Th ratio (Prettyman et al., 2012, 2013, 2015) (Table 13.3). Analysis of epithermal neutrons identified an equatorial region anomalously rich in hydrogen (Fig. 13.15); this area is interpreted as regolith contaminated with exogenic carbonaceous chondrite, which contains hydrous phyllosilicates (Prettyman et al., 2012).

The bulk composition of Vesta is chondritic, and its Fe/Mn ratio, oxygen-isotope composition, oxidation state (determined from HEDs), and its core size and density constraints (determined from Dawn data) can be modeled by a 3:1 mixture of H and CM

Table 13.5 Representative chemical analyses of achondrites

Element	Lew87051 *angrite*	Aubres *aubrite*	Sioux Co. *eucrite*	Shalka *diogenite*	Frankfort *howardite*
Na (mg/g)	0.174	0.706	3.3	0.024	1.01
Mg (mg/g)	117	232.7	42.9	155.5	126.3
Al (mg/g)	48.7	2.4	67.97	3.2	22.6
Si (mg/g)	188.9	273.7	229.2	241.1	237.9
P (mg/g)		2	0.39	0.01	0.15
K (μg/g)		517	295	13	81
Ca (mg/g)	77.2	2.77	74.0	5.2	26
Sc (μg/g)	36	6.5	34.5	9.9	20.5
Ti (mg/g)	4.4	0.185	4.08	0.37	1.36
Cr (mg/g)	1.09	0.241	1.98	16.5	7.94
Mn (mg/g)	1.9	1.49	4.34	4.28	4.34
Fe (mg/g)	148	4.1	144.7	126.5	139.1
Co (μg/g)	27.4	6	6.8	18	23.6
Ni (mg/g)	0.044	0.161	0.0017	0.005	0.125
Zn (μg/g)		3.61	5.3		2.17
Ga (μg/g)			1.27		0.51
Se (μg/g)		2.46			0.05
Rb (ng/g)		605	170		230
Sr (μg/g)	67		76		24
Zr (μg/g)			41		
La (ng/g)	2320		1820	9	980
Ce (ng/g)	6200	347	5190		2700
Nd (ng/g)		241	4320		1700
Sm (ng/g)	1480		1540	4	560
Eu (ng/g)	536	16.7	415	2	180
Tb (ng/g)	360	25.2	398		150
Yb (ng/g)	1520	101	1940	30	620
Lu (ng/g)	239	14.8	287	6	93
Hf (ng/g)	1170		1160		490
Ta (ng/g)	110		147		40
W (ng/g)			43		126
Os (ng/g)		0.843			
Ir (ng/g)		0.822			4.1
Au (ng/g)		4.27			3.4
Th (ng/g)	220		287		
U (ng/g)		3.17	74		38

Data sources given by Mittlefehldt (2014).

chondrites (Toplis et al., 2013). The average densities of Vesta's crust (2.97 g/cm^3) and mantle (3.16 g/cm^3) were derived by modeling the gravity field determined by careful tracking of the spacecraft's orbit and velocity (Konopliv et al., 2014). Vesta has a substantial metallic core, but core size and density cannot be determined independently from gravity data. The core radius is estimated at 110 km for an assumed density of 7.10–7.80 g/cm^3, as in iron meteorites (Russell et al., 2012). The core mass fraction (18%) is consistent with estimates (15–20%) from models of siderophile element depletion in HEDs. The exploration of

Table 13.6 Average chemical compositions of major iron meteorite groups

Element	IAB	IIAB	IIIAB	IVA	IVB
Ni (mg/g)	95.0	56.5	83.3	85.1	171
S (mg/g)		(170)	(120)	(30)	(0)
Co (mg/g)	4.9	4.6	5.1	4.0	7.6
Cu (μg/g)	234	133	156	137	<9
Ir (μg/g)	2.0	12.5 (1.3)	3.2 (5.0)	1.8 (1.8)	18.0 (22)
Au (μg/g)	1.75	0.71 (1.0)	1.12 (0.7)	1.55 (1.6)	0.14 (0.15)
Ga (μg/g)	63.6	58.6	19.8	2.14	0.23
Ge (μg/g)	247	174	39.1	0.12	0.055
Re (ng/g)	260	1780 (250)	439 (200)	230 (150)	2150 (3500)

Data sources given in Benedix et al. (2014). Values in parentheses are calculated initial liquid compositions of the core; for elements like Ir and Re with solid/liquid distribution coefficients that are far from unity, meteorite composition may be different from that of the initial liquid core.

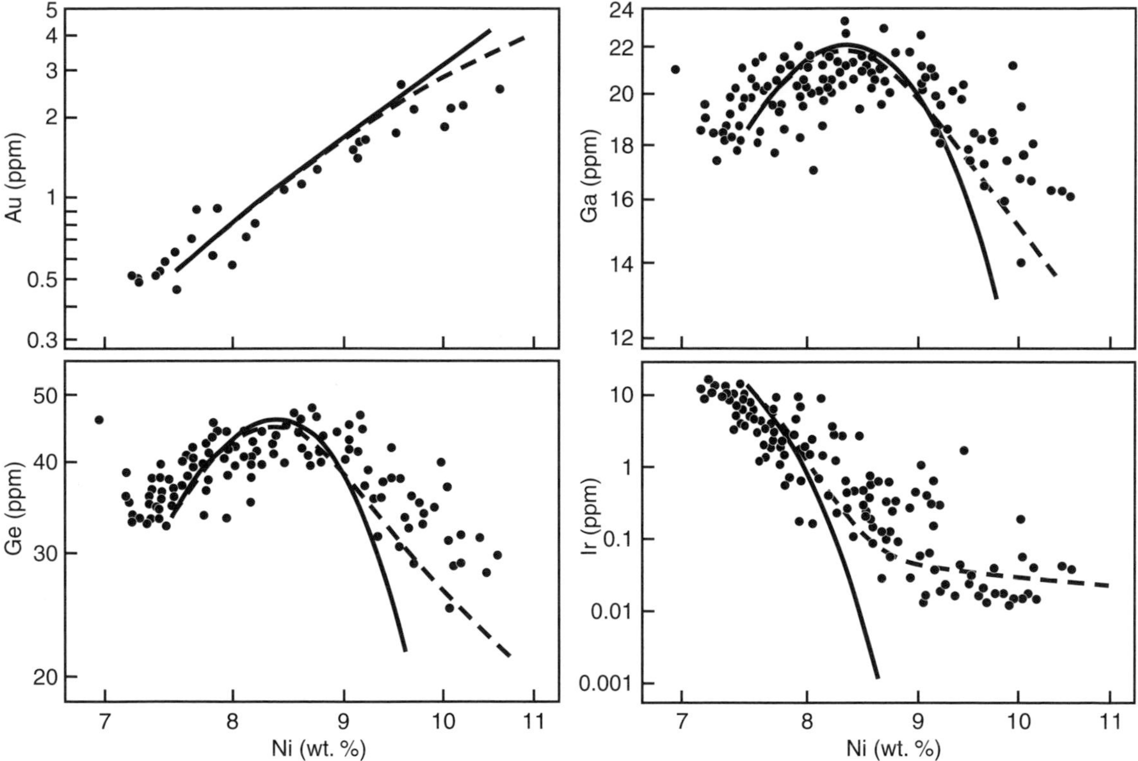

Figure 13.13 Elemental abundances in group IIIAB iron meteorites (dots), compared to calculated fractional crystallization trends that assume complete separation of crystals and melt (solid curves) and incomplete separation of those phases (dashed curves). Modified from Haack and McCoy (2004).

Vesta thus reveals a differentiated body composed of magmatic achondritic and iron meteorite materials. A compositional model for Vesta, providing geologic context for HEDs, is illustrated in Figure 13.16. Numerous thermal models for Vesta, based on an early magma ocean, have been used to explain the asteroid's differentiation (e.g. Mandler and Elkins-Tanton, 2013).

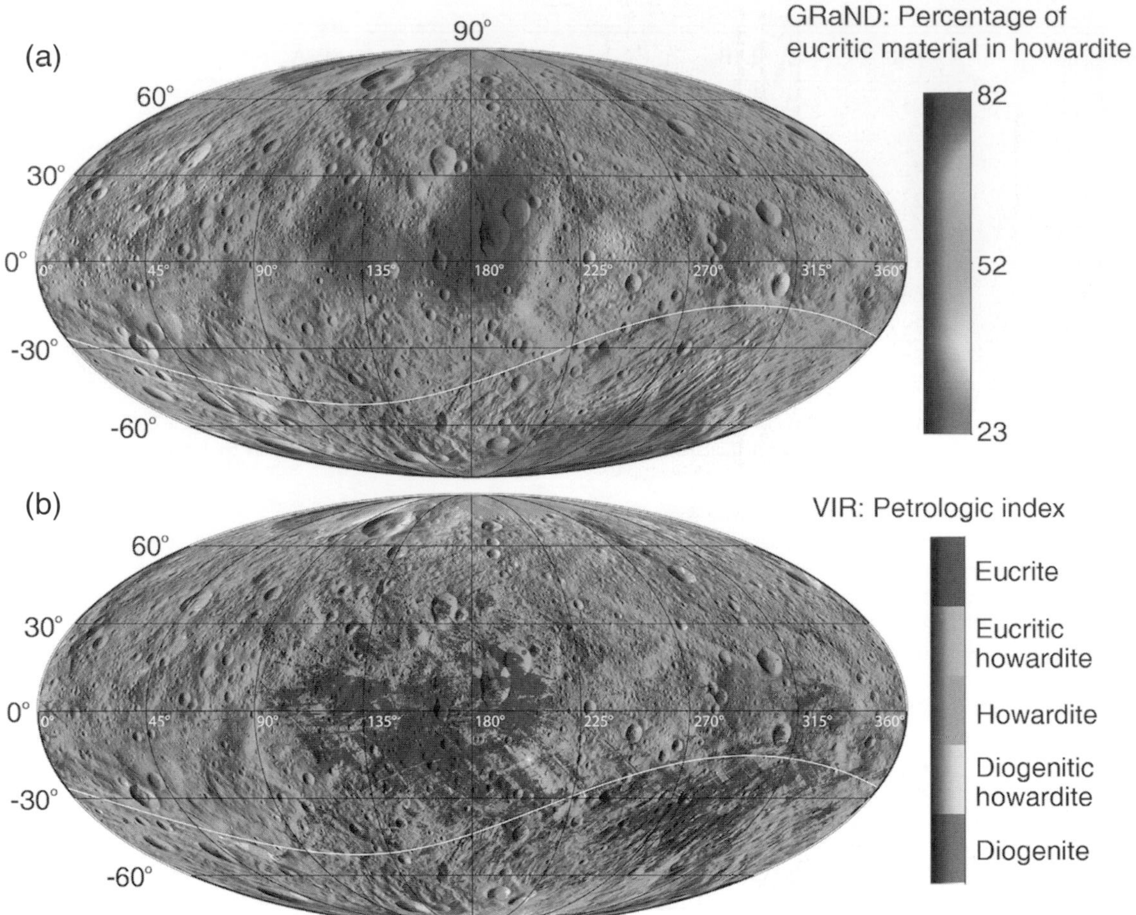

Figure 13.14 Compositional maps of Vesta in terms of proportion of eucrite (relative to diogenite and howardite), based on data from two Dawn instruments: (a) GRaND and (b) VIR. Methods used to construct the maps were described by Prettyman et al. (2013) and Ammannito et al. (2013), respectively. The white lines in (a) and (b) indicate the outline of the Rheasilvia basin at the south pole. From Raymond et al. (2017), with permission. *For the colour version, refer to the plate section.*

Summary

Carbonaceous and noncarbonaceous chondrites have approximately solar abundances of refractory elements, but are notably depleted in volatile elements to varying degrees. Variations in the elemental abundances of chondrites can result from nebular sorting of components, depletion of volatiles through evaporation or condensation, and changes in oxidation state. Because they are not differentiated, chondritic meteorites provide estimates of the bulk compositions of their parent planetesimals. Some large chondritic bodies, like Ceres, show partial differentiation – separation of water and rock.

Primitive achondrites are residues from the extraction of limited amounts of melt; their major elements remain unchanged, but trace elements that partition into melts are commonly depleted. Magmatic achondrites and iron meteorites show a wide variety of bulk compositions resulting from fractional crystallization, magma mixing, incorporation of solid materials into melts or trapping of melts between growing crystals, and liquid immiscibility. Asteroid Vesta has a chondritic bulk composition but is composed of achondrites (HEDs) and a large metal core (unsampled), and thus illustrates the complexities that attend differentiation.

Figure 13.15 Map of hydrogen, expressed as water equivalents, on Vesta, based on GRaND epithermal neutron absorption data. The solid white lines indicate the outlines of Rheasilvia and a smaller, superimposed basin, and the dashed white lines are contours of high water abundance defining an area impacted by a carbonaceous chondrite body that contained hydrous minerals. Adapted from Prettyman et al. (2012). *For the colour version, refer to the plate section.*

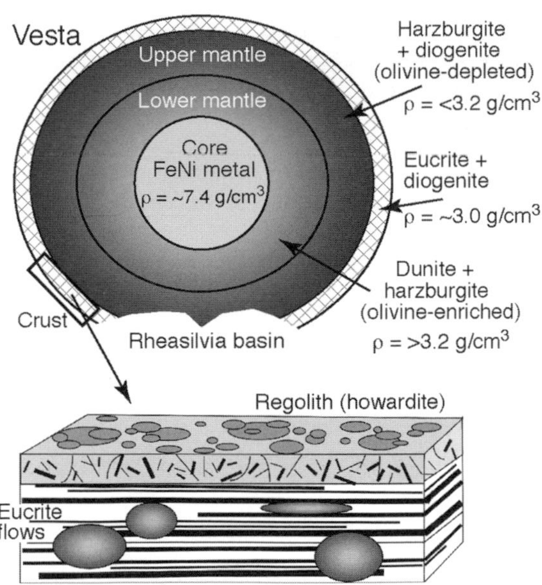

Figure 13.16 Compositional model of Vesta, based on data from the Dawn spacecraft. The surface regolith is mostly howardite – pulverized mixtures of eucrite and diogenite. The crust, <20 km thick, is basaltic, dominated by eucrite flows, and has been intruded by plutons containing diogenite. The mantle is composed of olivine + pyroxene (harzburgite) grading downward to greater proportions of olivine (dunite). A massive core of FeNi metal has a diameter almost half that of the asteroid.

Questions

1. What kinds of bulk chemical variations are exhibited by the various classes of chondrites?
2. Briefly explain how the various kinds of spacecraft instruments used to measure chemical composition work. What elements can be measured?
3. Compare the chemical compositions of IDPs to carbonaceous chondrites.
4. What processes can lead to chemical variations in achondrites and irons? How can we determine exactly what processes occurred in a given set of samples?
5. When a chondrite begins to melt, what is the composition of the first melt and how does it evolve with further melting?

Suggestions for Further Reading

Brearley A. J., and Jones R. H. (1998) Chondritic meteorites. In *Planetary Materials*, Papike J. J, editor, *Reviews in Mineralogy*, **36**, pp. 3-1 to 3-398, Mineralogical Society of America, Washington. A comprehensive review of the compositions of chondrites.

Mittlefehldt D. W. (2014) Achondrites. In *Treatise on Geochemistry, 2nd Edition, Vol. 1: Meteorites and Cosmochemical Processes*, Davis A. M., editor, pp. 235–266, Elsevier, Oxford. A superb review of achondritic meteorites, containing many high-quality analyses.

Nittler L. R., McCoy T. J., Clark P. E., et al. (2004) Bulk element compositions of meteorites: A guide for interpreting remote-sensing geochemical measurements of planets and asteroids. *Antarctic Meteorite Research*, 17, 231–251. A comprehensive summary of meteorite compositions, compiled for comparison with spacecraft remote-sensing data.

Other References

Alexander C. M. O'D., Bowden R., Fogel M. L., and Howard K. T. (2015) Carbonate abundances and isotopic compositions in chondrites. *Meteoritics & Planetary Science*, **50**, 810–833.

Ammannito E., De Sanctis M. C., Capaccioni F., et al. (2013) Vestan lithologies mapped by the visual and infrared spectrometer on Dawn. *Meteoritics & Planetary Science*, **48**, 2185–2198.

Benedix G. K., Haack H., and McCoy T. J. (2014) Iron and stony-iron meteorites. In *Treatise on Geochemistry, 2nd Edition, Vol. 1: Meteorites and Cosmochemical Processes*, Davis A. M., editor, pp. 267–285, Elsevier, Oxford.

Bradley J. P. (2014) Early solar nebula grains – interplanetary dust particles. In *Treatise on Geochemistry, 2nd edition, Vol. 1: Meteorites and Cosmochemical Processes*, Davis A. M., editor, pp. 287–308, Elsevier, Oxford.

Castillo-Rogez J., Neveu M., McSween H. Y., et al. (2018) Insights into Ceres' evolution from surface composition. *Meteoritics & Planetary Science*, **53**, 1820–1843.

Chabot N. L., and Haack H. (2006) Evolution of asteroidal cores. In *Meteorites and the Early Solar System II*, Lauretta D. S., and McSween, H. Y., editors, pp. 747–771, University of Arizona Press, Tucson.

Collinet M., and Grove T. L. (2020) Formation of primitive achondrites by partial melting of alkali-undepleted planetesimals in the inner solar system. *Geochimica et Cosmochimica Acta*, **277**, 358–376.

Day J. M. D., Ash R. D., Liu Y., et al. (2009) Early formation of evolved asteroid crust. *Nature*, **457**, 179–182.

De Sanctis M. C., Ammannito E., Carrozzo G., et al. (2018) Ceres' global and localized mineralogical composition determined by Dawn's Visible and Infrared Spectrometer (VIR). *Meteoritics & Planetary Science*, **53**, 1844–1865.

Ermakov A. I., Fu R. R., Castillo-Rogez J. C., et al. (2017) Constraints on Ceres' internal structure and evolution from its shape and gravity measured by the Dawn spacecraft. *Journal of Geophysical Research*, **122**, 2267–2293.

Evans L. G., Starr R. D., Bruckner J., et al. (2001) Elemental composition from gamma-ray spectroscopy of the NEAR-Shoemaker landing site on 433 Eros. *Meteoritics & Planetary Science*, **36**, 1639–1660.

Flynn G. J., and Sutton S. R. (1992) Trace elements on chondritic stratospheric particles: Zinc depletion as a possible indicator of atmospheric entry heating. *Proceedings of the Lunar and Planetary Science Conference*, **22**, 171–184.

Flynn G. J., Bleuet P., Borg J., et al. (2006) Elemental compositions of comet 81P/Wild2 samples collected by Stardust. *Science*, **314**, 1731–1735.

Haack H., and McCoy T. J. (2004) Iron and stony-iron meteorites. In *Treatise on Geochemistry, Vol. 1: Meteorites, Comets, and Planets*, Davis, A. M., editor, pp. 325–345, Elsevier, Oxford.

Jarosewich E. (1990) Chemical analyses of meteorites: A compilation of stony and iron meteorite analyses. *Meteoritics*, **25**, 323–337.

Konopliv A. S., Asmar S. W., Park R. S., et al. (2014) The Vesta gravity field, spin pole and rotation period, landmark positions and ephemeris from the Dawn tracking and optical data. *Icarus*, **240**, 103–117.

Krot A. N., Keil K., Scott E. R. D., et al. (2014) Classification of meteorites and their genetic relationships. In *Treatise on Geochemistry, 2nd edition, Vol. 1: Meteorites and Cosmochemical Processes*, Davis A. M., editor, pp. 1–63, Elsevier, Oxford.

Lauretta D. S., DellGiustina D. N., Bennett C. A., et al. (2019) The unexpected surface of asteroid (101955) Bennu. *Nature*, **568**, 55–60.

Lawler M. E., Brownlee D. E., Temple S., and Wheelock M. M. (1989). Iron, magnesium, and silicon in dust from comet Halley. *Icarus*, **80**, 225–242.

Lawrence D. J., Peplowski P. N., Beck A. W., et al. (2018) Compositional variability on the surface of 1 Ceres revealed through GRaND measurements of high-energy gamma rays. *Meteoritics & Planetary Science*, **53**, 1805–1819,

Lodders K., and Fegley B. Jr. (1998) *The Planetary Scientist's Companion*, Oxford University Press, New York, 371 pp.

Mandler B. E., and Elkins-Tanton L. T. (2013) The origin of eucrites, diogenites, and olivine diogenites: Magma ocean crystallization and shallow magma chamber processes on Vesta. *Meteoritics & Planetary Science*, **48**, 2333–2349.

McCoy T. J., plus 16 coauthors (2001) The composition of 433 Eros: A mineralogical-chemical synthesis. *Meteoritics & Planetary Science*, **36**, 1661–1672.

McKeegan K. D., plus 45 coauthors (2006) Isotopic compositions of cometary matter returned by Stardust. *Science*, **314**, 1724–1728.

McSween H. Y., Binzel R. P., De Sanctis M. C., et al. (2013). Dawn; the Vesta-HED connection; and the geologic context for eucrites, diogenites, and howardites. *Meteoritics & Planetary Science*, **48**, 2090–2114.

McSween H. Y., Emery J. P., Rivkin A. S., et al. (2018) Carbonaceous chondrites as analogs for the composition and alteration of Ceres. *Meteoritics & Planetary Science*, **53**, 1793–1804.

Mittlefehldt D. W., McCoy T. J., Goodrich C. A., and Kracher A. (1998) Non-chondritic meteorites from asteroidal bodies. In *Planetary Materials,* Papike J. J., editor, *Reviews in Mineralogy*, **36**, pp. 4-1 to 4-195, Mineralogical Society of America, Washington.

Mittlefehldt D. W., Herrin J. S., Quinn J. E., et al. (2013) Composition and petrology of HED polymict breccias: The regolith of (4) Vesta. *Meteoritics & Planetary Science*, **48**, 2105–2134.

Nittler L. R., and 15 coauthors (2001) X-ray fluorescence measurements of the surface elemental composition of asteroid 433 Eros. *Meteoritics & Planetary Science*, **36**, 1673–1695.

Ogliore R. C., Nagashima K., Huss G. R., et al. (2015) Oxygen isotopic composition of coarse- and fine-grained material from comet 81P/Wild 2. *Geochimica et Cosmochimica Acta*, **166**, 74–91.

Okada T., Shirai K., Yamanoto Y., et al. (2006) X-ray fluorescence spectrometry of asteroid Itokawa by Hayabusa. *Science*, **312**, 1338–1341.

Park R. S., Konopliv A. S., Bills B. G., et al. (2016) A partially differentiated interior for (1) Ceres deduced from its gravity field and shape. *Nature*, **537**, 515–517.

Prettyman T. H., Mittlefehldt D. W., Yamashita N., et al. (2012) Elemental mapping by Dawn reveals exogenic H in Vesta's regolith. *Science*, **338**, 242–246.

Prettyman T. H., Mittlefehldt D. W., Yamashita N., et al. (2013) Neutron absorption constraints on the composition of 4 Vesta. *Meteoritics & Planetary Science*, **48**, 2211–2236.

Prettyman T. H., Yamashita N., Reedy R. C., et al. (2015) Concentrations of potassium and thorium within Vesta's regolith. *Icarus*, **258**, 39–52.

Prettyman T. H., Yamashita N., Toplis M. J., et al. (2017) Extensive water ice within Ceres' aqueously altered regolith: Evidence from nuclear spectroscopy. *Science*, **355**, 55–59.

Prettyman T. H., Yamashita N., Ammannito E., et al. (2018) Elemental composition and mineralogy of Vesta and Ceres: Distribution and origins of hydrogen-bearing species. *Icarus*, **318**, 42–55.

Raymond C. A., Russell C. T., and McSween H. Y. (2017) Dawn at Vesta: Paradigms and paradoxes. In *Planetesimals*, Elkins-Tanton L., and Weiss B., editors, pp. 321–340, Cambridge University Press, Tucson.

Russell C. T., Raymond C. A., Coradini A., et al. (2012) Dawn at Vesta: Testing the protoplanetary paradigm. *Science*, **336**, 684–686.

Schramm L. S., Brownlee D. E., and Wheelock M. M. (1989) Major element composition of stratospheric micrometeorites. *Meteoritics*, **24**, 99–112.

Toplis M. J., Mizzon H., Monnereau M., et al. (2013) Chondritic models of 4 Vesta: Implications for geochemical and geophysical properties. *Meteoritics & Planetary Science*, **48**, 2300–2315.

Yurimoto H., Abe K., Abe M., et al. (2011) Oxygen isotopic compositions of asteroidal materials returned from Itokawa by the Hayabusa mission. *Science*, **333**, 1116–1119.

14 Geochemical Exploration

The Moon and Mars as Case Studies

Overview

The exploration of other planets increasingly involves combining the detailed chemical analysis of samples (in the laboratory or *in situ* by landers or rovers) with chemical mapping by orbiting spacecraft to provide geologic context. In this chapter, we illustrate this approach to exploration by reviewing what has been learned about the Moon and Mars.

Lunar surface materials (Apollo and Luna returned samples and lunar meteorites) are classified into three major geochemical end members: anorthosite, mare basalt, and KREEP. These components are clearly associated with various geochemically mapped terrains of different ages on the lunar surface. The composition of the lunar interior is inferred from the geochemical characteristics of basalts that formed by mantle melting. Geochemistry constrains the Moon's origin by giant impact and its differentiation via an early magma ocean.

Martian meteorites and rover analyses of surface rocks suggest that the martian crust is predominantly basaltic – a conclusion supported by orbital measurements – and sedimentary rocks derived from volcanic protoliths. Basalts of different ages have different compositions, suggesting magmatic evolution over time. The martian mantle is highly oxidized and compositionally heterogeneous. Since its original differentiation, mantle source regions with distinct trace-element and radiogenic isotope abundances have remained geochemically isolated, although they have been periodically melted to produce basalts. Water, once important in producing clays and sulfates on ancient Mars, has now retreated into the subsurface or been lost to space.

14.1 Why the Moon and Mars?

The strategy for planetary exploration usually follows a proscribed sequence: global characterization and regional geologic mapping from flyby or orbiter missions, surface analysis and local geologic mapping by landers, rovers, or astronauts, and return of samples for analysis in terrestrial laboratories. However, sample return is not always the last step, and data obtained from samples can be advantageously integrated into the science goals and planning for later orbital missions. The Moon and Mars are the only large bodies for which we have both samples for laboratory analysis and considerable chemical data from orbiting and landed spacecraft. Both bodies have experienced extensive melting and differentiation, as well as the development of surface *regoliths* by impacts or sedimentary processes. Following the distinction between cosmochemistry and planetary geochemistry made in Chapter 1, we refer to efforts to describe the chemical compositions of these bodies as "geochemistry." In the title of this chapter, we also use the term "exploration" to indicate that these studies involve spatial mapping of compositional differences and placing the analyzed areas and samples into their proper geologic context.

Scientific literature on the geochemistry of the Moon and Mars is voluminous, and we can only provide overviews of data acquisition and current understanding. Returned lunar samples, lunar meteorites, and martian meteorites (shergottites, nakhlites, and chassignites = SNCs, plus several unique meteories) were briefly described earlier in Chapter 6. Here we combine what has been learned from detailed analyses of lunar samples with the results of geochemical mapping by orbiting spacecraft. Similarly, we compare the results from laboratory analyses of martian meteorites with *in situ* geochemical analyses by Mars rovers, as well as geochemical mapping by Mars orbiters. Even though mineralogy and petrology play important roles in this story, we will focus on chemistry, in keeping with the theme of this book.

Although geochemical exploration generally defines the compositions of surface materials, we will also consider geochemical constraints on the compositions of the bulk crusts, as well as the unsampled mantles and cores of these bodies. Inferences about the unsampled interiors of the Moon and Mars can be drawn from, or are limited by, analyses of surface samples; trace-element and radiogenic isotope data are particularly useful in constraining mantle compositions, and depletions of siderophile and chalcophile elements provide insights into core compositions. Such measurements have so far been done only in laboratory settings. In the case of Mars, we will consider the geochemical changes wrought by alteration by fluids. We do not consider here how crust, mantle, and core compositions are combined to estimate the bulk compositions of planets – a topic to be considered in Chapter 15.

14.2 Global Geologic Context for Lunar Geochemistry

A gross subdivision of the Moon's surface is based on albedo. The bright regions are the *highlands*, which are clearly ancient, as revealed by the high densities of impact craters. The dark regions are younger *maria*, basaltic lavas that flooded the impact basins on the nearside of the Moon. Mare basalts are exposed over ~17% of the lunar surface and probably comprise only a small portion of the total crustal volume.

The global stratigraphic framework for the Moon is based on using the *ejecta* blankets of large impact basins and craters as marker horizons. These ejecta units, called "formations," are sufficiently widespread that their relative ages can be defined. Five major time intervals have been identified, as summarized in Figure 14.1 (Wilhelms,

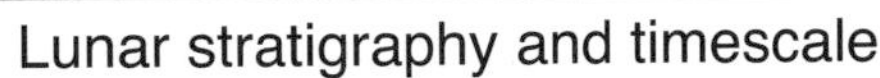

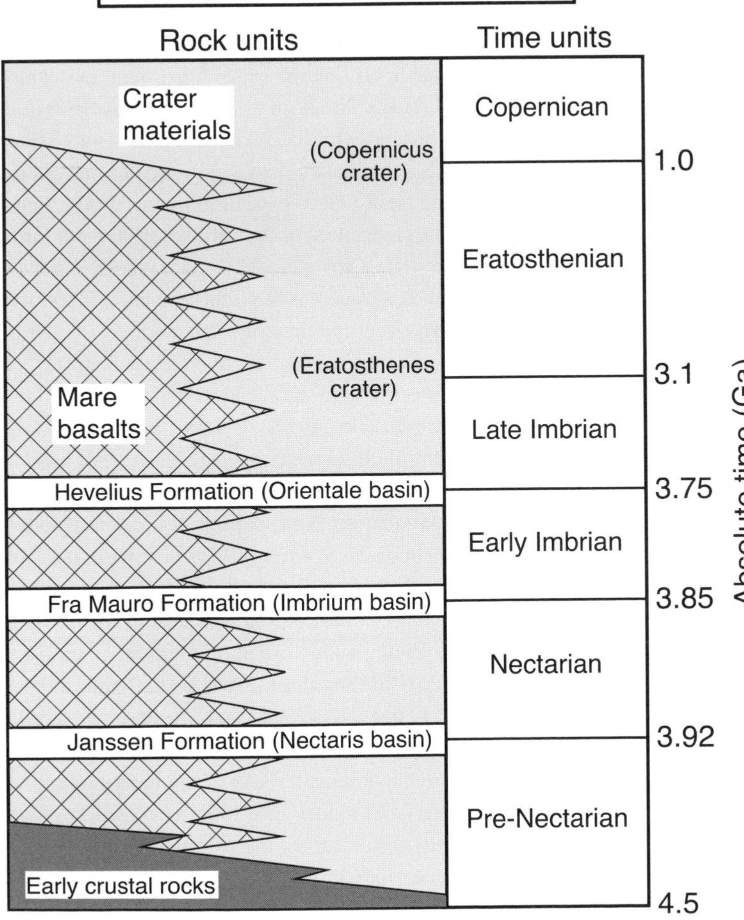

Figure 14.1 Schematic view of lunar stratigraphic and time units. The major rock and time periods are separated by formations representing the ejecta of huge impact basins. Estimates of the absolute ages of these boundaries are shown on the right.

1987). Assigning absolute ages to these stratigraphic formations is controversial and depends on the proper assignment of returned samples to specific major impact events and the measurement of their radiometric ages.

The pre-Nectarian Period began with the formation of the Moon, and includes the solidification of its crust during a long period of intense impact bombardment. A critical concept in understanding early lunar geologic history is the *magma ocean* model. It is generally believed (based on geochemical evidence described below) that much of the Moon, or perhaps nearly the entire Moon, was originally molten, resulting from its condensation from a vapor formed during a giant impact with the Earth. Fractional crystallization of such a magma ocean accounts for the feldspathic composition of the pre-Nectarian crust, and the complementary ultramafic composition represents the mantle.

The subsequent Nectarian Period was characterized by further bombardment that produced large basins and lasted only 0.07 Gyr. The Imbrian Period is subdivided into two intervals: the Early Imbrian Epoch included the formation of the Imbrium and Orientale basins, which had a profound effect on the present-day appearance of the Moon, and the Late Imbrian Epoch was marked by the cessation of giant impacts, which allowed the voluminous extrusion of basalts into the basins, forming the maria. The Eratosthenian Period, which lasted 2.1 Gyr, was characterized by significantly declining volcanism and impact flux. The Copernican Period, representing only the last 1.0 Gyr, marks the time during which rayed craters were preserved. Volcanism had ceased by this time.

The lunar surface is now covered by a thick regolith, formed by repetitive pulverization of crustal rocks by impacts. Its depth may be tens of kilometers in places. Rock and mineral fragments and impact-melted glasses have been chaotically mixed and churned over time, producing a global layer of debris. Although smaller craters tend to homogenize the regolith, the stratigraphic framework provided by ejecta from huge impact basins survives.

14.3 Geochemical Tools for Lunar Exploration

14.3.1 Instruments on Orbiting Spacecraft

The Lunar Prospector orbiter carried a gamma-ray/neutron spectrometer that made precise measurements of the concentration and distribution of thorium (Lawrence et al., 1998) and hydrogen (Feldman et al., 2008). Subsequent spectral deconvolutions (Prettyman et al., 2006) have enabled analyses of iron, titanium, potassium, magnesium, aluminum, calcium, and silicon. The principles of these analytical techniques were explained previously in Box 13.2.

The Clementine orbiter obtained high-resolution multispectral reflectance data. In comparing spectra with soil compositions at 39 locations sampled by Apollo astronauts, Blewett et al. (1997) determined a correlation between specific spectral features and chemical composition. This enabled Lucey et al. (1998) to develop an algorithm to estimate accurate FeO and TiO_2 abundances of mare basalts from Clementine spectra.

Global maps of FeO and thorium concentrations by Lunar Prospector (Jolliff et al., 2000), and of TiO_2 concentrations by Clementine (Giguere et al., 2000), especially, have allowed the determination of different compositional regions of the lunar crust, as well as new insights into chemical heterogeneities in the mantle.

14.3.2 Laboratory Analysis of Returned Lunar Samples and Lunar Meteorites

Six manned American Apollo missions returned 382 kg of rocks and soils from the nearside of the Moon. Three automated Soviet Luna missions also returned small amounts of soils. Most of these missions provided samples of the maria; only the Apollo 14, 15, and 16 missions sampled highlands materials and basin ejecta. Hiesinger and Head (2006) gave descriptions and outlined the geologic significances of the various Apollo and Luna landing sites. The Chang'e 5 mission returned ~2 g of lunar regolith from one of the youngest maria; it had not yet been analyzed at the time of publication.

The most informative geochemical analyses of highlands rocks come from rocks referred to as "pristine" (Warren and Wasson, 1977) to connote that they are coarse-grained plutonic rocks or large clasts (not brecciated mixtures of several rocks) and that there is no indication of contamination with siderophile elements (added by meteorite impact). Warren (1993) compiled a list of 89 pristine rocks and another 171 possibly pristine rocks from the Apollo collection. Several suites of pristine highlands rocks are distinguished: the ferroan anorthosite (FAN) suite, *KREEP* basalts, the magnesian suite, and the alkali suite. Ferroan (iron-rich) anorthosites, composed mostly of plagioclase with small amounts of pyroxene or olivine, are the dominant rocks of the lunar highlands. They have very old model ages (~4.5 Ga) and represent cumulate crust formed from the magma ocean. KREEP takes its name from high concentrations of potassium, rare earth elements (REEs), and phosphorus – all incompatible elements. Because incompatible elements are concentrated in residual liquids, KREEP is thought to represent the last dregs of the magma ocean. The few pristine KREEP rocks are basalts, and they may be melts of the KREEP source region in the mantle. Magnesian-suite rocks contain much more pyroxene and olivine and are easily distinguished by their high Mg/Fe ratios and high concentrations of incompatible elements – they have the unusual attribute that they represent relatively unfractionated magmas but have high trace-element abundances. They are generally thought to have sampled intrusions into the ancient FAN crust. Most models explain the magnesian suite as cumulates from KREEP basalt magmas, but assimilation might also account for their trace-element abundances. Their ages are generally younger than FAN rocks, implying later melting events. Alkali-suite rocks have sodium contents higher by a factor of four than those of ferroan anorthosites, and some small fragments have the compositions of granitic rocks. They may be derived by more extensive fractionation of KREEP basalt. The magnesian-suite and alkali-suite rocks are uncommon in the Apollo collection and absent from lunar meteorites, so they are unlikely to be major crustal lithologies. More details about the compositions of lunar crustal rocks can be found in Lucey et al. (2006), Shearer et al. (2006), and Wieczorek et al. (2006).

Mare basalts are lavas that erupted from fissures and filled large impact basins. Six of the nine missions to the Moon returned samples of basalts. The mare basalts from different sites have distinctive compositions and are classified based on their titanium contents, and to a lesser extent on their potassium contents (Fig. 14.2). A further subdivision is sometimes made based on Al_2O_3 contents. These compositional differences were inherited from their mantle sources. Volcanic glass beads, formed by fire fountains of hot lava erupting into the lunar vacuum, were found at several Apollo sites and eventually were shown to be a constituent of virtually every lunar soil. The glasses are ultramafic in composition and formed at high temperatures and great depths. The mare basalts and glasses have radiometric ages ranging from 3.85 to ~1.0 Ga.

Published geochemical studies of returned lunar samples are voluminous. Compilations of lunar rock analyses can be found in *The Lunar Sourcebook* (G. J. Taylor et al., 1991) and Papike et al. (1998).

Lunar meteorites (see review by Korotev et al., 2003) are mostly breccias of highlands crust (FAN) and regolith, although a few mare basalts are included in this collection. It is likely that the source craters for the meteorites are randomly distributed and thus include materials from

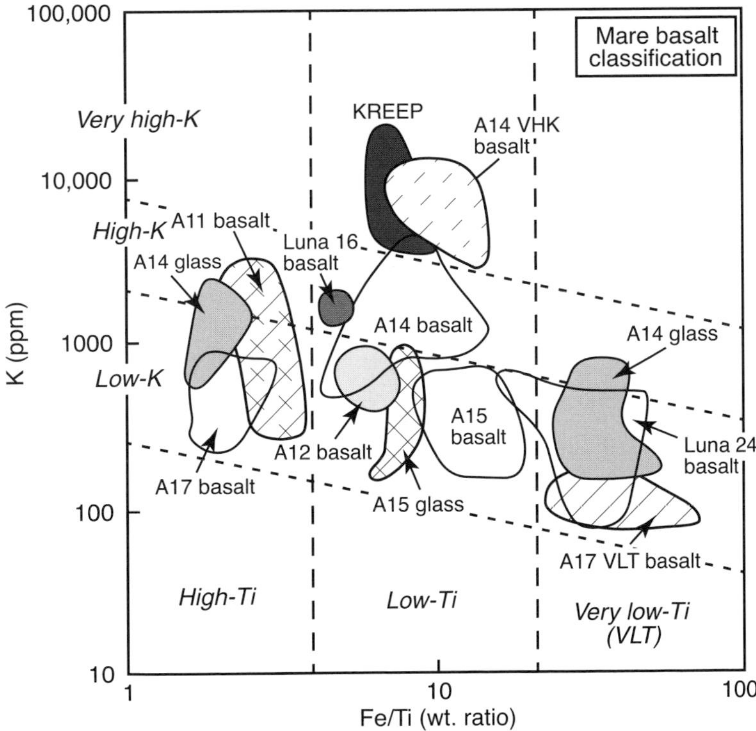

Figure 14.2 Lunar mare basalts from the various Apollo and Luna missions, classified by their potassium and iron/titanium ratios. After Lucey et al. (2006).

the lunar farside. As we will see, these meteorites provide a better estimate of the crustal composition than do the geographically biased samples returned by spacecraft.

14.4 Composition of the Lunar Crust

14.4.1 Sample Geochemistry

Samples returned by the Apollo and Luna missions can be readily distinguished based on their contents of FeO and thorium. This may seem like an unlikely choice of chemical components for classification, but they nicely discriminate rock types and are readily measured by remote sensing. The FeO and thorium contents of ferroan anorthosites, mare basalts, impact-melt breccias, and lunar meteorites are shown by various symbols in Figure 14.3. Anorthosites and basalts constitute two end members distinguished by FeO contents, and the impact-melt breccias extend upward toward a KREEP end member with high thorium and intermediate FeO contents. Besides thorium, the KREEP end member is obviously enriched in the other incompatible elements that define its name. The impact-melt breccias contain small clasts of KREEP basalt, which has not been sampled as large rocks.

The same ferroan anorthosite–mare basalt–KREEP components also define the compositions of lunar soils. The soils from each site contain different proportions of these end members. For example, Apollo 12 soils are mixtures of mare basalt and KREEP, whereas Apollo 15 soils contain all three components.

As noted earlier, lunar meteorites are mostly breccias of ferroan anorthosite and related early crustal rocks, although a few mare basalt meteorites are known. The lunar meteorites likely sample the whole lunar surface. The absence of KREEP-rich breccias, which are so common among Apollo samples collected from the nearside, in the lunar meteorite collection implies that KREEP-rich rocks cover only a small area of the Moon. In fact, the lunar highlands meteorites appear to provide a closer match to the average lunar crust than do the Apollo highlands samples (Fig. 14.4), as measured by geochemical mapping (see below).

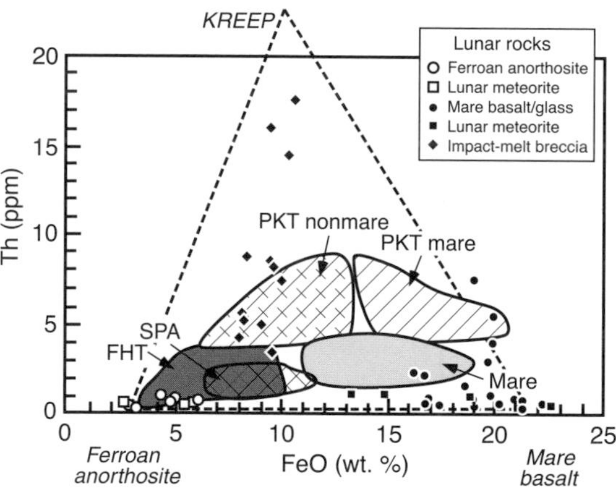

Figure 14.3 Analyses of thorium and iron in lunar rocks and lunar meteorites can be described by three compositional end members – ferroan anorthosite, KREEP, and mare basalt. The various lunar terranes defined by orbital measurements of thorium and FeO, illustrated by shaded and hatched fields, can be explained by mixtures of these components. FHT: Feldspathic Highlands Terrane, PKT: Procellarum KREEP Terrane, SPA: South Pole-Aiken Terrane. Modified from Jolliff et al. (2000).

14.4.2 Geochemical Mapping by Spacecraft

Orbital geochemical analyses since the Apollo and Luna sample return missions have revolutionized our understanding of the composition of the crust and mantle. As an example, Figure 14.5, which plots Al_2O_3 versus Mg/(Mg + Fe) from Lunar Prospector gamma-ray spectrometer (GRS) data in 5° bins, shows that the global data overlap major rock types in the Apollo collection. This comparison suggests that major geologic units have not been overlooked by biased sampling. However, all the Apollo samples were collected from one particular area on the lunar nearside, and lunar meteorite breccias commonly contain a lithology, magnesian anorthosite, that is not present in the returned samples (e.g. Roberts et al., 2019). These clasts plot in the FAN area of Figure 14.5 at Mg/(Mg + Fe) ratios of ~0.8. It is thus possible that the

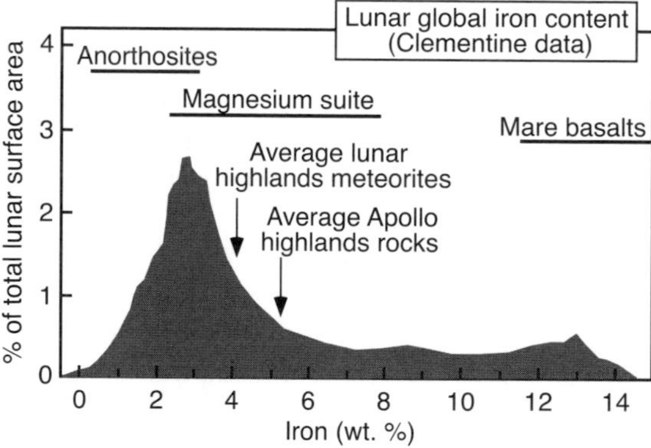

Figure 14.4 Distribution of iron on the lunar surface, as measured by the Clementine spacecraft, compared with the average iron contents (arrows) of Apollo highlands crustal rocks and lunar highlands meteorites. The meteorites, which presumably come from both the nearside and farside, more closely match the global iron peak. The compositional ranges for various lunar rock types are shown as horizontal bars.

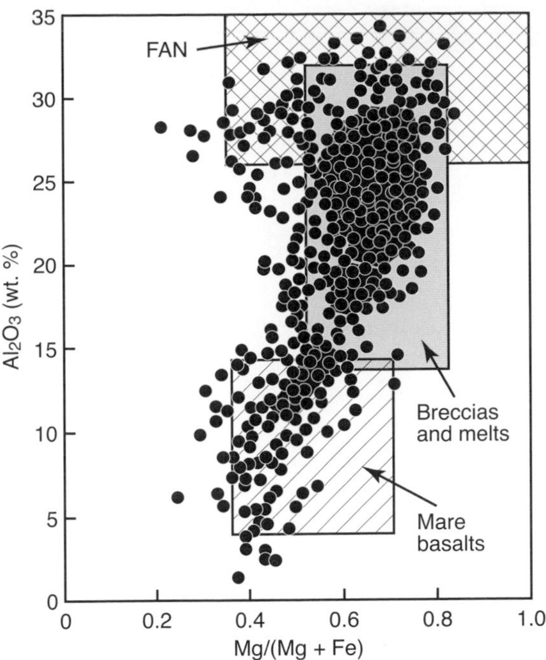

Figure 14.5 Lunar prospector measurements of Al_2O_3 and magnesium number Mg/(Mg + Fe) (black circles). These compositions are consistent with those of major lunar rock types from the Apollo collection (boxes). Modified from Prettyman et al. (2006).

highlands crust on the farside may contain feldspathic rocks more magnesian than ferroan anorthosite. Until more data are available, our understanding of lunar evolution continues to be dominated by Apollo samples.

Global geochemical maps of the concentrations of FeO and thorium (Fig. 14.6) have proved to be especially informative. Using those data, Jolliff et al. (2000) identified three important lunar terranes based on their geochemical characteristics. (Choice of the term "terrane" rather than "terrain" is intentional; in terrestrial usage, terranes are broad regions of the crust that share a common geologic history.)

The Feldspathic Highlands Terrane (FHT) is characterized by its relatively low FeO and thorium contents (Fig. 14.3), which are consistent with ferroan anorthosite. The FHT is the most extensive terrane and is concentrated on the lunar farside (Fig. 14.6). The Procellarum KREEP Terrane (PKT) occupies a large oval-shaped portion (Oceanus Procellarum) of the nearside (Fig. 14.6). The PKT has both light and dark areas, corresponding to highlands rocks and mare basalts. All of these materials are characterized by high contents of thorium (Fig. 14.3) and other incompatible elements (KREEP). PKT nonmare materials include rugged mountain ranges representing basin ring structures and Imbrium ejecta. Accordingly, most of these materials are breccias. The PKT is also more FeO rich than other highlands materials (Fig. 14.3), probably because of the inclusion of magnesian-suite rocks. The South Pole-Aitken Terrane (SPA) represents a gigantic (2600 km diameter) basin (Fig. 14.6). Its rocks have higher FeO contents than typical anorthositic crust (Fig. 14.3), probably because some lower crustal or mantle rocks are exposed.

All six Apollo missions landed near the boundaries between the two dominant terranes, the FHT and PKT. FHT and mare regions are fairly well represented in the Apollo collection and among lunar meteorites (Fig. 14.3). The Apollo 14 mission provided the best sampling of rocks from the PKT (both nonmare and mare). Although thorium-rich igneous rocks such as alkali anorthosite and granite occur among Apollo samples, no 5° areas have compositions dominated by these kinds of rocks. The SPA has apparently not been sampled by either the returned lunar samples or lunar meteorites. Lunar soils collected during the Apollo missions are not plotted in Figure 14.3, but Jolliff et al. (2000) showed that soils from the various sites define elongated, overlapping fields that collectively cover the terranes defined by orbital data. The only exception is Apollo 14 soils, which plot near the KREEP apex.

In discussing global geochemical mapping thus far, we have made no distinction between mare basalts, although we learned in an earlier section that they are commonly subdivided by their titanium contents. Titanium analyses of the nearside, where all the maria are, have been done using measurements based on Clementine reflectance spectra (Giguere et al., 2000) and Lunar Prospector neutron spectra (Elphic et al., 2002). Both studies produced similar maps showing variations attributed to different mare basalt compositions, although the TiO_2 concentrations from neutrons were lower than those from reflectance spectra (Fig. 14.7). More importantly, both global datasets demonstrated that there is a continuum between low-Ti and high-Ti basalt compositions, in contrast to the bimodal distribution of the compositions of returned lunar samples. Although samples analyzed in the laboratory are invaluable, they can be geographically (and hence compositionally) biased.

14.5 Compositions of the Lunar Mantle and Core

Even though the lunar crust has been thinned dramatically beneath large impact basins, no exposures of the mantle have been recognized. It is not surprising, then, that no

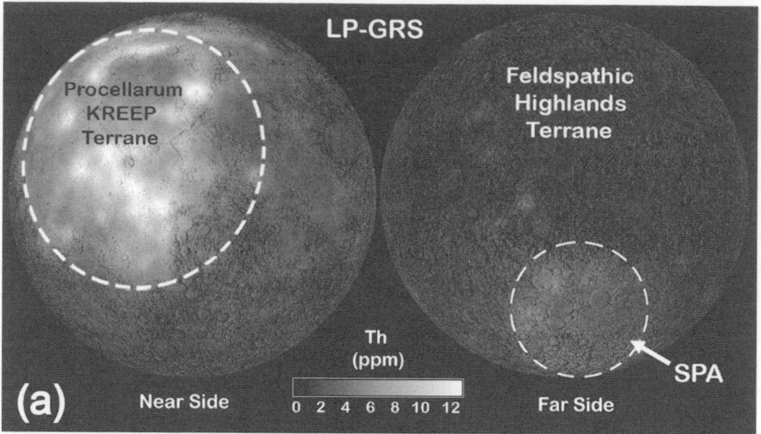

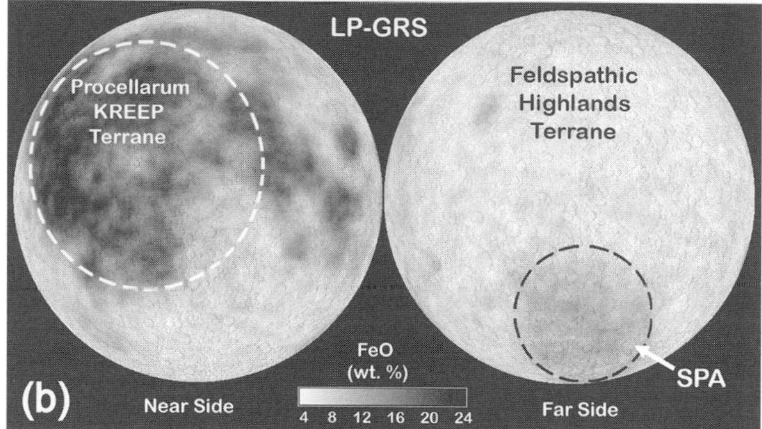

Figure 14.6 Geochemical maps of the nearside and farside of the Moon for (a) thorium and (b) FeO. Figures courtesy of B. L. Jolliff.

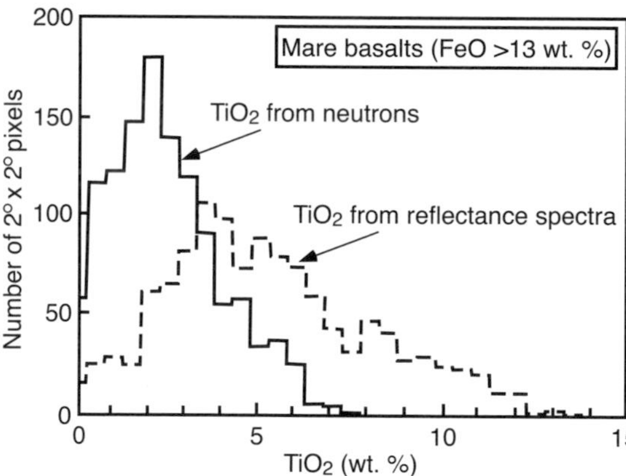

Figure 14.7 Comparison of titanium contents of lunar mare basalts, as measured from neutrons by the Lunar Prospector GRS and from reflectance spectra by the Clementine spacecraft. After Prettyman et al. (2006).

mantle rocks have been found among returned samples. Consequently, the composition of the lunar mantle must be determined indirectly, from the basaltic magmas that represent partial melts of mantle sources.

Lunar basalts and glasses have major-element compositions that require mantle sources that were not part of the primitive Moon but must be cumulates from the magma ocean (Shearer et al., 2006). Relative to terrestrial basalts, they have lower CaO, Al_2O_3, and Na_2O contents (all components of plagioclase) and higher FeO and MgO contents. The glasses are ultramafic in composition, and the mare basalts formed by fractionating ultramafic magmas.

The major-element compositions of these magmas determine what minerals crystallized as the magmas ascended and erupted. High-pressure experiments on these rocks can determine the pressures (depths) at which they are "multiply saturated," that is, in equilibrium with more than one mineral, as would be appropriate for a limited degree of melting. Such experiments indicate that the mantle sources consisted primarily of olivine and orthopyroxene, and suggest that lunar glasses may have originated at deeper levels than mare basalts. These experiments indicate considerable heterogeneity in mantle compositions, giving rise to magmas with varying titanium and aluminum contents.

REE patterns of mare basalts and crustal rocks (Fig. 14.8) provide strong evidence for the former existence of a magma ocean, as described below. In Chapter 11, we discussed how large, trivalent REE ions are excluded from most rock-forming minerals. Divalent europium, however, is an exception, in that it fits comfortably into the structure of plagioclase. The low oxidation state of the Moon explains the high ratio of divalent to trivalent europium. REE patterns for the FAN suite (highlands crust) and mare basalts (which mimic the patterns of their mantle source) are complementary (Fig. 14.8), since they both formed from the same magma ocean. KREEP, representing the last residual melt from crystallization of the magma ocean, is very enriched in REEs and has an even deeper negative europium anomaly (Fig. 14.8).

Besides potassium, REEs, and phosphorus, other trace elements that behave incompatibly in lunar basalts include strontium, barium, zirconium, niobium, tantalum, thorium, and uranium. The generally low incompatible-element concentrations in mare basalts indicate that they come from already depleted sources, as appropriate for mafic cumulates from the magma ocean. Some trace elements are difficult to fractionate, and differences in their abundances can signal roles for specific minerals that incorporate one or the other. For example, fractionation of hafnium from zirconium may reflect crystallization of ilmenite, and the depletion of heavy REEs relative to light REEs in some volcanic glasses may suggest garnet was left behind in the mantle source region.

Trace-element measurements in lunar basalts also indicate that the Moon is depleted in highly and moderately volatile elements (S. R. Taylor et al., 2006a; Ni et al., 2017). Estimates of some of the Moon's volatile element concentrations are compared with the Earth in Figure 14.9a. Lunar rocks contain no hydrous minerals,

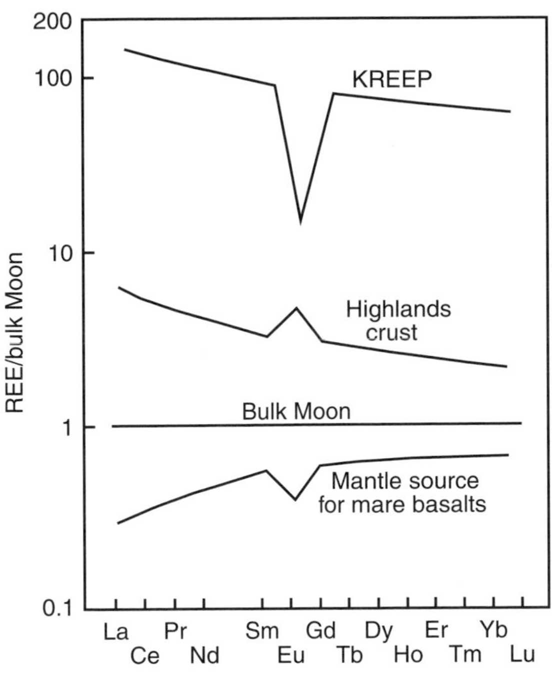

Figure 14.8 REE patterns, normalized to the bulk Moon composition, for various lunar materials. The highlands crust is enriched in REE over the bulk Moon because these rocks contain melts that concentrate incompatible elements, and the positive europium (Eu) anomaly indicates accumulation of plagioclase, which incorporates this element. The mantle should have a complementary REE pattern, depleted in abundance with a negative Eu anomaly, and this pattern was inherited by magmas generated from it, as represented by mare basalts. KREEP may represent the last dregs of the magma ocean, which would be extremely enriched in incompatible elements and exhibit a very large negative Eu anomaly because of the formation of the anorthositic crust.

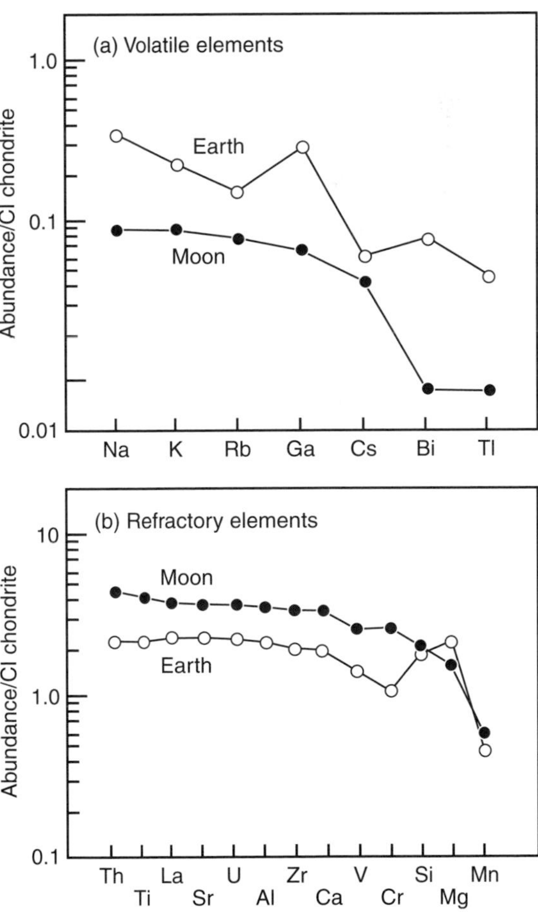

Figure 14.9 Comparison of the abundances of (a) volatile elements and (b) refractory elements, normalized to CI chondrites, for the Moon and the Earth. After S. R. Taylor et al. (2006a).

other than apatite, which originally led to the belief that it was bone dry. However, careful analysis of glass beads (Saal et al., 2013) and apatite (McCubbin et al., 2010) indicate that lunar magmas originally contained small amounts of water. The Moon may also be enriched in refractory elements (Fig. 14.9b). Volatile element depletion and refractory element enrichment are expected consequences of the giant impact origin and subsequent high-temperature accretion of the Moon, although volatile loss from the magma ocean may also have occurred.

Siderophile elements (cobalt, tungsten, nickel, molybdenum, gold, palladium, rhenium, iridium) are also depleted in lunar basalts (Righter and Drake, 1996). The depletion in siderophile elements clearly points to segregation of metals, and seismic modeling of moonquakes indicates a small, partially molten metallic core (Weber et al., 2011).

14.6 Geochemical Evolution of the Moon

There is now consensus that the Moon formed following a collision of the early Earth with a Mars-sized impactor (Barr, 2016). The timing of the impact and subsequent assembly of the Moon are not tightly constrained, but models of radiogenic isotopes suggest an age of 45–50 Myr after solar system formation (Halliday, 2014). The oxygen-isotopic compositions of the Earth and Moon have long been thought to be identical (Wiechert et al., 2001), suggesting that the impactor (called Theia) and Earth must have had the same isotopic composition – a problem for the giant impact hypothesis. However, new data indicate slight differences in crustal and mantle lithologies, and the oxygen isotopes in mantle-derived rocks imply that Earth and Theia were distinct (Cano et al., 2020). Computer simulations indicate that the core of the impactor mostly accreted to the Earth, and the Moon formed mostly from silicate material in the impactor's mantle. This would account for the lack of a large lunar core. Vaporization and incomplete recondensation of material in Earth orbit could account for the Moon's depletion in volatile elements. Condensation occurred in an environment with higher pressure than the ambient nebula, possibly allowing liquids to form (Pahlevan and Stevenson, 2007). Accretion of the solid and liquid materials produced a Moon that was largely, or at least partially, molten – in effect, a magma ocean (S. R. Taylor et al., 2006b).

Conceptual crystallization models for a completely molten and a half-molten Moon are contrasted in Figure 14.10. In both models, olivine and orthopyroxene crystals formed early and sank to produce a cumulate mantle. Once fractionation of olivine and pyroxene had increased the Al_2O_3 content of the magma ocean, plagioclase crystallized and floated upward, forming a thick, feldspathic crust with a globally averaged thickness of 34–43 km (Wieczorek et al., 2013). These simplified models do not take into account complications from the complex dynamics of a molten ocean cooling by convection and radiation from its top (Shearer et al., 2006). Nor do they consider changing pressure as the depth to the

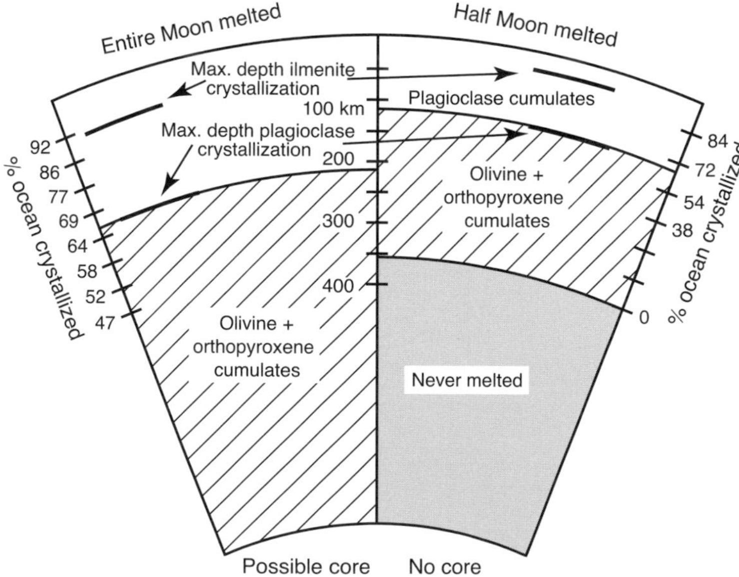

Figure 14.10 Two scenarios for the crystallization of the lunar magma ocean, involving whole-Moon melting and half-Moon melting. In both models, olivine and orthopyroxene cumulates sink and plagioclase floats, but the thickness of the plagioclase-rich crust is greater for a completely melted Moon. Ilmenite crystallizes late in both models, but mixing of dense, ilmenite-rich rocks with early magnesium-rich cumulates could result from convection. After Ryder (1991).

bottom of the magma ocean decreases during crystallization, as determined in experiments by Lin et al. (2017). The amount of crystallization of the magma ocean before plagioclase begins crystallizing is shown by tick marks for both models in Figure 14.10, and experiments indicate that plagioclase crystallization is delayed in a hydrous magma ocean (Lin et al., 2020). Radiometric ages of FAN rocks range from 4.57 to 4.39 Ga (Borg et al., 2014), suggesting that solidification and cooling of the magma ocean required several hundred million years; however, thermal models imply faster solidification, and younger ages of FAN rocks may have been affected by prolonged cooling in the solid state. The last dregs of the magma ocean were very rich in iron, titanium, and incompatible trace elements (KREEP) and were sandwiched between the crystallized mantle and crust. Subsequent mixing of layers could have resulted from overturn of the dense KREEP layer or the rising of plumes of magnesium-rich mantle cumulates. The degree of melting would likely have affected the formation of a core, which can occur more readily in extensive melt.

The pronounced asymmetry of the lunar crust, with PKT dominating the nearside and FHT dominating the farside, is apparently a result of magma ocean crystallization. Tilted convection may have occurred because of small temperature differences between the hemispheres (Loper and Werner, 2002); such a convective pattern would force plagioclase accumulation to occur preferentially on the cooler farside. Some workers have advocated an anorthosite continent (FHT), with residual liquid (PKT) concentrated at its edge, and others have proposed that the PKT represents a mantle hotspot where thorium and other heat-producing elements were concentrated. The heat generated by their radioactive decay caused melting of the upper mantle and KREEP materials to produce plutons of the magnesian and alkali suites. The radiometric ages of the highlands crustal rocks range from 4.5 to 3.9 Ga.

The so-called *late heavy bombardment*, discussed more fully in Chapter 15, is envisioned as a period of intense impacts by large planetesimals, concentrated in the Nectarian and Early Imbrian periods. There is considerable disagreement about whether the bombardment was a punctuated event or whether it represents a progressively declining period of impacts by ever-larger planetesimals. In any case, the Procellarum and Imbrium impacts exposed subsurface rocks in the PKT. The South Pole-Aiken terrain (SPAT) formed during this time and represents not only the largest impact basin on the Moon, but also the biggest basin the solar system. It exposes lower crust and perhaps mantle materials.

Following the formation of the giant impact basins, mare basalts formed by partial melting of the mantle, especially in the PKT region, where heat-producing radioactive elements were concentrated. Beginning at about 4.0–3.9 Ga, these magmas filled impact basins on the nearside for perhaps 2.8 Gyr (see Chapter 10). Most mare basalt activity occurred during the Late Imbrian and decreased significantly during the Eratosthenian Period.

Schematic cross sections of the various lunar terranes are illustrated in Figure 14.11 (Jolliff et al., 2006). These sketches illustrate current thinking about the complexities of the crust and mantle in these regions.

14.7 Global Geologic Context for Mars Geochemistry

Mars has a global dichotomy that separates the southern highlands, which stand well above the average surface elevation, and the northern lowlands. The north–south dichotomy is also expressed through differences in crater density (reflecting differences in age, with the northern plains being younger than the highlands) and in crustal thickness (with the lowlands crust being thinner than the highlands). Orbital spectra suggest that the highlands and lowlands are predominantly basaltic, although numerous deposits of sediments, mostly derived from basalts, occur locally.

The Tharsis bulge is a huge volcanic center located near the dichotomy boundary. The Tharsis hot spot formed early and has been volcanically and tectonically active for most of Mars' history. It is the locus of Mars' towering shield volcanoes. Several other regions, including a smaller bulge called Elysium, represent additional loci of volcanism.

The global stratigraphic system for assigning ages to various parts of the martian surface is based on the density of impact craters. The following time periods have been defined: pre-Noachian (the earliest, ranging from the planet's formation at ~4.5 to 4.1 Ga), Noachian (between ~4.1 and ~3.7 Ga), Hesperian (between ~3.7 and ~3.0 Ga), and Amazonian (postdating ~3.0 Ga). To first order, the southern highlands are dominated by Noachian rocks with some Hesperian cover, while in the northern lowlands, the Noachian basement is almost completely buried by Hesperian-age materials. The surface rocks of Tharsis and Elysium are Amazonian.

A comprehensive source for details about Mars geology is *The Surface of Mars* (Carr, 2006). Solomon et al. (2005) provided an excellent review of the geologic evolution of Mars, and various chapters in *The Martian Surface: Composition, Mineralogy, and Physical*

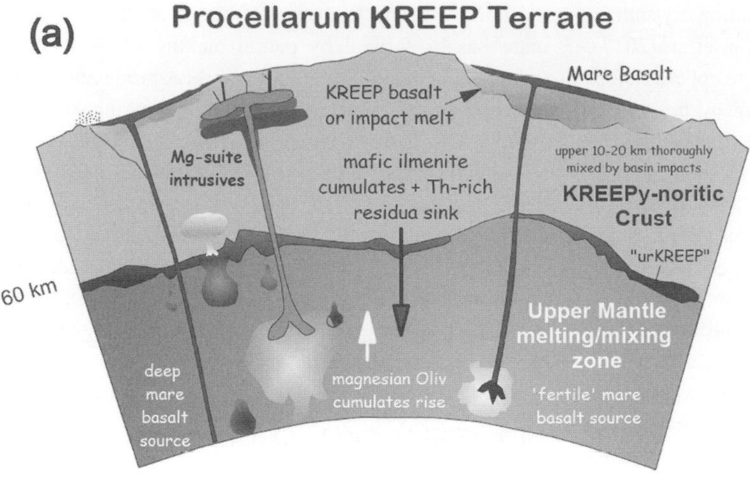

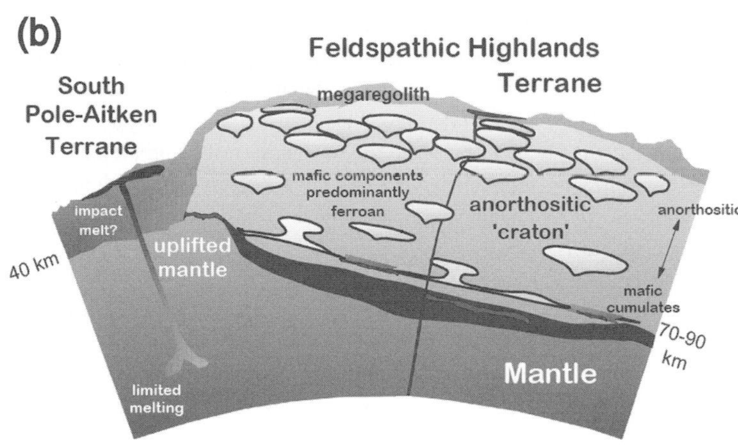

Figure 14.11 Sketches of cross sections of lunar terranes, illustrating processes that produced these terranes. (a) Procellarum KREEP Terrane, and (b) Feldspathic Highlands Terrane and South Pole-Aitken Terrane. Figures courtesy of B. L. Jolliff.

Properties (Bell, 2008) and the Mars chapter in *Treatise in Geochemistry* (McSween and McLennan, 2014) give summaries of Mars' geochemistry. The composition of the martian atmosphere was already discussed in Chapter 11 and will not be considered here.

14.8 Geochemical Tools for Mars Exploration

14.8.1 Instruments on Orbiting Spacecraft

Gamma-ray/neutron spectrometers have been flown on the Soviet Phobos 2 and United States Mars Odyssey orbiters. Because of their high orbiting altitudes, these instruments have larger footprints than Lunar Prospector data, and only the GRS on Mars Odyssey provided useful geochemical data. The principles of this instrument are as already described in Box 13.2. GRS provided analyses of hydrogen, silicon, iron, calcium, potassium, chlorine, aluminum, sulfur, and thorium, as well as low-resolution geochemical maps of global element distributions for most of these elements (Boynton et al., 2008; Feldman et al., 2008).

Mars Global Surveyor carried a *Thermal Emission Spectrometer* (TES), which provided thermal infrared (TIR) spectra. Lattice vibrations in many minerals absorb thermal energy, so the emitted radiation is missing those bands. Although designed as a mineral identification tool, several groups (Hamilton et al., 2001; Wyatt et al., 2001) have explored the use of these data in calculating bulk chemical compositions. Because TIR spectra for minerals combine linearly, a TES spectrum can be deconvolved into its constituent minerals, giving information on their relative proportions. Combining mineral proportions with chemical compositions of those minerals in the spectral

Box 14.1 Constraining the Composition and Depth of the Lunar Magma Ocean

The novel hypothesis that the Moon once had a magma ocean began with the recognition that its crust was composed of plagioclase (i.e. anorthosite rock). Wood et al. (1970) noted feldspar grains in the first soil samples returned by Apollo 11 and hypothesized that these grains floated in a magma ocean. The formation of a thick feldspathic crust, estimated to average perhaps 50 km thick, requires a great deal of parental mafic magma. Geochemistry – specifically Al_2O_3 content – has been used to test the plausibility of various proposed magma ocean compositions and the depth of the magma ocean, as summarized by Shearer et al. (2006). The choice of alumina derives from the aluminum-rich composition of calcic plagioclase, $CaAl_2Si_2O_8$, in the crust. We noted above that the degree of melting to form the lunar magma ocean is uncertain, as illustrated by the whole-Moon and half-Moon melting scenarios in Figure 14.10.

S. R. Taylor (1982) advocated a bulk Moon composition that was refractory rich, containing 6 wt. % Al_2O_3. Once magma ocean differentiation had occurred, the mantle and crust must have had complementary Al_2O_3 contents. Lines showing the complementary Al_2O_3 abundances in the crust (of varying thickness) and mantle for magma oceans (MO) representing different degrees of melting (whole-Moon [WM] melting and 500 km-deep melting) are illustrated in Figure 14.12. Using mass-balance calculations, Taylor attempted to match estimates for the density (2.93 g/cm^3) and thickness (73.4 km) of the lunar crust. From this calculation, he estimated that the Al_2O_3 content of the bulk crust was 26 wt. %. He further assumed that as much as 20% of the crust was composed of basalt, with an Al_2O_3 content of 10 wt. %. (This is certainly an overestimation, since maria occur only on the nearside, and the crustal thickness used is nearly twice the current estimate.) Removing that basalt from the crust increases Al_2O_3 to 30 wt. % and decreases thickness to 58 km. This volume would be composed of FAN and plutonic (magnesium-suite and alkali-suite) rocks.

Another bulk Moon model, called the Lunar Primitive Upper Mantle (LPUM), has 4 wt. % Al_2O_3 (Hart and Zindler, 1986). Analogous lines for whole-Moon and 500 km-deep magma oceans to produce crust of varying thicknesses for the LPUM model are also shown in Figure 14.12.

The 26–30 wt. % Al_2O_3 estimate for the lunar crust, depending on the amount of mare basalt it contains, is shown as a shaded vertical bar in Figure 14.12. Both whole-Moon and 500 km-deep melting of the Taylor model can satisfy this range of crustal compositions, but 500 km-deep melting of the LPUM model is problematic. Other evidence cited by Shearer et al. (2006) favors the LPUM model, which suggests that the lunar magma ocean was greater than 500 km deep.

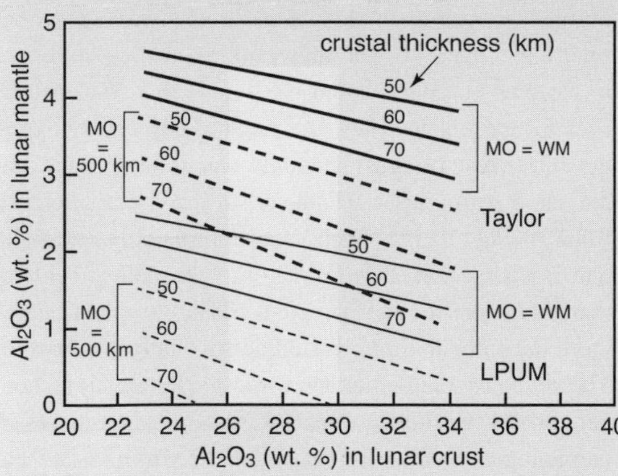

Figure 14.12 Calculated Al_2O_3 contents of the lunar crust and mantle are complementary, as shown by parallel lines representing different thicknesses for the crust. Calculations for two models for the bulk Al_2O_3 content of the Moon are shown: Taylor (1982) and LPUM (Hart and Zindler, 1986); lines for both models represent magma oceans formed by whole-Moon melting or melting only to 500 km deep. The shaded vertical bar defines a range of Al_2O_3 contents of the lunar crust, depending on how much mare basalt is present. Modified from Shearer et al. (2006).

library gives the bulk chemical composition. Wyatt and his colleagues compared the chemical compositions of a variety of terrestrial lavas analyzed in the laboratory with those calculated from TES spectra and concluded that the derived element abundances were generally within 10% of analyzed values. Various TES-derived estimates of Mars' surface chemistry, based on mineral deconvolutions using different spectral libraries, were compiled by McSween et al. (2003). A subsequent spectral study identified 11 compositionally different regions on Mars

(Rogers and Christensen, 2007), and Rogers and Hamilton (2014) used TES and GRS data to evaluate compositional variations among units of different age. However, martian terrains covered with even thin coatings of fine-grained dust emit black-body radiation and are not decipherable in terms of mineralogy or chemistry.

Models of visible/near-infrared (VNIR) spectra from the OMEGA instrument on the Mars Express orbiter can likewise provide mineral modes that have been recast into chemistry. These data have also been used to calculate major-element abundances for 11 regions on Mars (Poulet et al., 2009).

A major difference between the GRS- and the TES- and OMEGA-derived chemical datasets is that γ-rays penetrate and measure to 20–30 cm deep, whereas TIR and VNIR spectra sample only the outermost micron of surface grains. McSween and McLennan (2014) compared contents of FeO* (total iron expressed as FeO) and SiO_2, two of the best-determined oxides analyzed by the three instruments, and found little agreement (Fig. 14.13). As we will see, surface alteration of rocks and soils has modified their surface compositions to such an extent that the TES and OMEGA data do not give bulk sample chemistry, although they still provide valuable information on chemical alteration.

14.8.2 Instruments on Landers and Rovers

The Viking landers provided X-ray fluorescence data for a limited number of martian soils at two landing sites. These data were compiled by Clark et al. (1982).

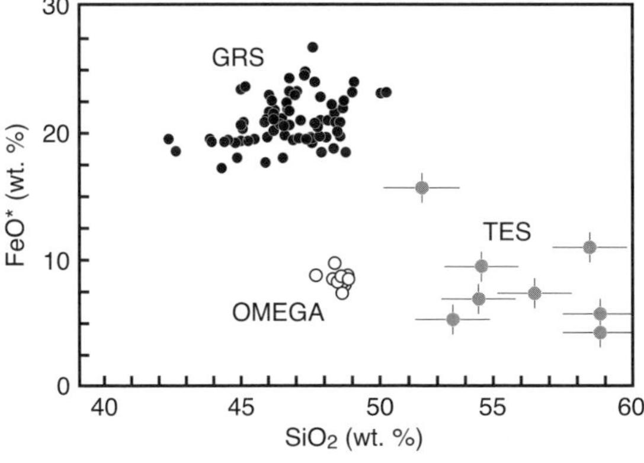

Figure 14.13 Comparison of FeO* (total iron expressed as FeO) and silica abundances of the Martian surface, derived from various datasets from orbiting spacecraft: analyzed directly by γ-ray spectra (GRS), or calculated from mineral modes determined from TIR spectra (TES) or VNIR spectra (OMEGA). Modified from McSween and McLennan (2014).

The Mars Pathfinder rover carried an Alpha Proton X-ray Spectrometer (APXS), and the two Mars Exploration Rovers (MER; Spirit and Opportunity) and the Mars Science Laboratory rover (MSL; Curiosity) carried *Alpha Particle X-ray Spectrometers* (also called APXS, but in this case more precise versions of the Pathfinder instrument, though without the ability to monitor protons for light-element analyses). The utility of the APXS instrument is described in Box 14.2. The MER rovers also carried Mossbauer spectrometers, which yielded information on iron oxidation state.

Mars Pathfinder analyzed six rocks and five soils (Wanke et al., 2001; Foley et al., 2008). The two MER rovers together have analyzed more than 250 rocks and 100 soils. The APXS on Curiosity has been used more sparingly, but still numerous analyses have been made. Compilations of MER geochemical analyses were given by Gellert et al. (2006), Ming et al. (2006), and Bruckner et al. (2008). MSL geochemical analyses are reported by McLennan et al. (2014) and Schmidt et al. (2014, 2018).

Wet chemistry experiments on the Phoenix lander extracted water-soluble ions (Ca^{2+}, Mg^{2+}, perchlorate, sulfate, chloride) from polar soil (Hecht et al., 2009). Carbon was below APXS detection limits (0.3 wt. %) in Mars Pathfinder soils, but organic molecules were identified by mass spectrometry in sedimentary rocks at the MSL landing site (Eigenbrode et al., 2018).

14.8.3 Laboratory Analyses of Martian Meteorites

Martian meteorites (McSween, 2015) were described briefly in Chapter 6, and their chronology was described in Chapter 10. We learned previously that Mars launch ages for these meteorites (the sums of cosmic-ray exposure and terrestrial ages) generally clump into groups with the same petrographic characteristics and classifications (refer to Fig. 10.16). Each group presumably represents one impact event at a launch site, so collectively the martian meteorites may sample perhaps 10 separate sites – more locations than those studied by landers and rovers. The young crystallization ages (most $\leq$1.3 Ga) of all but a handful of martian meteorites indicate that they are chronologically biased, probably reflecting the fact that impacts into young, coherent targets are better able to launch rocks to escape velocity. Most martian meteorites may be derived from Tharsis or Elysium. The old NWA 7034 (~4.5 Ga) and ALH 84001 (~4.1 Ga) meteorites represent the ancient crust of Mars.

Chemical and isotopic analyses of many of the meteorites are remarkably complete. Some compilations of bulk chemical analyses can be found in Treiman et al.

Box 14.2 APXS: A Geochemical Workhorse

The Alpha Particle X-ray Spectrometer has proved to be an extremely useful tool for determining the geochemistry of rocks and soils on the martian surface. On the Mars Pathfinder rover (Sojourner), MER rovers (Spirit and Opportunity), and MSL rover (Curiosity), the APXS instruments were mounted on robotic arms, allowing them to be placed against the target. Each instrument contained radioactive curium sources (Fig. 14.14) whose rapid decay produced energetic α-particles, which irradiated target rocks and soils. The resulting characteristic X-rays provided measurements of virtually all major- and minor-element abundances, and in a few cases, trace elements. The APXS is sensitive to elements ranging from sodium (atomic mass = 23) to bromine (atomic mass = 80). Examples of APXS spectra for a Mars rock are shown in Figure 14.14. The concentration of each element is determined from the corresponding peak area.

Many martian rocks are covered by airfall dust or coated by alteration rinds. Because the APXS analyzes only the outermost few microns (the actual depth varies for each element), a contaminated surface poses a common problem (also encountered in terrestrial rocks, but easily remedied by a hammer strike). In the case of Mars Pathfinder rocks, dust cover was estimated from sulfur content (which is concentrated in dust) and a dust-free composition was extrapolated (McSween et al., 1999). For MER rocks, the Rock Abrasion Tool (RAT) brushed away dust or abraded the weathering rind to expose fresh rock. For MSL rocks, a Dust Removal Tool was used with varying success, and a dust correction for analyses was made based on image-processing methods (Schmidt et al., 2018). Even so, APXS analyses may be somewhat compromised by matrix effects in unprepared samples; an analog study by Berger et al. (2020) suggests that rover-analyzed contents of aluminum and sulfur may be overestimated and magnesium underestimated.

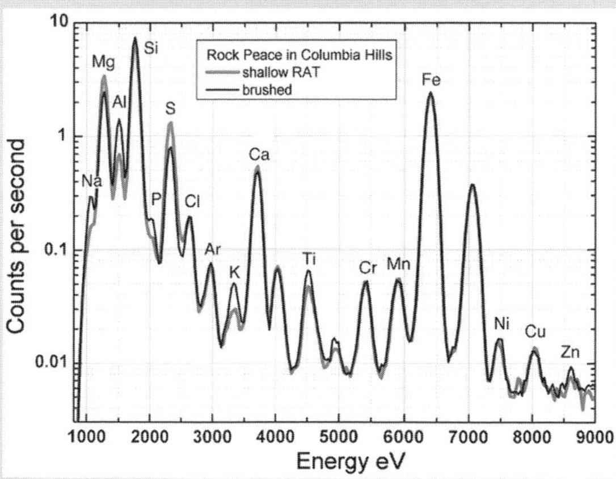

Figure 14.14 Image of an APXS showing a detector surrounded by an array of radioactive curium α-particle sources. The APXS spectra below are for a rock at the Spirit rover landing site that was brushed to remove surface dust and RAT-ground to remove any surficial weathering. *For the colour version, refer to the plate section.*

(1987), Warren and Kallemeyn (1997), Lodders (1998), Bridges and Warren (2006), and Udry et al. (2020). Although bulk martian meteorites contain very little H_2O, the initial contents of shergottite parent magmas have been estimated from analyses of apatite (McCubbin et al., 2016; Filiberto et al., 2019), and of nakhlite parent magmas from analyses of nominally anhydrous minerals (Peslier et al., 2019). Sources for data on stable isotopes of hydrogen, carbon, oxygen, nitrogen, sulfur, and noble gases in martian meteorites were given by McSween and McLennan (2014). Radiogenic isotopes and geochronology were discussed in Chapter 10, and analyses of organic matter were considered in Chapter 11.

14.9 Composition of the Martian Crust

The composition of Mars' surface materials can be assessed using laboratory analyses of martian meteorites, *in situ* APXS analyses from Mars rovers, and orbital geochemistry analyzed by GRS.

14.9.1 Analyses of Crustal Rocks

The rocks in Meridiani Planum analyzed by the Opportunity rover are sedimentary. To assess the composition of the crust, we will focus on the volcanic rocks in Gusev crater, analyzed by the Spirit rover, and in Gale crater, analyzed by the Curiosity rover. We will compare these compositions with those of martian meteorites and Bounce rock in Meridiani Planum, which is similar to shergottites (Zipfel et al., 2011). We will also consider orbital geochemical data obtained by GRS.

Figure 14.15 shows a plot of alkalis ($K_2O + Na_2O$) versus SiO_2, which is commonly used to chemically classify terrestrial volcanic rocks. Most martian meteorites and the geochemically similar Bounce rock plot in the basalt field. The Pathfinder dust-free rock composition is more silica rich and likely represents an alteration rind on the rocks. Igneous rock analyses from Gusev crater and Gale crater, as well as igneous clasts in the NWA 7034 meteorite breccia, are clearly more alkaline than the meteorites. GRS provided silica analyses (the width of the gray box represents the average silica abundance and standard deviation). GRS analyzed potassium but not sodium, so in constructing the vertical dimension of the box, we have assumed the average Na_2O/K_2O ratio of Gusev soils. The GRS analyses also plot in the basalt field, indicating that the martian crust is predominantly basaltic but appears to be more alkali rich than martian meteorites.

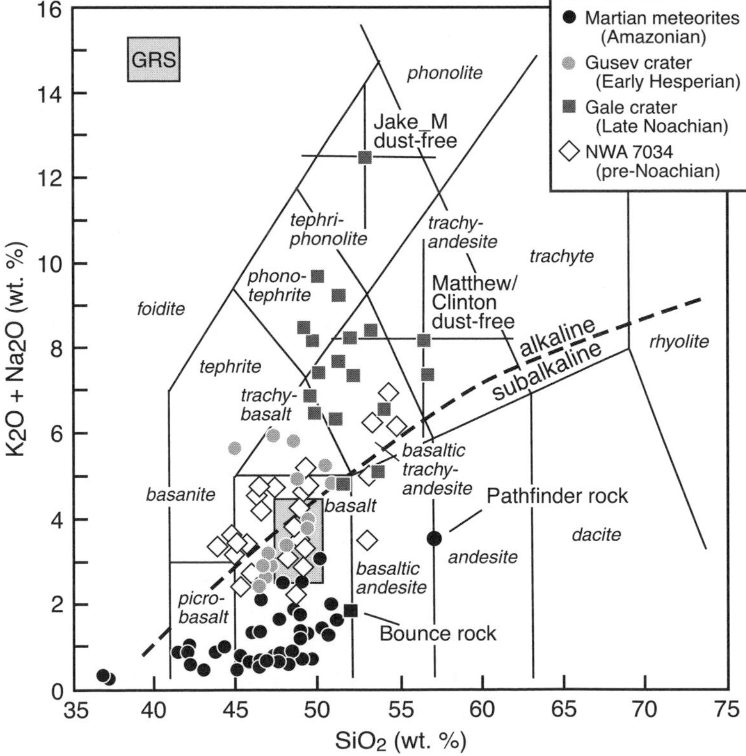

Figure 14.15 Alkalis versus silica diagram used for geochemical classification of volcanic rocks. Young martian meteorites and Bounce rock generally plot in the field of basalts, as does the average Mars Odyssey GRS analysis (gray box). Ancient rocks analyzed in Gusev and Gale craters by rovers, as well as clasts in the ancient NWA 7034 meteorite breccia, are more alkaline than the meteorites. Attempts to remove dust contamination from the analyses of Gale crater rocks are labeled "dust-free." The Mars Pathfinder dust-free rock plots near the andesite field, but this composition may be contaminated by an alteration rind. Data sources and figure (modified) from McSween et al. (2009).

The large variation in alkali contents in Figure 14.15 appears to be related to rock age. The ancient (pre-Noachian, Noachian, and Hesperian) samples have higher alkali contents than the younger (Amazonian) meteorites. The composition of the crust of Mars thus varies with age. GRS-measured potassium contents of surface units of varying age support the conclusion that older rocks are more alkali rich (Rogers and Hamilton, 2014). Silica content also appears to decrease with age (Fig. 14.15 and Baratoux et al., 2011); however, other proposed chemical changes based on GRS data, such as decreasing FeO* and increasing thorium in younger rocks, are not consistently supported by meteorite data.

A plot of Ca/Si versus Mg/Si ratios is illustrated in Figure 14.16. A similar diagram has been used previously for geochemical classification of martian meteorites (Ouri et al., 2003). The various classes of shergottites (enclosed by dashed ovals) exhibit an inverse correlation between these ratios, reflecting olivine fractionation (lower right arrow). The nakhlites have higher Ca/Si, reflecting augite fractionation (upper left arrow). Ancient rocks (Gale crater, Gusev crater, and NWA 7034) generally have lower Ca/Si ratios. For reference, GRS provided data on calcium and silicon, and the average Ca/Si ratio is 2.95 ± 0.5; GRS did not analyze magnesium.

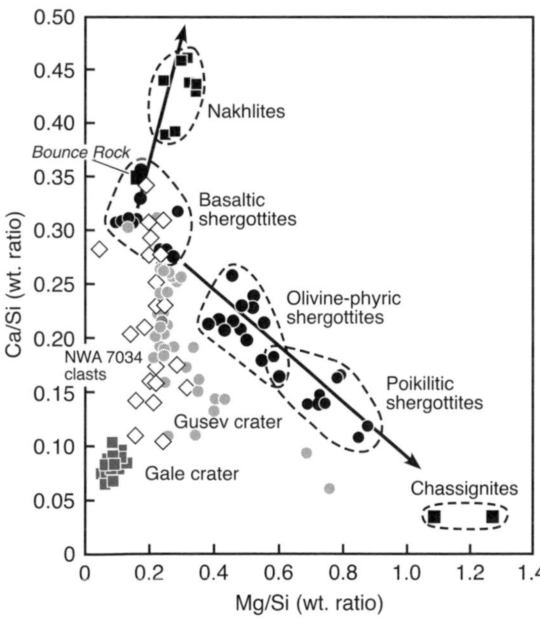

Figure 14.16 Weight ratios of Ca/Si versus Mg/Si in various Mars rocks and meteorites. Shergottites form a linear trend, reflecting olivine fractionation (arrow pointing to lower right), and nakhlites plot above the trend, reflecting high-Ca pyroxene (augite) fractionation (arrow pointing upward). Ancient rocks from Gusev and Gale craters and clasts in NWA 7034 plot at lower Ca/Si values. Modified from McSween et al. (2009).

The FeO*/MgO versus SiO_2 diagram (Fig. 14.17) is also useful in discriminating magma fractionation trends. Rocks plotting above the diagonal line show tholeiitic trends (shown as nearly vertical black arrows, exhibited by relatively anhydrous magmas), whereas those plotting below the line show calc-alkaline trends (gray arrows, reflecting hydrous melting and fractionation). Virtually all the meteorite and rover data plot in the tholeiitic field, supporting a conclusion that Mars does not exhibit calc-alkaline magmatism – a characteristic of subduction zone volcanism on Earth.

Several examples of global geochemical maps for specific elements analyzed by GRS (Boynton et al., 2007, 2008) are illustrated in Figure 14.18. Differences in the abundances of silicon and iron are relatively modest and consistent with a basaltic surface composition. Variations in thorium abundance are also shown. An assessment of the global K/Th ratio (G. J. Taylor et al., 2006) shows that the ratio has very little variability, consistent with a surface dominated by igneous processes. Potassium and thorium are both incompatible elements and thus should be fractionated together during melting and crystallization.

14.9.2 Water, Chemical Weathering, and Evaporites

Unlike the Moon, Mars has clearly had significant amounts of water on its surface, at least in the distant past. Estimating the abundance of water in the martian interior, which was presumably delivered to the surface by volcanic outgassing, has been controversial. Mars has commonly been assumed to be very dry, based on the low bulk water contents of martian meteorites (typically 50–150 ppm; Filiberto and Treiman, 2009). However, analyses of apatite in shergottites (McCubbin et al., 2012) suggest the magmas contained 730–2870 ppm H_2O. Because their parent magmas may have lost water on ascent and eruption, their initial water contents have been controversial. Based on the lack of saturation of volatile chlorine in shergottites, Filiberto et al. (2019) argued that shergottite magmas did not degas. Alternatively, oxidation of shergottite magmas during crystallization has been attributed to volatile degassing (Castle and Herd, 2018). In any case, the water content of martian meteorites represents only relatively recent magmas, and earlier magmas (and their mantle source regions) likely contained and outgassed greater amounts of water and other volatiles.

The presence of deuterium-enriched water in martian meteorites (Leshin et al., 1996; Leshin, 2000) indicates

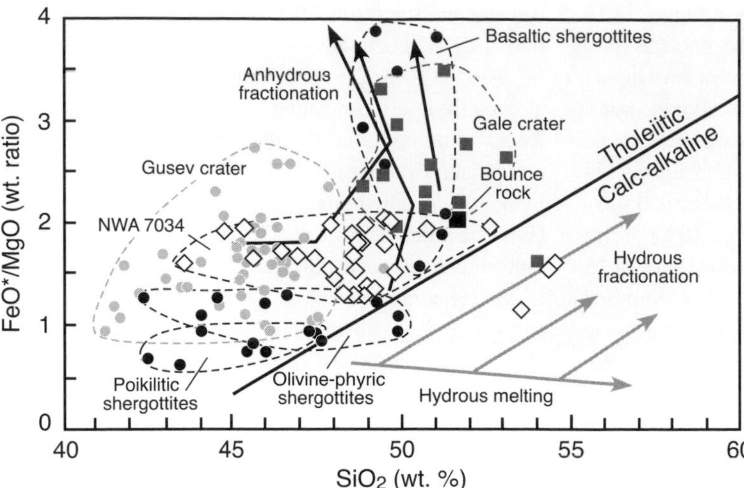

Figure 14.17 FeO*/MgO ratio versus silica in Mars rocks and meteorites. A diagonal line separates tholeiitic from calc-alkaline rocks; tholeiitic magmas are relatively anhydrous, and calc-alkaline magmas are hydrous. Trends for melting and fractionation are illustrated by arrows. Virtually all the martian rocks plot in the tholeiitic field. Modified from McSween et al. (2009).

that some water has been lost from the atmosphere, but the fate of most of the water that carved valleys and channels during the Noachian and early Hesperian is unclear. A GRS global map of hydrogen (Feldman et al., 2002), expressed as equivalent H_2O and previously illustrated in Figure 11.15, provides information on the locations of water in the subsurface at the present time. Hydrogen concentrations at high latitudes are thought to reflect the presence of ice in the form of permafrost, whereas hydrogen near the equator likely occurs as OH in hydrous minerals.

Sedimentary rocks have been studied at several locations by Mars rovers (Arvidson, 2016). Altered basaltic sandstones in Meridiani Planum, analyzed by the Opportunity rover (Ming et al., 2008), are cemented by salts and contain hematite concretions and secondary silica. APXS measurements indicate large, varying proportions of sulfur and chlorine and modest amounts of bromine. Sulfates and halides formed by evaporation of acidic waters. An extended diagenetic history, with multiple episodes of cementation and dissolution, indicates a chemically evolving groundwater system (McLennan et al., 2005). Sandstones and mudstones in Gale crater, analyzed by the Curiosity rover, represent paleolake deposits, also derived from basaltic protoliths. These rocks also contain amorphous material and are cemented by sulfate and chloride salts. Detrital material filing Gale crater has been dated at ~4.2 Ga, based on in situ K–Ar dating by Curiosity (Farley et al., 2014), as described in Chapter 10.

Rover-analyzed sedimentary and altered rocks provide an interesting insight into martian weathering reactions. Figure 14.19 is commonly used to summarize the effects of chemical weathering of basaltic rocks. Unlike weathering of basalt on the Earth, which leaches soluble CaO, Na_2O, and K_2O while enriching the weathering profile in insoluble clays along the Al_2O_3 join with FeO + MgO, weathered rocks on Mars form an array leading away from the FeO + MgO vertex. This difference might be controlled by pH; leaching experiments by Hurowitz et al. (2006) led them to propose that Mars weathering occurs under acidic conditions, whereas terrestrial weathering occurs under neutral to basic conditions. Alternatively, Fedo et al. (2015) demonstrated that hydrodynamic sorting of mineral grains could also account for martian soil compositions.

The occurrence of significant quantities of clay minerals has been documented by spectral observations made by instruments on the Mars Express and Mars Reconnaissance orbiters. The clays come in two forms: nontronite, which likely formed in hydrothermal systems, and saponite, which formed during chemical weathering on the surface. Bibring et al. (2005) noted that clays formed early in Mars history and that they were not stable during later periods. Acidic groundwaters, as documented at Meridiani Planum and inferred from other sulfate-bearing units elsewhere on Mars, would inhibit the formation of clays. High proportions of highly soluble sulfates and chlorides in soils imply an environmental change from wet to arid in later martian history. However, geologic evidence is most consistent with only episodically warm and wet conditions on early Mars.

14.10 Compositions of the Martian Mantle and Core

Geochemical models for the silicate portion of Mars (mantle plus crust, which is approximately equal to the mantle composition) are constrained by martian

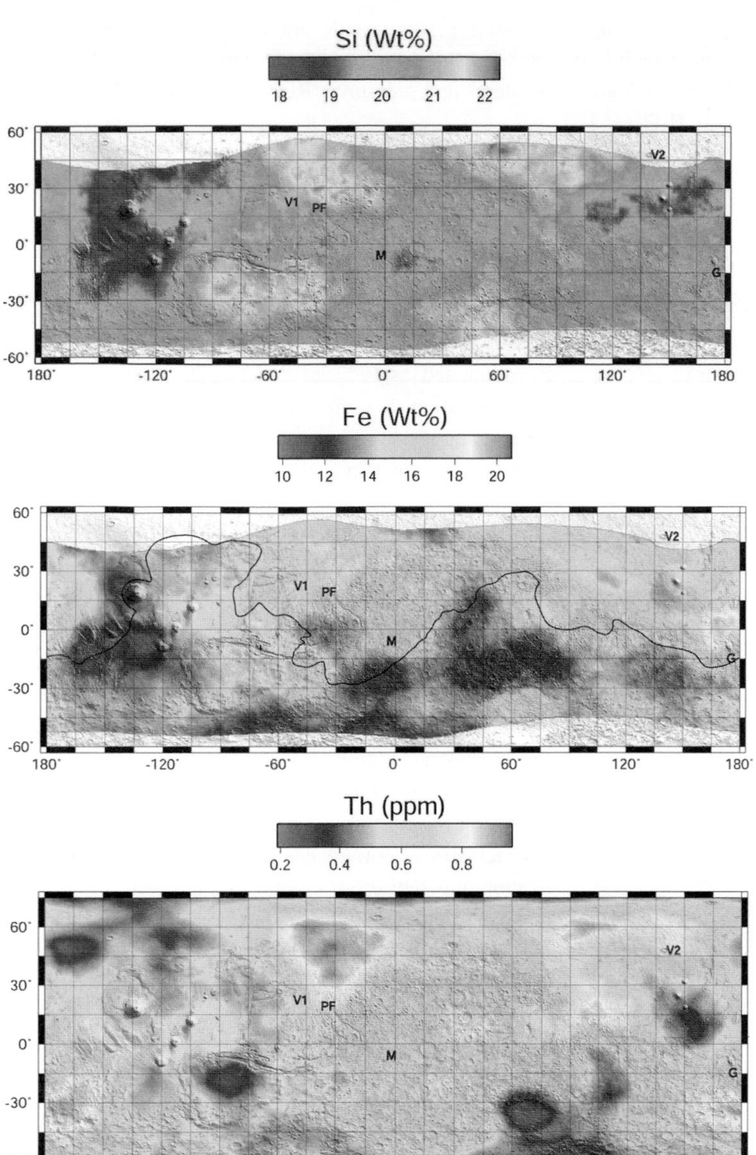

Figure 14.18 Global GRS compositional maps of Mars for (a) silicon, (b) iron, and (c) thorium. Letters represent spacecraft landing sites: V1 and V2 = Viking, PF = Mars Pathfinder, M = Merdiani (Opportunity), and G = Gusev (Spirit). The irregular line in (b) separates the southern highlands from the northern lowlands. From Boynton et al. (2008), with permission. *For the colour version, refer to the plate section.*

meteorites and spacecraft data. One early estimate (Wanke and Dreibus, 1988) was calculated from element ratios in the meteorites, and two others (Lodders and Fegley, 1997; Sanloup et al., 1999) are combinations of various chondrite types in proportions that reproduce the oxygen-isotope compositions of martian meteorites. G. J. Taylor (2013) refined the mantle composition by considering GRS data as well. Taylor's composition of the martian mantle is compared with the Earth's mantle in Figure 14.20. A common assumption in such models is that refractory element ratios are chondritic. Yoshizaki and McDonough (2020) took issue with that assumption and calculated a Mars composition based only on martian meteorites and spacecraft data. Their model and that of G. J. Taylor (2013) disagree on whether the Mars mantle is enriched or depleted in moderately volatile elements, but concur that it is more highly oxidized, so that Fe^{2+} is concentrated in the mantle, giving it a lower Mg/(Fe + Mg) ratio.

Correlated variations in the abundances of incompatible trace elements and radiogenic isotopes (Fig. 14.21), as well as the oxidation state in martian meteorites, are interpreted to indicate that the mantle is compositionally heterogeneous (Symes et al., 2008). The incompatible-element-enriched, highly radiogenic, oxidized mantle source also contains a higher abundance of H_2O (~55 ppm) than the depleted, less radiogenic, reduced mantle (~18 ppm), as calculated from apatite–melt exchange (McCubbin et al., 2016).

Estimates of Mars' core composition by the authors listed above suggest it is made of iron metal plus iron sulfide, the latter generally varying from 29 to 44 wt. %. However, Yoshizaki and McDonough (2020) favor substituting some oxygen and hydrogen for sulfur in the core. Depletions of siderophile (tungsten, phosphorus, cobalt, molybdenum, nickel) and chalcophile (indium, copper) elements in the mantle are consistent with equilibrium between sulfide, metal, and mantle silicate at high temperatures and pressures (Righter and Drake, 1996).

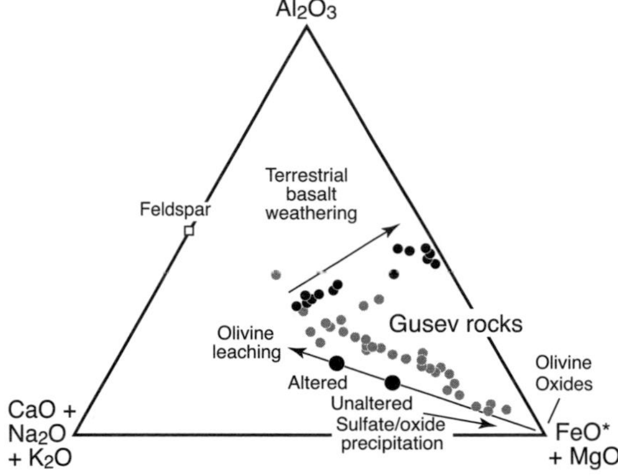

Figure 14.19 Molar chemical diagram showing different weathering trends for terrestrial basalts (small black circles) and martian basalts (gray circles), resulting from differences in the acidity of fluids. Experiments (large black circles) demonstrate that olivine is leached under acid conditions. Modified from Hurowitz et al. (2006).

14.11 Geochemical Evolution of Mars

Mars is more volatile rich than Earth (Filiberto et al., 2016), reflecting a higher proportion of accreted volatile-bearing planetesimals. It is also more highly oxidized (also a characteristic of volatile-rich meteorites like CI chondrites) than Earth, so that twice as much of its iron has remained as Fe^{2+} in the mantle, rather than being sequestered in the metallic core. Wanke and Dreibus (1988) suggested that iron oxidation occurred during accretion, as water in accreted planetesimals reacted with iron metal.

Both short-lived (^{182}Hf–^{182}W, ^{146}Sm–^{142}Nd) and long-lived (^{87}Rb–^{87}Sr, ^{147}Sm–^{143}Nd, U–Pb, ^{176}Lu–^{176}Hf, ^{187}Re–^{187}Os) radiogenic isotope systems in martian meteorites indicate early planetary differentiation, within 7–10 Myr after CAIs at 4567 Ma (e.g. Kruijer et al., 2017). It seems likely that Mars, like the Moon, had an early magma ocean (Elkins-Tanton et al., 2003) that crystallized within ~10–25 Myr, followed by ongoing crust formation until ~40 million years after solar system formation (Kruijer et al., 2017; Bouvier et al., 2018; see Chapter 10). Magma ocean models suggest that crystallization and settling of cumulates eventually produced a residual liquid rich in iron and incompatible elements near the top, and its high density caused mantle overturn (a similar situation is envisioned for the Moon). The overturned cumulates represent the depleted mantle, and rocks formed from the later liquid are the enriched mantle; local mixing of reservoirs would produce the intermediate

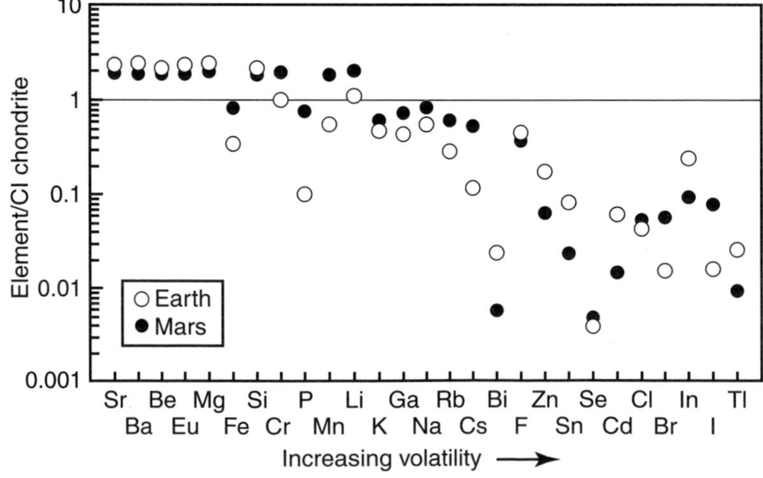

Figure 14.20 Mars bulk silicate (mantle + crust) composition, estimated from martian meteorites and GRS data, and compared to bulk silicate Earth. After G. J. Taylor (2013).

mantle source. This early fractionation of trace elements resulted in differences in radiogenic isotopes. Alternative models to account for mantle heterogeneity include partial melting that induces density differences and mixing of large, differentiated impactors into the mantle. Since its formation, the martian mantle has been geochemically isolated, because there is no plate tectonics to recycle crust back into the mantle.

Decompression melting associated with mantle overturn and continuing until ~40 million years after solar system formation (Kruijer et al., 2017) produced a crust comprising ~4% of the planetary volume – a much greater proportion of crust than on Earth. Although some models suggest that the early crust of Mars was andesitic, the crust we can measure is basaltic, with only small amounts of other rock compositions. Most volcanism was ancient, but some eruptions, especially in Tharsis and Elysium, extend into recent times. Martian basalts inherited the high ferrous iron contents, as well as trace-element contents, radiogenic isotope ratios, and redox states (Fig. 14.21) of their different mantle sources. Compositional variations among igneous rocks, especially in alkalis, suggest that magma compositions evolved with time. Mantle decompression melting may have occurred at different depths, reflecting global cooling and thickening of the lithosphere over time (Filiberto, 2017).

Early on, water existed at least periodically on the martian surface. The formation of clay minerals in the Noachian and the precipitation of layered deposits of sulfates and chlorides in the Hesperian were consequences of this water. Some fraction of the water evaporated and was lost to space, as indicated by the high D/H ratios in SNCs. Other water was apparently sequestered at the poles or underground as permafrost. The surface of Mars is cold and effectively dry now, limiting the chemical weathering of crustal rocks (Thorpe et al., 2021).

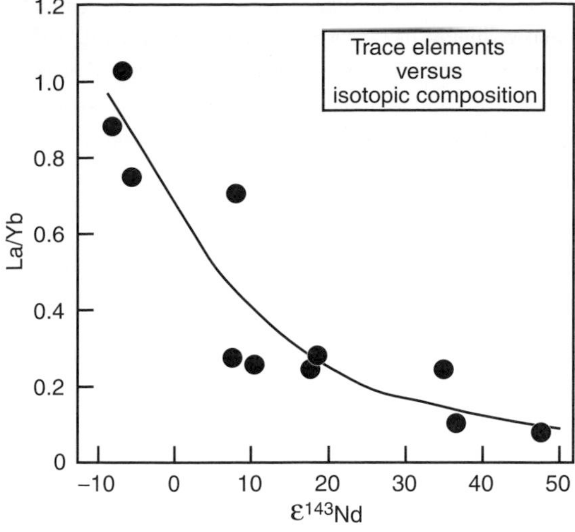

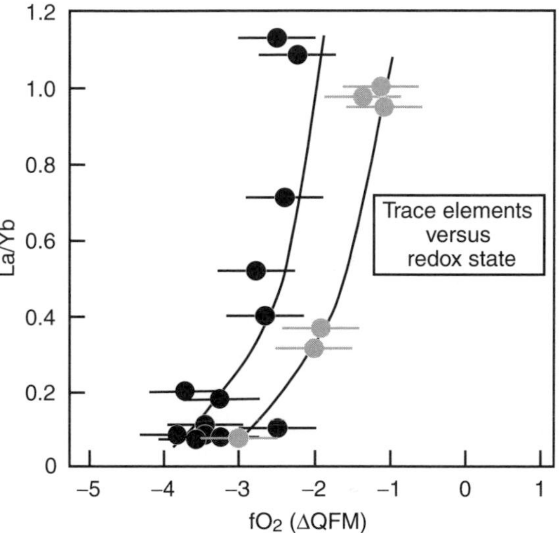

Figure 14.21 Enrichment or depletion in incompatible trace elements (represented by the La/Yb ratio) correlates with radiogenic isotope compositions (represented by ε^{143}Nd) and redox state (represented by difference in fO_2 relative to the quartz-fayalite-magnetite buffer (QFM) in martian shergottite meteorites. The black and gray circles in the lower diagram refer to enriched and depleted shergottites, respectively. These variations indicate that the mantle source regions are compositionally heterogeneous.

Summary

The Moon and Mars have very different compositions and have had very different geologic histories. However, the geochemical techniques applied to study both bodies are similar, and the results illustrate the power of chemical data in planetary exploration. Combinations of laboratory analyses of meteorites or returned samples, sometimes coupled with *in situ* analyses by rovers, provide the most detailed geochemical information. Geochemical mapping by orbiting spacecraft provide global or regional geologic context for those analyses.

The compositions of the crusts of the Moon and Mars are distinct – one is dominated by feldspathic cumulates from an early magma ocean and the other by basaltic lavas. Regional patterns or variations in trace elements, radiogenic isotopes, and oxidation state reflect differences in subjacent mantle compositions. The compositions of the mantles and cores of these bodies can be constrained by chemical analyses of mantle-derived basalts.

The interiors of both bodies have remained geochemically isolated, due to the absence of plate tectonics.

Having worked our way through the parts of the solar system for which we have significant cosmochemical or geochemical data, we will now see how what we have learned can constrain models for solar system formation.

Questions

1. Describe some analytical tools used on orbiting and landed spacecraft to analyze materials on the surface of the Moon and Mars.
2. Briefly describe the major groups of lunar rocks returned by Apollo astronauts and how they relate to groups determined from orbital measurements.
3. What are the major geochemical characteristics of the lunar mantle and crust?
4. How does the chemical composition of the martian crust differ from that of the Moon?
5. How has water affected the surface of Mars?
6. What are the major geochemical characteristics of the martian mantle and crust?

Suggestions for Further Reading

These publications provide excellent summaries of the geologic evolution of the Moon:

Wilhelms D. E. (1987) *The Geologic History of the Moon*, U.S. Geological Survey Professional Paper 1348.

Jolliff B. L., Wieczorek M. A., Shearer C. K., and Neal C. R., editors (2006) *New Views of the Moon, Reviews in Mineralogy & Geochemistry*, **60**, Mineralogical Society of America and Geochemical Society, Washington.

The following publications provide excellent summaries of the geologic and geochemical evolution of Mars:

Bell J. (2008) *The Martian Surface: Composition, Mineralogy, and Physical Properties*, Cambridge University Press, Cambridge, 636 pp.

McSween H. Y., and McLennan S. M. (2014) Mars. In Treatise on Geochemistry, 2nd edition, Vol. 2: Planets, Asteroids, Comets and the Solar System, Davis A. M., editor, pp. 251–300, Elsevier, Oxford.

Udry A., Howarth G. H., Herd C. D. K., et al. (2020) What martian meteorites reveal about the interior and surface of Mars. *Journal of Geophysical Research: Planets*, 125, e2020JE006523.

Taylor S. R., and McLennan S. M. (2009) *Planetary Crusts: Their Composition, Origin and Evolution*, Cambridge University Press, Cambridge, 378 pp. This book provides a thoughtful assessment of the origins and compositions of crusts on planets and the Moon.

Other References

Arvidson R. E. (2016) Aqueous history of Mars as inferred from landed measurements of rocks, soils, and water ice. *Journal of Geophysical Research, Planets*, **121**, 1602–1626.

Baratoux D., Toplis M. J., Monnereau M., and Gasnault O. (2011) Thermal history of Mars inferred from orbital geochemistry of volcanic provinces. *Nature*, **475**, 338–341.

Barr A. C. (2016) On the origin of Earth's Moon. *Journal of Geophysical Research, Planets*, **121**, 1563–1601.

Berger J. A., Schmidt M. E., Campbell J. L., et al. (2020) Particle induced X-ray emission spectrometry (PIXE) of Hawaiian volcanics: An analogue study to evaluate the APXS field analysis of geologic materials on Mars. *Icarus*, doi.org/10.1016/j.icarus.2020.113708.

Bibring J.-P., Langevin Y., Gendrin A., et al. (2005) Mars surface diversity as revealed by the OMEGA/Mars Express observations. *Science*, **307**, 1576–1581.

Blewett D. T., Lucey P. G., and Hawke B. R. (1997) Clementine images of the lunar sample-return stations: Refinement of FeO and TiO_2 mapping techniques. *Journal of Geophysical Research*, **102,** 16319–16325.

Borg L. E., Gaffney A. M., and Shearer C. K. (2014) A review of lunar chronology revealing a preponderance of 4.34-4.37 Ga ages. *Meteoritics & Planetary Science*, **50**, 715–732.

Bouvier L. C., Costa M. M., Connelly J. N., et al. (2018) Evidence for extremely rapid magma ocean crystallization and crust formation on Mars. *Nature*, **558**, 586–589.

Boynton W. V., Taylor G. J., Evans L. G., et al. (2007) Concentration of S, Si, Cl, K, Fe, and Th in the low- and mid-latitude regions of Mars. *Journal of Geophysical Research*, **112**, E12S99.

Boynton W. V., Taylor G. J., Karunatillake S., et al. (2008) Elemental abundances determined via the Mars Odyssey GRS. In *The Martian Surface: Composition, Mineralogy, and Physical Properties*, Bell J. F., editor, pp. 105–124, Cambridge University Press, Cambridge.

Bridges J. C., and Warren P. H. (2006) The SNC meteorites: Basaltic igneous processes on Mars. *Journal of the Geological Society of London*, **163**, 229–251.

Bruckner J., Dreibus G., Gellert R., et al. (2008) Mars Exploration Rovers: Chemical composition by the APXS. In *The Martian Surface: Composition, Mineralogy, and Physical Properties*, Bell J. F., editor, pp. 58–101, Cambridge University Press, Cambridge.

Cano E. J., Sharp Z. D., and Shearer C. K. (2020) Distinct oxygen isotope compositions of the Earth and Moon. *Nature Geoscience*, **13**, 270–274.

Carr M. (2006) *The Surface of Mars*, Cambridge University Press, Cambridge, 307 pp.

Castle N., and Herd C. D. K. (2018) Experimental investigation into the effects of oxidation during petrogenesis of the Tissint meteorite. *Meteoritics & Planetary Science*, **23**, 1–23.

Clark B. C., Baird A. K., Weldon R. J., et al. (1982) Chemical composition of Martian fines. *Journal of Geophysical Research*, **87**, 10050–10067.

Eigenbrode J. L., Summons R. E., Steele A., et al. (2018) Organic matter preserved in 3-billion-year-old mudstones at Gale crater, Mars. *Science*, **360**, 1096–1101.

Elkins-Tanton L. T., Parmentier E. M., and Hess P. C. (2003) Magma ocean fractional crystallization and cumulate overturn in terrestrial planets: Implications for Mars. *Meteoritics and Planetary Science*, **38**, 1753–1771.

Elphic R. C., Lawrence D. J., Feldman W. C., et al. (2002) Lunar Prospector neutron spectrometer constraints on TiO_2. *Journal of Geophysical Research*, **107** (E4), doi:10.1029/2000JE001460.

Farley K. A., Malespin C., Mahaffy P., et al. (2014) In situ radiometric and exposure age dating of the Martian surface. *Science*, **343**, doi:10.1126/science.1247166.

Fedo C. M., McGlynn I. O., and McSween H. Y. (2015) Grain size and hydrodynamic sorting controls on the composition of basaltic sediments: Implications for the interpretation of martian soils. *Earth & Planetary Science Letters*, 423, 67–77.

Feldman W. C., Boynton W. V., Tokar R. L., et al. (2002) Global distribution of neutrons from Mars: Results from Mars Odyssey. *Science*, **297**, 75–78.

Feldman W. C., Mellon M. T., Gasnault O., et al. (2008) Volatiles on Mars: Scientific results from the Mars Odyssey Neutron Spectrometer. In *The Martian Surface: Composition, Mineralogy, and Physical Properties*, Bell J. F., editor, pp. 125–148, Cambridge University Press, Cambridge.

Filiberto J. (2017) Geochemistry of martian basalts with constraints on magma genesis. *Chemical Geology*, **466**, 1–14.

Filiberto J., and Treiman A. H. (2009) Martian magmas contained abundant chlorite, but little water. *Geology*, **37**, 1087–1090.

Filiberto J., Baratoux D., Beaty D., et al. (2016) A review of volatiles in the martian interior. *Meteoritics & Planetary Science*, **51**, 1935–1958.

Filiberto J., McCubbin F. M., and Taylor G. J. (2019) Volatiles in martian magmas and the interior: Inputs of volatiles into the crust and atmosphere. In *Volatiles in the Martian Crust,* Filibert J., and Schwenzer S., editors., pp. 13–33, Elsevier, New York.

Foley C. N., Economou T. E., Clayton R. N., et al. (2008) Martian surface chemistry: APXS results from the Pathfinder landing site. In *The Martian Surface: Composition, Mineralogy, and Physical Properties*, Bell J. F., editor, pp. 35–57, Cambridge University Press, Cambridge.

Gellert R., Rieder R., Brückner J., et al. (2006), The Alpha Particle X-Ray Spectrometer (APXS): Results from Gusev crater and calibration report. *Journal of Geophysical Research*, **111**, E02S05.

Giguere T. A., Taylor G. J., Hawke B. R., and Lucey P. G. (2000) The titanium contents of lunar mare basalts. *Meteoritics & Planetary Science*, **35**, 193–200.

Halliday A. N. (2014) The origin and earliest history of the Earth. In *Treatise on Geochemistry, 2nd edition, Vol. 2: Planets, Asteroids, Comets and the Solar System*, Davis A. M., editor, pp. 149–211, Elsevier, New York.

Hamilton V. E., Wyatt M. B., McSween H. Y., and Christensen P. R. (2001) Analysis of terrestrial and Martian volcanic compositions using thermal emission spectroscopy: 2. Application to Martian surface spectra from the Mars Global Surveyor thermal emission spectrometer. *Journal of Geophysical Research*, **106**, 14733–14746.

Hecht M. H., Kounaves S. P., Quinn R. C., et al. (2009) Detection of perchlorate and soluble chemistry of the martian soil: Findings from the Phoenix Mars lander. *Science*, **235**, 64–67.

Hart S. R., and Zindler A. (1986) In search of a bulk-Earth composition. *Chemical Geology*, **57**, 247–267.

Heisinger H., and Head J. W. (2006) New views of lunar geoscience: An introduction and overview. In *New Views of the Moon*, Neal C. R., Jolliff B. L., Shearer C. K., and Wieczorek M. A., editors, *Reviews in Mineralogy and Geochemistry*, **60**, pp. 1–18, Mineralogical Society of America and Geochemical Society, Washington.

Hurowitz J. A., McLennan S. M., Tosca N. J., et al. (2006) In situ and experimental evidence for acidic weathering of rocks and soils on Mars. *Journal of Geophysical Research*, **111**, E02S19.

Jolliff B. L., Gillis J. J., Haskin L. A., et al. (2000) Major lunar crustal terranes: Surface expressions and crust-mantle origins. *Journal of Geophysical Research*, **105,** E2, 4197–4216.

Korotev R. L., Jolliff, B. L., Zeigler R. A., et al. (2003) Feldspathic lunar meteorites and their implications for compositional remote sensing of the lunar surface and the composition of the lunar crust. *Geochimica et Cosmochimica Acta*, **67**, 4895–4923.

Kruijer T. S., Kleine T., Borg L. E., et al. (2017) The early differentiation of Mars inferred from Hf-W chronometry. *Earth & Planetary Science Letters*, **474**, 3345–3354.

Lawrence D. J., Feldman W. C., Barraclough B. L., et al. (1998) Global elemental maps of the Moon: The Lunar Prospector gamma-ray spectrometer. *Science*, **281**, 1484–1489.

Leshin L. A. (2000) Insights into Martian water reservoirs from analysis of martian meteorite QUE94201. *Geophysical Research Letters*, **27**, 2017–2020.

Leshin L. A., Epstein S., and Stolper E. M. (1996) Hydrogen isotope geochemistry of SNC meteorites. *Geochimica et Cosmochimica Acta*, **60**, 2635–2650.

Lin Y., Tronche E. J., Steenstra E. S., and von Westrenen W. (2017) Experimental constraints on the solidification of a nominally dry lunar magma ocean. *Earth & Planetary Science Letters*, **471**, 104–116.

Lin Y., Hui H., Xia X., et al. (2020) Experimental constraints on the solidification of a hydrous lunar magma ocean. *Meteoritics & Planetary Science*, **55**, 207–230.

Lodders K. (1998) A survey of shergottite, nakhlite and chassigny meteorites whole-rock compositions. *Meteoritics & Planetary Science*, **33**, A183–A190.

Lodders K., and Fegley B. (1997) An oxygen isotope model for the composition of Mars. *Icarus*, **126**, 373–394.

Loper D. E., and Werner C.L. (2002) On lunar asymmetries 1. *Tilted convection and crustal asymmetry. Journal of Geophysical Research Planets*, **107**, E6, 5046.

Lucey P. G., Blewett D. T., and Hawke B. R. (1998) Mapping the FeO and TiO_2 content of the lunar surface with multispectral imagery. *Journal of Geophysical Research*, **103**, 3679–3699.

Lucey P. G., Korotev R. L., Gillis J. J., et al. (2006) Understanding the lunar surface and space-Moon interactions. In *New Views of the Moon*, Neal C. R., Jolliff B. L., Shearer C. K., and Wieczorek M. A., editors, *Reviews in Mineralogy and Geochemistry*, **60**, Mineralogical Society of America and Geochemical Society, Washington, pp. 83–219.

McCubbin F. M., Steele A., Hauri E. H., et al. (2010) Nominally hydrous magmatism on the Moon. *Proceedings of the National Academy of Sciences, USA*, **107**, 11223–11228.

McCubbin F. M., Hauri E. H., Elardo S. M., et al. (2012) Hydrous melting of the martian mantle produced both depleted and enriched shergottites. *Geology*, **40**, 683–686.

McCubbin F. M., Boyce J. W., Srinivasan P., et al. (2016) Heterogeneous distribution of H_2O in the martian interior: Implications for the abundance of H_2O in depleted and enriched mantle sources. *Meteoritics & Planetary Science*, **51**, 2036–2060.

McLennan S. M., Bell J. F., Calvin W. M., et al. (2005) Evidence for groundwater involvement in the provenance and diagenesis of the evaporite-bearing Burns formation, Meridiani Planum, Mars. *Earth & Planetary Science Letters*, **240**, 95–121.

McLennan S. M., Anderson R. B., Bell J. F., et al. (2014) Elemental chemistry of sedimentary rocks at Yellowknife By, Gale crater, Mars. *Science*, **343**, doi: 10.1126/science.1243480.

McSween H. Y. (2015) Petrology on Mars. *American Mineralogist*, **100**, 2380–2395.

McSween H. Y., and McLennan S. M. (2014) Mars. In *Treatise on Geochemistry, 2nd Edition, Vol. 2: Planets, Asteroids, Comets and the Solar System*, Davis A. M., editor, pp. 251–300, Elsevier, Oxford.

McSween H. Y., Murchie S. L., Crisp J. A., et al. (1999) Chemical, multispectral and textural constraints on the composition and origin of rocks at the Mars Pathfinder landing site. *Journal of Geophysical Research*, **104**, E4, 8679–8715.

McSween H. Y., Grove T. L., and Wyatt M. B. (2003) Constraints on the composition and petrogenesis of the martian crust. *Journal of Geophysical Research*, **108**, E12, 5135.

McSween H. Y., Taylor G. J., and Wyatt M. B. (2009) Elemental composition of the martian crust. *Science*, **324**, 736–739.

Ming D. W., Mittlefehldt D. W., Morris R. V, et al. (2006) Geochemical and mineralogical indicators for aqueous processes in the Columbia Hills of Gusev crater, Mars. *Journal of Geophysical Research*, **111**, E02S12.

Ming D. W., Morris R. V., and Clark B. C. (2008) Aqueous alteration on Mars. In *The Martian Surface: Composition, Mineralogy, and Physical Properties*, Bell J. F., editor, pp. 519–540, Cambridge University Press, Cambridge.

Ni P., Zhang Y., and Guan Y. (2017) Volatile loss during homogenization of lunar melt inclusions. *Earth & Planetary Science Letters*, **478**, 214–224.

Ouri Y., Shirari N., and Ebihara M. (2003) Chemical composition of Yamato (Y)980459 and Y000749: Neutron-induced prompt gamma-ray analysis study. *Antarctic Meteorite Research*, **16**, 80–93.

Pahlevan K., and Stevenson D. J. (2007) Equilibration in the aftermath of the lunar-forming giant impact. *Earth & Planetary Science Letters*, **262**, 438–449.

Papike J. J., Ryder G., and Shearer C. K. (1998) Lunar samples. In *Planetary Materials*, Papike J. J., editor, *Reviews in Mineralogy*, **36**, pp. 5-1 to 5-234, Mineralogical Society of America, Washington.

Peslier A. H., Hervig R., Yang S., et al. (2019) Determination of the water content and D/H ratio of the martian mantle by unraveling degassing and crystallization effects in nakhlites. *Geochimica et Cosmochimica Acta*, **266**, 382–415.

Poulet F., Mangold N., Platevoet B., et al. (2009) Quantitative compositional analysis of martian mafic regions using the Mex/OMEGA reflectance data. 2. Petrologic implications. *Icarus*, **201**, 84–101.

Prettyman T. H., Hagerty J. J., Elphic R. C., et al. (2006) Elemental composition of the lunar surface: Analysis of gamma ray spectroscopy data from Lunar Prospector. *Journal of Geophysical Research*, **111**, E12007.

Righter K., and Drake M. J. (1996) Core formation in the Earth's Moon, Mars and Vesta. *Icarus*, **124**, 513–529.

Roberts S. E., McCanta M. C., Jean M. M., et al. (2019) New lunar meteorite NWA 10986: A mingled impact melt breccia from the highlands – A complete cross section of the lunar crust. *Meteoritics & Planetary Science*, **54**, 3016–3035.

Rogers A. D., and Christensen P. R. (2007) Surface mineralogy of martian low-albedo regions from MGS-TES data: Implications for upper crustal evolution and surface alteration. *Journal of Geophysical Research*, **112**, E01003.

Rogers A. D., and Hamilton V. E. (2014) Compositional provinces of Mars from statistical analyses of TES, GRS, OMEGA, and CRISM data. *Journal of Geophysical. Research, Planets*, **120**, 62–91.

Ryder G. (1991) Lunar ferroan anorthosites and mare basalt sources: The mixed connection. *Journal of Geophysical Research*, **118**, 2065–2068.

Saal A. E., Hauri E. H., Van Orman J. A., and Rutherford M. J. (2013) Hydrogen isotopes in lunar volcanic glasses and melt inclusions reveal a carbonaceous chondrite heritage. *Science*, **340**, 1317–1320.

Sanloup C., Jambon A., and Gillet P. (1999) A simple chondritic model of Mars. *Earth & Planetary Science Letters*, **112**, 43–54.

Schmidt M. E., Campbell J. L., Gellert R., et al. (2014) Geochemical diversity in first rocks examined by the Curiosity rover in Gale crater; Evidence for and significance of an alkali and volatile-rich igneous source. *Journal of Geophysical Research, Planets*, **119**, 64–81.

Schmidt M. E., Perrett G. M., Bray S. L., et al. (2018) Dusty rocks in Gale crater: Assessing aerial coverage and separating dust and rock contributions in APXS analyses. *Journal of Geophysical Research, Planets*, **123**, 1649–1673.

Shearer C. K., Hess P. C., Wieczorek M. A., et al. (2006) Thermal and magmatic evolution of the Moon. In *New Views of the Moon*, Neal C. R., Jolliff B. L., Shearer C. K., and Wieczorek M. A., editors, *Reviews in Mineralogy and Geochemistry*, **60**, pp. 365–518, Mineralogical Society of America and Geochemical Society, Washington.

Solomon S. C., Aharonson O., Aurnoou J. M., et al. (2005) New perspectives on ancient Mars. *Science*, **307**, 1214–1220.

Symes S. J., Borg L. E. Shearer C. K., and Irving A. (2008) The age of the martian meteorite Northwest Africa 1195 and the differentiation history of the shergottites. *Geochimica et Cosmochimica Acta*, **72**, 1696–1710.

Taylor G. J. (2013) The bulk composition of Mars. *Chemie der Erde*, **73**, 401–420.

Taylor G. J., Warren P., Ryder G., et al. (1991) Lunar rocks. In *Lunar Sourcebook: A User's Guide to the Moon,* Heiken G. H., Vaniman D. T., and French B. M., editors, pp. 183–284, Cambridge University Press, Cambridge.

Taylor G. J., Stopar J. D., Boynton W. V., et al. (2006) Variations in K/Th on Mars. *Journal of Geophysical Research*, **111**, E03S06.

Taylor S. R. (1982) *Planetary Science: A Lunar Perspective.* Lunar & Planetary Institute, Houston, 481 pp.

Taylor S. R., Pieters C. M., and MacPherson G. J. (2006a) Earth-Moon system, planetary science, and lessons learned. In *New Views of the Moon*, Neal C. R., Jolliff B. L., Shearer C. K., and Wieczorek M. A., editors, *Reviews in Mineralogy and Geochemistry*, **60**, pp. 657–704, Mineralogical Society of America and Geochemical Society, Washington.

Taylor S. R., Taylor G. J., and Taylor L. A. (2006b) The Moon: A Taylor perspective. *Earth & Planetary Science Letters*, **70**, 5904–5918.

Thorpe M. T., Hurowitz J. A., and Siebach K. L. (2021) Source-to-sink terrestrial analogs for the paleoenvironment of Gale crater, Mars. *Journal of Geophysical Research, Planets*, 126, e2020JE006530.

Treiman A. H., Jones J. H., and Drake M. J. (1987) Core formation in the shergottite parent body and comparison with the Earth. *Journal of Geophysical Research*, **92**, E627–E632.

Wanke H., and Dreibus G. (1988) Chemical composition and accretion history of terrestrial planets. *Philosophical Transactions of the Royal Society of London*, **A325**, 545–557.

Wanke H., Bruckner J., Dreibus G., et al. (2001) Chemical composition of rocks and soils at the Pathfinder site. *Space Science Reviews*, **96**, 317–330.

Warren P. H. (1993) A concise compilation of petrologic information on possibly pristine nonmare Moon rocks. *American Mineralogist*, **78**, 360–376.

Warren P. H., and Wasson J. T. (1977) Pristine nonmare rocks and the nature of the lunar crust. *Proceedings of the Lunar Science Conference*, **8**, 2215–2235.

Warren P. H., and Kallemeyn G. W. (1997) Yamato-793605, EET 79001, and other presumed martian meteorites: Compositional clues to their origins. *Proceedings of NPR Symposium on Antarctic Meteorites*, **10**, 61–81.

Weber R. C., Lin P.-Y., Garnero E. J., et al. (2011) Seismic detection of the lunar core. *Science*, **331**, 309–312.

Wiechert U., Halliday A. N., Lee D.-C., et al. (2001) Oxygen isotopes and the Moon-forming giant impact. *Science*, **294**, 345–348.

Wieczorek M. A., Jolliff B. L., Khan A., et al. (2006) The constitution and structure of the lunar interior. In *New Views of the Moon*, Neal C. R., Jolliff B. L., Shearer C. K., and Wieczorek M. A., editors, *Reviews in Mineralogy and Geochemistry*, **60**, pp. 221–364, Mineralogical Society of America and Geochemical Society, Washington.

Wieczorek M. A., Neumann G. A., Nimmo F., et al. (2013) The crust of the Moon as seen by GRAIL. *Science*, **339**, 671–675.

Wood J. A., Dickey J. S., Marvin U. B., and Powell B. N. (1970) Lunar anorthosites. *Science*, **167**, 602–604.

Wyatt M. B., Hamilton V. E., McSween H. Y., et al. (2001) Analysis of terrestrial and martian volcanic compositions using thermal emission spectroscopy: 1. Determination of mineralogy, chemistry, and classification strategies. *Journal of Geophysical Research*, **106**, 14711–14732.

Yoshizaki T., and McDonough W. F. (2020) The composition of Mars. *Geochimica et Cosmochimica Acta*, **273**, 137–162.

Zipfel J., Schroder C., Jolliff B. L., et al. (2011) Bounce Rock – a shergottite-like basalt encountered at Meridiani Planum, Mars. *Meteoritics & Planetary Science*, **46**, 1–20.

15 Cosmochemical Models for the Formation and Evolution of Solar Systems

Overview

Cosmochemistry places important constraints on models for the Sun's accretion disk and for the formation and evolution of planets. We consider the thermal and chemical conditions under which meteorite components formed and the isotopic evidence in them for interaction of the nebula with the interstellar medium and perhaps a nearby supernova. We describe a pronounced dichotomy in nucleosynthetic isotopes in solar system materials and explore possible explanations for its origin. We explain how planetary bulk compositions are determined and used to reconstruct the formation of the terrestrial and giant planets and their subsequent differentiation. We also consider how the orbital and collisional evolution of planets may have redistributed materials formed in different thermal and compositional regimes within the solar system. Finally, we expand our perspective to include what little is known about the compositions of exoplanets and examine whether our solar system can be a template for the architecture of other planetary systems.

15.1 Constraining and Testing Models with Cosmochemistry

Understanding the formation of the solar system requires that we delve into processes for which there are no counterparts in terrestrial experience. Grand models for the solar nebula are mostly exercises in the physics of gravitational collapse and orbital mechanics. However, cosmochemistry imposes critical constraints on nebular chemistry, physical conditions, events, and chronology. In this chapter, we consider how astrophysical models for the formation and evolution of the Sun and its accretion disk can be informed by cosmochemistry. Theories about how the planets were assembled are likewise dominated by physical models of accretion, but cosmochemistry provides information about the compositions of precursor materials and the timescales for planet assembly and differentiation. Models for the orbital evolution and collisional histories of planets over time are also testable with cosmochemistry data. The bulk compositions and structures of planets around other stars, for which information is meager, can be inferred from cosmochemical reasoning.

15.2 From Gas and Dust to Stars and Our Sun

Stars form when dense regions of cold molecular clouds undergo gravitational collapse. In dense molecular cloud cores, the temperature is on the order of 10–20 K and the gas density is on the order of 10^6 particles/cm^3. Under these conditions, the gas phase consists mostly of molecular hydrogen, helium, and a few other highly volatile compounds, while the rest of the atoms are in solid phases as dust and ice. The dust includes mineral grains that formed in the atmospheres of dying stars, some of which we have identified as presolar grains, along with amorphous particles processed through cycles of evaporation by shock waves and ultraviolet light, and of recondensation under nonequilibrium conditions in interstellar space.

The process of star formation takes on the order of a few million years. This timescale is constrained both by astronomical observations and by the chronology of early solar system materials (meteorites). The stages of star formation were discussed in Chapter 10 and illustrated in Figure 10.7. Most stars – perhaps 70–90% – form in regions of cluster star formation, where thousands of stars form almost simultaneously. The star-forming region in Orion is a classic example of such an environment. During cluster star formation, the intense radiation generated by the first massive stars to form opens up a bubble of low-density hot gas around the stars. The radiation drives a shock front into the molecular cloud, which may trigger additional star formation. As the shock front overruns these newly forming systems, dense regions containing protostars are left behind, first as "pillars" of molecular gas still connected to the main cloud, and eventually as *evaporating gaseous globules* (EGGs). It is this sequence of events that accounts for the spectacular images we are able to capture of molecular clouds, such as that shown in Figure 15.1. The EGGs then evaporate, revealing the new stellar systems. Shortly thereafter, a nearby massive star may explode as a supernova, dispersing the surrounding cloud further and generally terminating star formation in the local region. But farther away, as the expanding supernova bubble slows down by piling up molecular cloud material ahead of the expanding shock front, new dense regions form that can initiate another round of star formation.

Based on the probabilities, we might infer that the Sun originated in such a setting of cluster star formation. Cosmochemistry provides ways to test this idea. As discussed in Chapter 10, short-lived radionuclides in the early solar system carry a record of nucleosynthesis in the galaxy and in the immediate environment where the solar system formed. The abundances of rapidly decaying ^{41}Ca, ^{36}Cl, ^{26}Al, and ^{60}Fe (Fig. 10.5) in the early solar

Figure 15.1 Hubble Space Telescope image of gas and dust in the star-forming region of the Carina Nebula. Two massive stars out of the field of view to the upper right and upper left are evaporating the molecular cloud, leaving pillars of gas pointing at the stars. Dense knots of cloud material are EGGs containing new stellar systems. Image courtesy of NASA and Hubble 20th Anniversary Team (STScI). *For the colour version, refer to the plate section.*

system were much too high to have been inherited from the interstellar medium. Production of these nuclides just before or during solar system formation is required. Even in the early days just after the discovery of ^{26}Al (Lee et al., 1977), a supernova was recognized as a strong candidate for its source. Cameron and Truran (1977) proposed a model in which the expanding shock front from a nearby supernova explosion compressed the gas and dust in the Sun's parent molecular cloud and triggered gravitational collapse, simultaneously seeding the solar system with newly synthesized material. The discoveries that calcium-aluminum-rich inclusions (CAIs) are ^{16}O rich compared to most solar system material (Clayton et al., 1973, 1977), and also contain isotopic anomalies in heavy elements (Wasserburg et al., 1977; McCulloch and Wasserburg, 1978a, 1978b), were thought to be explained by this model. Stellar sources of various kinds have been considered since that time, with supernovae and asymptotic giant branch (AGB) stars getting the most attention.

The discussion changed with the discovery that the short-lived radionuclide ^{10}Be was abundant in CAIs (McKeegan et al., 2000). Because ^{10}Be can only be produced by particle irradiation and not by stellar nucleosynthesis, serious attention was given to the idea that all short-lived nuclides might have been produced by irradiation in the nebula. This idea had been explored in the late 1970s to explain the presence of ^{26}Al, but it had been abandoned because the conditions in the solar nebula perceived at that time did not permit the high fluxes of high-energy particles required to make ^{26}Al, and the irradiation as modeled should also have produced huge isotopic anomalies in other elements that were not observed. But a model for the production of CAIs close to the Sun, known as the "X-wind model" (Shu et al., 1996; see below), provided a way to get the high particle fluxes required. Gounelle et al. (2001) formulated an energetic-particle irradiation model that reproduced the relative initial abundances of ^{10}Be, ^{26}Al, ^{41}Ca, and ^{53}Mn in CAIs. However, their calculations depended on special assumptions about target chemistry and rather extreme conditions for the energy and composition of solar cosmic rays.

The discussion changed again with a report that ^{60}Fe, which can only be produced efficiently in stars and not by irradiation, was more abundant in the early solar system than previously thought (Tachibana and Huss, 2003). However, the inferred initial abundance values of ^{60}Fe covered a very large range. An experimental problem with ratio bias in the secondary ion mass spectrometry data (Ogliore et al., 2011), and the realization that both iron and nickel are mobile in even the least metamorphosed chondrites (Telus et al., 2016), eventually led to lower inferred initial ratios, although ^{60}Fe was apparently present in the early solar system. At about the same time, new measurements increased the ^{60}Fe half-life from 1.5 Myr to 2.6 Myr (Rugel et al., 2009).

There is currently considerable debate about the origin of short-lived nuclides in the early solar system. The ^{60}Fe abundance was apparently not high enough to be a smoking gun for a supernova, and the longer half-life means that it could have come into the nebula from interstellar gas and dust, as is thought to have been true for at least part of the ^{53}Mn (half-life = 3.7 Myr). The ^{26}Al could have been provided by a supernova, but it could also have come in with the ejecta of a Wolf–Rayet star prior to its explosion as a supernova, or even from an AGB star. It is also likely that the ^{10}Be and some ^{36}Cl, ^{41}Ca, and perhaps even some ^{26}Al and ^{53}Mn were produced by particle irradiation in the early solar system. Sorting out the various contributions from these sources in order to create a comprehensive model for the short-lived radionuclides in the early solar system will require a lot more work.

The original supernova trigger model by Cameron and Truran (1977) proposed that the short-lived radionuclides were injected into the parent molecular cloud by the supernova that triggered the gravitational collapse of the solar system. Over the years, increasingly detailed models have tried to simulate this, without much success. The problem is that extremely hot gas, like that behind an expanding supernova shock, does not mix with cold molecular cloud gas. The two regimes interact across a shock front, and it is easier to assimilate the cold gas into the hot region than the other way around. For supernova ejecta to efficiently mix into a molecular cloud, the gas must cool, and the characteristic timescale for such cooling is a million years. Models have successfully gotten some supernova material into a collapsing cloud, but the range of conditions where this occurred was extremely limited.

The way to get around the mixing problem is to inject the short-lived nuclides in the form of dust directly into the already-formed accretion disk (Ouellette et al., 2005). The model builds on the picture described above, where the first massive stars trigger formation of additional stars and then uncover them by evaporating the surrounding molecular material, after which the massive star explodes as a supernova, pummeling the newly formed star systems with fast-moving gas and dust ejecta (Fig. 15.2). Once the protostar and accretion disk have formed, they are too dense to be evaporated by the supernova ejecta. Dust in

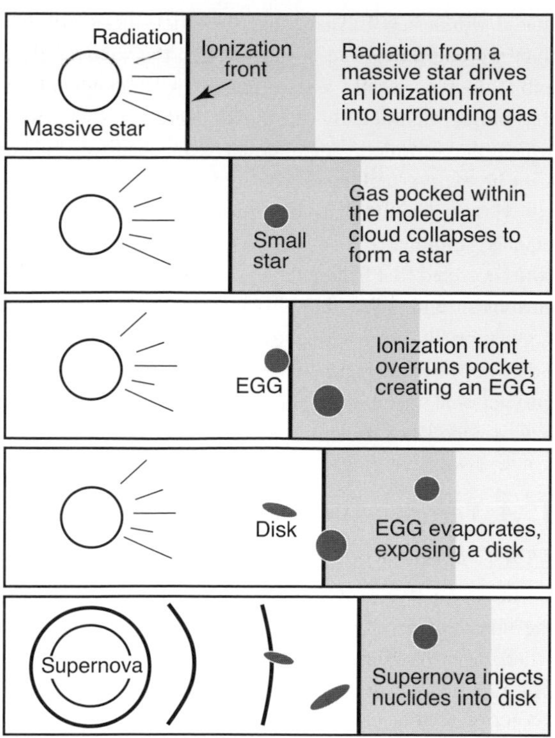
Radiation
Massive star
Ionization front
Radiation from a massive star drives an ionization front into surrounding gas
Small star
Gas pocked within the molecular cloud collapses to form a star
EGG
Ionization front overruns pocket, creating an EGG
Disk
EGG evaporates, exposing a disk
Supernova
Supernova injects nuclides into disk

Figure 15.2 Sequence of events (top to bottom) affecting low-mass stars like our Sun, formed in the vicinity of high-mass stars within molecular clouds. After Hester and Desch (2005).

the ejecta goes through the bounding shock and into the newly formed system. Although the dust evaporates in the bounding shock, it deposits its atoms into the accretion disk and thus seeds the disk with short-lived nuclides from the supernova. Numerical models seem to show that this works, but the probability of the right conditions occurring is low.

Another idea under investigation can be labeled "self-enrichment" of the molecular cloud. The idea uses the observation that in a giant molecular cloud, serial star formation may occur, where the massive stars in the first batch of stars to form explode as supernovae and trigger a second round of star formation, and then the massive stars from the second round explode and trigger yet another round of star formation. As each supernova explodes, it dumps short-lived nuclides into the interstellar medium that can be mixed into the surrounding molecular cloud on timescales of a few million years. If this happens repeatedly, the abundances of the short-lived nuclides will increase throughout the cloud. If the solar system formed late, then it would have inherited high abundances of the short-lived nuclides. While this idea is attractive, the predictions of the model in detail do not match the observations very well. Because the timescale for mixing newly synthesized material into a dense molecular cloud is at least a few million years, the abundances of ^{41}Ca and ^{36}Cl (half-life = 1–3 × 10^5 years) would be orders of magnitude lower than those inferred for the early solar system (see Chapter 10 and Fig. 10.5), even if mixing times were at the short end of the possible range. Although self-enrichment may not explain all of the short-lived radionuclides in the solar system, it does provide a mechanism for the formation of ^{53}Mn and ^{60}Fe (see above), without requiring an immediate supernova trigger.

It seems likely from statistical considerations that the Sun originated in a region of cluster star formation. Although the details have not been satisfactorily worked out, a supernova could have been involved in the origin of the solar system. One avenue of future work will be to investigate the potential contributions to stable isotopes from a late supernova. It is difficult to establish the level of such a contribution from observations because there are no labels on the atoms of stable isotopes produced in the last supernova as there are for the short-lived radioactive isotopes. It might be possible to infer the contribution by comparing solar system abundances with the composition of the galaxy in the solar neighborhood at ~4.57 Ga. This will require a major improvement in our knowledge of galactic abundances, but based on the amazing things we have learned to date, there is no reason to believe that we cannot eventually get the required information.

15.3 Formation of the Accretion Disk and Planets

The infant stars described in the previous section were initially embedded in optically thick, dusty nebulae. That is the first stage of stellar formation described previously in Chapter 10 and Figure 10.7. A new star will consume most of the mass of the surrounding nebula, and a small fraction that remains will acquire most of the system's angular momentum and flatten into a rotating *accretion disk.*

Dynamical models have identified several processes that facilitate accretion (Cuzzi and Weidenschilling, 2006). Settling of dust toward the midplane of the nebula would have increased the number density of particles at the midplane, facilitating grain growth. Nebular turbulence could have slowed or prevented settling and delayed grain growth. On the other hand, turbulence could also have brought more grains into contact, thereby facilitating grain growth. Eddies in the nebula could concentrate particles in a small volume of space, increasing the efficiency of grain growth. But these processes did not lead to growth of objects larger than about a meter, and these objects are just the right size for gas drag to cause them to spiral into the Sun.

For a long time, it was not clear how meter-sized objects got together to form larger bodies. More recently, the concept of *pebble accretion* has been developed, in which gas drag reduces the velocity of centimeter- to meter-sized objects and allows them to be captured more easily by massive bodies (Lambrechts and Johansen, 2012). This process should have worked most efficiently in the outer solar system, where both icy and rocky pebbles were present, facilitating the accretion of giant planets. However, pebble accretion should also have worked in the inner solar system (Morbidelli et al., 2015). Once bodies reached the sizes of planetesimals (1–10 km), they were no longer controlled by nebular gas drag. Thereafter, they grew by mutual collisions and became bound by gravity. Eventually, a few dominant bodies would have emerged, large enough that their gravity focused more collisions onto them. This would have resulted in runaway growth, as the most massive bodies gobbled up the smaller ones (Morbidelli et al., 2012). Smaller objects not accreted into these larger bodies would eventually be gravitationally scattered by close encounters, and would perhaps be lost from the solar system. Bodies remaining in the asteroid belt escaped this fate, but models suggest that as much as 99% of the original mass of the Main belt has been removed, probably through gravitational interactions with Jupiter.

The formation of the terrestrial planets from planetary embryos (100–1000 km in size) has been analyzed by numerical models (Morbidelli and Raymond, 2016), starting with a few dozen objects in nearly circular orbits. Gravitational interactions cause them to be perturbed into elliptical orbits that cross each other, allowing further collision and accretion. The timescale for planet formation in such "standard" models is typically a few tens of millions to 100 million years.

Jupiter and Saturn – the gas giants – must have formed in a way that allowed them to accrete nebula gas directly. Uranus and Neptune – the ice giants – accreted gases that had already condensed as ices. There are two competing theories for the formation of the giant planets. In the *disk fragmentation model*, a gravitationally unstable accretion disk fragments directly into self-gravitating clumps of gas and dust that then contract to become giant gaseous protoplanets (Cameron, 1978; Boss, 2004). Models indicate that disk fragmentation only occurs in the outer parts of accretion disks, typically beyond 50 AU (Paardekooper and Johansen, 2018). The attractiveness of this model is that it permits the rapid formation of Jupiter and Saturn. The main drawbacks are that it works best in disks that are much more massive than the one thought to have produced the solar system, and it does not easily lead to planets with compositions that differ significantly from the accretion disk from which they formed. An alternative mechanism, called *core accretion*, facilitated by pebble accretion, is generally viewed as more plausible for having formed the giant planets (Lunine, 2014). This idea assumes that rocky cores accreted first from pebbles, boulders, and then planetesimals, and once they became large enough, they gravitationally attracted nebular gas or ices. Core accretion was once thought to be significantly slower than gravitational instability, but pebble accretion seems to have solved this problem.

15.4 Temperatures in the Accretion Disk

15.4.1 A Hot Nebula?

By the late 1960s, it was almost universally accepted that the inner solar system went through a hot nebula stage where all pre-existing solids had evaporated. This conclusion was based on modeling by Cameron (1962) and others that showed very high temperatures in the disk, and on the conclusion from isotope measurements that all the matter to which scientists had access had the same isotopic compositions (previously identified isotopic anomalies had all been shown to be experimental artifacts). From the late 1960s to the early 1990s, most cosmochemical data were interpreted in terms of a hot solar nebula. However, even in the early days, there were hints that this model was not strictly correct. Reynolds and Turner (1964) identified isotopic anomalies in xenon that could not be explained by known processes operating on a homogeneous composition in the solar system. Black and Pepin (1969) identified a neon component consisting of nearly pure ^{22}Ne, which they called Ne-E, that they attributed to the decay of short-lived ^{22}Na (half-life = 2.6 years). Because of this short half-life, they proposed that Ne-E was probably located in a carrier that condensed in the atmosphere of a dying star where ^{22}Na was produced. Then came the discovery of mass-independent oxygen-isotope variations in CAIs (Clayton et al., 1973, 1977). In the years that followed, isotopic anomalies that could not be explained in the context of a hot solar nebula turned up with increasing regularity. By the mid-1980s, new nebula models showed that the accretion disk at the radius of the asteroid belt was probably never hot enough to vaporize all solids. Then in the late 1980s, presolar grains were discovered in meteorites (Lewis et al., 1987), and by 1990 they were known to be present in all classes of primitive chondrites (Huss, 1990). The hot solar nebula

Box 15.1 ALMA and VLA Observations of Young Stellar Systems

The Atacama Large Millimeter/submillimeter Array (ALMA) is currently the world's largest radio telescope. Light at these wavelengths comes from frigid, dense regions of gas and dust where stars are forming. ALMA examines the spatial distribution of matter in accretion disks around other stars at scales as small as a few AU.

A series of 10 papers in 2018 described high-resolution ALMA images of 20 accretion disks around nearby stars (Andrews, 2018). These images upended formerly simplistic views of theoretical models of disks that featured smooth dust and gas distributions and planet accretion timescales of millions of years. Rings and gaps in accretion disks only a million years old (Fig. 15.3) have been interpreted as the orbital locations of large, unseen planets, likely similar in size to Neptune or Saturn, that are interacting with nebular dust. Such planets would have formed more quickly than predicted by the standard accretion models described above and are located far from their parent stars. The measured motions of gas provide further indirect evidence for planets. A planet would push gas and dust aside, opening a gap in the disk. ALMA observations indicate that gas above the gap then collapses into it, like a waterfall. However, the ring-shaped gaps in Figure 15.3 are well outside the orbits of planets in our solar system, even though the star is about the same size as the Sun – this complicates their interpretation.

The Very Large Array (VLA) is one of the most advanced radio telescopes in the world. VLA images are made at longer wavelengths than ALMA, providing complementary data on the inner structures of accretion disks. VLA and ALMA scientists have teamed up to create hundreds of images of the disks around young stars in the Orion Clouds (Tobin et al., 2020). Examples of disks imaged in this study are shown in Figure 15.4.

These incredible images of infant stars provide new constraints on the earliest stages of planet formation and suggest accretional timescales that are more rapid than most models would predict. Very young disks are seen to be similar in size to, but more massive than, older disks, implying that the host stars incorporate disk material over time. Larger planets might form earlier from more massive disks. The youngest protostars imaged are puffy and irregular in shape and do not have the flattened disks and well-developed polar outflows of Class 0 stars.

Figure 15.3 ALMA image of the accretion disk around HL Tauri, a Sun-like star, showing gaps interpreted to have been opened by unseen planets. *For the colour version, refer to the plate section.*

Box 15.1 (cont.)

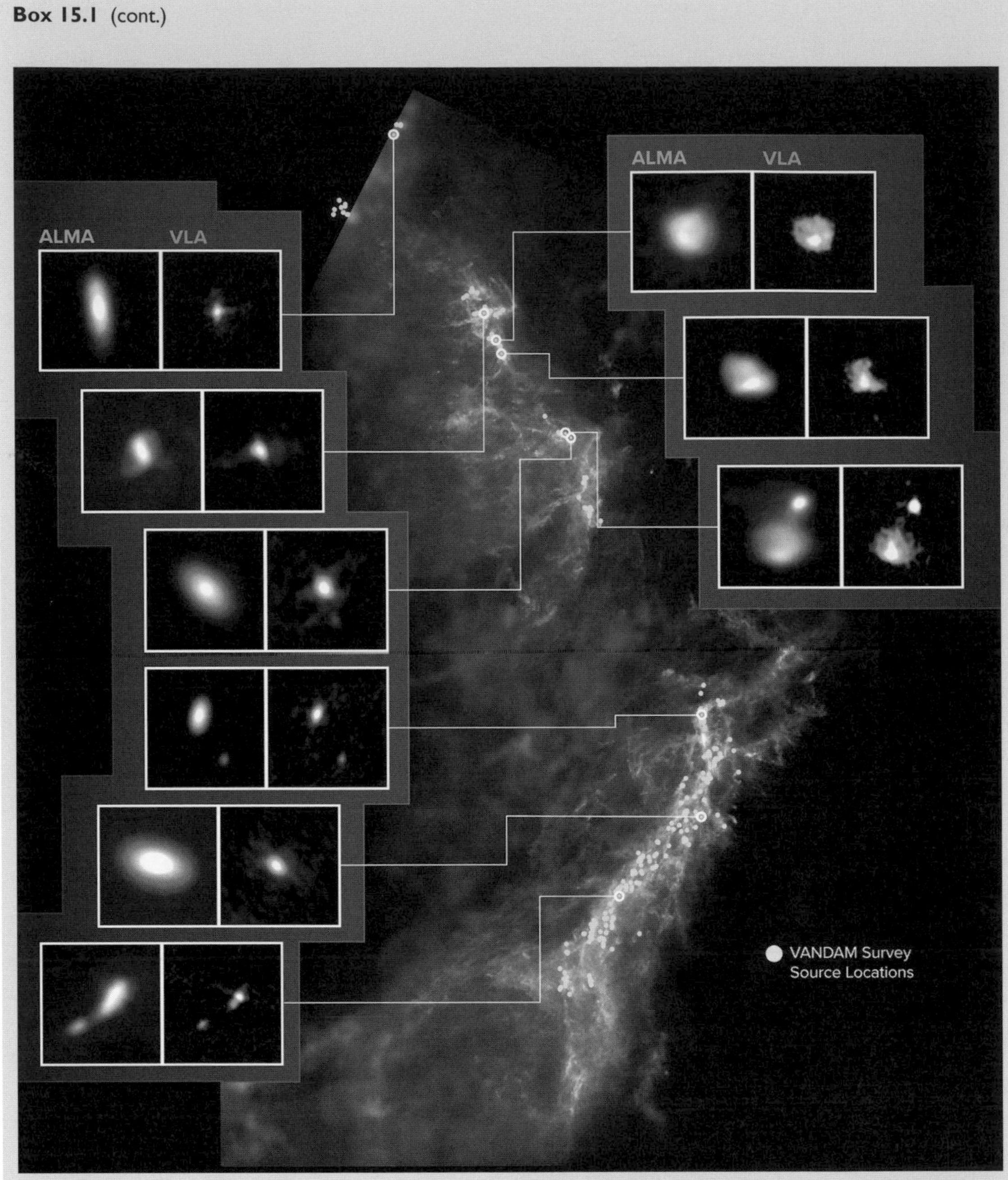

Figure 15.4 ALMA and VLA data complement each other. ALMA images the outer disk structure, visualized in blue, and VLA images the inner disk and star core (orange) in the Orion Molecular Clouds. Image courtesy of ALMA (ESO/NAOJ/NRAO), J. Tobin; VLA (NRAO/AUI/NSF), S. Dagnello; and Herschel/ESA. *For the colour version, refer to the plate section.*

model that had appeared to explain so much was in serious trouble.

Can a hot solar nebula be reconciled with the observations summarized above? Presolar grains are located in the matrix material of chondrites – material that was never melted. One possibility is that the presolar grains (and much of the matrix itself) might be a late addition to the region where chondrites formed. Perhaps they were added after the high temperatures that produced the volatility fractionations and the transient events that produced the chondrules had subsided. The main problem with this idea is that the presolar grains in each meteorite class record a heating event that closely tracks the event that depleted volatile elements in the bulk meteorites. While the two events for each class could have been separated in time (to allow the nebula to cool), it is more plausible that the presolar grains and the bulk compositions record the same events. Even if late addition did solve the problem with the presolar grains, it does not address the isotopic anomalies found in both high-temperature components of chondrites, such as CAIs, and in bulk meteorites, including differentiated bodies. In Chapter 8, we saw that nucleosynthetic anomalies are present in the majority of elements that make up chondrites and differentiated meteorites (Figs. 8.16 and 8.17). The bulk nebular material was clearly not homogenized on an atomic scale, so it does not appear that the hot solar nebula model can be reconciled with the observations.

How can we then account for the large number of observations that appear to be explained by a hot solar nebula? For example, what about the idea that all of the materials to which we have access seem to have formed from the same well-mixed isotopic reservoir? Although many objects have very similar isotopic compositions, we know that when examined in sufficient detail, there are small isotopic differences between most materials. Everything that happens to matter after it is ejected from stars works to homogenize its composition. Dust and gas from a wide variety of sources are mixed by turbulence, stellar winds, and supernova explosions. If dust grains are evaporated by supernova shocks, the atoms are scattered and likely recondense along with atoms with completely different histories as amorphous material or ices. By the time the solar system formed, its precursor materials were likely very well mixed, although individual grains that had not evaporated would still be isotopically and chemically anomalous. Within the solar system, any process that melts a representative sample of this well-mixed material will give very similar results. Overall, the observations are consistent with the formation of most solar system objects out of processed – but not completely evaporated – molecular cloud material. The close similarity between the isotopic compositions of most objects does not require a hot solar nebula.

Let us consider another example. The bulk chemical compositions of high-temperature materials in chondrites – that is, the CAIs, amoeboid olivine aggregates (AOAs), and chondrules – appear to follow approximately the calculated path for solids arising from equilibrium condensation (Fig. 15.5) (MacPherson et al., 2005). As we discussed

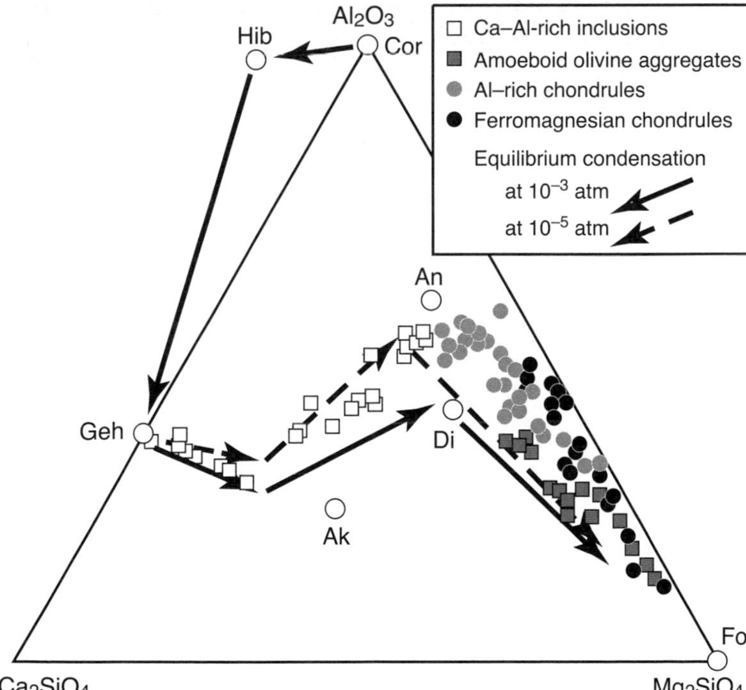

Figure 15.5 Bulk chemical compositions of CAIs, AOAs, and chondrules, compared to trends for calculated total condensed solids formed by equilibrium condensation of a hot solar gas at two different pressures. Condensed minerals are corundum (Cor), hibonite (Hib), gehlenite (Geh), akermanite (Ak), anorthite (An), diopside (Di), and forsterite (Fo). Modified from MacPherson et al. (2005).

in Chapter 7, equilibrium condensation is an end member of a continuum of volatility-based gas–solid fractionation processes. The equilibrium calculations tell us, to a first approximation, what the outcome of all of these processes is likely to look like. For example, moving from an equilibrium condensation regime to a regime where kinetic fractionations dominate is not going to make a CAI composition volatile. It might raise the temperature for complete evaporation of moderately volatile elements, but it will not make them less volatile than refractory elements. So general agreement with the prediction of equilibrium calculations is to be expected, almost independent of the details of the actual process that took place. Also, the precursors of CAIs, AOAs, and chondrules could have formed in a wide variety of places, both within and outside of the solar system, under conditions that ranged from near equilibrium (in a stellar atmosphere or during a transient high-temperature event in the nebula, for example) to very far from equilibrium. Thus, Figure 15.5 does not provide a compelling argument that equilibrium condensation in a hot solar nebula controlled the compositions of these objects. We already discussed in Chapter 7 how heating of molecular cloud dust could explain the differences in volatile depletions and mixtures of presolar grains among chondrite classes. Other observations once thought to require equilibrium condensation have similar explanations. So, while there is clear evidence for transient high-temperature events affecting inner solar system materials, there is no compelling evidence recorded in the meteorites or terrestrial planets of a global hot stage when all pre-existing solids were evaporated.

So how hot was the solar nebula? Current models indicate that during the main accretion stage of the Sun, temperatures at the midplane of the accretion disk were likely hot enough to vaporize silicates within a few AU of the star (e.g. Humayun and Cassen, 2000), but this high-temperature phase was short-lived and, as matter moved through the disk and accreted onto the Sun, the temperature in the disk dropped. But not all materials processed in this hot zone were necessarily swept into the Sun. Some of the earliest-formed CAIs are thought to have formed near the Sun and were carried outward (Yang and Ciesla, 2012), eventually to be incorporated into carbonaceous chondrite asteroids. Viscous radial expansion of the disk was fastest early on, when molecular cloud material was added close to the proto-Sun. Most of the nebula, however, experienced low ambient temperatures, consistent with the widespread survival of isotopic anomalies in meteorite components and bulk meteorites and of presolar grains in primitive chondrites.

15.4.2 Localized Heating: Formation of Chondrules and CAIs

Chondrules and chondrule fragments comprise the major portion of most chondrites, the most abundant type of meteorites. Levy (1988) estimated that the present-day asteroid belt contains ~10^{24} g of chondrules, and the primordial nebula may have contained much more. Clearly, chondrule-forming events must have been widespread. The origin of chondrules remains an unresolved problem in cosmochemistry. Chondrules formed in the Sun's accretion disk through some sort of transient flash-heating event(s). Some CAIs apparently also were melted in the disk. What was the process (or processes) that melted the chondrules and CAIs? Whatever it was, it dominated the disk for at least a few million years.

When Henry Sorby first studied chondrites with the petrographic microscope, he correctly inferred that chondrules had crystallized from molten droplets. In 1877, he wrote, "melted globules with well-defined outlines could not be formed in a mass of rock pressing against them on all sides, and I therefore argue that some at least of the constituent particles of meteorites were originally detached glassy globules, like drops of fiery rain." Sorby thought that the molten droplets that crystallized to form chondrules were likely pieces of the Sun, and the idea that the droplets condensed from hot gas was popular for a long time.

We now know that the vast majority of chondrules formed by the melting of solid precursors in some sort of flash heating event (see Grossman et al., 1988 or Krot et al., 2018 for reviews of chondrule properties). Evidence includes relict grains that did not melt during chondrule formation and the porphyritic textures of most chondrules, which require the presence of residual solid nuclei to prompt the formation of large crystals. The source of the heat that melted chondrules is among the most controversial questions in cosmochemistry. Temperatures were not uniformly high enough to melt silicates on a nebular scale, and the cooling time for a hot nebula is many orders of magnitude too long to explain quenched chondrule textures. A localized heating process is required. Proposed mechanisms have included impact, lightning, solar heating, and shock waves, among others.

The formation of chondrules by impact was investigated after samples returned from the Moon were found to contain round, glassy droplets of a size similar to chondrules. The idea lost favor because the production efficiency of glassy droplets in an impact (at least on the Moon) is very low, so a concentration mechanism would also be required. More recent versions of the impact model consider collisions between planetesimals. One model proposes that chondrules formed by impacts between growing planetary embryos, with shock heating producing jets of molten material that form droplets that cooled to become chondrules (Johnson et al., 2018). An alternative version considers impacts between molten planetesimals, where melt ejected into space cooled to form chondrules (e.g. Sanders and Scott, 2018). Neither of these models has gained wide acceptance as the mechanism for formation of the vast majority of chondrules. However, a subset of chondrules in the rare CB and CH chondrites might have formed through impact. These chondrules all have the same unusual composition, the same texture, and the same oxygen-isotope composition. They are hypothesized to have been formed by condensation of liquid droplets from the vapor plume produced by the impact of two large asteroids late in nebular history (Krot et al., 2005).

Lightning received considerable attention in the 1990s (Love et al., 1995; Desch and Cuzzi, 2000). A lightning discharge potentially carries plenty of energy. The heating timescale is rapid and the cooling timescale depends on ambient conditions but could match that of chondrules. Lightning is expected to have occurred in a turbulent nebula. The event is localized, but perhaps too localized to produce huge numbers of chondrules. The idea has lost favor because the conditions necessary to produce sufficiently energetic lightning bolts are limited and may not match average nebular conditions, and because the efficiency of such a process is expected to be low.

Solar heating has been discussed in a number of contexts. For example, Huss (1988) suggested that the intense radiation of the FU Orionis eruptions known to occur in young stellar objects could have both melted dust aggregates to form chondrules and helped them to accrete rapidly to form chondrites. Liffman and Brown (1995) suggested that liquid droplets melted close to the Sun by solar radiation could be lifted by bipolar outflows and deposited in the asteroid belt. They suggested that traveling through the thin gas above the disk could sort the droplets according to size, thus explaining why chondrite classes have unique and narrow size ranges for their chondrules. Neither these nor other similar models have gained much of a following.

A model that incorporated both the lightning concept and the energy of the Sun is the *X-wind model* proposed by Shu et al. (1996, 2001). In this model, the magnetic fields of the young Sun and the surrounding disk interacted, producing a gap between Sun and disk (Fig. 15.6). When materials spiraling inward reached the inner edge of the disk (called the "X-region"), they took one of two paths.

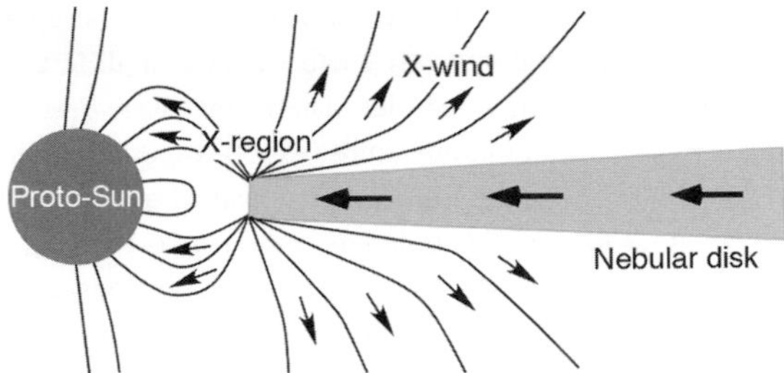

Figure 15.6 Sketch of the X-wind model, showing a gap between the proto-Sun and the inner edge of the nebular disk caused by interfering magnetic fields. Materials migrating inward in the disk reach the X-region, where they either accrete onto the star or are ejected outward above the disk (X-wind), only to be accreted back onto the disk farther out. After Shu et al. (1996).

Rapidly orbiting materials were flung outward along magnetic field lines (producing the X-wind), whereas less rapidly moving materials fell onto the star (Fig. 15.6). Intense solar heating in the X-region was proposed as the source of heat to melt CAIs and chondrules, which were then ejected out of the nebular plane and reaccreted onto the nebula disk at greater solar distance. A modification of the model added the production of short-lived radionuclides by irradiation in the reconnection ring (Gounelle et al., 2001). The model has lost favor in recent years. Desch et al. (2010) offered an extensive critique of the model, demonstrating that many of its predictions do not match observations. However, the idea that CAIs formed near the Sun and were transported to the region where meteorites formed has survived.

The passage of shock waves through the nebula potentially has the power to melt dust aggregates and form chondrules (e.g. Desch and Connolly, 2002; updated by Morris and Boley, 2018). In the *shock wave model*, the dust aggregates are heated both by gas drag as they pass through the shock and by radiation from the hot gas behind the shock. Shocks could be generated either by gravitational instabilities in the outer accretion disk or by the passage of giant protoplanets through the disk. This is currently the most popular mechanism for heating chondrules.

None of these models has been universally accepted, though, and all of them have problems (Connolly and Jones, 2016). It is possible that we have not yet conceived of the true mechanism that formed chondrules, but whatever it was, it operated pervasively, at least in the area where chondrites accreted.

The formation of CAIs also requires transient high-temperature events in the early solar system. However, CAIs formed at an earlier time (see Chapter 10) and probably in a different place than chondrules. The coarse-grained igneous type B and compact type

A CAIs, which appear to have experienced several episodes of (often incomplete) melting, have a different thermal history than chondrules. These CAIs were heated to a lower maximum temperature (~1400–1450°C versus 1450–1850°C for chondrules), but remained at elevated temperatures longer and cooled more slowly (1–10 versus 100–1000°C/hour) than chondrules. Both igneous and non-igneous CAIs formed at lower pressures than chondrules and in a more reducing environment, with an oxygen fugacity close to that of solar gas. Unaltered CAIs also have ^{16}O-rich compositions, close to the isotopic composition of the Sun (McKeegan et al., 2011). The ambient temperature where most CAIs formed was above the condensation temperature of forsterite, as shown by forsterite-rich condensates on the outside of many CAIs. AOAs represent the lowest ambient temperature recorded by refractory inclusions. CAIs also experienced intense particle irradiation, as shown by excesses of ^{10}B from the decay of short-lived ^{10}Be, and ^{53}Cr from the decay of ^{53}Mn over that inherited by the solar system. These observations indicate that CAIs formed close to the Sun, rather than in the asteroid belt where chondrules likely formed, and were scattered outward into the region where chondrites accreted. This outward migration could have been due to turbulent diffusion in the nebula (Ciesla, 2007) or to entrainment in bipolar outflows originating near the Sun.

The mechanisms of CAI formation are not at all well understood. The original idea that they are condensates from a hot solar nebula cannot explain most CAIs. Igneous CAIs, which contain many isotopic anomalies, formed by the melting of pre-existing refractory dust aggregates. Much of the precursor dust cannot be nebular condensate because it preserved isotope anomalies. Condensation was probably involved in producing components of fine-grained CAIs and AOAs and in forming rims on CAIs. All of this processing most likely took place near the Sun, and the resulting products were scattered outward. Some CAIs show evidence of having been processed through chondrule-forming events. CAIs have been found inside a few chondrules, and some CAIs have been remelted under the conditions experienced by chondrules. Considerable work is still required to sort out the details of CAI formation.

To summarize, chondrules and CAIs formed by transient heating events that processed a large fraction of the matter in the accretion disk. These heating events appear to have overprinted the thermal processing that produced the volatile element depletions among chondrites. Although models for chondrule formation are generally more detailed, there is no consensus on the formation mechanisms for either chondrules or CAIs. In the current view, CAIs formed in a localized distinct region, mostly likely near the proto-Sun, within the first 0.3 Myr of solar system history, whereas chondrule formation occurred farther outboard, likely in the region of the asteroid belt, over 1.5–5 Myr.

15.5 Compositional Variations within the Accretion Disk

The outward transition from the rocky terrestrial planets to the gas- and ice-giant planets indicates a radial compositional difference in volatile contents. The transition point in the nebula – the *snowline* – was actually multiple lines with different radial distances for each condensing gas species, and the position of any particular snowline was likely not stationary over time.

Variations in asteroid spectral classification with heliocentric distance in the Main belt (see Fig. 12.12) and farther outward also imply variations in the composition and oxidation state of accreted materials. It is currently unclear whether the radial distributions of asteroid classes are a primordial feature of the nebula or represent the implantation of materials into the Main belt from elsewhere by later gravitational perturbations (described below).

Processing of materials in the solar nebula contributed to the homogenization of primary nucleosynthetic signatures. Other processes operated either to preserve chemical or isotopic differences or to fractionate various components within the nebula. In Chapter 8, we learned that planets and meteorites contain small but resolvable nucleosynthetic anomalies in isotopes such as ^{50}Ti, ^{54}Cr, ^{58}Ni, and ^{95}Mo, relative to other stable isotopes of the same elements. In particular, many isotope systems reveal an isotopic dichotomy between carbonaceous chondrites and related differentiated meteorites (CCs) and noncarbonaceous chondrites and related differentiated meteorites (NCs) (Warren, 2011; see Fig. 8.17). This *isotopic dichotomy* is striking and its origin is the subject of lively debate. Among the solar system bodies for which we have samples, the Earth, the Moon, Mars, Vesta, ordinary and enstatite chondrites, numerous achondrites, and some irons constitute the noncarbonaceous (NC) group, whereas carbonaceous chondrites and a few relatively oxidized achondrites and irons constitute the carbonaceous (CC) group. The currently favored explanation for the dichotomy is that the nebula was divided into two spatially and isotopically distinct reservoirs located inside and outside the orbit of Jupiter (Fig. 15.7). Because materials that are more highly oxidized and volatile

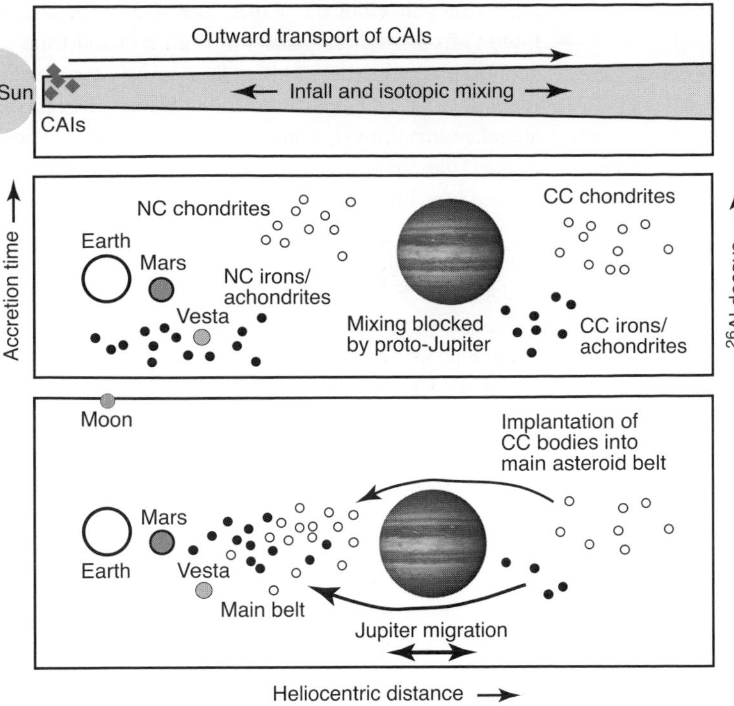

Figure 15.7 A model for the evolution of bodies in the early solar system. The top panel illustrates the earliest stage, when dust and gas carrying nucleosynthetic isotope signatures were falling into the disk and mixing. The first solids, CAIs, formed near the Sun and were transported outward to the locations where carbonaceous chondrites accreted. The middle panel shows the locations of noncarbonaceous (NC) bodies inside the orbit of Jupiter and carbonaceous (CC) bodies outboard the orbit of Jupiter. In this panel only, the vertical axis is time; earlier-accreted objects had more live ^{26}Al and thus experienced melting and differentiation. Mixing between the inner and outer regions of the disk was blocked by the formation of proto-Jupiter. In the lower panel, CC asteroids in the outer disk were perturbed inward and implanted into the Main belt, perhaps by the migration of Jupiter and other giant planets. Modified from Kruijer et al. (2020).

enriched are commonly thought to have formed in the outer solar system, beyond the snowline, the CC group is inferred to have formed in the outer solar system (Scott et al., 2018; Kleine et al., 2020). The isotopic differences between the reservoirs are thought to have been inherited from the Sun's parent molecular cloud (Burkhardt et al., 2019). The two reservoirs were maintained for several million years due to the rapid formation of Jupiter, which prevented significant exchange of materials between the inner and outer solar system. Eventually, gravitational scattering, perhaps due to migration of the giant planets, caused CC asteroids to migrate inward into the Main belt (Fig. 15.7), accounting for the occurrence of NC and CC asteroids together (Kleine et al., 2020; Kruijer et al., 2020). This scenario has become popular enough that workers have attempted to establish the formation time of Jupiter based on the formation times of CC and NC objects (Kruijer et al., 2017).

Although this model has a considerable following, there is no evidence specifically requiring that the nucleosynthetic anomalies represent distinct physical reservoirs that lasted for several million years, or even that the CC reservoir was the outer solar system. The two groups could represent distinct processing pathways in the solar nebula. A particular set of processes could have arisen multiple times in different places and/or at different times in the nebula. Or, a particular location in the nebula could have experienced different styles of processing over the lifetime of the nebula. Material was constantly moving through the accretion disk, either inward to join the Sun or outward to conserve angular momentum. With this history, it seems possible that the isotopic dichotomy could have arisen due to different styles of nebular processing.

The chemical processing that generated carbonaceous and noncarbonaceous chondrites was distinctly different. In carbonaceous chondrites, chemical fractionation was primarily based on volatility, and these chondrites include both the least fractionated and the most highly fractionated chondrites known. Carbonaceous chondrites have the highest abundances of CAIs, thought to have formed near the Sun, and they tend to be oxidized. In contrast, although noncarbonaceous meteorites experienced some volatility-based fractionation, they show considerable variations in oxidation state and have experienced metal–silicate fractionation. Figure 7.11 shows the dichotomy in a purely chemical Si/Al versus Mg/Al plot. An alternative model might suggest that the isotopic dichotomy originated through two different types of nebular processing, with CC materials processed primarily by the removal of volatile elements, and NC materials processed by a combination of volatility fractionation, changes in oxidation state which changed the abundances and compositions of metal and silicate reservoirs, and metal–silicate fractionation.

There is currently no well-developed model to compete with the idea of Jupiter separating the NC and CC reservoirs. But that does not mean the currently favored model must necessarily be correct. It is worth remembering that both chemical and isotopic features of the NC and CC meteorites exhibit a dichotomy, perhaps suggesting that the chemical and physical dichotomies may have arisen through the same set of processes. The as-yet unresolved origin of the isotopic and chemical dichotomy among solar system materials will continue to be a major focus of cosmochemical research and dynamical modeling. Like the insights that presolar grains provide into astrochemistry and stellar nucleosynthesis, these isotope effects may provide new constraints on the dynamic and temporal structure of the early solar system.

15.6 How to Estimate Bulk Compositions of Planets

The formation of the terrestrial planets is constrained by their bulk chemical compositions, but determining the compositions of entire planets is challenging. Because planets are differentiated into crust, mantle, and core, there is no place on or within a planet that has the composition of the entire body. Before considering the formation of the terrestrial planets, let's review how we go about estimating their bulk compositions.

15.6.1 Constraints on Planet Compositions

To illustrate how a planetary bulk composition is estimated, let's begin with the Earth, about which we have the most information. For this analysis, we need estimates of the compositions of the crust, mantle, and core. Several widely accepted crust compositions have been determined by estimating the relative proportions of continental and oceanic crustal rocks and averaging their compositions. The crust comprises only a small fraction (~0.4%) of the Earth's mass, as judged from seismic data, but it is important to estimate its composition correctly as it contains a significant proportion of the planet's incompatible elements. Mantle compositions are obtained by analyzing mantle xenoliths brought up in lavas or mantle blocks exposed by tectonic forces. Many mantle samples are residues from partial melting that produced basaltic magma, and some proportion of basalt can be added to residue compositions to reconstruct the chemistry of the primitive mantle (before basaltic magma was extracted). The mantle composition is especially important because the mantle constitutes the largest proportion of the planet. The size of the core is inferred from the bulk density of the Earth and from its geophysical properties, while its chemistry is estimated from solar system elemental abundances (iron is so abundant that it must dominate the core) and knowledge of how siderophile elements partition between metal and silicate from high-pressure experiments. Estimates of the compositions of the Earth's crust, mantle, and core compiled from various sources have been tabulated by Lodders and Fegley (1998). The Earth's bulk composition, as derived from combining these compositions and relative proportions (Kargel and Lewis, 1993), is given in Table 15.1.

In estimating the bulk compositions of the other terrestrial planets, there are not nearly so many constraints. Determination of a planet's mass (obtained from its gravitational effect on the orbits of moons or nearby spacecraft) and volume (calculated from its diameter as measured by telescopes or spacecraft) enables the calculation of its *mean density*. A meaningful comparison of planet mean densities requires that we correct for the effects of self-compression due to gravity – massive planets are more internally compressed than less massive ones. The *uncompressed mean densities* (in g/cm^3) of the terrestrial planets are shown in Figure 15.8. To first order, these density variations reflect differences in the relative proportions of metal (core) and silicate (mantle plus crust), also illustrated in Figure 15.8. Because iron is by far the most abundant element in metal cores, these proportions depend on how iron is partitioned between its different oxidation states. The terrestrial planets show an inverse relationship between the mass fractions of

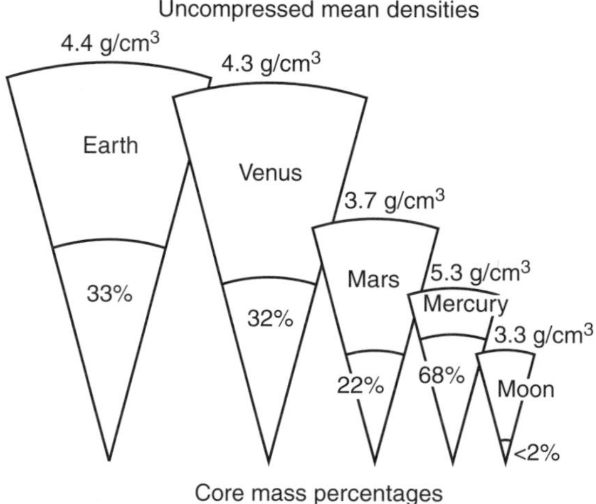

Figure 15.8 Uncompressed mean densities (numbers at top) and core fractions (expressed as volume percentages) of the terrestrial planets and the Moon (Lissauer and de Pater, 2013).

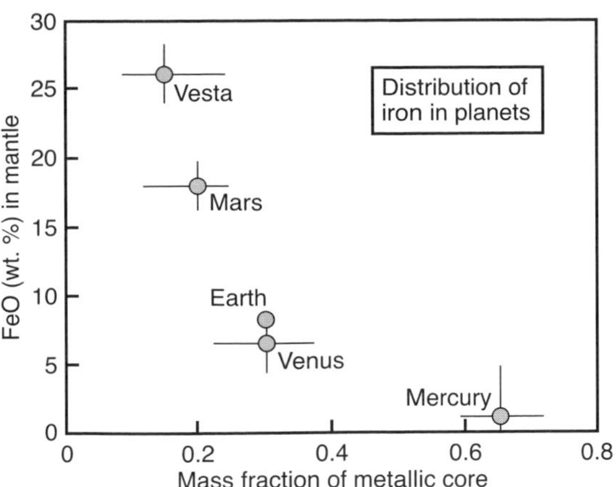

Figure 15.9 Mass fractions of cores (metallic iron) versus concentrations of FeO in mantles (oxidized iron) in the terrestrial planets and asteroid Vesta. After Righter et al. (2006).

metallic iron in cores and oxidized iron (FeO) in mantles (Fig. 15.9). A planet's *moment of inertia factor* is given by C/MR^2, where C is the moment of inertia along the rotational axis, M is mass, and R is mean radius. This factor, which can be obtained from the observed degree of flattening as the planet rotates, or from precise measurements of the precession of its rotational axis over time, provides information on how mass is radially distributed within its interior. Box 15.2 outlines how estimates of the bulk chemical compositions of planets are constrained by mean densities and moments of inertia.

The giant planets are composed mostly of hydrogen and helium. Uncompressed mean densities provide constraints on the proportion of rock to ice or gas, although the enormous internal pressures in some of these planets produce phase changes in hydrogen that complicate this determination (discussed below).

15.6.2 Models for Planet Bulk Chemistry

A number of different cosmochemical models have been devised to estimate the bulk compositions of planets for which we have very few, or more likely no samples.

The *equilibrium condensation model* assumes that solids thermally equilibrated with the surrounding nebular gas, and any uncondensed elements were somehow flushed from the system. Any planets accreted from these solids would then have compositions dictated by condensation theory (Saxena and Hrubiak, 2014). Because temperature and pressure decreased away from the Sun, the condensed solids would have varied with heliocentric distance. Figure 15.10 shows planets arranged along a presumed nebular temperature–pressure profile, with important

Box 15.2 Steps in Estimating and Testing a Planetary Bulk Composition

Planets are broadly chondritic in composition, and we can estimate the chemical composition of a planet from an accretional model using mixtures of chondritic components (explained more fully below). To a first approximation, we can ignore the crust and consider the planet to be composed of a silicate mantle and an iron metal core; alternatively, the crust and mantle can be combined into one silicate component. The mantle (or mantle plus crust) composition can be further constrained if we have igneous samples of the crust that were derived by melting the mantle. Testing our compositional model involves the following steps:

1. Calculate internal pressures as functions of depth, using the mass and radius of the planet. This calculation also requires that a temperature gradient be estimated or assumed.
2. Calculate the mass and radius of the metallic Fe (or Fe,Ni,S) core from the mean density. Subtract that much Fe (and Ni and S) from the bulk planet composition and assume the remaining Fe is FeO.
3. The remainder is the bulk silicate composition. Use that to estimate the major minerals present and their abundances as a function of depth (pressure and temperature), using mineral stabilities from experiments or thermodynamic calculations.
4. Knowing the densities of the various silicate phases and iron metal, calculate the planet's mean density from the volumetric proportions of mantle minerals and core at various depths. Calculate the moment of inertia factor using the mass, radius, and an assumed polar moment of inertia. Test the validity of the compositional model by comparing these calculated values with measured values for the planet.
5. Modestly change the bulk chemical composition (mostly involves changing the bulk Fe content or oxidation state) and iterate until the mean density and moment of inertia factor best match the measured values. The model can also be refined by adding additional layers, such as multiple mantle layers of different mineralogy and density, or a crustal layer.

state changed (probably independently) with time, and material moved inward and outward through that location. In such a dynamic system, only a general correlation between the properties of the final objects and heliocentric distance would be expected.

High-resolution chronometers (Wadhwa et al., 2006), especially that based on the ^{182}Hf–^{182}W system (Kleine et al., 2005), indicate that the earliest planetesimals were differentiated and that they formed 1–2 Myr before the chondrite parent bodies, as discussed in Chapter 10. Objects that accreted earlier would have contained more live ^{26}Al, and logically would have been heated to higher temperatures than objects that accreted later. Bottke et al. (2006) argued on dynamical grounds that the differentiated asteroids actually formed well inside the region of the present asteroid belt, probably within the terrestrial planet region. Accretion within this region was probably rapid, because of a higher density of planetesimals. Differentiated asteroids not swept up by planets subsequently drifted outward or were perturbed by encounters with planets into their present orbital locations in the Main belt. This dynamical shuffling of objects would also have blurred original correlations between properties of bodies and heliocentric distance.

The ratios of various elements and isotopes in the Earth are roughly chondritic, explaining why chondrites have commonly been considered as planetary building blocks. However, a detailed comparison of the Earth's composition with that of chondrites reveals some inconsistencies (Righter et al., 2006). For example, a plot of Mg/Si versus Al/Si, shown in Figure 15.11, reveals that no mixture of known chondrite types has the exact composition of the bulk silicate Earth (mantle plus crust), although carbonaceous chondrites provide the closest match. However, the $^{187}Os/^{188}Os$ ratio in carbonaceous chondrites is distinct from that measured for the Earth's mantle (Fig. 15.12); ordinary chondrites provide a better match for osmium isotopes. But neither carbonaceous nor ordinary chondrites can match the Earth's oxygen-isotope composition – instead, enstatite chondrites plot on the Earth's mass fractionation line; the mantle isotopic signatures of calcium, titanium, neodymium, chromium, nickel, molybdenum, and ruthenium are also most similar to enstatite chondrites (Dauphas, 2017). Thus, none of these chondrite types, or mixtures of them, consistently match the composition of the Earth, although enstatite chondrites come closest and may have been the major component.

Perhaps this is not surprising, given that recent chronological data (Chapter 10) have raised questions about the approach of using chondrites as building blocks.

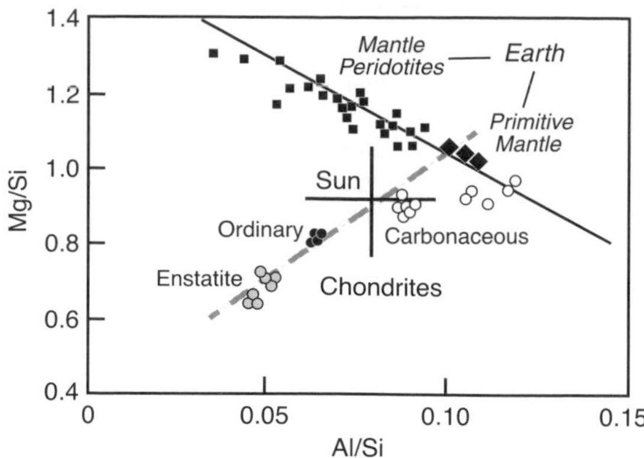

Figure 15.11 Weight ratios of Mg/Si and Al/Si for mantle rocks (peridotites) and estimates of the primitive mantle composition of the Earth compared with various groups of chondrites and the Sun (solar photosphere). No mixture of chondrite types provides an exact match to the primitive mantle composition, although some carbonaceous chondrites provide the closest match. Modified from Righter et al. (2006).

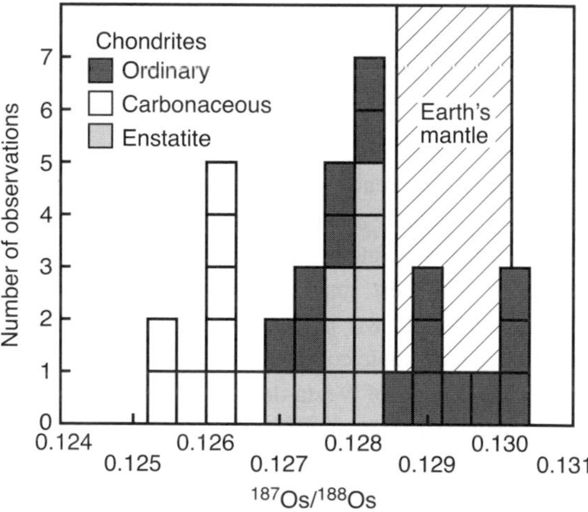

Figure 15.12 The osmium isotopic ratio of the bulk silicate Earth overlaps measurements of ordinary chondrites but is distinct from other chondrite groups. Adapted from Righter et al. (2006).

We have already noted that differentiated planetesimals formed earlier than chondritic planetesimals, and differentiated bodies probably dominated the terrestrial planet region. These bodies would have had different elemental fractionation patterns relative to the bulk solar system composition than objects formed farther from the Sun. Dynamical models suggest that the feeding zones for large bodies during much of their accretion were relatively narrow, implying that the majority of the Earth accreted from planetesimals from the same region of the solar nebula, not from the asteroid belt. It then follows that differentiated planetesimals, not chondrites, were building blocks for the terrestrial planets. This idea has much to recommend it, but we should note that the bulk compositions of differentiated planetesimals were probably chondritic in terms of many element ratios. The differences between achondrites and chondrites are primarily in the abundances of the moderately volatile elements, with differentiated materials being significantly more depleted than chondrites. The Earth, Mars, and probably the other terrestrial planets share this chemical characteristic (Yoshizaki and McDonough, 2020), likely inherited from their planetesimal building blocks.

15.7.2 Volatiles in the Terrestrial Planets

If the terrestrial planets were assembled from differentiated, volatile-depleted planetesimals located well inside the snowline, how did these planets obtain their water and other volatiles? During the early stages of accretion, models suggest that the feeding zones for planets were fairly narrow. However, during the later stages of accretion, feeding zones may have widened substantially and extensive mixing of nebular materials occurred. Alternatively, volatile-rich planetesimals in the outer accretion disk could have been scattered inward by migrating giant planets. Competing models suggest that the Earth accreted most of its water before the Moon-forming giant impact (Greenwood et al., 2018) or much later (Albarede, 2009). In either case, the composition of accreting materials likely varied with time, as planets grew larger. Even though the terrestrial planets formed mostly from already differentiated planetesimals, they probably incorporated some chondritic material with higher volatile element contents as veneers later in their assembly.

The Earth's oceans reveal an abundance of water that corresponds to ~1/1000th of the planet's mass, and rather uncertain estimates of the total planetary water content are perhaps 18 times higher (Peslier et al., 2017). Mars, too, once had liquid water that sculpted its surface, and water ice still resides at its poles and in its subsurface at high latitudes. The high D/H ratio in the atmosphere of Venus suggests that it once may have contained water in similar abundance to the Earth. Even Mercury, baking in the Sun's glare, appears to have water ice at its poles. The amounts of water in the terrestrial planets are modest, relative to the amounts of water in gas- and ice-rich planets in the outer solar system, but the importance of water for planetary habitability demands that we consider how the inner planets got their water.

Gravitational stirring of icy planetesimals by the giant planets could have sent many comets and asteroids careening into the inner solar system, providing a mechanism for late addition of water to the terrestrial planets. Comets impacting the Earth and the other terrestrial planets would have delivered water as ice (Delsemme, 1999), whereas the accretion of already altered carbonaceous chondrite asteroids would have delivered water in the form of hydroxyl-bearing minerals (Alexander et al., 2018) or organic matter. This water could have been liberated by modest heating (Nakano et al., 2020).

Noble gases may provide a constraint on the source of water and other volatiles. The abundance pattern of noble gases in planetary atmospheres resembles that of chondrites, perhaps arguing against comets. However, there are some differences, especially in the abundance of xenon. Relative to solar system abundances, krypton is more depleted than xenon in chondrites, but in the planets, krypton and xenon are present in essentially solar relative abundances (see Fig. 11.5). Measurement of cometary xenon isotopes by the Rosetta spacecraft indicates that comets provided ~22% of the xenon in the Earth's atmosphere (Marty et al., 2017). A counter-argument is that the Ar/H_2O ratio in comets (if the few available measurements are accurate and representative) limits the cometary addition of volatiles to the Earth to only ~1%.

Another relevant observation is that the D/H ratio of the Earth's oceans is roughly half that measured in comets, but is similar to that for carbonaceous chondrites (cf. Fig. 11.14). For this reason, Sarafian et al. (2014) calculated that no more than 10% of our planet's H_2O was accreted as comet ice – a conclusion consistent with dynamical constraints (Morbidelli et al., 2000). Despite the importance that D/H measurements have been given in recent publications, fractionation of these isotopes when ice is sublimated has been demonstrated experimentally. Consequently, the measurements of D/H in comet comae may not accurately reflect the D/H ratio in comet ice. We have also seen that not all comets are alike, so some as-yet-unrecognized comet reservoir might have an Earth-like hydrogen isotopic composition. O'Brien et al. (2018) further noted that the delivery efficiency of comets is very low, providing further support for delivery of water via carbonaceous chondrite accretion.

It is also possible that neither carbonaceous chondrites nor comets provided most of the water to the inner planets. Another hypothesis is that absorption of water onto dust particles in the accretion disk might account for the Earth's oceans (Drake, 2005). As already mentioned, the amount of water required to explain Earth's water is not large on a per-gram basis. Regardless of whether comets, carbonaceous asteroids, or nebular dust were the source of our planet's oceans, such models imply that water likely came from more distant regions of the nebular disk. A rival hypothesis is that water was indigenous to the inner solar system planetesimals that accreted to form the terrestrial planets. Piani et al. (2020) found that enstatite chondrites contain enough hydrogen (equivalent up to half a weight % H_2O) to account for the Earth's oceans. This idea has problems too – E chondrites are very reduced, so there must be an extra source of oxygen, and the isotopic compositions of hydrogen and nitrogen in the meteorites must be adjusted by addition of some carbonaceous chondrite.

15.7.3 Planetary Differentiation

Differentiation is among the most important geochemical processes that affected planets. We have already discussed the differentiation of planetesimals in Chapter 12; here we will consider differentiation in large planets, which differs because of the longer duration and the effects of significantly higher pressures. Specifically, we will focus on the Earth, since we know most about it. A complete discussion of the Earth's differentiation is beyond the scope of this book, and readers are referred to reviews of this topic (e.g. Tolstikhin and Kramers, 2008; Halliday and Wood, 2009) for more detailed information.

The timescale for Earth's differentiation was already discussed in Chapter 10. Tungsten isotopes indicate that the Earth's core had probably formed by ~30 Myr after CAIs (Kleine and Walker, 2017). Differentiation must have been complete before the giant impact that formed the Moon, currently constrained to have occurred ~30–60 Myr after the solar system formed (Avice and Marty, 2014).

The Earth's core is composed of iron–nickel alloy, with an inner solid core surrounded by a molten outer core. A mismatch between the inferred density of the outer core and that predicted for iron–nickel metal at high pressure suggests that some light element(s) must dilute the iron in the molten core. Some possibilities are oxygen, sulfur, silicon, and hydrogen – all elements with high cosmic abundances that can alloy with iron at very high pressures. Core formation required that the Earth was sufficiently molten that metal droplets could coalesce and sink through the silicate mantle. There is a growing consensus that the early Earth was largely melted, forming a magma ocean, although no direct evidence for that is apparent. A Moon-forming impact would have produced planet-wide melting, but even without such a catastrophic collision, short-lived radionuclides could have caused widespread melting in the early stages of accretion. In fact, magma oceans might have been commonplace on all planets. The accretion of already differentiated planetesimals would also have increased the efficiency of core separation, since large masses of metal sink more readily than small droplets.

Knowledge of the behavior of siderophile elements in iron meteorites is not necessarily a good guide to their behavior deep within the Earth, because the pressure effects inside asteroids are negligible. Experiments indicate that element partitioning between metal and silicate melts varies, depending on pressure. Siderophile-element abundances in the mantle do not match predictions from experiments at low pressures and moderate temperatures (Fig. 15.13). Righter (2011) and Carlson et al. (2014) reviewed partitioning behavior at high pressures and suggested that fractionation of metals near the bottom of a magma ocean might account for the observed partitioning. Chabot et al. (2005) found that variations in redox state might allow partitioning experiments for various elements to be reconciled. Another interesting observation is that the highly siderophile elements (elements on the right side of Fig. 15.13) show relatively constant

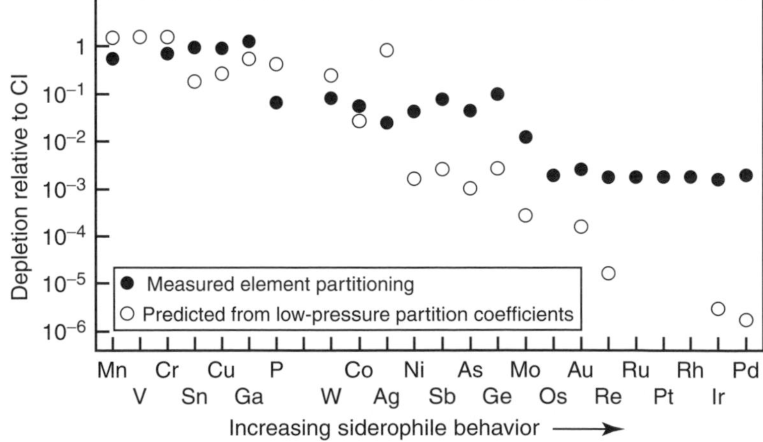

Figure 15.13 Chondrite-normalized abundances of siderophile elements in the Earth's mantle. The measured concentrations do not match those expected from low-pressure metal–silicate partition coefficients determined by experiments. Modified from Tolstikhin and Kramers (2008).

depletions, unlike the depletions expected from experiments. Some workers have suggested that equilibrium between silicate melts and metal alloys did not occur during core formation, and others hypothesize that highly siderophile elements were added as a chondritic veneer after core separation. Regardless of the exact partitioning behavior and conditions, the Earth's mantle has clearly been depleted in siderophile elements relative to chondritic abundances.

Crystallization of the magma ocean would have allowed fractionation of high-pressure silicate and oxide minerals that dominate the Earth's deep interior. The olivine-rich (silica-poor) upper mantle may owe its non-chondritic composition to fractional crystallization of bridgemanite, the high-pressure form of $MgSiO_3$. Crystallization and settling of bridgemanite would produce a complementary, silica-rich lower mantle (Agee and Walker, 1988). However, the strong and uniform fractionation of some element ratios that would be expected from crystallization of a magma ocean is not apparent on Earth. Perhaps our planet's magma ocean was vigorously stirred by convection, obliterating such geochemical effects.

Formation of the crust required repeated episodes of mantle melting, which concentrated incompatible elements to a remarkable degree. The continental crust, especially, represents the geochemical complement to the upper mantle (Fig. 15.14). The composition of the continents is nearly equivalent to andesite, the common magma type produced in subduction zones (Taylor and McLennan, 2009), implying that subduction is largely responsible for that material. The oceanic crust is basaltic in composition. Subduction recycles oceanic crust back into the mantle. In the process, partial melting produces andesitic volcanism that concentrates incompatible elements in the granitic continental crust that remains at the surface. Crust formation was probably accompanied by outgassing, as mantle-derived magmas carried dissolved gases to the surface.

Differentiation of other terrestrial planets must have varied in important ways from that of the Earth, because of differences in chemistry and conditions. For example, in Chapter 14, we learned that the crusts of the Moon and Mars are anorthosite and basalt, respectively – both very different from the crust of the Earth. Neither has experienced recycling of crust back into the mantle, because of the absence of plate tectonics, and neither has sufficient water to help drive the repeated melting events that produced the incompatible-element-rich continental crust (Taylor and McLennan, 1995). The mantles of the Moon and Mars are compositionally different from that of the Earth, although all are ultramafic. Except for these bodies, our understanding of planetary differentiation is rather unconstrained and details are speculative.

The oxidation state of precursor materials may direct the evolutionary path of planets and planetesimals as they

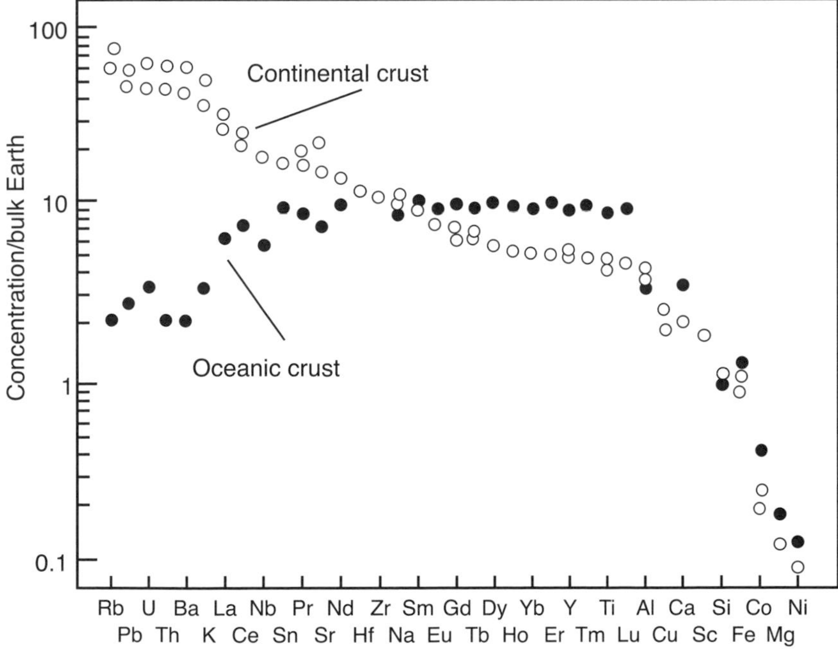

Figure 15.14 Comparison of element concentrations in the continental crust and oceanic crust. Elements are ordered from left to right by increasing compatibility, so the greatest differences are among the most incompatible elements. The highly and moderately incompatible elements in both crustal components are enriched relative to bulk Earth, but the concentrations of the most highly incompatible elements show complementary abundance patterns. Oceanic basalts are derived from the upper mantle, which has been significantly depleted in incompatible elements by the extraction of magmas that made the continental crust over billions of years. *For the colour version, refer to the plate section.*

melt and differentiate. Bodies composed of reduced materials differentiate to form cores of iron–nickel metal, mantles of magnesium-rich peridotite, and basaltic crusts. In contrast, studies of oxidized achondrites indicate that such bodies differentiate to produce cores dominated by FeNi sulfide, ferroan peridotite mantles, and more-felsic crusts (Crossley et al., 2020). Redox state can also affect the thickness of a planet's crust. More highly oxidized bodies have less iron sequestered in cores, and the more Fe^{2+}-rich mantles melt more readily, thus producing greater volumes of crust (Dyck et al., 2021).

Despite the many variations in differentiation behavior, we can make some general statements about the geochemistry of differentiated planets. The planetesimals from which planets accreted had compositions determined largely by element volatility. Once assembled into a planet and heated, the partitioning of elements into cores and mantles was governed by their siderophile or lithophile affinities, as well as oxidation state. Further differentiation of mantles to form crusts was controlled by compatible or incompatible behavior of elements.

15.8 Compositions and Differentiation of the Giant Planets

The giant planets were assembled from materials in the solar nebula, which had a bulk composition as illustrated previously in Figure 11.1. The "metals" component, comprising all elements except hydrogen and helium, can be expressed in terms of ices (water, methane, and ammonia) and rock (everything else), as illustrated in the right side of that figure. Let's see how these abundances compare to the bulk compositions of the giant planets.

To the accuracy of the measurement of molecular weights for the giant planets, only hydrogen and helium have significant abundances. The relative proportions of these elements, expressed as the molar fraction He/H, are 0.136 ± 0.003 for Jupiter, 0.135 ± 0.025 for Saturn, 0.152 ± 0.033 for Uranus, and 0.190 ± 0.032 for Neptune (Lodders, 2010; Lunine, 2014). The present-day solar He/H ratio is nearly identical to that for Jupiter, but the ratio at the time the planets formed and corrected for helium sedimentation in the Sun (taken as the nebula ratio) was ~15% higher.

The proportions of nebular gas incorporated into the giant planets, relative to accreted solids (ice plus rock), appear to decrease monotonically from Jupiter outward to Neptune. At one time it was thought that the giant planets had solar compositions and thus had formed directly from nebular gases. However, the abundances of elements heavier than hydrogen and helium are generally enriched relative to solar abundances. The Galileo probe, which was sent down into the atmosphere of Jupiter, found that heavy noble gases, expressed as [noble gas]/H_2 mixing ratios, are 2.9 for argon, 2.4 for krypton, and 2.5 for xenon, indicating that they are enriched by about a factor of ~2.5 relative to solar abundances (Lodders, 2010). (Unlike the other noble gases, neon is depleted, as explained below.) The total abundances of heavy elements in Jupiter and Saturn are estimated to be considerably greater than would be appropriate for bodies of solar composition. The overabundances of heavy elements support a core accretion model.

The outer layers of the atmospheres of Jupiter and Saturn show depletions in helium relative to hydrogen, but those of Uranus and Neptune do not (Table 15.3). This pattern may be explained by considering the equations of state of hydrogen and helium at the high pressures that obtain inside giant planets. A possible phase diagram for hydrogen is illustrated in Figure 15.15. At high pressures appropriate for the deep interiors of massive Jupiter and Saturn, hydrogen changes from molecular to metallic form. (In this case, "metallic" refers to a material in which electrons flow freely.) The presence of electrically conducting metallic hydrogen, along with planetary rotation, provides a natural dynamo mechanism to create the magnetic fields of those planets. Once the stability field for metallic hydrogen is reached, hydrogen and helium become immiscible, causing drops of helium to form and rain downward. Neon is more soluble in helium than

Table 15.3 Abundances of major elements and molecules in the atmospheres of the giant planets, relative to solar abundances (Lunine, 2014)

	Jupiter	Saturn	Uranus	Neptune
Helium	0.8	0.8	1	1.3
Methane, CH_4	2.5–3.5	7	30–60	30–60
Ammonia, NH_3	3–4	0.5–3	>1?	20–40 as N_2
Water, H_2O	>2			
Phosphine, PH_3	>0.2	6		
H-sulfide, H_2S	2.2–2.9			
Arsine, AsH_3	0.6–3	2–8		
Germane, GeH_4	0.1	6		
Neon	0.2			
Argon	2.0–3.0			
Krypton	2.2–3.2			
Xenon	1.8–3.0			

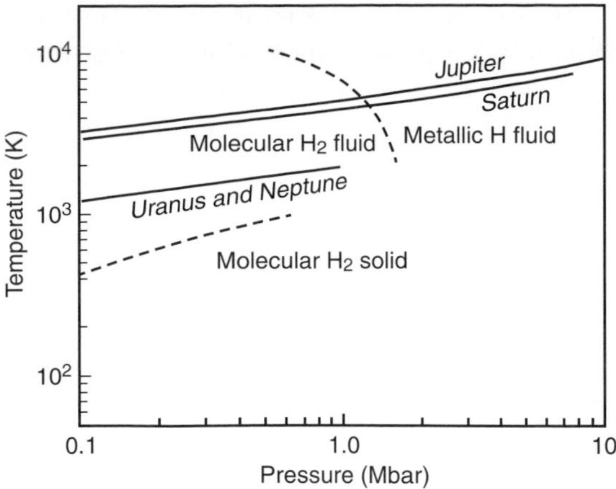

Figure 15.15 Phase diagram for hydrogen at high pressures and temperatures, showing the conditions under which hydrogen changes from molecular (H_2) to metallic (H^+). Helium becomes immiscible in metallic hydrogen fluid. Adiabats for Jupiter and Saturn cross the molecular/metallic hydrogen transition, but the Uranus and Neptune adiabats do not reach such high pressures. Modified from Lissauer and de Pater (2013).

hydrogen, so it too becomes depleted in the atmosphere (Table 15.3). This process, somewhat analogous to core separation in the terrestrial planets, fractionated hydrogen and helium (and neon) in Jupiter and Saturn. Uranus and Neptune are smaller planets, with interior pressures that do not reach those required for the formation of metallic hydrogen. Consequently, hydrogen and helium remain miscible throughout and their outer layers do not show helium depletions (Table 15.3).

Models of the interiors of the giant planets depend on assumed temperature–pressure–density relationships that are not very well constrained. Models for the gas giants, Jupiter and Saturn, feature concentric layers (from the outside inward) of molecular hydrogen, metallic hydrogen, and ice, perhaps with small cores of rock. However, theoretical models and studies of Jupiter's gravity by the Juno spacecraft suggest that the planet's core may be diffuse, with heavy elements mixed into the surrounding envelope (Helled and Stevenson, 2017; Mazevet et al., 2019). It is also possible that the solid cores of giant planets nucleate and grow over time as the planets cool. Models for the ice giants, Uranus and Neptune, are similar, except that there is no metallic hydrogen, the interior layers of ice (likely supercritical fluid at depth) are thicker, and the rocky cores are relatively larger. Based on mean densities, Jupiter contains 11–42 Earth masses of elements heavier than hydrogen and helium, and Saturn <22 Earth masses.

The interior compositions of giant planets are speculative, but we can actually measure the chemical compositions of their atmospheres (Table 15.3). However, detailed measurements from spacecraft only exist for Jupiter. Because hydrogen dominates, the other elements occur in reduced molecular forms – carbon as CH_4, nitrogen as NH_3, oxygen as H_2O, sulfur as H_2S, arsenic as AsH_3, phosphorus as PH_3, and so forth. However, less-reduced molecules like CO have also been detected, perhaps dredged up from more highly oxidized regions in the interior or produced by photochemistry in the atmosphere. Photochemical reactions also produce some hydrocarbons such as ethane (C_2H_6) and acetylene (C_2H_2). The atmospheres of the giant planets are highly reducing because of their high hydrogen abundances, so complex organics are recycled back to methane on timescales that are short relative to the age of the planets. Consequently, organic molecules in the giant planets cannot provide information on their sources, nor do they increase in complexity over time. Like the heavy noble gases, carbon and nitrogen are enriched in the atmosphere (Table 15.3), further indicating that Jupiter does not have solar composition.

Isotopic abundances for hydrogen have been measured in giant planet atmospheres, as shown in Figure 11.14. The D/H ratios in Jupiter and Saturn are similar to those in the Sun, but lower than those in the Earth's oceans or in comets. D/H ratios in Uranus and Neptune fall between the Jupiter-Saturn values and comet values. The decrease in the amount of gas accreted relative to ice plus rock from Jupiter to Neptune appears to be reflected in the varying D/H ratios, assuming that the accreted ice component had a hydrogen isotopic composition similar to comets.

15.9 Orbital and Collisional Evolution of the Solar System

No description of our solar system's formation would be complete without a discussion of the profound changes wrought by its orbital evolution and collisional history. Although these physical processes may not seem to be related to cosmochemistry, they have changed the spatial distribution of planets and small bodies of differing compositions within the solar system, and in some cases, even altered the bulk compositions of large bodies. Understanding these processes can help us appreciate how some of the cosmochemical conundrums and complexities of the solar system arose.

15.9.1 Giant Impact Models for the Moon and Mercury

First let's consider major collisional events. One such impact is thought to have affected the early Earth. As mentioned in Chapter 14, a glancing blow to the Earth from a Mars-sized body (Theia) is thought to have produced a huge volume of fragments and vaporized materials that went into orbit around the Earth and rapidly reaccreted to form the Moon. Cosmochemical evidence in support of this giant impact hypothesis includes the Moon's bulk depletion in iron (estimates of its abundance vary between 8% and 12%) relative to the Earth (31% iron). Models suggest that the metallic core of the impactor would have accreted to the Earth, and the Moon would have been formed primarily from silicates from the impactor and dislodged from the Earth's mantle. The similar oxygen-isotope composition of the Earth and Moon has been frequently cited as evidence for a close relationship between the colliding bodies; however, as discussed in Box 8.2, it is not clear why the proto-Earth and Theia should have shared a common isotopic composition. In any case, newer oxygen-isotope measurements indicate slightly different compositions for the Earth and Moon (Cano et al., 2020).

Collisions can either result in net accretion, as demonstrated by the assembly of planets from planetesimals, or disruption, as in the case of the formation of the Moon. Even accretional collisions can redistribute materials far and wide, altering and perhaps blurring the distributions of compositionally distinct surface materials and exposing subsurface compositions.

Another example of a massive impact with cosmochemical consequences may be Mercury. The abnormally massive core of that planet may result from a large collision that stripped off mantle material and left a planet with nonchondritic abundances of siderophile versus lithophile elements (Brown and Elkins-Tanton, 2009).

15.9.2 Migrations of the Giant Planets

The *Grand Tack model* (Walsh et al., 2011) posits the inward and then outward drift of Jupiter and Saturn during the nebula phase. The model's name alludes to a sailing maneuver that permits a vessel to sail against the wind. Gravitational interactions of these massive planets with planetesimals would have scrambled the orbits of the smaller bodies and ejected many of them from their formation locations in the accretion disk. This migration has also been hypothesized to have caused the inward scattering of carbonaceous chondrite bodies, implanting them in the outer Main belt and allowing the late accretion of volatile-bearing veneers onto the growing terrestrial planets.

The *Nice model* (named for the location in France where it was originally formulated in a trio of papers published simultaneously; see Gomes et al., 2005) describes later wanderings of giant planets. It suggests that as the orbits of Jupiter and Saturn evolved, due to gravitational interactions between them and a massive disk of planetesimals, they passed through a state in which Jupiter orbited the Sun exactly twice for each orbit of Saturn (a 1:2 mean motion resonance). This resonance caused their orbits to become eccentric and brought them closer to Uranus and Neptune. This later migration of the giant planets would have disrupted the orbits of bodies in the asteroid belt, sending them careening into the inner solar system. Orbital migration of the giant planets may likewise have stirred the legions of icy planetesimals in the outer solar system into a frenzy, scattering those that originated near Jupiter and Saturn outward to form the Oort cloud and causing the mean radius of the Kuiper belt to increase. Perhaps the dark, organic-rich bodies of the outer asteroid belt may even be interlopers that were scattered inward by migrating giant planets, complicating the meaning of the snowline.

In any case, if the locations and distributions of objects in the present solar system are not original, this would have profound implications for any inferences drawn from the chemical compositions, radioisotope chronologies, and geologic histories of planets and asteroids now at different heliocentric distances. Planetary dynamics calculations are providing a new paradigm for how the early solar system evolved and that, in turn, affects how the discoveries of cosmochemistry are interpreted.

15.9.3 A Late Heavy Bombardment?

The huge basins that occur on the lunar surface have been described as a result of the *late heavy bombardment*. U–Pb analyses of Apollo lunar samples yield a Concordia plot array with intercepts at 4.4 and 3.9 Ga, and Tera et al. (1974) interpreted the younger age as being due to thermal resetting during a terminal cataclysm. Support for such an event comes from Rb/Sr and $^{40}Ar/^{39}Ar$ ages for lunar rocks (Bogard, 1995), which show peaks at ~3.9 Ga (Fig. 15.16), and similar peaks in the $^{40}Ar/^{39}Ar$ ages of HED meteorites from Vesta. There has been considerable debate questioning whether this bombardment was actually a punctuated event of global scale on the Moon (Hopkins and Mojzsis, 2015; Bottke and Norman, 2017). The Apollo and Luna samples all come from the Moon's nearside – most derived from the two largest proximal basins – and their locations represent only a few percent of the surface area. Lunar meteorites, which likely represent broader sampling, show no 3.9 Ga spike

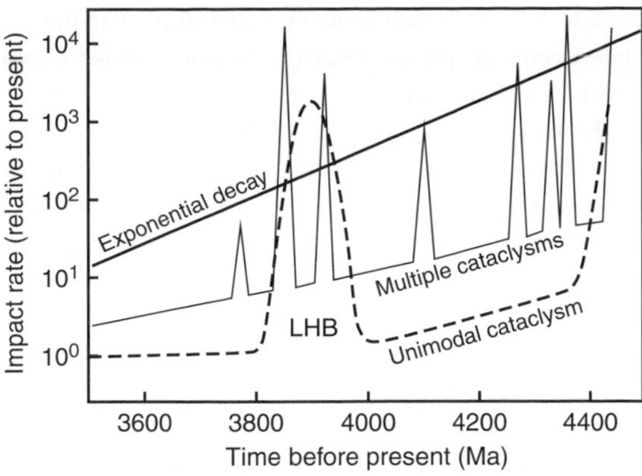

Figure 15.16 A representation of various lunar bombardment chronologies. The dashed line represents a spike in giant impacts at ~3.9 Ga (the late heavy bombardment, LHB), whereas the multiple-cataclysms model illustrates several bombardment episodes. The solid line labeled "exponential decay" represents the progressive exhaustion of impactors over time and is calibrated with crater counts and ages from Apollo landing sites. Modified from Hopkins and Mojzsis (2015).

(Cohen et al., 2005). Countering that is the observation that Mars and Mercury show evidence for large impacts of similar age, and globally distributed spherule beds on Earth suggest bombardment of our planet as well. However, crater-counting ages for the other planets are based on lunar chronology, which may introduce an apparent concordance. Another possibility is that the late heavy bombardment represents the exponential waning of solar-system-wide, protracted accretion (Fig. 15.16).

The Nice model, described above, would have scattered objects in the asteroid belt into the inner solar system to impact the terrestrial planets (Tsiganis et al., 2005; Minton and Malhotra, 2011). Such an event could have left a record in the form of the craters that we assign to the late heavy bombardment. Alternatively, there may have been multiple spikes in impact basin formation (Fig. 15.16), perhaps resulting from the terminal phase of planetesimal accretion and from giant planet migration (Bottke and Norman, 2017).

15.10 Inferring the Compositions of Exoplanets

Over the last several decades, more than 4000 planets around other stars (*exoplanets*) have been confirmed. Originally detected from the wobbles of stars caused by nearby orbiting planets, most newer discoveries have been made by stellar partial eclipses as exoplanets transited in front of stars, as viewed by the Kepler orbiting space telescope (Batalha, 2014). Systems of multiple planets are commonplace. Planetary orbits are usually coplanar, consistent with their formation in accretion disks (like our solar system's ecliptic), but some systems are misaligned (i.e. their orbital planes are at large angles with respect to the stellar rotation plane), and some planets even orbit around stellar poles. Exoplanets come in a bewildering variety of sizes, but most are smaller than Neptune and larger than Earth – a size absent from our solar system (Jontof-Hutter, 2019). Many occur in extreme and unstable orbits, supporting the idea of planetary migrations. In fact, gas giants orbiting very close to their parent stars ("hot Jupiters") require inward migration. Our solar system, with its small, rocky planets inside the snowline and giant, volatile-rich planets farther from the Sun, apparently is not a universal stellar-system architecture (He et al., 2019).

Kepler directly measured the diameters of exoplanets, as well as indirectly determined their masses in multiplanet systems by analyzing how their gravity fields disturb otherwise elliptical orbits. Measurements of sizes and masses for ~200 exoplanets are compared with models for planets consisting of pure hydrogen, water, rock (assumed to be olivine), and iron metal in Figure 15.17. The data are biased towards larger planets, which are easier to measure. The planets in our solar system are also plotted, and their similarities support the idea that they may be compositionally analogous.

From Figure 15.17, we can see that the smallest exoplanets, ranging up to 1.5 times Earth's radius, generally have mean densities (>5 g/cm^3) consistent with rocky bodies with metal cores. Their densities increase with size, explainable by gravitational compression (Marcy et al., 2014). Many larger "sub-Neptunes," with radii 1.5–4 times that of Earth, have densities that resemble Uranus and Neptune (1.27 and 1.63 g/cm^3, respectively), suggesting ices and gases surrounding rocky cores. However, planets in this size range show a great variety of densities, implying diverse compositions (Jontoff-Hutter, 2019). Even larger exoplanets are gas giants like Jupiter and Saturn, with mean densities near 1.0 g/cm^3. Inferences about the compositions and internal structures of exoplanets of necessity rely heavily on cosmochemical reasoning and knowledge of planets in our own solar system (Spiegel et al., 2014).

We previously noted that astronomers refer to elements heavier than hydrogen and helium as "metals." Exoplanet studies reveal that giant planets are more likely to occur around stars with high metallicity. This observation is consistent with the core accretion model of planet formation, whereby planet cores form from solids in the

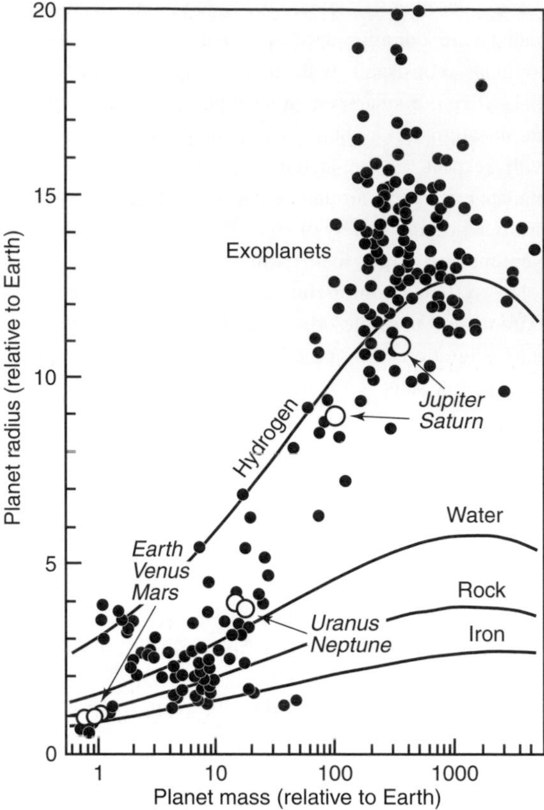

Figure 15.17 Exoplanet radius versus mass, compared with density models for planets composed of hydrogen, water, rock (olivine), and iron metal. The planets of our solar system are labeled. Adapted from McSween et al. (2019).

accretion disk, followed by runaway growth and direct capture of nebular gas. Smaller, rocky planets are found around stars with a range of metallicity.

Elements in the atmospheres of exoplanets can sometimes be identified from the difference between the combined spectra of the planet and its star versus the spectrum of the star alone. Although these spectra are not very detailed, they have allowed sodium and molecules of H_2O and CO to be detected (e.g. de Wit et al., 2016).

Most of the exoplanets discovered so far are around main-sequence stars. In Chapter 3, we saw that once such stars exhaust the hydrogen fuel in their cores, they expand enormously, expel their outer layers in strong stellar winds, and eventually become white dwarfs. In the process, they shatter close-by planets. The remnants of the planetary system are ground into debris disks, surrounding what are known as "polluted" white dwarfs. This may be the ultimate fate of our solar system. Manser et al. (2019) found a planetesimal orbiting within the debris disk around a white dwarf. Its estimated high density is consistent with a naked metal core, and further studies of such bizarre exoplanets may provide new insights into planetary differentiation.

Kepler ceased operations in 2013. Future exoplanet discoveries are expected from the Transiting Exoplanet Survey Satellite (TESS), the Characterizing Exoplanet Satellite (CHEOPS), both already launched, and the James Webb Space Telescope.

Summary

Our general understanding of how the solar system came to be comes from a variety of different sources. Astronomical observations and theoretical modeling provide a broad picture of how stars form. The abundances of elements and short-lived radionuclides in chondritic meteorites tell us about the environment in which the solar system formed. Based on the above considerations, the Sun likely formed in a region of cluster star formation in association with one or more nearby supernovae.

The formation and evolution of our solar system's early accretion disk can be modeled theoretically, but the models depend on constraints from astronomical observations and cosmochemistry. It is now recognized that the material that comprises asteroids (and meteorites) did not experience global temperatures in the nebula high enough to evaporate all pre-existing solids. The chemical properties of solar system objects (e.g. planets, asteroids, comets) reflect thermal and physical processing in the Sun's accretion disk. Isotopic anomalies observed in meteorites were primarily inherited from the Sun's parent molecular cloud, and we observe them either because we measure presolar grains that survived to carry the isotopic composition of the ejecta of dying stars into the solar system, or because solar system processes have separated different isotopic components, leaving isotopically unrepresentative samples for us to measure. The compositions of planets and meteorites so far analyzed exhibit a pronounced dichotomy in nucleosynthetic isotopes and other chemical parameters. The distinct compositions of non-carbonaceous and carbonaceous bodies may reflect the physical separation of reservoirs in the inner and outer solar system, or differences in processing pathways at different times or in different locations within the disk.

CAIs were the earliest-formed solar system solids. They apparently formed close to the Sun and were transported to where the meteorites formed by either viscous spreading of the disk or by bipolar outflows. Melting of the ubiquitous chondrules requires rapid heating and cooling events that postdate the general chemical fractionations observed in meteorites. The nature of the heating mechanism that formed chondrules has not been definitively identified.

The formation of the terrestrial planets is constrained by knowledge of their bulk chemical compositions. Obtaining these data is challenging because planetary differentiation ensures that no samples have the composition of the whole. The Earth's composition has been estimated by integrating analyses of crust and mantle samples with inferences about the composition of the core, as constrained by seismic data. For planets other than the Earth, we rely on cosmochemical models constrained by measurements of bulk density and moment of inertia. Although planet bulk compositions are commonly modeled using chondrites as building blocks, the early chronology of achondrites and irons suggests that the planetesimals that accreted to form the terrestrial planets were already differentiated. Relative to chondrites, differentiated meteorites are depleted in moderately volatile elements – a characteristic shared by the terrestrial planets. Late accretion of volatile-rich materials (chondritic asteroids or comets) may be required to account for water in the terrestrial planets. Differentiation of the planets involved complex geochemical fractionations under varying conditions, leading to a variety of core, mantle, and crust compositions. Separation of cores and mantles involved pressure-dependent partitioning of siderophile, chalcophile, and lithophile elements, and formation of crusts concentrated incompatible elements.

The giant planets are composed mostly of hydrogen and helium, along with smaller amounts of heavier elements, accreted as gas, ices, and dust. Their bulk compositions are not well constrained, but limited analyses of atmospheres and inferences from bulk densities suggest that all of them have lower H/He ratios than the nebula gas and are enriched in heavy elements relative to cosmic abundances. The gas giants formed rapidly and early, allowing them to accrete nebular gas directly. Fractionation in gas giants has occurred because high pressures produced metallic hydrogen, in which helium became immiscible and separated. The gas giants may have rocky cores or internally dispersed concentrations of heavy elements. The ice giants formed later after gases had condensed to ices. Ice giants are not massive enough to form metallic hydrogen, but they apparently contain rocky cores.

Orbital migrations of the giant planets, such as described by the Grand Tack and Nice models, may have led to the ejection of most asteroids from the Main belt, the implantation of carbonaceous asteroids into the Main belt, the scattering of icy planetesimals into the outer reaches of the solar system, and the bombardment of planets by large impactors. Gravitational interactions with massive planets may thus have altered primordial cosmochemical distributions with heliocentric distance.

Exoplanets show great diversity in composition, size, and distribution, but rocky planets, ice giants, and gas giants appear to be represented in other planetary systems. Understanding exoplanets draws heavily from cosmochemistry and from the exploration of planets in our solar system.

Questions

1. What is the evidence that our Sun formed in the vicinity of other stars and was influenced by a nearby supernova?
2. Discuss the evolution of ideas about temperatures in the solar nebula, and how that relates to the formation of CAIs.
3. What constraints do we have on the bulk chemical compositions of planets?
4. Describe the building blocks that accreted to form the terrestrial planets, and explain how that may relate to their observed volatile element depletions.
5. How do the chemical compositions and internal structures of the giant planets differ from those of the terrestrial planets and from each other?
6. What aspects of cosmochemistry are used to infer the compositions and internal structures of exoplanets?

Suggestions for Further Reading

Halliday A. N. (2014) The origin and earliest history of the Earth. In *Treatise on Geochemistry, 2nd edition, Vol. 2: Planets, Asteroids, Comets and the Solar System*, Davis A. M., editor, pp. 149–211, Elsevier, Oxford. A thoughtful summary of a very large topic.

Hester J. J., and Desch S. J. (2005) Understanding our origins: Star formation in H II region environments. In *Chondrites and the Protoplanetary Disk*, ASP Conference Series, **341**, Krot A. N., Scott E. R. D., and Reipurth A., editors, pp. 107–130, Astronomical Society of the Pacific, San Francisco. A clear review of astronomical observations that constrain models for solar system formation.

Taylor S. R. (1992) *Solar System Evolution: A New Perspective.* Cambridge University Press, Cambridge, 307 pp. A thorough treatise that covers ideas about the origin and evolution of the terrestrial planets.

Other References

Agee C. B., and Walker D. (1988) Mass balance and phase density constraints on early differentiation of chondritic mantle. *Earth & Planetary Science Letters*, **90**, 144–156.

Albarède F. (2009) Late accretion of volatiles to Earth. *Nature*, **461**, 1227–1233.

Alexander C. M. O'D., McKeegan K. D., and Altweg K. (2018) Water reservoirs in small planetary bodies: Meteorites, asteroids, and comets. *Space Science Reviews*, **214,** 36.

Andrews S. (2018) Focus on DSHAP results. *Astrophysical Journal Letters*, **869**, L41–L50 (introduction to a series of 10 papers).

Avice G., and Marty B. (2014) The iodine-plutonium-xenon age of the Moon-Earth system revisited. *Philosophical Transactions of the Royal Society A*, **372**, 20130260.

Barshay S. S., and Lewis J. S. (1976) Chemistry of primitive solar material. *Annual Reviews of Astronomy & Astrophysics*, **14**, 81–94.

Basaltic Volcanism Study Project (1981) *Basaltic Volcanism on the Terrestrial Planets*. Pergamon Press, New York, 1286 pp.

Batalha N. M. (2014) Exploring exoplanet populations with NASA'S Kepler mission. *Proceedings of the National Academy of Sciences, USA*, **111**, 647–654.

Black D. C., and Pepin R. O. (1969) Trapped neon in meteorites, II. *Earth & Planetary Science Letters*, **6**, 395–405.

Bogard D. D. (1995) Impact ages of meteorites: A synthesis. *Meteoritics*, **30**, 244–268.

Boss A. P. (2004) Convective cooling of protoplanetary disks and rapid giant planet formation. *Astrophysical Journal*, **610**, 456–463.

Bottke W. F., Nesvorny D., Grimm R. E., et al. (2006) Iron meteorites as remnants of planetesimals formed in the terrestrial planet region. *Nature*, **439**, 821–824.

Bottke W. F, and Norman M. D. (2017) The Late Heavy Bombardment. *Annual Reviews of Earth & Planetary Sciences*, **45**, 619–647.

Brown S. M., and Elkins-Tanton L. T. (2009) Compositions of Mercury's earliest crust from magma ocean models. *Earth & Planetary Science Letters*, **286**, 446–455.

Burkhardt C., Dauphas N., Hans U., et al. (2019) Elemental and isotopic variability in solar system materials by mixing and processing of primordial disk reservoirs. *Geochimica et Cosmochimica Acta*, **261**, 145–170.

Cameron A. G. W. (1962) The formation of the sun and planets. *Icarus*, **1**, 13–69.

Cameron A. G. W. (1978) Physics of the primitive solar accretion disk. *Moon & Planets*, **18**, 5–40.

Cameron A. G. W., and Truran J. W. (1977) The supernova trigger for formation of the solar system. *Icarus*, **30**, 447–461.

Cano E. J., Sharp Z. D., and Shearer C. K. (2020) Distinct oxygen isotope compositions of the Earth and Moon. *Nature Geoscience*, **13**, 270–274.

Carlson R. W., Garnero E., Harrison T. M., et al. (2014) How did early Earth become our modern world? *Annual Reviews of Earth & Planetary Sciences*, **42**, 151–178.

Chabot N. L., Draper D. S., and Agee C. B. (2005) Conditions of core formation in the Earth: Constraints from nickel and cobalt partitioning. *Geochimica et Cosmochimica Acta*, **69**, 2141–2151.

Ciesla F. J. (2007) Outward transport of high-temperature materials around the midplane of the solar nebula. *Science*, **318**, 613–615.

Clayton R. N., Grossman L., and Mayeda T. K. (1973) A component of primitive nuclear composition in carbonaceous meteorites. *Science*, **182**, 485–487.

Clayton R. N., Onuma N., Grossman L., and Mayeda T. K. (1977) Distribution of the pre-solar component in Allende and other carbonaceous chondrites. *Earth & Planetary Science Letters*, **34**, 209–224.

Cohen B. A., Swindle T. D., and Kring D. A. (2005) Geochemical and $^{40}Ar/^{39}Ar$ geochronology of impact-melt clasts in feldspathic lunar meteorites: Implications for lunar bombardment history. *Meteoritics & Planetary Science*, **40**, 755–777.

Connolly H. C. Jr., and Jones R. H. (2016) Chondrules: The canonical and noncanonical views. *Journal of Geophysical Research, Planets*, **121**, 1885–1899.

Crossley S. D., Ash R. D., Sunshine J. M., et al. (2020) Sulfide-dominated partial melting pathways in brachinites. *Meteoritics & Planetary Science*, **55**, 2021–2043.

Cuzzi J. N., and Weidenschilling S. J. (2006) Particle-gas dynamics and primary accretion. In *Meteorites and the Early Solar System II*, Lauretta D. S. and McSween H. Y., editors, pp. 353–381, University of Arizona Press, Tucson.

Dauphas N. (2017) The isotopic nature of the Earth's accreting material through time. *Nature*, **541**, 521–524.

Desch S. J., and Cuzzi J. N. (2000) The generation of lightning in the solar nebula. *Icarus*, 143, 87–105.

Desch S. J., and Connolly H. C. (2002) A model of the thermal processing of particles in solar nebula shocks: Application to the cooling rates of chondrules. *Meteoritics & Planetary Science*, **37**, 183 207.

Desch S. J., Morris M. A., Connolly H. C., and Boss A. P. (2010) A critical examination of the X-wind model of chondrule and CAI formation and radionuclide production. *Astrophysical Journal*, **725**, 692–711.

Delsemme A. H. (1999) The deuterium enrichment observed in recent comets is consistent with the cometary origin of seawater. *Planetary & Space Science*, **47**, 125–131.

de Wit J., Wakeford H.R., Gillon M., et al. (2016) A combined transmission spectrum of the Earth-sized exoplanets TRAPPIST-1 b and c. *Nature*, **537**, 69–72.

Drake M. J. (2005) Origin of water in the terrestrial planets. *Meteoritics & Planetary Science*, **40**, 519–527

Dyck B., Wade J., and Palin R. (2021) The effect of core formation on surface composition and planetary habitability. *Astrophysics Journal Letters*, arxiv.org/pdf/2104.10612.pdf.

Fegley B. F., and Cameron A. G. W. (1987) A vaporization model for iron/silicate fractionation in the Mercury protoplanet. *Earth & Planetary Science Letters*, **82**, 207–222.

Goettel K. A. (1988) Present bounds on the bulk composition of Mercury: Implications for planetary formation processes. In *Mercury*, Vilas F., Chapman C. R., and Matthews M. S., editors, pp. 613–621, University of Arizona Press, Tucson.

Gomes R., Levison H. F., Tsiganis K., and Morbidelli A. (2005) Origin of the cataclysmic late heavy bombardment period of the terrestrial planets. *Nature*, **435**, 466–469.

Gounelle M., Shu F. H., Shang H., et al. (2001) Extinct radioactivities and protosolar cosmic-rays: Self-shielding and light elements. *Astrophysical Journal*, **548**, 1051–1070.

Greenwood R. C., Barrat J.-A., Miller M. F., et al. (2018) Oxygen isotopic evidence for accretion of Earth's water before a high-energy Moon-forming giant impact. *Science Advances*, **4**, eaao5928.

Grossman J. N., Rubin A. E., Nagahara H., and King E. A. (1988) Properties of chondrules. In *Meteorites and the Early Solar System*, Kerridge J. M., and Mathews M. S., editors, pp. 619–659, University of Arizona, Tucson.

Halliday A. N., and Wood B. J. (2009) How did Earth accrete? *Science*, **325**, 44–45.

He M. Y., Ford E. B, and Ragozzine D. (2019) Architectures of exoplanetary systems – I. A clustered forward model for exoplanetary systems around Kepler's FGK stars. *Monthly Notices of the Royal Astronomical Society*, **490**, 4575–4605.

Helled R., and Stevenson D. (2017) The fuzziness of giant planets' cores. *Astrophysical Journal Letters*, **840**, L4.

Hopkins M. D., and Mojzsis S. J. (2015) A protracted timeline for lunar bombardment from mineral chemistry, Ti thermometry and U-Pb geochronology of Apollo 14 melt breccia zircons. *Contributions to Mineralogy & Petrology*, **169**, 1–18.

Humayan M., and Cassen P. (2000) Processes determining the volatile abundances of the meteorites and terrestrial planets. In *Origin of the Earth and Moon*, Canup R. M., and Righter K., editors, pp. 3–23, University of Arizona Press, Tucson.

Huss G. R. (1988) The role of presolar dust in the formation of the solar system. *Earth, Moon & Planets*, **40**, 165–211.

Huss G. R. (1990) Ubiquitous interstellar diamond and silicon carbide in primitive chondrites: Abundances reflect metamorphism. *Nature*, **347**, 159–162.

Johnson B. C., Ciesla F. J., Dullemond C. P., and Melosh H. H. (2018) Formation of chondrules by planetesimal collisions. In *Chondrules, Records of Protoplanetary Disk Processes*, Russell S. S., Connolly H. C. Jr., and Krot A. N., editors, pp. 343–360, Cambridge University Press, Cambridge.

Jontoff-Hutter D. (2019) The compositional diversity of low-mass exoplanets. *Annual Reviews of Earth & Planetary Sciences*, **47**, 141–171.

Kargel J. S., and Lewis J. S. (1993) The composition and early evolution of Earth. *Icarus*, **105**, 1–25.

Kleine T., and Walker R. J. (2017) Tungsten isotopes in planets. *Annual Reviews of Earth & Planetary Science*, **45**, 389–417.

Kleine T., Mezger K., Palme H., et al. (2005) Early core formation in asteroids and late accretion of chondrite parent bodies: Evidence from ^{182}Hf-^{182}W in CAIs, metal-rich chondrites and iron meteorites. *Geochimica et Cosmochimica Acta*, **69**, 5805–5818.

Kleine T., Budde G., Burkhardt C., et al. (2020) The non-carbonaceous – carbonaceous meteorite dichotomy. *Space Science Reviews*, **216**, 55, doi.org/10.1007/s11214-020-00675-w.

Krot A. N., Amelin Y., Cassen P., and Meibom A. (2005) Young chondrules in CV chondrites from a giant impact in the early solar system. *Nature*, **436**, 989–992.

Krot A. N., Nagashima K., Libourel G., and Miller K. (2018) Multiple mechanisms of transient heating events in the protoplanetary disk. In *Chondrules, Records of Protoplanetary Disk Processes*, Russell S. S., Connolly H. C. Jr., and Krot A. N., editors, pp. 11–56, Cambridge University Press, Cambridge.

Kruijer T. S., Burkhardt C., Budde G., and Kleine T. (2017) Age of Jupiter inferred from distinct genetics and formation times of meteorites. *Proceedings of the National Academy of Sciences, USA*, **114**, 6712–6716.

Kruijer T. S., Kleine T., and Borg L. R. (2020) The great isotopic dichotomy of the early solar system. *Nature Astronomy*, **4**, 32–40.

Lambrechts M., and Johansen A. (2012) Growth of gas-giant cores by pebble accretion. *Astronomy & Astrophysics*, **544**, A32.

Lee T., Papanastassiou D. A., and Wasserburg G. J. (1977) Aluminum-26 in the early solar system: Fossil or fuel? *Astrophysical Journal Letters*, **211**, L107–L110.

Levy E. H. (1988) Energetics of chondrule formation. In *Meteorites and the Early Solar System*, Kerridge J. F., and Matthews M. S., editors, pp. 697–711, University of Arizona Press, Tucson.

Lewis R. S., Tang M., Wacker J. F., et al. (1987) Interstellar diamonds in meteorites. *Nature*, **326**, 160–162.

Liffman K., and Brown M. (1995) The motion and size sorting of particles ejected from a protostellar accretion disk. *Icarus*, **116**, 275–290.

Lissauer J. J., and de Pater, I. (2013) *Fundamental Planetary Science: Physics, Chemistry and Habitability*. Cambridge University Press, Cambridge, 583 pp.

Lodders K. (2010) Atmospheric chemistry of the gas giant planets. *Geochemical News*, **142**, geochemsoc.org/publications/geochemicalnews/gn142.

Lodders K., and Fegley B. F. (1998) The Planetary Scientist's Companion. Oxford University Press, New York, 371 pp.

Love S. G., Keil K., and Scott E. R. D. (1995) Electrical discharge heating of chondrules in the solar nebula. *Icarus*, **115**, 97–108.

Lunine J. I. (2014) Giant planets. In *Treatise in Geochemistry, 2nd edition, Vol. 2: Planets, Asteroids, Comets and the Solar System*, Davis A. M., editor, pp. 301–311, Elsevier, Amsterdam.

MacPherson G. J., Simon S. B., Davis A. M., et al. (2005) Calcium-aluminum-rich inclusions; Major unanswered questions. In *Chondrites and the Protoplanetary Disk*, ASP Conference Series, **341**, Krot A. N., Scott E. R. D., and Reipurth A., editors, pp. 225–250, Astronomical Society of the Pacific, San Francisco.

Manser C. J., Gansicke B. T., Eggl S., et al. (2019) A planetesimal orbiting within the debris disk around a white dwarf star. *Science*, **364**, 66–69.

Marcy G. W., Weiss L. M., Petigura E. K., et al. (2014) Occurrence and core-envelope structure of 1-4x Earth-size planets around Sun-like stars. *Proceedings of the National Academy of Sciences USA*, **111**, 655–660.

Marty B., Altwegg K., Balsiger H., et al. (2017) Xenon isotopes in 67P/Churyumov-Gerasimenko show that comets contributed to Earth's atmosphere. *Science*, **356**, 1069–1073.

Mazevet S., Musella R., and Guyot F. (2019) The fate of planetary cores in giant and ice-giant planets. *Astronomy & Astrophysics*, **631**, L4.

McCulloch M. T., and Wasserburg G. J. (1978a) Barium and neodymium isotopic anomalies in the Allende meteorite. *Astrophysical Journal Letters*, **220,** L15–L19.

McCulloch M. T., and Wasserburg G. J. (1978b) More anomalies from the Allende meteorite: Samarium. *Geophysical Research Letters*, **5**, 599–602.

McKeegan K. D., Chaussidon M., and Robert F. (2000) Incorporation of short-lived ^{10}Be in a calcium-aluminum-rich inclusion from the Allende meteorite. *Science,* **289**, 1334–1337.

McKeegan K. D., Kallio A. P. A., Heber V. S., et al. (2011) The oxygen isotopic composition of the Sun inferred from captured solar wind. *Science*, **332**, 1528–1532.

McSween H. Y., Moersch J. E., Burr D. M., et al. (2019) *Planetary Geoscience*. Cambridge University Press, Cambridge, 334 pp.

Minton D. A., and Malhotra R. (2011) Secular resonance sweeping of the main asteroid belt during planet migration. *Astrophysical Journal*, **732**, 53–64.

Morbidelli A., Chambers J., Lunine J. I., et al. (2000) Source regions and timescales for the delivery of water to the Earth. *Meteoritics & Planetary Science*, **35**, 1309–1320.

Morbidelli A., and Raymond S. N. (2016) Challenges in planet formation. *Journal of Geophysical Research, Planets*, **121**, 1962–1980.

Morbidelli A., Lunine J. I., O'Brien D. P., et al. (2012) Building terrestrial planets. *Annual Reviews of Earth & Planetary Sciences*, **40**, 251–275.

Morbidelli A., Lambrechts M., Jacobson S., and Bitsch B. (2015) The great dichotomy of the solar system: Small terrestrial embryos and massive giant planet cores. *Icarus*, **258**, 418–429.

Morris M. A., and Boley A. C. (2018) Formation of chondrules by shock waves. In *Chondrules, Records of Protoplanetary Disk Processes*, Russell S. S., Connolly H. C. Jr., and Krot A. N., editors, pp. 375–399, Cambridge University Press, Cambridge.

Nakano H., Hirakawa H., Natsubara Y., et al. (2020) Precometary organic matter: A hidden reservoir of water inside the snow line. *Scientific Reports*, **10**, doi:10.1038/s41598–020-64815-6.

O'Brien D. P., Izidoro A., Jacobson S. A., et al. (2018) The delivery of water during terrestrial planet formation. *Space Science Reviews*, **214**, 47, doi.org/10.1007/s11214–018-0475-8.

Ogliore R. C., Huss G. R., and Nagashima K. (2011) Ratio estimation in SIMS analysis. *Nuclear Instruments & Methods in Physics Research B*, **269**, 1910–1918.

Ouellette N., Desch S. J., Hester J. J., and Leshin L. A. (2005) A nearby supernova injected short-lived radionuclides into our protoplanetary disk. In *Chondrites and the Protoplanetary Disk*, ASP Conference Series, **341**, Krot A. N., Scott E. R. D., and Reipurth A., editors, pp. 527–538, Astronomical Society of the Pacific, San Francisco.

Paardekooper S-J., and Johansen A. (2018) Giant planet formation and migration. *Space Science Reviews*, **214**, 38, doi.org/10.1007/s11214–018-0472-y.

Peslier A. H., Schonbacher M., Busemann H., and Karato S.-I. (2017) Water in the Earth's interior: Distribution and origin. *Space Science Reviews*, **212**, 743–810.

Piani L., Marroucchi Y., Rigaudier T., et al. (2020) Earth's water may have been inherited from material similar to enstatite chondrite meteorites. *Science*, **369**, 1110–1113.

Reynolds J. H., and Turner G. (1964) Rare gases in the chondrite Renazzo. *Journal of Geophysical Research*, **69**, 3263–3281.

Righter K. (2011) Prediction of metal-silicate partition coefficients for siderophile elements: An update and assessment of PT conditions for metal-silicate equilibrium during accretion of the Earth. *Earth & Planetary Science Letters*, **304**, 158–167.

Righter K., Drake M. J., and Scott E. R. D. (2006) Compositional relationships between meteorites and terrestrial planets. In *Meteorites and the Early Solar System II*, Lauretta D. S., and McSween H. Y., editors, pp. 803–828, University of Arizona Press, Tucson.

Rugel B., Faestermann T., Knie K., et al. (2009) New measurements of the ^{60}Fe half-life. *Physical Review Letters*, **103**, 072502.

Sanders I. S., and Scott E. R D. (2018) Making chondrules by splashing molten planetesimals: The dirty impact plume model. In *Chondrules, Records of Protoplanetary Disk Processes*, Russell S., Connolly H. C. Jr., and Krot A. N., editors, pp. 361–374, Cambridge University Press, Cambridge.

Sarafian A. R., Nielsen S. G., Marschall H. R., et al. (2014) Early accretion of water in the inner solar system from a carbonaceous chondrite-like source. *Science*, **346**, 623–626.

Saxena S. K., and Hrubiak R. (2014) Mapping the nebular condensates and the chemical composition of the terrestrial planets. *Earth & Planetary Science Letters*, **393**, 113–110.

Scott E. R. D., Krot A. D., and Sanders I. S. (2018) Isotopic dichotomy among meteorites and its bearing on the protoplanetary disk. *Astrophysical Journal*, **854**, 164–176.

Shu F. H., Shang H., and Lee T. (1996) Toward an astrophysical theory of chondrites. *Science*, **271**, 1545–1552.

Shu F. H., Shang H., Gounelle M., et al. (2001) The origin of chondrules and refractory inclusions in chondritic meteorites. *Astrophysical Journal*, **548**, 1029–1050.

Spiegel D. S., Fortney J. J., and Sotin C. (2014) Structure of exoplanets. *Proceedings of the National Academy of Sciences, USA*, **111**, 622–627.

Tachibana S., and Huss G. R. (2003) The initial abundance of ^{60}Fe in the solar system. *Astrophysical Journal Letters*, **588**, L41–L44.

Taylor S. R., and McLennan S. M. (1995) The geochemical evolution of continental crust. *Reviews of Geophysics*, **33**, 241–265.

Taylor S. R., and McLennan S. M. (2009) *Planetary Crusts: Their Composition, Origin and Evolution*. Cambridge University Press, Cambridge, 378 pp.

Telus M., Huss G. R., Ogliore R. C., et al. (2016) Mobility of iron and nickel at low temperatures: Implications for ^{60}Fe-^{60}Ni systematics of chondrules from unequilibrated ordinary chondrites. *Geochimica et Cosmochimica Acta*, **178**, 87–105.

Tera F., Papanastassiou D. A., and Wasserberg G. J. (1974) Isotopic evidence for a terminal lunar cataclysm. *Earth & Planetary Science*, **22**, 1–21.

Tobin J., Sheehan P. D., Megeath S. T., et al. (2020) The VLA/ALMA Nascent Disk and Multiplicity (VANDAM) Survey of Orion protostars. II. A statistical characterization of class 0 and class 1 protostellar disks. *Astrophysical Journal*, **890**, doi.org/10.3847/1538-4357/ab659e.

Tolstikhin I., and Kramers J. (2008) *The Evolution of Matter from the Big Bang to the Present Day Earth*. Cambridge University Press, Cambridge, 521 pp.

Toplis M. J., Mizzon H., Monnereau M., et al. (2013) Chondritic models of 4 Vesta: implications for geochemical and geophysical properties. *Meteoritics & Planetary Science*, **48**, 2300–2315.

Tsiganis K., Gomes R., Morbidelli A., and Levison H. F. (2005) Origin of the orbital architecture of the giant planets of the solar system. *Nature*, **435**, 459–461.

Wadhwa M., Srinivasan G., and Carlson R. W. (2006) Timescales of planetesimal differentiation in the early solar system. In *Meteorites and the Early Solar System II*, Lauretta D. S., and McSween H. Y., editors, pp. 715–732, University of Arizona Press, Tucson.

Walsh K. J., Morbidelli A., Raymond S. N., et al. (2011) Sculpting the inner solar system by gas-driven orbital migration of Jupiter. *Nature*, **475**, 206–209.

Wänke H. (1981) Constitution of terrestrial planets. *Philosophical Transactions of the Royal Society of London*, **A325**, 545–557.

Warren P. H. (2011) Stable-isotopic anomalies and the accretionary assemblage of the Earth and Mars: A subordinate role for carbonaceous chondrites. *Earth & Planetary Science Letters*, **311**, 93–100.

Wasserburg G. J., Lee T., and Papanastassiou D. A. (1977) Correlated O and Mg isotopic anomalies in Allende inclusions II: Magnesium. *Geophysical Research Letters*, **4**, 299–302.

Yang L., and Ciesla F. J. (2012) The effects of disk building on the distributions of refractory materials in the solar nebula. *Meteoritics & Planetary Science*, **47**, 99–119.

Yoshizaki T., and McDonough W. F. (2020) The composition of Mars. *Geochimica et Cosmochimica Acta*, **273**, 137–162.

Appendix: Some Analytical Techniques Commonly Used in Cosmochemistry

The purpose of this appendix is to present some additional information on various analytical techniques and methods that are used in cosmochemistry. This is not an exhaustive review. The goal is to give the reader an idea of the techniques that are available and their capabilities and limitations. References to more-detailed discussions are provided.

A.1 Chemical Compositions of Bulk Samples

There are a variety of methods for determining the bulk chemical composition of a sample. Each has its strengths and weaknesses. The choice of which technique to use depends on the elements of interest, the methods to which you have access, and the funding that is available for the analysis, among other things. The minimum sample requirement of the technique is an important consideration, but it can be very misleading. The true minimum sample size is the size that adequately accounts for the chemical and mineralogical complexity of the sample.

A.1.1 Wet Chemical Analysis

The traditional method of determining the bulk chemical composition of meteorites is wet chemical analysis. The process is time-consuming and requires skill and patience to obtain quality results. The best meteorite data were obtained by Hugo Wiik (e.g. Wiik, 1956) and Eugene Jarosewich (e.g. Jarosewich, 1990). The procedure is as follows. The sample is first crushed and divided into two fractions by passing it through a 100-mesh sieve. The coarse fraction is primarily metal, while the fine fraction consists primarily of silicate. This division permits accurate assignment of iron to metal and oxide. Six different aliquots of the homogenized sample are analyzed to determine different subsets of elements. A typical analysis gives concentrations of 16 elements, along with the amount of iron present as metal and the amount of bound water. A detailed discussion of the methodology is provided by Jarosewich (1990).

The format for reporting chemical data comes from wet chemical analyses. Major- and minor-element abundances are normally reported as wt. % oxides, although oxygen is not actually measured and oxides are calculated based on stoichiometry. Units of weight are the "natural" units for chemical analysis because the amount of each element is determined by weighing the samples. Atomic % element would probably make more sense but is not commonly used. Weight % oxide must be converted to atomic % element in order to calculate the mineral structural formula, and many modern analytical techniques count atoms rather than weigh the samples. Unfortunately, because of the weight % convention, a cosmochemist must become adept at converting published analyses to atomic % element.

A.1.2 X-Ray Fluorescence

X-ray fluorescence (XRF) uses secondary (fluorescence) X-rays emitted from a material than has been bombarded with high-energy X-rays. The sample is ground into a fine powder, homogenized, and pressed into a pellet. The pellet is irradiated by a monochromatic X-ray beam. Atoms in the sample absorb some of the X-rays, causing electrons to be ejected. The ejection of an electron causes other electrons to drop to lower energy levels to fill the vacancy, and fluorescence X-rays are emitted. The X-ray energies are characteristic of the elements from which they were emitted. The secondary X-rays are either detected simultaneously with one of the various types of energy-dispersive, solid-state detectors, including lithium-drifted silicon, lithium-drifted germanium, and PIN-diode detectors, or sequentially with wavelength-dispersive spectrometers. This method can, in theory, measure elements heavier than beryllium, but in practice, elements lighter than sodium are difficult to quantify unless background corrections and comprehensive inter-element absorption corrections are made. XRF was popular in the 1960s and 1970s for analyzing bulk compositions of meteorites.

A.1.3 Neutron Activation Analysis

During the late 1960s and early 1970s, neutron activation analysis provided a new way to measure bulk chemical composition. Neutron activation analysis utilizes (n,γ) reactions to identify elements. A sample is placed in a nuclear reactor, where thermal neutrons are captured by atoms in the sample. Some of the resulting atoms are radioactive and emit characteristic γ-rays when they decay. The energies of these γ-rays are measured to determine abundances. Approximately 35 elements are routinely measured by neutron activation analysis. A number of others produce radioactive isotopes that emit γ-rays, but the half-lives are too short to be useful. Unfortunately, silicon is one of these elements. Other elements do not produce γ-ray-emitting isotopes when irradiated with neutrons. There are two methods of using neutron activation to determine bulk compositions: instrumental neutron activation analysis (INAA) and radiochemical neutron activation analysis (RNAA).

In INAA, a rock or mineral sample is irradiated in the reactor. The irradiated sample is removed from the reactor, and the dangerous radioactivities are allowed to decay. Then the sample is placed into a counter and the γ-rays emitted by each element in the sample are counted. A variety of counters are used, including scintillation counters, gas-ionization counters, or semiconductor counters. For the most precise results, background counts in the detectors produced by electronic noise, cosmic rays, and other radioactive decays must be eliminated. The technique is very sensitive, and samples as small as a few tens of milligrams can be measured.

In RNAA, the sample is irradiated as for INAA. But after irradiation, the sample is chemically processed to separate elements with γ-rays whose energies cannot be resolved by the detectors and to concentrate the elements of interest to improve sensitivity. Typically, RNAA involves larger samples in order to facilitate chemical processing.

The extreme sensitivity of neutron activation analysis can be both an advantage and a disadvantage. The obvious advantage is that one can obtain data on small and precious samples. However, a danger is that one might measure a sample that is too small to be representative of the material of interest. For example, 50 mg samples of chondrites do not adequately sample all of the components in the rock, so either larger samples must be measured, or multiple samples must be measured and the results combined to get a good measurement.

A.2 Petrology, Mineralogy, Mineral Chemistry, and Mineral Structure

A.2.1 Optical Microscopy

The polarizing microscope is an important tool for investigating the mineralogy of a sample and the textural relationships between minerals. The microscope can be used with either transmitted light or reflected light. The light sources are designed to illuminate the sample from either directly below or directly above, using the microscope lens to focus the light.

In order to use transmitted light, samples are prepared as standard thin sections, 30 μm thick, thin enough for light to be transmitted through rock samples (see below). The light is first passed through a polarizer to generate plane-polarized light. The minerals in the sample rotate the direction of polarization of the light to varying degrees. When a second polarizer, oriented at 90 degrees to the first, is inserted between the sample and the observer, the minerals appear with different colors. These "interference colors," also known as "birefringence," are used, along with other properties such as shape, cleavage, twinning, etc., to identify the minerals. Transmitted light is used to identify silicates, carbonates, and some oxides. The textural relationships are also easy to study. Transmitted-light microscopy is not able to investigate materials with grain sizes smaller than the 30 μm thickness of the thin section. This is because the grain boundaries scatter the light. Material with a mean grain size of less than a few microns will appear opaque in transmitted light due to this scattering.

Reflected light can be used either on polished thin sections or on thicker polished sections. Reflected light is used primarily for identifying opaque minerals such as metals, sulfides, and some oxides. Each of these minerals has a unique appearance and properties in reflected light.

A.2.2 Electron Beam Techniques

When a high-energy electron beam strikes a sample, the electrons undergo elastic and inelastic scattering that generate a number of signals that can be used for imaging and for investigating the chemical and structural properties of a sample. The signals originate at different places in the sample, as illustrated schematically in Figure A.1. *Auger electrons* are generated when the primary electron beam hits and ejects an electron from an inner shell of an atom of the sample. The vacancy is filled by an electron from an outer shell, a process that releases energy that is carried away by either an auger electron or an X-ray photon (see below). Auger electrons have energies characteristic of the element from which they are emitted and originate within 1–3 nm of the surface. Electrons generated deeper in the sample are scattered by sample atoms and lose energy. *Secondary electrons*, the most commonly used imaging signal, are generated in a volume just below the surface when an incident electron passes near enough to an atom to impart some of its energy to one or more of the outer electrons. Some of these electrons leave the sample at low energy, becoming secondary electrons. Secondary electrons sample the outer ~10 nm of the sample. *Backscattered electrons* are generated over a somewhat larger volume when an incoming electron impacts an atom directly and is scattered backward in the direction that it came from with little or no loss of energy (an elastic collision). Backscattered electrons are generated from a larger region than secondary electrons (Fig. A.1). *Characteristic X-rays* are generated when an electron from the primary electron beam ejects an electron from an atom in the sample. In order for another electron to fill the vacancy, it must lose energy, which is removed by an X-ray with an energy that is characteristic of the element from which it was emitted. The volume of the sample from which characteristic X-rays (and to a lesser extent backscattered electrons) are generated is a function of the accelerating voltage of the primary electron beam, with higher accelerating voltages resulting in a larger interaction volume in the sample. *Cathodoluminescence* (CL) is visible light that is generated when the electron beam excites the outermost electrons in an atom to a higher state. When the electron drops back into its original orbital, it emits a photon that can be detected. The volume from which CL is generated is larger than those of the other signals (Fig. A.1).

Modern instruments use all of these signals to extract information about a sample. These include the scanning electron microscope (SEM), the transmission electron microscope (TEM), the electron microprobe (EMP), and the auger nanoprobe. Cathodoluminescence detectors are often added to modern SEMs and EMPs. For many of these techniques, a conductive coating (carbon, gold) is put on the sample to permit the charge imparted to the sample by the electron beam to be dissipated.

A.2.2.1 *Scanning Electron Microscope*

An SEM consists of a source that produces a beam of electrons accelerated to high energy (accelerating voltages can range from ~5 to ~30 keV), a column of electrostatic lenses and apertures to focus the beam, and one or more detectors. In standard operation, a focused beam of high-energy electrons is rastered across a rectangular area of a sample and an image is constructed from signals generated in the sample using a computer. Information about the position of the primary beam as a function of time is used to generate the image. SEMs use secondary electrons and backscattered electrons for imaging. Secondary electrons produce higher-quality images because they originate closer to the surface of the sample and because more of them are produced per incoming primary electron. Secondary-electron images are used to obtain morphological information about a sample. Backscattered electrons provide information on the mean atomic weight of the sample. In a backscattered-electron image, the brighter areas have higher mean atomic weight, and an experienced analyst can often determine the mineralogy of a sample from a backscattered-electron image.

The spatial resolution of an SEM is governed by a combination of the size of the primary electron beam and the size of the interaction volume. For secondary electrons, the size of the electron beam dominates. Two types of electron guns are in current use: the thermionic gun and

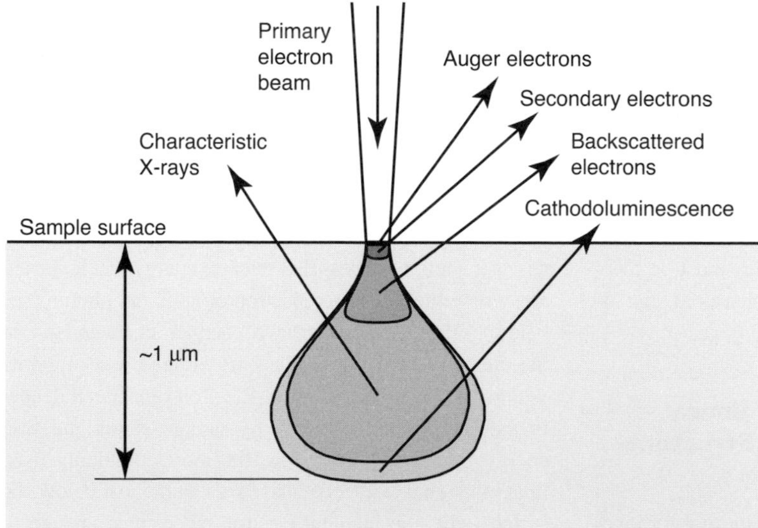

Figure A.1 When a high-energy electron beam strikes a sample, a series of interactions occurs that produce several types of radiation that can be used for analysis of the sample. Different signals come from different parts of the interaction volume (see text). At the sample surface, auger electrons are produced. At slightly greater average depth, secondary electrons are generated, followed at greater depth by backscattered electrons. Characteristic X-rays are generated over a volume of ~1 cubic μm (the exact volume depends in the impact energy of the electrons). Auger electrons are also produced at this depth, but they do not leave the sample because of interactions and scattering with other atoms. Cathodoluminescence requires somewhat less energy to produce and thus can be generated from a slightly larger volume than X-rays can.

the field emission gun (FEG). The thermionic gun is most common. It is easier and cheaper to operate and maintain than the FEG. Electrons are emitted from a heated filament and then accelerated toward an anode. The electrons emerging from the anode are focused into a beam by a set of electrostatic lenses. The FEG uses a very strong electric field to extract electrons from a metal filament. It operates at a much lower temperature than the thermionic gun and it is a much brighter source. However, it requires a much better vacuum and tends to be harder to operate. The field emission source permits a much smaller primary electron beam to be generated, resulting in much higher spatial resolution on the samples. A modern field emission SEM can image features down to ~10 nm in size.

SEMs are often equipped with an energy-dispersive X-ray spectrometer (EDS) for semi-quantitative and quantitative chemical analysis. Historically, these detectors have been lithium-doped silicon detectors that must be operated at liquid nitrogen temperature. The newer silicon-drift detectors can be operated with electronic Peltier cooling and thus do not need liquid nitrogen. The electronics recognize the energy of each incoming X-ray and produce an X-ray spectrum from which the composition of the sample can be determined. Modern instruments come with software that estimates the composition without the use of standards, using sophisticated computer software. The resulting compositions are reasonably accurate. Precise chemical compositions require careful standardization. If precise compositions are required, the EMP is a better instrument to use. Chemical information can be obtained from a spot or from an area, and compositional maps can be generated.

SEMs can be equipped with a CL detector. CL provides information on the mineralogy of a sample. An electron backscatter diffraction (EBSD) detector can also be mounted on an SEM to permit investigation of the structure of minerals. To use an EBSD detector, the sample is carefully polished to produce a damage-free surface. It is then mounted in the SEM at a high tilt angle so that the primary electron beam arrives almost parallel to the sample surface. The electrons are scattered by the atoms in the sample based on their positions, and the crystal structure of the mineral can be determined. The principal advantage of this technique over classical X-ray diffraction is that the petrographic context of the sample is maintained. Crystal orientation maps can be generated with an EBSD detector.

A.2.2.2 *Transmission Electron Microscope*

Transmission electron microscopy uses high-energy electrons to probe the structure and composition of materials at high spatial resolution. Electrons are accelerated to 60–400 keV, focused to a very small spot, and directed onto the sample to be studied. The sample must be thinned to electron-beam transparency; the thickness depends on the "electron density" of the material, but <200 nm is a good guide and 50 nm or less is best for high-resolution TEM (HRTEM) and electron energy-loss spectroscopy (EELS). TEMs come in two varieties: conventional (CTEM) and scanning (STEM). CTEM uses a fixed, stationary probe on the sample, whereas STEM rasters a fine probe across the area of interest. The spatial resolution of the TEM is limited primarily by spherical aberration, but software correction of such aberration for the HRTEM gives spatial resolutions of <0.1 nm, allowing individual atomic columns to be investigated.

The electrons passing through the thin sample comprise three types. 1) The fraction that does not interact with the sample as it passes through gives information on thickness and mean atomic weight. Thicker samples and areas of higher mean atomic weight will appear darker on the imaging screen. The contrast in the images is caused by absorption of electrons in the sample. At lower magnifications, this so-called "bright-field image" gives information on the shapes and morphologies of samples being studied. 2) Elastic scattering gives rise to diffraction effects and is the basis for the atomic-scale information in HRTEM images and selected area electron diffraction (SAED) patterns. 3) Inelastic scattering is where the incident monochromatic electron beam loses energy through interaction with the electrons in the solid. Inelastically scattered electrons form the basis of EELS, where the incident monochromatic electrons lose energy through interaction with core electrons; for example, interaction of a 100,000 eV electron with a carbon 1s electron (with a binding energy of ~290 eV) will reduce the final energy of the transmitted electron to <99,710 eV. The incident electron beam can also "interact" with the atoms in the sample, generating backscattered, secondary, and auger electrons, and X-rays. One of the great advantages of TEM is the ability to generate images, structure-specific HRTEM and SAED data, chemical and bonding EELS data, and energy-dispersive X-ray spectra from the same nanometer-sized region. For a detailed review of TEM instrumentation and techniques, see Williams and Carter (2009).

The TEM has found wide use in studying presolar grains, IDPs, and Stardust samples, and in analyzing the fine-grained alteration minerals in carbonaceous chondrites.

A.2.2.3 *Electron Microprobe*

An EMP nondestructively analyzes the chemical compositions of small volumes of minerals or glasses (Reed,

1993; Goldstein et al., 2003). The electron beam is generated in the same way as in an SEM. However, for EMP analysis, the beam is focused on a single spot and the characteristic X-rays are counted for a few seconds to a few minutes using wavelength-dispersive spectrometers. Each wavelength-dispersive spectrometer consists of one of several types of crystals that are used to diffract the X-rays of interest (using the principle of Bragg diffraction) into a gas-filled counter. A wavelength-dispersive spectrometer is better able to separate X-rays of different energy and is more sensitive than the energy-dispersive detector commonly found in SEMs, but each spectrometer can only measure one element at a time. An EMP typically has several spectrometers to permit analysis of multiple elements simultaneously. Major, minor, and many trace elements can be measured with a modern electron probe. Some EMPs are also equipped with energy-dispersive detectors to help with choosing points for analysis and otherwise to support the wavelength-dispersive analysis.

EMPs are used in spot mode to measure the chemical compositions of individual minerals. Mineral grains down to a few microns across are routinely analyzed. The chemical composition of the sample is determined by comparing the measured X-ray intensities with those from standards of known composition. Sample counts must be corrected for matrix effects (absorption and fluorescence). The spatial resolution of the EMP is governed by the interaction volume between the electron beam and the sample (Fig. A.1). An electron probe can also be operated in scanning mode to make X-ray maps of a sample. You will often see false-color images of a sample, where three elements are plotted in different colors. Such maps allow rapid identification of specific minerals. EMP analysis is the standard tool for characterizing the minerals in meteorites and lunar samples.

A.2.2.4 *Auger Nanoprobe*

Auger electron spectroscopy (AES) uses the auger electrons, which are generated at the surface of the sample by the impacting primary electron beam (Fig. A.1). Primary electrons eject an electron from an inner shell of an atom in the sample. An electron from an outer shell fills the vacancy and either another electron, the auger electron, or an X-ray photon carries away the excess energy. The energy of the auger electron is governed by the energy levels of the electrons and is thus characteristic of the element. The detector uses an electrostatic analyzer to steer electrons of a specific energy into the detectors. Because only electrons of a single energy can be measured at one time, an analysis can take a long time.

Auger electrons are generated in the top 1–3 nm of the sample, and the spatial resolution is controlled entirely by the size of the primary electron beam. An auger nanoprobe has a very small primary electron beam and the resulting spatial resolution is a few tens of nanometers. Auger spectroscopy cannot detect hydrogen or helium but is sensitive to all other elements. It is most sensitive to elements of low atomic number. The analysis must be done at high vacuum because the auger electrons come out with low energy, and because auger spectroscopy is a surface technique, the sample must be free of adsorbed oxygen and the carbon coat must be removed at the point of analysis. Modern auger nanoprobes are equipped with an ion beam (typically using argon) to sputter clean the surface before analysis.

As with the EMP, the chemical composition is determined through comparison with standards. Corrections for interactions with different elements are also necessary. However, the standardization and correction procedures for AES are much less mature than those for the electron probe. In cosmochemistry, the auger nanoprobe is used primarily to determine the chemical compositions of presolar grains. It is ideal for this application because it is a surface technique and has the same spatial resolution as the NanoSIMS (see below), which is used to identify presolar grains *in situ* in meteorite samples and IDPs.

A.2.3 Ion Beam Techniques

A.2.3.1 *Proton-Induced X-Ray Emission*

Proton-induced X-ray emission (PIXE) is a technique that uses an MeV proton beam to induce inner-shell electrons to be ejected from atoms in the sample. As outer-shell electrons fill the vacancies, characteristic X-rays are emitted that can be used to determine the elemental composition of a sample. Only elements heavier than fluorine can be detected due to absorption of lower-energy X-rays in the window between the sample chamber and the X-ray detector. An advantage of PIXE over electron-beam techniques is that there is less charging of the sample from the incoming beam and less emission of secondary and auger electrons from the sample. Another is the speed of analysis and the fact that samples can be analyzed without special preparation. A disadvantage for cosmochemistry is that the technique is not as well quantified as electron-beam techniques. PIXE has not been widely used in cosmochemistry.

A.2.3.2 *Focused Ion Beam Instruments*

A focused ion beam (FIB) instrument combines an electron column and an ion column, mounted at an angle to each other so that both focus on the sample at the same location. The electron beam is designed to navigate around the sample and to position the target of interest precisely under the ion beam. A FIB instrument typically uses a gallium ion beam. A FIB can be used to extract grains of interest from a sample and to prepare electron-transparent sections for TEM measurements. Alternatively, it can be used to deposit material on a sample, either to protect the sample from damage or to construct nanostructures. Many FIB instruments are equipped with a micromanipulator. The micromanipulator needle can be affixed in the vacuum system to the sample section that is being excavated by the ion beam using the deposition feature of the FIB so that the sample is not lost. The sample can then be moved to a holder, where it can be attached, again by ion deposition. This is observed in real time using the electron beam. In most modern FIB instruments, the electron-beam portion is a high-performance SEM, often with a field emission source, secondary and backscattered electron detectors, and EDS capability. This combination of a high-performance SEM with a versatile ion beam instrument makes a FIB instrument a very powerful tool for sample analysis.

A.2.3.3 *Ion Microprobe*

An ion microprobe (or secondary ion mass spectrometer, SIMS) uses a beam of charged particles (ions) to sputter material from a sample. Sputtered (secondary) ions are extracted into a mass spectrometer and measured. Ion microprobes can provide detailed chemical and isotopic information about a sample, while retaining the petrographic context. They can measure elements that make both positive and negative secondary ions. There are several types of ion microprobes with different types of mass spectrometers. The major tool for cosmochemistry uses a magnetic-sector mass spectrometer to measure the sputtered ions. These instruments are used to measure isotopic compositions of elements from hydrogen through the first row of transition metals. They are also used to measure trace-element abundances with sensitivities down to tens of ppb. Several types of magnetic-sector ion probes are commercially available. Another type of ion probe uses a time-of-flight mass spectrometer. These instruments are primarily used to measure trace elements.

A.2.4 Other Techniques for Determining Chemical Composition and Mineral Structure

A.2.4.1 *Inductively Coupled Plasma-Atomic Emission Spectroscopy*

Emission spectroscopy utilizes the characteristic line emission from atoms as their electrons drop from the excited to the ground state. The earliest version of emission spectroscopy as applied to chemistry was the flame test, where samples of elements placed in a Bunsen burner will change the flame to different colors (sodium turns it yellow, calcium turns it red, copper turns it green). The modern version of emission spectroscopy for the chemistry laboratory is inductively coupled plasma-atomic emission spectroscopy (ICP-AES). In this technique, rocks are dissolved in acid or vaporized with a laser, and then the sample liquid or gas is mixed with argon gas and turned into a plasma (ionized gas) by a radio frequency generator. The excited atoms in the plasma emit their characteristic energies that are measured either sequentially with a monochromator and photomultiplier tube, or simultaneously with a polychromator. The technique can analyze 60 elements in minutes.

A.2.4.2 *X-Ray Diffraction*

X-ray diffraction (XRD) uses the elastic scattering of X-rays from structures that have long-range order to give structural information. Diffracted waves from different atoms in the structure can interfere with each other and the resulting intensity distribution is directly related to the atomic distances in the structure. For a given set of lattice planes with an interplane distance of d, the condition for a diffraction peak is $2d \sin \theta = \lambda$, where λ is the wavelength of the incident X-rays. This relationship is known as Bragg's law.

Two types of XRD measurements are typically made on cosmochemical samples. Powder diffraction, where a sample powder is mounted on the end of a fiber and placed in the X-ray beam, is most commonly used. The term "powder" in this usage means that the crystal lattice domains are randomly oriented. Powder diffraction is used to identify unknown materials and to determine their basic crystal structure. Single-crystal XRD is used to solve the complete structure of crystalline materials. For more details, see Cullity (1978). Micro XRD systems can determine the structures of individual crystals of a few tens to a few hundred microns.

A.2.4.3 *Synchrotron Techniques*

Several analytical techniques are based on X-rays emitted from high-energy synchrotron electrons. These have been used to study IDPs and presolar grains and are an important part of the analytical arsenal being used on the Stardust samples. There are several distinct synchrotron-based techniques.

Synchrotron XRD can be used to give structural information on single crystals and on powdered samples. The X-ray energy is tunable, which can minimize absorption and give access to a greater range of types and sizes of structures. Structures consisting of heavy elements are accessible. Beam intensity is high, facilitating rapid data acquisition and helping to minimize beam damage.

Synchrotron XRF uses a highly focused monochromatic X-ray beam to excite characteristic X-rays from the sample. The principle is the same as for the XRF systems discussed earlier. The synchrotron primary X-ray beam can be very intense and very small, permitting compositional information to be obtained at spatial scales as small as 50–100 nm. A typical analytical mode is to raster the sample under the X-ray beam to produce element maps of the sample. Phase-contrast maps can be generated from XRF maps, and three-dimensional tomography can also be done. This technique has been used to determine bulk compositions of IDPs and the material in Stardust tracks.

X-ray absorption spectroscopy (XAS) gives information about the atomic coordination and valence state of the atoms in a structure. XAS requires a high intensity, coherent X-ray beam that is tunable over a wide energy range. These requirements are met in a synchrotron source.

A scanning transmission X-ray microscope (STXM) uses a monochromatic X-ray beam focused to a very small spot to map a sample. The image is formed by scanning the sample through the focal spot of the beam in X and Y. A detector records the total transmitted intensity for each pixel. A map of the distribution of a single element can be made by setting the X-ray energy to the "absorption edge" of the element. An absorption edge is a sharp discontinuity in the absorption of X-rays as a function of energy that occurs when the energy of the X-ray photon corresponds to the energy of an electron shell of an atom. The absorption edge of each element is at a different X-ray energy. By changing the energy of the X-ray beam, maps of different elements can be obtained with a spatial resolution of 30–35 nm.

X-ray absorption near-edge spectroscopy (XANES) uses the STXM beam line. Instead of holding the X-ray wavelength constant and moving the sample, the sample stays fixed and the X-ray energy is varied around the absorption edge of the element of interest. The details of the shape of the absorption edge depend on the bonding environment of the atom. XANES has been used to investigate the carbon, nitrogen, and oxygen in the organic material of primitive meteorites and IDPs. Because the absorption edge for iron shifts slightly in energy as a function of oxidation state, XANES is also a good way to determine the oxidation state of a sample.

Synchrotron techniques are nominally nondestructive, although X-ray beams carry a lot of energy and can disrupt the structures of some materials.

A.3 Mass Spectrometry

Mass spectrometers are used primarily as tools for measuring isotopic compositions, although some kinds can also be used to determine elemental abundances. Mass spectrometers have three basic components: (1) a means of ionizing the sample, (2) a mass analyzer that separates atoms based on their masses, and (3) a detector. Most of the time, the mass spectrometer is identified by its source, although the mass analyzer can also be identified. In the next few paragraphs, we will describe the various sources, the different mass analyzers, and the detectors, and then describe the most common configurations used in cosmochemistry. For more details, see Gross (2017).

A.3.1 Ion Sources

There are a number of different ion sources in common use for mass spectrometry.

A.3.1.1 *Gas Source*

Gas-source mass spectrometers work with elements in the gaseous state. The elements of interest are extracted from the samples either by chemical means or by heating, purifying, and introducing them into the mass spectrometer source as a gas. The gas is ionized by a beam of electrons passing through the sample gas with an energy of ~70 eV. The electrons in the beam knock an electron out of a gas atom, creating an ion that can be accelerated by a high-voltage plate into the mass spectrometer. For this reason, a gas source is also called an "electron-impact source." Gas-source mass spectrometers are used primarily to measure hydrogen, carbon, nitrogen, oxygen, and noble gases. Gas-source mass spectrometers can also be coupled to a laser that vaporizes the sample. The vaporized sample is transferred as a gas to the mass spectrometer and measured. This technique permits measurements of samples, while retaining the petrographic context. The spot size is typically a few tens of microns across.

A.3.1.2 *Thermal Ionization*

Thermal-ionization mass spectrometers (TIMS) use a hot filament to ionize the sample. In most applications, the element of interest is first purified using wet chemistry and then is loaded onto a source filament, often along with another substance that makes ionization easier and a more stable function of temperature. The filament is heated, and as the sample evaporates, it is ionized. In some cases, a small mineral grain can be directly loaded onto the filament, but the sample must be inherently free of interfering elements and compounds. Both positive and negative ions can be created by thermal ionization, depending on the electronegativity of the element to be measured. TIMS instruments are used to measure a wide variety of rock-forming and trace elements.

A.3.1.3 *Inductively Coupled Plasma*

Inductively coupled plasma (ICP) sources use a hot plasma to ionize the sample. The plasma is generated by electromagnetic induction. The plasma source operates at several thousand degrees and the sample is fully ionized. The ions are then extracted into the mass spectrometer. The ICP source is very stable and mass spectrometers using this technique can produce high-precision, high-accuracy results. However, relatively large amounts of sample are required. The ICP source can also be coupled to a laser that vaporizes the sample. The gas is then transferred to the ICP source, ionized, and admitted to the mass spectrometer. This approach retains the petrologic context of the sample, but does not produce the same level of precision as the standard technique.

A.3.1.4 *Sputtering with High-Energy Ions*

An ion microprobe (or secondary ion mass spectrometer, SIMS) is equipped with an ion gun that focuses a beam of charged particles to sputter atoms from a sample surface and ionize them for mass spectrometry. The ion beam can be aimed very precisely at the sample of interest, preserving the petrologic context. Depending on the instrument, the primary ion beam can be focused to a spot as small as 100 nm across, or can be defocused to a larger area. The sputtered (secondary) ions are pulled into a mass spectrometer and analyzed. Ion guns can be configured to emit positive or negative secondary ions, and the choice of primary ions can enhance the yield of secondary ions. A Cs^+ primary ion beam gives a high yield of negative secondary ions, such as O^-, S^-, or CN^-. An O^- primary ion beam favors production of positive ions, such as Mg^+, Ti^+, or Ca^+. Recently, a new oxygen source (the Hyperion-II) has become available which can produce a bright, stable primary oxygen beam that significantly outperforms the previous duoplasmatron oxygen source. Other elements, such as gallium and argon, can also be used in an ion gun, but these elements do not enhance the secondary ion yield in the way that O^- and Cs^+ do. An ion beam sputtering a sample produces every species imaginable, including neutral atoms, positive and negative ions of various charges, molecular species, etc. For this reason, the sputter source is typically coupled to a high-performance mass spectrometer.

A.3.1.5 *Laser Resonance*

Laser-resonance mass spectrometers use tuned laser beams to ionize only the element(s) of interest from a gas or a plume of atoms vaporized from a sample. In most cases, two lasers are used for ionization, with the first one raising an outer-shell electron to an excited state and the second one adding enough additional energy to ionize the atom. Because the allowed electron energy states in an atom are discrete and unique to each element, laser-resonance ionization mass spectrometry (RIMS) can be extremely selective. The key is to find a suitable ionization scheme for the element of interest.

A.3.2 Mass Analyzers

The mass analyzer is the heart of the mass spectrometer. Several types of mass analyzers are commonly used and several other types may see use in the future. The performance of a mass spectrometer is typically discussed in terms of its mass-resolving power. Mass-resolving power is defined as $m/\Delta m$, where m is the mass of the species and Δm is the width of the peak that can separate that mass from an adjacent mass. Implicit in the mass-resolving power is a definition of what it means to separate a mass peak from an adjacent peak. A commonly used definition is that the signal decreases to 10% of the maximum between the two peaks (the 10% height). Other definitions include the 50% height or full width at half maximum. In addition, a given mass-resolving power at low mass will give a lower mass-resolving power at high mass, because the fractional difference in mass between two isotopes is smaller at higher mass. So, in order to interpret a number given for mass-resolving power, one must understand the definition that is being used and the mass range of the elements being measured.

A.3.2.1 *Magnetic Sector*

Magnetic-sector mass spectrometers are the most common type used in cosmochemistry. In these devices, ions are separated based on their mass-to-charge ratio. The equation governing the magnetic-sector mass analyzer is:

$$\frac{m}{q} = \frac{B^2 r^2}{2V} \tag{A.1}$$

where m is the mass of the ion, q is its charge, B is the magnetic field, r is the radius of curvature of the magnetic sector, and V is the accelerating voltage. At a constant accelerating voltage, each value for the magnetic field corresponds to a unique mass-to-charge ratio. The radius of the magnetic sector governs the mass dispersion of the instrument and the mass-resolving power. In general, bigger instruments have higher mass-resolving power. The mass-resolving power can be increased by adding slits before and after the magnetic sector to limit the size of the beam going into the magnet and to block masses that are off axis from reaching the detector. The performance of a magnetic-sector instrument can be improved by coupling the magnetic sector with an electrostatic sector, which filters the ions based on their energy.

A.3.2.2 *Time-of-Flight*

Time-of-flight analyzers use a pulsed electric field to accelerate ions to the same kinetic energy and then measure the time it takes for the ions to reach the detector. The velocity of the ions, and thus the time of flight to the detector, depends only on the mass-to-charge ratio:

$$t = L\sqrt{\frac{m}{q} \times \frac{1}{2V}} \quad \text{A.2}$$

where t is the time of flight, L is the length of the flight path, m is the mass of the ion, q is its charge, and V is the accelerating voltage. A pulsed ion source is required for a time-of-flight instrument. The mass-resolving power is governed by the length of the flight path and the speed of the electronics at the detector end. Most time-of-flight mass spectrometers have an electrostatic mirror called a "reflectron" in the flight path so that the detector can be placed near the source and the overall length of the instrument can be minimized. Time-of-flight instruments are used in a wide variety of applications, including on spacecraft.

A.3.2.3 *Quadrupole Mass Analyzers*

Quadrupole mass analyzers consist of four rods that produce a radio frequency quadrupole field. Only a single mass-to-charge ratio has a stable trajectory through the mass analyzer, but the electric potentials can be swept rapidly, either continuously or in discrete steps to measure a variety of masses. Quadrupole mass spectrometers are used in a wide variety of applications, but are not widely used in cosmochemistry.

A.3.2.4 *Quadrupole Ion Traps*

Quadrupole ion traps are closely related to quadrupole mass analyzers, but in the ion trap, the fields are manipulated so that the ions of interest are trapped, rather than passed through to a detector. Instruments of this type are being investigated for spaceflight applications.

A.3.3 Detectors

There are two basic types of detectors used to measure ion signals: current detectors and ion counters. Each type has a different implementation. A third type of detector is an imaging detector. In some SIMS instruments, the mass spectrometer is also an ion microscope, which transmits a stigmatic image of the sample to a detector plane.

A.3.3.1 *Faraday Cup*

A Faraday cup is a current detector consisting of a conductive cup designed to catch charged particles in a vacuum. The charged particles transfer their charge to the metal cup, which is connected to a current-measuring device. In most systems, measurements are made by a high-precision voltmeter that measures the voltage developed across a resistor from the conducting lead to ground. In most systems, the Faraday cup has a repeller plate in front of the cup to prevent secondary electrons from leaving the cup and changing the reading.

A.3.3.2 *Daly Detector*

A Daly detector is an ion detector that consists of a metal "doorknob," a scintillator (typically a phosphor screen), and a photomultiplier. Incoming ions hit the doorknob and release secondary electrons. The electrons are attracted to the scintillator by a high voltage (~20 keV), where they are converted to photons. The photons are detected by the photomultiplier. An advantage of the Daly detector is that the photomultiplier tube can be placed outside the vacuum system of the mass spectrometer, separated by a window. This extends the life of the detector and permits use of a higher voltage between the doorknob and the scintillator, potentially improving sensitivity for high-mass ions.

A.3.3.3 *Electron Multiplier*

An electron multiplier is an ion detector that converts a single incoming ion into a cascade of 10^4–10^8 electrons, depending on the device. Electron multipliers typically used in cosmochemistry are discrete dynode multipliers, consisting of a series of plates or dynodes. An incoming ion strikes the first dynode, known as the "conversion dynode," and ejects several secondary electrons. The number of electrons ejected is a function of the velocity and mass of the incoming ion. These electrons are accelerated into the multiplier, where they strike the second dynode, with each electron ejecting one to three more electrons. Each of these is again accelerated to strike the third dynode. This process continues until a single incoming ion has generated a pulse of 10^6 to 10^7 electrons. This pulse then passes through a discriminator that rejects smaller noise pulses, shapes the pulse, and then sends it to a counter. While the system is amplifying and counting each pulse, it cannot respond to another incoming ion. This is known as the "dead time." A typical electron multiplier has a dead time on the order 5–10 nanoseconds, but because the length of the dead time depends on the size of the electron pulse, the electronics are set up to constrain the dead time to a fixed length, longer than the longest electron pulse. Typical values of the overall system dead time range from ~18 to as much as 60 nanoseconds, depending on the details of the counting system. The dead time is accounted for in the data reduction, but if it is too large, the uncertainty in the dead time can begin to contribute to the measurement uncertainty. The maximum practical count rate for an electron multiplier system is around 10^6 counts per second.

Electron multipliers can also be operated in analog mode as current detectors. In this mode, they have a lower gain and measure higher signals than in pulse-counting mode. This was the normal way to run an electron multiplier until fast digital electronics came along. In analog mode, the electron multiplier output may not be linear with input signal, so calibration curves must be established to get accurate data. In GCMS systems (see below), electron multipliers are typically run in analog mode.

A.3.3.4 *Imaging Detectors*

Some mass spectrometers, such as the IMS series CAMECA ion microprobes, are also ion microscopes. They can project a stigmatic ion image of the sample surface onto a two-dimensional imaging detector. This capability was originally intended for use in tuning the instrument (a tremendous advantage), but after a few years, investigators began to try to make quantitative ion maps of their samples. The original imaging detector on the CAMECA ion microprobes was a channel plate and phosphorescent screen, with the output captured on a video camera. This system was adapted by the Washington University group to produce semi-quantitative isotope-ratio maps that were used to search for presolar grains (which exhibit highly unusual isotopic compositions).

Recently, a Japanese group under the direction of Hisayoshi Yurimoto has developed a solid-state imaging detector for the CAMECA ion microprobes called SCAPS (Nagashima et al., 2001). This CMOS-type detector has a sensitivity of about three ions and is linear over a range of ~5×10^4. Adaptive electronics that read individual pixels as they reach saturation can increase that range significantly. This detector has been used to identify presolar grains in primitive chondrites and to investigate oxygen-isotope variations within CAI minerals. This detector is much more robust than an electron multiplier, can measure high currents without dead time, and yet can also measure extremely low ion signals. This detector and the next generation of solid-state detectors will certainly have a greater role as ion detectors in the future.

A.3.4 Mass Spectrometer Systems Used in Cosmochemistry

A.3.4.1 *Stable-Isotope Mass Spectrometers*

The instruments used to measure isotopic compositions of hydrogen, carbon, nitrogen, oxygen, and sometimes sulfur are gas-source machines with simple magnetic-sector mass spectrometers. Both single- and multiple-detector systems can be used; modern instruments are multidetector instruments. Faraday cups are the most common detectors. To make a measurement, the element of interest must be extracted from the sample in an extraction line and converted to a chemical compound that is suitable for measurement. The mass spectrometer typically has relatively low mass-resolving power, so the sample must be effectively cleaned up before measurement. The mass spectrometer is designed to compare the sample to a standard. The source has two inlets – one for the standard and one for the sample. First, the standard is introduced into the source and measured, then the inlet is switched to the sample and it is measured. By switching back and forth between sample and standard, instrumental effects can be minimized. The results are typically reported as parts per thousand (permil, ‰) deviation relative to a standard composition. Oxygen analyses can be done with laser fluorination, where a laser is used to extract the gas from a specific portion of a sample in the presence of fluorine. The purified gas is then introduced into the mass spectrometer and measured. The sensitivity of stable-isotope mass spectrometers has been improved with the introduction of a gas chromatograph inlet system. These systems use a carrier gas to bring the sample to the mass spectrometer, and the gas chromatograph helps to purify the sample. These systems are orders of magnitude more sensitive than conventional stable-isotope mass spectrometers.

A.3.4.2 *Noble Gas Mass Spectrometers*

The mass spectrometers used to measure noble gases are also gas-source machines, typically with magnetic-sector mass analyzers. However, unlike the stable-isotope machines, noble gas machines are run in static mode. Noble gas mass spectrometers have low mass-resolving power (~180 to ~600), so gases must be purified before measurement to eliminate interferences. Electron-multiplier detectors are common for many applications, but Faraday cups are also used in some situations. The exact protocol for measuring the noble gases depends on the goal of the study. We give an example protocol here. The noble gases are released from the sample by heating or oxidation. Chemically active gases are removed by chemical reactions with various types of "getters." The noble gases are then frozen onto a cold finger at liquid nitrogen temperature, which quantitatively removes argon, krypton, and xenon from the gas phase. The helium and neon are then admitted to the mass spectrometer, the valve is closed, and the gases are measured. At the end of the measurement, the helium and neon are pumped from the mass spectrometer. The cold finger is warmed up somewhat to release argon, which is cleaned up with getters and then admitted to the mass spectrometer for measurement. The mass spectrometer is again pumped out and krypton is released from the cold figure and measured. After krypton is pumped out of the spectrometer, xenon is released from the cold finger, introduced into the mass spectrometer and measured. Protocols are designed to minimize noble gas blank and interferences from various compounds, and to insure high sensitivity. In recent years, laser extraction systems have been added to noble gas mass spectrometers so that gases can be selectively extracted from just the grains of interest.

A.3.4.3 *Thermal-Ionization Mass Spectrometers*

Thermal ionization mass spectrometry (TIMS) combines a hot-filament source with a magnetic-sector mass spectrometer. The mass spectrometers are operated at low to moderate mass-resolving power. Many elements can be measured with TIMS. Special care is taken to purify the samples using ion-exchange columns (see below). Samples are loaded onto the filaments along with an emitter, and a typical run may take several hours. Modern systems have multiple collectors so that several isotopes can be measured simultaneously. High-precision measurements are made with Faraday cup detectors, but low-abundance isotopes can be measured on electron multipliers. Modern machines are capable of precisions of 0.1–0.01 permil.

A.3.4.4 *Inductively Coupled Plasma Mass Spectrometers*

Inductively coupled plasma mass spectrometry (ICPMS) represented a major advance in capability compared to TIMS, by expanding the list of elements that could be measured and by driving improvements in measurement precision. The plasma source makes most of the Periodic Table accessible to measurement, so several radioactive-isotope systems that used to be impractical to use for chronology are now routinely used (see Chapters 9 and 10). Most machines used for cosmochemistry are magnetic-sector instruments, often including an electrostatic sector to minimize the spread in energy. Both single-collector machines and multicollector machines are used. Single-collector machines coupled to a laser source are used to measure trace elements. Multicollector machines are used for high-precision isotope work. Lasers have also been coupled to multicollector machines for isotopic measurements that require petrographic information on measured spots. The ICP source is very stable, permitting precise correction for instrumental mass fractionation.

A.3.4.5 *Magnetic-Sector Ion Microprobes*

Ion microprobes are becoming increasingly important in isotopic analysis of extraterrestrial materials as spot size decreases and the precision and accuracy of the measurements improve. The first of the commercially available ion microprobes used in cosmochemistry were the CAMECA IMS 3f–7f series machines, which became available in the mid-1980s. These multipurpose instruments are able to measure the isotopic compositions of most elements of interest to cosmochemistry and can also be used to measure trace-element abundances. Their main drawback is that the relatively small mass spectrometer can only be operated at mass-resolving powers below about 9000, and at this mass-resolving power, the transmission of the mass spectrometer is very low.

To remedy this problem, the large-geometry SHRIMP machines, made by Anutech, an arm of the Australian National University, and the CAMECA IMS 1270–1300 series were developed. These instruments have much larger magnetic-sector mass spectrometers and much higher sensitivity at high mass-resolving power. The initial application for large-geometry ion probes was U–Pb dating, and this is still a major thrust of the work done on these machines. But recent improvements in ion optics, electronic stability, and automation have made the CAMECA IMS 1280–1300 instruments into highly versatile instruments capable of high-precision isotope-ratio measurements that rival the capabilities of the traditional mass spectrometers of the 1970s. This may not sound like much, but in the 1970s, scientists generally believed that ion microprobes would never be quantitative instruments. The high-precision measurements coupled with the retention of the petrographic context make these ion probes very popular, even though they are very expensive.

The large-geometry CAMECA ion probes are ion microscopes as well as ion microprobes. That means that they can transmit direct ion images to an imaging detector. This capability was originally used for tuning (a huge help), but it has now been adapted for semiquantitative and quantitative ion imaging. Direct ion imaging does not depend on the size of the primary ion beam for its spatial resolution. The new solid-state SCAPS detector has a spatial resolution of less than a micron, close to that of the NanoSIMS (see below).

Another new development in ion microprobes is the NanoSIMS, which was originally designed as an imaging tool for biological applications, but which has now been adapted to cosmochemistry. The primary ion beam can be focused to less than 100 nm, about three times smaller than the IMS 1280–1300. NanoSIMS is a true multicollector instrument and has high transmission at high mass-resolving power. It was originally brought into cosmochemistry to study presolar grains, but its applications have expanded to include stable-isotope work and radiochronology. Its main weakness is that it cannot generate high enough ion currents to measure trace elements in natural samples.

A.3.4.6 *Time-of-Flight Ion Microprobes*

The time-of-flight ion microprobe (TOF SIMS) has unique capabilities not found in other mass spectrometers. A pulsed ion beam, typically either cesium or gallium, ejects atoms and molecules from the sample. Ionized species are accelerated down the flight tube and the time of arrival in the detector is recorded, giving the mass of the species (see discussion of time-of-flight mass analyzers above). TOF SIMS instruments used in cosmochemistry have spatial resolutions of <1 micron. They are used to determine elemental abundances in IDPs and Stardust samples. The spatial distribution of elements within a small sample can also be determined. TOF SIMS instruments can get good data with very little consumption of the sample.

A.3.4.7 *Resonance Ionization Mass Spectrometry*

Resonance ionization mass spectrometry (RIMS) has finally arrived as a productive experimental technique. Two instruments are now producing high-quality data: CHILI (the Chicago Instrument for Laser Ionization at the University of Chicago) and LION (Laser Ionization of Neutrals at the Lawrence Livermore National Laboratory). The technique as implemented in these two instruments can maintain the petrologic context of the sample. To measure the sample, an ion gun or a desorption laser is fired at the sample, removing a few atomic layers and generating a cloud of gaseous atoms and ions. The ions are ejected from the cloud by a pulsed electromagnetic field. Then the lasers are fired into the cloud to ionize only the elements of interest. In a two-laser scheme, one laser raises an outer-shell electron to an excited state and the second adds enough additional energy to ionize the atom. The ions are extracted into a time-of-flight mass spectrometer and measured. Ionization schemes have been developed for iron, nickel, strontium, zirconium, molybdenum, ruthenium, and other elements. Laser-resonance ionization is very selective and efficient. The current instruments can put several tens of percent of the atoms extracted from the sample into the detector, making RIMS at least an order of magnitude more efficient than SIMS. Measurements have initially concentrated on *s*-process elements in presolar grains (Stephan et al., 2019), but RIMS has also shown great potential for measurements of short-lived radionuclides such as ^{60}Fe (Trappisch et al., 2018).

A.3.4.8 *Gas Chromatography Mass Spectrometry*

Gas chromatography mass spectrometry (GCMS) systems are a mainstay of organic chemical analysis in cosmochemistry. These systems are used even more widely in other fields, such as environmental monitoring, criminal forensics, security, and monitoring food and beverage quality. A GCMS system consists of two parts: a gas chromatograph, which separates molecules based on chemical properties, and a mass spectrometer, which measures the chemicals separated by the gas chromatograph by breaking them into ionized fragments and measuring their mass-to-charge ratio. The combined system permits a much finer degree of substance identification than either system used alone. A GCMS system flew on the Mars Viking landers as part of the instrument package to search for evidence of life.

A.4 Atom Probe

Atom probe tomography (APT) is a nanoscale-materials analysis technique that provides three-dimensional spatial and compositional information about a sample (e.g. Kelly and Larson, 2012). An atom probe combines a field ion microscope with a mass spectrometer and has single-atom detection capability. The field ion microscope uses a very strong electric field to pull atoms from a needle-like tip. The extracted atoms are transferred into a time-of-flight mass spectrometer. The output is a three-dimensional reconstruction of the atom positions in the sample tip. In metallic samples, the depth resolution is ~0.3 nm and the lateral resolution is ~0.5 nm. In insulating or semiconducting samples, the resolution is ~ 1 nm. The atom probe can detect all elements and can identify isotopes. The atom probe measures essentially every atom in the sample. Chemical sensitivity is ~10 ppm, controlled primarily by the number of atoms in the sample tip. Sample preparation consists of preparing a conical tip with a radius of ~100 nm. This can be done with a FIB instrument. APT is thus a destructive technique. APT is the only technique that has the possibility of measuring the carbon isotopic composition of a single 2000-atom presolar diamond.

A.5 Spectroscopic Methods

A.5.1 Raman Spectroscopy

Raman spectroscopy is a nondestructive technique that is used in cosmochemistry for the identification of minerals and to evaluate the bonding and composition of organic molecules. The technique does not require special sample preparation; raw rock samples, polished sections, fine-grained powders, and liquids can be analyzed. Raman spectroscopy is the basis for the Sherloc instrument on the Mars 2020 Perseverance rover and for several instruments that are being developed for upcoming NASA missions.

As we discussed for electrons, protons, and X-rays, the photons of visible and infrared light interact with molecules in a sample to induce transitions between electron energy states. Most photons are elastically scattered, and the emitted photon has the same wavelength or energy as the absorbed photon. Inelastic scattering of photons by molecules produces the Raman effect. These photons comprise a very small fraction of the incident photons (~ 1 in 10^7). In Raman scattering, the energies of the incident and scattered photons are different. The difference is equal to the energy of a vibration of the scattering molecule. Thus, Raman spectroscopy measures the vibrational energies of molecules induced by the inelastic scattering of light. A laser provides a monochromatic light source that is much more powerful than ambient light. The inelastic scattering by the different materials in the sample produces characteristic shifts in wavelength that can be measured by a spectrometer. Different wavelengths of laser light can be used to better couple with the elements in the sample or to minimize fluorescence. Infrared spectroscopy, also sensitive to the vibrational state of a molecule, yields complementary information.

A confocal Raman microscope couples a Raman spectrometer with a confocal microscope. The confocal microscope uses a point illumination and an appropriately placed pinhole in front of the detector to eliminate out-of-focus signals. The laser light is passed through the pinhole to create the point illumination and the return Raman-scattered light is passed to the spectrometer. This significantly lowers the background signal in comparison to a classic system and permits acquisition of Raman data from very small regions both on the sample surface and within a transparent sample. When the microscope is equipped with an automated stage, Raman maps of natural samples can be obtained that have a spatial resolution of close to 1 μm. For a detailed discussion of Raman spectroscopy, see Smith and Dent (2005).

A.5.2 Fourier-Transform Infrared Spectroscopy

The infrared region of the electromagnetic spectrum has longer wavelengths than visible light. Minerals and organic materials absorb characteristic frequencies of infrared light, controlled by the bonding between atoms. The visible-near-infrared region (VIS-NIR, 0.4–1 μm) is important for iron oxide identification. The NIR region (0.8–2.5 μm; 12,500–4000 cm^{-1}) is the region of hydroxyl, organic, and carbonate bonds, and the mid-infrared region (2.5–25 μm; 4000–400 cm^{-1}) has absorption features related to OH bonds, tectosilicates, oxides, and sulfide minerals.

In Fourier-transform infrared spectroscopy (FTIR), a broadband beam of light shines into an interferometer and then onto the sample. The interferometer consists of a set of mirrors that is moved by a motor. Each wavelength of light is periodically blocked and transmitted as the mirrors move, due to wave interference. The collected data is then transformed by Fourier transform from intensity as a function of displacement of the mirrors to intensity as a function of wavenumbers (units of cm^{-1}). The result is a spectrum of the signal as a function of a series of discrete wavelengths. Data can be collected in both transmission and absorption mode. In recent years, nanoFTIR has become a mature technology. This technique incorporates an atomic-force microscope and can provide data at a spatial resolution of 10–20 nanometers (e.g. Nan et al., 2020).

FTIR analysis is widely used in cosmochemistry, particularly in studies of organic materials, water content, and aqueous alteration. FTIR analysis of meteorites provides ground truth for spacecraft-based infrared analysis of asteroids (e.g. Beck et al., 2014).

A.5.3 Cavity Ring-Down Spectroscopy

Cavity ring-down spectroscopy (CRDS) is a form of laser absorption spectroscopy. A laser pulse is admitted into a highly reflective detection cavity. The intensity of the pulse decreases as an exponential function of time due to scattering by the sample within the cell and reflectivity losses. The signal of interest is the decay rate of the laser pulse rather than the absolute absorbance. The time required for the intensity of the light to fall to 1/e of the initial intensity is called the "ring-down time," which is a function of the sample within the cavity. Advantages of CRDS include: (1) it is not affected by fluctuations in laser intensity because it measures the decay rate of the laser signal, and (2) it is very sensitive due to the very long path length of the light within the reflective chamber (e.g. Berden et al., 2010). Commercial CRDS systems are small, portable, and relatively inexpensive and are capable of making measurements of isotopic ratios of carbon, oxygen, and hydrogen with accuracy and precision comparable to gas-source mass spectrometry.

A.6 Flight Instruments for Spacecraft Missions

A variety of different instruments have been put on spacecraft to determine the chemical compositions of planetary objects. Several of the instruments discussed above have been adapted as flight instruments. The technical requirements for instruments on orbiters are different from those on landers or rovers, but some techniques have been adapted to both.

A.6.1 X-Ray Fluorescence

Ground-based XRF, described above, uses a monochromatic X-ray beam to generate secondary X-rays with energies characteristic of the elements in the sample. XRF can be used from orbit to get compositional information from a planetary or asteroidal surface. In this application, the Sun provides the primary X-rays that generate the fluorescence spectrum from the target surface. The solar X-ray source is not very strong, so this technique is typically limited to a few low-atomic-weight rock-forming elements. XRF instruments were used by the NEAR Shoemaker, Hayabusa, Hayabusa2, and OSIRIS-REx spacecraft to analyze asteroid surfaces.

XRF has also been used on landers. XRF was part of the instrument package on the Viking landers that operated on Mars in 1976. In this case, the sample was picked up and placed inside the lander, where it was irradiated by two X-ray sources and the secondary X-rays were detected by gas proportional counters. The CheMin instrument on the Curiosity Mars rover is primarily an XRD instrument, but it has XRF capability. The PIXL instrument on the Mars 2020 Perseverance rover is a microfocus XRF instrument capable of gathering high-spatial-resolution data on the martian soil.

A.6.2 Alpha-Particle X-Ray Spectrometer

The alpha-particle X-ray spectrometer (APXS) technique is similar to, but much more sensitive than, traditional XRF. It combines two standard terrestrial methods, proton-induced X-ray emission (PIXE, see above) and XRF to determine surface chemistry. All Mars rover missions to date (except Mars 2020) have carried APXS instruments for chemical analysis of rocks and soils (see Fig. 14.14). The Pragyan rover on the Indian Chandrayaan-2 lunar mission also carried an APXS instrument. The source consists of radioactive ^{244}Cu, which decays with a short half-life to produce α-particles, which then irradiate the sample. Secondary X-rays characteristic of specific elements are then released and measured by a silicon-drift detector. The Mars Pathfinder APXS also measured the backscattered α-particles for detection of light elements, but the Mars Exploration Rovers measured only the X-rays. The instrument on the Curiosity rover determined major-, minor-, and trace-element abundances in the range from sodium to bromine.

A.6.3 Gamma-Ray and Neutron Spectrometers

Gamma-ray and neutron spectrometers are used from orbit to get compositional information about planetary surfaces. The principles of these techniques are described in Box 13.2. Both kinds of radiation, produced in the target by cosmic rays, are not directional, so the instruments sense them coming from the target as well as from the spacecraft itself (which is why, when possible, the instruments are mounted on a boom). Although γ-rays and neutrons are measured by separate instruments, they are commonly used together and the data are complementary. The measurements depend critically on counting statistics, which are functions of integration time and distance from the source. For normal spacecraft orbital altitudes, integration times for most detected elements are measured in months. Elements vary considerably in detection sensitivity and abundance, and only a few elements can usually be measured.

A neutron spectrometer separates detected neutrons into different energy ranges (thermal, epithermal, fast). Proportional counters containing ^{3}He are commonly used for thermal and epithermal neutron detection, whereas some other material, like a scintillator made of borated plastic, is used for fast neutrons. Neutrons are sensitive to light elements, especially hydrogen, or elements with especially high neutron capture cross sections.

There are a number of possible sensor options for a γ-ray spectrometer. These include a germanium sensor or scintillators made of various synthetic materials. Elements that are routinely analyzed with γ-rays include silicon, iron, titanium, magnesium, calcium, and aluminum, plus the radioactive elements potassium and thorium (uranium concentrations are usually too low).

Gamma-ray and neutron spectrometers were carried on the Lunar Reconnaissance Orbiter and the Lunar Prospector orbiter to the Moon, the Mars Odyssey orbiter to Mars, the MESSENGER orbiter to Mercury, and the Dawn mission to asteroids Vesta and Ceres.

A.6.4 Laser-Induced Breakdown Spectroscopy

Laser-induced breakdown spectroscopy (LIBS) is a type of atomic emission spectroscopy. It uses a high-energy laser pulse to atomize and excite the sample. The laser is focused onto a small area, and when it is discharged, it ablates a few atomic layers of the sample, generating a plasma plume with temperatures in excess of 100,000 K. Initially the plasma emits a continuum of radiation, which does not contain useful information. But as the plasma expands and cools, it passes through a temperature regime where the characteristic emission lines of the elements can be observed. The detector is gated only to measure when the emission lines are present. A LIBS instrument is part of the ChemCAM payload on the Curiosity rover, and part of the SuperCam payload for the Perseverance rover.

A.6.5 Mössbauer Spectrometer

Mössbauer spectroscopy is a technique based on the resonant emission and absorption of γ-rays in solids. A sample is exposed to γ-radiation, and a detector measures the intensity of the beam transmitted through the sample. The atoms in the γ-ray source must be of the same isotope as the atoms in the sample absorbing them. In this case, the emitted γ-rays will not lose energy to recoil and thus will have the right energy to be absorbed by the target atoms. The minor differences are attributable to the chemical environment of the target atoms, which is what we wish to measure. In cosmochemistry, ^{57}Fe is the isotope normally studied by Mössbauer spectroscopy. Three kinds of nuclear interactions (isomer shift, quadrupole splitting, and hyperfine splitting) are observed. These three parameters can often be used to identify a particular iron-bearing mineral, and this is also an important method for determining iron oxidation states. Mössbauer instruments were included on the Mars Exploration Rovers (Spirit and Opportunity).

A.6.6 Mass Spectrometers

Mass spectrometers have flown on a wide variety of spacecraft missions. The first mass spectrometers to fly were single-focusing magnetic-sector instruments. Mass spectrometers were deployed on booms attached to the Apollo 15 and 16 service modules on the Apollo 17 lander to measure the lunar atmosphere, but most of the measured gas probably came from the spacecraft. The Viking missions to Mars (1976) had two types of mass spectrometers. Neutral mass spectrometers mounted to the aeroshells that protected the Viking landers during atmospheric entry successfully measured the composition of the martian atmosphere. The Viking missions also carried GCMS instruments to search for organic molecules, but the results were ambiguous. Neutral mass spectrometers carried on atmospheric probes from the Pioneer Venus mission (1978) provided data on the composition of the Venusian atmosphere. A double-focusing mass spectrometer was included in the science payload of the Rosetta Mission, launched to comet 67P/Churyumov–Gerasimenko in 2004.

Quadrupole mass spectrometers have also been widely used. The Pioneer Venus Orbiter carried the first planetary quadrupole mass spectrometer, which measured the upper atmospheric gas densities. Subsequent instruments incorporating the same basic design were flown on the Galileo probe to study the Jovian atmosphere (1995), the Huygens probe to study Titan's atmosphere (2005), the LADEE neutral mass spectrometer to measure ions eroded from the lunar surface (2013–2014), and the MAVEN Neutral Gas and Ion Mass Spectrometer to characterize atmospheric gravity waves in the martian thermosphere (2014), among others. The Curiosity rover currently operating on Mars carries a sample-analysis package consisting of a modern quadrupole mass spectrometer that can be operated independently or with a gas chromatograph to make a GCMS instrument.

As electronics have improved, time-of-flight instruments have become more popular. Two such instruments were used to study comet Halley in 1987. The Russian Vega 2 spacecraft carried the PUMA instrument, and the European Space Agency's Giotto spacecraft carried the PIA instrument. Taking advantage of the high encounter speeds between spacecraft and comet, these time-of-flight mass spectrometers relied on collisions between the comet dust and a plate at the entrance to the mass spectrometer to provide the ions to be measured. The Rosetta mission also included time-of-flight mass spectrometers. The Cassini mission, which studied Saturn and its moons from 2004 to 2017, included a Cosmic Dust Analyzer, which was built around a time-of-flight mass spectrometer. Future missions will carry even more capable mass spectrometers. Europa Clipper, to be launched in 2025, will carry MASPEX, a time-of-flight mass spectrometer that employs multiple bounces to lengthen the flight path within the instrument and increase mass-resolving power.

For a full listing of missions and details of the various types of mass spectrometers used, see Arevalo et al. (2020).

A.7 Sample Preparation

A.7.1 Thin Section Preparation

Preparing thin sections of rocks and minerals is not conceptually difficult, but it is challenging to make high-quality thin sections. Thin sections are made either on standard rectangular glass slides (the dimensions differ in the United States and in Europe) or on one-inch-round glass slides. The latter are used if electron-probe or ion-microprobe analysis is planned because of the sample requirements of the instruments. First, a suitably sized chip of the rock is prepared. It is ground flat on one side and polished to a high polish. The polished surface is then glued to the glass slide using a suitable epoxy that has the same refractive index as glass. Sometime the glass is "frosted" first by grinding with a coarse grit to give the epoxy a rough surface to which to bond. The glass with the sample on it is put into a precision saw and the rock is cut off parallel to the surface of the glass within half a millimeter or so. The sample is then ground to near the official thickness of 30 μm and polished with various grades of diamond polish to achieve both the standard thickness and a high polish (1 μm diamond polish or better is the final step). The thickness is determined by checking the birefringence of known minerals in the section. In terrestrial samples, quartz is ideal for this purpose, but in meteorites, olivine is often used. A top-quality thin section has a uniform thickness across the section, a very high polish, and a very low degree of "plucking" of softer and more fragile materials. Although thin sections used for petrography in a petrographic microscope often have cover slips on them, if the sections are to be used for analytical work, a cover slip cannot be used. Cover slips are seldom seen on thin sections in cosmochemistry research.

A.7.2 Sample Preparation for EBSD

The EBSD technique measures the top few monolayers of a sample to determine its crystal structure. However, most polishing techniques leave a damaged layer on the samples. Even 1 or 0.25 μm diamond polish is not sufficient. For EBSD work, a final polish with colloidal silica (grain size = 0.05 μm) is often required to remove the damaged surface layers.

A.7.3 Sample Preparation for the TEM

The TEM has relatively strict sample requirements. As discussed above, the sample must be thinned to electron-beam transparency, typically less than 200 nm, and in some cases as thin as 50 nm. In addition, the sample must be small enough to fit into the TEM, on the order of <3 mm wide. There are several techniques used to thin samples to electron transparency. Ion milling has traditionally been used. An ion mill uses a beam of argon accelerated to high energy to sputter-thin the sample. The main drawback of this technique is that the thinned areas are on the edges of holes in the sample, and it is often impossible to choose a specific spot to make thin. Another way to obtain electron-transparent samples is to use a microtome. This device uses a very sharp and strong knife to cut slices from the sample of interest. Often the sample is mounted in epoxy prior to slicing in order to provide a substrate to control and keep track of the sample. A microtome can produce 50–200 nm slices. The slices are then mounted on copper-grids to transfer to the TEM. Microtome sections of minerals tend to fracture as the section is removed from the main sample.

A new way of preparing an electron-transparent section of small grains, such as presolar grains from meteorites, is the FIB lift-out technique. This technique permits one to prepare a section of a specific micron-sized grain. First, a mask of platinum is laid down over the grain of interest. The FIB then uses a tightly focused gallium beam to dig a trench on either side of the sample, leaving the thin, electron-transparent section behind. A micromanipulator needle is attached to the platinum mask. The sample is cut free, and the micromanipulator lifts the sample out and transfers it to a TEM grid. This technique is very time-consuming, but it is the only way to get a specific grain into the TEM for analysis. Note that a 100 nm-thick section can be produced with a 500 nm ion beam because the FIB removes the material from around the material of interest. Curf loss (the material lost to cutting) is quite high, and in fact, a single section is all one can get from a micron-sized grain.

A.7.4 Preparing Aerogel "Keystones"

The return of the Stardust mission with its treasure trove of comet dust posed a major challenge in sample preparation. The sample particles are dispersed along tracks in the aerogel, a material that is notoriously hard to work with. Andrew Westphal and his colleagues came up with a very clever way to extract complete particle tracks from the aerogel. This system uses a high-quality binocular microscope and a programmable micromanipulator. A very fine glass needle is prepared and given the required shape. It is inserted into a clamp in the arm of the micromanipulator. The micromanipulator is programmed to poke a series of closely spaced holes in the aerogel to cut out the track. It turns out that poking with a fine needle breaks the individual strands of the aerogel silica in a controlled way and leaves a clean trench around the track. A small silica "pickle fork" is inserted into the remaining aerogel around the track to serve as a handle. The aerogel "keystone" with the track inside is then lifted out of the aerogel block, ready for analysis (Westphal et al., 2004). The keystoning technique is one of the main ways of processing the Stardust samples for analysis.

A.7.5 Preparation of Samples for TIMS and ICPMS

Because the mass spectrometers used for TIMS and ICPMS measurements typically do not have the mass-resolving power required to separate the elements of interest from isobaric and molecular interferences, careful chemical separations must be carried out. Details of chemical procedures used by various laboratories are available in the published literature, although specialized knowledge derived from experience is typically necessary to implement the published procedures. To a first approximation, the initial step in all of these procedures is to get the sample into solution. Various acids, bases, and organic chemicals are used, sometimes in controlled environments. In some cases, in order to get everything into solution, high temperatures or high-pressure bombs must be used. This chemistry must be very clean so as not to introduce contaminants into the sample. Ultrapure water, acids, and bases are produced and used for this chemistry, which is typically done in a specialized clean laboratory.

Once the sample is fully into solution, it is then necessary to separate the components. Distillation is one means of separating volatile components from refractory ones. Ion-exchange columns are also important tools. Very effective ion-exchange resins are commercially available. Ion-exchange resins take advantage of reversible chemical reactions. For example, a resin with hydrogen ions available will exchange those ions for magnesium ions from solution according to the following reaction:

$$2(\text{R-SO}_3\text{H}) + \text{Mg}^{2+} = (\text{R-SO}_3)_2\text{Mg} + 2\text{H}^+ \qquad \text{A.3}$$

The R indicates the organic portion of the resin. Synthetic resins are produced as small porous beads, which are packed into a glass tube to make an ion-exchange column. The sample solution is poured into the top of the column and moves slowly down the column. The column is set up to have many more active sites than there are atoms of the element of interest in the solution, so that all of the atoms are absorbed in the column. The magnesium in the above example can be recovered from the resin by increasing the concentration of H_3O^+ in the solution (i.e. by passing an acid solution through the column). A wide variety of resins are available, including resins to attract positive ions and negative ions. A well-designed chemical procedure coupled with the right ion-exchange resins can produce clean chemical separations with high recovery (>90%).

An important method for determining the amount of an element is known as "isotope dilution." A spike, consisting of a known amount of a single isotope, is added to the sample. All the isotopes are then measured. The resulting spectrum consists of all of the isotopes of the target element in their natural relative abundances, plus the extra atoms of the spike isotope. The mixture can be deconvolved mathematically to give the abundance of the target element. One can use a double spike to determine the instrumental mass fractionation and thereby determine the intrinsic mass fractionation in the sample. In this case, a known mixture of two highly purified isotopes is added to the sample. For isotope dilution to work properly, care must be taken to insure that the spike is completely chemically equilibrated with the sample.

A.8 Details of Radiometric Dating Systems that Use Neutron Activation

A.8.1 ^{40}Ar–^{39}Ar dating

The ^{40}Ar–^{39}Ar system was developed to avoid the necessity of making a separate measurement of the potassium content, which must be done on a separate aliquot of the sample. In ^{40}Ar–^{39}Ar dating, the sample is irradiated with fast neutrons, and a portion of the ^{39}K is converted to ^{39}Ar*. The sample is then placed in an ultra-high-vacuum system and the argon is extracted by heating the sample in a series of increasing temperature steps. The argon released at each step is purified and admitted to a mass spectrometer for measurement. With this technique, the ^{39}Ar* from the irradiation and the ^{40}Ar* from the decay of ^{40}K are directly associated, minimizing the corrections for unrelated ^{40}Ar. By extracting the argon at different temperature steps, the argon in different minerals can be extracted separately, giving information about the thermal history of the rock and allowing isotopic disturbance of the sample to be recognized.

In order to obtain a date from the measurement of an irradiated sample, it is necessary to know the proportion of ^{39}K that was converted to ^{39}Ar. The number of ^{39}Ar* atoms formed by irradiation of the sample is given by

$$^{39}Ar^* = {}^{39}K\Delta T \int \phi(\varepsilon)\sigma(\varepsilon)\mathrm{d}\varepsilon \qquad \text{A.4}$$

where ^{39}K is the number of ^{39}K atoms in the sample, ΔT is the duration of the irradiation, $\phi(\varepsilon)$ is the neutron flux density at energy ε, and $\sigma(\varepsilon)$ is the capture cross section of ^{39}K for neutrons of energy ε. The integration is carried out over the entire neutron energy spectrum. However, the neutron flux density and capture cross sections for ^{39}K are not well known. To calculate the date from the irradiated sample, it is useful to introduce a parameter J, defined as:

$$J = \frac{\lambda_e}{\lambda} \frac{{}^{39}K \Delta T}{{}^{40}K} \int \phi(\varepsilon)\sigma(\varepsilon) d\varepsilon \qquad \text{A.5}$$

where λ_e is the decay constant for electron capture and λ is the combined decay constant. The value of J can be determined by including a sample of known age in the irradiation as a flux monitor. Because the decay constant of ^{40}K is known, and the number of ^{40}Ar* and ^{39}Ar* atoms in the flux monitor can be measured, the value of J can be calculated as follows:

$$J = \frac{e^{\lambda t_m} - 1}{\left({}^{40}Ar^* / {}^{39}Ar^*\right)_m} \qquad \text{A.6}$$

where t_m is the known age of the flux monitor. The date for each step of the ^{40}Ar–^{39}Ar measurement can then be calculated from

$$t = \frac{1}{\lambda} \ln \left(\frac{{}^{40}Ar^*}{{}^{39}Ar^*} J + 1 \right) \qquad \text{A.7}$$

As with all radiometric dating systems, the ^{40}Ar–^{39}Ar system gives a valid age if several conditions are satisfied. In addition to the conditions given in Chapter 8, there are several conditions that apply specifically to the ^{40}K–^{40}Ar and ^{40}Ar–^{39}Ar systems and to extraterrestrial samples. 1) The abundance of ^{40}K in natural potassium is assumed to be constant. This assumption is required because ^{40}K is almost never measured directly and must be inferred from the total potassium content. Small shifts in the potassium isotopic abundances due to mass-dependent fractionation can occur, but these are limited to $<3\%$. In extraterrestrial samples, cosmic-ray-induced spallation reactions on ^{40}Ca can produce ^{40}K and alter the isotopic abundance. However, in most cases, the cosmic-ray exposure occurred late in the history of the object and the additional ^{40}K has not had time to significantly affect the abundance of ^{40}Ar*. In these cases, the additional ^{40}K can be ignored. 2) The radiogenic ^{40}Ar* measured in the sample must have been produced *in situ* by the decay of ^{40}K. 3) Suitable corrections can be made for nonradiogenic ^{40}Ar present in the object being dated. Argon-40 is the dominant argon isotope in the Earth's atmosphere. Atmospheric ^{40}Ar can be accounted for by measuring ^{36}Ar and subtracting an amount of ^{40}Ar equal to the $^{36}Ar_{measured} \times (^{40}Ar/^{36}Ar)_{air}$. Extraterrestrial samples may have nonradiogenic ^{40}Ar from other sources, but in most cases satisfactory corrections can still be made. 4) The sample must have remained a closed system since the event being dated. No addition or loss of potassium or ^{40}Ar* can have occurred. Many samples do not entirely meet this requirement, particularly if they have had a complex thermal history. But the ^{40}Ar–^{39}Ar technique permits the scientist to extract not only the formation age, but also additional information about the thermal history of many such objects. A good review of the ^{40}Ar–^{39}Ar technique is given by McDougall and Harrison (1988).

A.8.2 ^{129}I–^{129}Xe dating

Iodine-129 was the first short-lived radionuclide shown to have been present in the early solar system. It was difficult to prove that the excess ^{129}Xe found by Reynolds (1960a, 1960b) in meteorites was due to the decay of ^{129}I. A means had to be found to show that the excess ^{129}Xe was associated with iodine. Jeffery and Reynolds (1961) solved the problem by irradiating a meteorite sample with thermal neutrons in order to convert ^{127}I to ^{128}Xe. They were then able to show that the radiogenic ^{128}Xe* from the irradiation of ^{127}I, and the radiogenic ^{129}Xe* from the decay of ^{129}I were in the same crystallographic sites. The parent isotopes of ^{128}Xe* and ^{129}Xe* had both been iodine.

It has been difficult to establish the ^{129}I–^{129}Xe system as a chronometer in part because iodine is quite mobile in the environment. The neutron irradiation method has turned out to be the only practical way of distinguishing the primordial iodine in a sample as associating with the ^{129}Xe*. In addition, it is only necessary to measure the xenon isotopic abundances to derive the abundances of both the parent element and the daughter isotope. Even though xenon is present in very low abundances in most samples, it is easy to extract and purify for measurement in a gas-source, isotope-ratio mass spectrometer. By coupling neutron irradiation with stepped heating to extract the xenon, iodine that is in different sites from the radiogenic ^{129}Xe in the sample can be identified and excluded from the final result.

If the iodine in a system is isotopically homogeneous, then a suite of objects forming at the same time from that system will have the same ^{129}I/^{127}I ratio. If the system remains closed, then ^{129}Xe* from the decay of ^{129}I will remain in association with ^{127}I. In an iodine-bearing phase, the measured ^{129}Xe consists of radiogenic ^{129}Xe* from the decay of ^{129}I and the ^{129}Xe that was originally trapped in the sample ($^{129}Xe_t$). In the same way, the measured ^{128}Xe consists of ^{128}Xe* produced by irradiation from the ^{127}I and the $^{128}Xe_t$ originally trapped in the sample. Thus, both ^{129}Xe and ^{128}Xe will be two-component mixtures between a trapped component and an iodine-derived component, and, if the assumptions of radiometric dating have been met, measured samples of the same age will plot on a single line. The xenon isotope ratios are calculated with respect to either ^{132}Xe, the most abundant isotope, or to ^{130}Xe, an isotope shielded from production by fission. In the case where the xenon isotopes are normalized to ^{132}Xe, the equation of the line, the isochron on a ^{129}Xe/^{132}Xe versus ^{128}Xe/^{132}Xe plot such as Figure 9.26, is:

$$\frac{^{129}\mathrm{Xe}}{^{132}\mathrm{Xe}} = M \times \frac{^{128}\mathrm{Xe}}{^{132}\mathrm{Xe}} + \left[\left(\frac{^{129}\mathrm{Xe}}{^{132}\mathrm{Xe}}\right)_t - M \times \left(\frac{^{128}\mathrm{Xe}}{^{132}\mathrm{Xe}}\right)_t\right] \quad \text{A.8}$$

where M is the slope (^{129}Xe*/^{128}Xe*), and the term in brackets is the trapped component represented by the end point of the line at the lower left on Figure 9.26 (analogous plots and equations can be created if ^{130}Xe is used for normalization). The trapped component is typically the dominant noble gas component in chondrites, known as Xe-P1 or Xe-Q. Although the measurements must be corrected for blank in the mass spectrometer and for small amounts of adsorbed air, these corrections are usually very small and the composition of atmospheric xenon has similar isotopic ratios to the trapped component, so uncertainties in the corrections do not affect the slope of the isochron.

It turns out to be quite difficult to determine the absolute ^{127}I abundance through the neutron irradiation method. This is because typical iodine-bearing minerals have iodine concentrations many orders of magnitude larger than the meteorite samples of interest, which results in nonlinear activation behavior in the reactor (e.g. Hohenberg and Pravdivtseva, 2008). Thus, the absolute ^{129}I/^{127}I ratio for the early solar system is not particularly well known. In practice, this does not affect ^{129}I–^{129}Xe dating because the Shallowater aubrite is used as a flux monitor. In this meteorite, the iodine is sited in the enstatite and thus is tightly bound, and the igneous history of the meteorite means that all samples of the meteorite have the same ^{129}Xe/^{127}I ratio. A piece of Shallowater is included in every irradiation, so the initial ^{129}I/^{127}I ratios of the unknown samples are determined relative to the ratio for Shallowater (Fig. 9.26). Over the last few years, the absolute age of the Shallowater meteorite of 4563.3 ± 0.4 Ma has been determined through use of several isotope chronometers. This has placed the ^{129}I–^{129}Xe ages determined relative to Shallowater on an absolute scale.

References and Further Reading

Arevalo R., Ni Z., and Danell R. M. (2020) Mass spectrometry and planetary exploration: A brief review and future projection. *Journal of Mass Spectrometry*, **55**, e4454.

Beck P., Garenne A., Quirico E., et al. (2014) Transmission infrared spectra (2–25 μm of carbonaceous chondrites (CI, CM, CV-CK, CR, C ungrouped): Mineralogy, water, and asteroidal processes. *Icarus*, **229**, 263–277.

Berden G., Peeters R., and Meijer G. (2010) Cavity ring-down spectroscopy: Experimental schemes and applications. *International Reviews in Physical Chemistry*, **19**, 565–607.

Cullity B. D. (1978) *Elements of X-ray Diffraction.* Addison-Wesley, Reading, Massachusetts, 555 pp.

Goldstein J., Newbury D., Joy D., et al. (2003) *Scanning Electron Microscopy and X-ray Microanalysis*. Springer, New York, 689 pp.

Gross J. H. (2017) *Mass Spectrometry – A Textbook, 3rd edition.* Springer-Verlag, Berlin, 968 pp.

Hohenberg C. M., and Pravdivtseva O. V. (2008) I-Xe dating: From adolescence to maturity. *Chemie der Erde*, **68**, 339–351.

Jarosewich E. (1990) Chemical analyses of meteorites: A compilation of stony and iron meteorite analyses. *Meteoritics*, **25**, 323–337.

Jeffery P. M., and Reynolds J. H. (1961) Origin of excess Xe^{129} in stone meteorites. *Journal of Geophysical Research*, **66**, 3582–3583.

Kelly T. F., and Larson D. J. (2012) Atom probe tomography 2012. *Annual Reviews of Materials Research*, **42**, 1–31.

McDougall I., and Harrison M. T. (1988) *Geochronology and Thermochronology by the ^{40}Ar/^{39}Ar method.* Oxford Monographs on Geology and Geophysics No. 9, Oxford University Press, New York, 212 pp.

Nagashima K., Kunihiro T., Takayanagi I., et al. (2001) Output characteristics of stacked CMOS-type active pixel sensor for charged particles. *Surface and Interface Analysis*, **31**, 131–137.

Nan C., Yue W., Tao L., and Yang X. (2020) Fourier transform infrared nano-spectroscopy: Mechanism and applications. *Applied Spectroscopy Reviews*, doi.org/10.1080/05704928.2020.1830789.

Reed S. J. B. (1993) *Electron Microprobe Analysis.* Cambridge University Press, Cambridge, 322 pp.

Reynolds J. H. (1960a) Determination of the age of the elements. *Physical Reviews Letters*, **4**, 8–10.

Reynolds J. H. (1960b) Isotopic composition of xenon from enstatite chondrites. *Zeitschrift für Naturforschung*, **15a**, 1112–1114.

Smith E., and Dent G. (2005) *Modern Raman Spectroscopy, A Practical Approach.* John Wiley and Sons, West Sussex, England, 210 pp.

Stephan T., Trappitsch R., Hoppe P., et al. (2019) Molybdenum isotopes in presolar silicon carbide grains: Details of *s*-process nucleosynthesis in parent stars and implications for *r*- and *p*-process. *Astrophysical Journal*, **877**, 101.

Trappisch R., Boehnke P., Stephan T., et al. (2018) New constraints on the abundance of ^{60}Fe in the early solar system. *Astrophysical Journal*, **857**, L15.

Wiik H. B. (1956) The chemical composition of some stony meteorites. *Geochimica et Cosmochimica Acta*, **9**, 279–289.

Westphal A. J., Snead C., Butterworth A., et al. (2004) Aerogel keystones: Extraction of complete hypervelocity impact events from aerogel collectors. *Meteoritics and Planetary Science*, **39**, 1375–1386.

Williams D. B., and Carter C. B. (2009) *Transmission Electron Microscopy: A Textbook for the Material Sciences*. Springer, New York, 832 pp.

Glossary

Absolute age – radiometric age that gives the time relative to today

Achondrites – igneous meteorites that have undergone melting and differentiation; these may be solidified melts or cumulates formed by accumulation of crystals from a magma

Accretion disk – a flattened disk of gas and dust surrounding an infant star, from which planets can accrete

Actinium series – decay series starting with ^{235}U and ending in ^{208}Pb, consisting of seven alpha decays and four beta decays

Agglutinate – impact-melt products consisting of small particles of minerals or rocks bonded together by glass

Albedo – the measured reflectivity of an object

Alpha (α) decay – the decay mode of a radioactive isotope where an α-particle, which consists of two protons and two neutrons (a ^{4}He nucleus), is emitted, lowering the atomic number by two and the atomic mass by four

Alpha particle (α-particle) – a ^{4}He nucleus containing two protons and two neutrons

Aliphatic – organic molecules in which carbon atoms are linked into chains

Amino acids – organic compounds containing both the amino and carboxylic acid functional groups

Amoeboid olivine aggregates (AOAs) – irregularly shaped refractory inclusions composed primarily of forsterite and iron–nickel metal, with lesser amounts of other minerals found in CAIs

Aromatic – organic compounds in which carbon atoms are linked to form rings

Asteroid family – fragments of a disrupted, larger body that share similar orbital characteristics

Asymptotic giant branch (AGB) – a region on the Hertzsprung–Russell diagram that lies above and roughly parallel to the red giant region; it is occupied by stars that have finished burning helium in the core and are experiencing helium-shell flashes and third dredge-up mixing events

Atmophile – gaseous compounds at the temperatures of the Earth's surface and above; they are typically sited in the atmospheres of terrestrial planets and are depleted relative to their cosmic abundances

Atom – the smallest unit of ordinary matter that constitutes a chemical element; it consists of a nucleus of protons and neutrons surrounded by a cloud of electrons

Atomic mass unit (amu) – used to describe the mass of an atom, equal to 1/12th of the mass of a ^{12}C atom (also called a *unified atomic mass unit*)

Atomic number (Z) – the number of protons in the nucleus of an atom

Atomic weight of an element – the abundance-weighted average value of masses of the various stable isotopes of an element

Baryogenesis – the hypothetical process in the early universe that produced the excess of matter (baryons) over antimatter (antibaryons) in the observed universe

Baryon – a subatomic particle composed of three quarks that has a mass equal to or greater than that of a proton; protons and neutrons are baryons

Beta (β^-) decay – the decay mode of a radioactive isotope where a neutron emits an electron and becomes a proton, increasing the atomic number by one

Beta particle (β-particle) – energetic electron emitted by radioactive decay

Black-body spectrum – the spectrum of electromagnetic radiation emitted by an idealized, opaque, nonreflective body (a black body) that is in thermodynamic equilibrium with its surroundings. The range of wavelengths in the emitted spectrum is inversely proportional to the temperature of the body

Black dwarf – a white dwarf that has cooled to the point that it is no longer luminous

Blocking temperature – the temperature at which diffusion of an element in a mineral effectively stops; the blocking temperature is different for every element

Blue-shift – in astronomy, the displacement of the electromagnetic spectrum to shorter wavelengths (higher frequency), caused by a celestial object moving toward the observer

Boundary diffusion – diffusion of atoms along grain boundaries in a polycrystalline solid

Breccias – rocks formed from broken fragments of pre-existing rocks; these can be monomict (clasts of the same chondrite group or achondrite rock type) or polymict (clasts of different chondrite groups or achondrite rock types)

Calcium-aluminum-rich inclusions (CAIs) – refractory inclusions composed of minerals that condense or melt at high temperatures (hibonite, corundum, perovskite, melilite, spinel, diopside, anorthite, forsterite); their ancient radiometric ages indicate that they were the first-formed solids in the solar system

Carbon burning – occurs only in stars more massive than 8–10 $M_{\odot}$; converts ^{12}C into ^{16}O, ^{20}Ne, ^{24}Mg, and ^{23}Na

Carbonaceous chondrite fission xenon (CCFXe) – a component enriched in heavy xenon isotopes, the origin of which was attributed to fission of an unknown heavy element

Carbonaceous chondrites – highly variable group of chondrites traditionally thought to have high abundances of organic matter

Centaur asteroids – planetesimals orbiting between the giant planets

Chandrasekhar mass – a limit giving the largest stellar mass that can be supported against gravitational contraction by electron degeneracy; for matter with equal numbers of protons and neutrons, the Chandrasekhar mass is ~1.44 $M_{\odot}$

Chalcophile – elements that prefer to bond with sulfur

Chirality – an organic molecule that cannot be superposed on its mirror image by any combination of rotations and translations is said to be chiral; this property is called chirality

Chondrite mixing model – a way to estimate a planet's bulk composition by mixing various classes of chondritic meteorites

Chondrites – meteorites that have a nearly average solar system elemental composition, except for depletion of the most volatile elements

Chondritic porous IDPs (CP IDPs) – fluffy-textured interplanetary dust particles consisting primarily of anhydrous silicates; thought to be derived from comets

Chondritic smooth IDPs (CS IDPs) – compact interplanetary dust particles that often contain hydrous silicates; thought to come from asteroids

Chondrules – quenched molten droplets of ultramafic composition that are the dominant component in most chondritic meteorites

Circumstellar condensate – a presolar grain that condensed in the atmosphere or ejecta of a dying star (commonly called *stardust*)

Clathrate hydrates – crystalline water-based solid physically resembling ice, in which liquid or gas molecules are trapped inside cages of hydrogen-bonded, frozen water molecules

Compatible element – an element that substitutes readily in a common rock-forming mineral because of similar ionic size and charge

Compound-specific isotope analysis – the analysis of stable isotopes in individual organic species

Concordia – locus of points on a $^{206}Pb/^{238}U$ versus $^{207}Pb/^{235}U$ plot that describes concordant decay of the two radioactive uranium isotopes

Condensation – formation of solids or liquids from a cooling gas

Condensation sequence – the calculated sequence of solid phases formed from a cooling gas at a particular pressure, typically done in terms of a gas of solar composition

Condensation temperature 50% – the temperature at which half of an element has condensed from gas to solid; a common measure of element volatility

Contact binary – two small objects physically connected by mutual gravitational attraction

Core accretion model – a model for forming the giant planets, whereby rocky cores accreted first, and once they became massive enough, they gravitationally attracted nebular gas or ices

Cosmic composition (or abundances) – the abundance of elements in the universe, often incorrectly said to be equivalent to solar system abundances

Cosmic microwave background – in Big Bang cosmology, the electromagnetic radiation that derives from a very early stage of the universe, also known as "relic radiation"; this radiation has been red-shifted until it lies in the microwave region of the electromagnetic spectrum

Cosmic ray – high-energy particles, mostly protons and α-particles, that originate both in the Sun (solar cosmic rays) and in interstellar space (galactic cosmic rays), which typically travel at near the speed of light

Cosmic-ray exposure age – the time a meteoroid spends in space or on the surface of an airless body where it is exposed to cosmic rays, determined from the production of cosmogenic nuclides

Cosmic inflation – a theory of the exponential expansion of space during an extremely brief period almost immediately after the Big Bang; cosmic inflation appears to explain much of the large-scale structure of the universe

Cosmochemistry – study of the chemical compositions of the universe and its constituents and the processes that produced those compositions

Cosmogenic nuclide – isotope formed by interaction with cosmic rays

Coupled diffusion – diffusion involving two atoms, both of which must move in order for diffusion to occur

Covalent bond – a covalent bond forms when electrons are shared between two atoms; usually each atom contributes an electron to form a pair of shared electrons

Cumulate – an igneous rock formed by the accumulation of crystals by either settling or floating in a magma

Dalton (Da) – unit used to describe the mass of an atom, defined as 1/12th of the mass of a ^{12}C atom (also called a *unified atomic mass unit (u) or Dalton*)

Dark energy – a mysterious form of energy postulated to explain the apparent acceleration of the expansion of the universe over the last few billion years

Dark matter – a form of matter that cannot be detected by electromagnetic radiation and is thought to account for approximately 85% of the matter in the universe

Differentiation – change in chemical composition, as in differentiation of a planet into core, mantle, and crust, or as in differentiation of a magma by fractional crystallization

Disk fragmentation model – a model for forming giant planets, whereby a gravitationally unstable accretion disk fragments into self-gravitating clumps of gas and dust that then contract to become planets

Distribution coefficient (D) – describes how an element is partitioned between a mineral and a coexisting melt under equilibrium conditions; defined as concentration of element in mineral/concentration of element in liquid; can vary with melt composition, temperature, and pressure

Doppler shift – the change in frequency of a wave as experienced by an observer who is moving relative to the wave source, or by a stationary observer of a moving wave sources

Dwarf planet – a solar system object smaller than a planet but massive enough to adopt a nearly spherical shape but not to clear out other bodies in its orbit

Ejecta – material ejected from impact craters

Electron – a stable subatomic particle with a negative electrical charge and a mass of 1/1836th that of a proton

Electron degeneracy – a consequence of the Pauli exclusion principle, which says that two identical half-integer-spin particles (in this case electrons) cannot occupy the same quantum state; for extremely dense states of matter at sufficiently low temperature, such as in normal white dwarf stars, electrons are forced into their lowest available energy states (electron-degenerate matter); electron-degenerate matter with a mass less than the Chandrasekhar mass (~1.44 $M_{\odot}$) cannot be compressed further

Enantiomers – one of two stereoisomers that are mirror images of each other and are not superimposable; enantiomers rotate polarized light in different directions

Enstatite (E) chondrites – a class of chondrites distinguished by their highly reduced minerals

Equilibrium – a state in which opposing forces or influences are balanced

Equilibrium condensation model – in the context of planet formation, a model that assumes that the accreted solids thermally equilibrated with the enclosing nebular gas; a more general definition assumes that the solids in the solar system condensed from a gas of solar composition

Evaporating gaseous globules (EGGs) – gravitationally bound cores of gas and dust, a few hundred AU in size, usually, but not always containing a protostar, that have been revealed and are being evaporated by the ultraviolet flux from nearby O and B stars.

Evaporation – transformation of a solid (in the nebula) or liquid (in the nebula or on a parent-body surface) to a gas

Exoplanets – planets outside our solar system; most orbit other stars

Extinct radionuclide – term sometimes used for a short-lived radionuclide with a half-life short enough that the original atoms of the nuclide in the solar system have almost entirely decayed away

Ferroan anorthosites (FAN) – the dominant rocks of the lunar highlands, composed mostly of plagioclase with small amounts of pyroxene and olivine; ferroan anorthosites represent the cumulate crust formed in the lunar magma ocean

First dredge-up – convective mixing event in a star at the end of core hydrogen burning that mixes the products of partial hydrogen burning from just outside the core throughout the envelope

Fission – the decay mode of a heavy radioactive isotope where an atom splits into two or more smaller particles; fission can happen spontaneously or can be induced by capturing a neutron

Fractionation – separation of elements or isotopes from one another by any process; also, shorthand for fractional crystallization of a magma

Fractional crystallization – separation of crystals from a cooling magma, resulting in differentiation

Functional groups – in organic chemistry, groups of elements bonded to a carbon backbone which are usually reactive

Fusion – nuclear reaction where two or more nuclei combine into a new heavier nucleus plus subatomic particles

Gas giants – Jupiter and Saturn, whose compositions are dominated by hydrogen and helium in gaseous or metallic forms

Geochemical affinity – the tendency of an element to occur in silicates or oxides (*lithophile*), to combine with iron into metal alloys (*siderophile*), to react with sulfur to form sulfides (*chalcophile*), or to form gases (*atmophile*)

Geochemistry – study of the chemical composition of the Earth; the term can be extended to other planets (*planetary geochemistry*) that have experienced geologic processing (e.g. melting and differentiation)

Geothermometer – a natural phenomenon that provides information about temperature; commonly, geothermometers use chemical or isotopic exchange reactions between coexisting minerals to quantify temperature

Glass with embedded metal and sulfide (GEMS) – micron- to submicron-sized particles that are common constituents of interplanetary dust particles

Gluon – an elementary particle that acts as the exchange particle for the strong force between quarks

Grand Tack model – model that posits an early migration of the giant planets inward and then outward again during the nebula phase; such a scenario would have scrambled the orbits of planetesimals and depleted the asteroid belt

Granularity (of the Sun's surface) – term used to describe areas of convective upwelling and downwelling of material of different temperatures at the Sun's surface

Half-life – the time it takes for half of the atoms of a radioactive nuclide to decay

Helium burning – combination of nuclear reactions that converts three α-particles into a ^{12}C atom (also called the triple-alpha process)

Helioseismology – the study of oscillations in the Sun

Hertzsprung–Russell (H–R) diagram – plot of stellar luminosity (variously plotted as luminosity, absolute bolometric magnitude, apparent visual magnitude, etc.) versus effective surface temperature (variously plotted as temperature, color index B–V, or spectral class)

Heterogeneous accretion model – planet formation model in which sequentially accreted materials have different compositions

High-mass stars – stars of >~8 $M_\odot$; these stars are massive enough to burn all of their nuclear fuel, producing an iron core; with no more fuel, the star collapses and explodes as a supernova

Highlands – ancient mountainous terrains on the Moon, composed mostly of anorthosite

Hubble constant – describes how fast the universe is expanding at different distances from a particular point in space; the farther away the object, the faster it is receding; numerical values for the Hubble constant range from 67.4 to 73.4 km/s/Mpc

Hydrocarbons – organic compounds consisting only of carbon and hydrogen

Hydrogen bond – a weak bond between two molecules resulting from an electrostatic attraction between a proton in one molecule and an electronegative atom in the other; hydrogen bonding is responsible for many of the properties of water

Hydrogen burning – the conversion of four hydrogen nuclei (protons) into a ^4He atom (two protons plus two neutrons) in the interior of stars; in stars somewhat more massive than our Sun, catalyzed hydrogen burning using the CNO cycle transforms hydrogen to helium

Hydrostatic equilibrium – in astronomy, the balance between thermal pressure (outward) and the pressure caused by gravity (inward) that maintains a star in a steady state and allows it to shine; the thermal pressure is generated by nuclear reactions during most of a star's life

Ice giants – Uranus and Neptune, giant planets containing large amounts of ices

Incompatible element – an element that does not fit comfortably into the crystal structures of rock-forming minerals because of its large size and/or high ionic charge

Initial mass function (IMF) – the distribution of stellar masses in a single star-forming event in a given volume in the galaxy

Insoluble organic matter (IOM) – complex, high-molecular-weight compounds that cannot be separated with solvents

Interdiffusion – the situation in which two different components are diffusing in such a way as to homogenize the material; concentration gradients in diffusion couples are produced by interdiffusion

Intermediate mass stars – stars of (~3 $M_\odot$ to ~8 $M_\odot$); these stars are hot enough to burn newly synthesized ^{12}C at the base of the stellar envelope, preventing them from becoming carbon stars; they end their lives as white dwarfs and do not explode as core-collapse supernovae

Internal isochron – an isochron constructed from measurements of radioisotopes in different minerals that formed together in the same rock

Interplanetary dust particle (IDP) – a small grain, generally less than a few hundred micrometers in size, in orbit around the Sun or having been recovered on Earth after being slowed down in the stratosphere; IDPs are responsible for the zodiacal light

Interstellar grains – presolar grains from meteorites and interplanetary dust particles that formed in interstellar space without direct association with a star

Interstellar medium (ISM) – the space between the stars

Interstitial diffusion – diffusion of atoms through interstitial sites rather than through lattice vacancies

Ion – an atom that has either lost or gained one or more electrons, giving it a positive or negative charge

Ionic bond – a chemical bond formed between two ions with opposite charges; one atom gives up one or more electrons to another atom, producing a filled outer electron shell and providing the charge that holds the compound together

Iron meteorites – meteorites composed primarily of FeNi metal (kamacite and taenite); most represent samples of metal cores in differentiated asteroids

Island of stability – in nuclear physics, a set of isotopes of super-heavy elements predicted to have considerably longer half-lives than other isotopes of similar mass due to their specific nuclear structure; on the Chart of the Nuclides, they would appear as an island separated from known naturally occurring nuclides

Isobar – each of two or more nuclides that have the same atomic mass (*A*)

Isomers – each of two or more organic compounds with the same chemical formula but a different geometric arrangement of atoms in the molecule and different properties

Isotope – any of two or more forms of the same element having the same number of protons (the same atomic number, *Z*) but different numbers of neutrons (different atomic masses, *A*)

Isotopic dichotomy – the separation of nebular and planetary materials into two groups (carbonaceous and noncarbonaceous) by distinctive nucleosynthetic isotopic compositions

Isotopologues – molecules that differ only in their isotopic composition

Kinetics – a branch of chemistry concerned with studying the rates of reactions

Kirkwood gaps – gaps in heliocentric distance and inclination in the Main asteroid belt, caused by gravitational resonances with Jupiter

KREEP – lunar rock type rich in potassium (K), rare earth elements (REEs), and phosphorus (P), all of which are incompatible; KREEP may represent the last dregs of the lunar magma ocean

Kuiper belt – the belt of icy planetesimals extending outward from beyond Neptune to ~50 AU; also called the Edgeworth–Kuiper belt

Kuiper belt objects (KBOs) – icy bodies located in the Kuiper belt

Late heavy bombardment – a period of intense impacts on the Moon (and other planetary bodies) by large planetesimals, thought to have occurred between ~4.1 and ~3.8 Ga, significantly after the main accretion of the Moon; an alternative model proposes that the late heavy bombardment actually represents the tail end of the accretionary process

Lepton – any of six types of subatomic particles that respond only to the electromagnetic force, weak force, and gravitational force and are not affected by the strong forces; they can carry one unit of electrical charge or can be neutral; leptons include the electron, muon, and tau and their antiparticles

Lithophile – rock-loving elements that predominantly bond with oxygen and preferentially partition into silicate or oxide minerals, which dominate the crusts and mantles of the terrestrial planets and asteroids

Long-lived radionuclide – a radioactive nuclide with a half-life long enough that some of its primordial solar system atoms have survived until today

Low-mass stars – stars with masses $<\sim 3\ M_\odot$; low-mass stars typically become carbon stars as ^{12}C from helium burning is dredged up into the envelope during the asymptotic giant branch phase

Magic numbers – in nuclear physics, the number of nucleons (either protons or neutrons) that fill a shell in the atomic nucleus; the numbers 2, 8, 20, 28, 50, 82, and 126 are considered magic

Main asteroid belt – belt of planetesimals located between the orbits of Mars and Jupiter

Main sequence – the continuous and distinctive band of stars on a Hertzsprung–Russell diagram extending from the upper left to the lower right; most stars spend the majority of their life on the main sequence as they burn hydrogen to helium in their cores

Magma ocean – a deep, globe-encircling body of magma, formed by the melting of a significant portion of a planet, moon, or asteroid early in its history

Maria (singular: Mare; Latin for "seas") – large expanses of basaltic lava filling impact basins on the near side of the Moon

Martian meteorites (SNCs) – shergottites (basalts and gabbros), nakhlites (clinopyroxenites), chassignites (dunites), and several unique cumulates or breccias from Mars

Mass defect – the difference between the mass of a nucleus and the masses of the individual protons and neutrons that make it up; the missing mass is the energy that was released when the nucleus was formed

Mass-dependent isotope variations – differences caused by fractionation of isotopes according to their relative masses, usually accomplished by geologic processes

Mass excess – the difference between the mass of a nuclide and the mass of the most stable nuclide of the same atomic mass (*A*)

Mass-independent isotope variations – differences caused by fractionation of isotopes that is independent of their masses, usually accomplished by nucleosynthetic processes

Mass number (A) – the number of protons plus neutrons in the nucleus of an atom

Matrix – the fine-grained portion of chondrites that fills the space between chondrules, chondrule fragments, metal, and troilite grains

Mean density – the average density (mass/volume) for a solid body; *uncompressed mean density* has been corrected for self-compression in large planets

Mean life – the average life expectancy of a radioactive atom, equal to the half-life divided by the natural log of two (~0.693)

Metal – in astronomy, all elements heavier than hydrogen and helium; in conventional usage, lustrous material, usually alloyed with iron, that conducts electricity and heat

Metallic bond – a type of chemical bond formed between positively charged atoms in which the valence electrons are shared among a lattice of cations

Metallicity (Z) – in astronomy, the mass fraction of elements other than hydrogen and helium; typically denoted by the symbol *Z*, with *X* and *Y* referring to the proportions of hydrogen and helium, respectively

Metal–silicate fractionation – the separation of metal from silicate observed in the compositions of meteorites and planets; could have resulted from core formation, by gravitational setting of metal in a quiescent nebula or aerodynamic sorting in turbulent nebular eddies, or by glancing impacts that stripped mantle and crust from differentiated bodies

Meteor shower – a shower of tiny meteors that occurs when the Earth passes through a meteor stream

Meteor stream – trail of small debris left in the wake of a comet

Micrometeorites – small particles ranging from ~50 micrometers to 2 millimeters that have survived passage through the atmosphere to arrive on the Earth's surface; many were melted during atmospheric passage and are called ablation spheres

Mineral – naturally occurring solid substance with a unique crystal structure and a specific chemical composition or limited range of composition

Model isochron – an isochron constructed from one or more measurements with very similar parent/daughter element ratios and an assumed value for the initial isotopic composition of the system

Model age – The time of various fractionation events in planetary bodies can be inferred from models of the behavior of radionuclides during and after the event. Models of the ^{87}Rb–^{87}Sr and U-Pb systems follow the evolution of isotopic reservoirs through proposed fractionation events. In the ^{147}Sm–^{142}Nd, ^{176}Lu–^{176}Hf, and ^{187}Re–^{187}Os, systems model ages are calculated relative to the chondritic uniform reservoir (CHUR). Differences in parent/daughter ratio cause evolution of the daughter isotopic composition of a rock at a different rate than CHUR. The model age, for a single stage model, is the time when isotopic composition of the rock matches that of CHUR. This gives the time of separate of the rock reservoir from CHUR.

Molecular clouds – cold, relatively dense clouds of interstellar gas molecules and ice-coated dust in the interstellar medium

Moment of inertia factor – a dimensionless quantity that characterizes the radial distribution of mass inside a planet

Near-Earth objects (NEOs) – comets and asteroids with elliptical orbits that approach or cross the Earth's orbit

Ne-E – neon component consisting of pure or nearly pure ^{22}Ne; two types are known: Ne-E(L) is released at low temperatures and is carried in presolar graphite, and Ne-E(H) is released at high temperatures and is carried in presolar silicon carbide

Neon burning – occurs in massive stars and converts ^{20}Ne to ^{24}Mg; lasts ~1 year in a 25 $M_\odot$ star

Neutron – a subatomic particle of about the same mass as a proton but without an electrical charge, found in the nucleus of all atoms except ordinary hydrogen

Neutron-capture cross section – a means of expressing the probability that a nucleus will capture a neutron

Nice model – a dynamic model in which migration of the giant planets disrupted the orbits of planetesimals; such a scenario could have led to the late heavy bombardment of the terrestrial planets and the scattering of icy bodies into the Oort cloud

Noble gases – unreactive elements with completely filled electron shells

Nova – or "new star," occurs when a white dwarf in a close binary system accretes hydrogen from its companion until enough hydrogen is accreted to ignite the hydrogen in a fusion runaway; the system brightens until the fuel is exhausted and then settles down; in many cases, the white dwarf accretes more hydrogen and ignites another fusion runaway

Nuclear noble gas component – a specific combination of noble gas isotopes produced by nuclear transmutation, such as decay of naturally occurring radionuclides, spontaneous fission of radionuclides, or spallation by cosmic rays

Nuclear binding energy – the energy that holds a nucleus together; it can be equated with the mass defect through Einstein's equation: $E = mc^2$

Nucleosynthesis – collective noun describing all of the processes that generated the elements and isotopes; typically, it refers to processes happening in stars

Nuclide – a term to describe an atom with any specific combination of protons and neutrons

Oddo–Harkins rule – elements with even atomic numbers are more abundant than adjacent elements with odd atomic numbers

Onion-shell asteroid – thermally metamorphosed planetesimal in which more severely metamorphosed (higher petrologic type) materials are located in the center of the body

Oort cloud – spherical volume of icy planetesimals extending outward from 2,000 to 200,000 AU

Ordinary chondrites – the most common type of chondrites, subdivided into H, L, and LL groups based on content of metallic iron

Organic molecules – compounds composed of carbon and hydrogen, usually combined with other atoms such as oxygen, nitrogen, sulfur, or phosphorus; the term does not necessarily imply a relationship with organisms

Oxygen burning – occurs in massive stars and produces silicon, sulfur, argon, calcium, chlorine, potassium, and other elements up to scandium; lasts ~6 months in a 25 $M_{\odot}$ star

p-process – nucleosynthesis process responsible for ~35 elements bypassed by neutron-capture nucleosynthesis; the *p*-process was originally envisioned to be a proton-capture process, but it is in fact a photodisintegration process

Pauli exclusion principle – quantum mechanical principle which states that two or more identical fermions cannot occupy the same quantum state within a quantum system simultaneously

Parsec – an astronomical measure of distance equal to ~3.24 light years (3.086×10^{13} km)

Partial melting – in a rock, melting of the minerals with the lowest melting points, producing a melt that has a different composition than the parent rock

Pebble accretion – a mechanism in which small particles ("pebbles") accrete onto large bodies (planetesimals or planetary embryos) in a gas-rich environment, proposed to facilitate rapid formation of gas-giant planets

Peridotite – ultramafic rock composed of olivine and pyroxenes; the principal lithology of the mantles of rocky planets

Petrologic type – classification for chondrites based on degree of thermal metamorphism or aqueous alteration

Planetary geochemistry – study of the chemical compositions of planets and planetary materials that have undergone geologic processing

Planetary nebula – an expanding shell of gas and dust ejected from a low- to intermediate-mass star at the end of its lifetime

Planetary noble gases – a fractionated noble gases component in meteorites, characterized by much more extreme depletions of light noble gases than heavy noble gases relative to solar system abundances, similar to the atmospheres of the terrestrial planets; planetary gases in chondritic meteorites have been shown to consist of several subcomponents, some of which are carried in presolar grains

Polar hydrocarbons – organic compounds containing nitrogen, oxygen, or sulfur replacing carbon in ring structures or attached to them; since these elements are more electronegative than carbon, C–N, C–O, and C–S bonds are polar

Polycyclic aromatic hydrocarbons (PAHs) – organic compounds composed of multiple aromatic rings

Positron – a subatomic particle with the same mass as an electron and an equal but positive charge

Positron (β^+) decay – the decay mode of a radioactive isotope where a proton emits a positron and becomes a neutron, decreasing the atomic number by one

Present-day mass function (PDMF) – the relative abundances of stars of different masses in the galaxy today

Presolar grains – small particles in meteorites that predate the solar system and formed around other stars

Primary isotope – in stellar nucleosynthesis, species whose yields are independent of the initial metallicity of the star; they can be produced in stars consisting initially of only hydrogen and helium. Examples are ^{16}O, ^{28}Si, and ^{56}Fe

Primitive achondrites – meteorites that are solid residues from which partial melts have been extracted

Proton – a stable subatomic particle occurring in the nucleus of all atoms with a positive electric charge equal to that of an electron but of opposite sign

Proton–proton chains – several sets of reactions that produce ^4He atoms from four ^1H atoms

Protostar – a very young star that is still accreting gas and dust from its parent molecular cloud, prior to the onset of hydrogen burning

Quark – a type of elementary particle; quarks combine to form hadrons, the most stable of which are protons and neutrons

r-process – nucleosynthesis through the capture of neutrons at a rate that is very much faster than the time required for an unstable nucleus to β decay

Racemic mixtures – mixtures of chiral molecules that have D- and L-enantiomers in equal proportions, often taken as a signature of abiotic origin

Radioactive isotopes – isotopes that are unstable and spontaneously decay to other isotopes

Radiogenic – isotopes formed by radioactive decay

Radiometric dating – determining the age of a sample using radioactive isotopes

Radiometric date – the time of origin as given by the radiometric dating equation

Radiometric age – a radiometric date becomes an age when four criteria are met: 1) a specific event to start the clock, 2) the isotopic system must have remained closed, 3) the value of the decay constant must be constant and known, and 4) the initial abundance of the daughter element must be known

Red-shift – in astronomy, the displacement of the electromagnetic spectrum to longer wavelengths (lower frequency), caused by a celestial object moving away from the observer

Refractory elements – elements that condense from gas to solid at high temperatures, typically above ~1300 K

Refractory inclusions – inclusions in chondrites consisting of minerals predicted to condense at high temperatures; see *calcium-aluminum-rich* inclusions and *amoeboid olivine aggregates*

Relative ages – give the sequence of events, such as which object formed first; must be calibrated to an absolute chronometer to give absolute ages

Regolith – unconsolidated surface material, usually formed by impact pulverization

Rubble-pile asteroid – a planetesimal that has been disrupted by impact, with fragments subsequently reaccreted in a haphazard way

Rumuruti (R) chondrites – chondrite group characterized by its high oxidation state

Saturated hydrocarbons – organic compounds of hydrogen and carbon that contain only single bonds; called "saturated" because all possible bond sites are occupied by hydrogen–carbon bonds

s-process – nucleosynthesis through the capture of neutrons at a rate that is slower than the time typically required for an unstable nucleus to β decay

Second dredge-up – convective mixing event that occurs at the end of core helium burning; in stars of >4–5 $M_\odot$, products of partial hydrogen burning are mixed into the envelope

Secondary isotope – in stellar nucleosynthesis, secondary isotopes are synthesized from pre-existing metals (elements heavier than helium); thus, the synthesis of a secondary isotope is proportional to the initial metallicity of the star; examples are ^{17}O, ^{18}O, ^{29}Si, and ^{30}Si

Self-diffusion – diffusion in a one-component material; atoms and vacancies exchange places and the direction of flow of atoms is opposite to that of the vacancies

Shock wave model – a model for forming chondrules, whereby dust aggregates are heated as they pass through nebular shock waves generated by gravitational instabilities in the outer accretion disk, or by passage of giant protoplanets through the disk

Short-lived radionuclide – radioactive isotope with a half-life sufficiently short that atoms present when the solar system formed have completely decayed away; in cosmochemistry, short-lived nuclides are considered to be those with half-lives <100 million years

Siderophile – iron-loving elements; elements that concentrate in the metal phase; in differentiated bodies, they are found primarily in the metallic core

Silicon burning – the ultimate stage of quiescent nucleosynthesis, consisting of photodisintegration and other processes that produce an iron core; lasts about a day in a 25 $M_\odot$ star

Snow line (or frost line) – heliocentric distance corresponding to the condensation temperature of water ice in the solar nebula; other ices had different snow lines governed by their condensation temperatures

Solar nebula – rotating, disk-shaped cocoon of gas and dust that enveloped the early Sun

Solar noble gases – a noble gas component in meteorites characterized by relatively slight depletions in light gases relative to heavy gases compared to the solar compositions; this component has been shown to be trapped solar wind

Solar system abundances – the average abundances of elements in the solar system, approximated by the composition of the Sun; the term "cosmic abundances" is sometimes used interchangeably

Solar wind – charged particles (mostly protons, but also other ions) ejected from the Sun's upper atmosphere

Soluble organic compounds – organic compounds that can be extracted with solvents

Space weathering – a process that modifies the optical properties of the surface of an airless body exposed to the space environment

Spallation – fragmentation of atomic nuclei into secondary particles due to collision with a cosmic ray

Spectral reddening – in a visible-light spectrum of an object, an increase in slope for wavelengths longer than 0.55 μm, typically due to the presence of iron–nickel metal or complex organic molecules

Spectroscopy – measuring the properties of an object by separating the light reflected or emitted by the object into its component wavelengths

Stardust – presolar grain that formed in the atmosphere or ejecta of a dying star (also called a *circumstellar condensate*); also the name of a NASA mission that returned dust from comet Wild2

Stony irons – meteorites that consist of roughly equal parts of FeNi metal and silicates; these include pallasites and mesosiderites

Supergiant star – a very massive star that is burning nuclear fuel in two or more shells and is losing mass at a tremendous rate; supergiants typically end their lives in supernova explosions

Supernova – or "super new star", is a powerful stellar explosion; a Type II supernova marks the collapse of the core of a massive star at the end of its life; the remnants of core-collapse supernova can be either neutron stars or black holes. Another type of supernova (Type Ia) occurs in C-O-Ne white dwarf that accretes enough mass from a binary companion to exceed the Chandrasekhar mass, triggering explosive carbon burning that completely destroys the star

Terrestrial age – the amount of time a meteorite has been on Earth, determined by measuring certain cosmogenic nuclides

Thermodynamics – a branch of physical science dealing with the relations between heat and other forms of energy; used in cosmochemistry to predict the stability of phases under specific equilibrium conditions

Thermodynamic equilibrium – the condition or state of a thermodynamic system in which the properties do not change with time; the system can be changed to another condition only through effects from outside the system

Third dredge-up – series of convective mixing events in low- and intermediate-mass stars triggered by ignition of the helium shell outside of the core; material from the helium shell and from the intershell region between the hydrogen and helium shell are mixed throughout the envelope

Tholins – complex organic compounds formed by interactions with ultraviolet radiation and cosmic rays

Trace elements – elements that are present in magmas, rocks, and minerals only in very low abundances, typically less than ~0.1%; these elements do not form their own minerals and each atom must be accommodated in the structures of minerals by replacing a major element

Trapped noble gas components – noble gas components that are trapped in the structures of meteoritic minerals; trapped components include trapped solar wind (the solar component) and all of the various planetary components, including those trapped in presolar grains

Triple-alpha process – set of nuclear reactions by which three ^{4}He nuclei are converted into ^{12}C

Trojan asteroids – planetesimals orbiting at Lagrange points 60 degrees ahead of and behind Jupiter

Type A CAIs – fine-grained refractory inclusions with mineralogy and texture consistent with formation by nebular condensation

Type B CAIs – coarse-grained refractory inclusions with textures and element partitioning indicating formation by crystallization of a liquid; these could represent remelted condensates or residues from partial evaporation

Unified atomic mass unit (u) – unit used to describe the mass of an atom, defined as 1/12th of the mass of a ^{12}C atom (also called an *atomic mass unit or Dalton*)

Unequilibrated chondrites – the most primitive (least metamorphosed) chondritic meteorites, characterized by wide ranges in the compositions of olivine and pyroxene

Unsaturated hydrocarbons – compounds of hydrogen and carbon that contain double and triple carbon–hydrogen bonds; these compounds are called "unsaturated" because there is room to add additional hydrogen atoms

Uranium series – decay series starting with ^{238}U and ending in ^{206}Pb, consisting of eight alpha decays and six beta decays

Valence electron – an electron in the outer shell of an atom that can participate in the formation of a chemical bond

Valley of β Stability – in nuclear physics, the region of the Chart of the Nuclides where stable isotopes plot; these isotopes have the lowest masses among isotopes of a given atomic number (*A*); unstable isotopes have higher masses and are less tightly bound

van der Waals forces – weak, short-range, electrostatic, attractive forces between uncharged molecules, arising from the interaction of permanent or transient electric dipole moments

Volatile elements – elements that condense from gas to solid at low temperatures; can be highly volatile, condensing below ~640 K, or moderately volatile, condensing between ~1300 and ~640 K

Volatility – the tendency for elements to exist in gaseous versus condensed (solid) states

Water-equivalent hydrogen (WEH) – abundance of H expressed as an equivalent concentration of H_2O

Xe-HL – xenon component characterized by large excesses of light and heavy isotopes compared to solar system xenon that is carried in presolar diamonds

Xe-S – xenon component found in presolar silicon carbide that consists only of isotopes produced by the *s*-process

Xe-X – early name for Xe-HL, a xenon component enriched in light and heavy isotopes compared to solar system xenon-114

X-ray – a form of electromagnetic radiation with high energy and very short wavelength (0.02–3 nanometers)

X-wind model – a model for the formation of chondrules and CAIs based on heating close to the Sun, followed by ejection outward along magnetic field lines

White dwarf – the end state of a low- to intermediate-mass star that has exhausted its nuclear fuel and is cooling down

Whole-rock isochron – an isochron constructed from radioisotope analyses of bulk samples of different rocks thought to have formed from the same source material at the same time

Wolf–Rayet star – the final burst of activity for a very massive star before it exhausts its nuclear fuel and explodes as a core-collapse supernova; this stage is characterized by massive stellar winds that eventually eject most of the stellar envelope

Zero-point energy – the lowest possible energy that a quantum-mechanical system can have. It is higher than the lowest point on a classical potential energy curve

Index